ENGINEERED NANOPARTICLES

Dedication

This book is dedicated to the memory of my niece Sony, who left us at a young age.

Ashok K. Singh

ENGINEERED NANOPARTICLES

STRUCTURE, PROPERTIES AND MECHANISMS OF TOXICITY

ASHOK K. SINGH, PHD

Associate Professor, Department of Veterinary Population Medicine, University of Minnesota, Minneapolis, MN, USA

AMSTERDAM • BOSTON • HEIDELBERG • LONDON
NEW YORK • OXFORD • PARIS • SAN DIEGO
SAN FRANCISCO • SINGAPORE • SYDNEY • TOKYO

Academic Press is an imprint of Elsevier

Academic Press is an imprint of Elsevier
125 London Wall, London EC2Y 5AS, UK
525 B Street, Suite 1800, San Diego, CA 92101-4495, USA
225 Wyman Street, Waltham, MA 02451, USA
The Boulevard, Langford Lane, Kidlington, Oxford OX5 1GB, UK

Disclaimer
Every effort has been made to cite all of the work adapted from other researchers. Please inform the author (singh001@umn.edu) if any such work is included in the book, but not cited in the references. We will make every effort to acknowledge the omission.

ISBN: 978-0-12-801406-6

British Library Cataloguing-in-Publication Data
A catalogue record for this book is available from the British Library

Library of Congress Cataloging-in-Publication Data
A catalog record for this book is available from the Library of Congress

For information on all Academic Press publications
visit our website at http://store.elsevier.com/

Publisher: Mica Haley
Acquisition Editor: Erin Hill-Parks
Editorial Project Manager: Molly McLaughlin
Project Manager: Edward Taylor
Designer: Victoria Pearson Esser

Typeset by TNQ Books and Journals
www.tnq.co.in

Printed and bound in the United States of America

Contents

Foreword

A rare Roman cage cup at the British Museum in London has been a source of delight to the eyes and wonderment to the mind for nearly 17 centuries. When viewed *en face*, its jade green glass depicts the scene of a Thracian king ensnared by vines. However, when backlit, the dichroic glass of the eponymous Lycurgus cup assumes a ruby red coloration. An explanation for this dazzling effect can be found roughly a mile away at the Faraday Museum, where a container of ruby red liquid containing minute colloidal gold particles sits on display. Colloidal gold and silver are embedded in the glass of the Lycurgus cup. Michael Faraday and his scientific successors in the nineteenth and twentieth centuries discovered that these gold nanoparticles in solution possess intriguing optical properties. They are capable of altering the wavelength of transmitted light in a size-dependent fashion. The precise mechanism underlying this phenomenon is clearly described in this excellent monograph authored by my colleague, Professor Ashok Singh, who is a distinguished chemist, toxicologist, and educator.

The growing field of nanoengineering was sparked by the advent of the electron microscope in the 1930s and inspired by physicist Richard Feynman's 1959 Caltech lecture entitled "There's Plenty of Room at the Bottom". Numerous materials have since been created, which possess distinctive properties at nanometric sizes (i.e., ranging from 1 to 100 nanometers in any single dimension). These novel materials, manufactured in all shapes and nanometric dimensions, have already made a significant, overarching impact on many essential aspects of our daily lives. Nanoscale materials have become incorporated into the processed foods that we consume, cosmetic and clothing products we wear, medications we take, vehicles that move us daily from place to place, and the computational devices to which we are constantly and intimately tethered in the early twenty-first century—and we are just at the beginning stages of this new technology. There is more to come, and the fruits of this endeavor are destined to transform humankind.

The few examples that follow represent only a small fraction of efforts in this area of inquiry and design. In the medical arena alone, oncologists have long been anticipating the development of a biocompatible nanoscale platform that will detect and then chemically treat localized tumors within the body, a so-called theranostics strategy. Virus-like nanoparticles and nanoengineered formulations of drugs are being designed to achieve high drug concentrations in localized body regions while reducing the risk of systemic adverse effects. Finally, the engineering of controlled biomolecular motors that can shuttle cargo around inside cells offer promise as a key step in the development of nanomachines, which can improve human health and wellness.

The first chapters of this book provide the reader with a fundamental understanding of the physicochemical properties of nanoparticles and their current and potential applications.

Professor Singh writes in a simple, direct prose style that will satisfy an expert, yet not entangle a casual reader with a basic scientific background in unexplained technical jargon and complexities.

Where there is great promise, there may be also pitfalls and even perils. Since the beginning of the twenty-first century, there has been increasing interest and concern about the safety of nanoparticles. As Professor Singh emphasizes throughout this book, nanomaterials with a large surface area to volume ratio may possess chemical reactivity and toxic properties that are not observed when they are present in larger sizes. Because they are very small, nanoparticles may become readily airborne and capable of being inhaled into the lungs. A fraction of an inhaled amount of nanoparticulate material may reach the lowest level of the lungs, the alveoli, and after traversing a single layer of alveolar epithelial cells, enter the bloodstream.

As nanotoxicology was becoming a recognized discipline, the toxic effects of airborne engineered nanoparticles in the respiratory system were found to be analogous those of known "incidental" ultrafine particles from polluted air. Since then, the toxicological properties of nanomaterials entering by other routes have been examined, including ingestion and skin contact. Regardless of how they enter the body, these materials are in general eliminated quite slowly and can persist in the body. Indeed, multiple exposures to nanoparticles may result in their accumulation within a host organism, augmenting the potential for adverse effects. Moreover, engineered nanoparticles are raising environmental concerns as they have been detected in soil, water and air, although the fates, behaviors, and ecological impacts of many nanomaterials in these settings are not yet fully understood.

Similar to other exciting and rapidly emerging technologies, the processes of design, creation, and production of new materials often precede those concerned with hazard protection and containment. In the latter half of this book, Professor Singh provides guidance on how to deploy the steadily growing knowledge base of toxicokinetic and toxicodynamic information on nanoparticles to formulate thoughtful risk assessments and effective containment strategies for safeguarding people, other organisms, and the environment.

The Roman admiral and natural philosopher Pliny the Elder perished in the eruption of Mount Vesuvius in 79 AD, presumably after inhaling nanoparticles of volcanic ash. To him is the following quote attributed: "In these matters, the only certainty is that nothing is certain." This adage could be applied to the current, early phase of nanotechnology. Compilations of existing knowledge presented with precision and clarity, as represented in this monograph, will support future developments in this exciting area, which lies at the intersection of materials science, bioengineering, and chemical biology.

David R. Brown, Ph.D.
Professor of Pharmacology
Past Chair, University of Minnesota
Institutional Biosafety Committee
Vice Chair, Department of Veterinary
and Biomedical Sciences
University of Minnesota
College of Veterinary Medicine
St. Paul, Minnesota, USA

CHAPTER

1

Introduction to Nanoparticles and Nanotoxicology

OUTLINE

1. INTRODUCTION TO NANOPARTICLES

In 1966, the science-fiction film *Fantastic Voyage*, directed by Richard Fleischer and produced by 20th Century Fox, was released. In the movie, agent Charles Grant, pilot Captain Bill Owens, Dr Michaels, surgeon Dr Peter Duval, and his assistant Cora Peterson are placed aboard a submarine that is then miniaturized and injected into a patient suffering from a life-threatening brain clot. Their goal was to travel to and then fix the clot. After an amazing journey across various organs, they reach their destination and achieve the goal.

Engineered Nanoparticles
http://dx.doi.org/10.1016/B978-0-12-801406-6.00001-7

How did the scientists shrink the submarine and its inhabitants? By reducing Plank's constant (you will read about Plank's constant in Chapter 3) that increased the speed of light and reduced the graininess of the universe. According to the movie, the submarine was reduced to the size of a bacterium (1 μm in diameter); therefore, its inhabitants may have been reduced to smaller size, probably in the nanometer range. I viewed this movie in the early 1970s and was amazed by the special effects, especially the scenes depicting the submarine attacked by the immune cells (Figure 1(B)). I never expected the plot to become a reality. However, 50 years after the movie's release, although nanosized humans are still a fantasy, nanosized drug-loaded submarines that travel within the body may become a reality. Scientists have developed devices that can travel into the body and fix diseased cells (Sailor and Park, 2012; Peer et al., 2007). The submarine in *Fantastic Voyage* may become a reality when nanomotors are controlled inside a living cell and literally fight the disease. This is the miracle of nanotechnology.

In 2002, Michael Crichton published his novel *Prey*, which tells the story of a mechanical plague that occurred when a cloud of self-sustaining and self-reproducing nanoparticles with collective intelligence (remember the "Borg" of *Star Trek*, Figure 1(C)) escaped from the laboratory. The nanocloud (*cloud*—Does this word sound familiar?) learns from its experiences and becomes more deadly with each passing hour. The novel describes the desperate efforts of a

FIGURE 1 (A) In the movie *Fantastic Voyage*, the miniaturized submarine and its occupants are injected into the scientist's body. (B) The body's defense cells attach the submarine. (C) The "Borg" of *Star Trek*—a collection of species that turned into a cyberorganism functioning collectively.

handful of scientists to stop the cloud as time is running out for the entire humanity. Although *Prey* is fiction, it brings up many of the issues relevant to this topic, especially the conflict between commerce and public safety.

How much risk can we, as humans, take to make our life better? Many earlier technologies, such as use of fossil fuel for energy (the first industrial revolution), the development of pharmaceutical drugs and insect eradicators (the chemical revolution), and/or the construction of nuclear power plants (the nuclear revolution), were incorporated in household and commercial products because of the perception that these technologies were safe to society and the environment without addressing safety concerns or developing regulations. We all are aware of the extensive pollution and ensuing health and environmental consequences that these "revolutions" created—and we are still dealing with their effects. Is nanotechnology heading in the same direction? The technology has been extensively integrated in our daily lives (food, cosmetics, medicine, electronics, and energy) despite the lack of adequate safety information and regulations. Therefore, the overall aim of this book is to present basic information regarding the structure, beneficial effects, and toxicity of nanoparticles, which can be used to make regulatory decisions.

1.1 Historical Aspects

Nanotechnology, although considered to be a recent phenomenon, is also evident in ancient civilizations. The Celtic-red enamels dating from 400 to 100 BC contain copper and cuprous oxide (cuprite Cu_2O) nanoparticles (Brun et al., 1991), while most of the red-tesserae used in Roman mosaics were made of glass containing copper nanocrystals (Brun et al., 1991; Colomban et al., 2003; Ricciardi et al., 2009). The Roman artisans achieved unusual color changes by adding noble metal-bearing material to glass prior to being molten. The middle age (1066–1485 AD) saw an emergence of glazed ceramics with striking optical effects obtained from metallic nanoparticles (Caiger-Smith, 1991). Ancient Indian (Ayurveda) and Chinese medicines used gold bhasma (Yadav et al., 2012) and soluble gold (http://www.zhengjian.org/sci/sci/home/newscontent.asp?ID=11330), respectively, for therapeutic purposes; these substances have been shown to contain gold nanoparticles mixed with larger particles. Ancient civilizations, however, did not understand the unique properties and potential of their preparations as we understand them now. Recent technological advancements have revolutionized the synthesis, characterization, and applications of nanoparticles, which are gradually becoming an integral part of society.

In 1959, Richard Feynman (http://metamodern.com/2009/12/29/theres-plenty-of-room-at-the-bottom%E2%80%9D-feynman-1959/) suggested the possibility of building machines small enough to manufacture objects with atomic precision. He also predicted that information could be stored with amazing density. The term *nanotechnology* was coined by Norio Taniguchi (1974), but it was used unknowingly by Eric Drexler in his 1986 book *Engines of Creation: The Coming Era of Nanotechnology*. The technology is growing and diversifying rapidly, as shown in Table 1.

1.2 Nanotechnology

In general, nanotechnology deals with the fabrication and control of nanoparticles less than 100 nm in at least one dimension. Nanoparticles exhibit unique physicochemical properties that are absent in their bulk (>500 nm) counterparts. An exception to the 100-nm rule is solid-lipid nanoparticles that exhibit the unique nanoparticle-related properties at diameters greater than 100 nm (Carla et al., 2011). Because of their unique properties, nanoparticles are being used in diverse applications such as the following.

TABLE 1 History of Nanotechnology

Year	Items
1981	Scanning tunneling electron microscope that could process single atoms.
1985	Discovery of fullerene, 60 carbon atoms in a circle (C60).
1992	Discovery of carbon nanotubes that are stronger than steel; can be used in drug delivery, energy storage, and power transmission.
1993	Discovery of quantum dots.
2000	Construction of passive nanoparticles for applications as nano fuel cells and in products of daily use, including cosmetics.
2005	Construction of active nanoparticles for target-directed drugs and other adaptive structures.
Future	Nanosystems, hierarchical nanoarchitectures, atomic devices, nano DNA-based computers, diagnostic robots, etc.

- Medical (targeted drug delivery, imaging, and personalized medicine) and cosmetic (makeup and sunscreens) applications (Patel et al., 2011; Pathak and Thassu, 2011; Raj et al., 2012; Smijs and Pavel, 2011; Wen-Tso, 2006; West and Halas, 2000)
- Efficient energy-storage devices using hybrids of carbon nanotubes and oxide nanoparticles (Gruner et al., 2006), as the development of high-temperature, heat-transfer nanofluids may allow storage of thermal energy (Wong and DeLeon, 2010)
- Efficient removal of pollutants such as metals (cadmium, copper, lead, mercury, nickel, zinc), nutrients (phosphate, ammonia, nitrate, and nitrite), cyanide, organics, algae (cyanobacterial toxins), viruses, bacteria, parasites, and antibiotics from water in waste treatment plants (Tiwari et al., 2008)
- Construction of implantable devices to treat neural disorders using carbon-nanotube-protein hybrids (Andersen et al., 2004; Green and Hersam, 2011)
- Clothing, sporting equipment, food packaging, dietary supplements, etc.

These developments may represent the next revolution—the nano-industrial revolution—which will significantly alter society. However, the pace of nanotechnology commercialization is much faster than the assessment of their safety, thus posing a significant health risk to the general population. A key hurdle in determining the health risk of nanoparticles is their structural heterogeneity.

1.3 Atoms, Nanoparticles, and Bulk Materials

Atoms (less than 0.1 nm) are the smallest unit taking part in chemical reactions. They are transformed into bulk materials (greater than 500 nm) through formation of clusters (approximately 1 nm) followed by small nanoparticles (1–100 nm) and large nanoparticles (greater than 100 nm), as shown below:

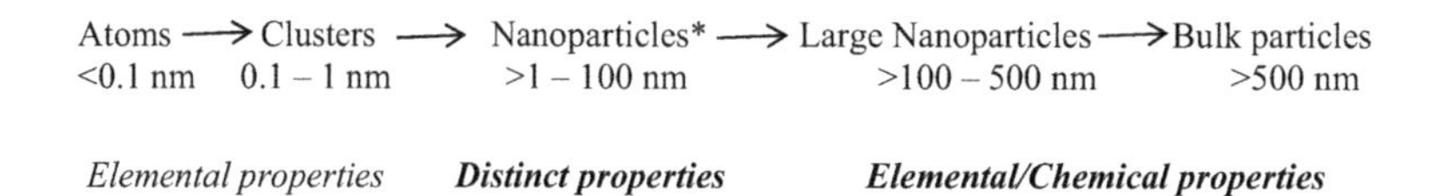

Clusters ranging from 0.1 to 1.0 nm in diameter possess elemental characteristics. As the clusters grow from 1 to 100 nm, they get transformed into nanoparticles possessing distinct physicochemical properties (described earlier) that are absent in their bulk counterparts. The relationship between the number of atoms in a nanoparticle and the percentage of atoms at the

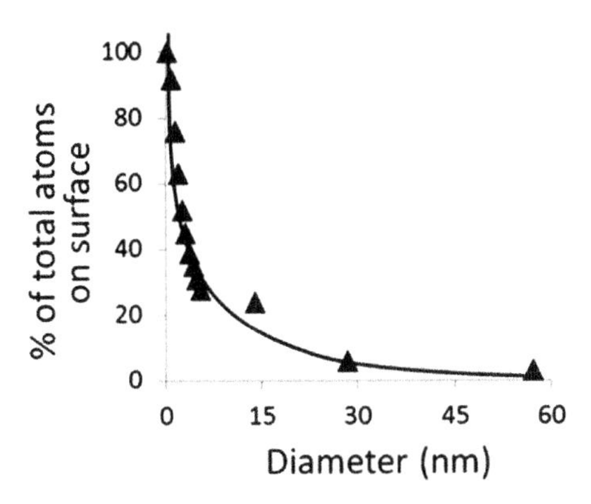

FIGURE 2 Relationship between diameter and percentage of atoms present at the surface. There is an inverse relationship between the two indices. An increase in nanoparticle size is associated with a decrease in the percentage of total atoms at the surface. All of the atoms were surface atoms in nanoparticles having 1 nm (12–25 atoms) diameter, while only 5% of atoms were surface atoms in nanoparticles having 5 nm (about 2000 atoms) diameter. In bulk particles, the share of the surface atom is <0.1%.

surface is shown in Figure 2 and Table 2. There is a direct relationship between the total number of atoms in a nanoparticle and the number of surface atoms or the particles' diameter (Table 2). There is an inverse relationship between the diameter (nm) and the percentage of atoms at the surface (Figure 2).

As nanoparticles become smaller, the proportion of atoms on the surface increases. In the nanometer range, a size reduction results in a drop in melting temperature and an increase in reactivity, as well as the dominance of surface atoms over the core atoms (Gupta et al., 2014; Lia et al., 2013). A transition from classical mechanics to quantum mechanics occurs when free electrons in nanoparticles start to behave like a wave (described in Chapter 3).

1.4 Classification of Nanoparticles

As shown in Figure 3, nanoparticles can be classified according to their dimensions, origin, application, chemistry, and counterpart types and applications.

TABLE 2 Mathematical Relationship between Total Atoms, Diameter, Surface Atoms, and Percentage of Surface Atoms in Gold Nanoparticles

Shell	Diameter (nm)	Total Atoms	Surface Atoms	Surface Atoms (%)
1	0.288	1	1	100
2	0.864	13	12	92
3	1.44	55	42	76
4	2.01	147	92	63
5	2.59	309	162	52
6	3.16	561	252	45
7	3.74	923	362	39
8	4.32	1415	492	35
9	4.89	2057	642	31
10	5.47	2869	812	28
25	14.1	4.9×10^4	5083	24
50	28.5	4.04×10^5	2.40×10^4	6
100	57.3	3.28×10^6	8.80×10^4	3

1.4.1 Dimension-Based Classification

Nanoparticles exist as zero-dimensional (0D), one-dimensional (1D), two-dimensional (2D), and three-dimensional (3D) particles. The 0D nanoparticles, such as nanospheres and nanoclusters, are less than 100 nm in all dimensions (Figure 4). The 1D nanomaterials, such as nanotubes, nanorods, and nanofibers, are less than 100 nm in at least one dimension. The 2D nanomaterials are films (graphene, molybdenum disulfide, and germanane (a single-layer crystal composed of germanium)) with less than 100 nm thickness. The 3D nanomaterials are more than 100 nm in all dimensions (Figure 4).

1.4.2 Natural or Anthropogenic Nanoparticles

Natural nanoparticles originate from forest fires, volcanic eruptions, lightning, etc. (Angelucci

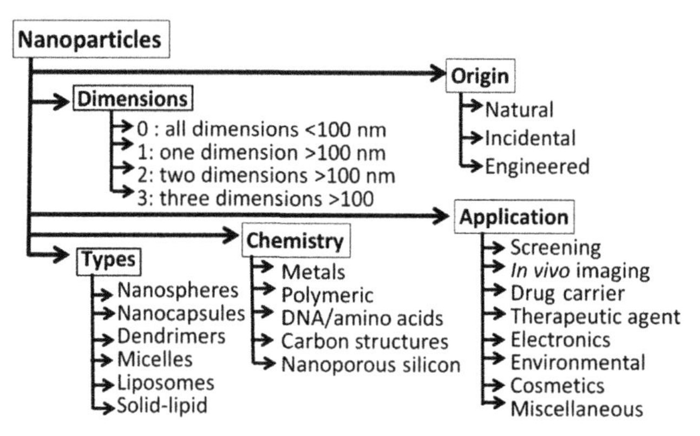

FIGURE 3 Nanoparticle classification.

et al., 2010; Buzea et al., 2007). They have been an integral part of the environment since the origin of the planet (http://sustainablenano.com/2013/03/25/nanoparticles-are-all-around-us/). Anthropogenic nanoparticles fall into two general categories: incidental and engineered nanoparticles. Incidental nanoparticles are heterogeneous in size and shape; they are generated by burning fossil fuel (gasoline, diesel, coal, and propane), large-scale mining, incinerating forests for agriculture, etc. (Buzea et al., 2007; Kittelson, 1998). Engineered nanoparticles are specifically designed particles having precisely controlled sizes, shapes, and compositions. They may even contain multiple layers (e.g., a gold nanoparticle covered in drug-loaded porous silica nanoparticles coated with specifically chosen antibodies). Engineered nanoparticles are becoming more complex with each passing year.

1.4.3 Classification of Nanoparticles According to Their Chemistry

Chemically, nanoparticles consist of metals/metal oxides, DNA and other biological materials, carbon, polymers, and clays. In addition to the common size-related properties, nanoparticles retain their chemical characteristics, which may be helpful in the selection of appropriate nanoparticles for a particular use. Some important groups of nanoparticles include the following:

- *Metal nanoparticles* (gold, copper, silicon, iron, etc.) are widely used in catalysis, electronics, sensors, photonics, environmental remedies, and medicine. Because of surface plasmon resonance and paramagnetic properties, metal nanoparticles find unique applications in medical and electronic technology

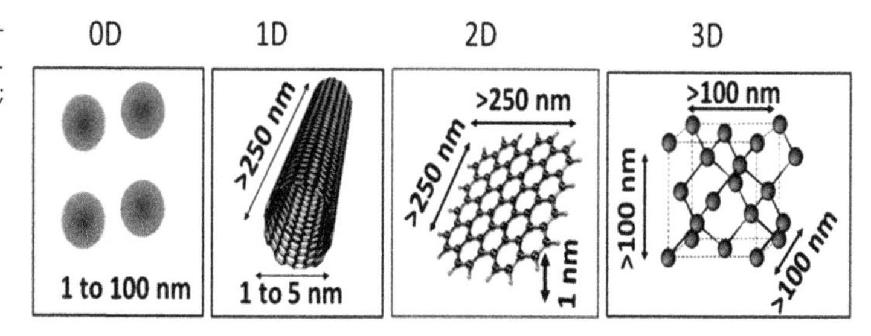

FIGURE 4 Types of nanoparticles determined by the size of their structural elements. 0D, zero-dimensional; 1D, one-dimensional; 2D, two-dimensional; 3D, three-dimensional.

(http://www.mpikg.mpg.de/886767/MetalNanoparticles.pdf). Porous silicon nanoparticles contain microscopic reservoirs that can hold and protect sensitive drugs in a pH-sensitive manner. Acidic pH disrupts the drug-nanoparticle binding, thus releasing the drug load. Functionalizing the surface with target molecules provides target-selective delivery of the nanoparticles.

- *Polymeric nanoparticles* are prepared from either synthetic polymers such as poly(2-hydroxy ethyl methacrylate), poly(*N*-vinyl pyrrolidone), poly(methyl methacrylate), poly(vinyl alcohol), poly(acrylic acid), polyacrylamide, or natural polymers such as gums (e.g., acacia, guar, etc.), chitosan, gelatin, and sodium alginate. In recent years, biodegradable polymeric nanoparticles have attracted considerable attention as potential drug delivery devices.
- *Biochemical nanoparticles* such as DNA, proteins, and poly-amino acids such as poly-L-lysine and poly-L-serine are synthetized from biological precursors. DNA nanoparticles are three strands of DNA with a lipid and functional molecule attached to its ends. In water solutions, the combination of hydrophilic DNA and lipophilic lipids causes the units to self-assemble into hollow spheres consisting of multiple layers of DNA, lipids, and cargo.
- *Carbon nanotubes* (*CNTs*) are formed from rolled-up graphite sheets. Depending on the direction of hexagons, carbon nanotubes can exhibit metallic or semiconductor properties. CNTs are twice as strong as steel but weigh many times less. In 1996, a new form of carbon—the Buckminster fullerene—was discovered; it looks like a nanometer-sized soccer ball made from 60 carbon atoms (Thess et al., 1996).
- *Nanoclays* are layers of mineral silicate nanoparticles. Organically modified or hybrid organic-inorganic nanomaterials have potential uses in polymer nanocomposites and as rheological modifiers, gas absorbents, and drug delivery carriers.

1.4.4 Isotropic and Anisotropic Nanoparticles

Isotropic nanoparticles include nanocapsules (Figure 5(1)), nanospheres (Figure 5(2)), dendrimers (Figure 5(3)), liposomes (Figure 5(4)), spheres (solid), capsules, and liposomes, whose physical and chemical properties are not dimensional. Nanospheres (solid) and nanocapsules (hollow) are polymeric nanoparticles consisting of a shell and a space, in which desired substances may be loaded and protected from the environment. Dendrimers are artificially manufactured branched nanoparticles comprised of many smaller ones linked together, built up from branched units called monomers. Liposomes consist of an outer single or multilamellar membrane and an inner liquid core. Liposomes consisting of natural or synthetic phospholipids are similar to those in cellular plasma membranes. Because of this similarity, liposomes are utilized by the cells. Micelles are similar to liposomes but they do not have an inner liquid

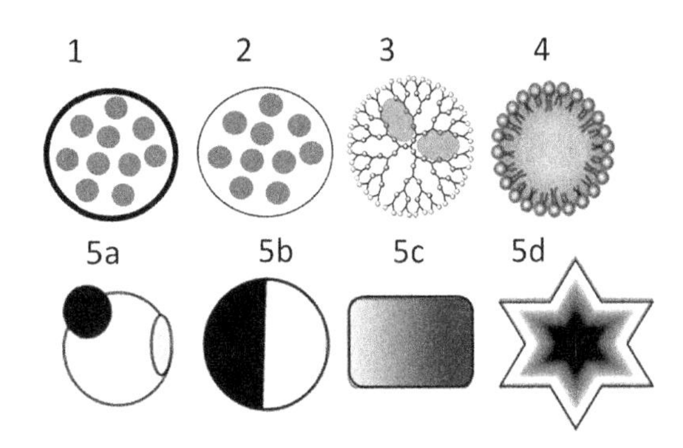

FIGURE 5 Examples of different forms of nanoparticles. (1) capsules, (2) sphere, (3) dendrimers, (4) liposomes, and (5) noble-metal anisotropic nanoparticles. (5a) Patchy particles that have one or more patches and exhibit strong surface anisotropy. (5b) A Janus particle defined as a particle with two faces or a particle with one patch covering half of the particle surface (Wurm and Kilbinger, 2009). (5c) Nanorods Janus particle. (5d) Nanoclusters as an example of patch particles.

compartment (Yang et al., 2004). A solid-lipid nanoparticle is typically spherical with an average diameter of 50–1000 nm. Solid-lipid nanoparticles possess a solid lipid core matrix.

Anisotropic nanomaterials, shown in Figure 5(5a–5d), were first described by Banholzer (2011) and Casagrande and Veyssie (1988). They exhibited direction- and dimension-dependent physicochemical properties (dendrimers can be constructed to be anisotropic). Multifunctional anisotropic nanoparticles have attracted increasing attention because of their promising properties for applications in biotechnology, nanotechnology, electronics, and clean/reusable energy (Grosse and Delgado, 2010; Perro et al., 2005; Rhodes et al., 2009; Walther and Muller, 2008; Wood et al., 2010).

1.4.5 Nanoparticle Classification Based on Application

Nanoparticles are currently being applied in medicine, environmental remediation, cosmetics, electronics, and energy-storage industries. Medicinal uses include screening, in vivo imaging, drug carriers, and treatment (Dykman and Khlebtsov, 2011; Salata, 2004). Nanoparticle applications in environmental remediation (Henn and Waddill, 2006; Masciangioli and Zhang, 2003; Wang and Zhang, 1997) include the clean up of oil spills (photocatalytic copper tungsten oxide nanoparticles), the destruction of volatile organic pollutants in air (gold nanoparticles embedded in a porous manganese oxide), and the removal of metals from water samples. Iron nanoparticles remove carbon tetrachloride from ground water. Iron oxide nanoparticles are used to clean up arsenic from water wells (Zhang, 2003). Nanoparticle applications in energy and electronics include low-cost electrodes for fuel cells, energy storage (http://www.understandingnano.com/nanoparticles.html, Lianga and Zhi, 2009), and catalysts such as a platinum-cobalt hybrid for fuel cells that produce 12 times more catalytic activity than pure platinum. Construction of a memory field-effect transistor (combining gold nanoparticles with organic molecules) can function in a way similar to synapses in the nervous system. Silicon nanoparticle-coated anodes of lithium-ion batteries can increase battery power and reduce recharge time. Nanoparticles are used in cosmetic products (Bertrand et al., 2013; Patel et al., 2011; Raj et al., 2012) including deodorant, soap, toothpaste, shampoo, hair conditioner, antiwrinkle cream, moisturizer, foundation, face powder, lipstick, blush, eye shadow, nail polish, perfume, and after-shave lotion.

2. INTRODUCTION TO NANOTOXICOLOGY

Toxicology is study of the nature, effects, and detection of toxins and the treatment of toxicosis. Toxins are natural or synthetic chemicals capable of causing harm or disease when introduced into the body. In general, there are three basic laws of toxicology:

- *The dose makes the poison* (Paracelsus theory). In general, dose is defined as the mass of a chemical per unit of body weight, such as g/kg body weight. From Paracelsus's time to the present, the mass-based dose has been used to determine a chemical's beneficial effects and toxicity. However, as we will see later, the mass-based definition of dose may not be entirely applicable to nanoparticles. This is because, in addition to the mass, the size, shape, and surface functionality may also play a significant role in determining a nanoparticle's beneficial effects and toxicity. Therefore, the concept of a dose based on size, surface area, or surface reactivity will be introduced later.
- *The biological actions of chemicals are specific to each chemical.* In the sixteenth century, Ambriose Paré recognized that the toxic effects of a chemical are specific to the chemical's structure. He postulated that each

chemical may exhibit unique toxicity. This postulation is the fundamental basis for structure-activity relationship studies.

- *Humans are animals*. Therefore, protection against the toxicity of agents would be impossible without the ability to study the effect of toxins on laboratory animals.

2.1 Dose–Response Relationship for Bulk Particles

Paracelsus (1493–1541), known as the father of toxicology, established the first law of toxicology. He stated "All substances are poisons: there is none which is not a poison. The right dose (mass based) differentiates a poison and a non-poison." Later, the Royal Society of Chemistry revised Paracelsus's statement to make sure that the pharmaceutical industry is not hurt by calling all chemicals "poison." The revised statement was, "While there is no such thing as a safe chemical, it must be realized there is no chemical that cannot be used safely by limiting the dose or exposure. Poisons can be safely used and be of benefit to society when used appropriately." Both statements emphasize dose as the hallmark of toxicology and the dose–response relationship as one of the three laws of toxicology. The dose–response relationship describes the change in effect on an organism (individual or population) caused by varying the doses after a certain exposure time (Figure 6). Many regulatory decisions have been made based on the indices calculated from the dose–response curve (see Chapter 5).

The dose–response curve brings up the following important issues:

- The concept of a threshold dose is considered, below which no adverse effects occur from exposure to the chemical. This can be attributed to the living organisms' defenses against many toxic agents. Cells, especially in the liver and kidneys, break down chemicals into nontoxic substances that are eliminated from the body in urine and feces. Therefore, at a dose below the threshold, organisms can take some toxic insult and still remain healthy.
- The shape of the dose–response curves can offer some information about the toxic potency of chemicals. A steep curve that begins to climb even at a small dose suggests high potency, while a relatively flat curve may represent a poorly toxic chemical. It is important to keep in mind that the more potent the chemical is, the less it takes to cause toxicity.
- After a maximum response, a further increase in dose does not increase the response. The maximum response is the efficacy of the toxin.
- The no-observed-effect level (NOEL) is the highest exposure level at which there are no significant increases in the severity of adverse effects between the exposed population and its appropriate control.

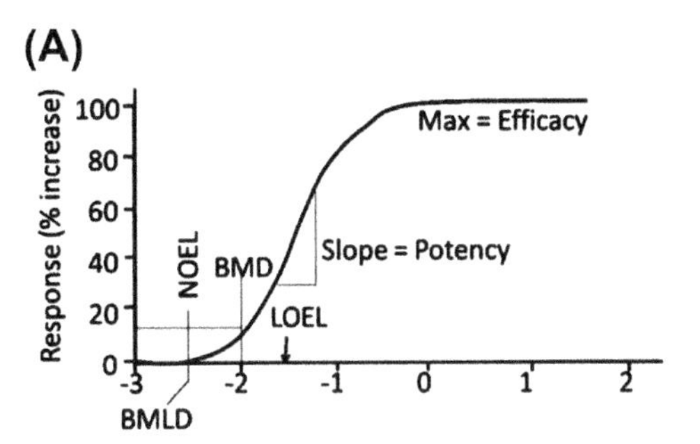

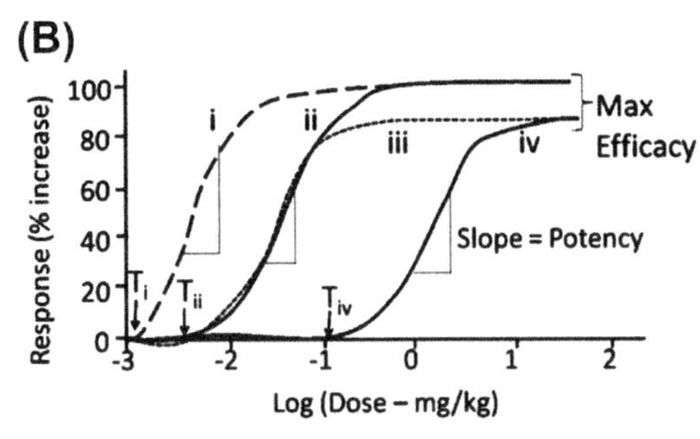

FIGURE 6 Sketches of hypothetical dose–response curves (not experimental data used, for demonstration only) showing the factors that influence toxicokinetics of bulk and nanoparticles (SA: surface area, V: volume). Curves i to iv are for toxins i to iv, respectively. Ti to Tiv are threshold level for toxins i to iv, respectively. Max: maximum The details are described in the text.

- The lowest-observed-effect level (LOEL) is the lowest exposure level at which a significant increase in frequency or severity of adverse effects between the exposed population and its appropriate control group occur.
- The benchmark dose (BMD) is a dose of a toxin that, when ingested, produces a predetermined change (benchmark response, BMR) in the response rate of an adverse-effect relative to the background response rate of this effect. Below Minimal Detection Limit (BMDL) is a lower one-sided confidence limit on the BMD that is determined by modeling the dose–response data, while NOEL or LOEL is determined from the curve (EPA, 2012).

For bulk particles, the dose–response relationship plays a key role in characterizing the toxicity of a chemical and performing risk assessment. The size and shape of the toxin are not compounding factors.

2.2 Is the Mass-Based Dose–Response Relevant to Nanotoxicology?

As discussed earlier, advances in chemistry and engineering have created a new technology (nanotechnology) involving the development of nanometer-sized products that are being increasingly used in commercial and medicinal products. Because the general public may not understand this new class of chemicals, the use of nanoparticles has either gone unnoticed or generated unnecessary fear in individuals regarding health risks. Although such fears are not entirely unfounded, often they are based on misinformation. This may be because investigation of the health effects often lags far behind the commercial advances in nanotechnology. The research data generated over the past decade have provided much needed information regarding the physicochemical properties, beneficial effects, and adverse effects of nanoparticles. The differences between bulk and nanosized particles may be attributed to certain physicochemical properties of nanoparticles that are not found in traditional bulk chemicals, such as the following:

1. A decrease in the size of a particle to less than 100 nm increased their surface area to volume ratio (Figure 7(A)), thereby increasing the influence of the surface atoms (which are more reactive due to the presence of unpaired electrons) over the core atoms (which are relatively less reactive due to the bonded atoms). Below 10 nm, surface atoms dominate the core atoms and determine the particle's properties, which may be completely different from the bulk properties. An example is shown in Figure 7(B) for gold nanoparticles, which lose their characteristic color at the nanometer size range. In addition to optical properties, the size has considerable influence on the viscosity and dielectric dispersion of gold nanoparticles.

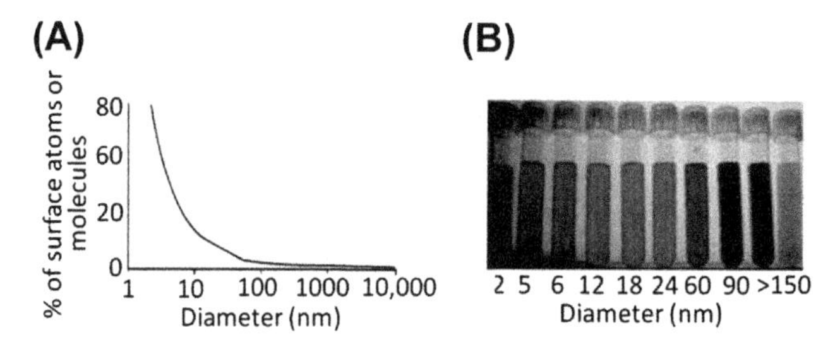

FIGURE 7 Influence of size on nanoparticle surface atoms (A) and optical properties. (B) As the size decreases, the percentage of atoms at the surface increase, which is associated with changes in the gold color.

2. Functionalization (conjugation) of the nanoparticle surface with certain organic chemicals (polyethylene glycol (PEG) and poly-L-lysine (PLL) or poly-L-glycane (PLG) alter their reactivity.
3. Nanosizing below 100 nm facilitates movement of particulates across cellular and intracellular barriers.
4. Nanoparticles, in a size- or surface-dependent manner, interact with, and sometimes hybridize with, subcellular structures such as microtubules and DNA.
5. An alteration in the nanoparticles' size or surface functionality may also alter their beneficial and toxicological responses.

The observation that, at a given dose, the size and/or surface of nanoparticles may significantly affect their bioactivity is a departure from the classic dose–response based risk assessment that is commonly used for bulk particles. This brings up the themes of size-response, surface area-response, and shape-response in the determination of the safety of nanoparticles. Because of a deviation from exclusively mass-based dose to mass-, size-, surface-, and reactivity-based dose, the mass-based dose–response curve may not accurately characterize nanoparticle toxicity. For example, a 1 mg/kg dose of a 10-nm nanoparticle may exhibit greater toxicity than a 1 mg/kg dose of a chemically identical 50-nm nanoparticle. However, the toxic effects of a 1 mg/kg dose of a bulk particle will not be modulated by either size or shape. Thus, the bulk particle data may not accurately predict the risks associated with nanoparticle (less than 100 nm).

Despite these discrepancies, the bulk particle data collected using mass-based dose–response curves are commonly used to assess the toxicity of commercial nanoparticles. These observations suggest that an adequate physicochemical characterization of nanoparticles is necessary prior to undertaking experiments for toxicity assessments. Warheit et al. (2007, 2008) has recommended that, at a minimum, toxicologists should characterize the following (prioritized) physicochemical properties prior to conducting hazard studies with nanoparticles:

- Particle size and size distribution (wet state) and surface area (dry state) in the relevant media being used
- Crystal structure/crystallinity
- Aggregation status in the relevant media
- Composition/surface coatings
- Surface reactivity
- Method of nanomaterial synthesis and/or preparation including postsynthetic modifications (e.g., neutralization of ultrafine TiO_2 particles)
- Purity of sample

An important problem encountered when studying the properties of nanoparticles is that the physicochemical properties of nanoparticles are commonly measured in cell-free media or in vitro/cell culture conditions. However, when administered into an organism, the physicochemical characteristics of nanoparticles are likely to change because of the particles' interaction with the bodily components. Unfortunately, the interaction will be highly heterogeneous depending on size and shape.

2.3 Redefining the Dose

The classic definition of dose is mass/unit body weight (such as g/kg) or moles/unit body weight (such as 1 mM/kg). When dealing with nanoparticles, dose in terms of mass may not relate to the toxicity. Dose in terms of size (diameter/kg or surface area/volume ratio/kg) is at least as important as the mass-dose, if not more important. Many studies have examined the dose as mass (μg), number, surface area (nm^2), surface-to-volume ratio, and surface reactivity in characterizing the toxicity of nanoparticles (Hedwig et al., 2014; van Ravenzwaay et al., 2009; Oberdörster et al., 2005, 2007;

Wittmaack, 2007). Although there is a general agreement that the dose in terms of mass may not be an accurate predictor of nanoparticle toxicity, some disagreements exist regarding the applicability of particle number, size, surface area, or reactivity in predicting nanoparticles' toxicity. This disagreement may be because of the diversities in the structure and function of nanoparticles and their surface functionalization.

2.4 Exposure of Humans and Animals to Nanoparticles

The principal routes of occupational and environmental exposure are through the skin, gastrointestinal tract, and respiratory tract (Figure 8). Two minor routes are injection (medicines) and implants containing nanoparticles. Dermal exposure is the main exposure mode for many metal nanoparticles present in cosmetics (Schulz et al., 2002). Ingestion is the main source of exposure for nanoparticles intended for medical imaging, therapeutics, and food. Airborne nanoparticles (workplace or environmental) enter the body via the respiratory tract (Charron and Harrison, 2003).

2.5 Nanoparticles in the Environment

Nanoparticles enter the environment either intentionally (direct use in waste and environmental treatment) or accidently (industrial products and wastes tend to end up in waterways such as drainage ditches, rivers, lakes, estuaries, and coastal waters). In the environment, aerosol and volatile nanoparticles enter the atmosphere; ionized and polar particles dissolve in water; and hydrophobic particles settle down and bind the sediment soil (Smita et al., 2012; Oberdörster et al., 2004). Suspended sediment particles in water sequester and transport nanoparticles away from the source (Darlington et al., 2009). The hydrodynamic and morphological characteristics of bodies of water and coastal zones determine the distribution of bound nanoparticles. Environmental nanoparticles undergo chemical and photochemical changes (Gorham et al., 2012; Lahiria et al., 2006) that may alter their distribution in air, water, and soil, as well as their toxicity (Figure 9). Potential routes by which nanoparticles enter aquatic organisms are direct ingestion and/or entry across epithelial boundaries such as gills, olfactory organs, or body walls. Uptake of nanoparticles into the aquatic biota is a major concern because they may bioaccumulate in food chains (Judy et al., 2011).

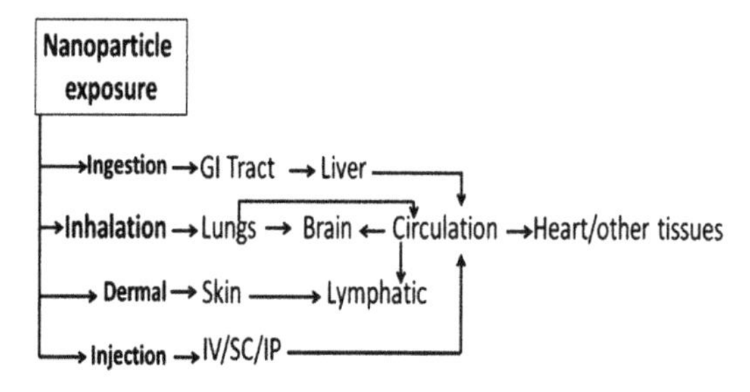

FIGURE 8 A flow diagram showing the modes of nanoparticle exposure.

2.6 Fate and Toxicity of Nanoparticles

Humans and animals are exposed to toxins via dermal exposure, ingestion, and/or inhalation.

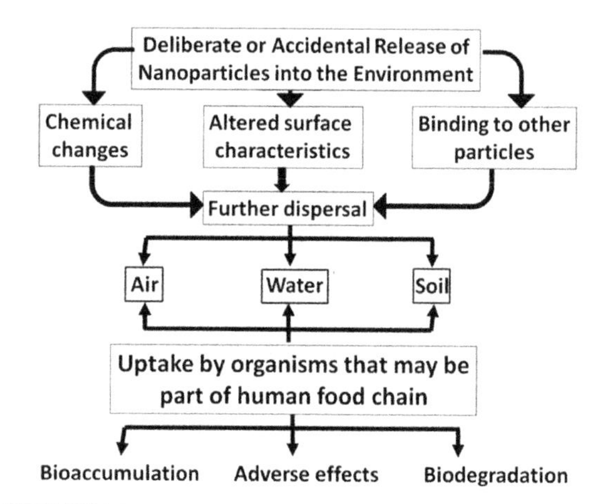

FIGURE 9 A flow diagram showing the release and fate of nanoparticles in the environment.

Toxin concentrations in the systemic bloodstream depend upon the route of exposure and transfer of the ingested toxin through the membrane barriers, metabolism and excretion, and nonspecific distribution of toxins into different tissues. A very small fraction of the toxin may reach the target site (Figure 10). The toxin–target interaction initiates a series of mechanisms, as shown in Table 3, and ensuing cytotoxicity (necrosis, DNA damage, and membrane toxicity) (Ho et al., 2011).

2.6.1 *Necrosis*

Nanoparticle exposure may cause necrotic damage (death of cells or tissues through injury or disease, especially in a localized area of the body) in the liver and other peripheral organs (Abdelhalim and Jarrar, 2011, 2012; Loh et al., 2010; Pan et al., 2009; Wilhelmi et al., 2013). Liver necrosis is an indication of injured hepatocytes due to direct nanoparticle toxicity. Necrosis occurs when cells are overwhelmed with the residues resulting from metabolic and structural disturbances. These alterations are size-dependent, with smaller ones inducing more damage than larger ones.

2.6.2 *Membrane Toxicity*

Loh et al. (2010) have shown that necrotic cell death is caused by cell membrane damage and resultant enzyme leakage. It has long been recognized that cationic nanoparticles induce cell

TABLE 3 Toxicological Pathways Responsible for Nanoparticle Toxicity

Toxicological Pathways	Nanoparticles
Membrane damage/leakage/thinning	Cationic NPs
Protein binding, unfolding responses, loss of function, fibrillation	Metal oxide NPs, polystyrene dendrimer, carbon nanotubes
DNA cleavage/mutation	Ag NPs
Mitochondrial damage, electron transfer, adenosine triphosphate synthesis, apoptosis	Si NPs, Cationic NPs, ultra fine particles
Lysosomal damage, proton pump activity, lysis, frustrated phagocytosis	CNTs, UFPs, Cationic NPs
Inflammation: signaling cascade, cytokines, chemokines, adhesion	Metal oxide NPs, CNTs
Fibrogenesis, tissue remodeling injury	CNTs
Blood platelet, vascular endothelial and clotting abnormalities	SiO_2
Oxidative stress injury, Glutathione depletion, lipid peroxidation, membrane oxidation, protein oxidation	UFPs, CNTs, metal oxide NPs, cationic NPs

CNT, carbon nanotube; NP, nanoparticle.

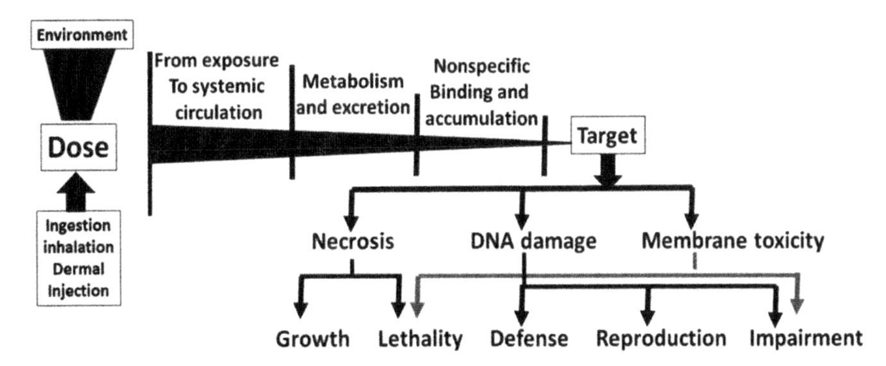

FIGURE 10 Relationship of the exposure dose and the dose reach the target size with various indices of adverse effects.

membrane permeability by forming nanoscale pores (Noronha et al., 2000) in supported lipid bilayers. Earlier studies (Chen et al., 2009; Yang et al., 2007; Fatima et al., 1999) have described pore formation by the direct effects of cationic nanoparticles and/or nanoparticles containing amphiphilic phenylene ethynylene antimicrobial oligomer (AMO-3), a small molecule that makes well-defined 3- to 4-nm holes in lipid bilayers. In most cases, the pores are sealed within 1–2 min.

2.6.3 DNA Cleavage

Metal nanoparticles, especially gold and silver nanoparticles, bind with double-stranded DNA, resulting in compromised DNA replication. It is not clear whether silver nanoparticles directly interact with DNA polymerases which, in the presence of silver nanoparticles, cause replication of error rates in vitro that could account for a significant fraction of in vivo mutations. Transition metal nanoparticles may show nuclease-like activity that promotes DNA cleavage by direct strand scission or base modification (Jose et al., 2011; Burrows and Muller, 1998), possibly by generation of oxidative-free radicals (Lopez et al., 2010, Shen et al., 2009; Lipovsky et al., 2009).

3. CONCLUSIONS

Nanoparticles, because of their outstanding characteristics and properties, have allowed the development of novel medical, electronic, and industrial products. Research and commercial application of nanoparticles has increased exponentially in the last several years, thus generating greater interest and safety and health aspects of nanoparticles. In addition, the unique characteristics and properties, especially exceptionally high surface area/volume ratio and surface activity, further complicate their safety and health issues.

Physicochemical properties, beneficial effects, and the toxicity of bulk particles (greater than 500 nm) depend solely on the dose (in terms of mass/body weight), independent of their shape, size, and mode of synthesis. Therefore, assessment of the dose (mass/kg)-response relationship characterizes the pharmacokinetics, toxicokinetics, and risk assessments of bulk particles in humans and animals (Goodman and Gilman, 2011). Unfortunately, the bulk particles' toxicity data cannot be applied to the nanoparticles, possibly because nanoparticles possess many novel properties not present in bulk particles. Nanoparticles exhibit (1) high surface area-to-volume ratio, (2) surface atoms influencing the particles' properties, and (3) size- and shape-dependent physicochemical properties, quantum confinement, (semiconductors), surface plasmon resonance (some metals), and superparamagnetism (magnetic particles) (Khanna and Verma, 2013; Nel et al., 2006; Chatterjee et al., 2003; Ramhan et al., 2003). Therefore, surface area and/or diameter may influence a nanoparticle's, but not bulk particle's, beneficial effects and toxicity (Gliga et al., 2014). In addition, the physicochemical properties of nanoparticles are also dependent on the procedure used for their synthesis (Ahmadi et al., 2011). This suggests that an understanding of the nanoparticles' structure and synthesis is needed to understand their beneficial and toxic effects.

One of the key aspects of determining public or environmental safety of a toxin is the *precautionary principle*, described as follows:

1. Whenever there is a possibility of harm to the environment or human health, some precautionary measures should be taken, even if the cause–effect relationships are not fully established.
2. Because the precautionary principle deals with risks that may not be credible, have unknown outcomes, and have poorly known probabilities, the unqualified possibility is sufficient to trigger an implementation of the precautionary principle. If the risks are well established and scientifically proven, then the preventive principle should be applied.

3. Possible effects that threaten the lives of future generations should be explicitly considered.
4. Interventions should be proportional to the chosen level of protection and the magnitude of possible harm.

The EU member states have accepted precaution as a general principle of environmental policy and have been undergoing implementation exercises. Several other countries, including Hungary and Brazil, have adopted precaution as a guiding principle. The United States has not adopted precaution as an explicit basis for environmental policy, even though it has ratified the Rio Declaration on Environment and Development, which obliges nations to exercise the precautionary principle. Unfortunately, the entire discussion is based on the safety of bulk particles. Although nanoparticles are not included in these studies, an assumption is made that the precaution principles developed for bulk particles will apply for nanoparticles. Generally, the US Environmental Protection Agency (EPA) does not consider a nanoscale version of a chemical to be a different chemical substance from the bulk version; thus, the nanoscale version would not be listed separately on the Toxic Substances Control Act Inventory. If a nanoscale substance is considered chemically the same as a bulk substance, then it would not require any special notice from the EPA. Because of this, the US cosmetic industry does not mention nanoparticles in their content list but list only the metals or metal oxides. The EPA's definition of nanoparticles forms the basis of the perception that the toxicity parameters established for bulk particles will be applicable to the nanoparticles. However, this is not the case—many nontoxic bulk metals become highly toxic in their nanoparticle form.

The toxic effects of bulk particles are defined by their dose–response relationships. Toxicologists, hygienists, public health professionals, and to a small extent, the general public are aware of the traditional physicochemical and biological properties of bulk hazardous substances. Yet, most people—including professionals—lack a good understanding of nanoparticle hazards. This is because toxicological risks from nanoparticles exposure are thought to be associated not only with dose but also with their size, shape, and surface functionalization.

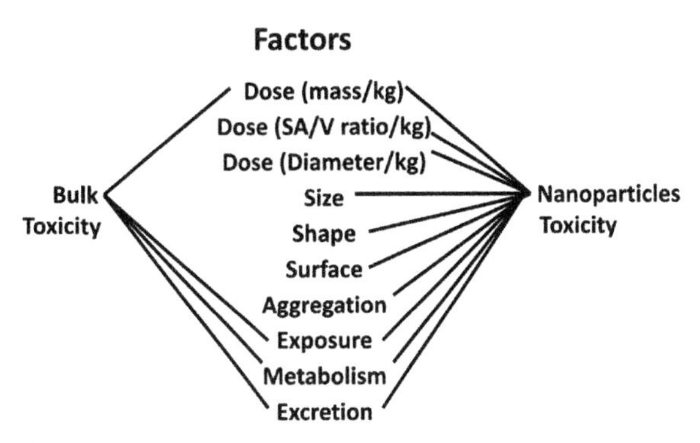

FIGURE 11 A list of factors modulating toxicity in bulk particles and nanoparticles.

Because of the complex confounding factors, as shown in Figure 11, the establishment of toxicokinetic models (including biodistribution predictions, data extrapolation, and property–biodistribution relationships) for nanoparticles is much more difficult than for bulk particles. In the following chapters, the structure, synthesis, and physicochemical properties of nanoparticles, as well as their relevance to the toxicity of nanoparticles, will be discussed.

References

Abdelhalim, M.A.K., Jarrar, B.M., 2011. Gold nanoparticle induced cloudy swelling to hydropic degeneration cytoplasmic hyaline vacuolation, polymorphism, binucleation, karyopyknosis, karyolysis, karyorrhexis and necrosis in the liver. Lipids Health Dis. 10, 166.

Abdelhalim, M.A.K., Jarrar, B.M., 2012. Histological alterations in the liver of rats induced by different gold nanoparticle sizes, doses and exposure duration. J. Nanobiotech. 10, 5.

Ahmadi, M., Ghasemi, M.R., Rafsanjani, H.H., 2011. Study of different parameters in TiO_2 nanoparticles formation. J. Mater. Sci. Eng. 5, 87–93.

Andersen, R.A., Burdick, J.W., Musallam, S., Pesaran, B., Cham, J.G., 2004. Cognitive neural prosthetics. Trends Cogn. Sci. 8, 486–493.

Angelucci, G., Bedin, K., Tirler, W., Donega, M., 2010. Ultra-fine particles in emissions of a municipal solid waste incinerator and wood. In: Proceedings Venice, Third International Symposium on Energy from Biomass and Waste. Venice, Italy, November 8–11, 2010.

Banholzer, M.J., 2011. Anisotropic Nanomaterials: Synthesis, Optical and Magnetic Properties, and Applications. ProQuest, UMI Dissertation Publishing.

Bertrand, N., Wu, J., Xu, X., Kamaly, N., Farokhzad, O.C., 2013. Cancer nanotechnology: the impact of passive and active targeting in the era of modern cancer biology. Adv. Drug Deliv. Rev. 66C, 2–25.

Braakhuis, H.M., Park, M.V.D.Z., Gosens, I., De Jong, W.H., Cassee, F.R., 2014. Physicochemical characteristics of nanomaterials that affect pulmonary inflammation. Part. Fibre Toxicol. 2014, 18.

Brun, N., Mazerolles, L., Pernot, M., 1991. Microstructure of opaque red glass containing copper. J. Mater. Sci. Lett. 10, 1418–1420.

Burrows, C.J., Muller, J.G., 1998. Oxidative nucleobase modifications leading to strand scission. Chem. Rev. 98, 1109–1152. http://www.jnanobiotechnology.com/sfx_links?ui=1477-3155-9-9&bibl=B13.

Buzea, C., Blandino, I.I.P., Robbie, K., 2007. Nanomaterials and nanoparticles: sources and toxicity. Biointerphases 2, MR17–MR172.

Caiger-Smith, A., 1991. Lustre Pottery. Technique, Tradition and Innovation in Islam and the Western World. New Amsterdam Books, New York, USA, ISBN 1-56131-030-1.

Carla, V., Filomena, A., Almeida, A.J., et al., 2011. The size of solid lipid nanoparticles: an interpretation from experimental design. Colloids Surf. B 84, 117–130.

Casagrande, C., Veyssié, M., 1988. Janus beads-realization and 1st observation of interfacial properties. C. R. Acad. Sci. 306, 1423–1425.

Charron, A., Harrison, R.M., 2003. Primary particle formation from vehicle emissions during exhaust dilution in the roadside atmosphere. Atmos. Environ. 37, 4109–4119.

Chatterjee, J., Haik, Y., Che, C.-J., 2003. Size dependent magnetic properties of iron oxide nanoparticles. J. Magnetism Magn. Mater. 257, 113–118.

Chen, J., Hessler, J.A., Putchakayala, K., Panama, B.K., Khan, D.P., Hong, S., Mullen, D.G., DiMaggio, S.C., Som, A., Tew, G.N., Lopatin, A.N., Baker Jr., J.R., Holl, M.M.B., Orr, B.G., 2009. Cationic nanoparticles induce nanoscale disruption in living cell plasma membranes. J. Phys. Chem. B 113, 11179–11185.

Colomban, P., March, G., Mazerolles, L., Karmous, T., Ayed, N., Ennabli, A., Slim, H., 2003. Raman identification of materials used for jewellery and mosaics in Ifriqiya. J. Raman Spectrosc. 34, 205–213.

Darlington, T.K., Neigh, A.M., Spencer, M.T., Nguyen, O.T., Oldenburg, S.J., 2009. Nanoparticle characteristics affecting environmental fate and transport through soil. Environ. Toxicol. Chem. 28, 1191–1199.

Dykman, L.A., Khlebtsov, N.G., 2011. Gold nanoparticles in biology and medicine: recent advances and prospects. Acta Naturae 3, 34–55.

EPA, 2012. Benchmark Dose Technical Guidance. EPA/100/R-12/001.

Fatima, Z., Aamrani, E.L., Kumar, A., Sastre, A.M., 1999. Kinetic modelling of the active transport of copper(II) across liquid membranes using thiourea derivatives immobilized on microporous hydrophobic supports. New J. Chem. 23, 517–523.

Gliga, A.R., Skoglund, S., Wallinder, I.O., Fadeel, B., Karlsson, H.L., 2014. Size-dependent cytotoxicity of silver nanoparticles in human lung cells: the role of cellular uptake, agglomeration and Ag release. Part. Fibre Toxicol. 2014, 11.

Goodman, L.S., Gilman, F., 2011. The Pharmacological Basis of Therapeutics, twelfth ed. McGraw-Hill, New York, ISBN 978-0-07-162442-8.

Gorham, J.M., MacCuspie, R.I., Klein, K.L., Fairbrother, D.H., Holbroo, R.D., 2012. UV-induced photochemical transformations of citrate-capped silver nanoparticle suspensions. J. Nanopart. Res. 14, 1139.

Green, A.A., Hersam, M.C., 2011. Properties and application of double-walled carbon nanotubes sorted by outer-wall electronic type. ACS Nano 5, 1459–1467.

Grosse, C., Delgado, A.V., 2010. Dielectric dispersion in aqueous colloidal systems. Curr. Opin. Colloid Interface Sci. 15, 145–159.

Gruner, G., Hu, L., Hecht, D., 2006. Graphene Film as Transparent and Electrically Conducting Material, Issued, US 7449133.

Gupta, A., Main, B.J., Taylor, B.L., Gupta, M., Whitworth, C.A., Cady, C., Freeman, J.W., El-Amin 3rd, S.F., 2014. In vitro evaluation of three-dimensional single-walled carbon nanotube composites for bone tissue engineering. J. Biomed. Mater. Res. A. http://dx.doi.org/10.1002/jbm.a.35088.

Henn, K.W., Waddill, D.W., 2006. Utilization of nanoscale zero-valent iron for source remediation–a case study. Remed. J. 16, 57–77.

Ho, M., Wu, K.Y., Chein, H.M., Chen, L.C., Cheng, T.J., 2011. Pulmonary toxicity of inhaled nanoscale and fine zinc oxide particles: mass and surface area as an exposure metric. Inhal Toxicol. 23, 947–956.

Jose, G.P., Santra, S., Mandal, S.K., Sengupta, T.K., 2011. Singlet oxygen mediated DNA degradation by copper nanoparticles: potential towards cytotoxic effect on cancer cells. J. Nanobiotechnol. 9, 9.

Judy, J.D., Unrine, J.M., Bertsch, P.M., 2011. Evidence for biomagnification of gold nanoparticles within a terrestrial food chain. Environ. Sci. Technol. 45, 776–778.

Khanna, L., Verma, N.K., 2013. Size-dependent magnetic properties of calcium ferrite nanoparticles. J. Magnetism Magn. Mater. 336, 1–7.

Kittelson, D.B., 1998. Engines and nanoparticles: a review. J. Aerosol Sci. 29, 575–588.

Lahiria, D., Subramanian, V., Bunker, B.A., Kamat, P.V., 2006. Probing photochemical transformations at TiO_2/Pt and TiO_2/Ir interfaces using X-ray absorption spectroscopy. J. Phys. Chem. 124, 204720.

Lia, H., Hana, P.D., Zhangb, X.B., Lib, M., 2013. Size-dependent melting point of nanoparticles based on bond number calculation. Mater. Chem. Phys. 137, 1007–1011.

Lianga, M., Zhi, L., 2009. Graphene-based electrode materials for rechargeable lithium batteries. J. Mater. Chem. 9, 5871–5878.

Lipovsky, A., Tzitrinovich, Z., Friedmann, H., Applerot, G., Gedanken, A., Lubart, R., 2009. EPR study of visible light-induced ROS generation by nanoparticles of ZnO. J. Phys. Chem. C 113, 15997–16001. http://www.jnanobio technology.com/sfx_links?ui=1477-3155-9-9&bibl=B16.

Loh, J.W., Yeoh, G., Saunders, M., Lim, L.-T., 2010. Uptake and cytotoxicity of chitosan nanoparticles in human liver cells. Toxicol. Appl. Pharmacol. 249, 148–157.

Lopez, T., Figueras, F., Manjarrez, J., Bustos, J., Alvarez, M., Silvestre-Albero, J., Rodriguez-Reinto, F., Martínez-Ferre, A., Martínez, E., 2010. Catalytic nanomedicine: a new field in antitumor treatment using supported platinum nanoparticles. In vitro DNA degradation and in vivo tests with C6 animal model on Wistar rats. Eur. J. Med. Chem. 2010 (45), 1982–1990.

Masciangioli, T., Zhang, W.-X., 2003. Nanotechnology could substantially enhance environmental quality and sustainability through pollution prevention, treatment, and remediation. Environ. Sci. Technol. 102A–108A.

Nel, A., Xia, T., Li, N., 2006. Toxic potential of materials at the nanolevel. Science 311, 622–627.

Noronha, F.S.M., Cruz, J.S., Beirao, P.S.L., Horta, M.F., 2000. Macrophage damage by *Leishmania amazonensis* cytolysin: evidence of pore formation on cell membrane. Infect. Immun. 68, 4578–4584.

Oberdörster, G., Oberdörster, E., Oberdörster, J., 2005. Nanotoxicology: an emerging discipline evolving from studies of ultrafine particles. Environ. Health Perspect. 113, 823–839.

Oberdörster, G., Sharp, Z., Atudorei, V., Elder, A., Gelein, R., Kreyling, W., Cox, C., 2004. Translocation of inhaled ultrafine particles to the brain. Inhal. Toxicol. 16, 437–445.

Oberdörster, G., Oberdörster, E., Oberdörster, J., 2007. Concepts of nanoparticle dose metric and response metric. Environ. Health Perspect. 115, A290–A294.

Pan, Y., Leifert, A., Ruau, D., Neuss, S., Bornemann, J., Schmid, G., Brandau, W., Simon, U., Jahnen-Dechent, W., 2009. Gold nanoparticles of diameter 1.4 nm trigger necrosis by oxidative stress and mitochondrial damage. Small 5, 2067–2076.

Patel, A., Prajapatil, P., Boghral, R., 2011. Overview on application of nanoparticles in cosmetics. Asian J. Pharm. Sci. Clin. Res. 1, 40–55.

Pathak, Y., Thassu, D., 2011. Drug Delivery Nanoparticles Formulation and Characterization, vol. 2. Taylor & Francis Group. Gruner, R. Aligned Carbon Nanotube–Oxide Nanoparticle Composites as Electrodes in Energy Storage Devices. UCLA, Tech ID: 22250/UC Case 2011–565–0.

Peer, D., Karp, J.M., Hong, S., Farokhzad, O.C., Rimona Margalit, R., Langer, R., 2007. Nanocarriers as an emerging platform for cancer therapy. Nat. Nanotechnol. 2, 751–760.

Perro, A., Reculusa, S., Ravaine, S., Bourgeat-Lami, E., Duguet, E., 2005. Design and synthesis of Janus micro- and nanoparticles. J. Mater. Chem. 15, 3745–3760.

Rahman, I.A., Vejayakumaran, P., Sipaut, C.S., Ismail, J., Chee, C.K., 2003. Size-dependent physicochemical and optical properties of silica nanoparticles. J. Magnetism Magn. Mater. 257, 113–118.

Raj, S., Jose, S., Sumod, U.S., Sabitha, M., 2012. Nanotechnology in cosmetics: opportunities and challenges. J. Pharm. Bioallied Sci. 4, 186–193.

Rhodes, R., Asghar, S., Krakow, R., Horie, M., Wang, Z., Turner, M.L., et al., 2009. Hybrid polymer solar cells: from the role colloid science could play in bringing deployment closer to a study of factors affecting the stability of non-aqueous ZnO dispersions. Colloids Surf. A Physicochem. Eng. Asp. 343, 50–56.

Ricciardi, P., Colomban, P., Tournié, A., Macchiarola, M., Ayed, N., 2009. A non-invasive study of Roman Age mosaic glass tesserae by means of Raman spectroscopy. J. Archaeol. Sci. 36, 2551–2559.

van Ravenzwaay, B., Landsiedel, R., Fabian, E., Burkhardt, S., Strauss, V., Ma-Hock, L., 2009. Comparing fate and effects of three particles of different surface properties: nano-TiO(2), pigmentary TiO(2) and quartz. Toxicol. Lett. 2009 (186), 152–159.

Sailor, M.J., Park, J.H., 2012. Hybrid nanoparticles for detection and treatment of cancer. Adv. Mater. 24, 3779–3802.

Salata, O.V., 2004. Applications of nanoparticles in biology and medicine. J. Nanobiotechnol. 2, 3.

Schulz, J., Hohenberg, H., Pflucker, F., Gartner, E., Will, T., Pfeiffer, S., Wepf, R., Wendel, V., Gers-Barlag, H., Wittern, K.-P., 2002. Distribution of sunscreens on skin. Adv. Drug Deliv. Rev. 54 (Suppl. 1), S157–S163.

Shen, Q., Nie, Z., Guo, M., Zhong, C.J., Lin, B., Li, W., Yao, S., 2009. Simple and rapid colorimetric sensing of enzymatic cleavage and oxidative damage of single-stranded DNA with unmodified gold nanoparticles as indicator. Chem. Commun. 28, 929–931. http://www.jnanobiotechnology.com/sfx_links?ui=1477-3155-9-9&bibl=B15.

Smijs, T.G., Pavel, S., 2011. Titanium dioxide and zinc oxide nanoparticles in sunscreens: focus on their safety and effectiveness. Nanotechnol. Sci. Appl. 2011, 95–112.

Smita, S., Gupta, S.K., Bartonova, A., Dusinska, M., Gutleb, A.C., Rahman, Q., 2012. Nanoparticles in the environment: assessment using the causal diagram approach. Environ. Health 11 (Suppl. 1), S13.

Taniguchi, N., 1974. On the Basic Concept of Nano-Technology. Proc. Intl. Conf. Prod. Eng. Tkyo. Part II, Japan Society of Precision Engineering.

Thess, A., Lee, R., Nikolaev, P., Dai, H., Petit, P., Robert, J., Xu, C., Lee, Y.H., Kim, S.G., Rinzler, A.G., Colbert, D.T., Scuseria, G.E., Tománek, D., Fischer, J.E., Smalley, R.E., 1996. Crystalline ropes of metallic carbon nanotubes. Science 273, 483–487.

Tiwari, D.K., Behari, J., Sen, P., 2008. Application of nanoparticles in waste water treatment. World Appl. Sci. J. 3, 417–433.

TSCA Nanoscale Materials Inventory Paper: Public Comments with EPA Responses at 2 (12/20/07), EPA-HQ-OPPT-2004-0122-0098, available at: www.regulation.gov, cited in EPA, Nanoscale Materials Stewardship Program, 73 Fed. Reg. 4, 861 (January 28, 2008).

Walther, A., Muller, A.H.E., 2008. Janus particles. Soft Matter 4, 663–668.

Wang, C., Zhang, W., 1997. Nanoscale metal particles for dechlorination of PCE and PCBs. Environ. Sci. Technol. 31, 2154–2156.

Warheit, D.B., Webb, T.R., Colvin, V.L., Reed, K.L., Sayes, C.M., 2007. Pulmonary bioassay studies with nanoscale and fine-quartz particles in rats: toxicity is not dependent upon particle size but on surface characteristics. Toxicol. Sci. 2007 (95), 270–280.

Warheit, D.B., Sayes, C.M., Reed, K.L., Swain, K.A., 2008. Health effects related to nanoparticle exposures: environmental, health and safety considerations for assessing hazards and risks. Pharmacol. Ther. 120, 35–42.

Wen-Tso, L., 2006. Nanoparticles and their biological and environmental applications. J. Biosci. Bioeng. 102, 1–7.

West, J.L., Halas, N.J., 2000. Applications of nanotechnology to biotechnology—commentary. Curr. Opin. Biotechnol. 11, 215–217.

Wilhelmi, V., Fischer, U., Weighardt, H., Schulze-Osthoff, K., Nickel, C., Stahlmecke, B., Kuhlbusch, T.A., Scherbart, A.M., Esser, C., Schins, R.P., Albrecht, C., 2013. Zinc oxide nanoparticles induce necrosis and apoptosis in macrophages in a p47phox- and Nrf2-independent manner. PLoS One 8, e65704.

Wittmaack, K., 2007. In search of the most relevant parameter for quantifying lung inflammatory response to nanoparticle exposure: particle number, surface area, or what? Environ. Health Perspect. 115, 187–194.

Wong, K.V., DeLeon, O., 2010. Applications of nanofluids: current and future. Adv. Mech. Eng. Article ID 519659, 11 pages.

Wood, V., Panzer, M.J., Caruge, J.M., Halpert, J.E., Bawendi, M.G., Bulović, V., 2010. Air-stable operation of transparent, colloidal quantum dot based LEDs with a unipolar device architecture. Nano Lett. 10, 24–29.

Wurm, F., Kilbinger, A.F.M., 2009. Polymeric janus particles. Angew. Chem. Int. Ed. 48, 8412–8421. http://dx.doi.org/10.1002/anie.200901735.

Yadav, V., Makwana, M., Kamble, S., Qureshi, F., Sarmalkar, B., Salve, D., 2012. Different Au-content in Swarna bhasma preparations: evidence of lot-to-lot variations from different manufacturers. Adv. Appl. Sci. Res. 3, 3581–3586.

Yang, H., Morris, J.J., Lopina, S.T., 2004. Polyethylene glycol-polyamidoamine dendritic micelle as solubility enhancer and the effect of the length of polyethylene glycol arms on the solubility of pyrene in water. J. Colloid Interf. Sci. 273, 148–154.

Yang, L., Gordon, V.D., Mishra, A., Som, A., Purdy, K.R., Davis, M.A., Tew, G.N., Wong, G.C., 2007. Synthetic antimicrobial oligomers induce a composition-dependent topological transition in membranes. J. Am. Chem. Soc. 129, 12141–12147.

Zhang, W.-X., 2003. Nanoscale iron particles for environmental remediation: an overview. J. Nanopart. Res. 5, 323–332.

CHAPTER

2

Structure, Synthesis, and Application of Nanoparticles

OUTLINE

Engineered Nanoparticles
http://dx.doi.org/10.1016/B978-0-12-801406-6.00002-9

1. INTRODUCTION

Engineered nanoparticles are widely used in numerous industrial, environmental, electronic, and medical applications. Using selective functional groups, nanoparticles can be directed to an organ, tissue, or tumor using internal-signaling or an external cue (e.g., magnetic field, ultrasound, photons). However, nanoparticles can also nonspecifically bind to proteins, enzymes, antibodies, or nucleotides, resulting in modulation of the cell signaling process. For medicinal applications, nanoparticles must conform to a size smaller than 100 nm, have uniform physicochemical, and be nontoxic and biocompatible. Magnetic nanoparticles can be heated in alternating magnetic fields for use in hyperthermia-based tumor treatment. However, the nanoparticles' physicochemical properties, beneficial effects, and toxicity depend upon their size, surface properties, and mode of synthesis. A small change in size causes an exponential change in the properties and biological activities of nanoparticles. Therefore, a clear understanding of the nanoparticles' structure and synthesis is important to understand their structure and mode of synthesis. The aim of this chapter is to discuss the structure, synthesis, and application of different classes of nanoparticles. Each section is divided into two subsections that discuss (1) structure and synthesis and (2) applications.

2. METAL, SEMICONDUCTOR, AND QUANTUM DOT NANOPARTICLES

Metal, semiconductor, and quantum dot (QD) nanoparticles have been widely used as functional materials in physics, chemistry, and biology. They exhibit great chemical diversity with unique beneficial and adverse effects that may be related to their structure and synthetic procedure, as described in the following sections.

2.1 Structure and Synthesis

2.1.1 Metal Nanoparticles

2.1.1.1 STRUCTURE OF METAL NANOPARTICLES

Metals are located on the left side and the middle of the periodic table. Group IA and Group IIA metals are alkali metals, while group IB-VIIIB metals are transition metals (less reactive than alkali metals). The elements to the right of the transition metals are basic metals. One of the common properties of metals is their ability to donate electrons; thus, they are good electrical and heat conductors. They are also generally malleable and ductile. The valence electrons of transition metals are present in multiple orbitals, and they exhibit several common oxidation states. Iron (Fe), cobalt (Co), and nickel (Ni) generate a magnetic field (Chatterjee et al., 2003). Most metal nanoparticles are colloids suspended in a continuous dispersion phase. Because of the perpetual motion, nanoparticles in dispersion medium constantly collide with other nanoparticles and/or particles of the dispersion media. This improves dispersion of metal nanoparticles in a solution. When the light passes through a colloidal solution, its path becomes visible because of light scattering. The intensity of the scattered light depends on the difference between the refractive indices of the dispersed phase and the dispersion medium.

Structurally, the unique thermodynamic properties confer amorphous characteristics to the metal nanoparticles. Martin et al. (1991) and Martin (1996) have demonstrated the existence of anomalous stable conformations of metal nanoparticles called "magic number" nanoparticles (where the so-called magic number is defined as the discrete number of atoms that correspond to the formation of the energetically most favorable clusters). The ambient stability of Ag magic number clusters may be a function of size, where the smaller magic number clusters could be the most stable (Figure 1, Desireddy et al., 2013).

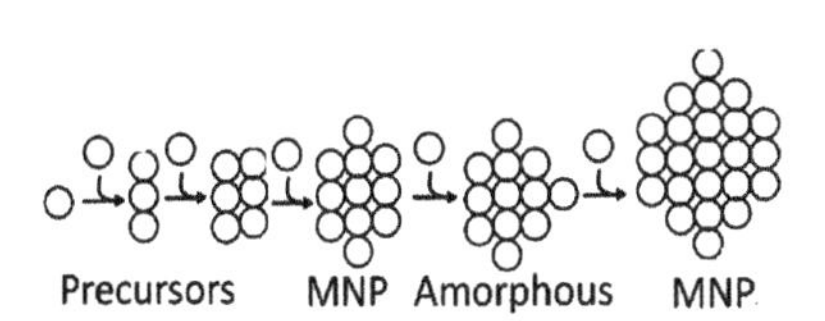

FIGURE 1 **The concept of magic number and nanoparticle stability.** Clusters are formed by the incorporation of monomers in to a central atom. Initially, they are unstable and amorphous. As they reach a magic number, they acquire a stable structure called "magic number" particles (MNPs). If monomer units are further added to the MNP, it again become amorphous and unstable until another magic number is reached. Calculation of the number of atoms/molecules in nanoparticles is described in Appendix 1.

2.1.1.2 SYNTHESIS OF METAL NANOPARTICLES

Metal nanoparticles are synthesized using top-down or bottom-up approaches (Chaki, 2002; Mafune et al., 2001; Raffi et al., 2007; Rosemary and Pradeep, 2003; Shukla and Seal, 1999). The bottom-up approach uses molecular components that are built-up into complex assemblies. The top-down approach uses microfabrication techniques where externally controlled tools cut, mill, and shape materials into the desired shape and size. Some of the commonly used physical and chemical methods are described here.

2.1.1.2.1 BOTTOM-UP METHODS Faraday et al. (1857) pioneered chemical synthesis of metal nanoparticles by reducing metal salts using a phosphorus and carbon disulfide mixture. At present, many reagents (e.g., citrate, hydrazine, $NaBH_4$, boranes, polyols, tetraoctylammonium bromide) are available for reduction of metal salts into colloids (Aslam et al., 2002; Chen et al., 2006b; Clavee et al., 1999; Kapoor et al., 2002; Panigrah et al., 2004; Shirtcliffe et al., 1999; Tan et al., 2004; Van Hyning and Zukoski, 1998; Wang et al., 2005a,b; Zhu et al., 2005). A number of green biosynthesis approaches using plant reducing agents have been developed for synthesis of magnetic nanoparticles (Iravani, 2011; Jana et al., 2001; Sharma et al., 2009). The hydroxyl and carbonyl groups present in plant extracts act as reducing as well as stabilizing agents.

In traditional reductive synthesis, the newly synthesized metal nanoparticles are highly reactive and aggregate rapidly into larger particles as well as undergo oxidation during synthesis (Datsyuka et al., 2008). To prevent this, the newly synthesized metal nanoparticles must be stabilized during or immediately following their synthesis. The Brust–Schiffrin method provides simultaneous synthesis and stabilization of metal nanoparticles (Figure 2) (Brust et al., 1994, 1995; Jorgensen et al., 2005; Lee et al., 2012; Li et al., 2011a,b; Zaluzhna et al., 2012). The method yields relatively high monodispersity and size control using a molar ratio of thiolated functional groups and metal salts (Perez-Dieste et al., 2003; He et al., 2004; Tao et al., 2008). The first step of the Brust–Schiffrin method is conversion of Au(III) into a [TOA][Au(III)X_4] complex by the surfactant tetraoctylammonium bromide (TOAB). The [TOA][Au(III)X_4] complex, being water insoluble, undergoes phase transfer from the aqueous solution to the toluene phase (Figure 2).

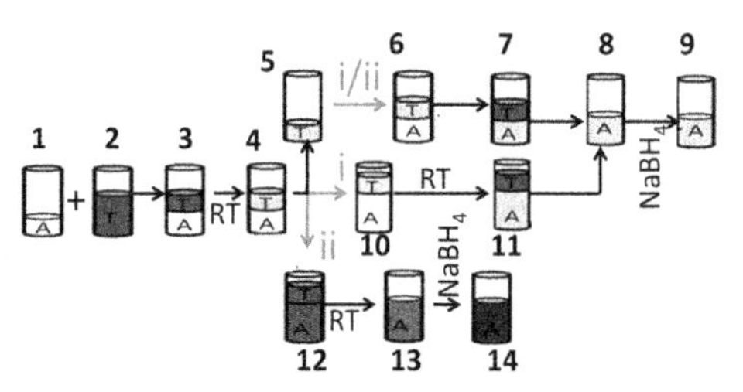

FIGURE 2 Synthesis of metal nanoparticles by using a one- or two-phase Brust–Schiffrin procedure. Nanoparticles are synthesized in three steps: (i) the common step for formation of toluene soluble precursor, (ii) one-phase synthesis, (iii) two-phase synthesis, and (iv) common reduction procedure. For the common step, metal salt (0.142 M, 0.14 ml, tube-1) and different concentrations (0–0.06 g) of tetra-*n*-octylammonium bromide (TOAB) (tube-2) are mixed and incubated at room temperature (RT) until the toluene layer becomes colorless (tube-4). Then, the mixture is split into three aliquots. For one-phase synthesis, aliquot-1 is centrifuged and the toluene phase containing a single precursor (for gold it is [TOA][AuX_4]) is collected (tube-5) and mixed with $C_{12}H_{25}SH$ (low concentration: <2 Eq or high concentration: ≥2 Eq) (tube-6). The tubes were mixed and incubated at room temperature. After centrifugation, the aqueous sample (tube-7) containing the soluble precursor [TOA][AuX_{n-1}] is collected (tube-8) and subjected to the reduction procedure described later (tube-8). For two-phase synthesis aliquot-2 was mixed with <2 Eq of $C_{12}H_{25}SH$ (i tube-10), incubated at room temperature, and processed as described for tube 7. The low concentration of thiol resulted in the formation of nanoparticles comparable to the one-phase synthesis. Aliquot-3 is mixed with high concentrations of (≥2 equivalent) of $C_{12}H_{25}SH$ (tube-12). This step resulted in the formation of a mixture of [TOA][AuX_4] and $[AuSR]_n$ that depends on the $C_{12}H_{25}SH$ concentration. For reduction, tubes 9 and 13 are stirred for 1 h followed by addition of 1 ml of the fresh reducing agent, $NaBH_4$ (10 mg/ml water) and the mixture was stirred for 3 h. The nanoparticles formed are precipitated with ethanol and then microfiltered. Nanoparticles prepared in aliquots 1 and 2 exhibited smaller variations (1.4 ± 0.16 nm) than nanoparticles prepared in aliquot 3 (2.6 ± 0.4 nm). (Procedure described in Brust et al., 1994).

Thereafter, the toluene phase is collected and used for one-phase synthesis, while the toluene-water mixture is used for two-phase synthesis of metal nanoparticles. Addition of the thiol ligand (dialkyl disulfide or dialkyl diselenide) partially reduces Au(III) and forms nanoparticle precursors. A single precursor is formed when the toluene phase is reduced (hence the term *one-phase synthesis*), while two precursors were formed when the toluene-water mixture was reduced (hence the term *two-phase synthesis*) with thiol ligands (Figure 2). The next step is complete reduction of Au cations to AuO using $NaBH_4$ aqueous solution, resulting in the formation of metal nanoparticles (Figure 2).

2.1.1.2.2 TOP-DOWN NANOFABRICATION
The top-down methods include nanoimprint, soft and step and flash lithography (Chou et al., 1996; Guo, 2007; Koo, 2008; Gilles et al., 2011; Prabhakaran et al., 2006; Kik et al., 2002), thermal embossing (Heyderman et al., 2000), particle replication in nonwetting templates (Maynor et al., 2006; Desimone et al., 2012), solvent molding-based fabrication (Martirosyan and Luss, 2009), ultraviolet (UV) embossing (Navaladian et al., 2008), and focused ion beam and nanodispension methods (Lian et al., 2005; Wang et al., 2005b). All of the listed methods exhibit nanometer resolution. Currently, nanoimprint lithography, including step flash-imprint lithography (S-FIL), particle replication in nonwetting templates, and solvent molding-based fabrication, are the most powerful methods for fabricating polymer nanocarriers of specific shape, size, and aspect ratios.

2.1.2 *Paramagnetic Metal Nanoparticles*

2.1.2.1 STRUCTURE OF PARAMAGNETIC NANOPARTICLES

Paramagnetic metal nanoparticles have an iron oxide core coated by inorganic materials (silica, gold), organic materials such as phospholipids, fatty acids, polysaccharides, peptides or other surfactants, and polymers (Laurent et al., 2008; Thorek et al., 2006). The unpaired electrons in paramagnetic nanoparticles, in the absence of a magnetic field, remain disorganized and do not exhibit magnetism. In applied magnetic fields, the electrons align in the direction of the field lines; thus, the nanoparticles gain magnetic moments (discussed in detail in Chapter 3). A repulsive force exists perpendicular to the aligned magnetized particles. The attractive force depends on the number of particles in the chain. The inducible magnetization allows them to be used for many applications, including as contrast-enhancing agents for magnetic resonance imaging (MRI), drug delivery systems, magnetic hyperthermia for local heat sources in the case of tumor therapy, and magnetically assisted transfection of cells (Banobre-Lopez et al., 2013; Besenius et al., 2012; Weinstein et al., 2010).

2.1.2.2 SYNTHESIS OF PARAMAGNETIC NANOPARTICLES

Hyeon (2003) and Murray et al. (1993) have described a simple procedure for synthesis of nanostructured T1 MRI contrast. Particle growth is controlled by Oswalt ripening (Mateo-Mateo et al., 2012) in which smaller nanoparticles dissolve and deposit on the bigger nanoparticles. Decreasing reaction temperature stopped the nanoparticle growth. Further size-selection processes can narrow down the particle size distribution below 5%. The iron oxide core provides excellent image contrast properties for MRI monitoring as well as for thermal ablation, providing a valuable tool for locating and destroying tumors.

2.1.3 Porous and Hollow Metal Nanoparticles

2.1.3.1 STRUCTURE OF POROUS AND HOLLOW METAL NANOPARTICLES

Metal nanoparticles with pores (micropores $\leq$2 nm, mesopores 2 to $\geq$50 nm, macropores >50 nm), or hollow metal nanoparticles (metal nanocapsules) have generated considerable interest over the past few years. These nanoparticles exhibit high loading capacity, functionalization for both inner and outer walls, high chemical and thermal stabilities, large surface area, permeability, and good biocompatibilities (Feng et al., 2008; Kim et al., 2002; Liu et al., 2004; Yuan et al., 2010). Porous nanoparticles have potential applications in catalysis, drug carriers, biosensors, prosthetic materials, gas adsorbents, and heavy metal ion adsorbents (Kim et al., 2002). Mesoporous silica nanoparticles are comprised of a honeycomb-like mesoporous structure that is able to encapsulate relatively large amounts of bioactive molecules. Unlike porous nanoparticles, hollow particles contain empty space that can be filled with active ingredients.

2.1.3.2 SYNTHESIS OF POROUS AND HOLLOW METAL NANOPARTICLES

2.1.3.2.1 POROUS SILICA In 1971, Chiola et al. (1971) filed a patent for preparation of low-density porous silica nanoparticles. Unfortunately, the method remained unnoticed for almost 20 years, until Beck et al. (1991) developed and patented an ordered mesoporous silica nanoparticle (MCM-41, an acronym for Mobil's Composition of Matter). The nanoparticle developed by Chiola et al. (1971) met all the criteria established for MCM-41. In 1992, Kresge et al. (1992) published a surfactant-template based synthesis of mesoporous silica materials. Thereafter, the synthesis and functionalization of nanoparticles for various applications, such as catalysis, medicine, and sensors, have been described. In general, porous nanoparticles are synthesized either (1) directly by condensation of oppositely charged metal and surfactant moieties or (2) indirectly by similarly charged metal and surfactant moieties and an oppositely charged halide ion Cl^- or Br^-, as shown in Figure 3 (Huo et al., 1994). A combination of different reagents may be useful in the

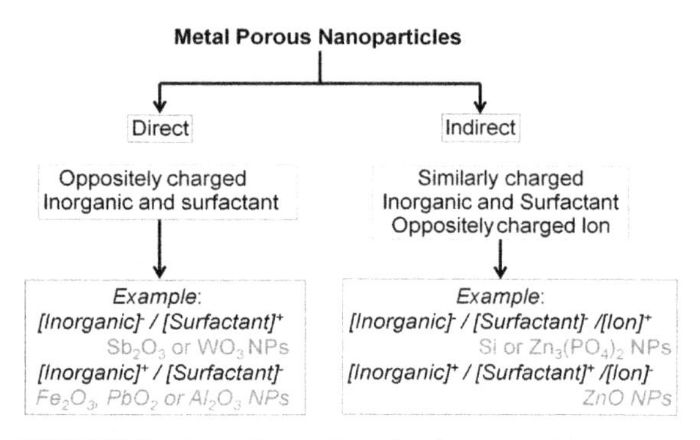

FIGURE 3 Overall procedures for the synthesis of porous nanoparticles.

formation of metal mesoporous nanoparticles, as shown in Figure 3 in red.

2.1.3.2.2 HOLLOW SILICA NANOCAPSULES

Specific templates, such as latex particles (Caruso et al., 1998; Yang et al., 2008), emulsion droplets (Fujiwara et al., 2004; Zoldesi and Imhof, 2005), and block copolymer micelles (Khanal et al., 2007), are commonly used for synthesis of hollow silica nanoparticles. Hu et al. (2013) devised a one-step method using water/oil inverse immersion and templates consisting of ammonia water stabilized in alkyl phenol polyoxyethyleneether, tetraethoxysilane, and cyclohexane. Lu et al. (2013a) described a method for synthesis of hollow nanoparticles in three simple steps:

1. A spherical SiO_2 template was synthesized by the hydrolysis of tetraethyl orthosilicate.
2. The TiO_2 shell was coated by hydrolysis of tetrabutyl titanate for formation of the SiO_2@TiO_2 nanosphere.
3. SiO_2@TiO_2 nanospheres were annealed at 500 °C followed by NaOH treatment that completely removed the inner SiO_2 core, leaving the well-defined anatase TiO_2 hollow sphere.

The hollow nanoparticle can be easily converted to a nanocup as needed (Lu et al., 2013a,b)

2.1.4 Semiconductor Nanocrystals and Quantum Dots

2.1.4.1 STRUCTURE OF SEMICONDUCTOR NANOCRYSTALS

There are three basic types of semiconductor nanoparticles:

1. *Intrinsic*, in which the number of electrons in the conduction band is equal to the number of "holes" in the Highest Occupied Molecular Orbital (e.g., Si, Ge, Ga, and SdS nanoparticles)
2. *N-type* doped, in which the number of electrons in the conduction band is greater than the number of holes in the HOMO (intrinsic nanoparticles doped with As or P)
3. *P-type* doped, in which the number of electrons in the conduction band is smaller than the number of holes in the HOMO (intrinsic nanoparticles doped with B, Ge, and Al)

In general, semiconductor nanoparticles can be classified according to their shape, including, but not limited to, ordered nanowires, nanorods, nanobelts, nanotubes, nanohelixes, and nanorings (Panda and Tseng, 2013; Panigrahi and Basak, 2011; Cho et al., 2010; Kang et al., 2009; Ra et al., 2009; Hsueh et al., 2007; Lao et al., 2006; Yi et al., 2005). Different types of semiconductor nanoparticles have potential applications in photodetectors, light-emitting devices, solar cells, biological labeling, medical diagnostics, etc. (Wang et al., 2007; Sun et al., 2003; Tessler et al., 2002; Colvin et al., 1994). These applications are facilitated by the nanoparticles' unique optical and electronic properties, which depend on the size, shape, and crystal phase (Li and Wang, 2003; Sun et al., 2003; Kim et al., 2006). Therefore, fabrication of semiconductor nanocrystals with well-defined shapes and desired crystal structures in a controlled manner is of key importance in designing and tailoring the properties of semiconductor nanocrystals. MnS nanoparticles are extensively used p-type semiconductors with wide band energy (3.7 eV), varied crystal structures, and promising novel magneto-optical properties (Puglisi et al., 2010; Tian et al., 2009; Zhang et al., 2008b; Kim et al., 2006; Joo et al., 2003; Tappero et al., 2001; Furdyna and Samarth, 1987). The n-doped nanoparticles are being used in the area of energy conversion and environmental cleanup, but their use in medicine is not fully known.

2.1.4.2 STRUCTURE OF QUANTUM DOTS

Quantum dots are semiconductor nanocrystals with a core–shell structure and a diameter that typically ranges from 2 to 10 nm (Medintz et al., 2005). The shell is usually ZnS, while the core usually composed of elements from groups II–VI, such as CdSe, CdS, or CdTe; groups III–V, such as InP or InAs; or groups IV–VI, such as PbSe (Cassette et al., 2013; Jamieson et al., 2007; Michalet et al., 2005). This core–shell structure confers QDs the status of an artificially fabricated atom in which charge carriers are confined in all three dimensions, similar to the electrons in real atoms. Quantum dots therefore exhibit properties (quantized energy levels and shell structures) described by the electron wave functions whose evolution is governed by the Schrödinger equation and Pauli's exclusion principle.

2.1.4.3 SYNTHESIS OF SEMICONDUCTOR NANOCRYSTALS AND QUANTUM DOTS

2.1.4.3.1 TPOP/TOP PROCEDURE FOR NANOCRYSTALS The commonly used procedure for synthesis of semiconductor nanoparticles (ZnO as an example) consists of two steps (Muray et al., 1993):

1. First is the synthesis of a single source precursor, such as *bis (benzoate) zinc* (reaction between benzoic acid and diethyl zinc) for ZnO or *bis (dialkylamidedithiocarbamate) zinc* (reaction between diethylamine and carbon disulfide and diethyl zinc) for ZnS nanoparticles.
2. The precursor is subjected to thermal decomposition in the presence of trioctylphosphineoxide (TPOP) and trioctylophosphine (TOP) for 1–20 nm nanoparticles. The TPOP/TOP procedure has been used extensively for various semiconductors.

The nanocrystals produced are hydrophobic in nature; thus, further functionalization may be needed to enhance their biocompatibility.

2.1.4.3.2 THE CO-PRECIPITATION METHOD FOR NANOCRYSTALS In co-precipitation, fractional precipitation of a specified ion results in the precipitation not only of the target ion but also of other ions existing side by side in the solution (Noboru et al., 1992). Some of the most commonly used substances used in co-precipitation are chlorides, hydroxides, carbonates, sulfates, and oxalates. Reactions (1) and (2) show the synthesis of ZnO nanoparticles using $ZnSO_4$ as the starting material. A carbonate buffer at pH 4.6 is added dropwise into the starting solution to produce a $Zn_4(CO_3^{2-})_3(OH)_2$ precipitate (reaction (1)):

$$4(Zn^{2+} + SO_4^{2-}) + 4(Na^+CO_3^-) + HOH \rightarrow Zn_4(CO_3^{2-})_3(OH)_2 + 4(Na^{2+} + SO_4^{2-}) \quad (1)$$

After addition of the buffer solution, the mixture is stirred for 2 h at room temperature. The precipitates are filtered, washed with distilled water, dried in an oven at 110 °C for 2 h, and heated (300 °C–1000 °C) in a muffle furnace. The basic zinc carbonate precipitate is decomposed in zinc oxide, white powder (reaction (2)):

$$Zn_4(CO_3^{2-})_3(OH)_2 \rightarrow 4ZnO + H_2O + 3CO_2 \quad (2)$$

Reaction (3) shows the synthesis of iron nanoparticles, Fe_3O_4, prepared by mixing ferric chloride and ferrous chloride in a 2:1 M ratio. The mixture is heated up to 50 °C for 10 min. Then, the solution is precipitated by ammonia solution (final pH >8.0) with continuous stirring with the magnetic stirrer at 50 °C. Black-colored particles of iron oxide are precipitated. These particles are then separated from the solution using a strong magnet and then washed many times with distilled water. The precipitated magnetite is black in color. The powder is then dried in hot air oven at 100 °C overnight.

$$Fe^{2+} + 2Fe_3{}^{+} + 8OH^{-} \rightarrow Fe_3O_4 + 4H_2O \quad (3)$$

A typical example of a co-precipitation procedure for synthesis of Cd nanoparticles begins by mixing the starting material $CdCl_2$, NH_4Cl, and thiourea in a 1:3:7 M ratio in two stages. In stage 1, $CdCl_2$ and NH_4Cl with a molar ratio 1:3 were dissolved in distilled water (pH 9.0 adjusted using ammonium solution). In stage 2, an aqueous solution of thiourea was prepared. Finally, the solution containing thiourea was injected into the reaction matrix dropwise. Particle size control in the deposited layer was attained by adding mercaptoethanol as a surfactant. The synthesis process was performed at three different temperatures: 35 °C, 50 °C, and 65 °C. The chemical reaction is shown below:

$$CdCl_2 + 2H_2O \rightarrow Cd(OH)_2 + 2HCl \quad (4)$$

$$\begin{aligned} &Cd(OH)_2 \rightarrow Cd_2{}^{+} + 2(OH)\text{-}2(NH)_2CS \\ &+ OH^{-} \rightarrow 2CH_2N_2 + H_2O + HS\text{-}HS^{-} \\ &+ OH^{-} \rightarrow H_2O + S^{2-} \end{aligned} \quad (5)$$

$$Cd2^{+} + S^{2-} \rightarrow CdS \quad (6)$$

Co-precipitation provides a simple method for synthesis of nanocrystals, provided their solubility curves are known. The method provides excellent solubility and size control.

2.1.4.3.3 SYNTHESIS OF QUANTUM DOTS The synthesis procedure for QDs varies according to their core components, buffers, and the synthesis solution used. Danek et al. (1994), Talapian et al. (2001), and Qu and Peng (2002) developed an organometallic method to produce highly fluorescent CdSe dots with a quantum efficiency of approximately 50%. Qu and Peng (2002) found that if the Cd:Se ratio was increased to 1:10, the fluorescent quantum yield increased to 80%. Wang et al. (2010) synthesized InP QDs from a high-temperature organic solution system, and then transferred the QDs into an aqueous phase through a ligand-exchange process using trichloro-*s*-triazine modified mPEG.

Huang et al. (2013) and Ma et al. (2009) have used alloyed bichalcogenide nanocrystals of type $ME_xE'_y$ (M = metal, E and E′ = different chalcogen atoms, $x + y \approx 1$) such as ZnSeS, CdSeS, and CdSeTe in a one-step synthesis of QDs. The photoluminescence can be tuned through the visible range by varying the alloy composition. The reactions involved are shown below. In these reactions, L is oleic acid and P is TOP, TOPS, or TOPSe. Reaction (7) forms CdSe nuclei, while reaction (8) forms CdS nuclei. Reaction (9) is the crystal growth to form $CdSe_xS_y$ nanocrystals.

$$Cd(L)_n + TOPSe \rightarrow (CdSe)(L)_m(P)_p \quad (7)$$

$$Cd(L)_n + TOPS \rightarrow (CdS)(L)_m(P)_p \quad (8)$$

$$\begin{aligned} &X(CdSe)(L)_m(P)_p \rightarrow Y(CdS)(L)_m(P)_p \\ &\rightarrow CdSe_xS_y(L)_m(P)_p \end{aligned} \quad (9)$$

As mentioned above, the selection of a method for the synthesis of QDs may depend upon the study's requirements and the solvents used.

2.1.5 Functionalization of Metal, Semiconductor, or Quantum Dot Nanoparticles

2.1.5.1 FUNCTIONALIZATION FOR IMPROVED DISPERSION AND DISSOLUTION

Bare metal nanoparticles agglomerate rapidly (Figure 4(A)), possibly due to attractive van der Waals (VDW) forces (Appendix 2) (Laaksonen et al., 2006), low zeta (ζ) potential, and/or the

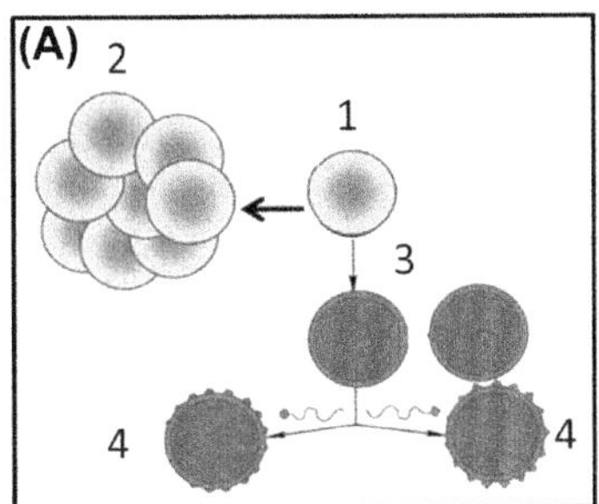

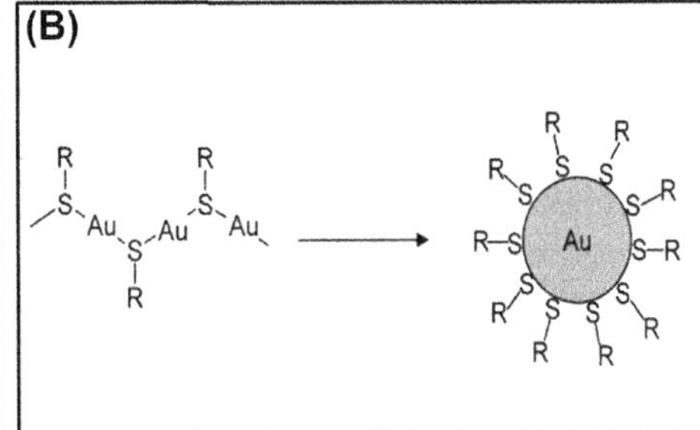

FIGURE 4 (A) Bare (nonfunctionalized) nanoparticles (1) are highly reactive and aggregate rapidly into large amorphous particles (2). PEGylation (3) added hydrophilicity and steric hindrance, resulting in the formation of dispersed particles (4). (B) Metal nanoparticles form strong and stable noncovalent bonds with thiol containing polymers or proteins.

pH (close to the nanoparticles' isoelectric points for "zero charge") of the dispersion liquid (Appendix 3). The ζ values greater than +25 mV or less than −25 mV typically indicate a stable colloid because of charge repulsion between two nanoparticles. An increase in ζ values stabilizes the nanoparticles, resulting in greater dispersion (Appendix 3). One of the procedures commonly used to stabilize colloidal nanoparticles is coupling the surface with a thiol group either during or after synthesis. Thiols form a noncovalent but stable bond with gold nanoparticles (Figure 4(B)) (Itoh et al., 2004). Studies have shown that mono-thiol ligands form relatively weak bonds with gold nanoparticles, especially at high temperatures, in the presence of competing thiols or oxidizing agents and in solvents where the ligand is extremely well solvated (Bhatt et al., 2011; Chompooeor et al., 2008; Klajn et al., 2007; Herdt et al., 2006; Hong et al., 2006; Kassam et al., 2006; Song et al., 2003; Song and Murray, 2012). A multivalent interaction between ligands with multiple thiol groups and the surface of gold nanoparticles may prevent ligand dissociation under extreme conditions (Kang and Taton, 2005). The coupling of the nanoparticle surface with an organic layer consisting of poly-L-lysine (PLL), poly-L-glycone (PLG), or polyethylene glycol (PEG) reduces the surface reactivity and improves dispersion in solution. Functionalized nanoparticles exhibit high bioavailability, evade the immune system, selectively release active ingredients on-site, and perform special functions such as biological switches.

PEG, being water soluble and nontoxic, has an advantage over PLL, which exhibits high cytotoxicity, and PLG, which exhibits poor water solubility. PEG functionalization (PEGylation) stabilizes the nanoparticles by adjusting their hydrophilic and hydrophobic properties and increasing their *zeta* potential and steric hindrance. PEG-functionalized nanoparticles, due to their ability to bind the cell membranes, can serve as excellent drug carriers. PEGs coupled with biomolecules such as lectin, lactose, and biotin have been used in cellular and intracellular targeting of biological materials. The stability and internalization of PEGylated nanoparticles may be affected by factors such as the molecular weight of PEG, the attached functional groups, the ligand, and the size of the nanoparticles used. In general, PEGylation has following advantages:

- High-molecular-weight PEGs may remain in the blood circulation for a longer period than low-molecular-weight PEGs ($t_{1/2}$ increased from 18 min to 1 day as the PEGs' molecular weight increased from 6000 to 190,000) (Lankveld et al., 2011).
- PEGylated nanoparticles accumulate in the tissues/organs, such as muscle, skin, bone,

and the liver, irrespective of the molecular weight.
- Small PEG, but not large PEG, tended to freely translocate from the circulation to extravascular tissues and vice versa.
- Urinary clearance decreased with increasing PEG molecular weight. Liver clearance increased with the increasing PEG molecular weight. PEG uptake by Kupffer cells was enhanced as the molecular weight became greater than 50,000 (Baumann et al., 2014).
- Attachment of large PEG moieties often reduces activity of the drug, and higher concentrations of the conjugate are necessary to achieve the required biological activity.

Permanent PEGylation is generally not applicable to small-molecule drugs because the bulky carrier usually prevents their binding to targets and cell penetration.

The beneficial effects of PEG ($-[CH_2-CH_2-O-]n-$) are attributed to its unusual structure and properties (Knop et al., 2010; Lin and Anseth, 2009). PEG is water soluble but also exhibits hydrophobic properties (Figure 5). Poly(butylene glycol), $-[CH_2-CH_2-CH_2-O-]n-$, having one more methylene (CH_2) group and poly(methylene oxide), $-[CH_2-O-]n-$, which has one less methyl group, are both hydrophobic and insoluble in water. Thus, there is something unique about PEG's $-CH_2-CH_2-O-$ unit. Possibly, the ethyl group yields sufficient hydrophobic properties to this hydrophilic molecule. PEG forms thin monolayers at the air–water interface that have distinct hydrophilic and hydrophobic regions. Water molecules bind to the ethylene oxide group via H-bonds to the $-O-$ group of polyethylene oxide. Another unique property of PEG is that its solubility in water depends on its molecular weight and, to a lesser extent, on the concentration. This is one of the least understood aspects of PEG's properties. While low-molecular-weight PEG generally induces cells or vesicles to adhere, high-molecular-weight PEG causes them to repel.

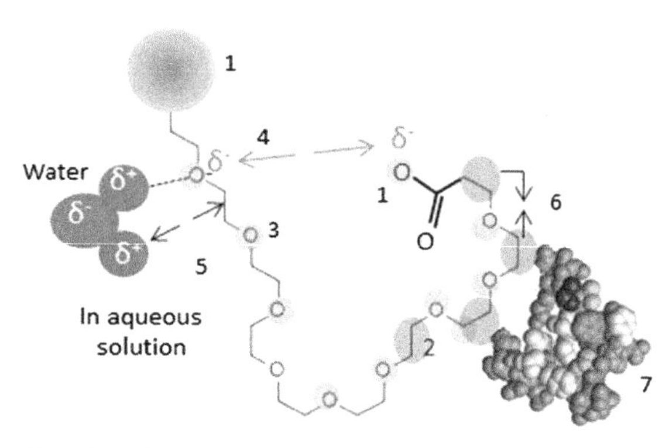

FIGURE 5 PEG's hydrophilic and hydrophobic sites. (1) metal nanoparticles, (2) hydrophobic sites, water with δ^- oxygen and δ^+ hydrogen that form hydrogen bonds with PEG, (3) PEG's δ^- sites, (4) hydrophilic repulsion, (5) hydrophobic-hydrophilic repulsion, (6) hydrophobic attraction, (7) hydrophobic interaction between PEG's and the protein's hydrophobic sites.

Bare metal nanoparticles, in addition to undergoing aggregation, also bind to biological proteins, especially plasma proteins. Protein adsorption on nanoparticles involves bond formation between proteins and surfaces, lateral diffusion on the surface, and conformational changes or rearrangements of adsorbed proteins (Ngadi et al., 2009). Driving forces for protein adsorption are hydrophobic interaction, electrostatic attraction, VDW, and hydrogen bonding. Aggregation and protein binding both facilitate their binding and subsequent engulfment into the macrophages (Saba, 1970; Owens and Peppas, 2006), and ensuing nanoparticles' hepatic clearance (Knop et al., 2006). This results in considerably lower residence time of bare nanoparticles in systemic blood, resulting in lower bioavailability (Figure 6). PEGs block the binding of proteins onto the nanoparticles' surface (Figure 7) by forming hydrogen bonds with water molecules (Figure 5), thus generating large repulsive forces. Blocking the adsorption of proteins on to nanoparticles also blocks their phagocytosis by macrophages, resulting in an increase in their circulation time. PEG chain length, conformation, and number density on the

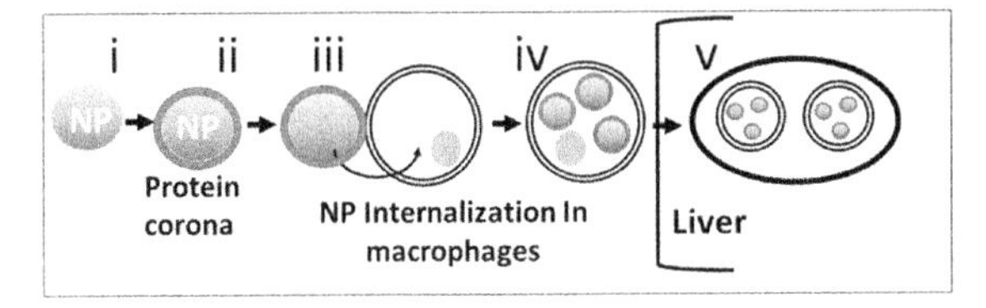

FIGURE 6 Bare metal nanoparticles (i) bind different proteins (ii) that facilitated their binding (iii) and internalization into the macrophages (iv). The structural changes in the macrophages facilitated their uptake in liver for further clearance (v).

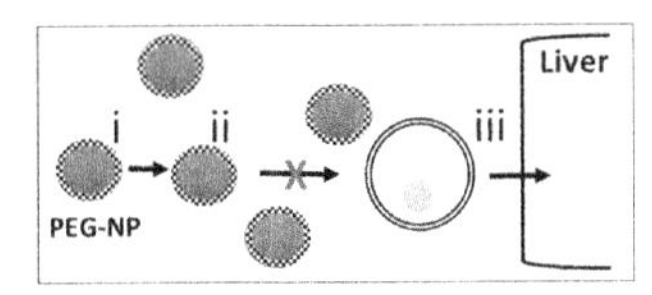

FIGURE 7 PEGylated metal PEGs (i) do not bind proteins, thus they are not engulfed by the macrophages (ii) or cleared via the liver (iii). This increases the half-life of nanoparticles.

surface are important factors for resisting protein adsorption. However, the hydrophobic sites in the PEG may produce attractive forces on proteins (Figure 5).

Macrophages are one of the first responders against foreign particles including the nanoparticles (others being bacteria, toxins, etc.). Studies have shown that intratumoral complement activation increases the concentration of complements 3 and 5 (C3 and C5, respectively) that may participate in cancer pathogenesis (Krishnan et al., 2010; Syrovets et al., 2012; Tabata and Ikada, 1990). Therefore, nanoparticle PEGylation may, in addition to suppressing phagocytosis, also suppress the complement concentrations, thus enhancing the potency of anticancer drugs. Contrary to the above observations, several studies have shown that the steric hindrance of the PEG chains (or other nonionic surfactants) on nanoparticles could not prevent activation of the complements (Al-Hanbali et al., 2006; Gbadamosi et al., 2002; Hamad et al., 2008; Moghimi, 2006; Moghimi and Moghimi, 2008; Moghimi and Murray, 1996; Moghimi and Szebeni, 2003; Reddy et al., 2007). More research may be needed to resolve this issue of the beneficial and harmful effects of PEGylated nanoparticles.

2.1.5.2 FUNCTIONALIZATION OF PEG FOR MEDICINAL/SCREENING APPLICATIONS

PEGylated metal nanoparticles, for performing specific functions, require specific end groups ($-NH_2$, $-COOH$, $-OH$, $-N_3$ and/or $-SH$) that interact, via either covalent or noncovalent bonding, with drugs, imaging dyes, or antibodies of specific proteins. For metal nanoparticles, PEGs bi-functionalized with $-SH$ groups at one end and $-NH_2$, $-COOH$, $-OH$ or $-SH$ groups at the other can be synthesized as described below (Figure 8).

2.1.5.3 ACID OR ENZYME CLEAVABLE LINKERS

Acid cleavable linkers remain stable at physiological pH but disintegrate at pH less than 6.0 (Aryal et al., 2009; Guarise et al., 2005; Santi et al., 2012). The enzyme-cleavable linkers contain an enzyme-selective substrate as a linker that is hydrolyzed by the enzyme. For cancer treatment, the cathepsin B substrates, such as Gly-Phe-Leu-Gly (GFLG), are used as the linker. Because cathepsin B is highly overexpressed in cancer cells, the cathepsin B substrate-linkers are disintegrated, resulting in an onsite release of drugs and/or dyes (Mai et al., 2000; Calderon et al., 2009). Figure 9 shows functionalization of gold nanoparticles with PEG (or other linker) coupled to a drug (Figure 9) or precoupled to an antibody (containing a fluorescence dye (red circle)-quencher (blue circle)) (Figure 10) via an enzyme-cleavable hydrazine linker. An important advantage of the acid-cleavable linker is that the released drug has full activity and is unencumbered by the bulky

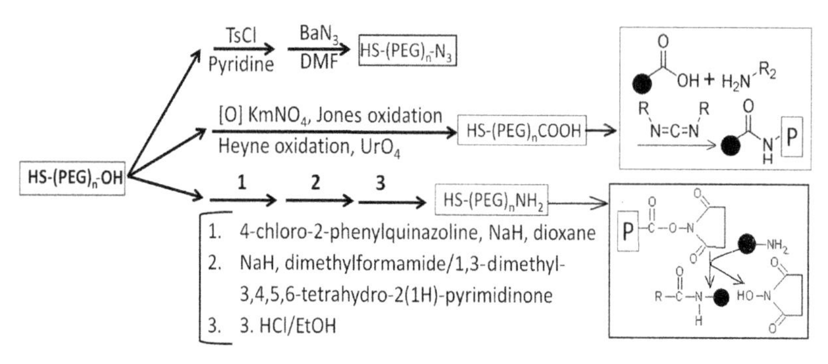

FIGURE 8 Conversion of hydroxyl-PEG into PEGs having other end groups (Barbara, 1968; Sun and Pelletier, 2007; Organic Chemistry Portal (http://www.organic-chemistry.org/namedreactions/)). However, the end groups are not highly reactive, except the –SH group that spontaneously interacts with metal colloidal nanoparticles and forms a stable bond (Love et al., 2005). Therefore, thiol-, amino-, and carboxylic-PEGs must be activated for further reactions using agents such as maleimide, NHS ester, and carbodiimide, respectively. P: protein and antibodies (Roberts et al., 2002).

macromolecular carrier. Studies have shown that the plasma and the extracellular fluid of normal tissues/cells have pH of around 7.4, while normal lysosomes and the extracellular fluid of cancer cells have pH of around <6 (Park et al., 1999; Wojtkowiak et al., 2012). The pH-sensitive probes remain intact at normal physiological pH but disintegrate at acidic pH, thus allowing onsite release of the drugs. One of the commonly used acid-cleavable linkers is hydrazine linker. Figure 9 shows the mechanism behind the pH-dependent release of ligands attached to the nanoparticle via a hydrazine linker. This approach allows the

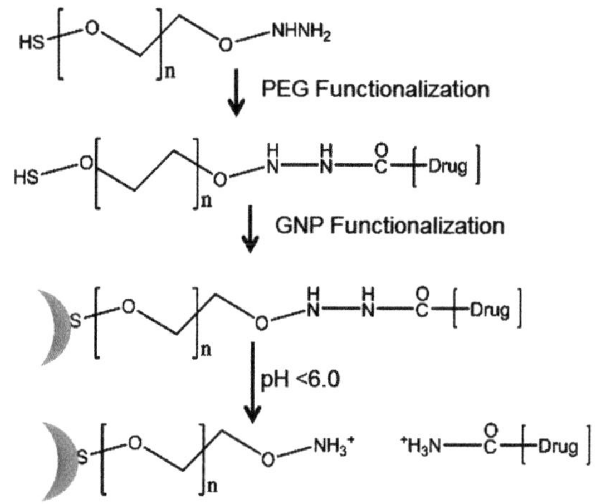

FIGURE 9 A simple procedure for synthesis of functionalized metal nanoparticles coupled to a drug via acid-cleavable linkage. At acidic pH, the linker dissolves resulting in the release of the loaded drug.

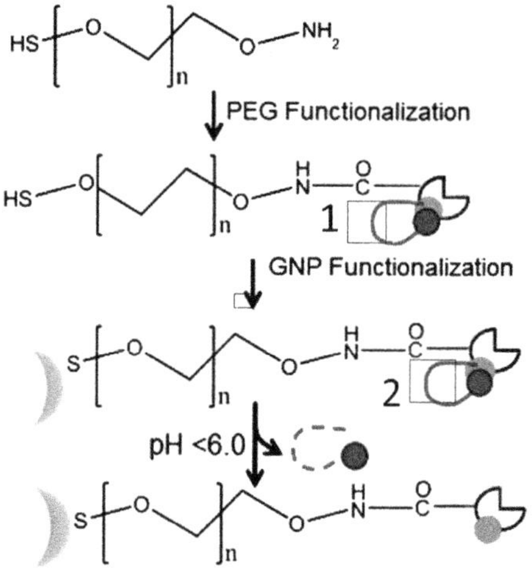

FIGURE 10 A simple procedure for synthesis of functionalized metal nanoparticles coupled to a fluorescent dye and a quencher linked via an enzyme-cleavable linker. (1) cathepsin substrate GFLG and (2) hydrolysis site is the G–F bond that releases the quencher from the probe resulting in fluorescing of the tumor.

establishment of the desired drug residence time in the body.

2.1.5.4 METAL NANOPARTICLES FUNCTIONALIZED WITH TUNABLE SWITCHES

Traditionally, metal nanoparticles are functionalized using thiolated ligands in a reducing environment. This procedure, although effective, may not be compatible with a wide range of intelligent functional groups, such as tunable switches, including a bifunctional stable electron donor–acceptor functionalization. Klajn et al. (2010a,b) devised novel functionalization of nanoparticles in which multiple components are not chemically bonded but cannot dissociate because of their topological linkage (Figure 11). There is much freedom of mechanical movement in the catenane rings to produce various co-conformers. These functional groups can play a key role in acting as switches to detect oxidative stress or other disorders.

Klajn et al. (2009a,b,c) functionalized metal (Au, Pt, Pd) nanoparticles using weakly protected "precursors" and dithiolanes into a redox-active mechanically interlocked molecules that retained their "switching" activity when attached to the nanoparticles. The oxidation potentials of the switches can be modulated by the properties of the metal nanoparticle surfaces. Many studies have reported DNA catenane systems with potential use in DNA topological labeling (Hua et al., 2013; Johann et al., 2013). To synthesize single-strand DNA catenanes, two DNA strands are synthesized and then cyclized using either enzymatic template-directed ligations or photo-crosslinking. Sannones and Sugiyama (2010, 2012) have developed a G-quadruplex method for the cyclization to avoid the introduction of modified bases. The DNA G-quadruplex is of great interest because of its roles in key biological processes, such as the maintenance of telomeres and regulation of gene transcription.

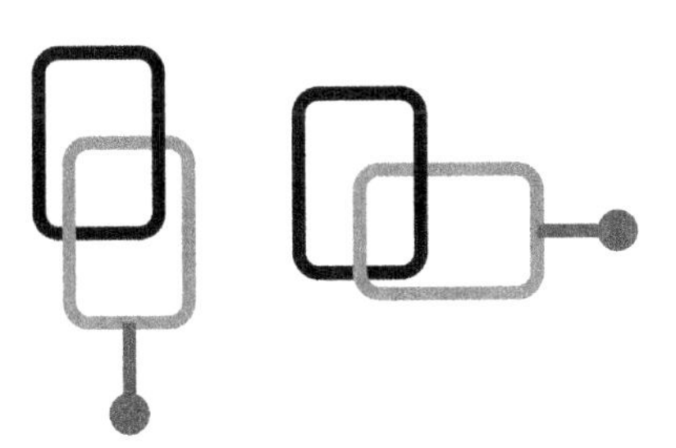

FIGURE 11 Structure of nanoparticles that can be used as tunable switches. These nanoparticles have multiple components that are not chemically bonded, but cannot dissociate because of their topological linkage. There is much freedom of mechanical movement in the catenane rings to produce various co-conformers (Lu et al., 2013). These functional groups can play a key role in acting as switches to detect oxidative stress or other disorders.

2.1.5.5 NONCOVALENT FUNCTIONALIZATION

Functional groups bind to nanoparticles via covalent bonds (shared electrons, single bond–sigma, double bonds–one sigma one π, triple bond–one sigma and two π) and relatively weak noncovalent interactions arising due to formation of hydrogen bonds, ionic interactions, and VDW interactions, such as London dispersion, dipole-related interactions, and π-electron-mediated interactions (Appendix 3).

2.1.6 Nanoparticle Characterization

Characterization refers to the study of materials' features such as its composition, structure, and various properties such as physical, electrical, magnetic, etc. Characterization is essential for nanoparticles ranging from 1 to 100 nm in diameter because their properties vary significantly with size and shape. Therefore, accurate measurement of a nanoparticle's size and shape is critical to its applications. The techniques commonly used for nanoparticle characterization are discussed in Chapter 4.

2.2 Applications of Metal Nanoparticles

2.2.1 Environmental Applications

Surface-modified nanoparticles are used for removal of arsenic and other toxic metals, as well as setting up a chemical reactive barrier to treat deep groundwater contaminants (Abid et al., 2013; EPA, 2003; Zhang, 2003). In some cases, such as groundwater remediation, bimetallic catalyst nanoparticles may be more effective than monometallic nanoparticles (Abid et al., 2013; EPA, 2003; Ponder et al., 2000, 2001; Schrick et al., 2002; Zhang and Elliott, 2006). Palladium-on-gold nanoparticles (Pd/Au NPs) are more effective than Pd-nanoparticles or Au-nanoparticles in catalyzing the dechlorination of trichloroethene in water at room temperature (Qian et al., 2014). Hydrophobic nanoparticles called submarines effectively removed oil from water sources (Guix et al., 2012). Modified magnetically enhanced separation technology separated the target contaminants from a sediment matrix or wastewater streams. These particles have a magnetic core, a shell that provides stability and protection from oxidation, and a surface to which contaminant-specific ligands are attached (http://web.mit.edu/cheme/news/seminars-11/Parekh.pdf). These observations suggest that metal nanoparticles may play a key role in environmental remediation.

2.2.2 *Biomedical Applications*

Metal nanoparticles are used in biomedical imaging, drug delivery, and high-throughput sensors.

2.2.2.1 IMAGING

Monomodal imaging using monofunctionalized nanoparticles (particles containing dyes, QDs, or guanidinium-functionalized nanoparticles and super-paramagnetic iron oxide) are common in diagnostic imaging. However, multifunctional imaging allows the combined diagnosis and therapy of cancer, thus visualizing the tumor to assess recovery in vivo. Huang and Hainfeld (2013) have described a biodegradable nanopolymer, poly (lactic-*co*-glycolic acid) (PLGA), functionalized with the following groups:

- Magnetic nanoparticles and doxorubicin for multimodal imaging and drug delivery
- Galactosylated chitosan (GC) that selectively targeted the hepatocarcinoma-selective glycoprotein receptors of hepatocarcinoma and allowed the nanoparticles' endocytosis within the carcinoma
- EGFP-N_2 (green fluorescence protein plasmid) that acted as contrast agent for fluorescence imaging

The engineered nanoparticles allowed multimodal resonance imaging, hyperthermia-induced cancer cell destruction, and selective degradation of the nanoparticles for on-site drug release (Banobre-Lopez et al., 2013). Cheng et al. (2013) developed a multimodal ultrasound-triggered phase transition nanoparticles called DiR-SPIO-ND (Figure 12) that exhibited significant ultrasound-triggered phase transition due to the vaporization of the pentafluoropentane (PFP) layer. The in vivo T_2-weighted images of MRI as well as the fluorescence images showed an enhancement in contrast in the liver and spleen of rodents after intravenous injection

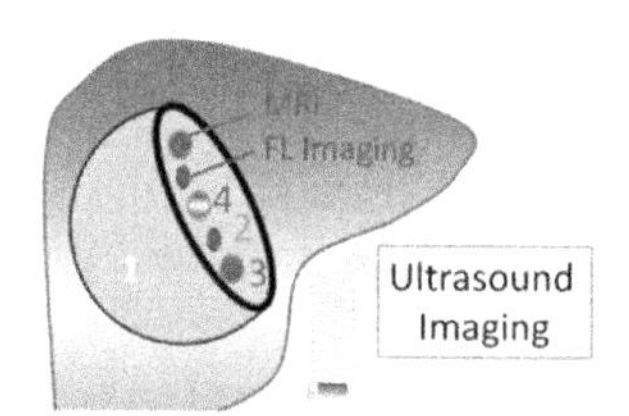

FIGURE 12 Structure of DiR-SPIO nanoparticle in which the core of the double emulsion pentafluoropentane (boiling point 29 °C) was encapsulated by the polymer shell of poly(lactic-*co*-glycolic acid) and poly(ethylene glycol) [PLGA-PEG-PLGA)PLGA-PEG-PLGA] (12-1) and polyvinyl alcohol (12-2). Hydrophobic super-paramagnetic iron oxide (SPIO, 12-3) nanoparticles were dispersed in the PLGA-PEG-PLGA layer, and DiR (near-infrared fluorescent dye 1,1′-dioctadecyl-tetramethyl indotricarbocyanine iodide) fluorescent dyes (12-4) in the PLGA-PEG-PLGA and PFP layers. DiR-SPIO-ND administration results in accumulation of this nanoparticle in liver. Upon ultrasound irradiation, the PVA moiety vaporizes, resulting in the release of the imaging agents and the drugs in vivo into the cancer cells. Carcinoma was imaged using fluorescence and MRI imaging in live subjects. Described by Cheng et al. (2013).

of DiR-SPIO-NDs. Furthermore, the ultrasound imaging in mice tumor as well as MRI and fluorescence imaging in liver of rats and mice showed that the DiR-SPIO-NDs had long-lasting contrast ability in vivo. This suggests that DiR-SPIO-NDs could potentially be a great MRI/fluorescence multimodal imaging contrast agent in the diagnosis of liver tissue diseases.

Earlier studies have described Gd^{3+}-doped upconversion nanoparticles (UCNPs) that converted near-infrared (NIR) radiations with lower energy into visible radiations with higher energy via a nonlinear optical process (Wang et al., 2011). Some of the commonly used UCNPs are the following:

1. Gd_2O_3: Thulium (Tm) + Ytterbium (Yb) that upconverts NIR wavelength to 455 nm
2. Gd_2O_3: Erbium (Er) + Yb that upconverts NIR to 564 nm
3. Er^{3+} in Gd_2O_3: Er + Yb that upconverts NIR to 661 nm (Bogdan et al., 2010; Guo et al., 2004; He et al., 2013; Naccache et al., 2009; Thanh-Dinh et al., 2010)

He et al. (2013) designed an upconversion nanoparticle coupled to a dual-functional PEG and two layers of poly(ethylenimine) (PEI) as a gene-delivery vector for the transfection of plasmid DNA encoding enhanced green fluorescent protein. When serum is added during transfection, it shows remarkably enhanced transfection efficiency, possibly due to enhanced receptor-mediated endocytosis via the binding of serum proteins (Figure 13).

The PEG-PEI_2 upconversion nanoparticles may be a promising gene vector that works well in the presence of serum proteins. The nanoparticles enter the cells via endocytosis and then deliver the plasmid into the nucleus, where the plasmid integrates into the cells' genome and expresses the encoded protein. The cells can be imaged via upconversion fluorescence imaging. Traditional fluorescence imaging using fluorescent dyes or QDs is a downconversion technique in which the emitted wavelength is of lower energy than the excitation wavelength. The NIR-to-visible upconversion methods have a number of advantages. Autofluorescence, a serious problem

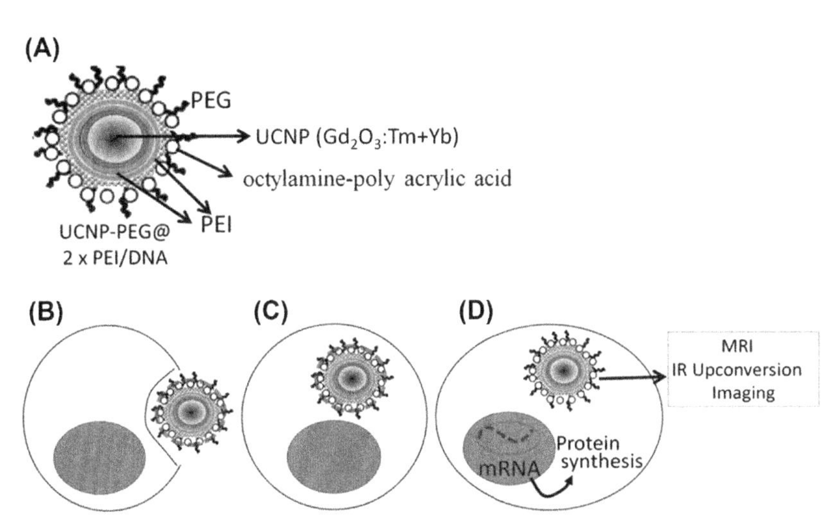

FIGURE 13 Gd-doped (example: Gd_2O_3:Tm + Yb) upconversion nanoparticle consisting of PEG, two layers of PEI and one layer of PA-PAA (A). UCNP enters the cells via endocytosis (B/C) where it releases the plasmid into the nucleus for induction of protein synthesis. The transfected cells can be imaged using IR upconversion imaging (D).

associated with the traditional imaging, is no longer an issue for upconversion imaging. This allows improved signal-to-noise ratios and imaging sensitivity (Yu et al. 2004a,b). NIR irradiation with better tissue penetration ability is used as the excitation source, facilitating in vivo imaging. However, UCNPs are resistant to photobleaching and have excellent photostability (Waynant et al., 2001; Yi et al., 2006).

2.2.3 *Drug Delivery*

The traditional allopathic drugs, although highly effective, have certain disadvantages, such as nonspecific binding to blood cells and proteins, rapid diffusion into the healthy tissues, short half-life in the bloodstream, and high overall clearance rate (Brown et al., 2011; Ehdaie, 2007). This reduces their efficacy in the body because relatively small amounts of the administered drug reach the target site. Distribution into healthy tissues may lead to severe side effects that often make many excellent drugs undesirable. Nanoparticles can be designed to circumvent the disadvantages of traditional drug formulation by improving their bioavailability, transporting them to the target site, and then releasing them as needed (Ghosh et al., 2013; Gibson et al., 2007; Singh et al., 2013; Zhao et al., 2013). The nanoparticles' performance as drug vectors depends on the size and surface functionalities of the particles, drug release rate, and particle disintegration. Figure 14 shows a hypothetical nanodrug that (1) simultaneously transports drugs to the disease site, releases the load on demand, and provides multimodal imaging; (2) improves bioavailability by enhancing aqueous solubility; (3) increases the residence time in the body by increasing half-life and/or decreasing their phagocytosis; (4) allows a reduction in the quantity of the drug required; and (5) does not cause any toxicity (Gupta et al., 2014; Irving, 2007). A transporter selective ligand may allow the drug carrier systems to pass through organ barriers, such as the blood–brain barrier (Abhilash, 2010).

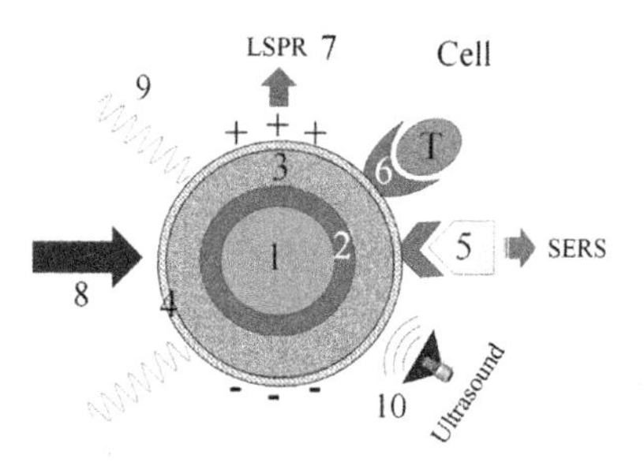

FIGURE 14 Multimodal functionalization of metal nanoparticles. The nanoparticle core (1) is hybridized with magnetic nanoparticles (2) for MRI. A silica nanoporous nanoparticles (Agbabiaka et al., 2013) is loaded with the drugs (3) in a pH, ultrasound, or infrared radiation (10) dependent manner. The porous structure is damaged either in acidic pH in lysosomes/tumor site or in the presence of ultrasound waves or infrared radiation, thus releasing the drugs. The surface is functionalized with PEG (4) coupled to a surface-enhanced Raman spectroscopy ligand or SERS ligand (5) and a transporter ligand (6) that, upon binding to a cell, facilitates the particle's internalization. When activated with photons (8), the particles may reflect light (9) or exhibit localized surface plasmon resonance (7) or LSPR. This nanoparticle will provide selective drug transport to the target site, on *demand* drug release and imaging for immediate verification.

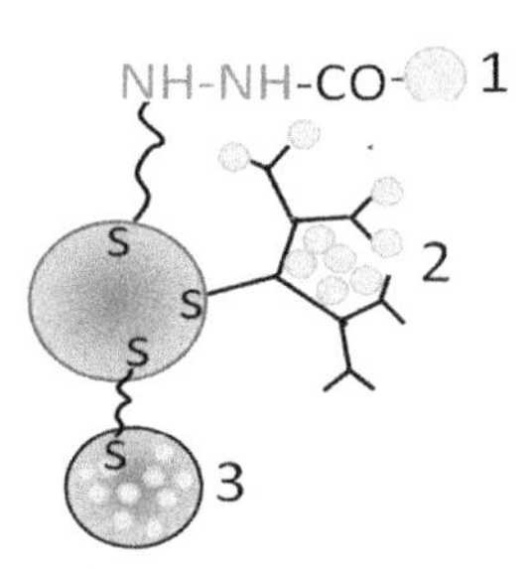

FIGURE 15 Drug loading of metal nanoparticles. Metal nanoparticles form a strong bond with the thiol-terminated chemicals. Thus, drug-loaded PEG (1), dendrimers (2), or gold capsules (3) can be attached to the nanoparticles via the thiol linkage.

One of the unique characteristics of certain transition metals, especially gold, is their strong affinity for thiol groups (Love et al., 2005; Weisbecker et al., 1996; Wolcott et al., 2006), which provides a simple means to load the drug-coupled PEG (Figure 15(1)) or drug-loaded dendrimers (Figure 15(2)) onto the metal nanoparticles. As an alternative approach, studies have described a photothermal modulated drug-delivery system consisting of hollow metal nanocapsules covered by a thermosensitive hydrogel matrix. These nanocapsules strongly absorb NIR light and release multiple amounts of many soluble materials held within the hydrogel matrix (Figure 15(3)).

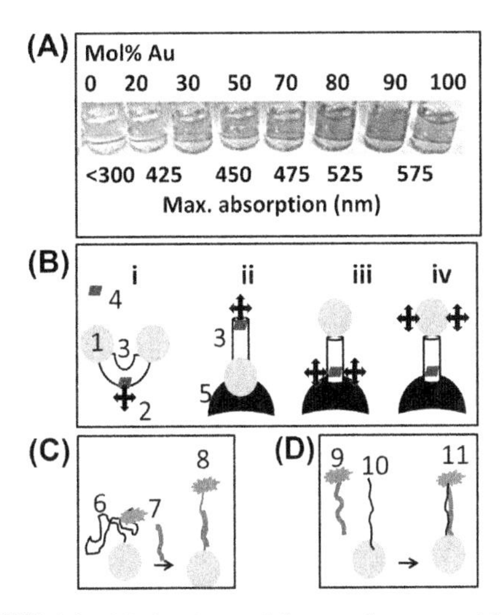

FIGURE 16 Mechanisms of the metal nanoparticle-based biosensing. (A) Application of localized surface plasmon resonance (LSPR) in biosensing. This method is based on the principle that an increase in the amount of metal shifts the LSPR-absorption band of metal nanoparticles to longer wavelengths (Baptista et al., 2011; Doria et al., 2010). Thus, quantification can be performed by scanning the light absorption curve. (B) Application of surface enhanced Raman spectroscopy in biosensing using nucleotides or antibodies as the probes. Nanoparticles are attached to the selective probes that recognize and bind to antigens or target-species containing Raman-active moiety. Different assay formats include (i) free (1: nanoparticles, 2: Raman-active dye, 3: linker, 4: analyte and 5: anchor base), (ii) anchored nanoparticles with Raman-active dye binding to the target, (iii) anchored nanoparticles with Raman-active dye binds to the anchor, and (iv) anchored nanoparticles with Raman-active dye binding to the nanoparticle (Vo-Dinh et al., 2005). (C) Fluorescence-based sensing of analytes. The nanoparticle fluorescence is quenched in the hairpin probe (6), but, upon binding with the target nucleotide (7), the probe unwinds, resulting in the development of fluorescence (8). (D) The target is labeled with a fluorescent dye (9) and fluorescence is measured. Upon binding the probe (10), fluorescence gets quenched (11), resulting in the loss of fluorescence intensity (Jennings et al., 2006; Griffin et al., 2009).

2.2.4 Metal Nanoparticle-Based Sensors

Metal nanoparticle-based biosensors exhibit high sensitivity for nucleic acids, proteins, antibodies, enzymes, and other biological molecules. The key detection procedures are localized surface plasmon resonance (LSPR), fluorescence enhancement/quenching, surface-enhanced Raman scattering (SERS), and electrochemical activity (Doria et al., 2010, 2012). The colorimetric approaches, due to their simplicity and portability, are one of the most promising methods for diagnosis at the point-of-care. Gold, silver, and platinum nanoparticles have been extensively used in biosensors. Figure 16(A–D) describes some of the commonly used approaches in screening by metal nanoparticle sensors.

3. CARBON NANOTUBES AND FULLERENES

3.1 Structure and Synthesis

3.1.1 Structure of CNTs and Fullerenes

CNTs and fullerenes are allotropes of carbon. Fullerenes are closed-caged structures consisting of hexagonal and pentagonal rings (Figure 17(A)). In C_{60} fullerene, each carbon is connected to three other carbon atoms by one double bond and two single bonds (Figure 17(B)). Electronically, this arrangement is referred to as sp^2-carbons because the three adjacent carbons are hybrids of the 2s orbital and the two 2p orbitals ($2p_x$ and $2p_y$). The remaining 2p orbital ($2p_z$) is responsible for the π-bond.

CNTs have diameters ranging from <1 to 4 nm and length-to-diameter aspect ratios up to 132,000,000:1. CNTs are one of the most massively manufactured nanoparticles (approximately 270 tons/year) worldwide (www.bccresearch.com/market-research/nanotechnology/carbon-nantubes-markets-technologies-nan024e.html). CNTs exhibit chemical and physical properties not found in the bulk allotropes of carbon: diamond, graphite, or amorphous carbon.

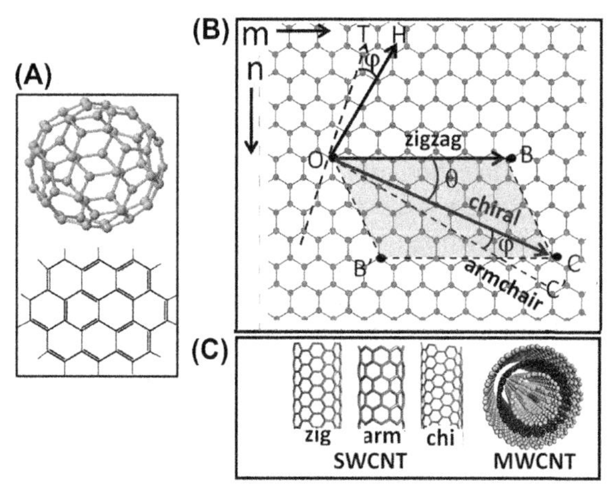

FIGURE 17 Structure or graphene, fullerenes, and carbon nanotubes. (A) Graphical representation (top) and chemical composition (bottom) of fullerenes. (B) Structure and folding of graphene (a one-atom thick planar sheet) consisting of sp^2-bonded carbon atoms. The carbon–carbon bond length in graphene is about 1.42 Å. Line OC is the chiral vector (Ch), line OH is the chiral translational axis (T), and θ is $<30°$ (but $>0°$) for chiral CNTs. The shaded rectangle is the unit cell for this CNT. The lines OT and OC′ represent Ch and T, respectively, and θ is 0° for armchair CNTs. Zigzag CNT's θ is 0°. Thus, folding patterns of the graphene sheets determine their metallic and semi-conductor nanotubes. (C) Structure of single-walled CNTs (SWCNTs) and multiple-walled CNTs (MWCNTs). Abbrevations: zig, zigzag; arm, armchair; chi, chiral SWCNTs.

CNTs have 100 times the tensile strength of steel, thermal conductivity comparable to diamond, and electrical conductivity greater than copper. These unique characteristics make CNTs a promising candidate for applications in microelectronics, medicine, hydrogen fuel cell technology, etc. However, consistent with earlier incidences of industrial pollutants including, but not limited to, freon, methyl-t-butyl ether flame retardants, and surfactants, there is high probability that unregulated production of nanoparticles, including CNTs, may lead to serious environmental problem and ensuing health issues.

There are three basic types of CNTs currently being produced: single-walled carbon nanotubes (SWCNTs), multiwalled carbon nanotubes (MWCNTs), and carbon nanowires (Harris, 2009; Saito et al., 1998; Weia and Srivastava, 2004; Avila et al., 2008). SWCNTs are rolled-up graphene sheets having an electronic makeup closely related to graphite (three-coordinated carbons in which three electrons are in sp^2 hybridization and one is delocalized) than to diamond (four-coordinated carbons with sp^3 hybridization). The intrinsic strength of the sp^2 carbon–carbon bonds may result in high mechanical strength and a very large Young's modulus, 50 times higher than that of steel (Chapter 3). As shown in Figure 17(B), the properties of SWCNTs are determined by the graphene sheet's folding patterns. The graphene sheet folding is represented by a pair of indices (n, m) that denote the number of unit vectors along the two directions in the honeycomb lattice of graphene (Ajayan, 1999; Moniruzzaman and Winey, 2006). If $m = 0$ (11, 0 in Figure 17(C)), the nanotubes are called zigzag nanotubes. If $n = m$, the nanotubes are called armchair nanotubes. The rest are called chiral nanotubes. The zigzag and armchair CNTs have θ values of 0° and 30°, respectively. The chiral CNTs have θ values greater than 0° but less than 30°. The CNT properties are determined by the chiral vector (Ch), charily angle (θ), structural defects, topology, diameter, surface area, and length. The armchair SWCNTs are always metallic (Dresselhaus et al., 1996). The zigzag and chiral CNTs, depending on whether or not they satisfy $n - m = 3I$ (where I is an integer), are metallic or semimetallic (Wildoer et al., 1998)). The CNTs diameter (D) and θ can be calculated using the following equation:

$$D = \left(\sqrt{3a_{c-c}(m^2 + nm + n)^2}\right)\Big/\pi \quad (10)$$

$$\theta = tan^{-1}\left[\frac{\sqrt{3}n}{2m + n}\right] \quad (11)$$

These equations suggest that the n and m vectors of nanoparticles can be used to calculate their diameter and chiral angle to determine whether they will have metallic or semiconductor properties.

The MWNTs are a group of concentric SWNTs separated by 0.35–0.40 nm and held together by VDW forces. The outer diameter of MWCNTs may range from several nanometers up to 100 nm. Unlike SWCNTs (Hirsch, 2002; Iijima, 2002), MWCNTs may be metallic–metallic (inner-outer), metallic–semiconducting, semiconducting–metallic, or semiconducting–semiconducting (Green and Hersam, 2011). Carbon nanowires consist of stacked cones defined by a finite cone angle. CNTs exhibit significantly higher surface-to-volume ratio (SA/V) that increases exponentially as particles get smaller without changing their shape (Warheit et al., 2004; Warheit, 2004). Since more surface area provides more reaction sites for the same volume, nanoparticles less than 100 nm react much faster than bulk particles. This suggests that chemical reactivity of structurally identical CNTs may depend on their diameter; a decrease in diameter increases their reactivity exponentially.

3.1.2 *CNT Synthesis (Gore and Sane, 2011)*

3.1.2.1 ARC-DISCHARGE METHOD (JUNG ET AL., 2002; LAI ET AL., 2001; TAN AND MIENO, 2010; XING ET AL., 2007)

In an arc-discharge method, the graphite anode and cathode terminals are placed in an inert environment (He or Ar at ~500 Torr) and a current of 100 A is delivered (Figure 18(A)). Generation of arc-induced plasma evaporates the carbon atoms in the graphite. The nanotubes grow from the surface of these terminals (http://www.appropedia.org/Synthesis_of_Carbon_Nanotubeshttp://www.appropedia.org/Synthesis_of_Carbon_Nanotubes). SWNTs are formed with the use of a metal catalyst such as iron or cobalt, while MWNTs can be formed without a catalyst.

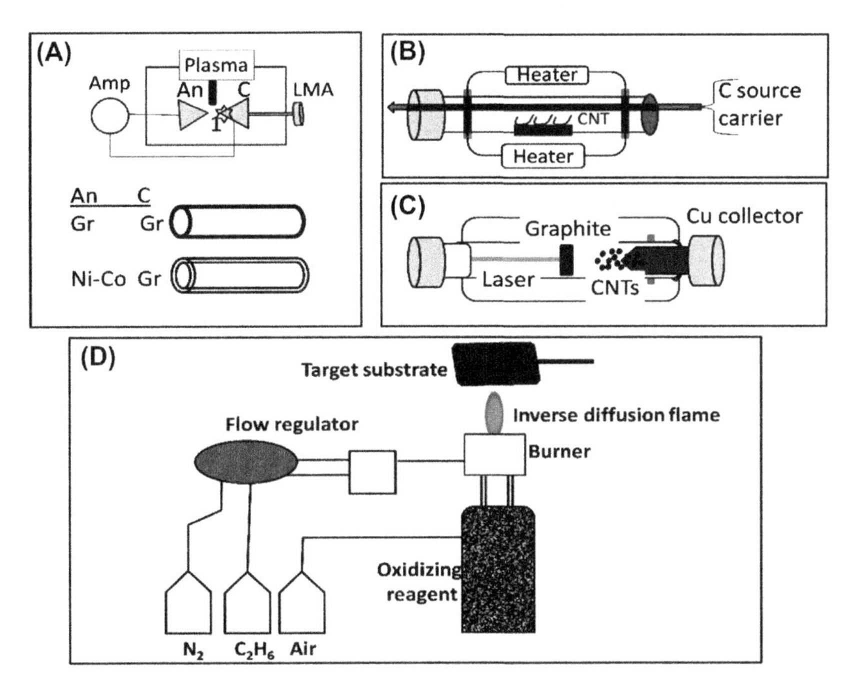

FIGURE 18 Schematic representation of methods used for carbon nanotube synthesis: (A) arc discharge (B) chemical vapor deposition (C) laser ablation (D) hydrocarbon flames. An, anode; c, cathode; Gr, graphite; Ni, nickel; Co, cobalt; Cu, copper.

3.1.2.2 LASER ABLATION METHOD (CHENA ET AL., 2005; GUO ET AL., 1995; THESS ET AL., 1996)

In this method, a laser beam evaporates carbon from graphite at high temperatures (approximately 1200 °C) under pure argon at 500 Torr. The nanotubes, mixed with undesired amorphous carbon, are collected on a cooled substrate at the end of the chamber. Arc-discharge and laser ablation methods have limited potential for scale-up (Figure 18(C)). The nanotubes produced are in an entangled form, and extensive purification is required to remove the amorphous carbon and fullerenes that are naturally produced in the process.

3.1.2.3 CHEMICAL VAPOR DEPOSITION (DANAFAR ET AL., 2011)

Chemical vapor deposition (CVD) is capable of mass producing defect-free CNTs at relatively low temperatures. A porous substrate, such as alumina or quartz, is subjected to electrochemical etching in hydrofluoric acid/methanol mixture. Nanotubes grow at a higher rate on a porous substrate. A catalyst (e.g., iron, nickel) is deposited onto the substrate by thermal evaporation (Figure 19(A) and (B)). An acid treatment followed by sonification removes the impurities. Catalyst (iron nanoparticles)-supported formation of CNTs occurs either by extrusion (Figure 19(A)), in which CNTs grow upward from the nanoparticles that remain attached to the substrate, or by lifting of the catalyst nanoparticles by growing CNT (Figure 19(B)). The catalyst particle diameter plays an important role in defining the synthesized carbon nanostructure (Figure 19(C)). Catalyst particles of the order of 1–15 nm diameter predominantly form SWCNTs, while particles with 20–50 nm diameter form MWCNTs; particles greater than 50 nm form a complex onion-like structure (Figure 19(C)) (Rao et al., 2001).

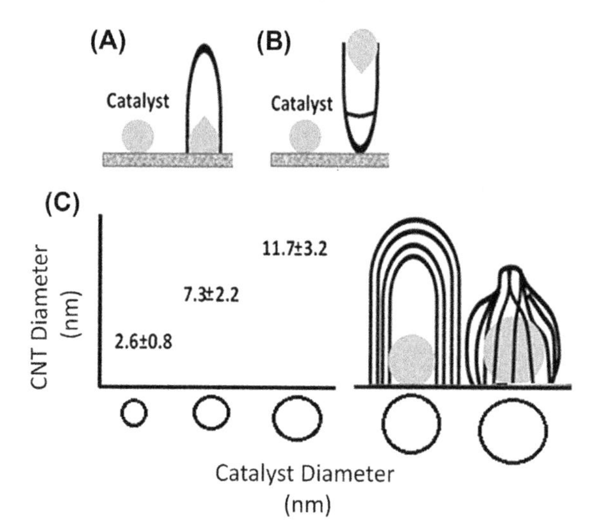

FIGURE 19 CNT growth strategies. (A) CNT growth via extrusion in which CNTs grow upward from the catalyst-nanoparticles (A) that remain attached to the substrate. (B) Lifting of the catalyst nanoparticles by growing CNT that remain attached to the base. (C) Control of CNT diameter: The catalyst diameter determines the CNT shape and diameter. For particles smaller than 20 nm, solid carbon atoms from a circular ring close to one of the diameters of the spherical particle. This accounts for the CNT diameter approximately corresponding to that of the catalyst particles. When the catalyst is about 50 nm, it forms MWCNTs, while catalysts 100 nm or greater form onion-like CNTs.

3.1.2.4 HYDROCARBON FLAMES (CHENG ET AL., 1998; EBBESEN AND AJAYAN, 1992)

Hydrocarbon flames provide a unique combination of the chemical and catalytic factors that are conducive to initiation and growth of CNTs (Figure 18(D)). Growth mechanisms similar to those observed in the CVD process govern the growth of nanotubes in flames. The choice of catalysts determines the structural properties of the carbon nanotubes.

3.1.2.5 CNT SYNTHESIS USING TWISTED GRAPHENE RIBBONS (NANO-TEST TUBE CHEMISTRY)

Although the seed-mediated synthesis of CNTs is commonly used in commercial applications, these methods result in considerable impurities and are not suitable for selective CNT synthesis, such as 5.0, 5.5, 11.5 (and so on) CNTs. To address the deficiencies, a nano-test tube chemistry method has been proposed (Lim et al., 2013; Chen et al., 2011; Khlobystov, 2011; Kitaure et al., 2009) in which molecules and atoms confined inside CNTs are used to synthesize selective CNTs via synthesis of nanoribbons (Figure 20). Jiang et al. (2011) showed that the graphene nanoribbons residing inside a CNT would impose geometric constraints, resulting in transformation of the nanoribbons into helical conformations with restricted chirality of the newly synthesized CNTs. Lim et al. (2013) showed that the chirality of CNTs can be controlled by carefully choosing the polycyclic aromatic hydrocarbon precursors. Miyata et al. (2010) developed the inner-shell extraction technique for extraction of inner CNTs from double-walled CNTs.

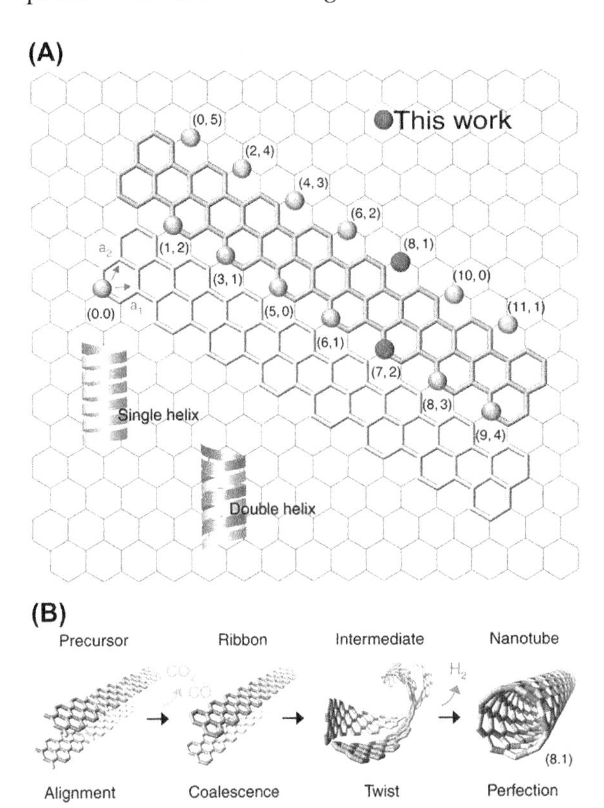

FIGURE 20 **Nano-test tube synthesis of carbon nanotubes.** (A) In a graphene background, chiral vectors for different (*n*, *m*) indices are shown obtained from the intertwining of helical perylene chains. (B) Twisting of graphene nanoribbons onto formation of CNTs. The precursor used was perylene-3,4,9,10-tetracarboxylic dianhydride monomers. Carbon is shown in blue (gray in print versions) and white and oxygen is shown in tallow. Thermal treatment removed the heteroatoms as CO and CO_2. *(Reprinted from Lim et al. (2013) with permission.)*

3.1.3 *Synthesis of Fullerenes*

In 1985, Kroto, Smalley, and Curl discovered fullerenes in carbon vapor from the laser ablation of graphite (Millon et al., 1992; Kroto et al., 1985), although the method resulted in microscopic quantities of fullerenes. Large quantities of fullerenes were produced by striking an electric arc between two graphite electrodes and then evaporating the graphite in low pressure (<100 Torr) under helium (Saidane et al., 2004). However, the method also generated carbonaceous soot impurities. A graphite arc-vaporization method is suitable for commercial formation of fullerenes (Haufler et al., 1990). Other methods of fullerene synthesis include ion sputtering and electron beam evaporation of graphite (Bunshah et al., 1992), vaporization of graphite using highly concentrated solar heating (Chibante et al., 1993; Fields et al., 1993), resistive heating of graphite (Hare et al., 1991), laser ablation of graphite (Wilson et al., 1993), inductive heating of graphite (Peters and Jansen, 1992), and carbon particle evaporation in a hybrid thermal plasma (Yoshie et al., 1992). A pyrolysis method uses naphthalene, a nongraphitic raw material, in an argon atmosphere at approximately 1000 °C (Taylor et al., 1993) to make fullerene.

3.1.4 *Carbon Nanotube Purification*

As synthetized, CNTs contain substantial impurities that can change their electronic and chemical properties. Crude nanotubes are not

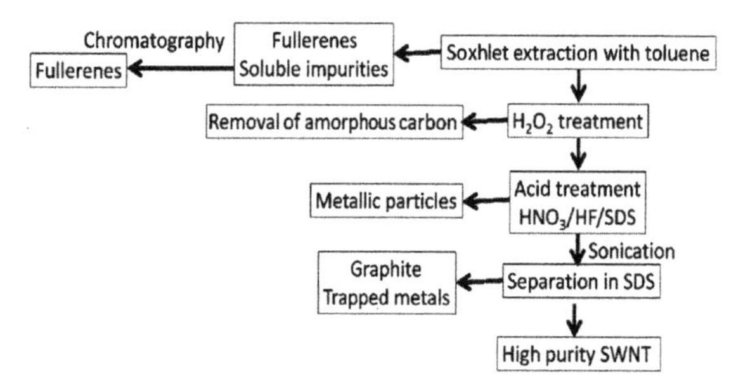

FIGURE 21 An integrated procedure for purification of carbon nanotubes.

suitable for biomedical applications. Therefore, the newly synthetized CNTs must be purified for further applications. An integrated method for purification of CNTs is shown in Figure 21. Use of strong acid and hydrogen peroxide is a problem in the purification of CNTs because the reagents may damage the structural integrity of the tubes (Rosolen et al., 2006). However, acid-induced defects in the nanotubes are often utilized in their functionalization.

3.1.5 *Structural Defects and Reactivity*

Pristine CNTs are homogenous hexagonal honeycomb structures that are chemically inert. However, when a graphene sheet is folded into CNTs, the curvature induces a band gap that confers CNTs either metallic or semiconductor properties (Popov, 2004). CNTs also have inherent structural defects that (1) modulate their mechanical, topological, and electronic properties and (2) have significant beneficial and/or toxicological consequences (Charlier, 2002; Fenoglio et al., 2008; Ouyang et al., 2002; Sammalkorpi, 2004; Mingo et al., 2008; Lee et al., 2010). In general, there are four types of defects reported in CNTs:

- *Vacancy or point defects* (Collins, 2009; Tunvir et al., 2007): Vacancy involves a missing or extra atom, resulting in dangling bonds capable of rehybridizing with another atom within the CNT or with an atom outside in the environment. Although vacancies occur in both graphene and CNTs, the defects are more complicated in CNTs because of circumferential and/or curvature-induced strain. The defect stabilities depend on their position within the CNT's lattice, diameter, and helicity (Carlsson and Scheffler, 2006).
- *Interstitials:* This defect is usually transient and occurs when a carbon atom is rendered free due to knockout damage. The reactive group may form a bond with other carbon atoms or environmental chemicals.
- *Bond rotation:* This defect may cause topological changes by replacing two hexagons with other forms, such as a pentagon and a heptagon. Because the bond angle strain in a pentagon is greater than that in a hexagon, replacement of two hexagons with a pentagon and a heptagon may increase overall molecular strain and ensuing changes in electronic properties and chemical reactivity. The electronic perturbations occurring due to a defect are not local but are spatially localized.
- *Stone-Wales defects (SWDs)*: This defect represents rotation of a C–C bond in hexagonal CNTs (Stone and Wales, 1986). SWDs can adsorb hydrogen and other foreign atoms; hence, CNTs containing this defect have a potential usage for energy storage.

Taken together, these observations suggest that the structural defects increase chemical reactivity and ensuing biological activity/toxicity of CNTs. Structurally defective CNTs can be used for (1) hydrogen storage, (2) miniature electron sources for *e*-beam and X-ray instruments and display applications, (3) detection of intracellular pH, and (4) screening of NO_2, NH_3, and CO (Cheng et al., 2001; De Jonge and Bonard, 2004; Dillon et al., 2002; Kong et al., 2000; Peng et al., 2001).

Since pristine SWCNTa are chemically inert, addition-based covalent binding of functional groups to nanotubes' surface is difficult. SWCNTs

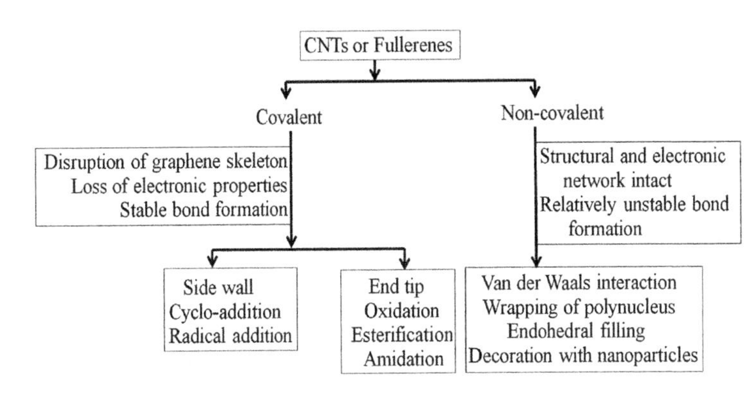

FIGURE 22 Overall approach for functionalization of CNTs and fullerenes.

have three main sources of reactivity: (1) the curvature-induced strain arising from sp^2 carbons, (2) the misalignment of the π orbitals, and (3) structural defects (Niyogi et al., 2002; Kuzmant, 2004). The most reactive place in any CNT is in the cap of the thinnest tube. Derivatization of the side wall required direct attachment of functional groups using either 1,3-dipolar cycloaddition (Poplawska et al., 2010; Tagmatarchis and Prato, 2004) or reactions with nitrenes, radicals, and carbenes (Holzinger et al., 2001).

3.1.6 CNT Functionalization

The medicinal application of CNTs has been severely limited due to their poor interfacial reactivity and aggregation due to VDW interactions. The commercial CNTs are heavily entangled and poorly dispersed bundles. Reactivity of fullerene molecules is strongly dependent on the curvature of the carbon framework. Their outer surface reactivity increases with an increase in curvature. The moderately curved CNTs are relatively less reactive than fullerene molecules. Structural defects increase their reactivity to organic molecules. Therefore, selective structural defects may be generated for further functionalization. In general, two basic approaches are commonly used to functionalize CNTs or fullerenes: covalent (chemical) and noncovalent (physical) functionalization, as shown in Figure 22.

3.1.6.1 COVALENT FUNCTIONALIZATION

Figure 23 shows an overall approach to covalently functionalize CNTs. Two important addition reactions of the CNT sidewall are fluoridation and aryl diazonium addition (Mickelson et al., 1998; Tasis et al., 2006). Fluoridation improves solubility and further functionalization (Kawasaki et al., 2004; Kelly et al., 1999). Diazonium reacts only with metallic CNTs; thus, this reaction can be used to separate metallic and semiconductor CNTs. In addition to fluoridation or diazonium addition, CNTs can also be functionalized using cycloaddition such as Diels–Alder reaction, carbene and nitrene addition, chlorination, bromination, hydrogenation, and azomethineylidation (Kumar et al., 2011). Many of these reactions require induction of defects in CNTs to generate free reactive functional groups such as carboxylic acid, ketone, alcohol, and ester groups that can be further functionalized via silanation, polymer grafting, esterification, thiolation, and biomolecule-addition (Ruelle et al., 2012; Martínez-Hernandez et al., 2010).

A significant disadvantage of covalent functionalization is that it leads to drastic changes in the CNT's electronic states that may extend beyond the vicinity of the addendum site (Park et al., 2006; Zhao et al., 2004). Monofunctionalization ($-F$, $-NH_2$, $-COOH$, etc.) distorts the Fermi level electronic state (Chapter 3), while bi-functionalization ($=NH$, $=O$, $=CCl_2$,

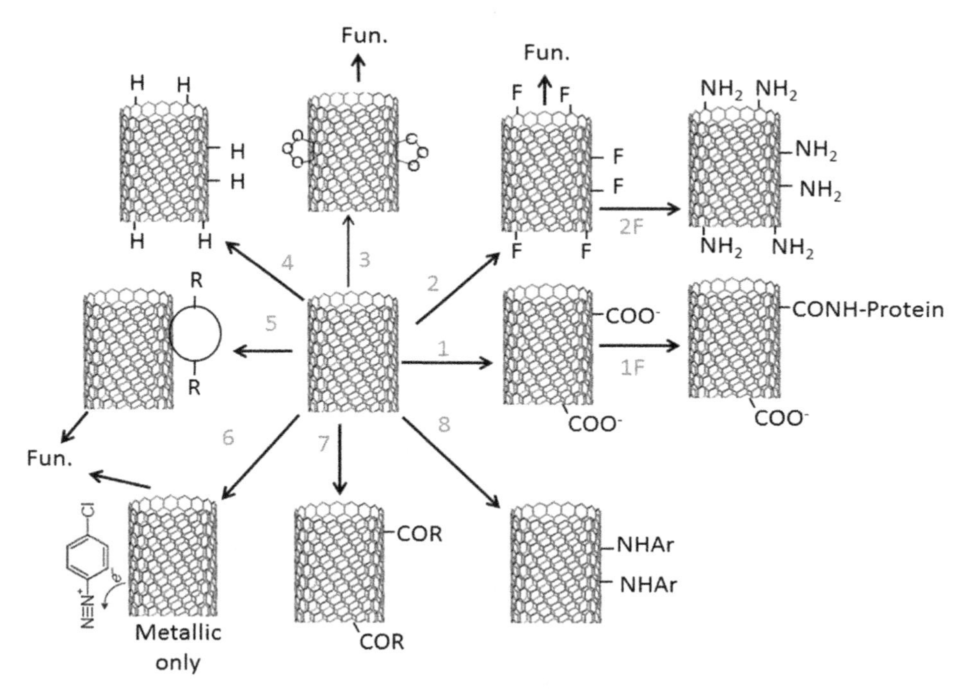

FIGURE 23 CNT functionalization reactions by acid oxidation (1) that can form amide bond with proteins and amino acids (1F), fluoridation (2) that can be transformed into an amidated CNTs (2F), oxidation (3), hydrogenation (4), cycloaddition (5), diazonium addition (6), esterification (7) and electrophilic oxidation (8).

etc.) moves the distortion away from the Fermi level (discussed in Chapter 3). Thus, mono-functionalization is more potent than bi-functionalization in distorting the metallic properties of the CNTs (Zhao et al., 2001). Unlike covalent functionalization, noncovalent functionalization does not alter the electronic state of CNTs (Zhao et al., 2004).

3.1.6.2 NONCOVALENT FUNCTIONALIZATION

The procedures for noncovalent functionalization of CNTs have been described earlier in the metal nanoparticle section. However, some of the procedures unique for noncovalent functionalization of CNTs are listed here.

3.1.6.3 SURFACE STABILIZATION

As described above, commercial CNTs exist in a highly aggregated form that cannot be integrated in biological and medicinal applications. Therefore, it is essential to disaggregate and disperse CNTs uniformly in appropriate media. One of the key approaches commonly used to decrease the nanotube agglomeration is ultrasonication of surfactant-stabilized CNT. Ultrasonication provides high shear that separates the CNT monomers that are stabilized by the surfactants (Figure 24). However, excessive sonication may fragment the CNT into smaller particles (Moore et al., 2003. O'Connell et al., 2001).

3.1.6.4 NANOPARTICLE SOLUBILIZATION

Qin et al. (2004) have described a two-step procedure using grafted poly(sodium 4-styrenesulfonate) that allows soft dissociation of aggregated CNTs. Islam et al. (2003) compared several different forms of surfactants and found that those containing a benzene ring are better dispersants than those lacking it (Figure 25).

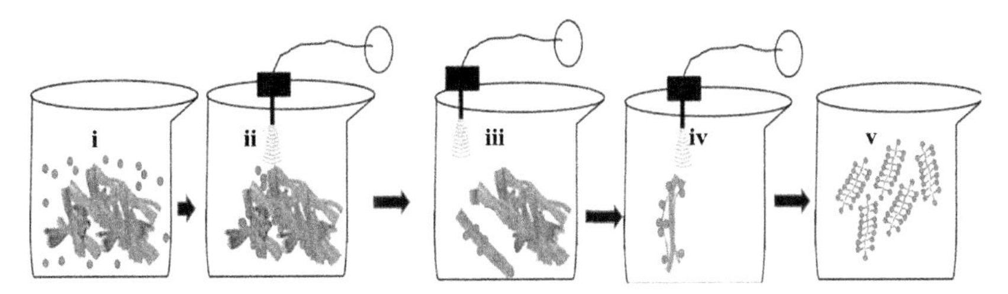

FIGURE 24 Dispersion and solubilization of CNTs using surfactants. Aggregated CNTs are suspended in aqueous solution containing an appropriate concentration of a surfactant (i). The suspensions is subjected to ultrasonication (ii). The shear force splits the non-covalent bonds and separate individual bundles (iii) that are collected (iv) and further solicited, resulting in the formation of uniformly dispersed nanotube (v).

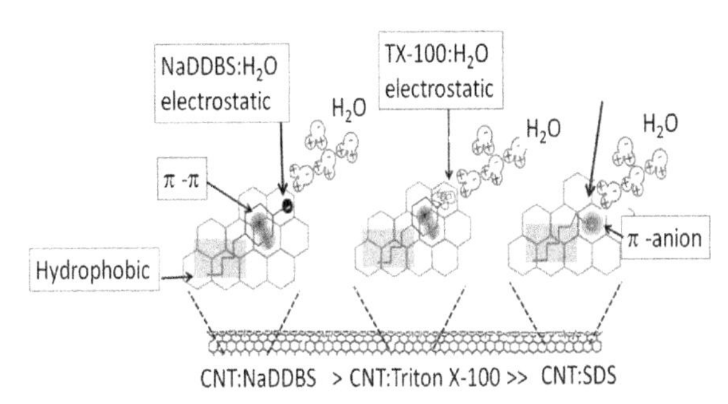

FIGURE 25 Absorption of different surfactants onto the nanotube surface (Islam et al., 2003). The results for sodium dodecyl benzenesulfonate (NaDDBS, $C_{12}H_{25}C_6H_4SO_3Na$), sodium dodecyl sulfate (SDS, $CH_3(CH_2)_{11}$ OSO_3Na) and t-octylphenoxypolyethoxyethanol (Triton-X 100, $C_8H_{17}C_6H_4$ $(OCH_2CH_2)_n$-OH) have been shown in this figure. The alkyl chain groups of the surfactant span along the length of the tube via hydrophobic interaction. The benzene moiety of the surfactants for $\pi-\pi$ interaction with the CNT, while the anionic site SO_3^- or SO_4^- dissolve in the aqueous solvents. SDS that lacks a benzene group interacts with the CNT via hydrophobic interaction and interacts with water via its anionic group. Thus, NaDDBS disperses better than TX100 that may be better than SDS.

3.1.6.5 ENDOHEDRAL FUNCTIONALIZATION

One of the unique aspects of CNT's noncovalent functionalization is endohedral (inner side of the tube) functionalization or filling in which particles are enclosed inside the CNTs for biomedical use. Unlike the graphite surfaces that have comparable reactivity, the external surface of CNTs is more reactive than the internal surface (Niyogi et al., 2002). This may be because the exterior C—C π-electrons may be more distorted than the interior surface; however, an increase in nanotube diameter reduces the distortion (Niyogi et al., 2002; Britzab et al., 2006). This relatively inert interior allows CNTs to store large quantities of particles without much CNT-particle interaction. CNTs could be filled with particles suspended in either gas or liquid (organic or aqueous). Gases, in a pressure- and temperature-dependent manner, preferred endohedral filling as opposed to surface binding (Brunauer and Emmett, 1938). Aqueous and polar-organic liquids (such as methanol) entered CNTs more effectively than nonpolar organic liquids. Joseph and Aluru (2008) and Mattia and Gogotsi (2008) have shown that despite the hydrophobic nature of nonfunctionalized CNTs, water can enter the nanotube at a very high rate that is optimal at 10 nm diameter, but decreased as (1) the nano-tube diameter increased and (2) the meniscus angle decreased (Figure 26).

Figure 27 describes adsorption of hydrophilic and hydrophobic particles to the nanotube interior. Exterior binding is preferred when CNT's diameter (D) is smaller than the particle's diameter. As the diameter increases ($D_{CNT} > D_{particle}$), the binding energy also increases that allows particles to enter the tube. A maximum binding energy occurs at an ideal diameter defined as $(d + 2) \times r_{vdw}$, where r_{vdw} is the thickness of

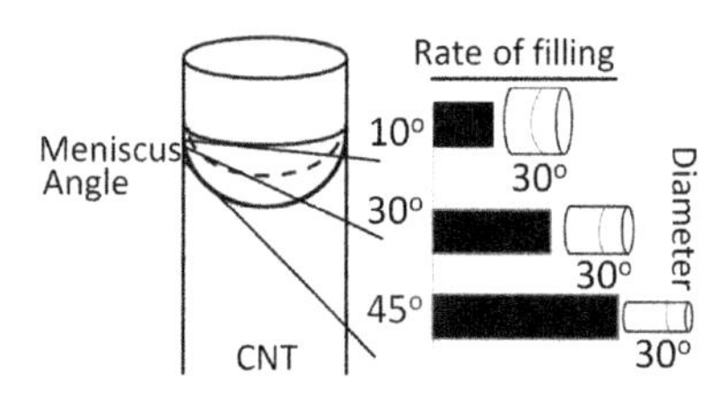

FIGURE 26 Behavior of water in CNTs. Although CNT is highly hydrophobic, water can enter the nanotube at a very high rate. Once inside the tube, the water meniscus is curved and the angle of the curve is inversely proportional to the tube's diameter.

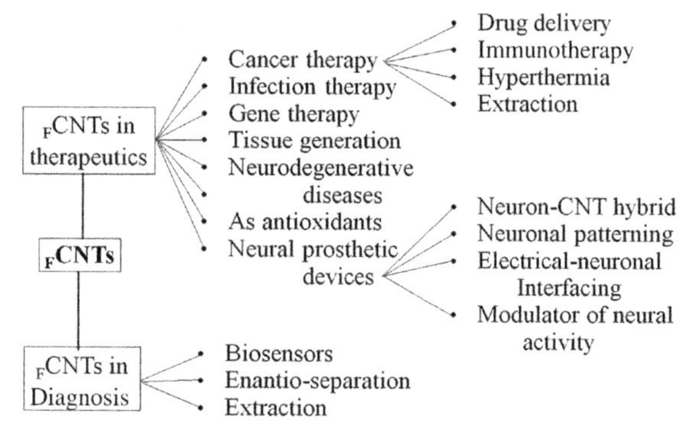

FIGURE 28 An overview of the application of functionalized CNTs ($_F$CNTs).

the nanoparticle's p-orbitals. At this diameter, particles exhibit the highest binding to the inner wall. Then, a gradual decrease in the particles' binding occurred at diameter greater than $(d + 2) \times r_{vdw}$. Accumulation of particles into a nanotube may alter the tube's mechanical and electronic properties, possibly due to formation of VDW bonds between the CNT and the nanotubes.

3.2 Biomedical Application of CNTs

CNTs, due to their high surface area and high containment volume, act as nanocarriers and nanocontainers for a wide variety of therapeutic and diagnostic applications (drugs, gene therapy, vaccines, antibodies, etc.). CNTs can be an excellent vehicle for drug delivery directly into the target cells without degradation by the enzymes present in systemic circulation. Other applications include gene therapy, tissue regeneration, biosensor diagnosis, enantiomer separation of chiral drugs, extraction and analysis of drugs and pollutants, and in neural prosthetic devices (Figure 28). In addition, CNTs synthesized using the plant extract are promising

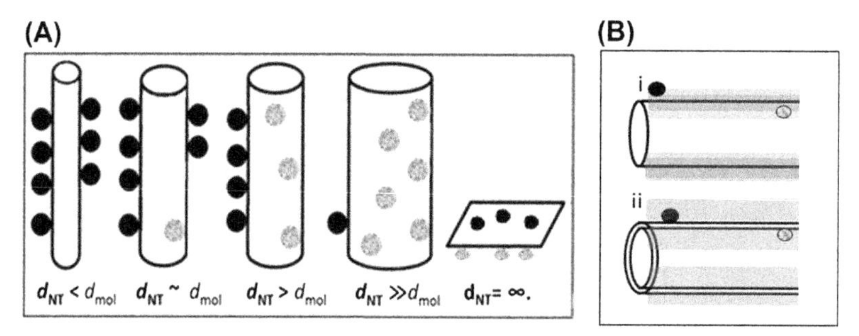

FIGURE 27 Binding of drugs or toxins to CNTs. (A) Binding of functional groups onto the exterior (exohedral) and/or interior (endohedral) of the SWCNT surface. Exohedral non-covalent functionalization of CNTs: A–C: non-covalent exohedral functionalization depends on van der Waals interaction between the particle and the nanotube as well as the tube length. Graphite represents $d = \infty$ in which both sides have comparable bonding energy. If the nanotube's diameter is less than that of the particles' VDW diameter (d_{vdw}), all particles will bind exohedrally. With an increase in diameter, the number of particles binding endohedrally increases. (B) Exohedral or endohedral binding of a molecule to SWCNT modulates properties of both surfaces (i), however, exohedral and endohedral binding of particles to double-walled carbon nanotubes (DWNTs) modulates the properties of their respective surfaces.

antioxidants (Sui et al., 2014; Paul and Samdarshi, 2011). In this section, some of the major biomedical applications of CNTs are described.

3.2.1 *Cancer Therapy*

CNTs can be functionalized into site-directed drug carriers capable of providing both diagnosis and treatment at the cellular level (Bhirde et al., 2009; Dhar et al., 2008; Kam et al., 2005; Lamprecht et al., 2009; Madani et al., 2012; Sahoo et al., 2011; Welsher et al., 2008). CNTs can also be filled with therapeutic drugs for site-directed release (Hilder and Hill, 2007a,b). As discussed previously, the traditional anticancer drugs exhibit high systemic toxicity, narrow therapeutic windows, drug resistance, and limited cellular penetration (Albini et al., 2010; Stuurman et al., 2013). CNTs can cross the cytoplasmic and nuclear membranes (Mu et al., 2009); thus, they can transport anticancer drugs in situ with intact concentration, resulting in higher efficacy against the tumor and stimulating the patient's immune system to attack the malignant tumor cells (Fadel and Fahmy, 2014). SWCNTs exhibit strong absorbance in the NIR region (NIR; 700–1100 nm) and then release significant amounts of heat (Burke et al., 2012; Chatterjee et al., 2010; Singh and Torti, 2013; Zhou et al., 2009), causing thermal ablation of tumor cells.

3.2.2 *Infection Therapy*

Functionalized CNTs, in addition to carrying drugs, have also been demonstrated to be able to act as carriers for antimicrobial agents (Jain et al., 2013; Afzal et al., 2013; Rosen and Elman, 2009). CNTs loaded with drugs such as amphotericin B reduced the antifungal toxicity about 40% as compared to the free drug (Prajapati et al., 2011).

3.2.3 *Gene Therapy*

Despite anticipated potentials, gene therapy has not succeeded in clinical settings, possibly because a safe transfection agent is not yet available. Viral reagents are not safe and the blood enzymes destroy the nonviral vectors. Studies have shown that the DNA-SWCNT transfection complex is protected from enzymatic cleavage and interference from nucleic acid binding proteins, thus exhibiting increased self-delivery capability of DNA (Bates and Kostarelos, 2013; Ojea-Jimenez et al., 2012; Siu et al., 2014). Stable complexes between plasmid DNA and cationic CNTs have demonstrated the enhancement of gene therapeutic capacity compared with naked DNA (Ahmed et al., 2009).

3.2.4 *Tissue Regeneration*

A composite nanomaterial consisting of SWCNTs and other organic nanopolymers, such as poly-L-lactide or poly-D,L-lactide-co-glycolide, has been shown to improve tissue regeneration (Jain, 2012; Mao et al., 2014), cell tracking and labeling, and sensing cellular behavior (Ferreira et al., 2008; Gul et al., 2010). CNTs have effectively enhanced bone tissue regenerations in mice and neurogenic cell differentiation by embryonic stem cells in vitro (Chen et al., 2006a; Hirata et al., 2013; Kabiri et al., 2012). Further research is needed to explore the beneficial effects of CNTs in tissue regeneration.

3.2.5 *Neurodegeneration Therapy*

CNTs, because of their 1D character and accessible external modifications, are able to cross the blood–brain barrier and deliver drugs to the brain. Many functionalized SWCNTs or MWSCNTs have been used as suitable delivery systems for treating neurodegenerative diseases or brain tumors (Fu et al., 2009; Li et al., 2011a; El-Jamal et al., 2011).

3.2.6 *Antioxidant*

It is well established that oxygen-free radicals play a central role in diseases pathogenesis and that antioxidants have significant health protective effects (Amiri et al., 2013). Recently, CNTs synthesized using plant extracts acted as free-radical scavengers with potential therapeutic potency (Oliveira et al., 2013).

3.2.7 *Neural Prosthetic Devices*

Functionalized CNTs may serve as a promising material for neuronal interfacing applications. The unique mechanical and electrical properties of CNTs make them excellent neuronal implants. Many reviews have described using CNTs in neurointerfacing applications, including cell adhesion, neuronal engineering, and multielectrode recordings (Jain et al., 2013a; Meredith et al., 2013; Parker et al., 2012; Zhang et al., 2013; Zhou et al., 2013).

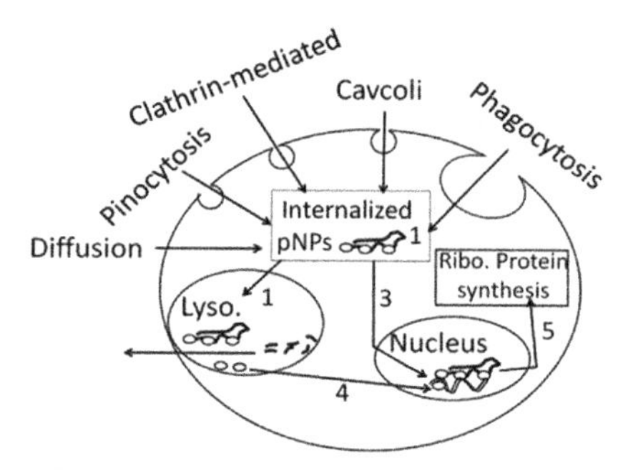

FIGURE 30 **Possible mechanisms underlying the intracellular uptake of nanopolymers (pNPs).** Nanopolymers enter the cells via simple diffusion, pinocytosis, clathrin-mediated uptake, cavcoli-mediated uptake, and phagocytosis (1). The internalized pNPs are taken up by the lysosomes (2) that release free drugs from pNPs coupled to the drugs via acid-cleavable bonds. Plasmid-coupled pNPs (3) enter the nucleus and induce protein synthesis (5). Free drugs may also enter the nucleus and modulate protein synthesis (5).

4. LINEAR NANOPOLYMERS

4.1 Structure and Synthesis

4.1.1 *Structure*

Polymeric nanoparticles (Figure 29) are one of the most investigated nanoparticles as drug delivery vehicles (Avgoustakis, 2004; Bender et al., 1996; Hans and Lowman, 2002; Kumari et al., 2010). Biodegradable nanopolymers can be functionalized to improve their biocompatibility and drug targeting to particular organs/tissues (Dunne et al., 2000). They can also be used as carriers of DNA in gene therapy, and in their ability to deliver proteins, peptides, and genes through oral administration (Panyam and Labhasetwar, 2012). Figure 30 shows possible mechanisms by which nanopolymers enter the cells and the intracellular fate of the nanopolymers.

4.1.2 *Synthesis*

Two basic procedures are used for synthesis of polymer nanoparticles: (1) the dispersion

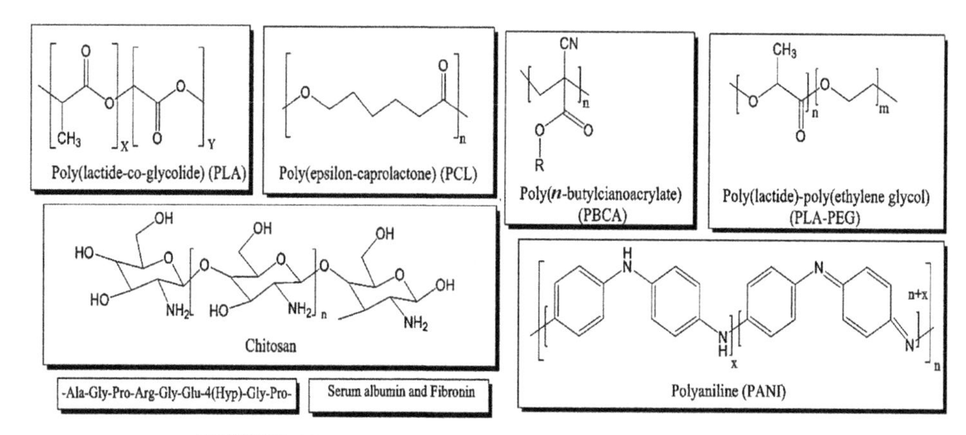

FIGURE 29 Structure of different classes of linear polymers.

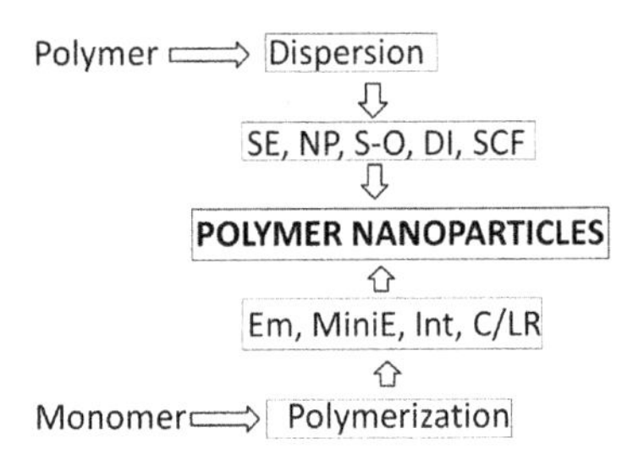

FIGURE 31 **Synthesis of polymeric nanoparticles.** SE, solvent evaporation; NP, nanoprotection; S–O, salting out; DI, dialysis; SCF, supercritical fluid; EmL, emission; Int, interfacial; CL/R, controlled living radical.

method, in which preformed polymers are used for synthesis; and (2) the polymerization method, in which synthesis occurs via direct polymerization of monomers (Nagavarma et al., 2012). Figure 31 lists important methods for the preparation of polymer nanoparticles.

1. *Dispersion methods*: In the solvent evaporation method, oil-in-water emulsions are subjected to either high-speed homogenization or ultrasonication, followed by evaporation of the solvent. In the nanoprecipitation method, polymers are suspended in aquatic and organic solvents to precipitate polymer nanoparticles. Salting-out involves dissolution of polymer and drug in an organic solvent (such as acetone) and then emulsification into an aqueous gel containing the salted-out electrolytes (magnesium chloride, calcium chloride, and magnesium acetate) or nonelectrolytes (sucrose) and a stabilizer (polyvinylpyrrolidone or hydroxylethylcellulose) (Reis et al., 2006). For dialysis, the polymer is dissolved in an organic solvent and then dialyzed using an appropriate cutoff membrane against a miscible nonsolvent. The displacement of the solvent inside the membrane results in the progressive formation of homogeneous suspensions of nanoparticles. In the supercritical fluid technology (Chernyak et al., 2001; Weber and Thies, 2002), the polymer is dissolved in a supercritical fluid, followed by a rapid expansion of the solution into an appropriate solution for nanosized particles.
2. *Monomer polymerization methods* (Ekman and Sjfholm, 1978; Lowe and Temple, 1994): In the emulsion method, a monomer is dissolved in a continuous aqueous solution. The polymerization is initiated when a monomer molecule collides with an initiator molecule that might be an ion or a free radical. Collisions among the monomer ions initiate chain growth. In the interfacial method, polymerization is carried out at the interface of two immiscible solvents, such as the monomers dissolved in an organic solvent and an oxidizing agent dissolved in an aqueous solution. In the microemulsion polymerization method, an aqueous solution containing the oxidizing agent and an organic solution containing the monomer are emulsified in the presence of a surfactant to prevent aggregation. In a controlled/living radical procedure, the radical-collision method is controlled. C/LR provides control over the molar mass, the molar mass distribution, the end functionalities, and the macromolecular architecture.

4.1.3 *Functionalization of Nanopolymers*

The overall strategies described earlier (metal nanoparticle section) are applicable for the functionalization of nanopolymers, except that nanopolymers exhibit no affinity for the thiol group. Therefore, covalent and noncovalent methods are used to functionalize nanopolymers with PEG.

4.2 Application of Polymer Nanoparticles

1. Cationic, but not anionic or nonionic, polymeric nanoparticles are internalized into the cells, suggesting the possibility of their application on gene therapy. The total number of particles is inversely proportional to their size.

2. Polymeric nanoparticles (1) increase the stability of volatile pharmaceutical agents; (2) offer protection of the drugs from the degradation enzyme of blood and liver, thus significantly improving their stability when given orally or administered by intravenous injection; and (3) deliver a higher concentration of pharmaceutical agent to a desired location. They also facilitate selective cancer therapy by delivering vaccines, antibiotics, and anticancer drugs at the target site.
3. Polymeric nanoparticles are the starting material for synthesis of other advanced nanoparticles, such as dendrimers, nanospheres, and nanocapsules.

5. DENDRIMER NANOPARTICLES

5.1 Structure and Synthesis

5.1.1 Structure

Dendrimers are a class of three-dimensional molecules having a high degree of structural symmetry, a density gradient, and defined number of terminal groups that may be chemically different from the inner groups. The dendrimer nanoparticles (Figure 32) are functional nanomaterials with unique electronic, optical, opto-electronic, magnetic, chemical, and biological properties (Astruc et al., 2010).

Dendrimers are repetitively branched molecules that radiate from a central unit called the core. The number of branching points when going from the core towards the surface is called the generation number (Figure 32). Dendrimers having one branching point when going from the center to the periphery are called first generation (G1) dendrimers. An increase in branching points increases the generation numbers. For example, G4 dendrimers have four branches radiating from the core. Dendrimers have three components: a central core (Figure 32, fourth generation, core (C)), an interior dendritic structure, the branches (G1–G4), and an exterior surface with functional surface groups. The branches can be homologous (G2 top, G3 and G4 top) or heterologous (G2 bottom and G4 bottom.). The synthetic methodology,

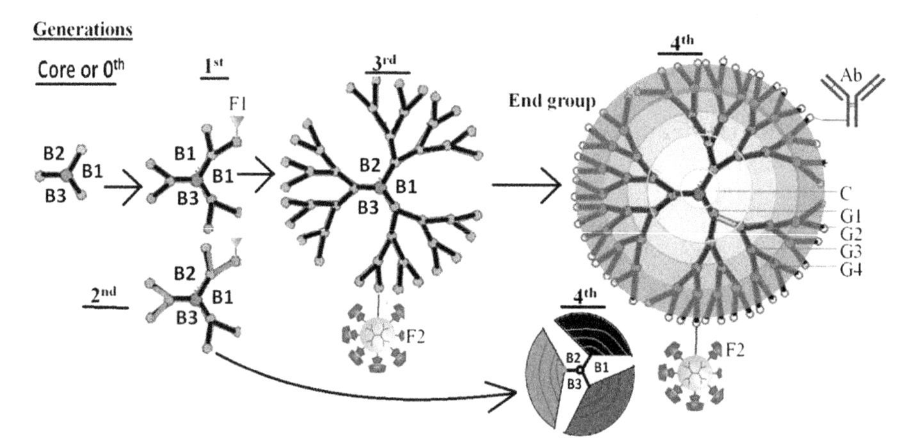

FIGURE 32 **Structures of different generations of dendrimers.** First generation dendrimers are simple molecules consisting of four atoms, one central and three branch atoms (B1, B2, B3). The second and higher generations (up to four generations are shown) contain increasing layers of atoms. The branches can be homologous (1st generation top) or heterologous (first generation bottom). An increase in the dendrimer generation results in the corresponding increase in their intra-dendrimer density. The outer layers of the dendrimer can be functionalized with different functional groups.

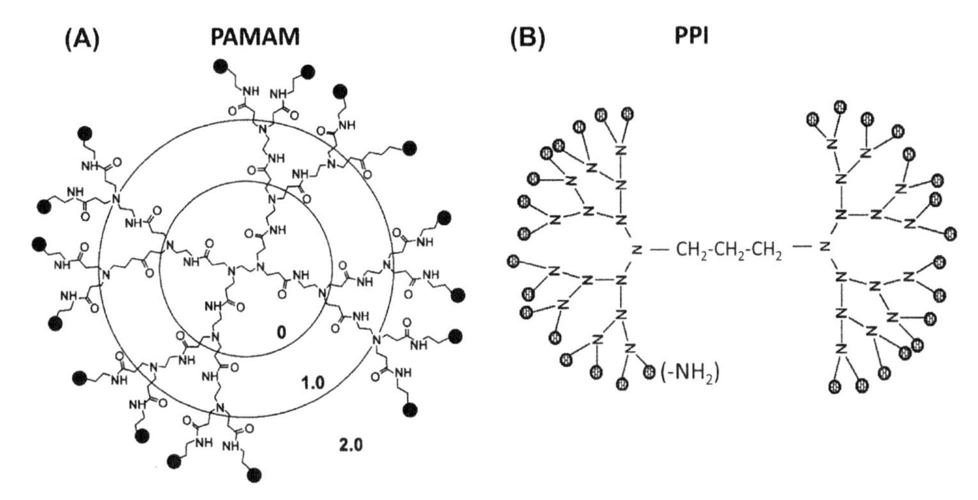

FIGURE 33 Structures of (A) PAMAM and (B) PPI dendrimers.

therefore, allows production of multifunctional dendrimers.

5.1.2 Types of Commonly Used Dendrimers (Zimmerman et al., 2001)

5.1.2.1 POLY (AMIDOAMINE) DENDRIMERS (PAMAM)

These dendrimers are synthesized from ammonia or ethylenediamine as the core reagent (Figure 33(A)). Products up to generation 10^7 have been synthetized. A modification of PAMAM is radially layered poly(amidoamine-organosilicon) (PAMAMOS) dendrimers that consist of hydrophilic, PAMAM, interiors, and hydrophobic organosilicon exteriors. Figure 33(B) shows poly(propyl imine) (PPI) dendrimers that are generally poly-alkyl amines having primary amines as end groups.

5.1.2.2 TECTO DENDRIMERS

These dendrimers are composed of a core dendrimer, surrounded by dendrimers designed to perform a specific function such as diseased cell recognition, diagnosis of the disease state, drug delivery, reporting location, or reporting outcomes of therapy. Multilingual dendrimers contain multiple copies of a particular functional group (Figure 34).

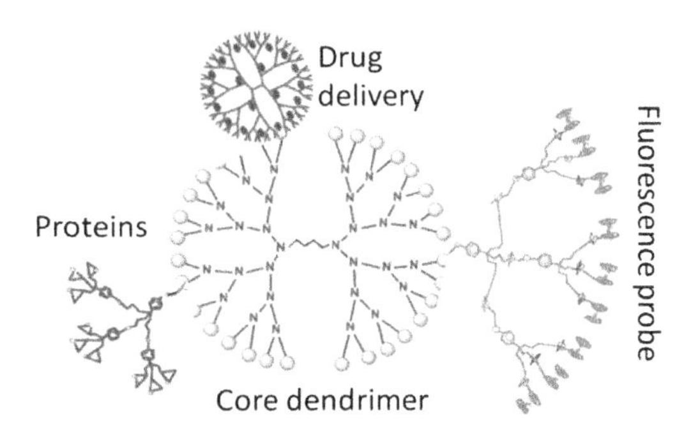

FIGURE 34 Structure of a Tecto dendrimer.

5.1.2.3 CHIRAL AND AMPHIPHILIC DENDRIMERS

Chiral dendrimers consist of constitutionally different but chemically similar branches to the chiral core (Figure 35(1)). Amphiphilic dendrimers (Figure 35(2)) are built with two segregated sites of a chain end—one half is electron donating and the other half is electron withdrawing.

5.1.2.4 POLYMERIC DENDRIMERS

Fréchet-type dendrimers (Figure 36(1)) are based on a poly-benzyl ether hyper-branched skeleton and carboxylic acid groups as surface groups. Figure 36(2) shows a different type of

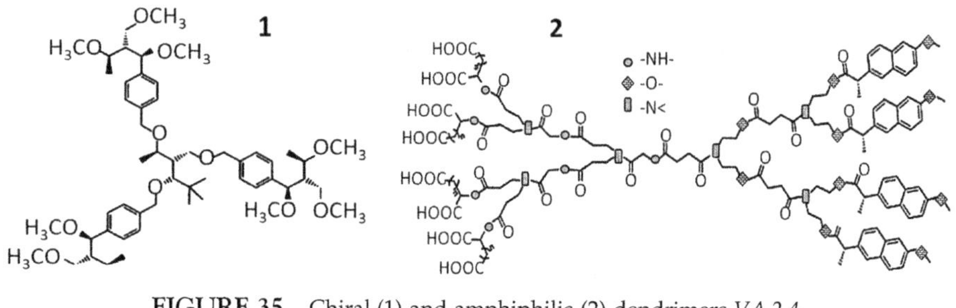

FIGURE 35 Chiral (1) and amphiphilic (2) dendrimers.*VA.2.4.*

fully aromatic dendrimer where the benzene rings are either linear or form an angle of 60° or 120° to each other. Therefore, rotation about the bonds between the rings is unable to change the structure of the molecule. Figure 36(3) shows PEG-based biodegradable dendrimers. These different types of dendrimers have unique properties that can be exploited for specific functions.

5.1.2.5 DNA-BASED DENDRIMERS

DNA is a helical polymer central to the genetic makeup of all organisms. Recently, polymeric DNA molecules have been designed that are important in nanomedicine including, but not limited to, disease screening and therapeutics (Roh et al., 2011). Synthetic DNA polymers generally belong to the following four topologies—linear, branched, dendritic

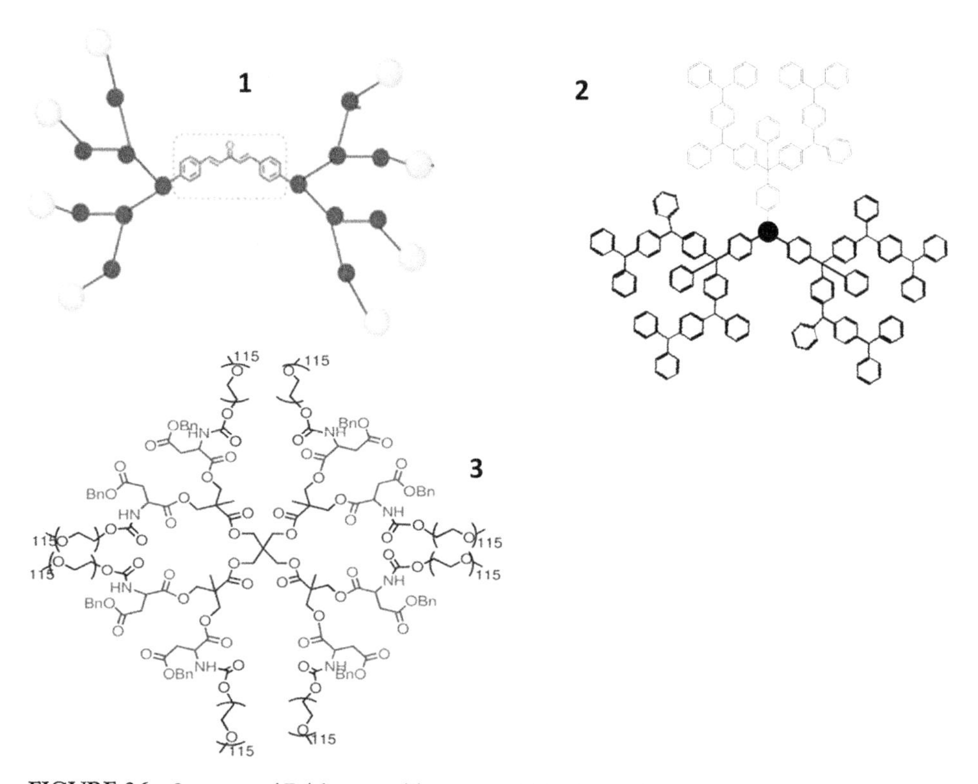

FIGURE 36 Structure of Fréchet-type (1) aromatic (2) PEG-based biodegradable (3) nanoparticles.

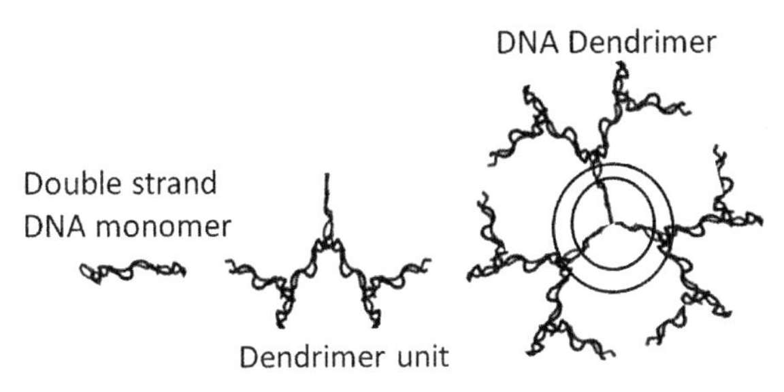

FIGURE 37 Topology of DNA-based dendrimer nanoparticles.

(Figure 37), and networked. Linear DNA nanopolymers are simple structures used extensively in drug delivery. The branched DNAs usually contain Y-shaped subunits and sticky ends that connect two to four units to make dendrimers. Branched DNA building blocks are also used to construct network DNA tiles and hydrogels.

Linear polymeric nanoDNAs form complexes with lipids, cationic polymers, and synthetic peptides that may get encapsulated within hydrogel nanoparticles (Liang et al., 2006). Linear polymeric nanoDNA is synthesized using monodispersed, single-stranded DNA (ssDNA) that (1) recognize and bind to complementary sequences and form a double-strand DNA (dsDNA)—a process known as hybridization (Li et al., 2004; Roh et al., 2011). They may self-anneal to form unique tertiary structures such as dendrimers that can bind specifically to macromolecules. The DNA dendrimers overcome some of the limitations of the linear DNA: they are inherently multivalent and designed to contain different functional moieties on different branches that are connected anisotropically. DNA dendrimers have the advantages of traditional chemical dendrimers such as tunable size, diverse architecture, and availability of multiple surface groups; however, unlike the chemical dendrimers, the "sticky ends" of DNA dendrimers allow construction of nanostructures with specific functional moieties. The dendrimeric DNA, however, lack rigidity in the structures, which prevents their self-assembly of ordered structures. This deficiency has been resolved by the development of networked DNA such as checkered-patterned DNA, hexagonal DNA, cross-shaped DNA, and star-patterned DNA lattices (He et al., 2005; Winfree et al., 1998a; Yan et al., 2003; Zhang et al., 2008a).

The networked DNA include double crossovers, triple crossovers, and paranemic crossover structures with increased rigidity and planar format (Fu and Seeman, 1993; LaBean et al., 2000; Winfree et al., 1998a). A combination of these structures form DNA tiles that can be used to form more complex patterns (He et al., 2005; Yan et al., 2003; Zhang et al., 2008a). Lou et al. (2008) and Um et al. (2006) have developed a simple method using branched DNAs having unique sticky ends that were used to construct dendritic DNAs (Lee et al., 2005, 2009). The peripheral Y-DNAs (with sticky ends complementary to the core Y-DNA) were ligated to the initial core to form the G1 dendrimeric DNA. Subsequent generations (G2, G3, G4, etc.) were produced by repeatedly ligating peripheral Y-DNAs to the sticky ends of the previous generation.

5.1.3 Dendrimer Synthesis

Synthesis of dendrimers having identical structural units involves repetitious alternation of a growth reaction and an activation reaction such as the Michael reaction, Williamson ether synthesis, solid-phase synthesis, organotransition-metal chemistry, organosilicon chemistry, organophosphorus chemistry, etc., depending upon the type of branching needed in the dendrimer. In general, four different approaches are used to synthesize dendrimers:

1. *Divergent* synthesis in which dendrimers grow outwards from the core, diverging into the space
2. *Convergent* synthesis, in which dendrimers are built from small molecules that end up at the surface of the sphere
3. *Hypercore* growth, in which lower generation dendrimers serve as a hypercore to which presynthesized dendrimers are attached

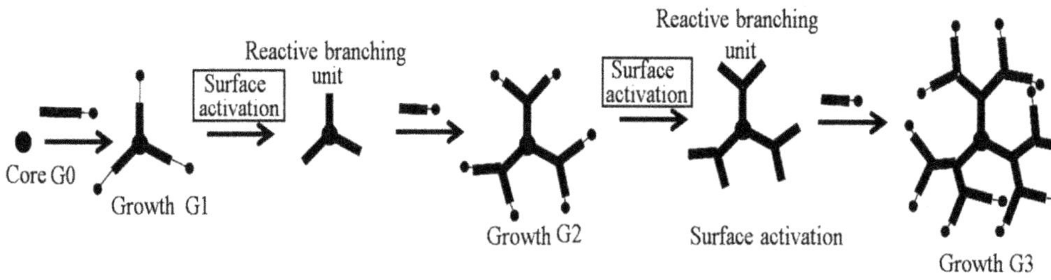

FIGURE 38 Divergent synthesis of dendrimers.

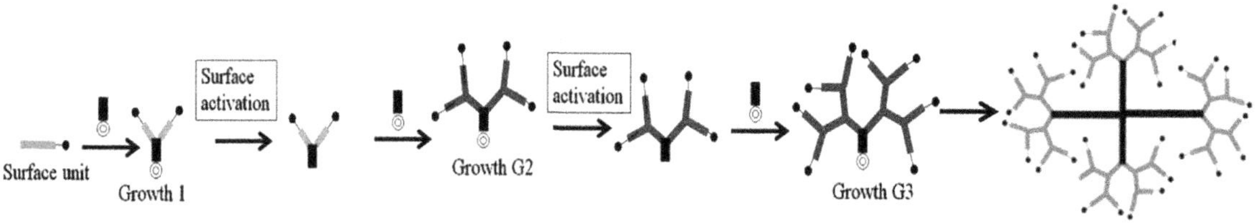

FIGURE 39 Convergent synthesis of dendrimers.

4. *Double exponential* growth, in which two different dendrimers (core and periphery) are activated and converged, resulting in the formation of higher generation dendrimers (Buhleier et al., 1978; Hawker et al., 1990; Newkome et al., 1985; Tomalia et al., 1985).

Divergent synthesis: In this method, the starting material is a reactive core (Figure 38). By addition of the reactive monomer units, generation G1 is grown. A G1 molecule is activated, which reacts with more monomers for the next generation (G2). The two steps can be repeated for further growth. This method allows production of large quantities of dendrimers because the molar mass of the dendrimer is doubled at each step. A disadvantage is that the incomplete growth and side reaction products are difficult to remove.

Convergent synthesis: This approach addresses the weaknesses of divergent syntheses. As shown in Figure 39, the growth begins at what will end up being the surface of the dendrimer. Then, growth works inwards by gradually linking surface units together. When the growing wedges reach a required size, they attach to a suitable core to give a complete dendrimer (Figure 39). In convergent growth, only two simultaneous reactions are required for any generation-adding step. The convergent methodology, however, suffers from low yields in the synthesis of large structures. Dendritic wedges of higher generations may encounter serious steric problems in the reactions of their branching points.

Hypercore synthesis: As shown in Figure 40, the first step of the hypercore and branched method involves pre-assembly of a surface unit and a core that form a branching monomer and hypercore, respectively. Hypercores and branched monomers allow synthesis of higher order dendrimers.

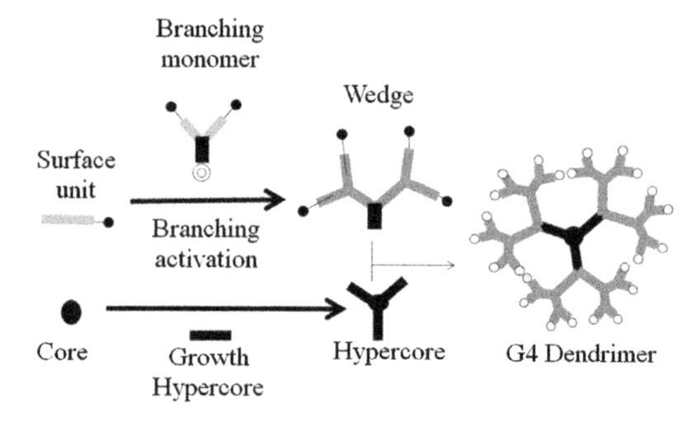

FIGURE 40 Hypercore synthesis of dendrimers.

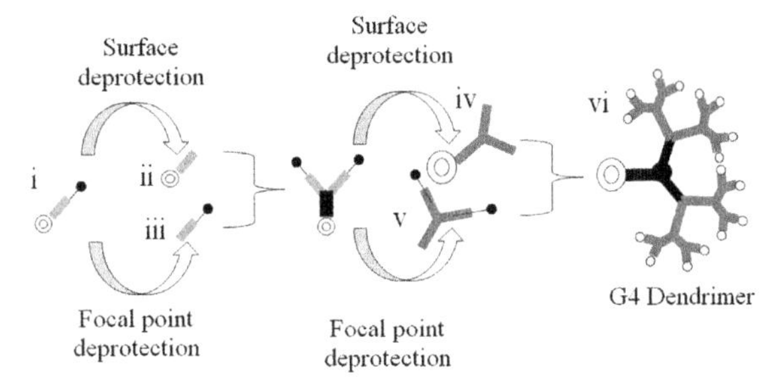

FIGURE 41 Double exponential synthesis of dendrimers.

Double exponential synthesis: Figure 41 shows the steps involved in double exponential growth. Basically, this method allows the preparation of monomers for both convergent and divergent growth from a single starting material. The monomers are then used to synthesize G4 dendrimers. This method allows large dendrimers in relatively few steps. The double exponential methodology provides a means whereby a dendritic fragment can be extended in either the convergent or the divergent direction as required.

5.1.4 Dendrimer Functionalization

Dendrimer (hydrophilic or hydrophobic) surfaces contain functional groups such as –OH, –COOH, –SH, $-NH_2$, –COR (ester), and $-NHNH_2$, which can be functionalized for an intended purpose. In general, dendrimers are functionalized in three stages:

Surface groups ⟶ Group Modification ⟶ Group activation ⟶ Functionalization
Example:
Dendrimer-Cl ⟶ *Dendrimer-NH_2* ⟶ *Dendrimer-NHS* ⟶ *Dendrimer-peptide*

Dendrimers are functionalized by either directly modifying the surface groups (Figure 42(A)) or coupling the surface groups with functionalized PEGs (Figure 42(B)). The advantages and disadvantages of PEG are described elsewhere in this chapter and summarized in Figure 42(C).

For functionalization, the end groups, such as carboxylic acids (–COOH) or amines (NH_2, RNH or R_2N), must be activated using appropriate activation reagents, such as aryl azide, hydrazine, *N*-hydroxysuccinamide, diazrine, or 1-ethyl-3-(3-dimethylaminopropyl)-carbodiimide (Roberts et al., 2002). Figure 43 shows the functionalization of a dendrimer for drug delivery and imaging simultaneously.

5.2 Application of Dendrimers

Dendrimers are tunable nanostructures that can be designed for specific functions including, but not limited to, host–guest chemistry, electrochemistry, photochemistry, pollution management, catalysis, and nanomedicine (Adronov and Frechet, 2000; Brady and Levy, 1995; Dykes, 2001; Hecht and Frechet, 2001; Langer et al., 1995; Miller et al., 1998; Service, 1995; Tully and Frechet, 2001; Twyman et al., 1999; Vogtle and Fisher, 1999; Wells and Crooks, 1996; Zhu and Shi, 2013). They are suitable for use (1) as

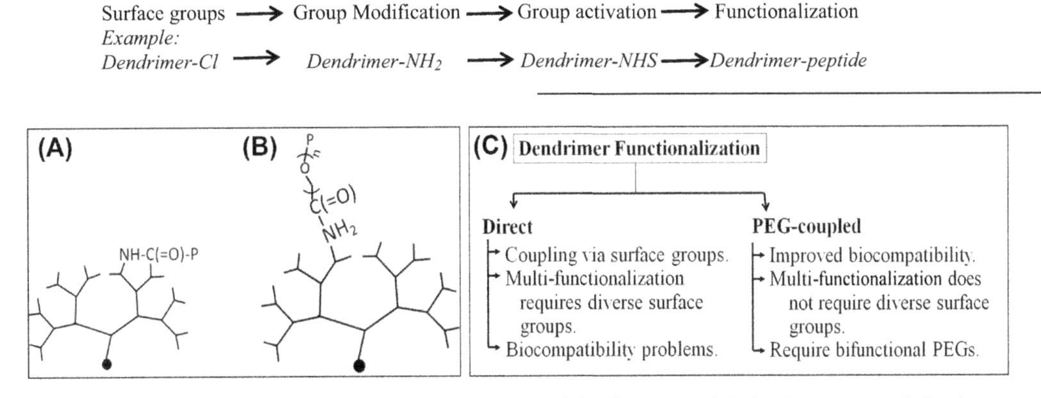

FIGURE 42 Direct (A) and PEG-coupled (B) functionalization of dendrimers and their advantages and disadvantages (C).

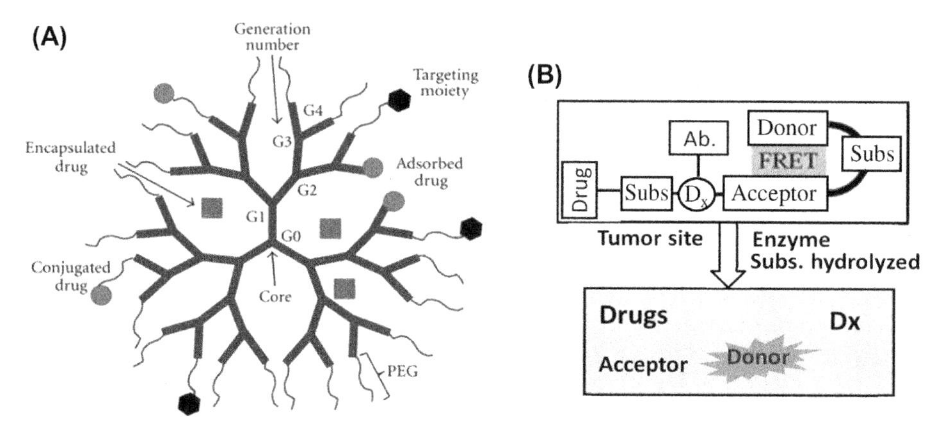

FIGURE 43 **Structure of functionalized dendrimers.** (A) Structure of a dendrimer containing (1) water-soluble and -insoluble drugs by adsorption to the surface (ionic interaction), (2) drugs encapsulated within hydrophobic micro-cavities inside branching clefts, (3) drugs conjugated (via cleavable or covalent linker) to the surface functional groups and (4) targeting moiety to steer the dendrimer to the desired site. (B) For cancer treatment, the dendrimer (Dx) will contain a cancer-specific targeting moiety that will steer the probe to the cancer site. In this example, a tumor-selective antibody has been used for targeting and a tumor-specific enzyme's substrate has been used for drug release and FRET-based imaging. The drugs may be linked to the dendrimer via an acid-cleavable bond. Dx is coupled to (1) an FRET system consisting of (acceptor-Subs (a cancer selective enzyme's substrate)-donor) connected to the Dx by a cancer-selective enzyme's substrate. In the absence of the enzyme, the fluorescence remains quenched. When the probe reaches the tumor cells, the substrate is hydrolyzed and the donor dissociates from the acceptor, resulting in fluorescing of the tumor.

therapeutic agents, (2) as targeted contrast agents for MRI and near-infrared imaging (NIRI), and (3) for controlled delivery of therapeutic agents. This section describes important applications of dendrimers.

5.2.1 Dendrimers as Therapeutic Agents

Although most of the literature deals with application of functionalized dendrimers in bioimaging and drug delivery (discussed later), some studies have reported direct medicinal effects of dendrimers. G4 PAMAM and PPI dendrimers eliminate prion infectivity from neuroblastoma ScN2a cells (Gajbhiye et al., 2009; Prusiner, 1982; Supattapone et al., 2001). G3 to G5 PAMAM dendrimers were capable of disrupting the previously formed Aβ aggregates in patients with Alzheimer's disease; higher generation dendrimers were more effective at lower concentrations than lower generation dendrimers (Heegaard et al., 2004; Klajnert et al., 2006; Patel et al., 2006; Selvaggini, 1993). Dendrimers functionalized with sialic acid suppressed growth of influenza virus (Landers et al., 2002). Sialic acid or polysulfated galactose-derivatized dendrimers interfere with fusion of HIV to the cell membrane, possibly via binding of the V3 loop within gp120 (Kensinger et al., 2004a,b). As shown in Figure 44(A), HIV surface-protein gp120 selectively binds to the cell membrane's silic acid residue and then fuses with the membrane, the initial step of viral infection. Silated or polysulfated galactose dendrimers compete with the cell membranes for gp120 binding; thus, the HIV-dendrimer complex results in unsuccessful infection (Figure 44(B)). Many metal-based dendrimers have been shown to catalyze O_2 into O free radical (Figure 44(C)), thus damaging the bacterial membranes (Chen and Cooper, 2002). Although promising, these applications are still in the experimental stage

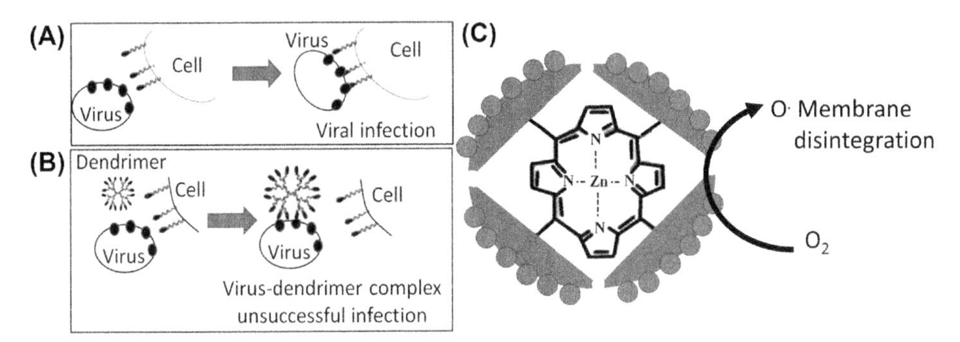

FIGURE 44 **Mechanisms of action of functionalized dendrimers.** (A) Interaction of HIV's gp120 with cell membrane that results in their internalization, the initial step of HIV infection. (B) Silated or polysulfated galactose dendrimers compete with the cell binding site and suppress the binding of gp120 to the cell membrane, thus suppressing their internalization. (C) Structure of a Zn-based dendrimer that can form superoxide free radicals, resulting in disintegration of bacterial membranes.

and more research is needed to make them available for clinical use.

5.2.2 *Dendrimers in Gene Therapy*

Gene therapy, although very promising, has not succeeded in the clinical setting for therapeutic purposes. This is because a safe method of gene transfection is not yet available. Current viral transfection agents are not safe and not approved for clinical applications. Recent studies have shown that polycationic PAMAM dendrimers may bind strongly to the negatively charged phosphate group of single-stranded (such as siRNAs) and double-stranded DNA strands and protect them from enzymatic degradation in the body (Gebhart and Kabanov, 2001; Stiriba et al., 2002; Sato et al., 2001). Because PAMAM forms compact polycations under physiological conditions, PAMAM dendrimers have been used as DNA delivery systems. As shown in Figure 45, DNA wraps around the dendronized cationic polymers over G9 generations (Goss et al., 2002).

A combination of electroporation (Wang et al., 2001) and β-cyclodextrin-(Roessler et al., 2001) coupled DNA-dendrimer or PAMAM dendrimer systems allowed superior gene expression relative to the electroporation alone (Arima et al., 2001; Wang et al., 2001). In addition, β-cyclodextrin improved the nanoparticle dispersion and yielded more monodispersed particles (Roessler et al., 2001). Poly(ethylene glycol) functionalization of G5-PAMAM dendrimers produced a 20-fold increase in transfection efficiency using plasmid DNA coding for a reporter protein β-galactosidase relative to partially degraded PAMAM dendrimers (Luo et al., 2002). These observations provide strong evidence for potential applications of nanoparticles in gene delivery (De Jong et al., 2001).

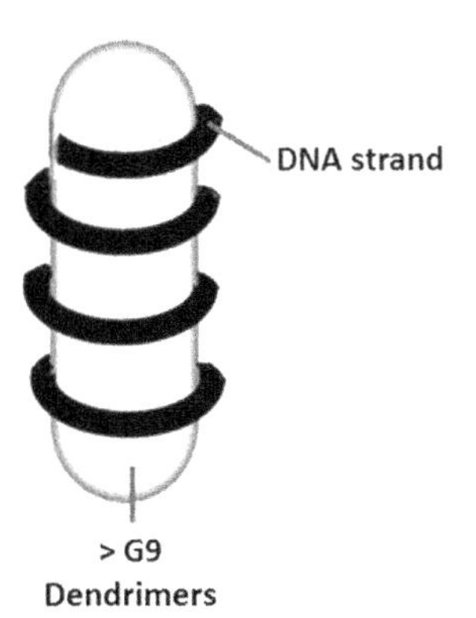

FIGURE 45 Structure of a dendrimer-DNA complex.

5.2.3 *In Vivo Imaging*

Dendrimers have been shown to be excellent contrast agents for MRI and X-ray computed tomography (Aime et al., 1999; Astruc et al., 2010, 2012; Kobayashi and Brechbiel, 2005; Langereis et al., 2007; Wolinsky and Grinstaff, 2008) and dye carriers for optical imaging (Kojima et al., 2011). The larger size of high-generation dendrimers offers potential to develop multipurpose agents that can act both as multimodal imaging agents and as delivery vectors (Longmire et al., 2008). Figure 46 shows a few examples of dendrimer-based probes for imaging.

5.2.4 *Dendrimers as Drug Carriers*

The unique structural and chemical properties of nanodendrimers make them ideal drug carriers. Dendrimers' empty internal cavities and open conformations allow encapsulation of hydrophobic or hydrophilic drugs (Jansen et al., 1994). In addition, a high density of surface groups permits conjugation of multiple drugs or dyes (Yang et al., 2004). Figure 47 describes

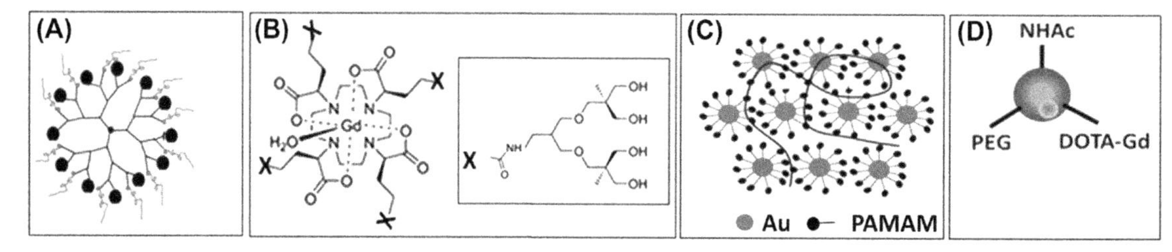

FIGURE 46 **Examples of nanoparticles used for imaging.** (A and B) PAMAM (G4) functionalized Gd at the end group (A) or part of the core (B). Polyethylene glycol (PEG)-surface modification is shown in red. PEG makes the dendrimer hydrophilic, thus reducing sequestration in the reticuloendothelial system (Luoa et al., 2011; Villaraza et al., 2010). (C) Structure of AuPAMAM nanoparticles non-covalently coupled to a plasmid for gene delivery (Figueroa et al., 2014). (D) Polyamine dendrimers with carbonyl end groups were coupled to quaterrylene dye that has absorption and emission wavelength close to 800 nm, the most preferred wavelength for NIR fluorescence imaging. Being highly hydrophobic, this dye aggregates extensively. The dendrimer–dye complex exhibits high biocompatibility and low aggregation.

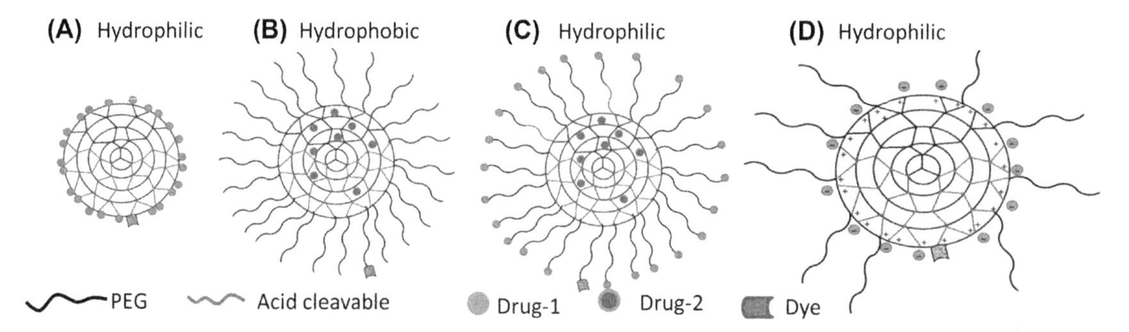

FIGURE 47 **Different strategies of drug loading in dendrimers.** (A) Hydrophilic drugs bind to the hydrophilic dendrimer surface. (B) Hydrophobic drugs fill in the intradendrimer space. (C) Hydrophilic drugs are attached to the PEG molecule. (D) Electrostatic attraction between anionic drugs and cationic dendrimer surface.

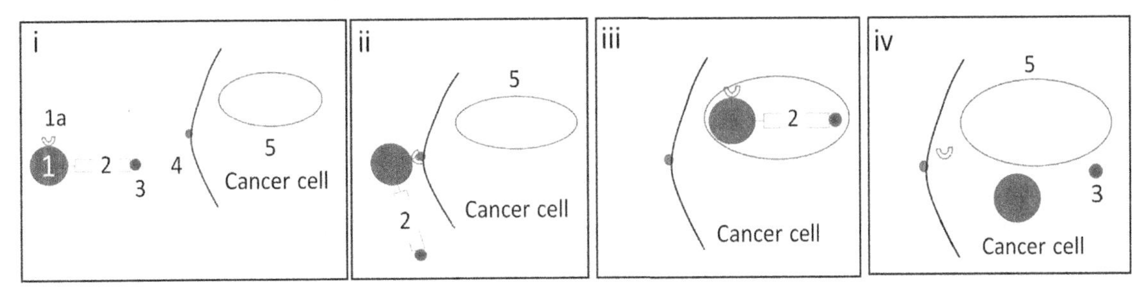

FIGURE 48 **Selective transport of anticancer drugs into the cancer cells.** (i) The dendrimer (1) was coupled to a cancer selective antibody or protein via a non-cleavable linker (1a) and the drug (3) via an acid cleavable hydrazine linker (2). The cancer cell membrane contains a selective antigen (4). The lysosome (5) pH is <5.0. (ii) The cancer selective protein of the probe in systemic blood binds to the cancer cell antigens and gets internalized into the lysosome (iii). The acidic pH of lysosome hydrolyzes the drug that is released into the cytoplasm of the cancer cells. The dendrimers and proteins are metabolized in body (iv).

possible dendrimer-drug interaction mechanisms. They covalently (Figure 47(A)) or noncovalently (Figure 47(D)) bind to the surface group of a hydrophilic dendrimer, covalently bind to PEG-functionalized hydrophobic dendrimers (Figure 47(B)), and/or encapsulate and bind to PEG (Figure 47(C)) (D'Emanuele and Attwood, 2005; Esfand and Tomalia, 2001; Gillies and Frechet, 2005).

Drug molecules are entrapped in the internal cavities of dendrimers via physical encapsulation, hydrophobic interaction, ionic interaction, or hydrogen bonding (Kojima et al., 2000). Many hydrophilic dendrimers such as PAMAM possess a positive charge at biological pH; thus, they bind negatively charged drugs via electrostatic interaction. Many acidic (containing −COOH) nonsteroidal anti-inflammatory, anticancer, and antibacterial drugs have been coupled with dendrimers by electrostatic interactions (Chauhan et al., 2003; Cheng and Xu, 2005; Gillies and Frechet, 2005; Gupta et al., 2006; Kolhe et al., 2003; Ma et al., 2006). A significant development in the release of anticancer drugs selectively in cancer cells is acid cleavable linkage for dendrimer-drug conjugates. The linkages remain intact in the systemic circulation (pH 7.4) but quickly degrade once internalized into the cancer cell (pH 5−6) and release the attached drug to produce the desired therapeutic activity (Bae et al., 2007) (Figure 48).

5.2.5 Unique Applications of DNA Dendrimers

Unique applications of DNA dendrimers include the following:

DNA as a template for self-assembly: DNA scaffolds are likely to play a promising role in the assembly of spatially addressable nanoscale components in future optoelectronic devices (Zhang et al., 2006; Park et al., 2005).

DNA as a guide for self-assembly: Real-time assembly of nanoscale components may be possible by virtue of DNA's molecular recognition capabilities (Mastroianni et al., 2009; Mirkin et al., 1996).

Medical diagnostics: DNA dendrimers provide a unique opportunity of constructing sensitive sensors and imaging devices (Cheng et al., 2009; Li et al., 2005, 2009).

DNA gel as a scaffold for cell-free protein production: Protein synthesis in biological systems is time-consuming and lacks commercial application. Studies have shown that DNA gel scaffolds can induce protein synthesis in cell-free medium on a commercial scale (Park et al., 2009).

6. CONCLUSIONS

Nanoparticles are structures between 1 and 100 nm in diameter at least in one dimension, having unique size- and shape-dependent physicochemical and biological properties not found in the corresponding bulk materials. Chemically, nanoparticles belong to the following groups: metals and metal oxides, carbon nanotubes, and polymers and dendrimers. However, despite their chemical variability, nanoparticles exhibit common physicochemical properties such as high surface-to-volume ratio, quantum confinement, surface plasmon resonance in some metal particles, and superparamagnetism in magnetic materials (discussed in detail in Chapter 3). The physical unity and chemical diversity of nanoparticles have allowed development of numerous novel function-oriented nanosystems.

7. APPENDICES

Appendix 1: Calculation of the Number of Atoms and Percentage of Surface Atoms in a Nanoparticle

For bulk particles, the number of atoms can be calculated using the following equation:

$$\text{Number of particles} = \text{Moles of substance} \times (6.022 \times 10^{23}) \quad (12)$$

Here, a mole (mass (g)/molecular weight) of any chemical represents 6.022×10^{23} molecules (Avogadro's number). However, Eqn (12) cannot be directly applied to determine nanoparticle numbers because nanoparticles are heterogenous mixtures of different size that, in suspension, may undergo aggregation and sedimentation. In the case of metal nanoparticles, molar concentration of the total metal can be calculated; thus, the number of metal atoms in one nanoparticle can be determined. However, this information may not be useful to determine the nanoparticles' application or toxicity. Determination of molarity may not be relevant to the study of nanoparticle toxicity. However, determination of nanoparticles' diameter, number of particles, percentage of surface atoms or molecules, or total surface area may be more relevant in nanoparticle research and risk assessment.

The size and distribution of nanoparticles can be measured using a transmission electron microscope that measures individual particles directly. The irregularity parameter is a simple concept that is employed to calculate the sphericity of particles. The specific surface area of the particles (summation of the areas of the exposed surfaces of the particles per unit mass (m^2/g) that is inversely related to the particle size) can be determined by measuring nitrogen adsorption (Brunauer, Emmett, and Teller (BET); (Brunauer et al., 1938). Reynolds et al. (2005) devised a viscosity/light scattering method for determining the weight of nanoparticle cores or total nanoparticles. The weight of nanoparticles can be determined according to the equation $h/h_o = 1 + 2.5\theta$, where h is nanoparticle suspension viscosity, h_o is the viscosity of solvent alone, and θ is the volume fraction of nanoparticles. The number of nanoparticles/volume (N) $= \theta/[4/3\pi(d/2)^3]$, where d is diameter. If the surface atoms forming a spherical shell have a thickness of d and the particles' outer radius is R, then its inner radius is $(R - d)$. The percentage of surface atoms can be calculated using the following equation: % of surface atoms $= ((R^3 - (R - d)^3)/R^3) \times 100$ (Figure 49).

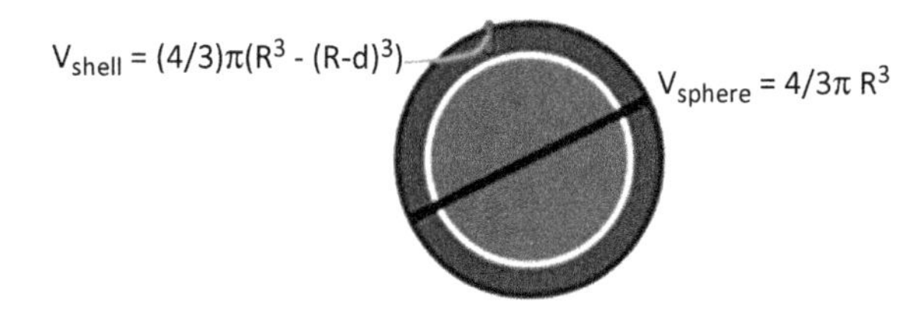

FIGURE 49 Calculation of percentage of atoms at a nanoparticle surface.

Appendix 2: Van der Waals Forces

Van der Waals forces are weak electrostatic forces that attract neutral molecules to one another. Particles in liquid or air vibrate and move constantly. Thus, they collide with other particles, including the media's particles such as water molecules—the process known as Brownian motion (Figure 50). When these neutral particles reach within a threshold distance, electrons from one particle are pulled towards the nucleus of the other particle, thus causing transient polarization (electron-rich domain: δ^- and electron-deficient domain: δ^+). The δ^+ side of one particle attracts the δ^- side of another (London dispersion interaction). Particle vibration or motion breaks the interparticle interaction and the particles separate.

VDW interactions have an *attractive interaction* between the atoms, which results from the induced dipoles, and a *repulsive interaction,* which results from overlap of the electron clouds of the two atoms, when they get too close to each other. The total energy of VDW interactions can be approximated by the Lennard-Jones expression (Figure 51).

The *π-electron interaction* occurs between two organic molecules having C–C π electrons or between an organic molecule having C–C π electrons and another polar or nonpolar

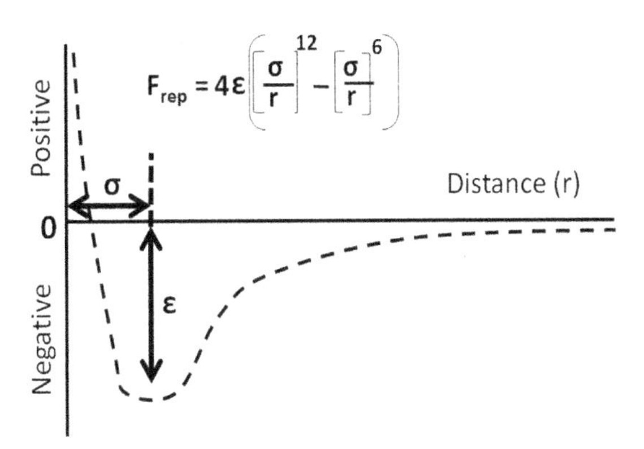

FIGURE 51 The van der Waals Potential is a function of the distance between the centers of two particles. When two nonbonding particles are in an infinite distance apart, the possibility of them coming together and interacting is minimal. As the distance of separation decreases, the probability of interaction increases. The particles come closer together until they reach a region of separation where the two particles become bound and their bonding potential energy decreases from zero to become a negative quantity. While the particles are bound, the distance between their centers will continue to decrease until the particles reach an equilibrium, which is specified by the separation distance at which the minimum potential energy is reached. Now, if we keep pushing the two bound particles together past their equilibrium distance, repulsion begins to occur, as particles are so close to each other that their electrons are forced to occupy each other's orbitals. Therefore, repulsion occurs as each particle attempts to retain the space in their respective orbitals. Despite the repulsive force between both particles, their bonding potential energy rises rapidly as the distance of separation between them decreases below the equilibrium distance.

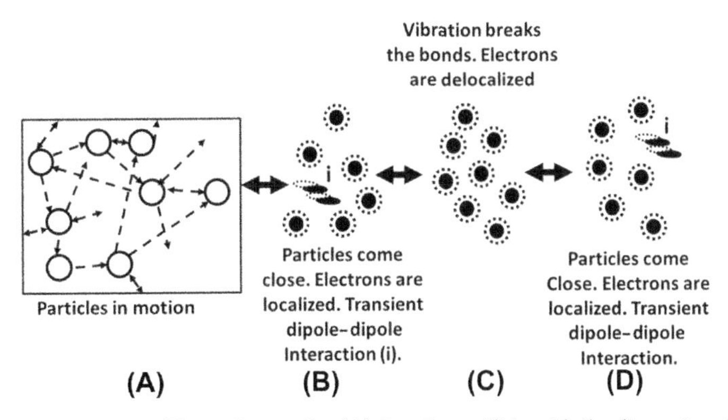

FIGURE 50 Nanoparticles in perpetual Brownian motion (A) thus they collide with the dispersion phase particles. Randomly, two particles come close so that their electrons localize (because of their electrostatic repulsing) and develop transient dipole–dipole interaction (B). Vibration breaks the weak bond and the electrons are delocalized (C). Simultaneously, other particles come close and form transient bonds (D). Thus, even the noble metals exhibit weak transient bond formations.

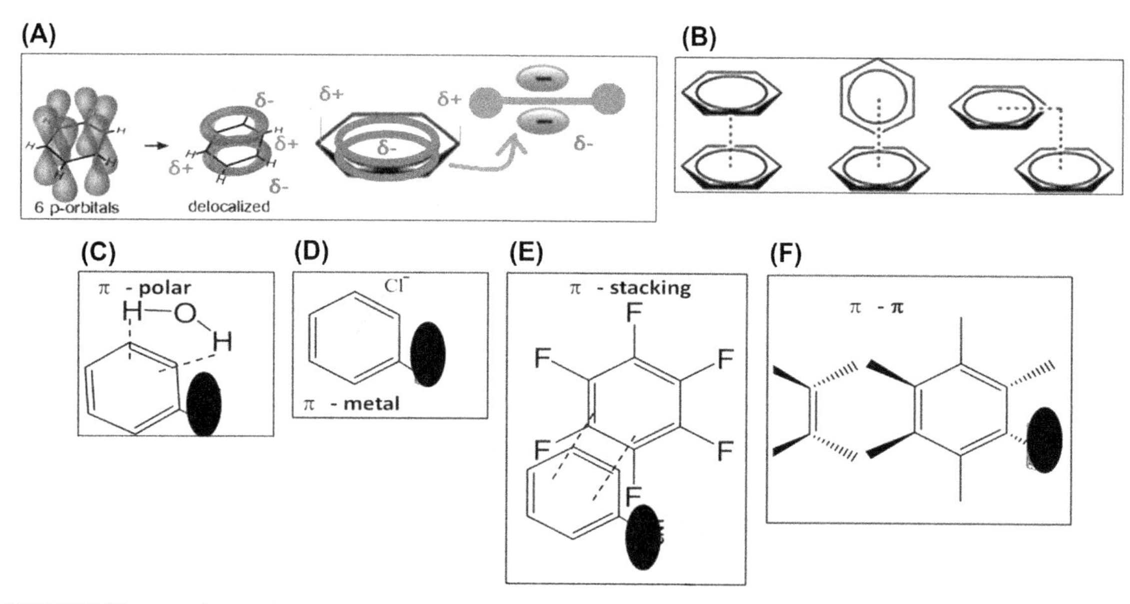

FIGURE 52 **Van der Waals interaction involving π electrons.** (A) Benzene ring contains six *p*-orbital electrons that forms two rings of delocalized charges (δ^- and δ^+). (B) Two benzene rings interact via $\pi-\pi$ bonds on different configurations. (C) Water and benzene rings form polar H-bonds between δ^+ of water's H atoms and δ^- of the benzene tings. (D) VDW interaction between a metal and an π electrons of benzene. (E) Stacking VDW bond formed between fluoro benzene and benzene. (F) Hydrophobic $\pi-\pi$ interaction between two hydrophobic groups.

molecule (Matthews et al., 2014). As shown in Figure 52(A), benzene ring contains six C–C π-electrons that, when delocalized, form two bands of electron-rich regions (δ^-) separated by a positively charged region (δ^+) due to the H atoms (Figure 52(A)). Two benzene rings form a π-π interaction in stacked, T-shaped, and parallel modes. Figure 52(B) shows an interaction between δ^+ and δ^- charges between two molecules. Figure 52(C–F) shows some of the examples of different types of π-electron interaction.

Appendix 3: Zeta Potential

Metal nanoparticles, depending on their chemical characteristics, possess either +ve or −ve charge that is modulated by the pH of the dispersion medium. The isoelectric points (IEP) for TiO_2, Fe_2O_3, Al_2O_3, ZnO, and CeO_2 nanoparticles have been shown to be 5.19, 4.24, 7.6, 8.5, and 7.6, respectively (Berg et al., 2009). This shows that, at physiological pH, TiO_2 and Fe_2O_3 will be negatively charged, ZnO will be positively charged, while Al_2O_3 and CeO_2 will be neutral. These nanoparticles, depending on their charge, attract oppositely charged ions, resulting in the formation of thin layers of ions to its surface (Figure 53(A)).

The attractive force between the nanoparticles and the ions is inversely related to the distance (Figure 53(A), inset). Studies have shown that the double layer of ions is stably adsorbed onto the nanoparticle and travel with it. The electrical potential at the boundary of the double layer is called the zeta (ζ) potential. The distance from the nanoparticle surface and the double-ionic layer surface is inversely related to the formation of aggregates: A smaller distance will result in aggregation (Figure 53(B)), while a larger distance will favor dispersion (Figure 53(C)). Zeta (ζ) potential is defined as the surface charge of nanoparticles in solution (colloids). Metal

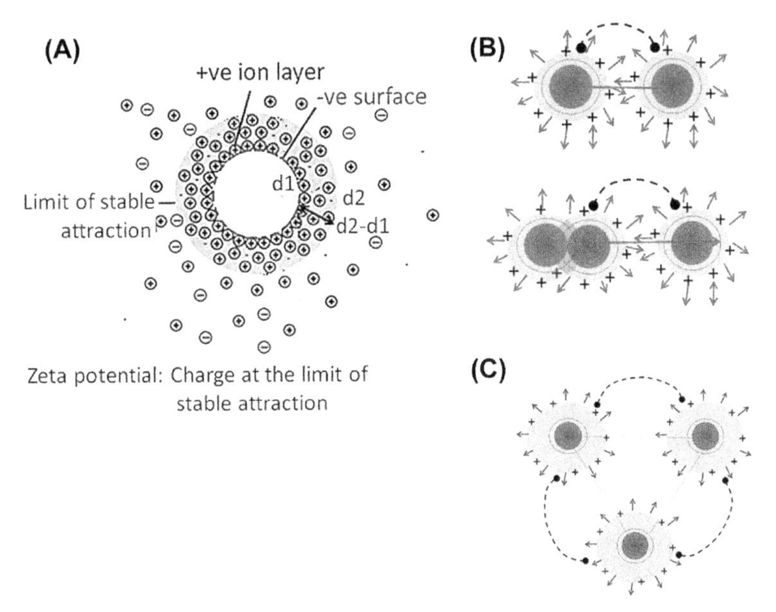

FIGURE 53 (A) Nanoparticle's surface (diameter *d*1) contain a bi-layer of positively charged ions that are permanently attached to and migrate with the particles (area in green (light gray in printed versions), limit of stable attraction, diameter *d*2). The positive charge prevents nanoparticles to aggregate. Zeta (ζ) potential, defined as the surface charge of nanoparticles in solution (colloids) at diameter *d*2. The *d*2-*d*1 difference determines interaction between two particles. (B) Smaller *d*2-*d*1 values may allow particles to come closer and aggregate (bottom). (C) Larger *d*1-*d*2 values prevent intra-particle interactions and promote dispersion.

nanoparticles have a negative surface charge that attracts a thin layer of positive ions on to its surface (Figure 53(C)). The electric potential at the boundary of the +ve ion layer stably attracted to the nanoparticles is known as the ζ potential of the particles. The values typically range from +100 mV to −100 mV. The magnitude of the ζ potential predicts colloidal stability. A ζ value greater than +25 mV or less than −25 mV typically indicates a stable colloid because of charge repulsion between two nanoparticles.

A nanoparticle coated with organic polymers, especially PEG, expands the ionic bilayer, thus increasing d2 (A1 non-PEG, B1 PEGylated). Thus, PEG may prevent nanoparticle aggregation. In non-PEG nanoparticles, an increase in d2-d1 increases the surface potential that peaks (ζ potential) at certain critical distance; thereafter, the potential decreases (54A2). In PEGylated nanoparticles, the ζ potential is maintained at wider distance range. Figure 54(A3) and (B3) show the relationship between aggregation of non-PEG (54A3) and PEGylated (54B3) nanoparticles (Fe_2O_3 (isoelectric point (IEP) 4.24) and ZnO (IEP 8) nanoparticles) and pH of the dispersion media. The ζ potential of a nanoparticle is highly dependent upon the dispersion phase pH. A change in pH also changes the ζ potential, which is zero at the isoelectric point. As shown in Figure 54(A3), non-PEG Fe_2O_3 nanoparticles aggregates at pH 4 and the ZnO nanoparticle aggregates at pH 8, the IEP for both particles. The PEGylated nanoparticles did not aggregate at any of the pH values tested. This confirms that nanoparticles highly aggregate at IEP.

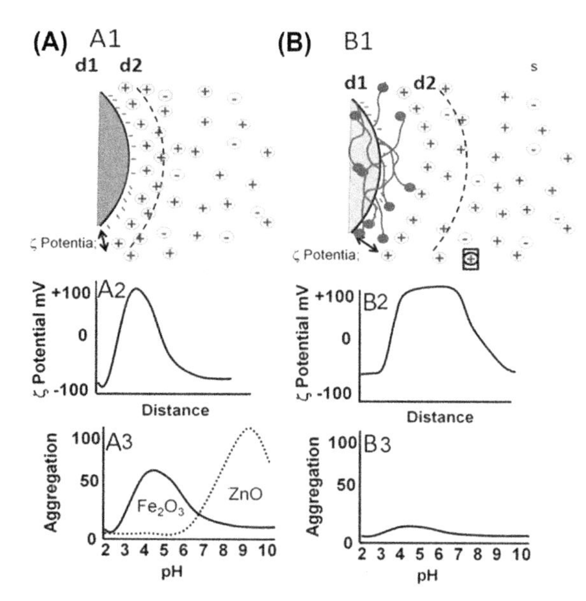

FIGURE 54 Zeta potential and aggregation of non-functionalized (A) and PEGylated (B) nanoparticles. (A1 and B1) Interaction between the nanoparticle and ions of the disperse phase. The limit of attraction (*d*2) is identified by a broken line. The *d*2-*d*1 values for PEGylated nanoparticles were larger than the values for non-functionalized nanoparticles. (A2 and B2) Theoretical plots of distance versus surface potential. The peak represents the particle's ζ potential. (A3 and B3) Theoretical plots showing relationship between aggregation and pH of the dispersion medium. Highest aggregation occurred at the particles' isoelectric potential when the ζ potential is zero.

References

Abhilash, M., 2010. Potential applications of nanoparticles. Int. J. Pharm. Bio Sci. 1, 1–12.

Abid, A.D., Kanematsu, M., Young, T.M., Kennedy, I.M., 2013. Arsenic removal from water using flame-synthesized iron oxide nanoparticles with variable oxidation states. Aerosol Sci. Technol. 47, 169–176.

Adronov, A., Frechet, J.M.J., 2000. Light harvesting dendrimers. Chem. Commun. 1701–1710.

Afzal, M.A., Kalmodia, S., Kesarwani, P., Basu, B., Balani, K., 2013. Bactericidal effect of silver-reinforced carbon nanotube and hydroxyapatite composites. J. Biomater. Appl. 27, 967–978.

Agbabiaka, A., Wiltfong, M., Park, C., 2013. Small angle X-ray scattering technique for the particle size distribution of nonporous nanoparticles. J. Nanopart. 2013, 640436.

Ahmed, M., Jiang, X., Deng, Z., Narain, R., 2009. Cationic glyco-functionalized single-walled carbon nanotubes as efficient gene delivery vehicles. Bioconjug. Chem. 20, 2017–2022.

Aime, S., Botta, M., Fasano, M., Terreno, E., 1999. Prototropic and water-exchange processes in aqueous solutions of Gd(III) chelates. Acc. Chem. Res. 32, 941–949.

Ajayan, P.M., 1999. Nanotubes from carbon. Chem. Rev. 99, 1787–1800.

Albini, A., Pennesi, G., Donatelli, F., Cammarota, R., De Flora, S., Noonan, D.M., 2010. Cardiotoxicity of anticancer drugs: the need for cardio-oncology and cardio-oncological prevention. J. Natl Cancer Inst. 102, 14–25.

Al-Hanbali, O., Rutt, K.J., Sarker, D.K., Hunter, A.C., Moghimi, S.M., 2006. Concentration dependent structural ordering of poloxamine 908 on polystyrene nanoparticles and their modulatory role on complement consumption. J. Nanosci. Nanotechnol. 6, 3126–3133.

Amiri, A., Memarpoor-Yazdi, M., Shanbedi, M., Eshghi, H., 2013. Influence of different amino acid groups on the free radical scavenging capability of multi walled carbon nanotubes. J. Biomed. Mater. Res. A 101, 2219–2228.

Arima, H., Kihara, F., Hirayama, F., Uekama, K., 2001. Enhancement of gene expression by polyamido-amine dendrimer conjugates with α-, β-, and γ-cyclodextrins. Bioconjug. Chem. 12, 476–484.

Aryal, S., Grailer, J.J., Pilla, S., Steeber, D.A., Gong, S., 2009. Doxorubicin conjugated gold nanoparticles as water-soluble and pH-responsive anticancer drug nanocarriers. J. Mater. Chem. 19, 7879–7884.

Aslam, M., Gopakumar, G., Shoba, T.L., Mulla, I.S., Vijayamohanan, K., Kulkarni, S.K., Vogel, W., 2002. Formation of Cu and Cu_2O nanoparticles by variation of the surface ligand. J. Colloid Interface Sci. 255, 79–90.

Astruc, D., Boisselier, E., Ornelas, C., 2010. Dendrimers designed for functions: from physical, photophysical, and supramolecular properties to applications in sensing, catalysis, molecular electronics, photonics, and nanomedicine. Chem. Rev. 110, 1857–1859.

Astruc, D., Liang, L., Rapakousiou, A., Ruiz, R., 2012. Click dendrimers and triazole-related aspects: catalysts, mechanism, synthesis, and functions. A bridge between dendritic architectures and nanomaterials. Acc. Chem. Res. 45, 630–640.

Avgoustakis, K., 2004. Pegylated poly(lactide) and poly(lactide-co-glycolide) nanoparticles: preparation, properties and possible applications in drug delivery. Curr. Drug Deliv. 1, 321–333.

Avila, A.F., Silveira, G., Lacerda, R., 2008. Molecular mechanics applied to single-walled carbon nanotubes. Mat. Res. 1, 325–333.

Bae, Y., Diezi, T.A., Zhao, A., Kwon, G.S., 2007. Mixed polymeric micelles for combination cancer chemotherapy through the concurrent delivery of multiple chemotherapeutic agents. J. Control. Release 122, 324–330.

Banobre-lopez, M., Teijeiro, A., Rivas, J., 2013. Magnetic nanoparticle-based hyperthermia for cancer treatment. Rep. Pract. Oncol. Radiother. 18, 397–400.

Baptista, P.V., Doria, G., Conde, J., 2011. Alloy metal nanoparticles for multicolor cancer diagnostics. Proc. SPIE 7909. http://dx.doi.org/10.1117/12.879025.

Barbara, E.C.B., 1968. In: Patai, S. (Ed.), The Chemistry of the Amino Group. Wiley-Interscience, London. http://www.organic-chemistry.org/synthesis/C2O/carboxylicacids/oxidationsalcohols. shtm.

Bates, K., Kostarelos, K., 2013. Carbon nanotubes as vectors for gene therapy: past achievements, present challenges and future goals. Adv. Drug Deliv. Rev. 65, 2023–2033.

Baumann, A., Tuerck, D., Prabhu, S., Dickmann, L., Sims, J., 2014. Pharmacokinetics, metabolism and distribution of PEGs and PEGylated proteins: quo vadis? Drug Discov. Today 19, 1623–1631.

Beck, J.S., Chu, C.T.W., Johnson, I.D., Kresge, C.T., Leonowicz, M.E., Roth, W.J., Vartuli, J.W., 1991. WO Patent 91/11390.

Bender, A.R., Von Briesen, H., Kreuter, J., Duncan, I.B., Rubsamen-Waigmann, H., 1996. Efficiency of nanoparticles as a carrier system for antiviral agents in human immunodeficiency virus-infected human monocytes/macrophages in vitro. Antimicrob. Agents Chemother. 40, 1467–1471.

Berg, J.M., Romoser, A., Banerjee, N., Zebda, R., Sayes, C.M., 2009. The relationship between pH and zeta potential of ~30 nm metal oxide nanoparticle suspensions relevant to in vitro toxicological evaluations. Nanotoxicology 3 (4), 276–283.

Besenius, P., Heynens, J.L.M., Straathof, R., Nieuwenhuizen, M.M.L., Bomans, P.H.H., Terreno, E., Aime, S., Strijkers, G.J., Nicolay, K., Meijer, E.W., 2012. Paramagnetic self-assembled nanoparticles as supramolecular MRI contrast agents. Contrast Media Mol. Imaging 7, I356–I361.

Bhatt, N., Huang, P.-J.J., Dave, N., Liu, J., 2011. Dissociation and degradation of thiol-modified DNA on gold nanoparticles in aqueous and organic solvents. Langmuir 27, 6132–6137.

Bhirde, A.A., Patel, Y., Gavard, J., Zhang, G., Sousa, A.A., Masedunskas, A., Leapman, R.D., Weigert, R., Gutkind, J.S., Rusling, J.F., 2009. Targeted killing of cancer cells in vivo and in vitro with EGF-directed carbon nanotube-based drug delivery. ACS Nano 3, 307–316.

Bogdan, N., Vetrone, F., Roy, R., Capobianco, J.A., 2010. Carbohydrate-coated lanthanide-doped upconverting nanoparticles for lectin recognition. J. Mater. Chem. 20, 7543–7550.

Brady, P.A., Levy, E.G., 1995. Inorganic and organometallic macromolecules: design and applications. Chem. Ind. (London) 18, 21.

Britzab, D.A., Khlobystov, A.N., 2006. Noncovalent interactions of molecules with single walled carbon nanotubes. Chem. Soc. Rev. 35, 637–659.

Brown, S.D., Nativo, P., Smith, J.-A., Stirling, D., Edwards, P.R., Venugopal, B., Flint, D.J., Plumb, J.A., Graham, D., Wheate, N.J., 2011. Gold nanoparticles for the improved anticancer drug delivery of the active component of oxaliplatin. J. Am. Chem. Soc. 132, 4678–4684.

Brunauer, S., Emmett, P.H., Teller, E., 1938. Adsorption of gases in multimolecular layers. J. Am. Chem. Soc. 60, 309–319.

Brust, M., Fink, J., Bethell, D., Schiffrin, D.J., Kiely, C., 1995. Synthesis and reactions of functionalised gold nanoparticles. J. Chem. Soc. Chem. Commun. 1995, 1655–1656.

Brust, M., Walker, M., Bethell, D., Schiffrin, D.J., Whyman, R., 1994. Synthesis of thiol-derivatised gold nanoparticles in a two-phase liquid–liquid system. J. Chem. Soc. Chem. Commun. 1994, 801–802.

Buhleier, E., Wehner, W., Vögtle, F., 1978. Cascade and nonskid-chain-like synthesis of molecular cavity topologies. In: Vögtle, H. (Ed.), Synthesis. Springer-Verlag, Berlin, pp. 155–158.

Bunshah, R.F., Jou, S., Prakash, S., Doerr, H.J., Isaacs, L., Wehrsig, A., Yeretzian, C., Cynn, H., Diedrich, F., 1992. Fullerene formation in sputtering and electron beam evaporation processes. J. Phys. Chem. 96, 6866–6869.

Burke, A.R., Singh, R.N., Carroll, D.L., Wood, J.C.S., D'Agostino Jr., R.B., Ajayan, P.M., Torti, F.M., Torti, S.V., 2012. The resistance of breast cancer stem cells to conventional hyperthermia and their sensitivity to nanoparticle-mediated photothermal therapy. Biomaterials 33, 2961–2970.

Calderon, M., Graeser, R., Kratz, F., Haag, R., 2009. Development of enzymatically cleavable prodrugs derived from dendritic polyglycerol. Bioorg. Med. Chem. Lett. 19, 3725–3728.

Carlsson, J.M., Scheffler, M., 2006. Structural, electronic, and chemical properties of nanoporous carbon. Phys. Rev. Lett. 96, 046806.

Caruso, F., Caruso, R.A., Mohwald, H., 1998. Nanoengineering of inorganic and hybrid hollow spheres by colloidal templating. Science 282, 1111–1114.

Cassette, E., Helle, M., Bezdetnaya, L., Marchal, E., Dubertret, B., Pons, T., 2013. Design of new quantum dot materials for deep tissue infrared imaging. Adv. Drug Deliv. Rev. 65, 719–731.

Chaki, N.K., Sundrik, S.G., Sonawane, H.R., Vijayamohanan, K., 2002. Single phase preparation of monodispersed silver nanoclusters using a unique electron transfer and cluster stabilising agent, triethylamine. J. Chem. Soc. Chem. Commun. 76.

Charlier, J.-C., 2002. Defects in carbon nanotubes. Acc. Chem. Res. 35, 1063–1069.

Chatterjee, D.K., Diagaradjane, P., Krishnan, S., 2010. Nanoparticle-mediated hyperthermia in cancer therapy. Ther. Deliv. 2, 1001–1014.

Chatterjee, J., Haik, Y., Chen, C.-J., 2003. Size dependent magnetic properties of iron oxide nanoparticles. J Magnetism Magnetic Materials 257, 113–118.

Chauhan, A.S., Sridevi, S., Chalasani, K.B., Jain, A.K., Jain, S.K., Jain, N.K., Diwan, P.V., 2003. Dendrimer mediated transdermal delivery: enhanced bioavailability of indomethacin. J. Control. Release 90, 335–343.

Chen, C.Z., Cooper, S.L., 2002. Interaction between dendrimer biocides and bacterial membranes. Biomaterials 23, 3359.

Chen, E.Y., Wang, Y.C., Mintz, A., Richards, A., Chen, C.S., Lu, D., Nguyen, T., Chin, W.C., 2006a. Activated charcoal composite biomaterial promotes human embryonic stem cell differentiation toward neuronal lineage. J. Biomed. Mater. Res. A 100, 2006–2017.

Chen, L., Zhang, D., Chen, J., Zhou, H., Wan, H., 2006b. The use of CTAB to control the size of copper nanoparticles and the concentration of alkylthiols on their surfaces. Mat. Sci. Eng. A 415, 156–161.

Chen, S., Lim, H.E., Miyata, Y., Kitaura, R., Bando, Y., Golberg, D., Shinohara, H., 2011. Transformation of ionic liquid into carbon nanotubes in confined nanospace. Chem. Commun. 47, 10368–10370.

Chena, C., Chenb, W., Zhanga, Y., 2005. Synthesis of carbon nano-tubes by pulsed laser ablation at normal pressure in metal nano-sol. Nanostructures 28, 121–127.

Cheng, E.J., Xing, Y.Z., Chen, P., Yang, Y., Sun, Y.W., et al., 2009. A pH-triggered, fast-responding DNA hydrogel. Angew. Chem. Int. Ed. 48, 7660–7663.

Cheng, H., Li, F., Su, G., Pan, H., He, L., Sun, X., et al., 1998. Large-scale and low-cost synthesis of single-walled carbon nanotubes by the catalytic pyrolysis of hydrocarbons. Appl. Phys. Lett. 72, 3282.

Cheng, H.-M., Yang, Q.-H., Liu, C., 2001. Hydrogen storage in carbon nanotubes. Carbon 39, 1447–1454.

Cheng, X., Li, H., Chen, Y., Luo, B., Liu, X., Liu, W., Xu, H., Yang, X., 2013. Ultrasound-triggered phase transition sensitive magnetic fluorescent nanodroplets as a multimodal imaging contrast agent in rat and mouse model. PLoS One 8, e85003.

Cheng, Y.Y., Xu, T.W., 2005. Solubility of nicotinic acid in polyamidoamine dendrimer solutions. Eur. J. Med. Chem. 40, 1384–1389.

Chernyak, Y., Henon, F., Harris, R.B., Gould, R.D., Franklin, R.K., Edwards, J.R., et al., 2001. Formation of perfluoro polyether coatings by the rapid expansion of supercritical solutions (RESS) process part1: experimental results. Ind. Eng. Chem. Res. 40, 6118–6126.

Chibante, L.P.F., Thess, A., Alford, J.M., Diener, M.D., Smalley, R.E., 1993. Solar generation of the fullerenes. J. Phys. Chem. 97, 8696–8700.

Chiola, V., Ritsko, J.E., Vanderpool, C.D., 1971. US Patent 3 556 725.

Cho, S., Jeong, H., Park, D.H., Jung, S.H., Kim, H.J., Lee, K.H., 2010. The effects of vitamin C on ZnO crystal formation. Cryst. Eng. Commun. 12, 968–976.

Chompoosor, A., Han, G., Rotello, V.M., 2008. Charge dependence of ligand release and monolayer stability of gold nanoparticles by biogenic thiols. Bioconjug. Chem. 19, 1342–1345.

Chou, S.Y., Krauss, P.R., Renstrom, P.J., 1996. Imprint lithography with 25-nanometer resolution. Science 272, 85–87.

Clavee, G.N., Jungbaur, M., Jackelen, A.L., 1999. Nanoscale materials synthesis. 1. Solvent effects on hydridoborate reduction of copper ions. Langmuir 15, 2322–2326.

Collins, P.G., 2009. Defects and disorders in carbon nanotibes. In: Narlikar, A.V., Fu, Y.Y. (Eds.), Oxford Handbook of Nanosciences and Technology: Frontiers and Advances. Oxford University Press, Oxford.

Colvin, V.L., Schlamp, M.C., Alivisatos, A.P., 1994. Light-emitting diodes made from cadmium selenide nanocrystals and a semiconducting polymer. Nature 370, 354–357.

D'Emanuele, A., Attwood, D., 2005. Dendrimer-drug interactions. Adv. Drug Deliv. Rev. 57, 2147–2162.

Danafar, F., Fakhru'l-Razi, A., Salleh, M.A.M., Biak, D.R.A., 2011. Influence of catalytic particle size on the performance of fluidized-bed chemical vapor deposition synthesis of carbon nanotubes. Chem. Eng. Res. Design 89, 214–223.

Danek, M., Jensen, K.F., Murray, C.B., Bawendi, M.G., 1994. Preparation of II–VI quantum dot composites by electrospray organometallic chemical vapor deposition. J. Crys. Growth 145, 714–720.

Datsyuka, V., Kalyvaa, M., Papagelisb, K., Partheniosa, J., Tasisb, D., Siokoua, A., Kallitsisa, I., Galiotisa, I., 2008. Chemical oxidation of multiwalled carbon nanotubes. Carbon 46, 833–840.

De Jong, G., Telenius, A., Vanderbyl, S., Meitz, A., Drayer, J., 2001. Efficient in-vitro transfer of a 60-Mb mammalian artificial chromosome into murine and hamster cells using cationic lipids and dendrimers. Chromosome Res. 9, 475–485.

De Jonge, N., Bonard, eM., 2004. Carbon nanotube electron sources and applications. Phil. Trans. R. Soc. Lond. A 362, 2239–2266.

Desimone, J.M., Wang, J.-Y., Wang, Y., 2012. Particle replication in non-wetting templates: a platform for engineering shape- and size-specific Janus particles. RSC Smart Materials No. 1. In: Jiang, S., Granick, S. (Eds.), Janus Particle Synthesis, Self-Assembly and Applications. The Royal Society of Chemistry, 2012, pp. 90–107. www.rsc.org.

Desireddy, A., Kumar, S., Guo, J., Bolan, M.D., Griffith, W.P., Bigioni, T.P., 2013. Temporal stability of magic-number metal clusters: beyond the shell closing model. Nanoscale 5, 2036–2044.

Dhar, S., Liu, Z., Lippard, S.J., 2008. Targeted single-wall carbon nanotube-mediated Pt(IV) prodrug delivery using folate as a homing device. J. Am. Chem. Soc. 130, 11467–11476.

Dillon, A.C., Gilbert, K.E.H., Parilla, P.A., Alleman, J.L., Hornyak, G.L., Jones, K.M., Heben, M.J., 2002. Hydrogen storage in carbon single-wall nanotubes. In: Proceedings of the 2002 U.S. DOE Hydrogen Program Review, NREL/CP-610–32405.

Doria, G., Conde, J., Veigas, B., Giestas, L., Almeida, C., Assunção, M., Rosa, J., Baptista, P.V., 2012. Noble metal nanoparticles for biosensing applications. Sensors 12, 1657–1687.

Doria, G., Larguinho, M., Dias, J.T., Pereira, E., Franco, R., Baptista, P.V., 2010. Gold-silver-alloy nanoprobes for one-pot multiplex DNA detection. Nanotechnology 21, 255101.

Dresselhaus, M.S., Dresselhaus, G., Eklund, P.C., 1996. Science of Fullerenes and Carbon Nanotubes. Academic, San Diego.

Dunne, M., Corrigan, O.I., Ramtoola, Z., 2000. Influence of particle size and dissolution conditions on the degradation properties of polylactide-co-glycolide particles. Biomaterials 21, 1659–1668.

Dykes, G.M., 2001. Dendrimers: a review of their appeal and applications. J. Chem. Technol. Biotechnol. 76, 903–918.

Ebbesen, T.W., Ajayan, P.M., 1992. Large-scale synthesis of carbon nanotubes. Nature 358, 220–222.

Ehdaie, B., 2007. Application of nanotechnology in cancer research: review of progress in the National Cancer Institute's alliance for nanotechnology. Int. J. Biol. Sci. 3, 108–110.

Ekman, B., Sjfholm, I., 1978. Improved stability of proteins immobilized in microparticles prepared by modified emulsion polymerization technique. J. Pharm. Sci. 67, 693–696.

El-Jamal, K.T., Bardi, G., Nunes, A., Guo, C., Bussy, C., Herrero, M.A., Bianco, A., Prato, M., Kostarelos, K., Pizzorusso, T., 2011. Functional motor recovery from brain ischemic insult by carbon nanotube-mediated siRNA silencing. PNAS 108, 10952–10957.

EPA (US Environmental Protection Agency), 2003. Technology Innovation Office, Permeable Reactive Barriers. http.//cli-in.org.

Esfand, R., Tomalia, D.A., 2001. Poly(amidoamine) (PAMAM) dendrimers: from biomimicry to drug delivery and biomedical applications. Drug Discov. Today 6, 427–436.

Fadel, T.R., Fahmy, T.M., 2014. Immunotherapy applications of carbon nanotubes: from design to safe applications. Trends Biotechnol. http://dx.doi.org/10.1016/j.tibtech.2014.02.005.

Faraday, F., 1857. Experimental relations of gold (and other metals) to light. Philos. Trans. 145–181.

Feng, Z., Li, Y., Niu, D., Li, L., Zhao, W., Chen, H., Li, L., Gao, J., Ruan, M., Shi, J., 2008. A facile route to hollow nanospheres of mesoporous silica with tunable size. Chem. Commun. 2008, 2629–2631.

Fenoglio, I., Greco, G., Tomatis, M., Muller, J., Raymundo-Piñero, E., Béguin, F., Fonseca, A., Nagy, J.B., Lison, D., Fubini, B., 2008. Structural defects play a major role in the acute lung toxicity of multiwall carbon nanotubes: physicochemical aspects. Chem. Res. Toxicol. 21, 1690–1697.

Ferreira, L., Karp, J.M., Nobre, L., Langer, R., 2008. New opportunities: the use of nanotechnologies to manipulate and track stem cells. Cell Stem Cell 3, 136–146.

Fields, C.L., Pitts, J.R., Hales, M.J., Bingham, C., Lewandowski, A., King, D.E., 1993. Formation of fullerenes in highly concentrated solar flux. J. Phys. Chem. 97, 8701–8702.

Figueroa, E.R., Lin, A.Y., Yan, J., Luo, L., Foster, A.E., Drezek, R.A., 2014. Optimization of PAMAM-gold nanoparticle conjugation for gene therapy. Biomaterials 35, 1725–1734.

Fu, T.J., Seeman, N.C., 1993. Biochemistry 32, 3211–3220.

Fu, Z., Luo, Y., Derreumaux, P., Wei, G., 2009. Induced beta-barrel formation of the Alzheimer's Abeta25-35 oligomers on carbon nanotube surfaces: implication for amyloid fibril inhibition. Biophys. J. 97, 1795–1803.

Fujiwara, M., Shiokawa, K., Tanaka, Y., Nakahara, Y., 2004. Preparation of hierarchical architectures of silica particles with hollow structure and nanoparticle shells: a material for the high reflectivity of UV and visible light. Chem. Mater. 16, 5420–5426.

Furdyna, J.K., Samarth, N.J., 1987. Magnetic properties of diluted magnetic semiconductors: a review. Appl. Phys. 61, 3526.

Gajbhiye, V., Palanirajan, V.K., Tekade, R.K., Jain, N.K., 2009. Dendrimers as therapeutic agents: a systematic review. J. Pharm. Pharmacol. 61, 989–1003.

Gbadamosi, K., Hunter, A.C., Moghimi, S.M., 2002. PEGylation of microspheres generates a heterogeneous population of particles with differential surface characteristics and biological performance. FEBS Lett. 532, 338–344.

Gebhart, C.L., Kabanov, A.V., 2001. Evaluation of polyplexes as gene-transfer agents. J. Control. Release 73, 401–416.

Ghosh, R., Singh, L.C., Shohet, J.M., et al., 2013. A gold nanoparticle platform for the delivery of functional microRNAs into cancer cells. Biomaterials 34, 807–816.

Gibson, J.D., Khanal, B.P., Zubarev, E.R., 2007. Paclitaxel-functionalized gold nanoparticles. J. Am. Chem. Soc. 129, 11653–11661.

Gilles, S., Kaulen, C., Pabst, M., Simon, U., Offenhausser, O., Mayer, D., 2011. Patterned self-assembly of gold nanoparticles on chemical templates fabricated by soft UV nanoimprint lithography. Nanotechnology 22, 295301 (7pp).

Gillies, E.R., Frechet, J.M.J., 2005. Dendrimers and dendritic polymers in drug delivery. Drug Discov. Today 10, 35–43.

Gore, J.P., Sane, A., 2011. Flame synthesis of carbon nanotubes. In: Yellampalli, S. (Ed.), Nanotechnology and Nanomaterials Carbon Nanotubes – Synthesis, Characterization, Applications. ISBN 978-953-307-497-9.

Goss, I., Shu, L., Schluter, A.D., Rabe, J.P., 2002. Molecular structure of single DNA complexes with positively charged dendronized polymers. J. Am. Chem. Soc. 124, 6860–6865.

Green, A.A., Hersam, M.C., 2011. Properties and application of double-walled carbon nanotubes sorted by outer-wall electronic type. ACS Nano 5, 1459–1467.

Griffin, J., Singh, A.K., Senapati, D., Rhodes, P., Mitchell, K., Robinson, B., Yu, E., Ray, P.C., 2009. Size- and distance-dependent nanoparticle surface-energy transfer (NSET) method for selective sensing of hepatitis C virus RNA. Chemistry 15, 342–351.

Guarise, C., Pasquato, L., Scrimin, P., 2005. Reversible aggregation/deaggregation of gold nanoparticles induced by a cleavable dithiol linker. Langmuir 21, 5537–5541.

Guix, M., Orozco, J., García, M., Gao, W., Sattayasamitsathit, S., Merkoçi, A., Escarpa, A., Wang, J., 2012. Superhydrophobic alkanethiol-coated microsubmarines for effective removal of oil. ACS Nano 6, 4445–4451.

Gul, H., Lu, W., Xu, P., Xing, J., Chen, J., 2010. Magnetic carbon nanotube labelling for haematopoietic stem/progenitor cell tracking. Nanotechnology 21, 155101.

Guo, H., Dong, N., Yin, M., Zhang, W., Lou, L., Xia, S., 2004. Visible upconversion in rare earth ion-doped Gd_2O_3 nanocrystals. Phys. Chem. B 108, 19205–19209.

Guo, L.J., 2007. Nanoimprint lithography: methods and material requirements. Adv. Mater. 19, 495–513.

Guo, T., Nikolaev, P., Thess, A., Colbert, D.T., Smalley, R.E., 1995. Catalytic growth of single-walled nanotubes by laser vaporization. Chem. Phys. Lett. 243, 49–54.

Gupta, A., Main, B.J., Taylor, B.L., Gupta, M., Whitworth, C.A., Cady, C., Freeman, J.W., El-Amin 3rd, S.F., 2014. In vitro evaluation of three-dimensional single-walled carbon nanotube composites for bone tissue engineering. J. Biomed. Mater. Res. A. http://dx.doi.org/10.1002/jbm.a.35088.

Gupta, U., Agashe, H.B., Asthana, A., Jain, N.K., 2006. Dendrimers: novel polymeric nanoarchitectures for solubility enhancement. Biomacromolecules 7, 649–658.

Hamad, A.C., Hunter, K.J., Liu, R.Z., Dai, H., Moghimi, S.M., 2008. Complement activation by PEGylated carbon nanotubes is independent of C1q and alternative pathway turnover. Mol. Immunol. 45, 3797–3803.

Hans, M.L., Lowman, L., 2002. Biodegradable nanoparticles for drug delivery and targeting. Current Opinion in Solid State and Materials Science 6, 319–327.

Hare, J.P., Kroto, H.W., Taylor, R., 1991. Preparation and UV/visible spectra of C60 and C70. Chem. Phys. Lett. 177, 394–398.

Harris, P.J.F., 2009. Ultrathin graphitic structures and carbon nanotubes in a purified synthetic graphite. J. Phys. Cond. Matter 21, 355009.

Haufler, R.E., Conceicao, J., Chibante, L.P.F., Chai, Y., Byrne, N.E., Flanagan, S., Haley, M.M., O'Brien, S.C., Pan, C., Xiao, Z., Billups, W.E., Ciufolini, M.A., Hauge, R.H., Margrave, J.L., Wilson, L.J., Curl, R.F., Smalley, R.E., 1990. Efficient production of C_{60} (Buckminsterfullerene), $C_{60}H_{36}$, and the solvated buckide ion. J. Phys. Chem. 94, 8634–8636.

Hawker, C.J., Frechet, J.M.J., 1990. Preparation of polymers with controlled molecular architecture. A new convergent approach to dendritic macromolecules. J. Am. Chem. Soc. 112, 7638–7647.

He, B., Tan, J.J., Liew, K.Y., Liu, H., 2004. Synthesis of size controlled Ag nanoparticles. J. Mol Catal. A Chem. 221, 121–126.

He, L., Feng, L., Cheng, L., Liu, Y., Li, Z., Peng, R., Li, Y., Guo, L., Liu, Z., 2013. Multilayer dual-polymer-coated upconversion nanoparticles for multimodal imaging and serum-enhanced genedelivery. ACS Appl. Mater. Interfaces 5, 10381–10388.

He, Y., Chen, Y., Liu, H., Ribbe, A.E., Mao, C., 2005. Sequence symmetry as a tool for designing DNA nanostructures. J. Am. Chem. Soc. 127, 12202–12203.

Hecht, S., Frechet, J.M.J., 2001. Dendritic encapsulation of function: applying Nature's site isolation principle from biomimetics to material science. Angew. Chem. Int. Ed. 40, 74.

Heegaard, P.M.H., Pedersen, H.G., Flink, J., Boas, U., 2004. Amyloid aggregates of the prion peptide PrP 106–126 are destabilised by oxidation and by action of dendrimers. FEBS Lett. 577, 127–133.

Herdt, A.R., Drawz, S.M., Kang, Y., Taton, T.A., 2006. DNA dissociation and degradation at gold nanoparticle surfaces. Colloids Surf. B 51, 130–139.

Heyderman, L.J., Schift, H., David, C., Gobrecht, J., Schweizer, T., 2000. Flow behaviour of thin polymer films used for hot embossing lithography. Microelectron. Eng. 54, 229–245.

Hilder, T.A., Hill, J.M., 2007a. Modelling the encapsulation of the anticancer drug cisplatin into carbon nanotubes. Nanotechnology 18, 275704.

Hilder, T.A., Hill, J.M., 2007b. Continuous versus discrete for interacting carbon nanostructures. J. Phys. A Math. Theor. 40, 3851–3868.

Hirata, E., Ménard-Moyon, C., Venturelli, E., Takita, H., Watari, F., Bianco, A., Yokoyama, A., 2013. Carbon nanotubes functionalized with fibroblast growth factor accelerate proliferation of bone marrow-derived stromal cells and bone formation. Nanotechnology 24, 435101.

Hirsch, A., 2002. Funktionalisierung von einwandigen Kohlenstoffnanorohern. Angew. Chem. Int. Ed. 114, 1933–1939.

Holzinger, M., Vostrowsky, O., Hirsch, A., Hennrich, F., Kappes, M., Weiss, R., Jellen, F., 2001. Sidewall functionalization of carbon nanotubes. Angew. Chem. Int. Ed. 40, 4002–4005.

Hong, R., Han, G., Fernández, J.M., Kim, B.-J., Forbes, N.S., Rotello, V.M., 2006. Glutathione-mediated delivery and release using monolayer protected nanoparticle carriers. J. Am. Chem. Soc. 128, 1078–1079.

Hsueh, T.J., Chen, Y.W., Chang, S.J., Wang, S.F., Hsu, C.L., Lin, Y.R., Lin, T.S., Chen, I.C., 2007. ZnO nanowire-based CO sensors prepared at various temperatures. J. Electrochem. Soc. 154, J393–J396.

Hu, H.-L., Yao, D.-X., Xia, Y.-F., Zuo, K.-H., Zeng, Y.-P., 2013. Porous Si_3N_4/SiC ceramics prepared via nitridation of Si powder with SiC addition. Int. J. Appl. Ceram. Technol. http://dx.doi.org/10.1111/ijac.12186.

Hua, M., Jiang, Y., Wu, B., Pan, B., Zhao, X., Zhang, Q., 2013. Fabrication of a new hydrous Zr(IV) oxide-based nanocomposite for enhanced Pb(II) and Cd(II) removal from waters. ACS Appl. Mater. Interfaces 5, 12135–12142.

Huang, H.S., Hainfeld, J.F., 2013. Intravenous magnetic nanoparticle cancer hyperthermia. Int. J. Nanomed. 8, 2521–2532.

Huang, L., Zhu, X., Publicover, N.G., Hunter Jr., K.W., Ahmadiantehrani, M., de Bettencourt-Dias, A., Bell, T.W., 2013. Cadmium- and zinc-alloyed Cu–In–S nanocrystals and their optical properties. J. Nanopart. Res. 15, 2056.

Huo, Q., Margolese, D.I., Ciesla, U., Demuth, D.G., Feng, P., Gier, T.E., Sieger, P., Firouzi, A., Chmelka, B.F., Schuth, F., Stucky, G.D., 1994. Organization of organic molecules with inorganic molecular species into nanocomposite biphase arrays. Chem. Mater. 6, 1176–1191.

Hyeon, T., 2003. Chemical synthesis of magnetic nanoparticles. ChemInform 34. http://dx.doi.org/10.1002/chin.200324224.

Iijima, S., 2002. Carbon nanotubes: past, present, and future. Phys. B 323, 1–5.

Iravani, S., 2011. Green synthesis of metal nanoparticles using plants. Green Chem. 13, 2638–2650.

Irving, B., 2007. Nanoparticle drug delivery systems. Inno. Pharm. Biotechnol. 24, 58–62.

Islam, M.F., Rojas, E., Bergey, D.M., Johnson, A.T., Yodh, A.G., 2003. High weight fraction surfactant solubilization of single-wall carbon nanotubes in water. Nano Lett. 3, 269–273.

Itoh, H., Naka, K., Chujo, Y., 2004. Synthesis of gold nanoparticles modified with ionic liquid based on the imidazolium cation. J. Am. Chem. Soc. 126, 3026–3027.

Jain, K.K., 2012. Advances in use of functionalized carbon nanotubes for drug design and discovery. Expert Opin. Drug Discov. 7, 1029–1037.

Jain, N.K., Mishra, V., Mehra, N.K., 2013a. Targeted drug delivery to macrophages. Expert Opin. Drug Deliv. 10, 353–367.

Jain, S., Sharma, A., Basu, B., 2013b. In vitro cytocompatibility assessment of amorphous carbon structures using neuroblastoma and Schwann cells. J. Biomed. Mater. Res. B Appl. Biomater. 101, 520–531.

Jamieson, T., Bakhshi, R., Petrova, D., Pocock, R., Imani, M., Seifalian, A.M., 2007. Biological applications of quantum dots. Biomaterials 28, 4717–4732.

Jana, N.R., Gearheart, L., Murphy, C.J., 2001. Evidence for seed-mediated nucleation in the chemical reduction of gold salts to gold nanoparticles. Chem. Mater. 13, 2313–2322.

Jansen, J.F.G.A., de Brabander-van den Berg, E.M.M., Meijer, E.W., 1994. Encapsulation of guest molecules into a dentritic box. Science 266, 1226–1229.

Jennings, T.L., Singh, M.P., Strouse, G.F., 2006. Fluorescent lifetime quenching near d = 1.5 nm gold nanoparticles: probing NSET validity. J. Am. Chem. Soc. 128, 5462–5467.

Jiang, Y., Li, H., Li, Y., Yu, H., Liew, K.M., He, Y., Liu, X., 2011. Helical encapsulation of graphene nanoribbon into carbon nanotube. ACS Nano 5, 2126–2133.

Johann, E., Alessandro, C., Zhiyuan, F., et al., 2013. Powering the programmed nanostructure and function of gold nanoparticles with catenated DNAmachines. Nat. Commun. 4, 2000.

Joo, J.H., Na, B., Yu, T., Yu, J.H., Kim, Y.W., Wu, F., Zhang, J.Z., Hyeon, T., 2003. Generalized and facile synthesis of semiconducting metal sulfide nanocrystals. J. Am. Chem. Soc. 125, 11100–11105.

Jorgensen, J.M., Erlacher, K., Pedersen, J.S., Gothelf, K.V., 2005. Preparation temperature dependence of size and polydispersity of alkylthiol monolayer protected gold clusters. Langmuir 21, 10320–10323.

Joseph, S., Aluru, R., 2008. Why are carbon nanotubes fast transporter of water? Nano Lett. 8, 452–458.

Jung, K.Y., Park, B.C., Song, W.Y., Eom, T.B., 2002. Measurement of 100-nm polystyrene sphere by transmission electron microscope. Powder Technol. 126, 255.

Kabiri, M., Soleimani, M., Shabani, I., Futrega, K., Ghaemi, N., Ahvaz, H.H., Elahi, E., Doran, M.R., 2012. Neural differentiation of mouse embryonic stem cells on conductive nanofiber scaffolds. Biotechnol. Lett. 34, 1357–1365.

Kam, N.W.S., O'Connell, M., Wisdom, J.A., Dai, H., 2005. Carbon nanotubes as multifunctional biological transporters and near-infrared agents for selective cancer cell destruction. Proc. Natl Acad. Sci. U.S.A. 102, 11600–11605.

Kang, D.S., Han, S.K., Kim, J.H., Yang, S.M., Kim, J.G., Hong, S.K., Kim, D., Kim, H., 2009. ZnO nanowires prepared by hydrothermal growth followed by chemical vapor deposition for gas sensors. J. Vac. Sci. Technol. B 27, 1667–1672.

Kang, Y., Taton, T.A., 2005. Core/shell gold nanoparticles by self-assembly and crosslinking of micellar, block-copolymer shells. Angew. Chem. Int. Ed. 44, 409–412.

Kapoor, S., Joshi, R., Mukherjee, T., 2002. Influence of I- anions on the formation and stabilization of copper nanoparticles. Chem. Phys. Lett. 354, 443–448.

Kassam, A., Bremner, G., Clark, B., Ulibarri, G., Lennox, R.B., 2006. Place exchange reactions of alkyl thiols on gold nanoparticles. J. Am. Chem. Soc. 128, 3476–3477.

Kawasaki, S., Komatsu, K., Okino, F., Tohara, H., Kataura, H., 2004. Fluorination of open- and closed-end single-walled carbon nanotubes. Phys. Chem. Chem. Phys. 6, 1769–1772.

Kelly, K.F., Chiang, I.W., Mickelson, E.T., Hauge, R.H., Margrave, J.L., Wang, X., Scuseria, G.E., Radloff, C., Halas, N., 1999. Insight into the mechanism of sidewall functionalization of single-walled nanotubes: an STM study. J. Chem. Phys. Lett. 313, 445–450.

Kensinger, R.D., Catalone, B.J., Krebs, F.C., Wigdahl, B., Schengrund, C.-L., 2004a. Novel polysulfated galactose-derivatized dendrimers as binding antagonists of human immunodeficiency virus type 1 infection. Antimicrob. Agents Chemother. 48, 1614–1623.

Kensinger, R.D., Yowler, B.C., Benesi, A.J., Schengrund, C.L., 2004b. Synthesis of novel, multivalent glycodendrimers as ligands for HIV-1 gp120. Bioconjug. Chem. 15, 349–358.

Khanal, A., Inoue, Y., Yada, M., Nakashima, K., 2007. Synthesis of silica hollow nanoparticles templated by polymeric micelle with core–shell–corona structure. J. Am. Chem. Soc. 129, 1534–1535.

Khlobystov, A.N., 2011. Carbon nanotubes: from nano test tube to nano-reactor. ACS Nano 5, 9306–9312.

Kik, P.G., Martin, A.L., Maier, S.A., Atwater, H.A., 2002. Metal Nanoparticle Arrays for Near Field Optical Lithography. http://daedalus.caltech.edu/publication/pubs/2002-kik-spie.pdf.

Kim, D.S., Lee, J.Y., Na, V.W., Yoon, S.W., Kim, S.Y., Park, J., Jo, Y., Jung, M.H., 2006. Synthesis and photoluminescence of Cd-doped α-MnS nanowires. J. Phys. Chem. B 110, 18262–18266.

Kim, S.W., Kim, M., Lee, W.Y., Hyeon, T., 2002. Fabrication of hollow palladium spheres and their successful application to the recyclable heterogeneous catalyst for Suzuki coupling reactions. J. Am. Chem. Soc. 124, 7642–7643.

Kitaura, R., Nakanishi, R., Saito, T., Yoshikawa, H., Awaga, K., Shinohara, H., 2009. High-yield synthesis of ultrathin metal nanowires in carbon nanotubes. Angew. Chem. Int. Ed. 48, 8298–8302.

Klajn, R., Bishop, K.J.M., Fialkowski, M., Paszewski, M., Campbell, C.J., Gray, T.P., Grzybowski, B.A., 2007. Plastic and moldable metals by self-assembly of sticky nanoparticle aggregates. Science 316, 261–264.

Klajn, R., Browne, K.P., Soh, S., Grzybowski, B.A., 2010a. Nanoparticles that "remember" temperature. Small 6, 1385–1387.

Klajn, R., Fang, L., Coskun, A., Olson, M.A., Wesson, P.J., Stoddart, J.F., Grzybowski, B.A., Metal, A., 2009a. Nanoparticles functionalized with molecular and supramolecular switches. J. Am. Chem. Soc. 131, 4233.

Klajn, R., Olson, M.A., Wesson, P.J., Fang, L., Coskun, A., Trabolsi, A., Soh, S., Stoddart, J.F., Grzybowski, B.A., 2009b. Dynamic hook-and-eye nanoparticle sponges. Nat. Chem. 1, 733–738.

Klajn, R., Stoddart, J.F., Grzybowski, B.A., 2010b. Nanoparticles functionalised with reversible molecular and supramolecular switches. Chem. Soc. Rev. 39, 2203–2237.

Klajn, R., Wesson, P.J., Bishop, K.J.M., Grzybowski, B.A.M., 2009c. Writing self-erasing images using metastable nanoparticle "inks". Angew. Chem. Int. Ed. 48, 6924.

Klajnert, B., Cladera, J., Bryszewska, M., 2006. Molecular interactions of dendrimers with amyloid peptides: pH dependence. Biomacromolecules 7, 2186–2191.

Knop, K., Hoogenboom, R., Fischer, D., Schubert, U.S., 2006. Poly(ethylene glycol) in drug delivery: pros and cons as well as potential alternatives. Angew. Chem. Int. Ed. 49, 6288–6308.

Knop, K., Hoogenboom, R., Fischer, D., Schubert, U.S., 2010. Poly(ethylene glycol) in drug delivery: pros and cons as well as potential alternatives. Angew. Chem. Int. Ed. 49, 6288–6308.

Kobayashi, H., Brechbiel, M.W., 2005. Nano-sized MRI contrast agents with dendrimer cores. Adv. Drug Deliv. Rev. 57, 2271–2286.

Kojima, C., Kono, K., Maruyama, K., Takagishi, T., 2000. Synthesis of polyamidoamine dendrimers having poly(ethylene glycol) grafts and their ability to encapsulate anticancer drugs. Bioconjug. Chem. 11, 910–917.

Kojima, C., Turkbey, B., Ogawa, M., Bernardo, M., Regino, C.A.S., Bryant Jr., L.H., Choyke, P.L., Kono, K., Kobayashi, H., 2011. Dendrimer-based MRI contrast agents: the effects of PEGylation on relaxivity and pharmacokinetics. Nanomedicine 7, 1001–1008.

Kolhe, P., Misra, E., Kannan, R.M., Kannan, S., LiehLai, M., 2003. Drug complexation, in vitro release and cellular entry of dendrimers and hyperbranched polymers. Int. J. Pharm. 259, 143–160.

Kong, J., Franklin, N., Zhou, C., Peng, S., Cho, J.J., Dai, H., 2000. Nanotube molecular wires as chemical sensors. Science 287, 622.

Koo, N., Plachetka, U., Otto, M., Bolten, J., Jeong, J.-H., Lee, E.-S., Kurz, H., 2008. The fabrication of a flexible mold for high-resolution soft ultraviolet nanoimprint lithography. Nanotechnology 19, 225304.

Kresge, C.T., Leonowicz, M.E., Roth, W.J., Vartuli, J.C., Beck, J.S., 1992. Ordered mesoporous molecular sieves synthesized by a liquid-crystal template mechanism. Nature 359, 710–712.

Krishnan, S., Diagaradjane, P., Cho, S., 2010. Nanoparticle-mediated thermal therapy: evolving strategies for prostate cancer therapy. Int. J. Hyperthermia 26, 775–789.

Kroto, H.W., Heath, J.R., O'Brien, S.C., Curl, R.F., Smalley, R.E., 1985. Cd buckrninsterfullerene. Nature 318, 162–163.

Kumar, I., Rana, S., Cho, J.W., 2011. Cycloaddition reactions: a controlled approach for carbon nanotube functionalization. Chem. A Eur. J. 17, 11092–11101.

Kumari, A., Yadav, K.S., Yadav, S.C., 2010. Biodegradable polymeric nanoparticles based drug delivery systems. Colloids Surf. B 75, 1–18.

Kuzmant, H., 2004. Functionaslization of carbon nanotubes. Synth. Met. 141, 113–122.

Laaksonen, T., Ahonen, P., Johans, C., Kontturi, K., 2006. Stability and electrostatics of mercaptoundecanoic acid-capped gold nanoparticles with varying counterion size. Chem. Phys. Chem. 7, 2143–2149.

LaBean, T.H., Yan, H., Kopatsch, J., Liu, F., Winfree, E., Reif, J.H., Seeman, N.C., 2000. Stepwise DNA self-assembly of fixed-size nanostructures. J. Am. Chem. Soc. 122, 1848–1860.

Lai, H.J., Lin, M.C.C., Yang, M.H., Li, A.K., 2001. Synthesis of carbon nanotubes using polycyclic aromatic hydrocarbons as carbon sources in an arc discharge. Mater. Sci. Eng. C 16, 23–26.

Lamprecht, C., Liashkovich, I., Neves, V., Danzberger, J., Heister, E., Rangl, M., Coley, H.M., McFadden, J., Flahaut, E., Gruber, H.J., Hinterdorfer, P., Kienberger, F., Ebner, A., 2009. AFM imaging of functionalized carbon nanotubes on biological membranes. Nanotechnology 20, 434001.

Landers, J.J., Cao, Z.Y., Lee, I., Piehler, L.T., Myc, P.P., Myc, A., Hamouda, T., Galecki, A.T., Baker, J.R., 2002. Prevention of influenza pneumonitis by sialic acid-conjugated dendritic polymers. J. Infect. Dis. 186, 1222–1230.

Lankveld, D.P., Rayavarapu, R.G., Krystek, P., Oomen, A.G., Verharen, H.W., van Leeuwen, T.G., De Jong, W.H., Manohar, S., 2011. Blood clearance and tissue distribution of PEGylated and non-PEGylated gold nanorods after intravenous administration in rats. Nanomedicine (Lond.) 6, 339–349.

Langer, R., 1995. Dendrimers II: architecture, nanostructure and supramolecular chemistry. Chem. Eng. Sci. 50, 4109.

Langereis, S., Dirksen, A., Hackeng, T.M., van Genderen, M.H.P., Meijer, E.W., 2007. Dendrimers and magnetic resonance imaging. New J. Chem. 31, 1152–1160.

Lao, C.S., Liu, J., Gao, P., Zhang, L., Davidovic, D., Tummala, R., Wang, Z.L., 2006. ZnO nanobelt/nanowire Schottky diodes formed by dielectrophoresis alignment across Au electrodes. Nano Lett. 6, 263.

Laurent, S., Forge, D., Port, M., Roch, A., Robic, C., Vander Elst, L., Muller, R.N., 2008. Magnetic iron oxide nanoparticles: synthesis, stabilization, vectorization, physicochemical characterizations, and biological applications. Chem. Rev. 108, 2064–2110.

Lee, C.C., MacKay, J.A., Fréchet, J.M., Szoka, F.C., 2005. Designing dendrimers for biological applications. Nat. Biotechnol. 23, 1517–1526.

Lee, G., Lee, H., Nam, K., Han, J.-H., Yang, J., Lee, S.W., Yoon, D.S., Eom, K., Kwon, T., 2012. Nanomechanical characterization of chemical interaction between gold nanoparticles and chemical functional groups. Nanoscale Res. Lett. 7, 608.

Lee, J.B., Roh, Y.H., Um, S.H., Funabashi, H., Cheng, W.L., et al., 2009. Multifunctional nanoarchitectures from DNAbased ABC monomers. Nat. Nanotechnol. 4, 430–436.

Lee, J.M., Park, J.S., Lee, S.H., Kim, H., Yoo, S., Kim, S.O., 2010. Selective electron- or hole-transport enhancement in bulk-heterojunction organic solar cells with N- or B-doped carbon nanotubes. Adv. Mater. 23, 629–633.

Li, G., Li, X., Zhang, S., 2009. Dendrimers-based DNA biosensors for highly sensitive electrochemical detection of DNA hybridization using reporter probe DNA modified with Au nanoparticles. Biosens. Bioelectron. 24, 3281–3287.

Li, H., Luo, Y., Derreumaux, P., Wei, G., 2011a. Carbon nanotube inhibits the formation of β-sheet-rich oligomers of the Alzheimer's amyloid-β(16-22) peptide. Biophys. J. 101, 2267–2276.

Li, J., Wang, L., 2003. Shape effects on electronic states of nanocrystals. Nano Lett. 3, 1357–1363.

Li, Y., Tseng, Y.D., Kwon, S.Y., d'Espaux, L., Bunch, J.S., McEuen, P.L., Luo, D., 2004. Controlled assembly of dendrimer-like DNA. Nat. Mater. 2004 (3), 38–42.

Li, Y., Zaluzhna, O., Xu, B., Gao, Y., Modest, J.M., Tong, Y.Y.J., 2011b. Mechanistic insights into the Brust–Schiffrin two-phase synthesis of organo-chalcogenate-protected metal nanoparticles. J. Am. Chem. Soc. 133, 2092–2095.

Li, Y.G., Cu, Y.T.H., Luo, D., 2005. Multiplexed detection of pathogen DNA with DNA-based fluorescence nanobarcodes. Nat. Biotechnol. 23, 885–889.

Lian, J., Zhou, W., Wang, L., Boatner, L.A., Rodney, C., Ewing, R.C., 2005. Focused ion beam-induced ripple and nanoparticle formation in $Cd_2Nb_2O_7$. Microsc. Microanal. 11 (Suppl. 2), 86–87.

Liang, X., Kuhn, H., Frank-Kamenetskii, M.D., 2006. Monitoring single-stranded DNA secondary structure formation by determining the topological state of DNA catenanes. Biophys. J. 90, 2877–2889.
Lim, H.E., Miyata, Y., Kitaure, R., Nishimura, Y., Nishimoto, Y., Irle, S., Warner, J.H., Kataura, H., Shinohara, H., 2013. Growth of carbon nanotubes via twisted graphene nanoribbons. Nat. Commun. 4, 2548.
Lin, C.C., Anseth, K.S., 2009. PEG hydrogels for the controlled release of biomolecules in regenerative medicine. Pharm. Res. 26, 631–643.
Liu, F.-K., Chang, Y.-C., Koa, F.-H., Chub, T.-C., 2004. Microwave rapid heating for the synthesis of gold nanorods. Mater. Lett. 58, 373–377.
Longmire, L., Choyke, P.L., Kobayashi, H., 2008. Dendrimer-based contrast agents for molecular imaging. Curr. Top. Med. Chem. 8, 1180–1186.
Lou, H., Komata, M., Katou, Y., Guan, Z., Reis, C.C., Budd, M., Shirahige, K., Campbell, J.L., 2008. Mrc1 and DNA polymerase epsilon function together in linking DNA replication and the S phase checkpoint. Mol. Cell 32, 106–117.
Love, J.C., Estroff, L.A., Kriebel, J.K., Nuzzo, R.G., Whitesides, G.M., 2005. Self-Assembled monolayers of thiolates on metals as a form of nanotechnology. Chem. Rev. 105, 1103–1170.
Lowe, P.J., Temple, C.S., 1994. Calcitonin and insulin in isobutylcyanoacrylate nanocapsules: protection against proteases and effect on intestinal absorption in rats. J. Pharm. Pharmacol. 46, 547–552.
Lu, C.-H., Cecconello, A., Elbaz, J., Credi, A., Willner, I.A., 2013b. Three-station DNA catenane rotary motor with controlled directionality. Nano Lett. 13, 2303–2308.
Lu, J., Zhang, P., Li, A., Su, F., Wang, T., Liu, Y., Gong, J., 2013a. Mesoporous anatase TiO_2 nanocups with plasmonic metal decoration for highly active visible-light photocatalysis. Chem. Commun. 49, 5817–5819.
Luo, D., Haverstick, K., Belcheva, N., Han, E., Saltzman, W.M., 2002. Poly(ethylene glycol)-conjugated PAMAM dendrimer for biocompatible high-efficiency DNA delivery. Macromolecules 35, 3456–3462.
Luoa, K., Liua, G., Shea, W., Wanga, G., Wanga, G., Hea, B., Aia, H., 2011. Gadolinium-labeled peptide dendrimers with controlled structures as potential magnetic resonance imaging contrast agents. Biomaterials 32, 7951–7960.
Ma, M.L., Cheng, Y.Y., Xu, Z.H., Xu, P., Qu, H.O., Fang, Y.J., Xu, T.W., Wen, L.P., 2006. Evaluation of polyamidoamine (PAMAM) dendrimers as drug carriers of anti-bacterial drugs using sulfamethoxazole (SMZ) as a model drug. Eur. J. Med. Chem. 42, 93–98.
Ma, W., Luthe, J.M., Zheng, H., Wu, Y., Alivisatos, A.P., 2009. Photovoltaic devices employing ternary PbSxSe1-x nanocrystals. Nano Lett. 9, 1699–1703.
Madani, S.Y., Tan, A., Dwek, M., Seifalian, A.M., 2012. Functionalization of single-walled carbon nanotubes and their binding to cancer cells. Int. J. Nanomed. 7, 905–914.
Mafune, F., Kohno, J., Takeda, Y., Kondow, T., 2001. Formation of gold nanoparticles by laser ablation in aqueous solution of surfactant. J. Phys. Chem. B 105 (22), 5114–5120.
Mai, J., Waisman, D.M., Sloane, B.F., 2000. Cell surface complex of cathepsin B/annexin II tetramer in malignant progression. Biochim. Biophys. Acta 1477, 215–230.
Mao, H., Cai, R., Kawazoe, N., Chen, G., 2014. Long-term stem cell labeling by collagen-functionalized single-walled carbon nanotubes. Nanoscale 6, 1552–1559.
Martin, T.P., Bergmann, T., Goehlich, H., Lange, T., 1991. Shell structure of clusters. J. Phys. Chem. 95, 6421–6429.
Martin, T.P., 1996. Shells of atoms. Phys. Rep. 273, 199–241.
Martínez-Hernandez, A.L., Velasco-Santos, C., Castano, V.M., 2010. Carbon Nanotubes Composites: Processing, Grafting and Mechanical and Thermal Properties. Curr. Nanosci. 6, 12–39.
Martirosyan, K.S., Luss, D., 2009. Fabrication of metal oxide nanoparticles by highly exothermic reactions. Chem. Eng. Technol. 32, 1376–1383.
Mastroianni, A.J., Claridge, S.A., Alivisatos, A.P., 2009. Pyramidal and chiral groupings of gold nanocrystals assembled using DNA scaffolds. J. Am. Chem. Soc. 131, 8455–8459.
Mateo-Mateo, C., Vázquez-Vázquez, C., Pérez-Lorenzo, M., Salgueiriño, V., Correa-Duarte, M.A., 2012. Ripening of platinum nanoparticles confined in a carbon nanotube/silica-templated cylindrical space. J. Nanomaterials Article ID 404159, 6 p. http://dx.doi.org.ezp1.lib.umn.edu/10.1155/2012/404159.
Matthews, R.P., Welton, T., Hunt, P.A., 2014. Competitive pi interactions and hydrogen bonding within imidazolium ionic liquids. Phys. Chem. Chem. Phys. 16, 3238–3253.
Mattia, D., Gogotsi, Y., 2008. Review: static and dynamic behavior of liquids inside carbon nanotubes. Microfluid. Nanofluid. 5, 289–305.
Maynor, B.W., Denison, G.M., Rolland, J.P., DeSimone, J.M., 2006. Scalable Fabrication of Polymeric and Organic Nanomaterials Using Particle Replication in Non-wetting Templates (PRINT) and Imprint Lithography, vol. 3. NSTI-Nanotech. ISBN 0-9767985-8-1. www.nsti.org.
Medintz, I.L., Uyeda, H.T., Goldman, E.R., Mattoussi, H., 2005. Quantum dot bioconjugates for imaging, labelling and sensing. Nat. Mater. 4, 435–446.
Meredith, J.R., Jin, C., Narayan, R.J., Aggarwal, R., 2013. Biomedical applications of carbon-nanotube composites. Front Biosci. (Elite Ed.) 5, 610–621.
Michaet, X., Pinaud, F., Bentolila, L., 2005. Quantum dots for live cells, in vivo imaging, and diagnostics. Science 307, 538–544.
Mickelson, E.T., Huffman, C.B., Rinzler, A.G., et al., 1998. Flurination of single-wall carbon nanotubes. Chem. Phys. Lett. 296, 18–19.

Miller, L.L., Kunugi, Y., Canavesi, A., Rigaunt, S., Moorefield, C.N., Newkome, G.R., 1998. Supramolecular photosensitive and electroactive materials. Chem. Mater. 10, 1751.

Millon, E., Weber, J.V., Theobald, J., Muller, J.F., 1992. Laser ablation of carbonaceous materials: a method to produce fullerenes. C.R. Acad. Sci. Paris 315, 947–953.

Mingo, N., Stewart, D.A., Broido, D.A., Strivastava, D., 2008. Phonon transmission through defects in carbon nanotubes from first principles. Phys. Rev. B 77, 033418.

Mirkin, C.A., Letsinger, R.L., Mucic, R.C., Storhoff, J.J., 1996. A DNA-based method for rationally assembling nanoparticles into macroscopic materials. Nature 382, 607–609.

Miyata, Y., Suzuki, M., Asada, Y., Kitaura, R., Shinohara, H., 2010. Solution-phase extraction of ultrathin inner shells from double-wall carbon nanotubes. ACS Nano 4, 5807–5812.

Moghimi, A.M., Szebeni, J., 2003. Stealth liposomes and long circulating nanoparticles: critical issues in pharmacokinetics, opsonization and protein-binding properties. Prog. Lipid Res. 42, 463–478.

Moghimi, S.M., Moghimi, M., 2008. Enhanced lymph node retention of subcutaneously injected IgG1–PEG2000–liposomes through pentameric IgM antibody–mediated vesicular aggregation. Biochim. Biophys. Acta 1778, 51–55.

Moghimi, S.M., Murray, J.C., 1996. Poloxamer-188 revisited: a potentially valuable immune modulator. J. Natl Cancer Inst. 88, 766–768.

Moghimi, S.M., 2006. Recent developments in polymeric nanoparticle engineering and their applications in experimental and clinical oncology. Anticancer Agents Med. Chem. 6, 553–561.

Moniruzzaman, M., Winey, K.I., 2006. Polymer nanocomposites containing carbon nanotubes. Macromolecules 39, 5194–5205.

Moore, V.C., Strano, M.S., Haroz, E.H., Hauge, R.H., Smalley, R.E., Schmidt, J., Talmon, Y., 2003. Individually suspended single-walled carbon nanotubes in various surfactants. Nano Lett. 3, 1379–1382.

Mu, Q., Broughton, D.L., Yan, B., 2009. Endosomal leakage and nuclear translocation of multiwalled carbon nanotubes: developing a model for cell uptake. Nano Lett. 9, 4370–4375.

Murray, C.B., Norris, D.J., Bawendi, M.G., 1993. Synthesis and characterization of nearly monodisperse Cde (E = S, Se, Te) semiconductor nanocrystallites. J. Am. Chem. Soc. 115, 8706–8715.

Naccache, R., Vetrone, F., Mahalingam, V., Cuccia, L.A., Capobianco, J.A., 2009. Controlled synthesis and water dispersibility of hexagonal phase NaGdF4:Ho^{3+}/Yb^{3+} nanoparticles. Chem. Mater. 21, 717–723.

Nagvarma, B., Hemant, V., Yadav, K.S., Ayaz, A., Vasudha, L.S., Shivkumar, H.G., 2012. Different techniques for preparation of polymeric nanoparticles – a review. Asian J. Pharm. Clin. Res. 5 (Suppl. 3), 16–23.

Navaladian, S., Viswanathan, B., Varadarajan, T.K., Viswanath, R.B., 2008. A rapid synthesis of oriented palladium nanoparticles by UV irradiation. Nanoscale Res. Lett. 4, 181–186.

Newkome, G.R., Yao, Z., Baker, G.R., Gupta, V.K., 1985. Cascade molecules: a new approach to micelles. J. Org. Chem. 50, 2003–2004.

Ngadi, N., Abrahamson, J., Fee, C., Morison, K., 2009. Are PEG molecules a universal protein repellent? World Acad. Sci. Eng. Technol. 3.

Niyogi, S., Hamon, M.A., Hu, H., Zhao, B., Bhowmik, P., Sen, R., Itkis, M.E., Haddon, R.C., 2002. Chemistry of single-walled carbon nanotubes. Acc. Chem. Res. 35, 1105–1113.

Noboru, I., Ozaki, Y., Kasu, S., James, M., 1992. Chemical manufacturing processes, superfine particle technology. Springer Verlag, New York.

O'Connell, M.J., Boul, P., Ericson, L.M., Huffman, C., Wang, Y.H., Haroz, E., Kuper, C., Tour, J., Ausman, K.D., Smalley, R.E., 2001. Reversible water-solubilization of single-walled carbon nanotubes by polymer wrapping. Chem. Phys. Lett. 342, 265–271.

Ojea-Jiménez, I., Tort, O., Lorenzo, J., Puntes, V.F., 2012. Engineered nonviral nanocarriers for intracellular gene delivery applications. Biomed. Mater. 7, 054106.

Oliveira, M.B., Valentim, I.B., de Vasconcelos, C.C., Omena, C.M., Bechara, E.J., da Costa, J.G., Freitas Mde, L., Sant'Ana, A.E., Goulart, M.O., 2013. *Cocos nucifera* Linn. (Palmae) husk fiber ethanolic extract: antioxidant capacity and electrochemical investigation. Comb. Chem. High Throughput Screen. 16, 121–129.

Ouyang, M., Huang, J.-L., Lieber, C.M., 2002. Fundamental electronic properties and applications of single-walled carbon nanotubes. Acc. Chem. Res. 35, 1018.

Owens, D., Peppas, N., 2006. Opsonization, biodistribution, and pharmacokinetics of polymeric nanoparticles. Int. J. Pharmaceut. 307, 93–102.

Panda, D., Tseng, T.-Y., 2013. One-dimensional ZnO nanostructures: fabrication, optoelectronic properties, and device applications. J. Mater. Sci. 48, 6849–6877.

Panigrah, S., Kundu, S., Ghosh, S.K., Nath, S., Pal, T., 2004. General method of synthesis for metal nanoparticles. J. Nanopart. Res. 6, 411–414.

Panigrahi, S., Basak, D., 2011. Core–shell TiO_2@ZnO nanorods for efficient ultraviolet photodetection. Nanoscale 3, 2336–2341.

Panyam, J., Labhasetwar, V., 2012. Biodegradable nanoparticles for drug and gene delivery to cells and tissue. Adv. Drug Deliv. Rev. 64 (Suppl.), 61–71.

Park, H., Zhao, J., Lu, J.P., 2006. Effects of sidewall functionalization on conducting properties of single wall carbon nanotubes. Nano Lett. 6, 916–919.

Park, H.J., Lyons, J.C., Ohtsubo, T., Song, C.W., 1999. Acidic environment causes apoptosis by increasing caspase activity. Br. J. Cancer 80, 1892–1897.

Park, N., Um, S.H., Funabashi, H., Xu, J.F., Luo, D., 2009. A cellfree protein-producing gel. Nat. Mater. 8, 432–437.

Park, S.H., Yin, P., Liu, Y., Reif, J.H., LaBean, T.H., Yan, H., 2005. Programmable DNA self-assemblies for nanoscale organization of ligands and proteins. Nano Lett. 5, 729–733.

Parker, R.A., Negi, S., Davis, T., Keefer, E.W., Wiggins, H., House, P.A., Greger, B., 2012. The use of a novel carbon nanotube coated microelectrode array for chronic intracortical recording and microstimulation. Conf. Proc. IEEE Eng. Med. Biol. Soc. http://dx.doi.org/10.1109/EMBC.2012.6346050.

Patel, D., Henry, J., Good, T., 2006. Attenuation of β-amyloid induced toxicity by sialic acid-conjugated dendritic polymers. Biochim. Biophys. Acta 1760, 1802–1809.

Paul, S., Samdarshi, S.K., 2011. A green precursor for carbon nanotube synthesis. New Carbon Mater. 26, 85–88.

Peng, S., O'Keeffe, J., Wei, C., Cho, K., 2001. Carbon nanotube chemical and mechanical sensors. In: Conference Paper for the 3rd International Workshop on Structural Health Monitoring, pp. 1–8.

Perez-Dieste, V., Castellini, O.M., Crain, J.M., Eriksson, M.A., Kirakosian, A., Lin, J.-L., McChesney, J.L., Himpsela, F.J., 2003. Thermal decomposition of surfactant coatings on Co and Ni nanocrystals. Appl. Phys. Lett. 83, 5053–5055.

Peters, G., Jansen, M., 1992. A new fullerene synthesis. Angew. Chem. Int. Ed. 31, 223–224.

Ponder, S., Darab, J.G., Bucher, J., Caulder, D., Craig, I., Davis, L., Edelstein, N., Lukens, W., Nitsche, H., Rao, L., Shuh, D.K., Mallouk, T.E., 2001. Surface chemistry and electrochemistry of supported zerovalent iron nanoparticles in the remediation of aqueous metal contaminants. Chem. Mater. 13, 479–486.

Ponder, S.M., Darab, J.G., Mallouk, T.E., 2000. Remediation of Cr(VI) and Pb(II) aqueous solutions using supported, nanoscale zero-valent iron. Environ. Sci. Technol. 34, 2564–2569.

Popławska, M., Żukowska, G.Z., Cudziło, S., Bystrzejewski, M., 2010. Chemical functionalization of carbon-encapsulated magnetic nanoparticles by 1,3-dipolar cycloaddition of nitrile oxide. Carbon 48, 1318–1320.

Popov, V.N., 2004. Carbon nanotubes: properties and application. Mater. Sci. Eng. R 43, 61–102.

Prabhakaran, K., Gotzinger, S., Shafi, K.V.P.M., Mazzei, A., Schietinger, S., Benson, O., 2006. Ultrafine luminescent structures through nanoparticle self-assembly. Nanotechnology 17, 3802–3805.

Prajapati, V.K., Awasthi, K., Gautam, S., Yadav, T.P., Rai, M., Srivastava, O.N., Sundar, S., 2011. Targeted killing of *Leishmania donovani* in vivo and in vitro with amphotericin B attached to functionalized carbon nanotubes. J. Antimicrob. Chemother. 66, 874–879.

Prusiner, S.B., 1982. Novel proteinaceous infectious particles cause scrape. Science 216, 136–144.

Puglisi, A., Mondini, S., Cenedese, S., Ferretti, A.M., Santo, N., Ponti, A., 2010. Monodisperse octahedral α-MnS and MnO nanoparticles by the decomposition of manganese oleate in the presence of sulfur. Chem. Mater. 22, 2804–2813.

Qian, H., Zhao, Z., Velazquez, J.C., Pretzer, L.A., Hecka, K.N., Wong, M.S., 2014. Supporting palladium metal on gold nanoparticles improves its catalysis for nitrite reduction. Nanoscale 6, 358–364.

Qin, S., Qin, D., Ford, W.T., Herrera, J.E., Resasco, D.E., Bachilo, S.M., Weisman, R.B., 2004. Solubilization and purification of Single-Wall carbon nanotubes in water by in situ radical polymerization of sodium 4-styrenesulfonate. Macromolecules 37, 3965–3967.

Qu, L., Peng, X., 2002. Control of photoluminescence properties of CdSe nanocrystals in growth. J. Am. Chem. Soc. 2002, 2049–2055.

Ra, H.W., Choi, D.H., Kim, S.H., Im, Y.H., 2009. Formation and characterization of ZnO/a-c core–shell nanowires. J. Phys. Chem. C 113, 3512–3516.

Raffi, M., Rumaiz, A.K., Hasan, M.M., Shah, S.I., 2007. Studies of the growth parameters for silver nanoparticle synthesis by inert gas condensation. J. Mater. Res. 22, 3378–3384.

Rao, C.N.R., Satishkumar, B.C., Govindaraj, A., Nath, M., 2001. Nanotubes. ChemPhysChem 2 (78105), 1439–1442.

Reddy, S.T., van der Vlies, A.J., Angeli, S.V., Randolph, G.J., O'Neill, C.P., Lee, L.K., Swartz, M.A., Hubbell, J.A., 2007. Exploiting lymphatic transport and complement activation in nanoparticle vaccines. Nat. Biotechnol. 25, 1159–1164.

Reis, C.P., Neufeld, R.J., Ribeiro, A.J., Veiga, F., 2006. Nanoencapsulation I. Methods for preparation of drug-loaded polymeric nanoparticles. Nanomedicine 2, 8–21.

Reynolds, F., O'loughlin, T., Weissleder, R., Josephson, L., 2005. Method of determining nanoparticle core weight. Anal. Chem. 77, 814–817.

Roberts, J., Bentley, M.D., Harris, J.M., 2002. Chemistry for peptide and protein PEGylation. Adv. Drug Deliv. Rev. 54, 459–476.

Roessler, B.J., Bielinska, A.U., Janczak, K., Lee, I., Baker, J.R., 2001. Substituted cyclodextrins interact with PAMAM dendrimer-DNA complexes and modify transfection efficiency. Biochem. Biophys. Res. Commun. 283, 124–129.

Roh, Y.H., Ruiz, R.C.H., Peng, S., Leea, J.B., Luo, D., 2011. Engineering DNA-based functional materials. Chem. Soc. Rev. 40, 5730–5744.

Rosemary, M.J., Pradeep, T., 2003. Solvothermal synthesis of silver nanoparticles from thiolates. J. Colloid Interface Sci. 268, 81–84.

Rosen, Y., Elman, N.M., 2009. Carbon nanotubes in drug delivery: focus on infectious diseases. Expert Opin. Drug Deliv. 6, 517–530.

Rosolen, J.M., Montoro, L.A., Matsubara, E.Y., Marchesin, M.S., Nascimento, L.F., Tronto, S., 2006. Step-by-step chemical purification of carbon nanotubes analyzed by high resolution electron microscopy. Carbon 44, 3293–3301.

Ruelle, B., Peeterbroeck, S., Godfroid, T., Bittencourt, C., Hecq, M., Snyders, R., Dubois, P., 2012. Selective grafting of primary amines onto carbon nanotubes via free-radical treatment in microwave plasma post-discharge. Polymers 4, 296–315.

Saba, T.M., 1970. Physiology and physiopathology of the reticuloendothelial system. Arch. Intern. Med. 126, 1031–1052.

Sahoo, N.G., Bao, H., Pan, Y., Pal, M., Kakran, M., Cheng, H.K.F., Li, L., Tan, L.P., 2011. Functionalized carbon nanomaterials as nanocarriers for loading and delivery of a poorly water-soluble anticancer drug: a comparative study. Chem. Commun. 47, 5235–5237.

Saidane, K., Razafinimanana, M., Lange, H., Huczko, A., Baltas, M., Gleizes, A., Meunier, J.L., 2004. Fullerene synthesis in the graphite electrode arc process: local plasma characteristics and correlation with yield. J. Phys. D Appl. Phys. 37, 232–239.

Saito, R., Dresselhaus, G., Dresselhaus, M.S., 1998. Physical Properties of Carbon Nanotubes. Imperial College press, London, UK.

Sammalkorpi, M., 2004. Molecular Dynamics Simulations of Strained and Defective Carbon Nanotubes (Doctoral dissertation). Helsinki University of Technology. ISBN: 951-22-7379-9.

Sannohe, Y., Sugiyama, H., 2010. Overview of Formation of G-quadruplex Structures, Current Protocols in Nucleic Acid Chemistry, Unit Number: Unit 17.2.

Sannohe, Y., Sugiyama, H., 2012. Single strand DNA catenane synthesis using the formation of G-quadruplex structure. Bioorg. Med. Chem. 20, 2030–2034.

Santi, D.V., Schneider, E.L., Reid, R., Robinson, L., Ashley, G.W., 2012. Predictable and tunable half-life extension of therapeutic agents by controlled chemical release from macromolecular conjugates. PNAS 109, 6211–6216.

Sato, N., Kobayashi, H., Saga, T., Nakamoto, Y., Ishimori, T., Togashi, K., Fujibayashi, Y., Konishi, J., Brechbiel, M.W., 2001. Tumor targeting and imaging of intraperitoneal tumors by use of antisense oligo-DNA complexed with dendrimers and/or avidin in mice. Clin. Cancer Res. 7, 3606–3612.

Schrick, B., Blough, J., Jones, A., Mallouk, T.E., 2002. Hydrodechlorination of trichloroethylene to hydrocarbons using bimetallic nickel-iron nanoparticles. Chem. Mater. 14, 5140–5147.

Selvaggini, C., DeGioia, L., Cantu, L., Ghibaudi, E., Diomede, L., Passerini, F., Forloni, G., Bugiani, O., Tagliavini, F., Salmona, M., 1993. Molecular characteristics of a proteaseresistant, amyloidogenic and neurotoxic peptide homologous to residues 106–126 of the prion protein. Biochem. Biophys. Res. Commun. 194, 1380–1386.

Service, R.F., 1995. Dendrimers: Dream Molecules Approach Real Applications. Science 267, 458–459.

Sharma, V.K., Yngard, R.A., Lin, Y., 2009. Silver nanoparticles: green synthesis and their antimicrobial activities. Adv. Colloid Interface 145, 83–96.

Shirtcliffe, N., Nickel, U., Schneider, S., 1999. Reproducible preparation of silver sols with small particle size using borohydride reduction: for use as nuclei for preparation of larger particles. J. Colloid Interface Sci. 211, 122–129.

Shukla, S., Seal, S., 1999. Cluster size effect observed for gold nanoparticles synthesized by sol-gel technique as studied by X-ray photoelectron spectroscopy. NanoStruct. Mater. 11, 1181–1193.

Singh, R., Torti, S.V., 2013. Carbon nanotubes in hyperthermia therapy. Adv. Drug Deliv. Rev. 65, 2045–2060.

Singh, K., Jiang, Y., Gupta, S., et al., 2013. Anti-inflammatory potency of nano-formulated puerarin and curcumin in rats subjected to the lipopolysaccharide-induced inflammation. J. Med. Food 16, 899–911.

Siu, K.S., Chen, D., Zheng, X., Zhang, X., Johnston, N., Liu, Y., Yuan, K., Koropatnick, J., Gillies, E.R., Min, W.P., 2014. Non-covalently functionalized single-walled carbon nanotube for topical siRNA delivery into melanoma. Biomaterials 35, 3435–3442.

Song, Y., Huang, T., Murray, R.W., 2003. Heterophase ligand exchange and metal transfer between monolayer protected clusters. J. Am. Chem. Soc. 125, 11694–11701.

Song, Y., Murray, R.W., 2012. Dynamics and extent of ligand exchange depend on electronic charge of metal nanoparticles. J. Am. Chem. Soc. 124, 7096–7102.

Stiriba, S.-E., Frey, H., Haag, R., 2002. Dendritic polymers in biomedical applications: from potential to clinical use in diagnostics and therapy. Angew. Chem. Int. Ed. 41, 1329–1334.

Stone, A.J., Wales, D.J., 1986. Theoretical studies of icosahedral C_{60} and some related structures. Chem. Phys. Lett. 128 (5–6), 501–503.

Stuurman, F.E., Nuijen, B., Beijnen, J.H., Schellens, J.H., 2013. Oral anticancer drugs: mechanisms of low bioavailability and strategies for improvement. Clin. Pharmacokinet. 52, 399–414.

Sun, B., Marx, E., Greenham, N., 2003. Photovoltaic devices using blends of branched CdSe nanoparticles and conjugated polymers. Nano Lett. 3, 961–963.

Sun, W., Pelletier, J.C., 2007. Efficient conversion of primary and secondary alcohols to primary amines. Tetrahedron Lett. 48, 7745–7746.

Supattapone, S., Wille, H., Uyechi, L., Safar, J., Tremblay, P., Szoka, F.C., Cohen, F.E., Prusiner, S.B., Scott, M.R., 2001. Branched polyamines cure prion-infected neuroblastoma cells. J. Virol. 75, 3453–3461.

Syrovets, T., Lunov, O., Simmet, T., 2012. Plasmin as a proinflammatory cell activator. J. Leukoc. Biol. 92, 509–519.

Tabata, Y., Ikada, Y., 1990. Phagocytosis of polymer microspheres by macrophages. New Polym. Mater. 107, 41.

Tagmatarchis, N., Prato, M., 2004. Functionalization of carbon nanotubes via 1,3-dipolar cycloadditions. J. Mater. Chem. 14, 437–439.

Talapin, D.V., Rogach, A.L., Kornowski, A., Haase, M., Weller, H., 2001. Highly luminescent monodisperse CdSe and CdSe/ZnS nanocrystals synthesized in a hexadecylamine-trioctylphosphine oxide-trioctylphospine mixture. Am. Chem. Soc. 2001, 207–211.

Tan, G., Mieno, T., 2010. Experimental and numerical studies of heat convection in the synthesis of single-walled carbon nanotubes by arc vaporization. Jpn. J. Appl. Phys. 49. ID 045102, 6 p.

Tan, W., Wang, K., He, X., Zhao, X.J., Drake, T., Wang, L., Bagwe, R.P., 2004. Bionanotechnology based on silica nanoparticles. Med. Res. Rev. 24, 621–638.

Tao, A.R., Habas, S., Yang, P., 2008. Shape control of colloidal metal nanocrystals. Small 4, 310–325.

Tappero, R., Wolfers, P., Lichanot, A., 2001. Electronic, magnetic structures and neutron diffraction in B_1 and B_3 phases of MnS: a density functional approach. Chem. Phys. Lett. 335, 449–457.

Tasis, D., Tagmatarchis, N., Bianco, A., et al., 2006. Chemistry of carbon nanotubes. Chem. Rev. 106, 1105–1136.

Taylor, R., Langley, G.J., Kroto, H.W., Walton, D.R.M., 1993. Formation of C_{60} by pyrolysis of naphthalene. Nature 366, 728–731.

Tessler, N., Medvedev, V., Kazes, M., Kan, S.H., Banin, U., 2002. Efficient near-infrared polymer nanocrystal light-emitting diodes. Science 295, 1506–1508.

Thanh-Dinh, N., Cao-Thang, D., Trong-On, D., 2010. Shape- and size-controlled synthesis of monoclinic ErOOH and cubic Er_2O_3 from micro- to nanostructures and their upconversion luminescence. ACS Nano 4, 2263–2273.

Thess, A., Lee, R., Nikolaev, P., Dai, H., Petit, P., Robert, J., Xu, C., Lee, Y.H., Kim, S.G., Rinzler, A.G., Colbert, D.T., Scuseria, G.E., Tománek, D., Fischer, J.E., Smalley, R.E., 1996. Crystalline ropes of metallic carbon nanotubes. Science 273, 483–487.

Thorek, D.L., Chen, A.K., Czupryna, J., Tsourkas, A., 2006. Superparamagnetic iron oxide nanoparticle probes for molecular imaging. Ann. Biomed. Eng. 34, 23–38.

Tian, L., Yep, L.Y., Ong, T.T., Yi, Y., Ding, J., Vitta, J.J., 2009. Synthesis of NiS and MnS nanocrystals from the molecular precursors $(TMEDA)M(SC\{O\}C_6H_5)_2$ ($M = Ni$, Mn). Cryst. Growth Des. 9, 352–357.

Tomalia, D.A., Baker, H., Dewald, J.R., Hall, M., Kallos, G., Martin, S., Roeck, J., Ryder, J., Smith, P., 1985. A new class of polymers: starburst-dendritic macromolecules. Polym. J. (Tokyo) 17, 117–132.

Tunvir, K., Kim, A., Nahm, S.H., 2007. Carbon nanotubes with vacancy defects and their mechanical properties. In: Proceedings of the International Conference on Mechanical Engineering 2007 (ICME2007) 29–31 December 2007, Dhaka, Bangladesh.

Tully, D.C., Frechet, J.M.J., 2001. Dendrimers at surfaces and interfaces: chemistry and applications. Chem. Commun. 6 (14), 1229–1239.

Twyman, L.J., Beezer, A.E., Esfand, R., Hardy, M.J., Mitchell, J.C., 1999. The synthesis of water soluble dendrimers, and their application as possible drug delivery systems. Tetrahedron Lett. 40, 1743–1746.

Um, S.H., Lee, J.B., Kwon, S.Y., Li, Y., Luo, D., 2006. Dendrimer-like Dan-based fluorescence nanobarcodes. Nat. Protoc. 1, 995–1000.

Van Hyning, D.L., Zukoski, C.F., 1998. Formation mechanisms and aggregation behavior of borohydride reduced silver particles. Langmuir 14, 7034–7046.

Villaraza, A.J.L., Bumb, A., Brechbiel, M.W., 2010. Macromolecules, dendrimers, and nanomaterials in magnetic resonance imaging: the interplay between size, function, and pharmacokinetics. Chem. Rev. 110, 2921–2959.

Vo-Dinh, T., Yan, F., Wabuyele, M., 2005. Surface-enhanced Raman scattering for medical diagnostics and biological imaging. J. Raman Spectrosc. 36, 640–647.

Vogtle, F., Fischer, M., 1999. Dendrimers: from design to application—a progress report. Angew. Chem. Int. Ed. 38, 884–905.

Wang, H., Qiao, X., Chen, J., Ding, S., 2005a. Preparation of silver nanoparticles by chemical reduction method. Colloids Surf. A 256, 111–115.

Wang, L., Boatner, L.A., Ewing, R.C., Lian, J., Zhou, J., 2005b. Focused ion beam-induced Ripple and nanoparticle formation in $Cd_2Nb_2O_7$. Microsc. Microanal. 11 (Suppl. 2), 86–87.

Wang, M., Abbinenib, G., Msb, C.A., Mao, C., Xu, S., 2011. Upconversion nanoparticles: synthesis, surface modification and biological applications. Nanomedicine 7, 710–729.

Wang, Q., Xu, Y., Zhao, X., Chang, Y., Liu, Y., Jiang, L., Sharma, J., Seo, D.-K., Yan, H., 2007. A facile one-step in situ functionalization of quantum dots with preserved photoluminescence for bioconjugation. J. Am. Chem. Soc. 129, 6380–6381.

Wang, Y., Bai, Y., Price, C., Boros, P., Qin, L., Bielinska, A.U., Kukowska-Latallo, J.F., Baker, J.R., Bromberg, J.S., 2001. Combination of electroporation and DNA/dendrimer complexes enhances gene transfer into murine cardiac transplants. Am. J. Transplant. 1, 334–338.

Wang, Y., Quek, C.H., Leong, K.W., Fang, J., 2010. Synthesis and cytotoxicity of luminescent InP quantum dots mater. In: Res. Soc. Symp. Proc. Materials Research Society 1241-XX02-04.

Warheit, D.B., Laurence, B.R., Reed, K.L., Roach, D.H., Reynolds, G.A.M., Webb, T.R., 2004. Comparative pulmonary toxicity assessment of single-wall carbon nanotubes in rats. Toxicol. Sci. 77, 117–125.

Warheit, D.B., 2004. Nanoparticles: health impacts? Mater. Today 7, 32–35.

Waynant, R.W., Ilev, I.K., Gannot, I., 2001. Mid-infrared laser applications in medicine and biology. Philos. Trans. R. Soc. Lond. Ser. A 359, 635–644.

Weber, M., Thies, M.C., 2002. Understanding the RESS process. In: Sun, Y.P. (Ed.), Supercritical Fluid Technology in Materials Science and Engineering: Synthesis, Properties, and Applications. Marcel Dekker, New York, pp. 387–437.

Weia, C., Srivastava, D., 2004. Nanomechanics of carbon nanofibers: structural and elastic properties. Appl. Phys. Lett. 85, 2208–2210.

Weinstein, J.S., Varallyay, C.G., Dosa, E., Gahramanov, S., Hamilton, B., Rooney, W.D., Muldoon, L.L., Neuwelt, E.A., 2010. Superparamagnetic iron oxide nanoparticles: diagnostic magnetic resonance imaging and potential therapeutic applications in neurooncology and central nervous system inflammatory pathologies, a review. J. Cereb. Blood Flow Metab. 30, 15–35.

Weisbecker, C.S., Merritt, M.V., Whitesides, G.M., 1996. Molecular self-assembly of aliphatic thiols on gold colloids. Langmuir 12, 3763–3772.

Wells, M., Crooks, R.M., 1996. Surface and colloid science. J. Am. Chem. Soc. 16, 3988–3989.

Welsher, K., Liu, Z., Daranciang, D., Dai, H., 2008. Selective probing and imaging of cells with single walled carbon nanotubes as near-infrared fluorescent molecules. Nano Lett. 8, 586–590.

Wildoer, J.W.G., Venema, L.C., Rinzler, A.G., Smalley, R.E., Dekker, C., 1998. Electronics of structure of atomically resolved nanotubes. Nature 391, 59–63.

Wilson, M.A., Pang, L.S.K., Quezada, R.A., Fisher, K.J., Dance, I.G., Willett, G.D., 1993. Fullerene production in alternative atmospheres. Carbon 31, 393–396.

Winfree, E., Liu, F., Wenzler, L.A., Seeman, N.C., 1998a. Quantitative analysis of molecular-level DNA crystal growth on a 2D surface. Nature 394, 539–544.

Wojtkowiak, J.W., Rothberg, J.M., Kumar, V., Schramm, K.J., Haller, E., Proemsey, J.B., Lloyd, Sloane, B.F., Gillies, R.J., 2012. Chronic autophagy is a cellular adaptation to tumor acidic pH microenvironments. Cancer Res. 72, 3938–3947.

Wolcott, A., Gerion, D., Visconte, M., Sun, J., Schwartzberg, A., Chen, S.W., Zhang, J.Z., 2006. Silica-coated CdTe quantum dots functionalized with thiols for bioconjugation to IgG proteins. J. Phys. Chem. B 110, 5779–5789.

Wolinsky, J.B., Grinstaff, M.W., 2008. Therapeutic and diagnostic applications of dendrimers for cancer treatment. Adv. Drug Deliv. Rev. 60, 1037–1055.

Xing, G., Jia, S.-L., Shi, Z.-Q., 2007. The production of carbon nano-materials by arc discharge under water or liquid nitrogen. New Carbon Mater. 22, 337–341.

Yan, H., Park, S.H., Finkelstein, G., Reif, J.H., LaBean, I.H., 2003. DNA-templated self-assembly of protein arrays and highly conductive nanowires. Science 301, 1882–1884.

Yang, H., Morris, J.J., Lopina, S.T., 2004. Polyethylene glycol-polyamidoamine dendritic micelle as solubility enhancer and the effect of the length of polyethylene glycol arms on the solubility of pyrene in water. J. Colloid Interface Sci. 273, 148–154.

Yang, J., Lind, J., Trogler, W., 2008. Synthesis of hollow silica and titania nanospheres. Chem. Mater. 20, 2875–2877.

Yi, G.C., Wang, C., Park II, W., 2005. ZnO nanorods: synthesis, characterization and applications. Semicond. Sci. Technol. 20, s22–s34.

Yi, G.S., Chow, G.M., 2006. Synthesis of hexagonal-phase NaYF4:Yb,Er and NaYF4:Yb,Tm nanocrystals with efficient up-conversion fluorescence. Adv. Funct. Mater. 16, 2324–2329.

Yoshie, K., Kasuya, S., Eguchi, K., Yoshida, T., 1992. Novel method for C_{60} synthesis: a thermal plasma at atmospheric pressure. Appl. Phys. Lett. 61, 2782–2783.

Yu, B., Shevchenko, V., Ponyavina, A.N., Rakhmanov, S.K., 2004a. Formation of silver and copper nanoparticles upon the reduction of their poorly soluble precursors in aqueous solution. Colloidal J. 66, 517–522.

Yu, M.X., Li, F.Y., Chen, Z.G., Hu, H., Zhan, C., Yang, H., Huang, C.H., 2004b. Laser scanning up-conversion luminescence microscopy for imaging cells labeled with rare-earth nanophosphors. Anal. Chem. 2009 (81), 930–935.

Yuan, J., Zhang, X., Qian, H., 2010. A novel approach to fabrication of superparamagnetite hollow silica/magnetic composite spheres. J. Magn. Magn. Mater. 322, 2172–2175.

Zaluzhna, O., Li, Y., Zangmeister, C., Allisonb, T.C., Tong, Y.Y.J., 2012. Mechanistic insights on one-phase vs. two-phase Brust–Schiffrin method synthesis of Au nanoparticles with dioctyl-diselenides. Chem. Commun. 48, 362–364.

Zhang, C., Su, M., He, Y., Zhao, X., Fang, P., Ribbe, A.E., Jiang, W., Mao, C., 2008a. Conformational flexibility facilitates self-assembly of complex DNA nanostructures. Proc. Natl Acad. Sci. U.S.A. 105, 10665–10669.

Zhang, H., Patel, P.R., Xie, Z., Swanson, S.D., Wang, X., Kotov, N.A., 2013. Tissue compliant neural implants from micro-fabricated carbon nanotube multilayer composite. ACS Nano 7, 7619–7629.

Zhang, J., Liu, Y., Ke, Y., Yan, H., 2006. Periodic square-like gold nanoparticle arrays templated by self-assembled 2D DNA nanogrids on a surface. Nano Lett. 6, 248–251.

Zhang, N., Yi, R., Wang, Z., Shi, R.R., Wang, H.D., Qiu, G.Z., Liu, X.H., 2008b. Hydrothermal synthesis and electrochemical properties of alpha-manganese sulfide submicrocrystals as an attractive electrode material for lithium-ion batteries. Mater. Chem. Phys. 111, 13–16.

Zhang, W.-X., Elliott, D.W., 2006. Applications of iron nanoparticles for groundwater remediation. Remediation J. 16, 7–21.

Zhang, W-x, 2003. Nanoscale iron particles for environmental remediation: an overview. J. Nanopart. Res. 5, 323–332.

Zhao, J., Chen, Z., Zhou, Z., Park, H., von Ragu Schleyer, P., Lu, J.P., 2001. Engineering the electronic structure of single-walled carbon nanotubes by chemical functionalization. ChemPhysChem 6, 598–601.

Zhao, J., Park, H., Han, J., Lu, J.P., 2004. Electronic properties of carbon nanotubes with covalent sidewall functionalization. Phys. Chem. B 108, 4227–4230.

Zhao, N., You, J., Zeng, Z., et al., 2013. An ultra pH-sensitive and aptamer-equipped nanoscale drug-delivery system for selective killing of tumor cells. Small 9, 3477–3484.

Zhou, F., Xing, D., Ou, Z., Wu, B., Resasco, D.E., Chen, W.R., 2009. Cancer photothermal therapy in the near-infrared region by using single-walled carbon nanotubes. J. Biomed. Opt. 14, 021009.

Zhou, H., Cheng, X., Rao, L., Li, T., Duan, Y.Y., 2013. Poly(3,4-ethylenedioxythiophene)/multi-wall carbon nanotube composite coatings for improving the stability of microelectrodes in neural prostheses applications. Acta Biomater. 9, 6439–6449.

Zhu, H., Zhang, C., Yin, Y., 2005. Novel synthesis of copper nanoparticles: influence of the synthesis conditions on the particle size. Nanotechnology 16, 3070–3083.

Zhu, J., Shi, X., 2013. Dendrimer-based nanodevices for targeted drug delivery applications. J. Mater. Chem. B 1, 4199–4211.

Zimmerman, S.C., Lawless, J.L., 2001. Supramolecular Chemistry of Dendrimers. In: Topics in Current Chemistry, vol. 217. Springer-Verlag, Berlin Heidelberg pp. 96–118.

Zoldesi, C., Imhof, A., 2005. Synthesis of monodisperse colloidal spheres, capsules, and microballoons by emulsion templating. Adv. Mater. 17, 924–928.

CHAPTER

3

Physicochemical, Electronic, and Mechanical Properties of Nanoparticles

OUTLINE

Engineered Nanoparticles
http://dx.doi.org/10.1016/B978-0-12-801406-6.00003-0

Nanoparticles are a group of structurally and chemically diverse particles smaller than 100 nm at least in one dimension. In comparison, the diameter of an atom is less than 0.1 nm, while the diameter of bulk particles is greater than 500 nm. As the nanoparticles get smaller, there is a gradual size-dependent transition in their physical (increase in hardness, strength, and ductility), optical (color change and surface plasmon), and surface-related (greater surface area to volume ratio) properties (Gupta et al., 2008; Lia et al., 2013). Below a certain size (usually <100 nm), the electronic properties of nanoparticles switch from classical mechanics to quantum mechanics (Rahman et al., 2003; Chatterjee et al., 2003). The aim of this chapter is to discuss the properties of nanoparticles. The chapter has been divided into two broad groups:

- Common size and/or surface-related properties that are independent of the structural and chemical differences (Section 1)
- Distinct structure- and chemistry-related properties, such as surface plasmon by metal nanoparticles, metallic and semiconductor properties in carbon nanotubes (CNTs), unique hydrophilic/hydrophobic properties in dendrimers, etc. (Section 2)

This unity and diversity in their physicochemical properties make nanoparticles useful for applications ranging from industrial to biomedical fields.

1. COMMON SIZE AND SURFACE-RELATED PROPERTIES

All nanoparticles, irrespective of their chemistry, exhibit comparable surface reactivity and thermodynamic, electronic, mechanical, and magnetic properties. This section provides a detailed review of the size-related effects of nanoparticles.

1.1 Surface Atoms

Nanoparticles exhibit exceptionally high surface area to volume (S/V) ratios compared to their bulk counterparts. The bulk particles (>500 nm) and mesoparticles (>100 nm but <500 nm) have less than 1% of total atoms on the surface (Figure 1(A)). As the size decreases, there is a decrease in surface area and volume, but a significant increase in the S/V ratio (Figure 1(B)). This is because surface area is $4\pi r^2$ and volume is $3/4\pi r^3$. The surface area to volume ratio will be $4\pi r^2/(3/4\pi r^3) = 3/r$. Because of this, an increase in radius reduces the S/V ratio, but increases the number of atoms at the surface (Figure 1(C)).

In 1.0-nm (at least in one dimension) nanoparticles, greater than 80% of the total atoms reside at the surface (Figure 1(A), right). Spherical nanoparticles of 1.8 nm diameter have equal numbers of core and surface atoms (or molecules), while nanoparticles <1 nm in diameter may have all of their atoms facing the surface (may be unstable). As the percentage of surface atoms increase

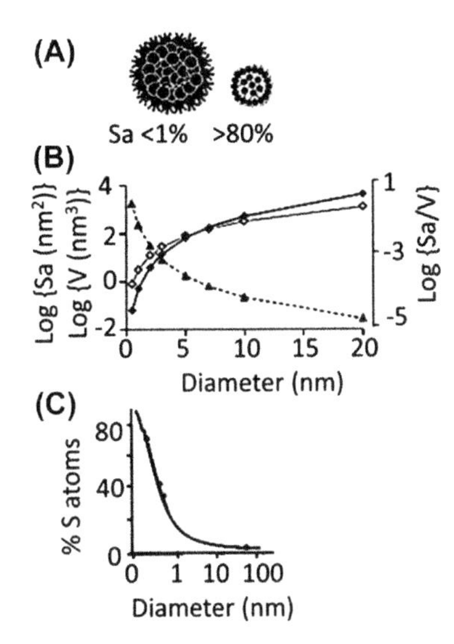

FIGURE 1 Nanoparticle size, surface to volume ratio, and surface atoms. (A) Surface and core atoms of bulk particles (left) and nanoparticles (right). (B) Relationship between nanoparticle diameter (nm) and surface area (unfilled diamond), volume (filled diamond), and surface area to volume ratio (triangle). (C) Relationship between diameter (nm) and the percentage of atoms at the surface. *Abbreviations*: Sa: surface area, S atom: surface atoms, V: volume.

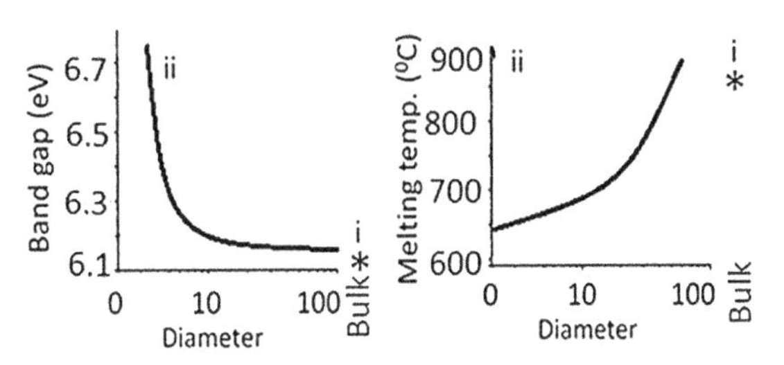

FIGURE 2 Effects of size reduction on band-gap energy and melting temperature. As the nanoparticles become smaller, their band gap increases (for semiconductors) and the melting point decreases (i: bulk, ii: nanoparticle). *The melting temperature plot was constructed using data from Xiong et al. (2011). The band-gap plot was constructed using data from Li and Li (2006). Reprinted with permission.*

relative to the core atoms, surface atoms begin to play a greater role in determining the nanoparticles' physicochemical, electronic, mechanical, and magnetic properties. Contrarily, surface atoms play a minuscule role in determining the properties of bulk particles. As shown in Figure 2, the melting temperature and the band-gap energy of nanoparticles but not bulk particles are dependent on their size. Thus, a reduction in size from the micro (bulk particles) to nano (nanoparticle) range enhances the relative importance of surface atoms (Boas and Heegaard, 2004). As the diameter increases, the nanoparticles' melting temperature decreases (Xiong et al., 2011), while the band-gap between the valence and the conduction bands increases (Gajewicz et al., 2011; Li and Li, 2006). Thus, many metals may become semiconductors at the nanometer level.

To understand the surface effects, it is important to understand the characteristics of the core and the surface atoms. The core atoms form stable covalent bonds with the nearest neighboring atoms. Because the paired electrons spin in opposite directions, they are stable and relatively nonreactive. The surface atoms contain nonbonded electrons, exhibiting greater uncompensated spin. Therefore, the surface atoms exhibit relatively low nearest neighbor (NN) numbers (Figure 3) and relatively great free energy, anisotropy, bond defects, surface strain, and narrowing of the band gap, which increase the density of electrons at the Fermi level that facilitates conduction of electrons (discussed later). These properties give the surface atoms many unique physicochemical and magnetic characteristics that are not present in the core atoms. In addition, the surface but not core atoms are in contact with oxygen, resulting in the formation of oxidized atoms, which further alters the properties of the surface atoms.

1.2 Size-Dependent Thermodynamic Properties

An understanding of the thermodynamic properties (melting temperature, enthalpy and entropy of melting, elastic moduli, specific heat capacity, etc.) of nanoparticles is necessary to improve their applications by minimizing the

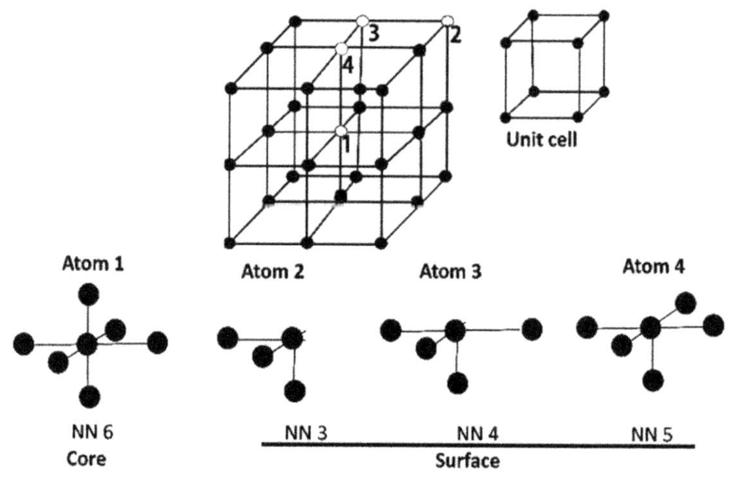

FIGURE 3 Three-dimensional shape of a crystal having a primitive cubic unit cell. The core atoms have six nearest neighbors (atom 1: NN = 6). The surface atoms 2, 3, and 4 have different NN numbers of 3, 4, and 5, respectively. This suggests that, depending on the location, different surface atoms may exhibit different electronic prop erties, commonly known as surface anisotropy. In addition, the surface and core atoms may exhibit different electronic properties, commonly known as structural anisotropy.

adverse effects. Many studies have shown a close relationship between the Gibbs free energy of a particle and its thermodynamic properties (Jia et al., 2009; Xiong et al., 2011; Nanda et al., 2003). The Gibbs free energy of bulk particles (G_b) is defined as G_b $(p, T) = \Delta H - T\Delta S$, where ΔH is the change in enthalpy, T is temperature, and ΔS is the entropy (calculated by the equation derived by Jia et al., 2009). Surface atoms have a minuscule role in determining the particle's G_b. However, as the particle size reaches the nanometer range, the surface free energy begins to dominate the core free energy; thus, the size becomes the key determinant of the nanoparticles' thermodynamic properties. The Gibbs free energy of nanoparticles (G_n) is sum of the Gibbs free energy of the surface (G_{surface}) and that of the core (G_{core}). For nanoparticles, G_{surface} is inversely related to diameter (D).

1.2.1 *Surface Free Energy*

Considering the importance of the surface in nanoparticle properties, it is important to understand the surface free energy of nanoparticles (γ_G) and its relationship with the nanoparticles' thermodynamic properties. Surface energy is defined as the energy required to produce a new surface by breaking the bonds of a nanoparticle core (Figure 4).

As shown in Eqn (1), γ_G is related to N_b that in the number of broken bonds, ε is the bond strength, and ρ is the number of atoms or molecules per unit surface area density.

$$\gamma_G = \frac{N_b(\varepsilon\rho)}{2} \tag{1}$$

The free energy of a surface is calculated using Eqn (2), where ΔH_s is the enthalpy of sublimation (energy needed to break all the bonds of a solid), number (#) of broken bonds can be determined using the nearest neighbor algorithm, Z is the coordination number (bonds/atom), ZN_A is the bonds/mol, and N/A is # of atoms/area.

$$\lambda = \frac{\Delta H_S\left(\frac{\#\ of\ broken\ bonds}{atom}\right)}{ZN_a}\frac{N}{A} \tag{2}$$

Table 1 shows examples of surface free energy of different nanoparticles and the corresponding bulk particles. In all cases, the free energy in bulk particles was lower than that in nanoparticles. Nanoparticles capped with functional groups or embedded in another particle exhibited lower free energy. This suggests that the free surface energy of nanoparticles also

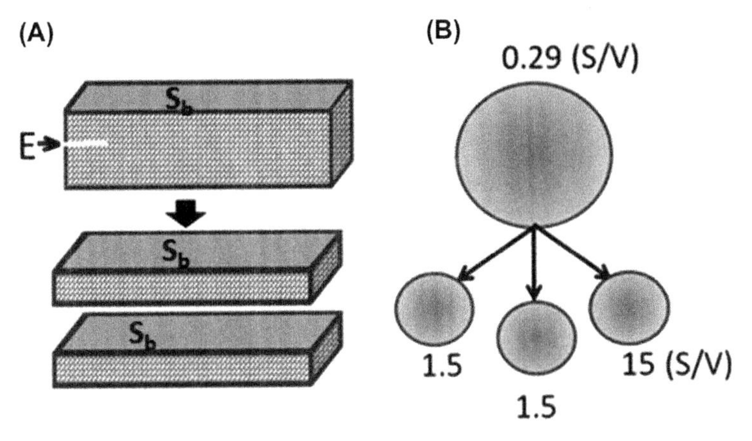

FIGURE 4 (A) Formation of a new surface by applying sufficient energy (known as surface free energy, γ) to split core bonds. (B) Splitting of a zero-dimensional nanoparticle (29 s/v) into three smaller ones (15 s/v) resulted in 55% increase in total surface area and ensuing increase in surface energy.

TABLE 1 Surface Free Energy (J/m^2) of Different Bulk Particles and Nanoparticles

	PdS Nanda et al. (2003)	CdS Goldstein et al. (1992)	PbS Nanda et al. (2003)	Ag Dubiel et al. (1997), Montano et al. (1984)
Bulk	1.8	1.51	1.47	1.06
NP bare	6.0	2.5	2.45	6.4
NP-capped		1.74		
Embedded				1.3–5.9

depends upon their functionalization and environment.

The size dependence of γ_G is explained by Eqn (3) (Xiong et al., 2011), in which γ_{G0} is the bulk free energy, d is the atom diameter, D is the nanoparticle diameter, x is the proportionality coefficient, k is a function of particle size constant for bulk change for N atoms, and E_0 is (cohesive energy for bulk (E_b)/coordination numbers for bulks).

$$\gamma_G = \gamma_{G0}\left(1 - \frac{1.45d}{D}\right) \text{ or } (\gamma_G - \gamma_{G0}) = -\gamma_0\left(\frac{1.45d}{D}\right) \quad (3)$$

$$\gamma_{G0} = \frac{13.21\, x\, k\, E_0}{\pi d^3} \quad (4)$$

These equations suggest that the surface free energy is related to $1/D$. For bulk particles $d << D$, then d/D approaches zero ($\rightarrow 0$). For nanoparticles <10 nm in diameter, $d \geq D$ and $d/D \neq 0$, an increase in their surface free energy occurs. The relationship between shape and γ_G is shown in Eqns (5–7) to describe the shape (Figure 5) dependence of surface energy.

Surface free energy (J/m^2) is shown by (Jia et al., 2009):

$$\gamma_{G\{100\}} = 4\varepsilon/a^2, \quad 1.85 \quad (5)$$

$$\gamma_{G\{110\}} = (5/2)^{0.5} - (\varepsilon/a^2) \quad 1.95 \quad (6)$$

$$\gamma_{G\{111\}} = 2(3)^{0.5} - (\varepsilon/a^2) \quad 1.75 \quad (7)$$

In these equations, a is the lattice constant. Each atom on surface {100} has four broken bonds, surface {110} has five broken bonds, and

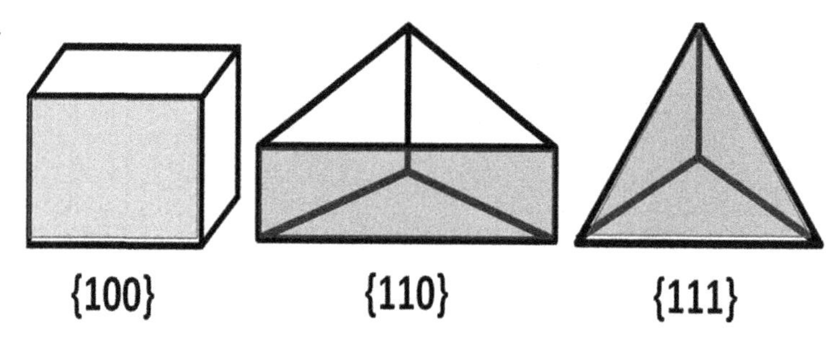

FIGURE 5 Structures of the {100}, {110}, and {111} crystals.

surface {111} has three broken bonds (shown in bold). Figure 6 shows a linear relationship between broken bonds (unpaired electrons) and the surface free energy of different face crystals. The close-packed {1 1 1} surface with the smallest number of broken bonds showed the lowest free energy, while the loosely packed {110} surface with the highest number of broken bonds showed the largest free energy.

1.2.2 *Thermodynamic Indices*

This section describes characteristics of and relationship among cohesive energy, melting temperature, elastic moduli, specific heat capacity, evaporation, Curie temperature, and Debye temperature.

Cohesive energy accounts for the strength of atomic bonds that is equal to the energy required to divide the nanoparticle into individual atoms (Sun et al., 2002). Consider a nanoparticle with diameter D composed of n atoms. If sufficient energy (E_n, also known as the cohesion energy) is applied, individual atoms will be separated and the cumulative surface area of n atoms will be $S = n.\ \pi d^2$ in which d is the individual atoms' diameter (Qi and Wang, 2002). E_n (cohesive

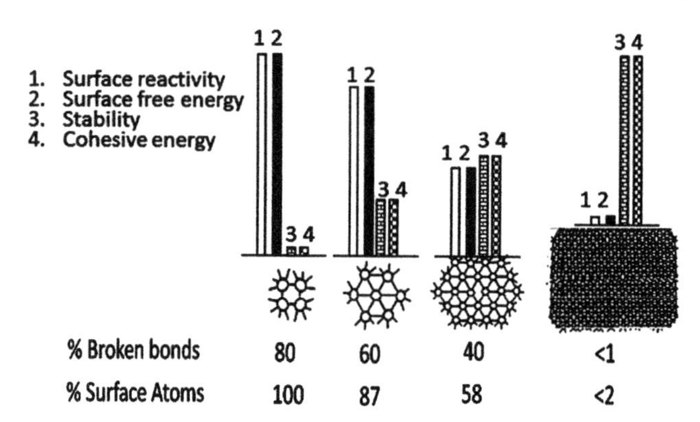

FIGURE 6 A hypothetical model for surface-related modulation of the particles' surface reactivity, surface free energy, stability, and cohesive energy. Almost all of the atoms in small clusters are at the surface (no core), approximately 90–40% of the atoms in nanoparticles face the surface, and less than 1% of the atoms in the bulk particles face the surface (defined as an interphase between two phases such as solid and gas or liquid). When surface is created, there is a disruption of intermolecular bonds that makes the surfaces intrinsically less energetically favorable than the core of a material. The surface free energy is defined as the excess energy at the surface of a material compared to the bulk. As clusters grow, there is a decrease in surface energy and an increase in cohesion energy (the difference between the energy per atom of a system of free atoms far apart from each other and the energy of a bonded atom in a crystal) that increases their stability. A relatively high cohesive energy is associated with the stability of the particle.

energy of nanoparticles) is related to the nanoparticle's diameter by Eqn (8), in which D is the diameter and E_n is the cohesive energy for nanoparticles, E_b is the cohesive energy for bulk particles, and d is the diameter of the atoms (Xiong et al., 2011; Miedema and Boom, 1978)

$$\frac{E_n}{E_b} = 1 - \frac{d}{D} \quad (8)$$

In Eqn (8), E_b can be calculated by the equation $E_b = k\pi\gamma_o d^2$. The value of d is dependent on the crystal type, as shown below (Figure 5; a is the lattice constant).

$$d_{\{100\}} = (3/\pi)^{1/3a} \quad (9)$$

$$d_{\{110\}} = (3/2\pi)^{1/3a} \quad (10)$$

$$d_{\{111\}} = \left(3\sqrt{a^2 \cdot c/2\pi}\right)^{1/3} \quad (11)$$

Figure 6 shows the possible relationship between cohesive energy and surface atoms, broken bonds (unpaired electrons), surface free energy, and surface reactivity of clusters (>90% atoms facing the surface), nanoparticles (90–40% atoms facing the surface), and bulk particles (<1% atoms facing the surface). The core is an ensemble of bound atoms or molecules, while the surface has unsaturated bonds. Clusters or nanoparticles may contain greater density of the unpaired electrons (broken bonds) than the bulk particles. An increase in the particle size increases the cohesive energy. In nanoparticles greater than 100 nm, cohesive energy reaches the bulk size constant value (Figure 6). Because cohesive energy is directly related to the bond strength of a core, an increase in cohesive energy is associated with an increase in intermolecular bonds that cannot be broken easily, while a decrease in cohesive energy is associated with a decrease in intermolecular bonds (such as in wax) that can be broken easily. The surface of the nanoparticles, because of their dangling bonds, may have lowest level of cohesive energy, possibly because of the inverse relationship between the surface area of a nanoparticle and its cohesion energy.

In general, cohesive energy depends directly on the coordination number (Shandiz et al., 2008); thus, it affects the thermal stability that modulates the melting temperature of particles. The melting temperature of nanoparticles has a significant impact on other thermodynamic properties: they all decrease with a decrease in size (Nanda et al., 2003; Kukowska-Latallo et al., 2012; Jia et al., 2009; Andrievski, 2009), as shown in Figure 7. The size effect is more pronounced at diameters 10 nm or lower. For nanoparticles over 50 nm in size, some of the thermodynamic values may reach the bulk value.

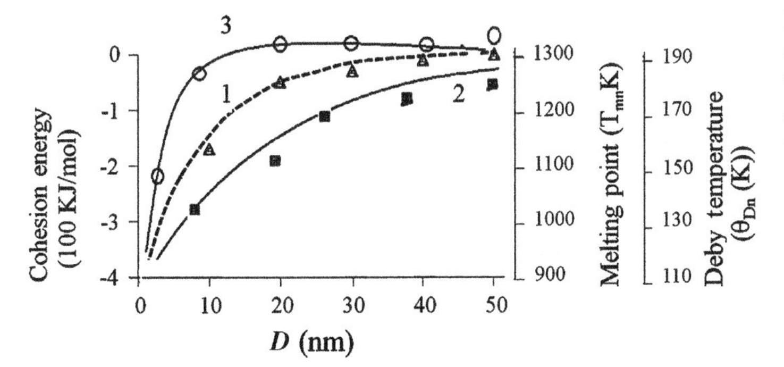

FIGURE 7 Relationship between the cohesive energy (plot 1) and melting point (plot 2) or Debye temperature (plot 3). *Data for this graph were derived from Xiong et al. (2011). Reprinted with permission.*

Xiong et al. (2011) has characterized the effects of size on the thermodynamic indices of nanoparticles, as shown below:

$$\frac{E_n}{E_b} = 1 - \frac{d}{D} \tag{12}$$

in which E_n is the cohesive energy for nanoparticles, E_b is the cohesive energy for bulk particles, and d is the diameter of the atoms.

$$\frac{T_{nm}}{T_{bm}} = 1 - \frac{P}{D}, \tag{13}$$

where T_{nm} and T_{bm} are the melting temperature for nanoparticles and bulk particles, respectively. D is diameter and $P = \frac{\pi N_A d^4 b}{\Delta S b}$.

$$\frac{T_{nv}}{T_{bv}} = 1 - \frac{P}{D},$$

where T_{nv} and T_{bv} are evaporation temperatures for nanoparticles and bulk particles, respectively. D is diameter and P is described above.

$$\frac{Q_{nD}}{Q_{bD}} = \left(1 - \frac{P}{D}\right)^{0.5}, \tag{14}$$

where Q_{nD} and Q_{bD} are Debye temperatures for nanoparticles and bulk particles, respectively. D is diameter and P is described above.

$$\frac{C_{np}}{C_{bp}} = 1 - \frac{K}{D}, \tag{15}$$

where K is -0.5 and C_{nP} and C_{bP} are specific heat capacities for nanoparticles and bulk particles, respectively. D is diameter.

These equations reveal that if D increases, then $1-(d/D)$ will increase; thus E_n will increase because E_b is constant. Ouyang et al. (2010) mathematically derived the following relationship: the cohesive energy is proportional to (the melting temperature)$^{1/2}$ and (the Debye temperature)$^{1/2}$. The cohesive energy of crystals reveals the strength of the chemical bonds; therefore, a decrease in the cohesive energy is associated with a decrease in the strength of the corresponding bond, which leads to a decrease in the melting point of nanoparticles.

1.3 Electronic Properties

1.3.1 Classic Theory of Atomic Structure

Atoms consist of negatively charged electrons orbiting the nucleus, which contains protons (positively charged) and neutrons (Figure 8(A)). According to Bohr's model, electrons are distributed in discrete energy bands called occupied orbitals (Bohr, 1923; Kramers and Helge, 1962; Housecroft and Sharpe, 2012). The outermost occupied orbital is the valence orbital or the highest occupied molecular orbital (HOMO). The number of electrons in valence orbital determines the chemical and electrical properties of an atom. In the occupied band, electrons (e^-) exist bound with a positively charged electron–hole, the mathematical opposite of electrons (Shockley, 1950; Figure 8(A), inset).

When an electron in HOMO is excited by absorbing energy (light or heat), it separates from the "hole" and jumps into the conduction band, resulting in the creation of a positively charged hole in the HOMO (Freund, 2002). The newly formed hole attracts neighboring electrons, resulting in propagation of a positively charged hole-current in the HOMO (Figure 8(B)), moving opposite to the direction of the electron current in the conduction bands (Nicolai et al., 2012). In metals, because the HOMO and LUMO overlap, electrons from both HOMO and LUMO can bind the hole.

1.3.2 Quantum Mechanical Theory

In simple terms, the basic rules of quantum theory of atomic structure are the following:

1. Electrons can be found in certain energy states called the density of states (DOS) in a occupied molecular orbital. Electrons cannot jump from the HOMO (valence band) to the lowest occupied molecular orbital (LOMO) unless a DOS is available in the conduction band.
2. Only one particle can occupy a particular state at any one time. This is the called the Pauli Exclusion Principle.

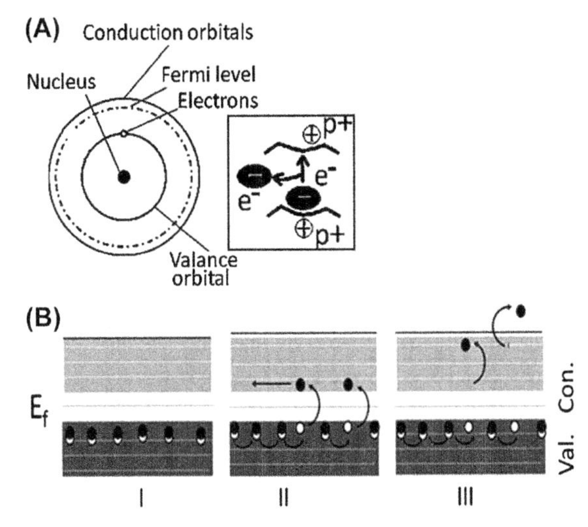

FIGURE 8 Structure of an atom. (A) According to Bohr's theory of atomic structure, atoms consist of a nucleus (protons and neutrons), and electrons encircling the nucleus in discrete orbits. The bands containing electrons are called the occupied molecular orbital. The outermost orbital is called valence orbital or the highest occupied molecular orbital (HOMO). In addition, atoms also contain conduction bands or unoccupied bands. The lowest (nearest to the HOMO) conduction band is called the lowest unoccupied molecular orbital or LUMO. The valence electrons in HOMO are bound with their mathematically opposite positively charged electron–hole (inset). When electrons absorb energy, they dissociate from the hole and either translocate into the conduction band (if sufficient energy is absorbed) or release energy and bind back to the hole (if the absorbed energy is not sufficient.) When electrons migrate to the conduction band, the HOMO contains a vacant hole that seeks electrons from neighboring atom. In semiconductors, there is an energy gap (band gap) between the HOMO and the LUMO ($E_{band\ gap} = E_{LUMO} - E_{HOMO}$). The valence electron must absorb $E_{band\ gap}$ equivalent energy to jump to the LUMO. The Fermi energy level is the critical energy an electron required to jump to the conduction band. In semiconductors, the Fermi energy is situated within the band gap; thus, valence electrons cannot jump into the conduction band without absorbing energy. In metal bands, there is no gap between HOMO and the LUMO and the Fermi level is situated within either HOMO or LUMO. Thus, valence electrons can migrate from HOMO to LUMO if electrons are being removed from the conduction orbital. (B) The electron and the electron–hole currents. When electrons jump from the HOMO into the LUMO (or higher conduction orbital), the hole seeks another electrons from neighboring atoms in the case of semiconductors. This results in the formation of a hole at another site that also seeks an electron. This generates a hole current.

3. Electrons exhibit both wave-like and particle-like behaviors (the wave-particle duality).
4. Classic physics describes light as a wave (electromagnetic radiation) with a set of frequency and amplitude. Einstein proposed that light travels in small quantized packets of energy called *photons* instead of a classic wave. For an electron to occupy a DOS, the photon's wavelength must match the wavelength of the atom's orbital.
5. The classic atomic structure is *deterministic*—that is, one can calculate both its location in the orbit and its velocity as it revolves around the orbit. The electron will always be there until some energy is absorbed or given off to change the orbit, all of which are totally certain. The quantum atomic structure is *probabilistic*—that is, it is not possible to predict where in its orbit an electron will be found. One can only determine the probability of finding an electron at certain points. The electron is therefore described in terms of its probability distribution, which is also known as probability density (Shrodinger's wave equation for matter: $H\psi = E\psi$).

In 1925, de Broglie (de Broglie, 1925) proposed particle-wave theory for electrons in which electrons, in addition to being particles, could also be thought of as a wave with a characteristic wavelength (de Broglie was awarded the Nobel Prize with Paul Dirac in 1929). Davison and Germer (1927) experimentally confirmed the particle-wave nature of electrons. Because electrons can exist as either a wave or a particle at a given instant of time, there is a fundamental limitation to finding an electron at a particular point in space (Heisenberg, 1925). The quantum mechanical model, therefore, shows that electrons are in random motion confined within the areas of electron probability distribution. In 1926, Erwin Schrödinger mathematically described the likelihood of finding an electron at a certain position and proposed an atom consisting of a nucleus surrounded by an electron cloud. The probability of finding the electron is greatest where the cloud is most dense and, conversely, the electron is less likely to be in a less dense area of the cloud, as shown in Figure 9(A) (Cutler, 1969). Figure 9(B) shows the shape of different orbitals.

As shown in Figure 10, only certain wavelengths of an electron wave will "fit" into a particular orbit. If the electron's wavelength is longer or shorter, then they will not fit in that orbit (wavelength exclusion). Thus, electrons may not exist at one single spot in its orbit, bringing up the idea of the DOS.

The DOS of a system describes the number of states at each energy level that are available for occupation by electrons. In metals, the LUMO, HOMO, and the Fermi level (Eg) overlap; thus, electrons nearest to the Fermi level cross to the conduction band. In semiconductors, there is an energy gap between the HOMO and the LUMO, with the Fermi level situated in between (at 0 K, Eg = E gap/2). The electrons in the HOMO must be energized to the Eg for its translocation from the HOMO to LUMO. In insulators, a large band gap prevents electron transfer (Figure 11). For bulk particles, the band structure is an intrinsic property of an atom that is independent of the particles' size or shape. For nanoparticles, the band structure is influenced by the particles' size and shape (Figure 11).

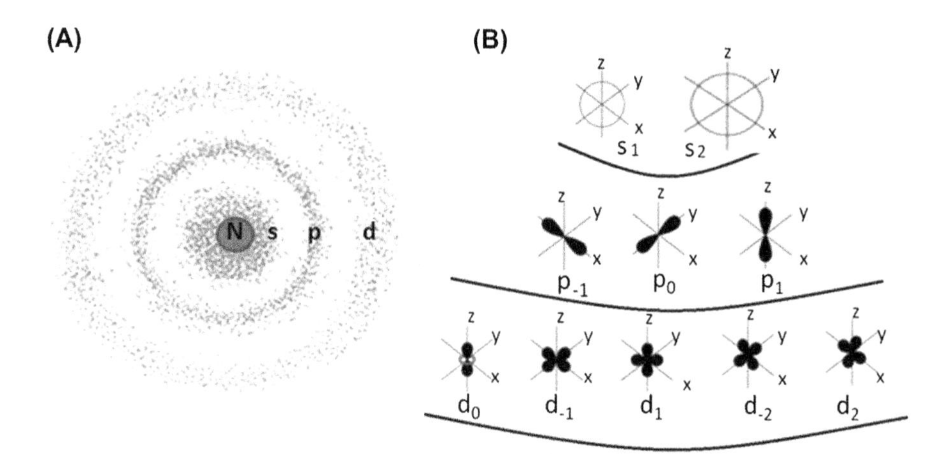

FIGURE 9 Electron cloud and shells. (A) Electron clouds. The darker region is where electrons are likely to be found. (B) Electron subshells. As shown in the top row of the figure, there are 2 s orbitals — one for 1s and the other for 2s. The s orbitals are spherical with the nucleus at the center. The 2s orbital is larger in diameter than the 1s orbital. The second row of the figure shows the shapes of the p orbitals, and the last row shows the shapes of the d orbital. The shapes get progressively more complex.

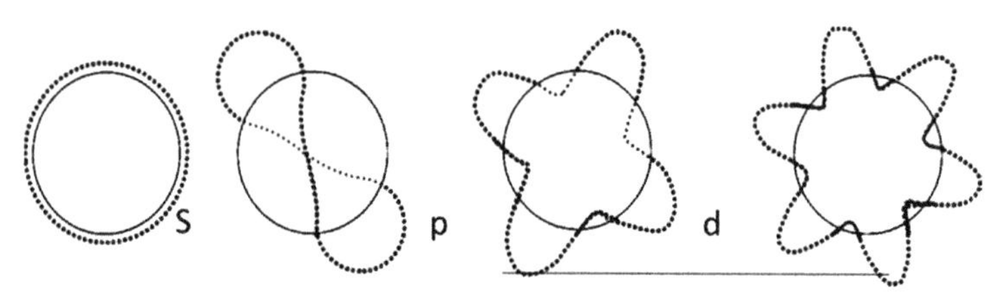

FIGURE 10 Different forms of standing wave in different electron orbitals.

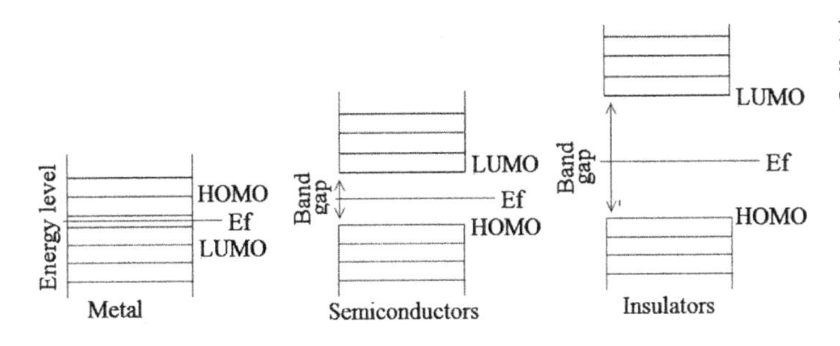

FIGURE 11 Electronic structure of metals, semiconductors, and insulators. Ef: Fermi energy.

1.3.3 *Unique Electronic Properties of Nanoparticles*

1.3.3.1 JELLIUM MODEL OF ELECTRONIC STRUCTURE

Figure 12 compares the electronic structure of an atom and a nanoparticle (Freund, 2002) based on the Jellium model (Sattler, 2002), in which nanoparticles can be envisioned as a single atom in which only the valence electrons are treated explicitly, while the positive charges are smeared out to a homogeneous background (Figure 12). The Jellium model allows evaluation of the valence and conduction electronic orbitals in nanoparticles. The HOMO and LUMO overlap in metals, while they are separated by an energy gap (band gap), characterized by a Fermi energy barrier in semiconductors and insulators (Figure 13). However, in semiconductors but not in insulators, electrons in the HOMO can absorb sufficient optical or heat energy to cross the Fermi energy barrier and jump into the conduction band, thus becoming free of its nucleus and moving freely within the atomic lattice (delocalized electron). Electron delocalization in the conduction band is crucial to the conduction process.

1.3.3.2 ELECTRON CONFINEMENT AND THE DENSITY-OF-STATE

Unlike the bulk particles, the band structure in the nanoparticles is strongly influenced by their size and shape. Smaller nanoparticles (less than 10 nm or smaller than the electron wavelength; Figure 14(A)) confine the motion of randomly moving electrons to a specific energy level (discreteness)—a process known as

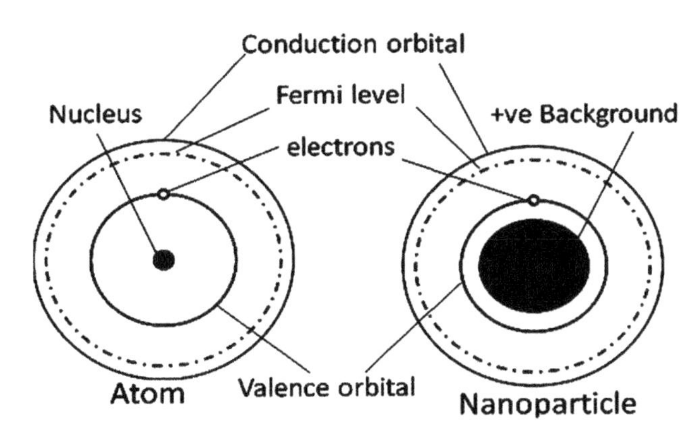

FIGURE 12 Comparative structure of an atom and a nanoparticle. The core of nanoparticles is envisioned as the positive charges smeared out to a homogeneous background and only the valence and conductance electrons are treated explicitly (the Jellium structure). Therefore, an atom and a nanoparticle may participate in chemical reactions.

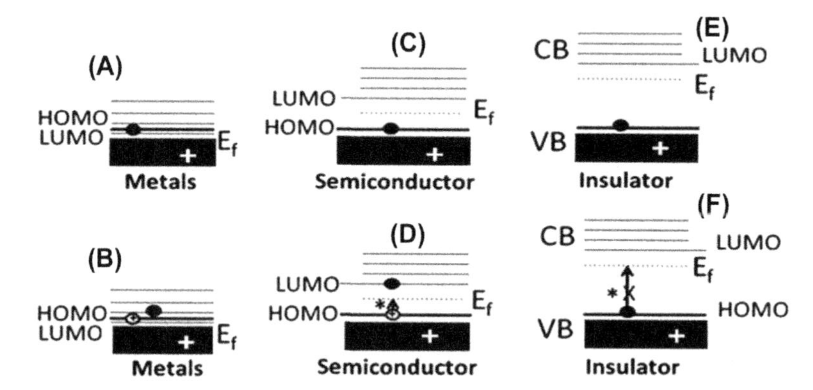

FIGURE 13 Electronic structure of metals, semiconductors, and insulators. As shown in Figure 12, the core of the nanoparticle is taken as a positively charged moiety that is balanced by the negatively charged valence electron in HOMO (A: metallic, C: semiconductor and E: insulator, Eg: Fermi energy). Metals lack band gap and Eg is situated within the LUMO, thus a level of energy that splits the electron–hole complex will allow the electrons to jump from the HOMO into the LUMO (B). Semiconductors have a band gap and the Eg is situated within the gap. This electron must absorb sufficient energy to split from the hole and then cross the Eg to jump into the LUMO (D). In insulators, electrons cannot absorb sufficient energy to cross from HOMO to LUMO (F).

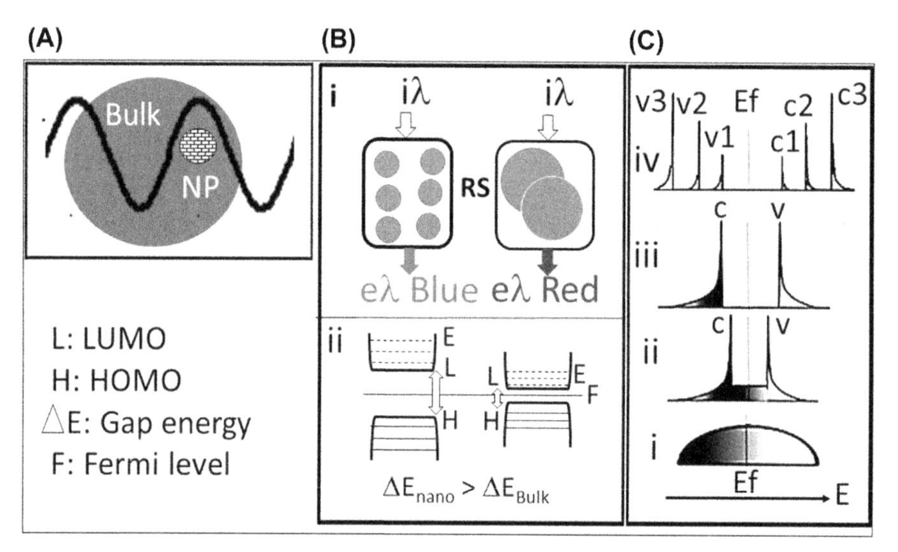

FIGURE 14 Nanoparticle size and electronic properties. (A) Nanoparticle size relative to the visible light wavelength. (B) A decrease in nanoparticle size is associated with an increase in the band-gap energy and a radiation shift from red to blue (decrease in emission wavelength). (C) Density of state for bulk metal (i), metallic nanoparticle (ii), semiconductor (iii) and a carbon nanotube with multiple DOSs (iv). Abbreviations: v1, v2, v3 are valence bands and c1, c2, d3 are corresponding conduction bands.

quantum confinement. As the nanoparticles become smaller, a decrease in confining dimensions makes the energy levels discrete, resulting in an increase or widening up of the band-gap energy (Figure 14(B)). Because the band-gap energy and wavelength are inversely related to each other, the wavelength decreases, resulting in the emission of blue radiation by the nanoparticles as opposed to red radiation emitted by the bulk particles (Figure 14(B)).

Figure 14(C) shows the DOS available for occupation by electrons. An increase in darkening is associated with an increase in the probability of finding an electron in that area. Figure 14(C-i) shows a continuous DOS for bulk metals in which Eg, HOMO, and LUMO overlap without any band gap, resulting in separation of electrons from their position and their migration to the LUMO. In the process, the valence orbital acquires a positive charge (the electron hole) that is mathematically opposed to the electronic charge. However, if the electron is not extracted from the conduction band, it may release photons and revert to the valence band. Figure 14(C-ii) shows discrete electronic structure of metallic nanoparticle (<100 nm) in which the DOS is quantized and the Fermi level overlaps with the HOMO. Figure 14(C-iv) shows a semiconductor carbon nanotube (CNT) exhibiting multiple DOS and a band gap where the Fermi level is situated in the band gap.

Although the size dependence of the band gap is well established (the band gap increases as the nanostructure size decreases), the shape dependence of band gap is not fully known. In earlier studies, Yu et al. (2003) have shown that the band gaps in one-dimensional (1D)-confined wells, two-dimensional (2D)-confined wires, and three-dimensional (3D)-confined dots evolved differently than the size effects. The theoretical relationships determined by Yu et al. (2003) are shown in Figure 15. The ΔEg (the increase in the effective band gap above the bulk value) depends linearly on 1/diameter^2 for wire and dots or 1/thickness^2 for wells. The ΔEg values for well, wire, and dot were 1.00, 1.17, 2.00, respectively. The differences in the slope are shown in Figure 15; the slope of a quantum-wire plot was 0.585 times that of a corresponding quantum-dot plot. The size-dependent changes showed the following relationship: dots (3D) > wire (2D) > wells (1D). Thus, losing dimension also reduced the effects of size on opening of the band gap. More research is needed to establish the effects of shape on band-gap energy.

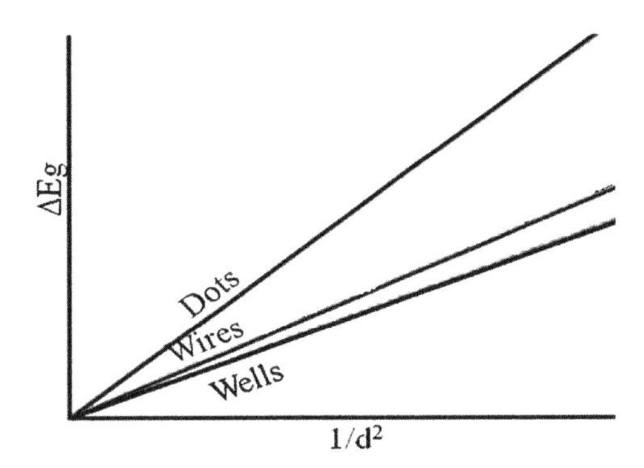

FIGURE 15 A sketch of the effects of shape on electronic band-gap in nanowells (1D), nanowires (2D), and quantum dots (3D) nanoparticles (d: diameter and ΔE_g: change in gap energy).

1.4 Optical Properties

Nanoparticles, along with exhibiting a size-dependent increase in electronic band gap, also exhibit a size-dependent increase in the optical band gap (Og)—a level of energy that activates the electron-hole complex but is not able to free the electron for conduction (Figure 16(ii)). The electron then returns to its original state by emitting photons in the process. The energy of the emission photon is equal to the Og energy.

In 1968, Tauc developed a model that related the Og values with optical energy (Eqn (16)):

$$-\alpha h\nu = A(h\nu - Og)^m \quad (16)$$

In this equation, α is the absorption coefficient given by $\alpha = 2.303 \log (T/d)$, where d is the thickness of the sample, T is the transmission, and $h\nu$ is the photon energy. A graph between $(\alpha h\nu)^2$ versus $h\nu$ is used to determine the optical band gap (Figure 17). The values of Og have been estimated by the extrapolation to the linear part of the curve to the Hv axis: $(\alpha h\nu)^2 \rightarrow 0$.

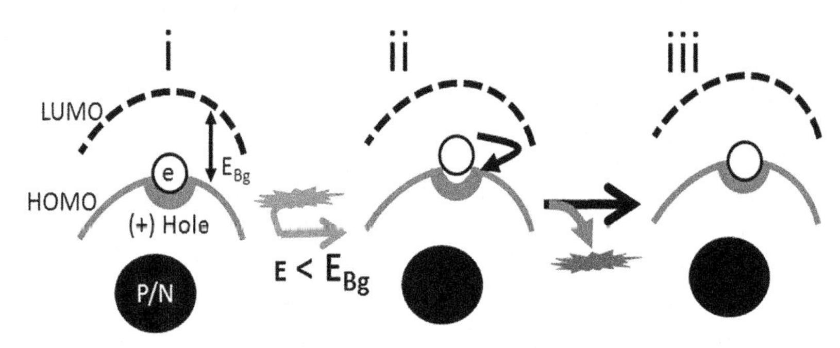

FIGURE 16 (i) An example of optical band gap and electron–hole complex. (ii) Optical band gap is a photon energy level that activated the electron–hole complex without freeing the electron to migrate into the conduction band. (iii) The complex returns to the resting state by releasing photons.

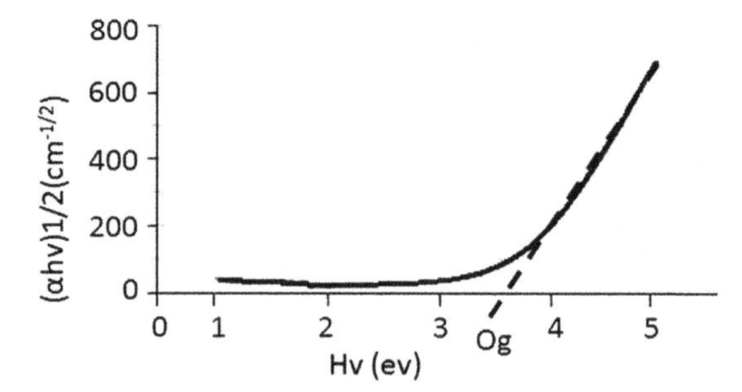

FIGURE 17 A sketch of photon energy (Hv) versus αhv^2-graph. Extrapolation of the linear plot yields the Og values.

A decrease in semiconductor particle size from the bulk level (>500 nm) to the nano level (<0 nm) resulted in a 200–450 nm absorption drop (Agostiano et al., 2000). Thus, as the particle size decreases, a shift in the absorption toward lower wavelengths occurs possibly because of a size-dependent increase in the Og. Absorption occurs at higher energies, resulting in a shift toward shorter wavelengths. Figure 18 shows the absorption wavelength spectra for different-sized semiconductor nanoparticles

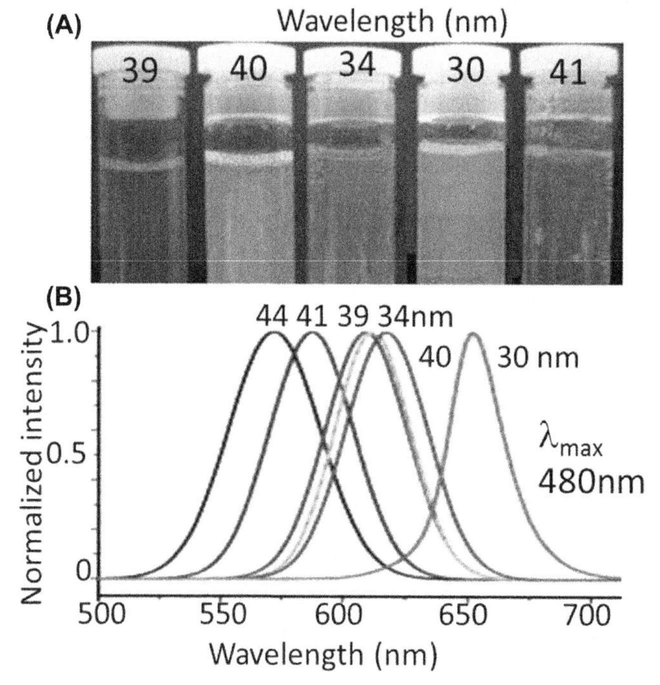

FIGURE 18 Size-dependent optical properties of CdSe semiconductor nanoparticles. (A) Photograph of photoluminescence under short-wavelength UV light. (B) Spectra of CdSe quantum dots (Peak ID: 1: 44, 2: 41, 3: 39, 4: 34, 5: 40, 6: 30 nm). *From Chu et al. (2012) with permission.*

(1.7–15 nm). Because of size-dependent optical properties, semiconductor nanoparticles can be applied for development of imaging and screening devices, where the free and bound particles may emit different light spectra.

1.5 Mechanical Properties

When the size of a nanoparticle approaches or become smaller than 0 nm, it acquires physical properties (fraction, hardness, elastic modulus, fracture toughness, scratch resistance, fatigue strength, etc.) that are different from its bulk counterpart (Sun et al., 2002; Chatterjee et al., 2003; Rahman et al., 2003; Qi and Wang, 2004; Schmid, 2004; Mohanraj and Chen, 2006; Gupta et al., 2008; Akbari et al., 2011; Lia et al., 2013; Guo et al., 2014). Mechanical properties have made important contributions to the fields of nanomedicine and nanobiology because they modulate the molecular forces that drive the molecular interaction, thermodynamic properties, and interface of the nanoparticle with the liquid, etc. Biological systems such as proteins and DNA create interfaces with the surrounding fluids that may govern their interactions with nanoparticles. The interaction of cell membranes with nanoparticles is governed by the mechanical properties, such as friction, adhesion, or elasticity, of both the cells and the materials because the cells dynamically react to the mechanical cues. Therefore, an understanding of the mechanical properties of nanoparticles is essential to bring modern nanomedicine from the bench to clinical applications. The objective of this section is to discuss the basic principles of the nanoparticles' mechanical properties, especially surface friction and tensile strength.

1.5.1 Surface Friction

1.5.1.1 BULK PARTICLES

The surface topology of bulk particles exhibits considerable roughness, depressions, and projections due to their molecular arrangements (Figure 19). An interlocking of the irregularities of two surfaces in contact causes friction (defined as the force resisting the relative motion of solid surfaces sliding against each other). If the two contacting surfaces possess interlocking structures (Figure 19(1)) or spacing (Figure 19(2)), then they can lock with each other periodically; this must be overcome to commence motion. If they have different atomic spacing, then the atoms of the top surface will not fit into the valleys (Figure 19(3)); thus, they will experience less friction (Dietzel et al., 2013).

Friction force (F_f) for bulk particles is defined by the equation $F_f = \mu F_n$, where F_f is the repulsive force linearly proportional in the opposite direction of the force applied ($F_{applied}$), μ is the ratio of the frictions of two surfaces, and F_n is the normal force (usually the force of gravity) on the particle mass. If the force applied is less than F_f, then the particle will remain static. If the force applied is greater than F_f, the particle will move (Figure 20(A)). In 1699, Guillaume Amontons proposed the two laws of friction:

1. *Friction is directly proportional to the applied load*, which is described by the equation $F_{max} = \mu F_n$. As an example, for a particle with 0.1 kg mass and 0.6 μm, F_f can be calculated

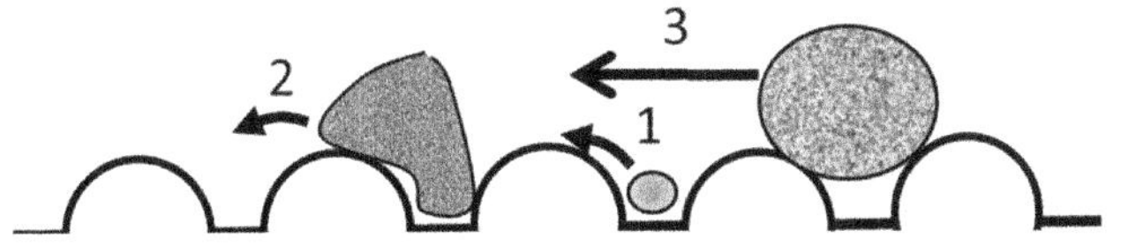

FIGURE 19 A sketch of surface topology, particle interlocking, and friction. The surface of bulk particles or nanoparticles is rough, consisting depressions and projections. If the two contacting surfaces possess a similar topology and molecular spacing, they interlock (1 and 2) with each other periodically. To commence motion, energy must be applied to overcome the interlocks such as pushing particles 1 and 2 out of the valley. If they have different atomic spacing, then the atoms of the top surface will not fit into the valleys (3), so they will experience less friction.

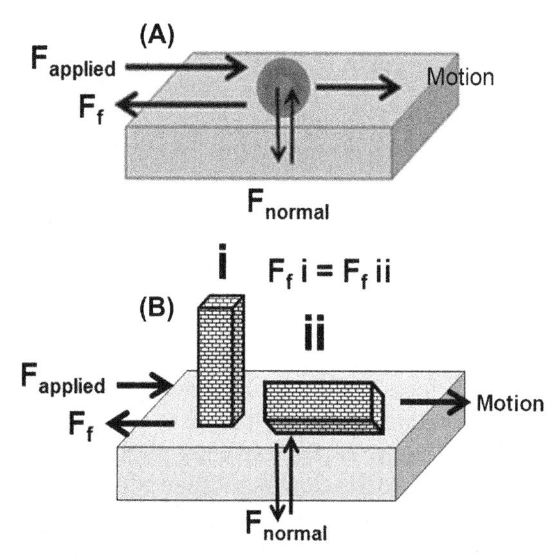

FIGURE 20 Mechanism of friction in bulk particles and nanoparticles. (A) Description of the friction-related forces in bulk particles. When two surfaces make contact, the following forces come into play. The resistive force, F_f, that is parallel, but opposite to the $F_{applied}$, the force that gravity applies on the particle's mass and α is the friction ratio of two surfaces. Particle will move if $F_{applied} > \alpha N$. (B) For bulk particles, the area of contact is not critical in determining F_f. Therefore, the brick placed vertically (i) or laterally (ii) will have the same F_f.

as follows: $F_f = ((0.1 \text{ kg} \times 9.8 \text{ m/s/s})) \times 0.6 = 0.59$ N. This suggests that a force >0.59 N must be applied to commence motion.

2. *Friction is independent of the area of contact.* Thus, a force greater than 0.59 N will be needed irrespective of the area of contact (Figure 20(B)).

1.5.1.2 NANOPARTICLES

Earlier studies have shown that many metal oxide nanoparticles or nanotubes, when mixed with lubricants, provided surface repair by reducing friction (Bakunin et al., 2004; Hernández Battez et al., 2008; Peng et al., 2010). A commercial firm Nanoprofix (http://www.nanoprofix.com/) has developed nanomaterial lubricant additives that, when mixed with oil and grease formulations, deliver a range of performance enhancements, including significant friction reduction, reduced wear of contacting surfaces, reconditioning of existing wear damage, enhanced energy and fuel efficiency, reduced operating temperature, etc. (www.digitaljournal.com/pr/1185283#ixzz30zs3z4K6). This superlubrication performance is attributed to the nanoparticles' nanometer size as well as replenishment of nanoparticles onto the contact, thus forming a transfer layer (Gnecco and Meyer, 2007; Dietzel et al., 2008; Guo et al., 2011). However, there is one basic flaw in this explanation. As described previously, for bulk particles, F_f is linearly proportional to the $F_{applied}$ and the area of contact is not a determinant of F_f (Guo et al., 2011; Khomenko and Prodanov 2010; Heim et al., 1999). For nanoparticles, the relationship between F_f and $F_{applied}$ appears to be extremely nonlinear (Urbakg et al., 2004). In addition, unlike for bulk particles, the area of contact (A_c) plays a critical role in the development of F_f (Figure 21(A)). The adhesion of small nanoparticles may depend not only on the A_c but also on the structure of the nanoparticles, particularly in the direction of F_{normal} to the surface (Figure 21(B)). Desanguliers (1734, reference not available) proposed that the contact between two surfaces is not whole-surface contact but a finite number of small *asperities*; although the surface may appear flat and smooth in bulk particles, it may be rough with structural deformations when viewed under a microscope (Figure 21(B)). Bulk particles may not interact with another bulk surface's nanoasperities (Figure 21(B-1)). Nanoparticles may interact with the bulk surface's nanoasperities in an area-dependent manner, with smaller nanoparticles (Figure 21(B-2)) exhibiting greater friction than larger ones (Figure 21(B-3)), via the following possible mechanisms:

1. *Absorption*, in which nanoparticles may form noncovalent bonds (van der Waals and hydrogen bonds) with the surface

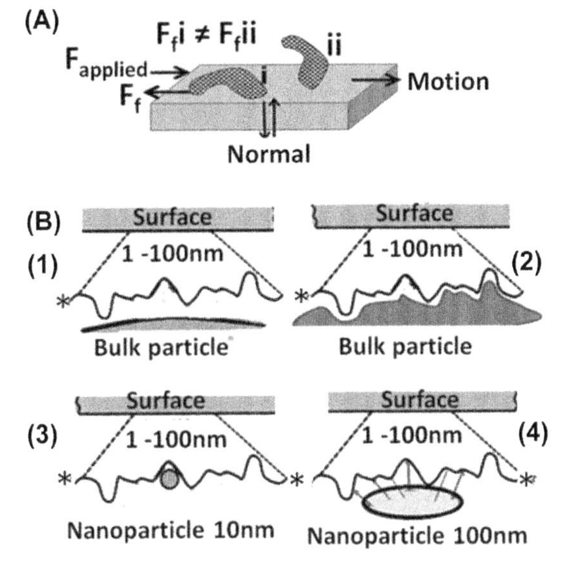

FIGURE 21 Friction forces in nanoparticles. (A) For nanoparticles, the area of contact is a critical determinant of F_f. Unlike for bulk particles, $F_f i \neq F_f ii$ for nanoparticles. (B) Even though the surface may appear smooth, at the microscopic level, the surface consists of roughness, deformities, and valleys. If a particle (such as a bulk particle) is larger than the surface deformations and the two surfaces are not complimentary, it will not experience much resistance (B-1). If the surface of the two particles are complementary, they will interlock (B-2). If a particle is smaller (such as nanoparticles less than 100 nm), it might interlock with the surface deformities and experience resistance (B-3). A certain amount of force must be applied to move the interlocked particles. Larger particles capable of forming bonds with the surface may also experience interlocking and require force to commence motion (B-4).

2. *Mechanical*, in which nanoparticles may form electrical double layers due to noncovalent electron sharing
3. *Topological indices*, in which nanoparticles may fit into the grooves; thus, more energy may be needed to push them out

In addition to metal oxide nanoparticles, carbon-based and inorganic nanotubes, either as pure solid lubricants or as additives to various fluid lubricants, may reduce friction and wear (Ritter et al., 2004; Greenberg et al., 2004; Rapoport et al. 2003a,b, 2005). Recently, Tevet et al. (2011) showed that an inorganic 2D nanotube combined with hollow polyhedral fullerene-like nanoparticles (called IF-WS^2) provided excellent lubricating properties (Figure 22). They explained this mechanism by superposition of rolling, sliding, and exfoliation-material transfer (third body).

Rolling is an important lubrication mechanism for IF-WS^2 in the relatively low range of normal stress (0.96 ± 0.38 GPa), sliding is relevant under slightly higher normal stress, and exfoliation of the nanoparticles is the dominant mechanism at the high end of normal stress (>1.2 GPa). The rolling mechanism, which leads to low friction and wear, requires spherical and dispersed nanoparticles and smooth contact surfaces (Tevet et al., 2011).

1.5.2 Tensile Properties

Tensile properties (the modulus of elasticity, elastic limit, elongation, proportional limit, reduction in area, tensile strength, yield point, yield strength, and other tensile properties) determine a particle's reaction to the forces being applied in tension. Out of these tests, tensile strength (the maximum stress that a material can withstand while being stretched or pulled before failing or breaking) is commonly used to measure a particle's strength by plotting the stress-strain curve (Figure 23).

The stress-strain curve is linear when large increase in stress causes small increase in strain. In the linear region, stress is proportional to strain and is called the *Young's modulus* that defines the properties of a material as it undergoes stress, deforms, and then returns to its original shape after the stress is removed. At some point, either the particle breaks (brittle materials, breaking strength) or the stress-strain curve deviates from the straight-line relationship and the strain increases faster than the stress. At this stage, some permanent deformation occurs and the material will not return to its original condition when the load is removed. There is

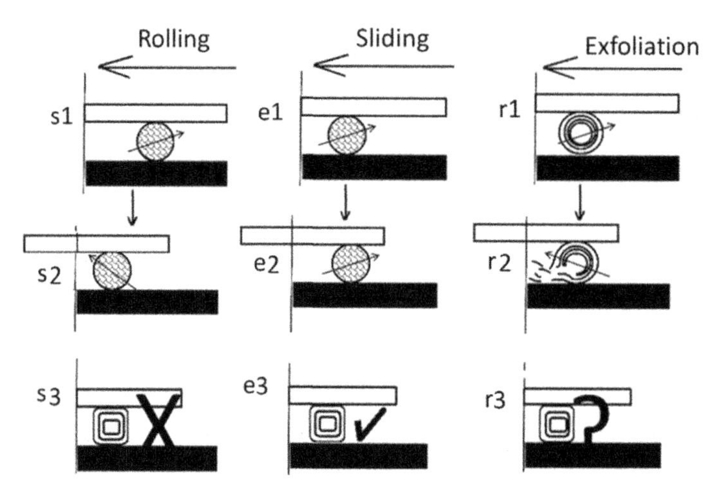

FIGURE 22 Proposed friction mechanisms of multilayered nanoparticles. **Rolling:** The nanoparticle acts as a ball bearing between the mating surfaces (Wu et al., 2009). The arrow shows position of the ball, s1 and s2 show the surface rolling and s3 suggests that rolling is highly dependent upon the nanoparticles' shape: a squire particle will not facilitate rolling. **Sliding:** If the nanoparticles (especially nanotubes) have low surface energy, they act as a separator providing low friction and facile shearing between the contact surfaces (Fischer, 1996; Fischer et al., 1997). Sliding depends upon the particles' surface energy, not the shape (s3). **Exfoliation:** Exfoliated layers from the multilayered nanoparticles are deposited on the asperities of the mating surfaces providing easy shearing (r1 to r3) (Rappaport et al., 2005). Possible role of shape in exfoliation is not fully established.

substantial, but not compelling, evidence that nanoparticles (especially metal nanoparticles and carbon nanotubes) increase both the linear and nonlinear phases of the stress-strain curve.

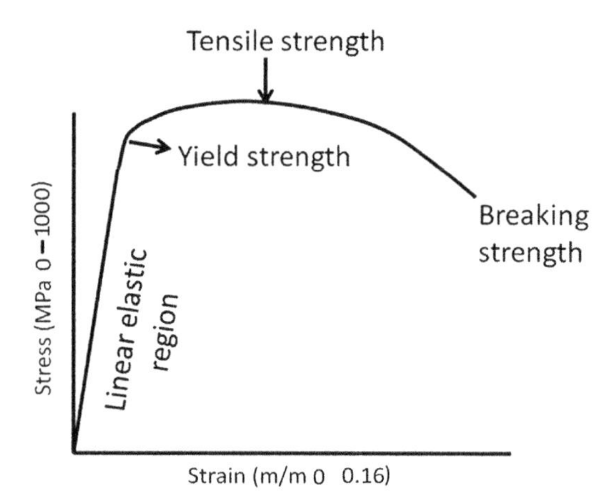

FIGURE 23 A sketch of the stress-strain relationship of metal nanoparticles. Details are described in the text.

The tensile strength and Young's modulus values for steel, single-walled carbon nanotubes (SWCNTs) and multiwalled carbon nanotubes (MWCNTs) are shown in Figure 24(i) and (ii). Both SWCNTs and MWCNTs, approximately 20 nm in diameter, exhibit several-fold greater tensile strength and Young's modulus values than that of stainless steel values (discussed in detail later). Figure 24(iii) and (iv) show tensile strength and Young's modulus values, respectively, for a nanoparticle mixture consisting of polymer/chitosan and $CaCO_3$ nanoparticles. The tensile strength increased as the percentage of nano-$CaCO_3$ increased from 0.5% to 1.0%, but decreased as the nano-$CaCO_3$ concentration increased further (Abdolmohammadi et al., 2012). As shown in Figure 24(iv), increasing the amount of nano-$CaCO_3$ in the mixture increased the Young's modulus. This may be due to an increase in rigidity of the nano-$CaCO_3$ and a strong interaction between the polymer and

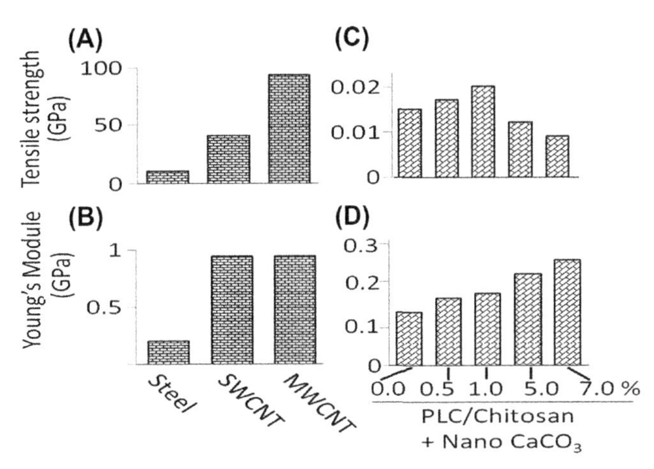

FIGURE 24 Tensile strength and Young's module of different nanoparticles. *(Data from Abdolmohammadi et al. (2012) with permission)* (open access, reuse allowed). Tensile strength (A) and Young's module (B) both are higher for CNTs than for steel. For chitosan nanoparticles, an increase in $CaCO_3$ proportion causes biphasic change in tensile strength (an increase for up to 1% and *n* decrease (C), but increased the Young's module values (D).

nano-$CaCO_3$ due to their large interfacial area (Chan et al., 2002; Chen et al., 2004). The enhanced composite modulus as a result of nanofiller loading has also been reported by several research groups (Reynaud et al., 2001; Zhang et al., 2004). These observations further support the hypothesis that the mechanical properties of nanoparticles deviate significantly from those of bulk particles.

2. PHYSICOCHEMICAL PROPERTIES OF SPECIFIC NANOPARTICLES

2.1 Dendrimers

As described in Chapter 2, dendrimers are hyperbranched (Figure 25) synthetic polymers that can be engineered into well-defined structures for various biological and pharmacological functions. Dendrimers contain a core group (C), branching generations (G1 to G4), and end groups that can be functionalized with functional groups such as an antibody (Ab) or another dendrimer containing a fluorescing group (Figure 25). During synthesis, as dendrimers grow in size, different generations begin to show distinct features that are amplified with increasing generations. These unique features (i) make dendrimers suitable for a wide range of biomedical applications and (ii) may be implicated in their toxicity. The following sections describe some of the unique properties of dendrimers.

2.1.1 *Intrinsic Viscosity (η)*

Intrinsic viscosity characterizes the frictional contribution of polymers in dilute solutions and facilitates determination of molecular weight, size, and topology of linear polymers, dendrimers, or biological molecules such as proteins and polysaccharides (Kukowska-Latallo et al., 1996; Bielinska et al., 1996). Figure 26 shows η and hydrodynamic (in solvent) radius (R_h) values for different-generation dendrimers (plot 1) and equivalent (in molecular weight) linear polymers (plot 2). The η values for linear polymers increase as their size increases. Conversely, dendrimers exhibited a biphasic response: the values increased as the dendrimer size increased from G0 to G4; then, a further increase in the dendrimer size decreased the η values (Frechet, 1994; Mourey et al., 1992). The hydrodynamic

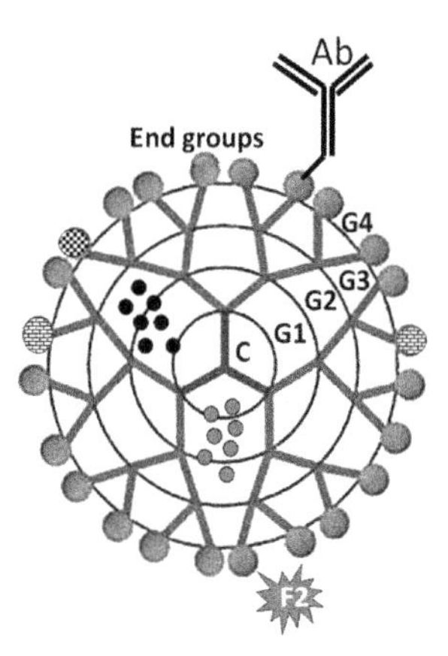

FIGURE 25 Structure of a functionalized G4 nanoparticle. Ab: antibody, F2: fluorescence probe, C: dendrimer core, G: generation. The end groups can be homogenous (all COOH or SH group) or heterogenous (a combination of COOH, SH or amino groups).

radius of the dendrimers increased linearly as their size increased (Klajnert and Bryszewska, 2001). Although mechanisms for the anomalous η values of dendrimers of different generations are not fully understood, this may be related to the discontinuous nature of steric hindrance that increases as the size increases from G0 to G4 and then decreases as the size increases further (Lu et al., 2010).

2.1.2 The Dendrite Box Concept

The branched structure of dendrimers contains empty and defined sized spaces surrounded by either hydrophilic or hydrophobic environments. These spaces can accept and store guest particles. The hydrophobic particles accumulate in the sites surrounded by a hydrophobic surface (black particles, Figure 25), while hydrophilic particles accumulate in sites surrounded by the hydrophilic environment (brown particles, Figure 25). Once entrapped, the particles are protected from the external environment. The entrapment space is known as the dendrimer box (Esfand and Tomalia, 2001). The very high number of functionalities located on the surface and the outer shell allows the dendrimers to participate in host–guest interactions and onsite release of the loaded particles.

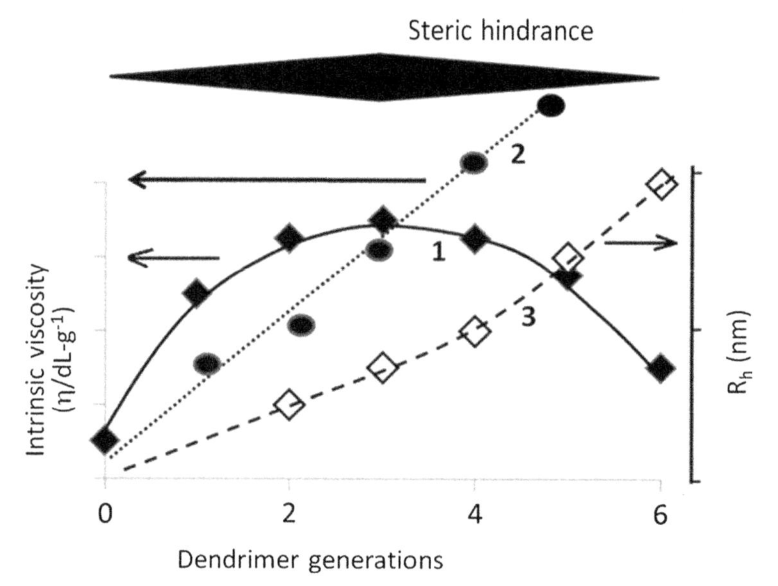

FIGURE 26 A sketch showing possible effects of dendrimer size (in terms of generations) on their intrinsic viscosity and hydrodynamic radius. Plot-1: linear polymers, plot-2: dendrimers and plot-3: hydrodynamic diameter (includes a thin electric dipole layer adhered to the dendrimer surface).

2.1.3 *Biomimicry*

One of the outstanding properties of dendrimers is their ability to mimic biological particles, especially globular proteins (Esfand and Tomalia, 2001; Koenig et al., 2001; Yue et al., 2012). Studies have shown the close resemblance of higher order dendrimers to many proteins, such as insulin (3G), cytochrome (4G), and hemoglobin (5G) (Esfand and Tomalia, 2001; Noriega-Luna et al., 2014). Dendrimers also mimic histone clusters; thus, they make stable complexes with the DNA and enhance gene expression (Fant et al., 2008; Dootz et al., 2011). Like proteins, dendrimers may respond to many external stimuli (solvent, pH, temperature, ionic strength) and adapt a tightly packed (resembling native proteins) or extended (resembling denatured proteins) conformation. However, there are some key differences between proteins and dendrimers.

1. Dendrimers normally have lower molecular density than comparable size proteins. Thus, they are less compact compared to the proteins.
2. The globular structure of proteins is due to the folding of the linear polypeptide chain into secondary and tertiary structures via intramolecular bonding (H-bonding, van der Waals interactions, ion pairing, and disulfide bonding), resulting in their compaction. In comparison, dendrimers acquire a globular shape by taking up smaller hydrodynamic volume.
3. Dendrimers incorporate a high degree of covalent bonding, which results in a less flexible structure than what is found in proteins.

2.1.4 *Solvent and pH-Dependent Folding of Dendrimers*

In 1983, De Gennes and Hervet showed that as the PAMAM dendrimers grow in size, the branching becomes increasingly crowded at the periphery, while remaining low density at the core (Figure 25). This gave an impression that dendrimers are rigid structures that, unlike proteins, are not capable of achieving multiple conformations. However, later studies (Maiti et al., 2005; Muller et al., 2002; Liu et al., 2009; Jansen et al., 1994) have shown the dendrimers to be relatively less rigid because of the high mobility of the surface atoms, which lack interaction between the surface and the core atoms. Therefore, the surface bonds can fold inward (resulting in a dense core) or outward (the opening of the dendrimers), depending upon the intermolecular charges as well as the van der Waals interactions.

Liu et al (2009), using G4 dendrimers (Figure 27), have shown the effects of pH on conformational changes (by measuring the density at various radii from the core) in the PAMAM dendrimer. At higher pH, dendrimers contained mostly nonionized, hydrophobic nitrogen moieties that facilitated low-energy van der Waals interactions between the surface and the core atoms, resulting in an inward folding of the surface atoms, thus increasing the core density (the highest density occurred at 1 nm). At neutral or lower pH, the nitrogen moieties were ionized, resulting in intermolecular repulsing and an ensuing opening of the dendrimers with dense outer shell (highest density occurred at 1.5–3 nm). This demonstrates an outward movement of dendrimer branches at acidic pH. In an earlier study, Jansen et al. (1994) constructed a dense-shell dendrimer by *boc*-modification of the surface groups that enhances formation of H-bonds between the surface groups.

Maiti et al. (2005) have investigated the effects of pH and solvents on dendrimer conformation by assessing the solvent-accessible areas, representing empty areas in the dendrimers (Figure 29). At high pH, the dendrimers exhibited only small, but restricted, areas and cavities. At neutral and low pH, the dendrimers open up, resulting in an increase in the empty surfaces. They also computed the void distribution of the inner

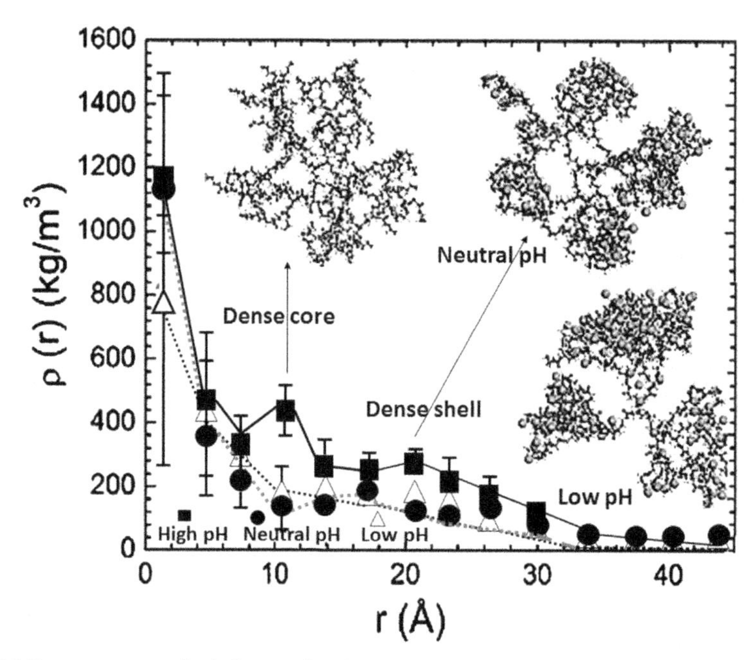

FIGURE 27 Radial (from core to surface) density distribution of G4-NH2 PAMAM dendrimer at various (black – acidic pH, red – neutral pH and purple – basic pH) pH values. The center of mass of the dendrimer was used as the reference. *Insets*: Schematic structure of dendrimers at different pH. At basic pH, the amino group of dendrimers loose a proton and become hydrophobic. Thus, the surface groups bend toward the hydrophobic core making the core denser. At acidic pH, the amino group acquire a proton and ensuing positive charge, resulting in the development of repulsive forces between the core and the surface groups. Therefore, the dendrimers open up. *Reprinted from Liu et al. (2009) with permission.*

cavities using alpha-shape and discrete flow methods. As shown in Figure 28, dendrimers in the absence of solvents and at a high pH exhibited small packets of empty areas. At a low pH, a single big cavity percolating the whole dendrimer structure and connecting the outer surface existed. At this pH, therefore, small molecules can diffuse in and out of the dendrimer interior.

2.1.5 Host–Guest Complex Formation

Dendrimers' unique dendritic topology allows their application in molecular caging; as controlled delivery agents, DNA transporters, and transfection agents; in the light-scattering polycation–dendrimer complex; and as potentiometric titration agents. These applications require an understanding of the nature of guest particle–dendrimer interaction. Earlier studies (Nanjwade et al., 2009; Gupta et al., 2007; Wang and Imae, 2004; Ottaviani et al., 1999; Kukowska-Latallo et al., 1996) have suggested the following criteria for guest particle–dendrimer interaction:

1. A critical effective dendritic charge (such as NH_3^+ or COO^- terminal groups) density is essential for complex formation. If the core is hydrophobic, then the ionic groups will be repelled (making the core more hollow), while the nonionic groups will be attracted to the core (making the core more crowded).
2. The chain's stiffness reduces its interaction with the dendrimer core.
3. A critical ionic strength facilitates complex formation. No complexation may occur above the critical ionic strength and the guest particles are released as the ionic strength is lowered.
4. Strong noncharged secondary interactions (e.g., hydrogen bonding) may maintain the complex once formed, thus hindering a salt or pH-triggered release.

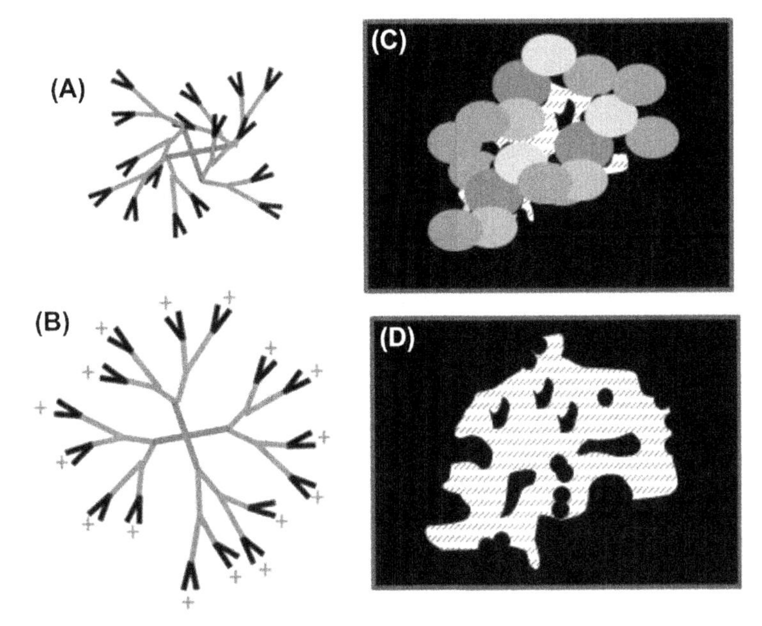

FIGURE 28 A sketch of the surface and core for G5 PAMAM dendrimer at high (<8) pH (A) and low (<6) pH (B). At high pH, the core was denser than the shell, while at lower pH the dendrimer opened up with larger cavities inside the particle. (C and D) show the inner channels and cavities for G5 dendrite under no solvent conditions (C) and low pH (D). In the absence of solvent, the dendrimer contains small pockets of empty spaces (various shades represent cavities of different sizes). At low pH (D), a single continuous channel with openings was observed. *Redrawn from Mati et al. (2005), with permission.*

5. The guest particle is encapsulated if it is small compared to the dendrimer. The reversibility of complex formation is significant in the context of controlled delivery.
6. If the dendrimer is smaller than the guest particle, a significant guest density will lie outside of the dendrimer. In case of a chain guest particle, it may penetrate the host dendrimer.

Figure 29 shows an example of dendrimer–guest interactions. When the dendrimer is much larger (Figure 29(i)) than the polymeric chain (Figure 29(ii)), the chain is completely internalized into the dendrimer (Figure 29(iii)). In case of large chains and small dendrimers, a novel "chain-walking" may occur (Figure 29(iv–vii)). In addition, the dimensions of both the dendrimer and chain may change upon complex formation (Welch and Muthukumar, 2000).

2.2 Properties of Carbon Nanotubes: Metals or Semiconductors

CNTs are formed by the folding of a hexagonal graphene sheet consisting of unit cells. Each cell contains two identical carbon atoms having two identical zero-energy electron states: one in which the electron resides on atom A, the other in which the electron resides on atom B (Appendix 1). The three electrons of each carbon atom are bonded with neighboring atoms, resulting in one free π electron for each atom, which can be found in an inverted-cone density of state, above or below the sheets' plain (Figure 30(A-ii)). The lattice structure of grapheme, made out of two interpenetrating triangular lattices a1 and a2, are the lattice unit vectors, while δ1, δ2, and δ3 are the nearest-neighbor vectors (Figure 30(A-iii)). Electrons can dislocate from one atom to another

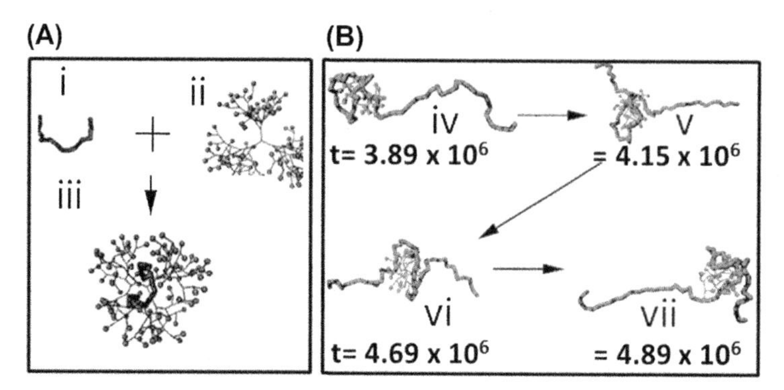

FIGURE 29 Examples of host (dendrimer)–guest (linear polymer) interaction. (A) When the dendrimer (host) is larger than the linear polymer (guest), the smaller particle is encapsulated by the larger one. (B) When the guest particle is smaller, it binds to the polymer (i) and walks along the chain (ii to iv). *Reprinted from Welch and Muthukumar (2000) with permission.*

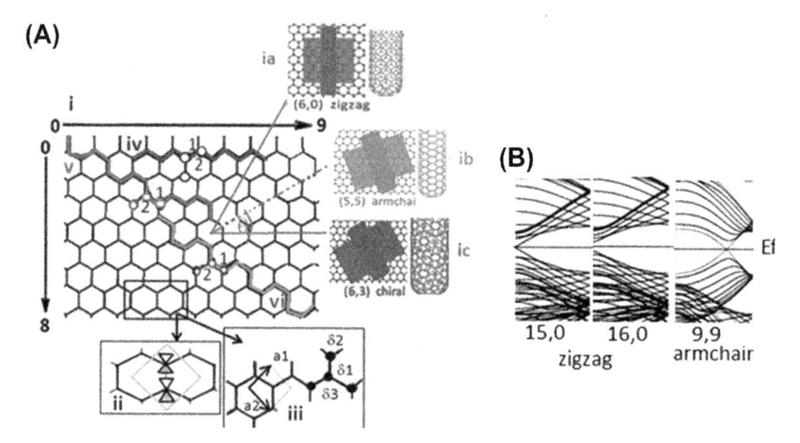

FIGURE 30 Structure and characteristics of graphene and CNTs. (A-i) Structure and folding of graphene sheet. b1 and b2 bonds (1 and 2, respectively) connect neighboring atoms. ia: folding of the graphene sheet for zigzag nanotubes, ib: folding of the graphene sheet for chiral nanotubes and ic: folding of the graphene sheet for armchair nanotubes. (A-ii) Graphene-unit cells containing two identical carbon atoms sharing one electron each (triangles show the density of state for each electron). (A-iii) The lattice structure of graphene is made out of two triangular lattices a1 and a2 that are the lattice unit vectors, while δ1, δ2, and δ3 are the nearest-neighbor vectors. (Aiv–vi) zigzag, armchair and chiral patterns in a graphene sheet. (B) Electronic structure (density of states) of 15,0 zigzag CNT (metallic), 16,0 CNT (semiconductor), and 9,9 armchair CNT(metallic). The (15,0) zigzag and (9,9) armchair CNTs do not have an energy gap (the HOMO and LUMO bands overlap) at the Fermi level (E_f), while (16,0) zigzag CNT has a band gap (HOMO and LUMO are separated by an energy gap) at the Fermi level.

within a cell. In addition, the electrons travel through the graphene sheet as if they carry no mass, as fast as just one hundredth that of the speed of light.

Figure 30(A-i) shows three different types of graphene sheet folding that result in the formation of three electronically different nanotubes. When $n = m$ (like 5,5 or 6,6), the nanotubes belong to the *armchair* family (Figure 30(ia)), whereas when $n = 0$ or $m = 0$, they are called *zigzag* tubes (Figure 30(A-ib)). All other combinations of n and m are called chiral nanotubes (Figure 30(A-ic)). The chiral angle ranges from 0° (zigzag) to 30° (armchair). The bond lengths b1 (Figure 30(A-1)) and b2 (Figure 30(A-2)) between three connected atoms are as follows:

- $b1(CNT_{zigzag}) > b1(graphene)$, $b2(CNT_{zigzag}) = b1(graphene)$.
- $b1(CNT_{armchair}) > b1(graphene)$, $b2(CNT_{armchair}) < b1(graphene)$.
- $b1(CNT_{charil}) = b1(grapheme)$, $b2(CNT_{charil}) = b1(grapheme)$.

Thus, the different patterns of graphene folding caused differential changes in b1 and b2 values that may be responsible for their differential electronic properties, as shown in Figure 30(B). Armchair CNTs exhibit metallic properties, while the zigzag and chiral CNTs can be semiconductors (containing a bang gap between HOMO and LUMO) if $n - m$ is a multiple of 3; otherwise they are metallic (Saito et al., 1998, 2000; Ajayan, 1999; Dresselhaus et al., 2001, 1996; Moniruzzaman and Winey, 2006; Tian et al., 2006). Band gaps of 0.4 to >1 eV have been reported for SWNTs. Recently Matsuda et al. (2010), using an improved technology, showed the presence of minor (<0.2 eV) gap energy in many of the zigzag and chiral CNTs that were previously classified as metallic CNTs with zero gap (Figure 31(A-a) and (A-b)). The energy gap was inversely related to the CNTs' diameter. Thus, as the diameter decreases, the electrons in the HOMO require relatively great energy to overcome the Fermi barrier.

Why do CNTs, depending upon their geometry, exhibit metallic or semiconductor properties? The following possibilities may answer this question:

1. The geometric differences allow CNTs to have certain unique intrinsic properties, such as a σ-π hybridizing effect and/or curvature stress that is inversely related to their diameter and has been shown to induce semiconductor–metal–semiconductor phase transitions in primary metallic SWCNTs (Ding et al., 2003).
2. As discussed in Chapter 2, CNTs possess various types of structural defects that modulate their electronic properties (Zhou et al., 2014). There has been sufficient evidence showing that vacancy defects may be an important factor in altering the CNTs' geometry and electronic properties (Zhang et al., 2010; Tien et al., 2008; Neophetou et al., 2007; Okada, 2007; Wang and Wang, 2006). Zeng et al. (2011) have shown that a variety of mono-vacancy, di-vacancy, tetra-vacancy, and hexa-vacancy defects differentially altered the electronic properties of semimetallic (12,0) CNTs, possibly due to the presence of mid-gap states originating from the defect, thereby enhancing conduction (Figure 31(B)). The pristine (12,0) SWNT, due to a curvature effect, has a small band gap of about 0.07 eV (Ouyang et al., 2001) with two DOSs in HOMO and LUMO bands. The defects of increasing severity (mono-to hexa-vacancy) did not cause a proportional increase in the conduction, but they cause different types of mid-gap DOSs. This suggests that the nanotube conductance is not a monotonic function of the defect size.
3. The electronic properties of CNTs, in addition to the vacancy defects, are also modulated by the adsorption of foreign materials (doping) such as catalyst impurities, hydrogen molecules, or polymers (Xia et al., 2013; Mananghava et al., 2012; Haruyama et al., 2011; Ayala et al., 2010; Zhao et al., 2004; Esfariani et al., 2003). The effects of hydrogen adsorption on CNTs are of current interest because of the tube's ability to store hydrogen. A number of studies have reported alterations in electronic properties of CNTs in response to hydrogen adsorption (Bianco et al., 2010; Jalili and Majidi, 2007; Wessley et al., 2007; Hou et al., 2003). Chen et al. (2012) have reported effects of partial hydrogenation and vacancy defects on the

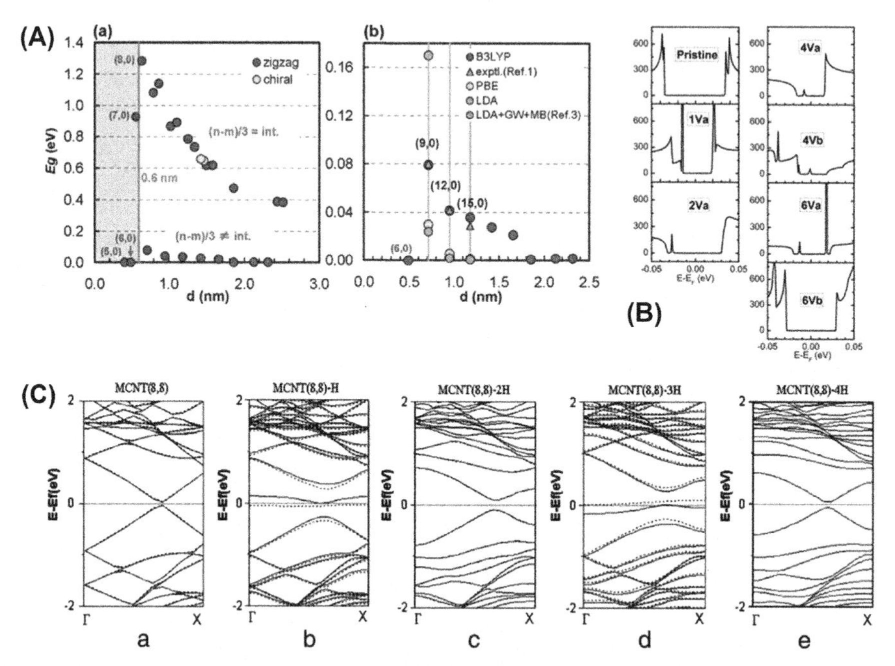

FIGURE 31 Metallic and semiconductor CNTs. (A—a) Effects of diameter on band-gap energy in zigzag CNTs where (n-m)/3 = I (semiconductor CNTs, top plot) or (n-m)/3 ≠I (metallic CNTs, bottom plot). An increase in diameter decreases the band-gap energy in semiconductors. (A—b) Some of the CNTs that were classified as zero-energy gap were reanalyzed by an improved method and found to contain minor diameter-dependent gap energy. *(Reprinted from Matsuda et al. (2010) with permission.)* (B) Sketches showing DOS for (12,0) CNTs subjected to mono-, di-, tetra-, and hexa-vacancy defects. The pristine CNT exhibited a minor gap energy of about 0.07 eV with two DOSs each at HOMO and LUMO bands, while CNTs with mono-vacancy defects caused a large DOS at both ends, di-vacancy basically abolished the pristine DOSs. The tetra-valence defect occurred in two forms, α and β, and both altered the DOS differently, α defect blocked the valence side, while the β defect blocked the conduction side of DOS with a minor DOS within the gap. The hexa-valence α defects caused an extremely high DOS at the conduction end, while the β defect caused only minor changes. This suggests that the CNT conductance is not a monotonic function of the defect size. *(Reprinted from Zeng et al. (2011) with permission.)* (C) Electronic characteristics of partially hydrogenated semi-metallic (8,8) SWCNTs ($SWCNT_{prestine}$ (a), $SWCNT_{H}$ (b), $SWCNT_{2H}$ (c), $SWCNT_{3H}$ (d) and $SWCNT_{4H}$ (e). In $SWCNT_{pristine}$, the states cross at the Fermi level, a typical metallic condition. In $SWCNT_{1H}$, the states split above and below the Fermi level, thus opening the gap and acquiring a magnetic moment of 0.483 μ_B. In $SWCNT_{2H}$, the band gap remains but magnetization is absent. In $SWCNT_{3H}$, the states lose symmetry, thus they have 0.501 μ_B magnetic moment. In $SWCNT_{4H}$, the band gap decreases and magnetization is lost. Thus, the effects of hydrogenation depends on the site at which H atoms bind. *Reprinted from Chen et al. (2012) with permission.*

electronic properties of metallic CNTs. They showed that the hydrogenation, depending upon the adsorption site, transitioned metallic CNTs in semiconductor CNTs (Figure 31(C)) lacking magnetic properties (some pristine metallic CNTs exhibited magnetic properties). The energy band structure of H-adsorbed metallic CNTs exhibits a spin-polarized flat band near the Fermi level.

These observations, taken together, suggest that vacancy defects and/or doping may allow tuning of CNTs' electronic and mechanical properties for developing nanomachines for different purposes. Conversely, defects and doping may

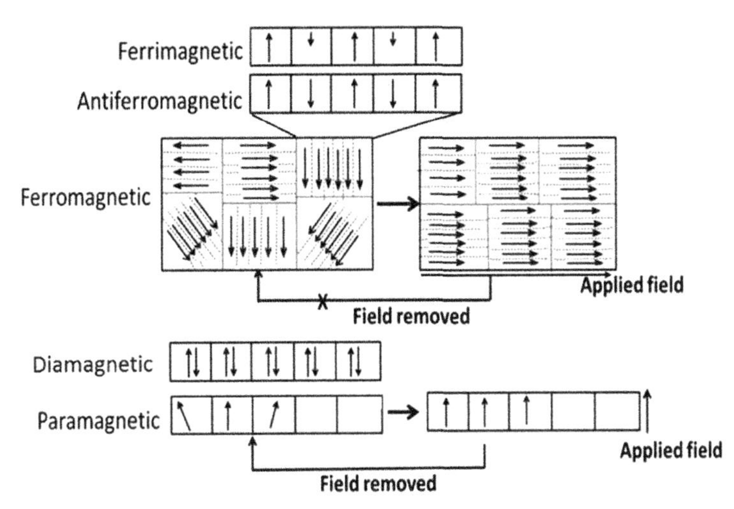

FIGURE 32 Structure of magnetic particles. The arrows indicate whether the electron spin is parallel (arrow up) or antiparallel (arrow down) to the direction of the applied force. *The details are discussed in the text and the following site: http://www.irm.umn.edu/hg2m/hg2m_b/hg2m_b.html.*

cause unique adverse effects not recognized at present.

2.3 Magnetic Nanoparticles

Recently, there has been a considerable increase in the application of magnetic nanoparticles, possibly due to their unusual size-dependent magnetism. The properties of magnetic nanoparticles having the same material and similar size show a strong dependence on their morphology, microstructure, or the nature of the matrix. As a consequence, there is a great variety of data, often contradictory, in the literature about magnetic nanoparticles. In this section, various key aspects of the magnetic nanoparticle properties have been discussed.

2.3.1 *Brief Overview of Magnetism*

As described in Appendix 2, the magnetic moment of electrons and intra-atomic interactions are the fundamental sources of magnetism. In general, magnetism can be classified as weak (diamagnetic or paramagnetic) or strong (ferromagnetic, antiferromagnetic, and ferromagnetic). Some examples of the different types of magnetism are shown in Figures 32 and 33 (http://www.irm.umn.edu/hg2m/hg2m_b/hg2m_b.html).

Diamagnetic particles do not have magnetic moments because all of their orbitals contain paired electrons with opposite spins. Paramagnetic particles contain unpaired electrons that, in the absence of a magnetic field, exhibit random coordination. In the presence of a magnetic field, the electrons spin parallel to the applied field, resulting in the development of transient magnetism that reverses when the force is removed. Ferromagnetic particles consist of a number of small domains (Stewart, 2008), each spontaneously magnetized to saturation. The boundaries between domains are called domain walls, in which the magnetization gradually changes from the direction on one side to that on the other (Grollier et al., 2011; Indira and Lakshmi, 2010). In the absence of an applied force, the electrons in different domains exhibit random orientation. In a magnetic field, all electrons spin parallel to the force. Bulk ferromagnetic particles remain magnetized even after the force has been removed (hysteresis effects, Appendix 2). In ferromagnetic particles, domains disappear

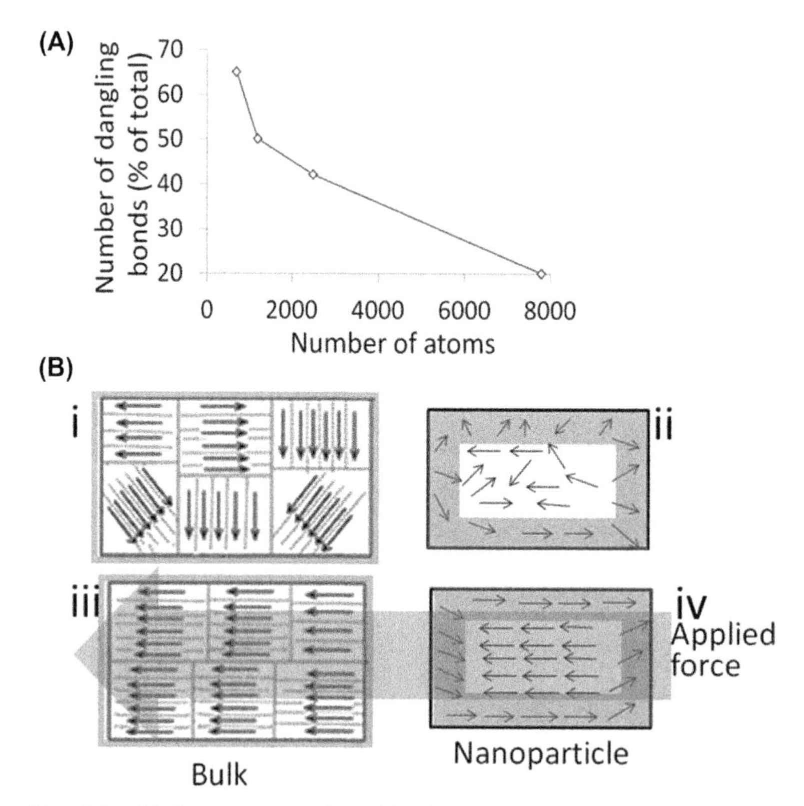

FIGURE 33 Possible relationship between the number of dangling (uncoupled single electron) bonds and the development of magnetization. (A) Percentage of dangling bonds (unpaired electrons) in nanoparticles of different size. As the nanoparticle size decreases, the particle's electronic anisotropy increases. (B) Magnetic properties of bulk particles and nanoparticles. The surface atoms of bulk ferromagnetic particles are minuscule compared to the core atoms that reside within specific domains. The surface atoms make a significant portion of the ferromagnetic nanoparticles. The proportion of the surface atoms to the core atoms increase exponentially as the particle size decreases (more than 80% of the atoms in 1 nm nanoparticles face the surface). Bulk particles contain domains (i), while nanoparticles lack domains (ii). When force is applied, (iii) the atoms in bulk particles spin parallel to the force applied, thus generating a magnetic moment, (iv) the surface and core atoms of a nanoparticle may spin differently, thus modulating the overall magnetic moment. The modulation may depend upon the size and shape of the nanoparticle.

over the Curie temperature (Tc) (Appendix 2); thus, the particles exhibit paramagnetic properties (Dobis et al., 2010). A coercive force (Hc) must be applied to reverse the magnetization to zero. The atoms in antiferromagnetic materials exhibit oppositely spinning electrons that cancel out the magnetic moments. In an applied field, the magnetic moments do not cancel out because they have different magnitudes, which result in a net spontaneous magnetic moment. The key factors that modulate the magnetic moment are chemical composition, type and degree of defectiveness of the particle lattice, size and shape, the morphology, particle interaction with the surrounding neighboring atoms, and the environment.

2.3.2 Comparison of Bulk Particles and Nanoparticles: The Concept of Critical Diameter

As shown in Table 1, size reduction alters the physicochemical, electronic, and magnetic properties of particles. However, for ferromagnetic particles, a critical diameter (D_c ranging from 15 to 35 nm, but in rare cases as high as 800 nm) exists, below which a particle undergoes a multiple-domain to single-domain transition, resulting in a large increase in coercivity. Thus, below a critical diameter, many novel properties appear in magnetic nanoparticles that are not found in bulk particles. The D_c has been defined by Eqn (17), in which A is the exchange constant, k_{eff} is the anisotropic constant, μ_0 is vacuum permeability, and M is the saturation magnetism:

$$D_c \approx 18\frac{\sqrt{A\,k_{eff}}}{\mu_0 M^2} \tag{17}$$

The following review articles have addressed this issue in detail: Issa et al. (1999, 2013), Lu et al. (2007), and Batlle and Labarate (2002).

2.3.3 Effects of Size Reduction on Magnetic Properties

The size-dependent effects of magnetic nanoparticles are classified as (i) the surface-mediated effects that include surface-to-volume ratio, differential electronic properties of surface and core atoms, surface oxidation, etc.; and (ii) the direct size-related effects that include thermodynamic disruption, electron confinement, and single- or multiple-domain structures.

2.3.3.1 MAGNETIC MOMENT: THE SURFACE AND CORE ATOMS

As shown in Figure 3, the core atoms of a particle form bonds with their nearest atoms. The paired electrons exhibit opposite spins that may cancel the magnetic moment of each electron. The surface electrons, however, exhibit uncompensated spin. The density of unpaired electrons increases as the particle size decreases (Figure 33(A)). This results in a high level of anisotropy between the surface and the core atoms in smaller nanoparticles. Therefore, the surface and core atoms may respond differently to an applied magnetic field (Figure 33(B)).

Surface atoms also modulate the properties of the inner layers (Figure 34). In Ni nanoparticles, the second layer may be magnetically dead, while in Fe and Co nanoparticles, the second layer acquires the properties of the surface atoms. In Fe nanoparticles, the fourth layer acquires antiferromagnetic properties. This suggests that the magnetic nanoparticles, depending on the parent metal, exhibit structural anisotropy that is amplified as the size decreases. This may result in size-dependent changes in the nanoparticles' magnetic properties. Studies have shown that magnetization (per atom) and the magnetic

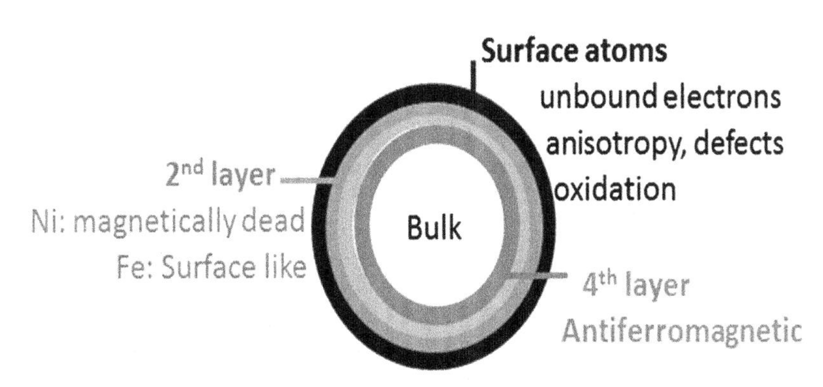

FIGURE 34 Structural anisotropy in magnetic nanoparticles.

anisotropy is much greater for nanoparticles than for bulk particles.

Morales et al. (1999) have shown that size reduction exponentially increases the surface area/volume ratio of nanoparticles, thus, increasing electronic anisotropy of the surface and ensuing magnetic changes.

2.3.3.2 DIRECT EFFECTS OF SIZE REDUCTION

In addition to altering the surface characteristics, size reduction also results in electron confinement and ferromagnetic to superparamagnetic transitions.

The quantum confinement effect: Quantum confinement is defined as a transition from continuous to discrete energy levels as the particle size becomes smaller than the electrons' wavelength, size of the magnetic domain (for ferromagnetic or paramagnetic particles), or wavelength of magnetic oscillations (for superparamagnetic properties). In general, a decrease in dimension results in a widening of the band gap, ferromagnetism to paramagnetism transition, and/or superparamagnetic to sub-superparamagnetic (electron freezing) transition in nanoparticles (Issa et al., 2013; Alberto, 2009; Kodama, 2999). Because of the confinement, many metals may acquire semiconductor properties. In the following paragraphs, possible roles of the spin state of electrons in a magnetic field, the Curie law, and Pauli paramagnetism in size-related magnetic changes are discussed.

- *The spin states of electrons*: Electrons, in addition to having negative charge, also exhibit magnetic moments with measurable north and south poles. The electronic magnetic moments possibly arise due to their angular moment around the nucleus and spin around its axis, although, in a quantum sense, electrons may not have an axis (Figure 35(A)). In general, the electrons can have two spins

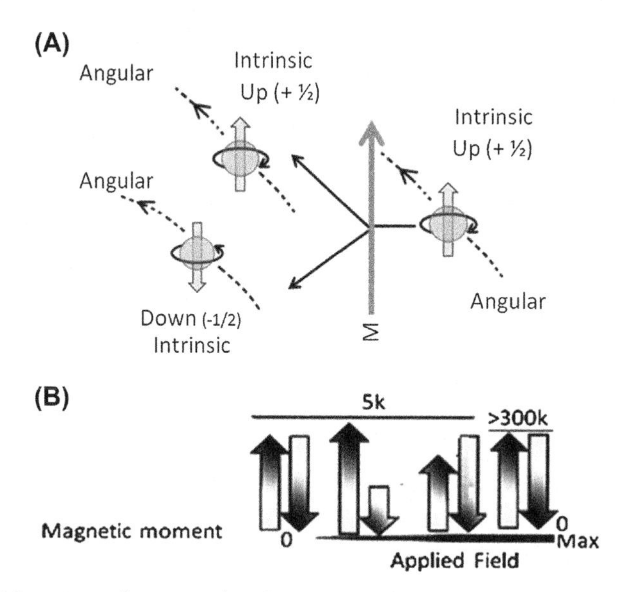

FIGURE 35 A sketch of the spin-up (low energy) and spin-down (high-energy) states, and their relevance to magnetism. (A) Magnetic nanoparticles, in the absence of magnetic field, exhibit net spin-up (+1/2) state. When a magnetic field is applied, electrons split into spin-up and spin-down states. The magnetic properties of the nanoparticles depend upon the ratio of the two states (Jun et al., 2008). (B) The density of spin-up and spin-down electrons changes as the strength of the applied magnetic field increases. In the absence of the field, the two states are equal. At lower force, spin-up is greater than spin-down electrons. At relatively high strength, spin-up decreases and spin-down increases, resulting in a net magnetic moment. When the temperature is increased to 300 K, even at highest field strength, the ratio equalizes with loss of magnetic moment.

called *up* (+) and *down* (−). The spins are given half integer values, either 1/2 (for up spin) or −1/2 (for down spin). When a magnetic field is applied, electrons spin either parallel to or antiparallel to the direction of the applied force. If the direction is parallel, the spin of the electron is termed *spin-down* or −1/2. If the direction is antiparallel, the spin of the electron is negative spin or *spin-up* or +1/2. According to the quantum theory, atomic orbitals have an angular momentum (designated I) (Atkins, 1974). The s-orbital has I = 0, p-orbital has I = 1, and so on. In a low-strength magnetic field, the electrons' energy states are separated into the magnetic quantum numbers that represents how much of the orbital momentum (ML) can be projected along the z axis: for ML = 1, the projection will be +h (h is Planck's constant); for ML = −1, the projection will be −h; for ML = 0, the projection = 0, etc.. At relatively high strength of the magnetic field, the ML splits into two states: $m_s + 1/2$ or $m_s - 1/2$. Therefore, if $m_s = 1/2$, the angular momentum along the z-axis (designated s_z) will be +h/2; if m_s is −1/2, the s_z will be −h/2 (Eastham and Kirkman, 2000). The electronic magnetic moment can be defined as $\mu = g\mu_B/2$, where g is 2.00232 and μ_B is the Bohr magneton. As shown in Figure 35(A), +1/2 and −1/2 represent electrons exhibiting antiparallel-spin (spin-up) and parallel-spin (spin-down), respectively, to the magnetic field. Figure 35(B) shows the effects of magnetic field and temperature on the two energy (spin-up and spin-down) fields. In the absence of a magnetic field, spin-up and the spin-down electrons are in equal proportion (Jun et al., 2008). At a lower magnetic field, spin-up electrons are greater than the spin-down electrons that may oppose the magnetic field (diamagnetic properties). At a high magnetic field, spin-up electrons are less than the spin-down electrons, which increases the magnetic moment. High temperatures neutralize the magnetic field by suppressing the spin-down electrons.

- *The Curie law* (Morton, 2012): The law states that the magnetic susceptibilities (M) of most paramagnetic substances are inversely proportional to the absolute temperatures in *K*. For metals having one unpaired electrons, magnetization (*M*) can be defined by Eqn (18), where *C* is the Curie constant ($C = (1/k_B)N^2_{\mu B}$), *T* is temperature in *K*, and *B* is magnetic induction:

$$M = C(B/T) \tag{18}$$

Equation (18) suggests that *M* is directly proportional to the magnetic induction, *B*, but is inversely proportional to the temperature (in *K*). Since the magnetic susceptibility (χ) is defined as $\chi = M/H$ (*H* is applied field), we propose that χ is inversely proportional to *T* as shown in Eqn (19), in which *N* is the number of magnetic atoms, κB is Boltzmann's constant, μ_B is the Bohr magneton (9.27400915e−24 J/T or A·m^2), and μ_0 is a constant ($1.2566370614 \times 10^{-6}$ Hm^{-1}).

$$\chi(Cirie) = N^2_{\mu B}\mu_0/\kappa B \cdot T \tag{19}$$

According to Eqn (19), an increase in temperature decreases a particle's magnetization. However, Curie's paramagnetism is related to the localized electrons in the metals; thus, it is not applicable to particles containing free conduction electrons.

- *Pauli's paramagnetism law* (Morton, 2012): Unlike Curie's law, Pauli's paramagnetism, however, is applicable to metals with free conduction electrons. Pauli's law states that free electrons spin up and down according to the applied magnetic field (Figure 36), yet the Fermi energies remain the same. Some spin-up electrons (antiparallel to the applied field) are converted into spin-down electrons (parallel to the applied force) (Figure 36). To understand Pauli's paramagnetism, let us consider the DOS for electrons that occupy

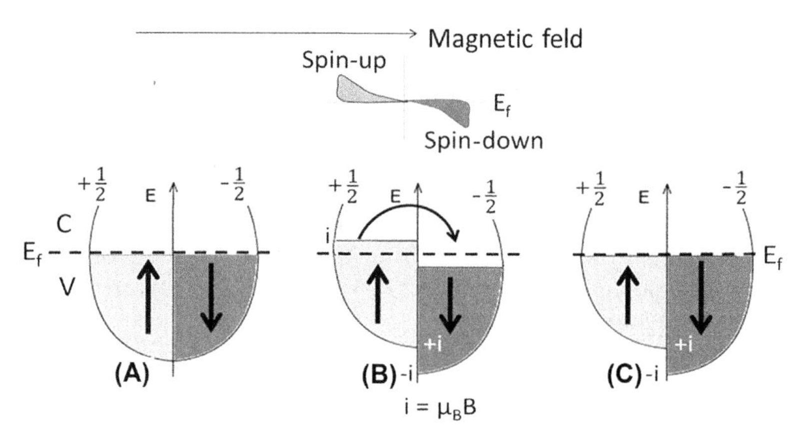

FIGURE 36 Pauli's paramagnetism law. In the absence of magnetic fields, the spin-up and spin-down states are comparable or up-spin may be greater than down-spin (A). In a magnetic field, the spin-up electrons flip into spin-down electrons (top). When force is applied, the valence spin-up electrons jump into the conduction band (B) and then flips into spin-down electrons (C), thus increasing electronic anisotropy and magnetic moment.

spin-up and spin-down states for up to Fermi energy (E_f) in metal particles without magnetic fields. The density of states would be the same for both states. When a magnetic field, B, is applied, these densities of states shift up and down by the energy of the spin interacting with the field ($\pm\mu_B B$). As a result, a number of the spin-up electrons must flip (Figure 36) to minimize the total energy according to Eqn (20), where gE_r is the DOS at the Fermi energy.

$$n(\text{flip}) = \mu_B B\left(\frac{gEr}{2}\right) \quad (20)$$

The susceptibility χ for Pauli paramagnetism can be determined using Eqn (21). In this equation, the Fermi energy is written in terms of the Fermi temperature ($E_r = kBT_f$).

$$\chi(\text{Pauli}) = \frac{2}{3}\frac{3N\mu_B^2\mu_0}{k_B\,T_f} \quad (21)$$

Equation (21) becomes similar to the Curie paramagnetic susceptibility, except the actual temperature T(K) is replaced with the Fermi temperature T_f.

- *Transition from ferromagnetism to superparamagnetism*: As described earlier, ferromagnetic bulk particles are multiple-domain particles in which each domain's local magnetization is saturated but not parallel to other domains' local magnetization. In the presence of a magnetic field, all domains exhibit parallel spin resulting in the development of a magnetic moment. As the particle size decreases, nanoparticles transition from a multiple-domain state to pseudo single-domain (mixture of multidomain and single-domain properties), single-domain paramagnetic (disordered atoms or electrons), single-domain superparamagnetic (spin reversal and loss of magnetic moment), single-domain sub-superparamagnetic (very high magnetic anisotropy—freezing behavior) states (Figure 37). The loss of the multidomain structure is one of the critical effects of the

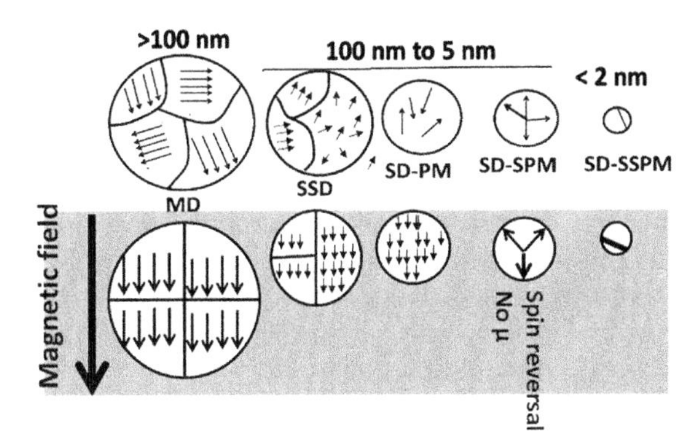

FIGURE 37 A sketch of the size-dependent alterations in nanoparticles magnetic properties. Bulk ferromagnetic particles are multi-domain (MD) particles. The electrons exhibit unidirectional spins in each domain, but the electrons in different domains may face in different directions. In the presence of magnetic fields, the domains (net individual electrons) rearrange such that all electrons are in the direction of the applied field. As the size decreases from 100 nm to 5 nm, it transitions into pseudo single-domain (SSD), single-domain paramagnetic (PM), and single-domain super-paramagnetic (SPM) particles, respectively. Further decrease in size transition nanoparticles in a sub-super-paramagnetic form (SSPM).

nanoparticle's size reduction at a temperature below the Curie point (at a temperature above the Curie point, bulk ferromagnetic particles lose atom coordination and become paramagnetic). The paramagnetic nanoparticles can be used in bioimaging, while the superparamagnetic nanoparticles can be used for separation processes in biochemistry.

- *Domain structure, hysteresis, and coercivity field* (Chen et al., 2013; Usov and Grebenshchikov, 2009; Bronstein et al., 2007; Ibusuki et al., 2001): Hysteresis is the dependence of the output of a ferromagnetic particle not only on its current input, but also on its history of past inputs. An alternate increase and decrease in the magnetization force yields a cyclic output, as shown in Figure 38(A). Starting from the zero field, an increase in H causes a nonlinear increase in the magnetic moment (plot point 1) until a saturation point is reached (plot point 2). When the magnetic field is removed, the magnetization undergoes an irreversible change and the magnetic moment is retained in the absence of the magnetic field (remanence field, plot point 3). A magnetic field in the opposite direction (−H) is needed to bring the magnetization to zero moment (coercivity field, plot point 4). A further increase in −H will result in saturation magnetism in the opposite direction (plot point 6). Then, if −H is removed, most of the reverse magnetization is retained (plot point 6) at zero magnetic field. However, the decrease in reverse magnetization took a

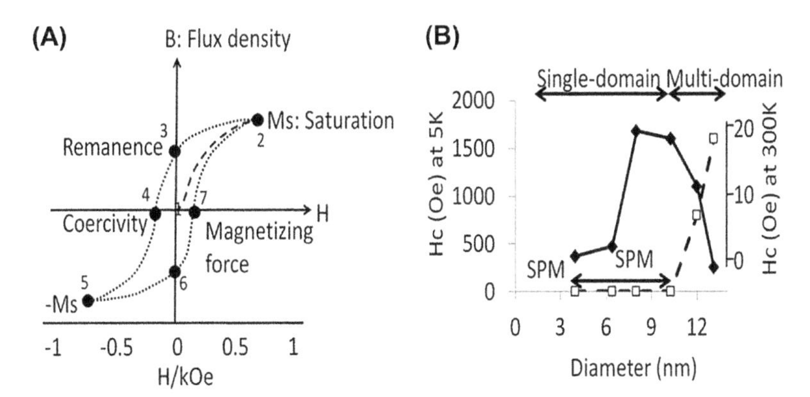

FIGURE 38 (A) Hysteresis curve for ferromagnetic bulk particle. (B) Effect of particle size and temperature (solid line 5 K and broken line 300 K) on coercivity field. *Details are discussed in the text by Chen et al. (2013).*

different paths. The cycle is complete when H strength is increased and the magnetic moment reached zero moment (plot point 7) and then to the magnetic saturation (plot point 2).

At a lower temperature (5 K), an increase in nanoparticle size caused biphasic effects on the coercivity field. An increase in particle size from 3.9 to 7 nm was associated with an increase in the coercivity field up to a critical diameter, then a further increase in the size decreased coercivity field (Figure 38(B)). At a higher temperature (300 K), the coercivity field was not observed for smaller particles (3.9–9.6 nm), but increased gradually as the particle size increased further. At 5 K, 3.9- to 9-nm nanoparticles possess a single-domain structure below a critical diameter. In single-domain nanoparticles, the atoms or electrons align unidirectionally and the coercivity field increases as the size increases.

3. CONCLUSIONS

There is considerable unity and diversity in the properties of structurally diverse nanoparticles. The unity stems from their size and surface-related electronic (quantum confinement, surface electron anisotropy, metal-to-semiconductor transition), magnetic (ferromagnetic and paramagnetic transition, development of superparamagnetism, lack of hysteresis), and optical (optical confinement, surface Plasmon) properties. The diversity stems from their distinct structure- and chemistry-related properties, such as metallic and semiconductor properties in CNTs, unique hydrophilic/hydrophobic properties in dendrimers, unique mechanical properties in metals, etc. This unity and diversity in physicochemical properties make nanoparticles useful for diverse applications ranging from industrial to medical applications. Recently, there has been growing interest in understanding the relationship between the physicochemical properties of nanoparticles and their potential risk to the environment and human health. This is because nanotechnology is creating many new materials with characteristics that are not always easily predicted from current knowledge. Within the nearly limitless diversity of these materials, it is not possible to characterize their toxicity using limited published data on bulk materials. Therefore, novel strategies are needed to define the risks posed by nanoparticles.

4. APPENDICES

Appendix 1: Unique Electronic Properties of Carbon Nanotubes

A1.1 Classic and Quantum Mechanical Theories of Atomic Structure

The classic theory: According to the classic theory, atoms consist of a nucleus (Figure 39A) and the following:

1. Positively charged protons and neutral neutrons that give the nucleus a net positive charge.
2. Negatively charged electrons (the number of electrons is identical to the number of protons) that orbit around the nucleus in energy-defined occupied molecular orbitals (OMOs; Rutherford theory). The electrons occupying the OMO closest to the nucleus have the highest energy. The electrons' energy decreases as their distance from the nucleus increases. The outermost orbital is called the valence orbital or highest OMO (HOMO). Electrons in HOMO have the lowest energy levels and they define the chemical properties of an atom. In occupied orbitals, the negatively charged electrons remain bound to a positively charged electron "hole" (Figure 39A, inset). The binding energy of an electron-hole complex is inversely related to its distance from the nucleus: the higher the distance, the lower the binding energy. The HOMO's electron–hole complex has lowest

binding energy; thus, it requires the lowest excitation energy for dissociation.

3. Figure (39A) also shows that each orbital (an example of HOMO is shown) consists of many subshells, called s, p, d, and f orbitals. Electrons fill the subshell orbitals in the following order: 1s, 2s, 2p, 3s, 3p, 4s, 3d, 4p, 5s, 4d, 5p, 6s, 4f, 5d, 6p, 7s, 5f, 6d, 7p.
4. In addition to the OMO, atoms also contain many unoccupied molecular orbitals (UMOs). The UMO nearest to the HOMO is called the lowest UMO or LUMO. The HOMO and LUMO may overlap (in metals (Figure 39A)) or be separated by an energy gap known as band gap or forbidden gap (in semiconductors (Figure 39B) and insulators Figure 39C). At the basal level, electrons are not present in the band gap. Atoms also contain an energy barrier called the *Fermi level* (red circle) that is situated within LUMO in metals (Figure 39A) but within the band gap in semiconductors (Figure 39B) or insulators (Figure 39C). An electron in the HOMO must absorb a critical level of energy equal to or more than the energy needed for it to separate from the hole and overcome the Fermi level to jump from HOMO to LUMO. If this happens, then the LUMO acquires an electron and the HOMO a positively charged hole.
5. In metals, the HOMO and LUMO overlap and the Fermi level is within the LUMO, the HOMO electrons freed from the hole can spontaneously jump from HOMO to LUMO. If electrons are not removed, it returns back to the HOMO hole and releases energy as photons in the process. In semiconductors, the Fermi level is situated in the band gap; this HOMO electron must absorb sufficient energy to split from the hole and then cross the Fermi level to jump into the LUMO or higher orbitals. In insulators, the Fermi level cannot be crossed by the electrons.

The quantum mechanical theory: According to the quantum mechanical theory, electrons exist as particles (defined mass) or waves (defined wavelength). Although electrons travel at a speed approaching 1/10th of the speed of light, many studies have proposed that they lack mass. Because of the particle-wave duality

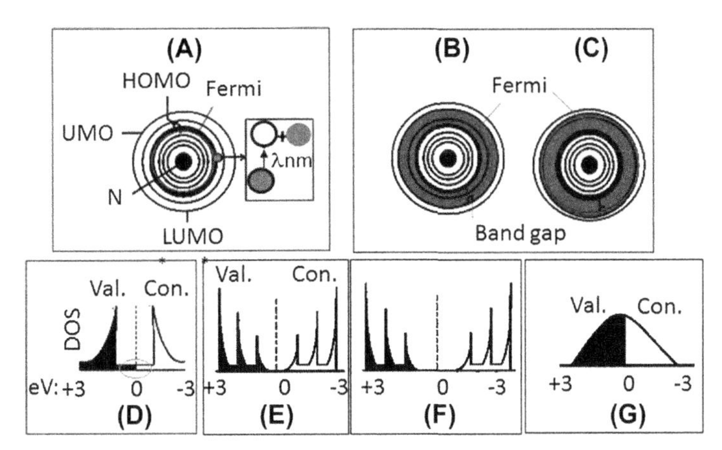

FIGURE 39 Classic and quantum-mechanical (QM) theory of the atomic structure. In classic theory, electrons are considered to be a particle encircling the nucleus in a fixed orbit. In QM theory, electrons have particle-wave duality. The details are described in the text. (A) metal nanoparticles, (B) semiconductors, (C) insulators, (D) density of state (DOS) for electrons in metal nanoparticles, (E) DOS for non-zero band-gap semiconductor nanoparticles, (F) DOS for zero band-gap graphene sheet, (G) DOS for insulator nanoparticles, and (H) DOS for bulk metals.

(electrons exhibit properties of waves [wavelength, interference, and diffraction] as well as particles), it is not possible to accurately predict the position of an electron at a given time. Instead, it is possible to statistically determine the probability of finding an electron in the density of state. The probability is related to the wave associated with the electron. The amplitude of the wave may determine the probability of finding an electron at a particular point (the larger the amplitude, the larger the probability; similarly, the smaller the amplitude, the smaller the probability). In fact, the probability is proportional to the square of the amplitude of the wave. Figure 39(D-F) shows the probability of finding an electron in carbon nanotubes (1D nanoparticle), while Figure 39(G) shows the probability of finding an electron (density of state) in a bulk particle. Figure 39(D) shows the density of state (DOS, statistical probability of finding an electron, shaded region) for metals. The HOMO and LUMO overlap and the Fermi level is within the LUMO. Figure 39(E) shows the DOS for nonzero band-gap semiconductor CNTs in which the HOMO contains several subshells of DOS and there is substantial, but achievable, band-gap between HOMO and LUMO. The Fermi level is within the gap. In a graphene sheet (a starting material for formation of CNTs), the HOMO contains several subshells of DOS, and there is a minute level (zero band gap is a misnomer) of energy gap between HOMO and LUMO. Graphene can absorb photons of all wavelengths; thus, they are used in solar cells. Figure 39(F) shows the DOS for insulators in which electron conduction is prohibited due to an unachievable band gap. Figure 39(G) shows the DOS for bulk metals in which the DOS is linear, not discrete as above for nanoparticles.

A1.2 Introduction to Carbon Nanotubes

The electronic structure of a nanoparticle can be explained using the *Jellium model*. The valence electrons of the metal atoms in a nanoparticle or metallic carbon nanotube are loosely bound so that they can move around everywhere and are not bound to a particular atom (Figure 40). The atoms without their outer electrons are positively charged ions. The free electrons move around jelly-like positive ions in the cluster. As can be seen in Figure 40, the Al_{13} cluster contains a 39-electron species that is one electron short of a stable magic number (energetically most stable nanoparticles have the following total number of electrons: 8, 20, 34, 40, 58). This suggests a

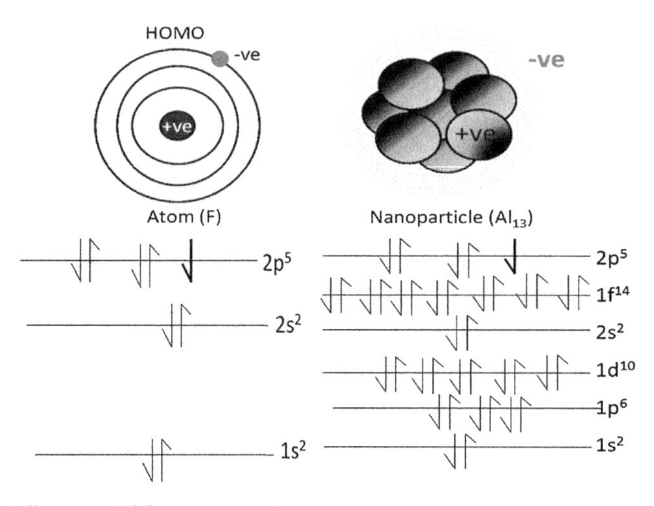

FIGURE 40 Jellium model for nanoparticles as pseudoatoms. The details are described in the text.

similarity between Al_{13} and the halogens (F electronic structure is shown). The electron affinity of Al_{13} is 3.6 eV, essentially the same as that of a chlorine atom. Taken together, these observations suggest that nanoparticles may act like an artificial atom in which the valence electrons may encircle a positively charged moiety.

Caution must be exercised when applying the Jellium model for carbon nanotubes. This is because the Jellium model assumes that atom localization does not play a significant role in describing the electronic structure of a nanoparticle. However, it is well established that diameter and chirality do have profound effects on the electronic properties of carbon nanotubes (armchair nanotubes are metallic; zigzag nanotubes with n-m as a nonzero multiple of three should be small gap semiconductors; and nanotubes not fitting in any of the groups are high-gap semiconductors). Complicated mathematical models developed to address these variabilities are outside the scope of this book.

A1.3 Density of State and Electron Conduction

The occupied molecular orbitals and conduction molecular orbitals contain DOS. In bulk particles, DOS is a continuous function of energy, while in CNTs (and other 1D nanoparticles), DOS is made of discontinuous spikes, ascending and then descending sharply—the Van hove singularities (Figure 39; v1, v2 and v3 are DOS for HOMO and c1, c2 and c3 for LUMO). Electrons migrate from one state of HOMO into the corresponding state of LUMO: v1 to c1 (11 migration), v2 to c2 (22 migration) or v3 to c3 (33 migration). Electron migration occurs in two stages: (1) dissociation of the electron-hole complex and (2) migration of dissociated electrons into the LUMO (Figure 41).

Electrons can dissociate from the hole by absorbing sufficient energy (light wave of a particular wavelength or magnetic field) to overcome the binding energy of the complex. The dissociated electron either migrates into the

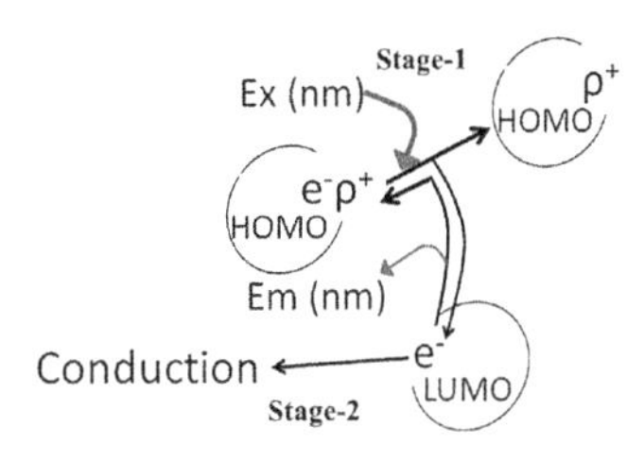

FIGURE 41 Possible mechanism for conduction of electrons from HOMO to LUMO.

LUMO or reverts to the positively charged hole in the HOMO by emitting photons. Migration of electrons from HOMO to LUMO is determined by (1) the presence (semiconductors) or absence (metals) of a band gap and (2) the Fermi level, whether it is situated within the HOMO or within the band gap.

In metals, HOMO and LUMO overlap, and the Fermi level is situated within the HOMO. Therefore, the dissociated electrons translocate spontaneously (without further energy input) from HOMO into LUMO (Figure 42). Then, either the electron is removed from the LUMO (conduction) or it will revert to the HOMO, bind to the hole, and release photons and/or weak phonons in the process.

In semiconductors having nonzero energy barriers or band gap (where the Fermi level is within the band gap), the dissociated electrons in HOMO must absorb sufficient energy (heat or photon) to overcome the Fermi barrier and move to LUMO. However, the excitation energy must match the band-gap energy for efficient absorption. Figure 43 shows the relationship between the wavelength (the amount of energy is inversely related to the wavelength of the radiation) of the excitation wave and electron conduction through a band-gap energy. No absorption occurs if the wave energy is smaller than the band-gap energy (wavelength restriction). If the wave energy matches the gap energy, energy is absorbed and the electron moves from HOMO to LUMO. If the wave energy is much higher than the gap energy, the electron moves

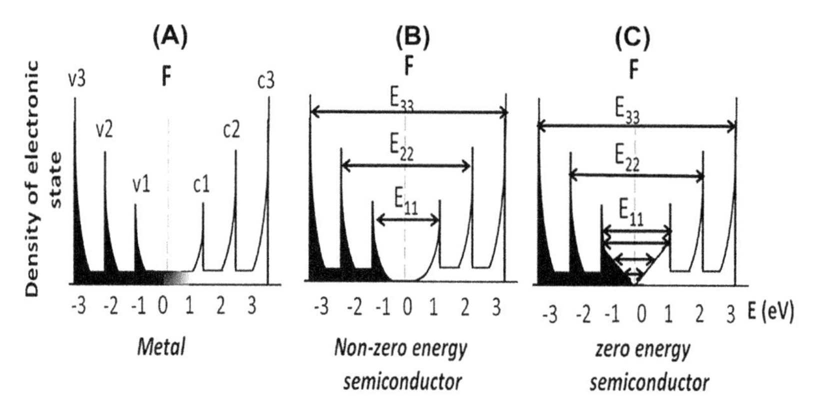

FIGURE 42 Quantum-mechanical model for valence and conduction bands in metals (A), nonzero band-gap energy for semiconductors (B) and zero band gap energy graphene (C). In semiconductors, electron conduction occurs only between corresponding DOSs (v1 to c1 (E11), v2 to c2 (E22), etc.). The v1 electron in zero band gap graphene may absorb energy of all wavelengths.

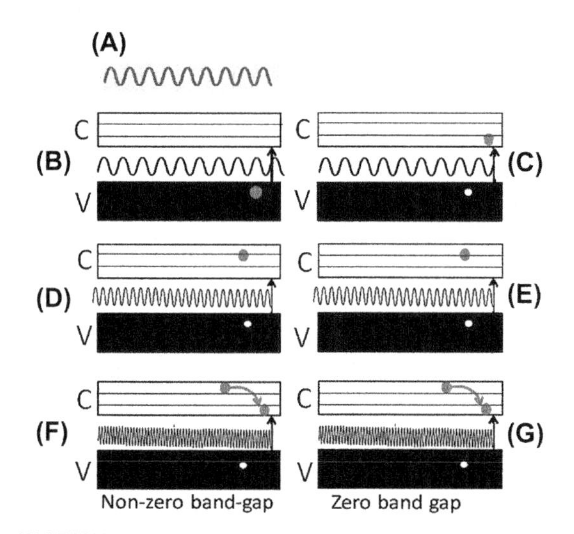

FIGURE 43 Diagrams showing possible relationship between the wavelength of the excitation wave and electron conduction through the band-gap energy. Electrons do not absorb any energy if the excitation wavelength (B) is larger than the band-gap energy (A) (wavelength restriction). If the excitation wavelength (C) matches the gap energy (i), energy is absorbed and the electron moves from HOMO to LUMO. If the wave energy is much higher than the gap energy, electron moves to a higher conduction orbital (D–G), but subsequently moves to a lower orbital by releasing phonons (F and G).

to a higher conduction orbital, but subsequently moves to a lower orbital by releasing phonons.

In semiconductors having zero energy barrier (Figure 42(C)), there is no wavelength restriction—that is, the electrons can absorb photons within the light spectrum from infrared to ultraviolet. The best example of a zero energy gap semiconductor is grapheme. The folding of graphene sheets results in a nonzero band gap in carbon nanotubes.

Appendix 2: Introduction to Magnetism

A magnetic field consists of imaginary lines of flux coming from moving or spinning electrically charged particles, such as spin and motion of an electron and spin of a proton in an electric circuit. The magnetic force (magnetism) is the force created by a magnetic field on other objects with magnetic fields.

To understand magnetic field, it is important to understand the basic structure of an atom. As described previously and in Figure 44, electrons orbit the nucleus (consisting of +ve charged protons and electrically neutral neutrons (N)) in distinct orbitals with fixed energy (two levels of electron orbitals are shown in Figure 44). The nucleus consists of positively

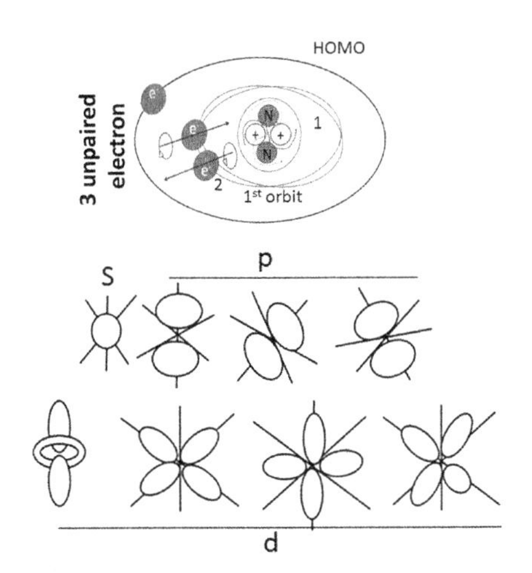

FIGURE 44 (Top) Electrons, protons, and magnetism. Protons and neutrons exhibit opposite spins, thus there is no net magnetic moment ((top)-1). Electron pairs in occupied orbitals (example first orbital) exhibit opposite spin, thus canceling the magnetic moments ((top)-2). The unpaired electrons in HOMO exhibit net magnetic moment ((top)-3). (Bottom) Electron movements in different suborbitals (s, p and d). The distinct electron routes generate some magnetic moment due to orbital moments.

charged protons and electrically neutral neutrons (Figure 44(1)). The electrons are negatively charged particles or waves (particle-wave duality) orbiting the nucleus in different orbitals (HOMO; Figure 44(3)) and the first orbital (Figure 44(2)). The protons and neutrons exhibit spin, while electrons exhibit spin as well as momentum of orbiting the nucleus. The orbital nearest to the nucleus contains two electrons exhibiting the opposite spin, thus canceling the magnetic momentum. However, the HOMO contains a single unpaired electron exhibiting net magnetic moment. Figure 44 shows that electrons are organized on different orbitals, each preceded by the n energy level in which it exists. The orbital 1s implies an s orbital with energy $n = 1$, the orbital 2p implies a p orbital with energy $n = 2$, and so on. If we were to fill an atom with electrons starting only from a central nucleus, the orbital will be filled in the following order: s, p, d, f. The s, p, d, and f orbitals can hold a maximum of 2, 6, 10, and 14 electrons, respectively.

A2.1 Neutrons, Protons, and Nuclear Spin

The positively charged protons and uncharged neutrons in nucleus spin around their own axis, thus producing a very weak magnetic field. Protons have a t_3 of $+1/2$ and neutrons have a t_3 of $-1/2$ spin energy, where t_3 is the magnetic quantum number associated with the spin quantum numbers (l, s, and j; described later). This suggests that protons and neutrons both have an intrinsic spin of $\pm 1/2$ that cancels out magnetic momentum. However, momentum may develop if there is an interaction between the electrons and the protons. Because every element has a specific number of protons, electrons, and neutrons, the nucleus of each element produces a magnetic field different from another element.

A2.2 Electron Spin, Orbital Moments, and Magnetism

Unlike protons that exhibit spin, electrons exhibit an orbit angular vector (Figure 45(L)) due to their orbiting around the nucleus not as particle, but a wave having a specific wavelength, similar to a standing wave and spin (Figure 45(S)). Thus, the electrons' motion is characterized by the quantum numbers associated with the angular momentum and spin quantum numbers (m, l, s, and j; described later).

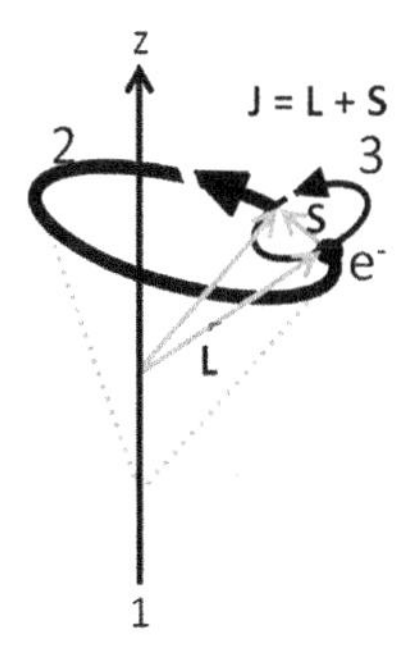

FIGURE 45 Spin (S) and orbital (L) moments of electrons. The total orbital moment (J) is sum of L and S.

The magnetic moment due to electron spin is called the spin magnetic moment (μ_s) defined by Eqn (22). In this equation, g is a constant ($g = 2.0023$ for the electron, $g = 5.586$ for the proton, and $g = -3.826$ for the neutron) and μ_B is the Bohr magneton:

$$\mu_S = \pm g\mu_B/2 \quad (22)$$

The orbital magnetic moment (μ_L) is characterized by Eqn (23), in which g is a constant, L is the angular momentum, and ħ is reduced Plank's constant:

$$\mu_L = \pm g_L\mu_B L/\hbar \quad (23)$$

The overall motion of protons and neutrons in an atom is characterized by their *total angular moment J*, where j_n is the angular momentum of individual electron and n is the number of electrons. In terms of quantum mechanics, $J = \sum$(L (angular momentum) + S (spin quantum momentum) for electrons and $J = \sum(S)$ for protons and neutrons that only have spin moments. In these equations, L is angular momentum quantum number, while l is the angular momentum of individual electron. Similarly, is the spin quantum number, while s total angular momentum of individual electron. If necessary, an axis-angle (like z-angle) correction can be made according to the equation $M_j = M_L + M_S$, where M_J is the quantum number J at z-axis. The total magnetic moment can be determined using Eqn (24) in which J is the total magnetic moment.

$$\mu_J = g_J\mu_B J/\hbar \quad (24)$$

A2.3 Quantum Characterization of the Atomic Subparticles

As discussed previously, protons and neutrons spin around their axis and, in addition, electrons orbit around the nucleus as a wave in an orbital. Orbitals describe a region of space in which there is a high probability of finding the electron. The atomic subparticles can be characterized by the following quantum numbers:

1. *Principal quantum number (n)* specifies the energy of an electron and the size of the orbital (1, 2, 3, 4.......... α). Electrons, depending on their energy (or wavelength), occupy a defined orbital. For a hydrogen atom with $n = 1$, the electron is in its *ground state.* If the electron is in the $n = 2$ orbital, it has absorbed energy and moved to an *excited state.* The total number of orbitals for a given n value is n^2.
2. *Angular momentum quantum number (l)* specifies the shape of an orbital with a particular principal quantum number ($l = 0$ is s-orbital, $l = 1$ is p orbital, $l = 2$ is d orbital, $l = 3$ *is f orbital*, etc.) The subshell with $n = 2$ and $l = 1$ is the 2*p* subshell; if $n = 3$ and $l = 0$, it is the 3*s* subshell, and so on.
3. *Magnetic quantum number (m_l)* specifies the orientation in space of an orbital of a given energy (n) and shape (l). This number divides the subshell into individual orbitals that hold the electrons. There are $2l + 1$ orbitals in each subshell.
4. *Spin quantum number (s)* specifies the orientation of the spin axis of an electron. An electron can spin in only one of two directions (sometimes called *up* and *down* or +1/2 and −1/2).

A2.4 Electrons and Magnetism

Pauli's Exclusion Principle states that no two electrons in the same atom can have identical values for all four of their quantum numbers. In other words, (1) no more than two electrons can occupy the same orbital and (2) two electrons in the same orbital must have opposite spins (Figure 46(i) and (ii)).

In general, opposing spins (proton +1/2 and neutron −1/2) of the proton and the neutron

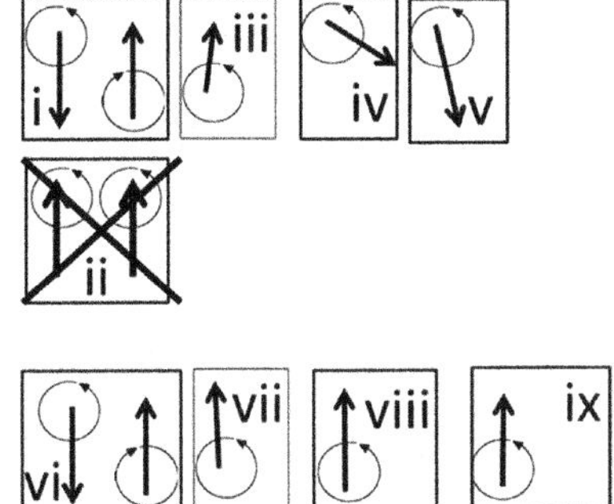

FIGURE 46 Electron spin and magnetic moment. (i) electron pairs with opposite spins cancel out magnetic moments, (ii) electron pair in an orbital cannot have parallel spins (Pauli's exclusion principle), (iii, iv and v): randomly oriented electrons do not result in net magnetism, (vii, viii and ix): electrons parallel to an applied field cause net magnetic moment.

cancel the magnetic field. In certain cases, such as hydrogen atom or certain isotopes, nuclear magnetic moment may be significant. Although electron spin generates magnetic momentum, the opposite spins of the two electrons in the same orbital cancel out their magnetic momentum with no residual magnetic momentum. Atoms with unpaired electrons spinning in the same direction contain net magnetic moments and are weakly attracted to magnets. The overall magnetic activity depends upon the alignment of their unpaired electrons. If they exhibit random movement (Figure 46(iii–v)), there will be no net magnetic moment. However, if the unpaired electrons are parallel to the same direction, the particles exhibit magnetic moment.

A2.5 Types of Magnetism

A magnet is a piece of a metal or an alloy that exhibits properties of magnetism, such as attracting other iron-containing objects or aligning itself to an external magnetic field. Magnetic materials are classified as (i) weak magnetic materials containing diatomic and paramagnetic materials and (ii) strong magnetic materials such as ferromagnetic, ferromagnetic, and antiferromagnetic particles (Figure 47).

In diamagnetic materials, atoms do not have any unpaired electrons, resulting in zero net magnetic moment (a quantity that determines

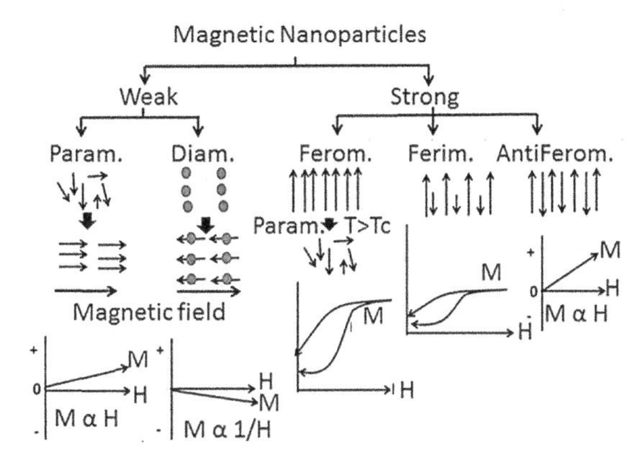

FIGURE 47 Classification of magnetic nanoparticles. In general, magnetic nanoparticles are classified as having weak (paramagnetic (Parm.) and diamagnetic (Diam.)) and strong (ferromagnetic (Ferom.), ferimagnetic (Ferim.) and antiferromagnetic (AntiFerom.)) moments.

- Paramagnetic nanoparticles (superconductors such as Ag, Au and water) exhibit random electronic spin in the absence of a magnetic field (H), but parallel spin in the direction of an applied magnetic field that generates magnetic moment (M) where M α H. The effects are reversed as the applied field is removed.
- Diamagnetic nanoparticles (Na, Al, Ca and U), in a magnetic field, exhibit electron spin opposite to the applied field (M α (1/H)), thus the strength of H decreases. The effects are reversed as the applied field is removed.
- Ferromagnetic nanoparticles (Fe, Co, Ni, Gd, SmCo nanoparticles) contain domains of electrons that align parallel to the applied field. The magnetic effects remain stable even after the applied field is removed. Ferromagnetic particles acquire paramagnetic properties when T > Curie temperature (Tc).
- The electron pairs of ferrimagnetic nanoparticles (Fe_3O_4, Yttrium granite, Gd, etc.) spin in the opposite direction, but, in magnetic field, the spin strength is stronger in one direction than the other. This results in relatively weak magnetization.
- The antiferromagnetic nanoparticles (Cr, FeMn, Ni) are similar to the ferromagnetic nanoparticles except the two spins are of equal strength. These nanoparticles act like paramagnetic nanoparticles.

the torque it will experience in an external magnetic field). These materials display a very weak response in an applied magnetic field. In paramagnetic materials, an atom has a net magnetic moment due to unpaired electrons, but magnetic domains are absent. When a magnetic field is applied, the magnetic moments of the atoms align along the direction of the applied magnetic field, forming a weak net magnetic moment. These materials do not retain magnetic moments when the magnetic field is removed.

The ferromagnetic atom has a net magnetic moment due to unpaired electrons. The unmagnetized bulk materials consist of domains, each containing large numbers of atoms whose magnetic moments are parallel, producing a net magnetic moment of the domain; however, atoms in different domains face in different directions, giving a zero net magnetic moment of the material. When the ferromagnetic material is placed in a magnetic field, the magnetic moments of the domains align along the direction of the applied magnetic field, forming a large net magnetic moment (Figure 48). A residual magnetic moment exists even after the magnetic field is removed.

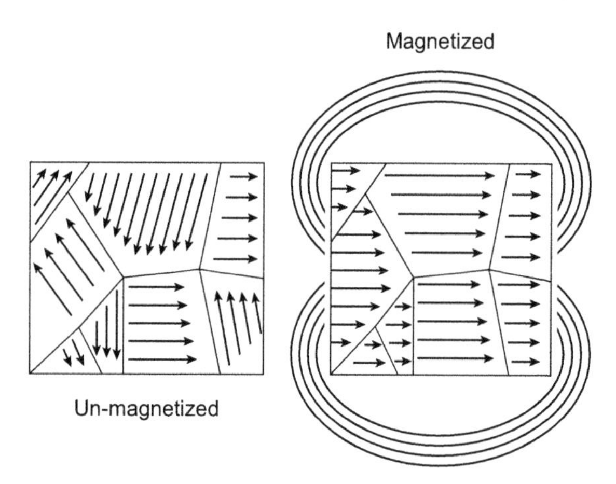

FIGURE 48 Response of ferromagnetic particles to a magnetic field. When the ferromagnetic material is placed in a magnetic field, the magnetic moments of the domains align along the direction of the applied magnetic field, thus forming a large net magnetic moment. A residual magnetic moment exists even after the magnetic field is removed.

Antiferromagnets are compounds of two different atoms that occupy different lattice positions and have magnetic moments that are equal in magnitude and opposite in direction; thus, the magnetic moments cancel out, resulting in zero net magnetic moment (Figure 47). However, in the presence of a magnetic field, a weak magnetic moment is observed. The atoms in ferromagnetic materials also reside in different lattice sites with antiparallel magnetic moments; however, unlike the antiferromagnetic materials, the magnetic moments do not cancel out because they have different magnitudes, which results in a net spontaneous magnetic moment.

The magnetic activity of a material depends on its intrinsic (properties such as composition, crystallographic structure, magnetic anisotropic energy, and vacancies and defects that are not dependent on the particles' microstructure) and extrinsic (properties such as size, shape, external magnetic field, temperature, etc. that are dependent on the particles' microstructure) characteristics. Some of the indices used to assess the intrinsic magnetic properties include the following:

- *Magnetic moment* (μ_B), a quantity that determines the torque it will experience in an external magnetic field
- *Curie temperature* (Tc) at which a material's permanent magnetism changes to induced magnetism
- *Saturation magnetisation* (Ms) at which an increase in external field does not cause further increase in magnetism
- *Exchange*, an interaction between electrons in different orbitals (similar spin direction) caused by Coulomb interactions
- *Magnetic order*, the atomic spin shown in Figure 47
- *Magnetocrystalline anisotropy* (*K*), a condition where it takes more energy to magnetize the particle in certain directions than in others

- *Hysteresis effect*, the lagging of an effect behind its cause, as when the change in magnetism of a body lags behind changes in the magnetic field. In bulk particles, a magnetic force (M (eμ/g)) proportional to the strength of the magnetic field is applied until a saturation state (magnetic saturation) is achieved. Once a saturation state is achieved, the applied magnetic field can be dropped to zero without any significant drop in the magnetic force. If an alternating magnetic field is applied to the material at a magnetic saturation state, its magnetization will trace out a loop, as shown in Figure 49. This loop is called a hysteresis loop. Hysteresis can be defined as the lack of retraceability of the magnetization curve, which may be related to the existence of magnetic domains in bulk ferromagnetic material. Once the magnetic domains are reoriented, it develops magnetic memory, which can be turned back again with some energy.

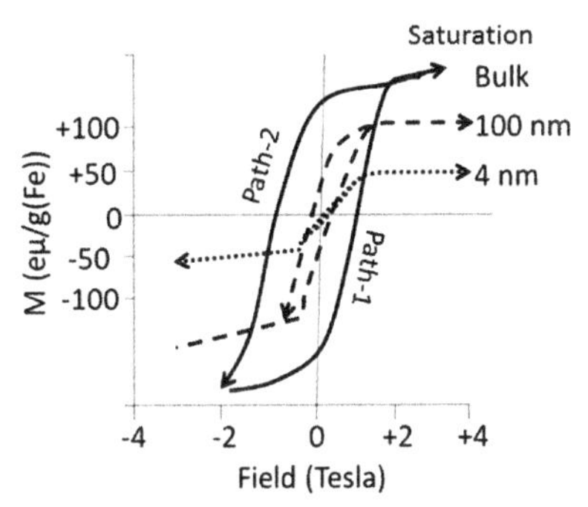

FIGURE 49 Bulk ferromagnetic particles retain the magnetic moment after the force has been removed, a process known as the *hysteresis effect*. M (eμ/g)) increases proportionally to the H values until a saturation state (magnetic saturation) is achieved. An alternating magnetic field is required to demagnetize the particles; however, the demagnetization takes a different path, thus forming a loop called a hysteresis loop. Hysteresis is defined as the lack of retraceability of the magnetization curve. At the nanorange, the particles gradually lose the hysteresis (4-nm particles exhibit greater loss than 100-nm particles).

The commonly used indices to assess the extrinsic properties that depend on size and shape are the following:

- *Coercivity (Hc) and remnant magnetisation (Mr)*: Hc measures the resistance of a ferromagnetic material to becoming demagnetized. *Mr* is the amount of magnetization retained by the magnetized nanoparticle at the zero driving field.
- *Magnetic susceptibility* (χ) is a dimensionless proportionality constant that indicates the degree of magnetization of a material in response to an applied magnetic field.

For bulk particles, the intrinsic and extrinsic properties are independent of changes in their size (from μm to >100 nm) or shape.

References

Abdolmohammadi, S., Siyamak, S., Ibrahim, N.A., Yunus, W.M.Z.W., Rahman, M.Z.A., Azizi, S., Fatehi, A., 2012. Enhancement of mechanical and thermal properties of polycaprolactone/chitosan blend by calcium carbonate nanoparticles. Int. J. Mol. Sci. 13, 4508–4522.

Agostiano, A., Catalano, M., Curri, M.L., Monica, M.D., Manna, L., Vasanelli, L., 2000. Synthesis and structural characterisation of CdS nanoparticles prepared in a four-components "water-in-oil" microemulsion. Micron 31, 253.

Ajayan, P.M., 1999. Nanotubes from carbon. Chem. Rev. 99, 1787–1800.

Akbari, B., Tavandashti, M.P., Zandrahimi, M., 2011. Particle size characterization of nanoparticles—a practical approach. Iran. J. Mater. Sci. Eng. 8, 48–56.

Alberto, P., 2009. Guimarães Principles of Nanomagnetism. Springer, Berlin/Heidelberg, Germany.

Andrievski, R.A., 2009. Size dependent effects in properties of nanostructured materials. Rev. Adv. Mater. Sci. 21, 107–133.

Atkins, P.W., 1974. Quanta: A Handbook of Concepts. Oxford University Press.

Ayala, P., Arenal, R., Rümmeli, M., Rubio, A., Pichler, T., 2010. The doping of carbon nanotubes with nitrogen and their potential applications. Carbon 48, 575–586.

Bakunin, V.M., Yu, A., Kuzmina, S.G.N., Parenago, O.P., 2004. Synthesis and application of inorganic nanoparticles as lubricant components – a review. J. Nanopart. Res. 6, 273–284.

Batlle, X., Labarta, A., 2002. Finite-size effects in fine particles: magnetic and transport properties. J. Phys. D Appl. Phys. 35, R15–R42.

Bianco, S., Giorcelli, M., Musso, S., Castellino, M., Agresti, F., Khandelwal, A., Lo Russo, S., Kumar, M., Ando, Y., Tagliaferro, A., 2010. Hydrogen adsorption in several types of carbon nanotubes. J. Nanosci. Nanotechnol. 10, 3860–3866.

Bielinska, A., Kukowska-Latallo, J.F., Johnson, J., Tomalia, D.A., Baker, J.R., 1996. Regulation of in vitro gene expression using antisense oligonucleotides or antisense expression plasmids transfected using starburst PAMAM dendrimers. Nucl. Acids Res. 1996 (24), 2176–2182.

Boas, U., Heegaard, P.M.H., 2004. Dendrimers in drug research. Chem. Soc. Rev. 33, 43–63.

Bohr, N., 1923. Structure of atoms. Nature 112, 29–42.

de Broglie, L., 1925. The Wave Nature of the Electron. The Nobel Foundation.

Bronstein, L.M., Huang, X., Retrum, J., Schmucker, A., Pink, M., Stein, B.D., Dragnea, B., 2007. Influence of iron oleate complex structure on iron oxide nanoparticle formation. Chem. Mater. 19, 3624–3632.

Chan, C.M., Wu, J., Li, J.X., Cheung, Y.K., 2002. Polypropylene/calcium carbonate nanocomposites. Polymer 43, 2981–2992.

Chatterjee, J., Haik, Y., Che, C.-J., 2003. Size dependent magnetic properties of iron oxide nanoparticles. J. Magn. Magn. Mater. 257, 113–118.

Chen, L.N., Wu, X.Z., Huang, W.R., Ma, S.S., Xu, H., 2012. Effects of partial hydrogenation and vacancy defects on the electronic properties of metallic carbon nanotubes. Solid State Commun. 152, 868–872.

Chen, N., Wan, C., Zhang, Y., Zhang, Y., 2004. Effect of nano-$CaCO_3$ on mechanical properties of PVC and PVC blendex blend. Polym. Test. 23, 169–174.

Chen, R., Christiansen, M.J., Anikeeva, P., 2013. Maximizing hysteretic losses in magnetic ferrite nanoparticles via model-driven synthesis and materials optimization. AcsNano 7, 8990–9000.

Chu, V.H., Nghiem, T.H.L., Le, T.H., Vu, D.L., Tran, H.N., Vu, T.K.L., 2012. Synthesis and optical properties of water soluble CdSe/CdS quantum dots for biological applications. *Adv. Nat. Sci.: Nanosci. Nanotechnol.* 3, 025017 (7pp).

Cutler, M., Mott, N.F., 1969. Observation of Anderson localization in an electron gas. Phys. Rev. 181, 1336.

Davison, C., Germer, T., 1927. The scattering of electrons by a single crystal of nickel. Nature 119, 558–560.

Dietzel, D., Ritter, C., Mönninghoff, T., Fuchs, H., Schirmeisen, A., Schwarz, U.D., 2008. Frictional duality observed during nanoparticle sliding. Phys. Rev. Lett. 101, 125505.

Dietzel, D., Feldmann, M., Schwarz, U.D., Fuchs, H., Schirmeisen, A., 2013. Scaling laws of structural lubricity. Phys. Rev. Lett. 111, 235502.

Ding, J.W., Yan, X.Y., Cao, J.X., Wang, D.L., Tang, Y., Yang, Q.B., 2003. Curvature and strain effects on electronic properties of single-wall carbon nanotubes. J. Phys. Condens. Matter 15, L439–L445.

Dobis, P., Bruestlova, J., Bartlova, M., 2010. Curie temperature in ferromagnetic materials and visualized magnetic domains. In: 3rd International Symposium for Engineering Education. University College Cork, Ireland.

Dootz, R., Toma, A.C., Pfohl, T., 2011. PAMAM6 dendrimers and DNA: pH deACpendent "beads-on-a-string" behavior revealed by small angle X-ray scattering. Soft Matter 7, 8343–8351.

Dresselhaus, M.S., Dresselhaus, G., Avouris, P., 2001. Carbon Nanotubes: Synthesis, Structure, Properties and Applications. Springer, Berlin Heidelberg. ISBN 9783540728641.

Dresselhaus, M.S., Dresselhaus, G., Eklund, P.C., 1996. Science of Fullerenes and Carbon Nanotubes. Academic Press, San Diego.

Dubiel, M., Hofmeister, H., Schurig, E., 1997. Compressive stresses in Ag nanoparticle doped glasses by ion implantation. Phys. Status Solidi (b) 203, R5–R6.

Eastham, D.A., Kirkman, I.W., 2000. Highly enhanced orbital magnetism on cobalt cluster surfaces. J. Phys. Condens. Matter 12, L525–L532.

Esfand, R., Tomalia, D.A., 2001. Poly(amidoamine) (PAMAM) dendrimers: from biomimicry to drug delivery and biomedical applications. Drug Discov. Today 6, 427–436.

Esfarjani, K., Chen, Z., Kawazoe, Y., 2003. Electronic properties of magnetically doped nanotubes. Bull. Mater. Sci. 26, 105–107.

Fant, K., Esbjörner, E.K., Lincoln, P., Nordén, B., 2008. DNA condensation by PAMAM dendrimers: self-assembly characteristics and effect on transcription. Biochemistry 47, 1732–1740.

Fischer, A., 1996. Well-founded selection of materials for improving wear resistance. Wear 194, 238–245.

Fischer, J.E., Dai, H., Thess, A., Lee, R., Hanjani, N.M., Dehaas, D.L., Smalley, R.E., 1997. Metallic resi stivity in crystalline ropes of single-wall carbon nanotubes. Phys. Rev. B 55, R4921.

Frechet, J.M.J., 1994. Functional polymers and dendrimers: reactivity, molecular architecture, and interfacial energy. Science 263, 1710–1715.

Freund, H.-J., 2002. Clusters and islands on oxides: from catalysis via electronics and magnetism to optics. Surf. Sci. 500, 271–299.

Gajewicz, A., Puzyn, T., Rasulev, B., Leszczynska, D., Leszczynski, J., 2011. Metal oxide nanoparticles: size-dependence of quantum-mechanical properties. Nanosci. Nanotechnol. Asia 1, 53–58.

Gnecco, E., Meyer, E., 2007. Fundamentals of Friction and Wear on the Nanoscale, first ed. Springer, Berlin.

Goldstein, A.N., Echer, C.M., Alivisatos, P.A., 1992. Melting in semiconductor nanocrystals. Science 256, 1425–1426.

Greenberg, R., Halperin, G., Etsion, I., Tenne, R., 2004. The effect of WS_2 nanoparticles on friction reduction in various lubrication regimes. Tribol. Lett. 17, 179–186.

Grollier, J., Chanthbouala, A., Matsumoto, R., Anane, A., Cros, V., Nguyen van Dau, F., Fert, A., 2011. Magnetic domain wall motion by spin transfer. Comptes Rendus Phys. 12, 309–317.

Guo, D., Xie, G., Luo, J., 2014. Mechanical properties of nanoparticles: basics and applications. J. Phys. D Appl. Phys. 47, 013001.

Guo, Y.-B., Wang, D.-G., Zhang, S.-W., 2011. Adhesion an friction of nanoparticles/polyelectrolyte multilayer films by AFM and micro-tribometer. Tribol. Int. 44 (7–8), 906–915.

Gupta, S.K., Talati, M., Jha, P.K., 2008. Shape and size dependent melting point temperature of nanoparticles. Mater. Sci. Forum 570, 132–137.

Gupta, U., Agashe, H., Jain, N.K., 2007. Polypropylene imine dendrimer mediated solubility enhancement: effect of pH and functional groups of hydrophobes. J. Pharm. Sci. 10, 358–367.

Haruyama, J., Matsudaira, M., Reppert, J., Rao, A., Koretsune, T., Saito, S., Sano, H., Iye, Y., 2011. Superconductivity in boron-doped carbon nanotubes. J. Supercond. Novel Magnet. 24, 111–120.

Heim, L.O., Jürgen, B., Preuss, M., Hans-Jürgen, B., 1999. Adhesion and friction forces between spherical micrometer-sized particles. Phys. Rev. Lett. 83, 3328–3331.

Heisenberg, W., 1925. Über quantentheoretishe Umdeutung kinematisher und mechanischer Beziehungen. Z. für Physik 33, 879–893.

Hernández Battez, A., González, R., Viesca, J.L., Fernández, J.E., Díaz Fernández, J.M., Machado, A., Chou, R., Riba, J., 2008. CuO, ZrO_2 and ZnO nanoparticles as antiwear additive in oil lubricants. Wear 265, 422–428.

Hou, P.-X., Xu, S.-T., Ying, Z., Yang, Q.-H., Liu, C., Cheng, H.-M., 2003. Hydrogen adsorption/desorption behavior of multi-walled carbon nanotubes with different diameters. Carbon 41, 2471–2476.

Housecroft, C., Sharpe, A.G., 2012. Inorganic Chemistry, fourth ed. Prentice Hall. ISBN 978 0273742753.

Ibusuki, T., Kojima, S., Kitakami, O., Shimada, Y., 2001. Magnetic anisotropy and behaviors of Fe nanoparticles. IEEE Trans. Magn. 37, 2223–2225.

Indira, T.K., Lakshmi, P.K., 2010. Magnetic nanoparticles, a review. Int. J. Pharm. Sci. Nanotechnol. 3, 1035–1042.

Issa, B., Obaidat, I.M., Albiss, B.A., Haik, Y., 2013. Magnetic nanoparticles: surface effects and properties related to biomedicine applications. Int. J. Mol. Sci. 14, 21266–21305.

Issa, B., Obaidat, I.M., Albiss, B.A., Haik, Y., Kodama, R.H., 1999. Magnetic nanoparticles. J. Magn. Magn. Mater. 200, 359–372.

Jalili, S., Majidi, R., 2007. The effect of atomic hydrogen adsorption on single-walled carbon nanotubes properties. J. Iran. Chem. Soc. 4, 431–437.

Jansen, J.F.G.A., de Brabander-van den Berg, E.M.M., Meijer, E.W., 1994. Encapsulation of guest molecules into a dendritic box. Science 266, 1226–1229.

Jia, M., Lai, Y., Tian, Z., Liu, Y., 2009. Calculation of the surface free energy of fcc copper nanoparticles. Model. Simul. Mater. Sci. Eng. 17, 015006.

Jun, Y.-W., Seo, J.-W., Cheon, J., 2008. Nanoscaling laws of magnetic nanoparticles and their applicabilities in biomedical sciences. Acc. Chem. Res. 41, 179–189.

Khomenko, A.V., Prodanov, N.V., 2010. Study of friction of Ag and Ni nanoparticles: an atomistic approach. J. Phys. Chem. C 114, 19958–19965.

Klajnert, B., Bryszewska, M., 2001. Dendrimers: properties and applications. Acta Biochim. Polinica 48, 199–208.

Kodama, R.H., 1999. Magnetic nanoparticles. J. Magn. Magn. Mater. 200, 359–372.

Koenig, S., Müller, L., Smith, D.K., 2001. Dendritic biomimicry: microenvironmental hydrogen-bonding effects on tryptophan fluorescence. Chemistry 7, 979–986.

Kramers, H., Helge, H.A., 1962. The Atomic and Bohr Theory of Its Structure: An Elementary Presentation. Gyldendal, London. ASIN B003F8LC24.

Kukowska-Latallo, J.F., Bielinska, A.U., Johnson, J., Spindler, R., Tomalia, D.A., Baker, J.R., 1996. Efficient transfer of genetic material into mammalian cells using Starburst polyamidoamine dendrimers. Proc. Natl. Acad. Sci. U.S.A. 93, 4897–4902.

Kukowska-Latallo, J.F., Bielinska, A.U., Johnson, J., Spindler, R., Tomalia, D.A., Baker Jr., J.R., Kumar, R., Kumar, M., 2012. Effects of size on cohesive energy, melting temperature and Deby temperature of nanomaterials. Ind. J. Pure Appl. Phys. 50, 329–334.

Li, M., Li, J.C., 2006. Size effects on the band-gap of semiconductor compounds. Mater. Lett. 60, 2526–2529.

Lia, H., Hana, P.D., Zhangb, X.B., Lib, M., 2013. Size-dependent melting point of nanoparticles based on bond number calculation. Mater. Chem. Phys. 137, 1007–1011.

Liu, Y., Bryantsev, V.S., Diallo, M.S., Goddard III, W.A., 2009. PAMAM dendrimers undergo pH responsive conformational changes without swelling. J. Am. Chem. Soc. 131, 2798–2799.

Lu, Y., Shi, T., An, L., Jinb, L., Wang, Z.-G., 2010. A simple model for the anomalous intrinsic viscosity of dendrimers. Soft Matter 6, 2619–2622.

Lu, A.H., Salabas, E.L., Schuth, F., 2007. Magnetic nanoparticles: synthesis, protection, functionalization, and application. Angew. Chem. Int. Ed. 46, 1222–1244.

Maiti, P.K., Cagin, T., Lin, S.-T., Goddard III, W.A., 2005. Effect of solvent and pH on the structure of PAMAM dendrimers. Macromolecules 38, 979–991.

Mananghaya, M., Rodulfo, E., Santos, G.N., Villagracia, A.R., Ladines, A., 2012. Theoretical investigation on single-wall carbon nanotubes doped with nitrogen, pyridine-like nitrogen defects, and transition metal atoms. J. Nanomater. 2012, 104891.

Matsuda, Y., Tahir-Kheli, J., Goddard III, W.A., 2010. Definitive band gaps for single-wall carbon nanotubes. J. Phys. Chem. Lett. 1, 2946–2950.

Miedema, A.R., Boom, R., 1978. Surface tension and electron density of pure liquid metals size effect on the cohesive energy of nanoparticle. Z. Metallkd 69, 183.

Mohanraj, V., Chen, Y., 2006. Nanoparticles—a review. J. Pharm. Res. 5, 561–573.

Moniruzzaman, M., Winey, K.I., 2006. Polymer nanocomposites containing carbon nanotubes. Macromolecules 39, 5194–5205.

Montano, P.A., Schulze, W., Tesche, B., Shenoy, G.K., Morrison, T.I., 1984. Extracting electrical energy from the vacuum by cohesion of charged foliated conductors. Phys. Rev. B 30, 1700–1702.

Morales, M.P., Veintemillas-Verdaguer, S., Montero, M.I., Serna, C.J., 1999. Surface and internal spin canting in ç-Fe_2O_3 nanoparticles. Chem. Mater. 1999 (11), 3058–3064.

Morton, J.J.L., 2012. Magnetic Properties of Materials. JJLM, Trinity. http://www.ucl.ac.uk/qsd/people/teaching/MPM-Part3.

Mourey, T.H., Turner, S.R., Rubenstein, M., Fréchet, J.M.J., Hawker, C.J., Wooley, K.L., 1992. Unique behaviour of dendritic macromolecules: intrinsic viscosity of polyether dendrimers. Macromolecules 25, 2401–2406.

Muller, T., Yablon, D.G., Karchner, R., Knapp, D., Kleinman, M.H., Fang, H., Durning, C.J., Tomalia, D.A., Turro, N.F., Flynn, G.W., 2002. AFM studies of high-generation PAMAM dendrimers at the liquid/solid interface. Langmuir 18, 7452–7455.

Nanda, K.K., Maisels, A., Kruis, F.E., Fissan, H., Stappert, S., 2003. Higher surface energy of free nanoparticles. Phys. Rev. Lett. 91, 106102.

Nanjwade, B.K., Bechra, H.M., Derkar, G.K., Manvi, F.V., Nanjwade, V.K., 2009. Dendrimers: emerging polymers for drug-delivery systems. Eur. J. Pharm. Sci. 38, 185–196.

Neophytou, N., Ahmed, S., Klimeck, G., 2007. Influence of vacancies on metallic nanotube transport properties. Appl. Phys. Lett. http://dx.doi.org/10.1063/1.2736295.

Nicolai, H.T., Kuik, M., Wetzelaer, G.A., de Boer, B., Campbell, C., Risko, C., Brédas, J.L., Blom, P.W., 2012. Unification of trap-limited electron transport in semiconducting polymers. Nat. Mater. 11, 882–887.

Noriega-Luna, B., Godínez, L.A., Rodríguez, F.J., Rodríguez, A., Zaldívar-Lelo de Larrea, G., Sosa-Ferreyra, C.F., Mercado-Curiel, R.F., Manríquez, J., Bustos, E., 2014. Applications of dendrimers in drug delivery agents, diagnosis, therapy, and detection. J. Nanomater. 2014 (507273), 19.

Okada, S., 2007. Energetics and electronic structure of carbon nanotubes with adatom-vacancy defects. Chem. Phys. Lett. 447, 263–267.

Ottaviani, M.F., Sacchi, B., Turro, N.J., Chen, W., Jockusch, S., Tomalia, D.A., 1999. An EPR study of the interaction between starburst dendrimers and polynucleotides. Macromolecules 32, 2275–2282.

Ouyang, G., Zhu, Z.M., Zhu, W.G., Sun, C.Q., 2010. Size dependent debye temperature and total mean square relative atomic displacement in nanosolids under high pressure and high temperature. J. Phys. Chem. C 114, 1805–1808.

Ouyang, M., Huang, J.-H., Cheung, C.L., Lieber, C.M., 2001. Energy gaps in "metallic" single-walled carbon nanotubes. Science 292, 702–705.

Peng, D.-X., Chen, C.-H., Kang, Y., Chang, Y.-P., Chang, S.-Y., 2010. Size effects of SiO_2 nanoparticles as oil additives on tribology of lubricant. Ind. Lubr. Tribol. 62, 111–120.

Qi, W.H., Wang, M.P., 2004. Size and shape dependent melting temperature of metallic nanoparticles. Mater. Chem. Phys. 88 (2–3), 280–284.

Qi, W.H., Wang, M.P., 2002. Size effect on the cohesive energy of nanoparticle. J. Mater. Sci. Letters 21, 1743–1745.

Rahman, I.A., Vejayakumaran, P., Sipaut, I.J., Chee, C.K., 2003. Size-dependent physicochemical and optical properties of silica nanoparticles. J. Magn. Magn. Mater. 257, 113–118.

Rapoport, L., Fleischer, N., Tenne, R., 2003a. Fullerene-like WS_2 nanoparticles: superior lubricants for harsh conditions. Adv. Mater. 15, 651–655.

Rapoport, L., Leshchinsky, V., Lapsker, I., Volovik, Y., Nepomnyashchy, O., Lvovsky, M., Popovitz-Biro, R.Y., 2003b. Tribological properties of WS_2 nanoparticles under mixed lubrication. Wear 255, 785b793.

Rapoport, L., Nepomnyashchy, O., Lapsker, I., Verdyan, A., Soifer, Y., Popovitz-Biro, R., Tenne, R., 2005. Friction and wear of fullerene-like WS_2 under severe contact conditions: friction of ceramic materials. Tribol. Lett. 19, 143–149.

Reynaud, E., Jouen, T., Gauthier, C., Vigier, G., Varlett, J., 2001. Nanofillers in polymeric matrix: a study on silica reinforced PA. Polymer 42, 8759–8768.

Ritter, C., Heyde, M., Stegemann, B., Rademann, K., 2004. Contact-area dependence of frictional forces: moving adsorbed antimony nanoparticles. Phys. Rev. B 71, 085405.

Saito, R., Dresselhaus, G., Dresselhaus, M.S., 1998. Physical Properties of Carbon Nanotubes. Imperial College Press, London, UK.

Saito, R., Dresselhaus, G., Dresselhaus, M.S., 2000. Trigonal warping effect of carbon nanotubes. Phys. Rev. B 61, 2981–2990.

Sattler, K., 2002. The energy gap of clusters, nanoparticles and quantum dots. In: Nalwa, H.S. (Ed.), Handbook of Thin Films Materials, Nanomaterials and Magnetic Thin Films, vol. 5. Academic Press.

Schmid, G., 2004. Nanoparticles: From Theory to Application. Wiley-VCH, Weinheim.

Shandiza, M.A., Safaeia, A., Sanjabia, S., Barberb, Z.N., 2008. Modeling the cohesive energy and melting point of nanoparticles by their average coordination number. Solid State Commun. 145, 432–437.

Shockley, W., 1950. Electrons and Holes in Semiconductors. Van Nostrand, Princeton, NJ.

Stewart, K.H., 2008. Ferromagnetic Domains. The University of California Press.

Sun, C.Q., Tay, B.K., Zeng, X.T., Li, S., Chen, T.P., Zhou, J., Bai, H.L., Jiang, E.Y., 2002. Bond-order–bond-length–bond-strength (bond-OLS) correlation mechanism for the shape-and-size dependence of a nanosolid. J. Phys. Condens. Matter 14, 7781.

Tauc, J., 1968. Optical properties and electronic structure of amorphous Ge and Si. Mater. Res. Bull. 3, 37–46.

Tevet, O., Von-Huth, P., Popovitz-Biro, R., Rosentsveig, R., Wagner, H.D., Tenne, R., 2011. Friction mechanism of individual multilayered nanoparticles. PNAS 108, 19901–19906.

Tian, F., Cui, D., Schwarz, H., Estrada, G.G., Kobayashi, H., 2006. Cytotoxicity of single-wall carbon nanotubes on human fibroblasts. Toxicol. In Vitro 20, 1202–1212.

Tien, L.G., Tsai, C.-H., Li, F.-Y., 2008. Influence of vacancy defect density on electrical properties of armchair single wall carbon nanotube. Diamond Relat. Mater. 17, 563–566.

Urbakh, M., Klafter, J., Gourdon, D., Israelachvili, J., 2004. The nonlinear nature of friction. Nature 430, 525–528.

Usov, N.A., Grebenshchikov, Y.B., 2009. Hysteresis loops of an assembly of superparamagnetic nanoparticles with uniaxial anisotropy. J. Appl. Phys. 106, 023917.

Wang, C., Wang, C.T., 2006. Geometry and electronic properties of single vacancies in achiral carbon nanotubes. Eur. Phys. J. B 54, 243–247.

Wang, D.J., Imae, T., 2004. Fluorescence emission from dendrimer & its pH dependence. J. Am. Chem. Soc. 126, 13204–13205.

Welch, P., Muthukumar, M., 2000. Dendrimer–polyelectrolyte complexation: a model guest–host system. Macromolecules 33, 6159–6167.

Wessely, O., Katsnelson, M.I., Nilsson, A., Nikitin, A., Ogasawara, H., Odelius, M., Sanyal, B., Eriksson, O., 2007. Dynamical core-hole screening in the X-ray absorption spectra of hydrogenated carbon nanotubes and graphene. Phys. Rev. B 76, 161402.

Wu, J.F., Zhai, W.S., Jie, G.F., 2009. Preparation and tribological properties of WS_2 nanoparticles modified by trioctylamine. Proc. Inst. Mech. Eng. 223, 695–703.

Xia, J., Shao CR,Wang, T., Zhang, J., Shao, Q.Y., 2013. First-principles investigation on B/N co-doping of ultra small diameter metallic single-walled carbon nanotubes. Chem. Phys. Lett. 579, 127–131.

Xiong, S., Qi, W., Cheng, Y., Huang, B., Wang, M., Li, Y., 2011. Universal relation for size dependent thermodynamic properties of metallic nanoparticles. Phys. Chem. Chem. Phys. 14, 10652–10660.

Yu, D., Wang, C., Guyot-Sionnest, P., 2003. N-type conducting CdSe nanocrystal solids. Science 300, 1277.

Yue, Y., Eun, J.S., Lee, M.-K., Seo, S.-Y., 2012. Synthesis and characterization of G5 PAMAM dendrimer containing daunorubicin for targeting cancer cells. Arch. Pharm. Res. 35, 343–349.

Zeng, H., Zhao, J., Hu, H., Leburton, J.-P., 2011. Atomic vacancy defects in the electronic properties of semi-metallic carbon nanotubes. J. Appl. Phys. 109, 083716.

Zhang, J.-M., Liang, R.-L., Xu, K.-W., 2010. Effect of uniaxial strain on the band gap of zigzag carbon nanotubes. Phys. B Condensed Matter 405, 1329–1334.

Zhang, Q.X., Yu, Z.Z., Xie, X.L., Mai, Y.W., 2004. Crystallization and impact energy of polypropylene/$CaCO_3$ nanocomposites with nonionic modifier. Polymer 45, 5985–5994.

Zhao, J., Park, H., Han, J., Lu, J.P., 2004. Electronic properties of carbon nanotubes with covalent sidewall functionalization. J. Phys. Chem. B 108, 4227–4230.

Zhou, Q.X., Wang, C.Y., Fu, Z.B., Tang, Y.J., Zhang, H., 2014. Effects of various defects on the electronic properties of single-walled carbon nanotubes: a first principle study. Front. Phys. 9, 200–209.

CHAPTER

4

Experimental Methodologies for the Characterization of Nanoparticles

OUTLINE

Engineered Nanoparticles
http://dx.doi.org/10.1016/B978-0-12-801406-6.00004-2

1. INTRODUCTION

Characterization is defined as the study of a material's features, such as chemical composition and various physical, chemical, electrical, and magnetic properties. Because the physicochemical properties of nanoparticles vary exponentially with variations in their size, the accurate characterization of nanoparticles is critical to their applications. The methods commonly used for nanoparticle characterization are listed in Figure 1. In general, nanoparticle properties can be divided into the following groups: surface, physicochemical, electronic, mechanical, magnetic, and thermodynamic properties.

The indices commonly measured for assessing a nanoparticle's chemical properties include its atomic and molecular composition, element analysis, and determination of surface functionalization. Surface is characterized by determination of surface area, volume, purity, electronic structure, functional groups, and the particle's shape, texture, and roughness. Assessment of physical properties includes the determination of dispersion and aggregation, size and distribution, and mechanical properties. Thermodynamic properties are assessed by measuring melting temperature, melting enthalpy, melting entropy, evaporation temperature, Curie temperature, Debye temperature, and specific heat capacity. Mechanical properties are assessed by measuring magnetic susceptibility, magnetic moment, and hysteresis.

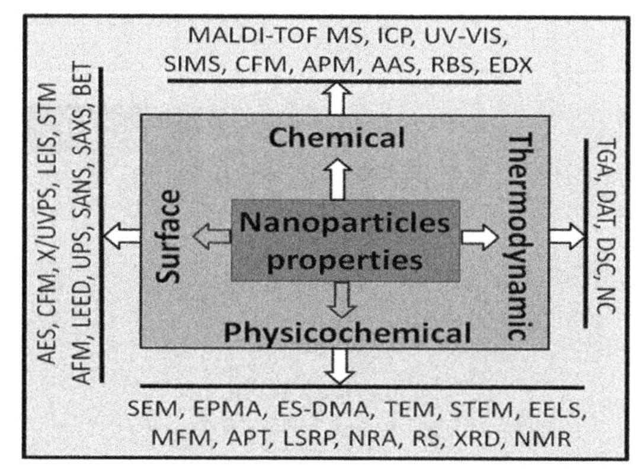

FIGURE 1 A list of methods commonly used for characterization of different properties of nanoparticles. AAS, atomic absorption spectroscopy; AES, auger electron spectroscopy; AFM, atomic force microscopy; APM, atomic probe microscope; APT, atomic probe tomography; BET, Brunauer–Emmett–Teller; CFM, chemical force microscopy; DSC, differential scanning colorimeter; DTA, differential thermal analysis; EDX, energy dispersion X-ray spectroscopy; EELS, electron energy loss spectroscopy; EPMA, electron probe microanalysis; ES-DMA, electrospray differential mobility analysis; FS, fluorescence spectroscopy; ICP, inductivity coupled plasma spectroscopy; LEED, low-energy electron diffraction; LSRP, localized surface plasmon resonance; MFM, magnetic force microscopy; NC, nanocalorimetry; NMR, nuclear magnetic resonance; NRA, nuclear reaction analysis; RBS, Rutherford back scattering; RS, Raman spectroscopy; SANS, small-angle neutron scattering; SAXS, small-angle X-ray scattering; SEM, scanning electron microscopy; STM, scanning tunneling microscope; STEM, scanning transmission electron microscopy; TEM, transmission electron microscopy; TGA, thermal gravimetric analysis; UPS, ultraviolet photoemission; UV/Vis, ultraviolet/visible spectrometry; XRD, X-ray diffraction; X/UV PS, X-ray or UV photoelectron spectroscopy.

The physicochemical, optical, and electronic properties of nanoparticles are dependent on their size and surface properties. A 2-nm particle may have different electronic and surface properties than a 10-nm particle. Surface functionalized nanoparticles may exhibit different properties than the corresponding bare particles. Therefore, their size measurement is essential to ensure reproducible beneficial and/or toxic effects. In this chapter, some of the important methods used for nanoparticle characterization are described.

2. NANOPARTICLE ENUMERATION, SIZE, AND SHAPE

2.1 Particle Counter

2.1.1 Conductivity Counters (Rosse and Loizeau, 2003; Beyer, 1987)

Conductivity (Coulter) counters consist of two chambers. Nanoparticles are suspended in an electrolyte solution and placed in the sample chamber, while an electrolyte alone is placed in a reference chamber. Then, the nanoparticle solution passes through the electrolyte solution. As the nanoparticle solution passes through the electrolyte solution, a change in conductivity occurs and is measured. The difference in conductivity is a function of the particle size.

2.1.2 Condensation Particle Counters

In this method, nanoparticles are drawn in an aerosol sample continuously through a heated saturator. Inside, alcohol or water is vaporized and the vapor diffuses into the sample stream. The aerosol sample saturated with the liquid vapor becomes supersaturated and ready to condense. Nanoparticles serve as a condensation area for the vapor; thus, they grow in size as their optical density is recorded.

The advantages of counting methods include measurement of true volume distribution and high resolution and a wide range of sample measurements. The disadvantages include a need for calibration, use of electrolyte as the medium (limited to hydrophilic particles), and low-particle concentration. In addition, errors may occur with porous particles, particles below the minimum detectable size go unnoticed, and there is difficulty with high-density materials.

2.2 Scanning Electron Microscopy

Scanning electron microscopy (SEM) with a secondary electron detector can visualize crystal shape, surface morphology, dispersed and agglomerated nanoparticles, and surface functionalizations (Lee et al., 1996; Montesinos et al., 1983). SEM can examine each particle, including the aggregate particles, individually; thus, the method is considered to be an absolute measurement of particle size. It can be coupled to image analysis computers for examination of each field for particle distribution. However, the depth of focus is only 0.5 μm at ×1000 and the diffraction effects increase with small particles, which causes blurring at the edges in the determination of particles.

Field emission scanning electron microscopy (FESEM) has narrower probing beams at low and high electron energy, so it provides improved spatial resolution while minimizing sample damage. The in-lens FESEM provides topographical information at magnifications of 250-1,000,000× with ion-free images. FESEM allows detection of small-area contamination spots at electron-accelerating voltages, compatibility with energy-dispersive X-ray spectroscopy, application of low kinetic energy electrons closer to the immediate material surface, and elimination of conducting coatings on insulating materials.

2.3 Laser Diffraction

When a laser beam is passed through nanoparticles, the photons are scattered (particle properties) or diffracted (wave properties), reflected, and/or transmitted. In this method, the diffracted light is detected over a range of angles in the forward direction (Lee and Groves, 1981).

The angles of diffraction are inversely related to the particle size. The advantages of laser diffraction include nonintrusive analysis, use of a low power laser beam, rapid assay time, precise and wide working range, absolute measurement without calibration, and versatility. However, it is expensive, all measurements are based on the volume measurement assuming spherical particles, and there must be a difference in the refractive indices of the particles and suspending medium.

2.4 Polarization Intensity Differential Scattering

Polarization intensity differential scattering (PIDS) is used in line with the conventional laser diffraction sample cell. The polarized light of three different wavelengths is scattered by the sample and recorded at six different angles. PIDS is claimed to be especially sensitive for scattering caused by the smallest particles (Bott and Hart, 1991).

2.5 Field-Flow Fractionation

In field-flow fractionation (FFF), particles of different sizes are separated in a field of solvent flow similar to liquid chromatography, but without a stationary phase. The lower channel wall is made of a frit covered with an ultrafiltration membrane, which is impermeable to the sample (Giddings, 1993). Because of a force field perpendicular to the channel flow, the particles concentrate according to their size in the direction of the lower accumulation wall. The particles closer to the accumulation wall are transported slower than those higher up in the channel, causing smaller particles to elute first. The mean square radius can be calculated from first principles, if the Rayleigh-Debye-Gans approximation holds (Wyatt, 1993). Sizes can be calculated without any concentration information from light scattering alone. However, the size and the refractive index increment can be used to calculate concentration (Wyatt Technology, 1998).

3. CHEMICAL ANALYSIS OF NANOPARTICLES

3.1 Atom Probe Field-Ion Microscopy

Atom probe field-ion microscopy (APFIM; Figure 2(A)) consists of a field-ion microscope,

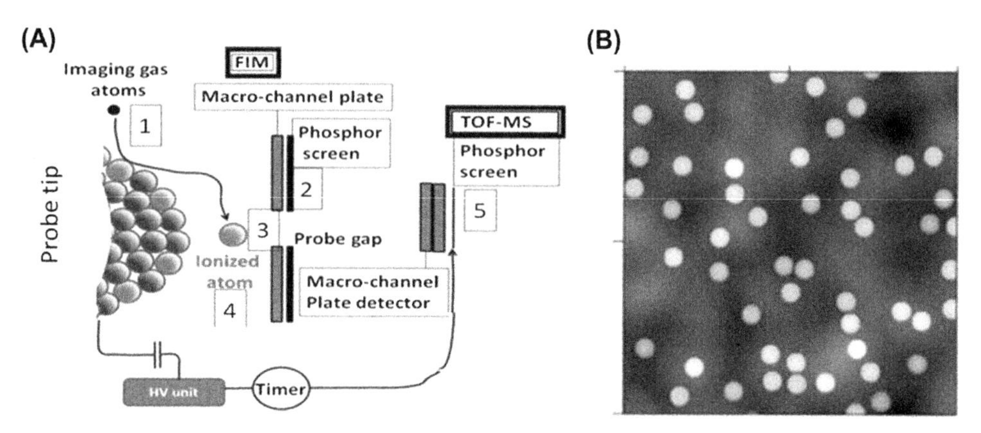

FIGURE 2 **Atomic probe field-ion microscopy (APFIM).** (A) APFIM consists of a combination of a field-ion microscope (FIM) and a detector (either time-of-flight mass spectrometer of a single-ion sensitivity and/or a phosphor screen). FIM contains a specimen tip, an imaging gas such as Ne or He (1), a microchannel plate (2) with a pore-gap (3), the imaging gas ionizes and desorbs the atoms (4) that passes through the pore-gap. The ions are detected through the time-of-flight detector (5). A unique feature of the atom probe compared with the other analytical instrument is its extremely high spatial resolution and the equal detection efficiency for light elements. (B) The phosphor image of individual particles with different configurations. Each and every ionized particle is visible. *Reprinted from Klapetek et al. (2011) with permission.*

a time-of-flight mass spectrometer (TOF-MS), and a power supply (Miller, 2000). When a high electric field is applied on the surface of a probe's tip, imaging gas (such as He) atoms are ionized and fluoresce the phosphor screen (Hono and Murayama, 1997; http://nims.jp/apfim/apfim/apfim.html). Higher electric fields also ionize the surface atoms that enter the time-of-flight (TOF) chamber via an aperture of 0.5–5 nm. The mass/charge ratio is measured by a mass spectrometer. Because atoms always evaporate from the surface, the spatial resolution of the method is in the direction of a mono-atomic layer. The method provides extremely high spatial resolution and element characterization. Figure 2(B) shows the distribution of nanoparticles scanned with no convolution.

3.2 Atomic Absorption Spectroscopy

Atomic absorption spectroscopy (AAS), in both flame and electrothermal modes, is one of the best methods for determination of the metal concentrations in various specimens that dissolved in acid (Kalbasi and Mosaddegh, 2012). Separation, preconcentration, and dissolution of samples are the vital steps in many procedures, especially in the case of low-metal concentrations. AAS is relatively sensitive to Pd and Rh (Scaccia and Goszczynska, 2004) and has been used for the determination of metal levels in nanoparticles (Kalbasi and Mosaddegh, 2012; Budiman et al., 2010).

3.3 Inductively Coupled Plasma Mass Spectrometry

Inductively coupled plasma mass spectrometry (ICP-MS; Figure 3) is a sensitive method for analysis and confirmation of metal ions with a high linear dynamic range (Allabashi et al., 2009). It is capable of analyzing almost all elements in the periodic table and can be applied to solutions, solids, and gases. A typical configuration of an ICP-MS is shown in Figure 3.

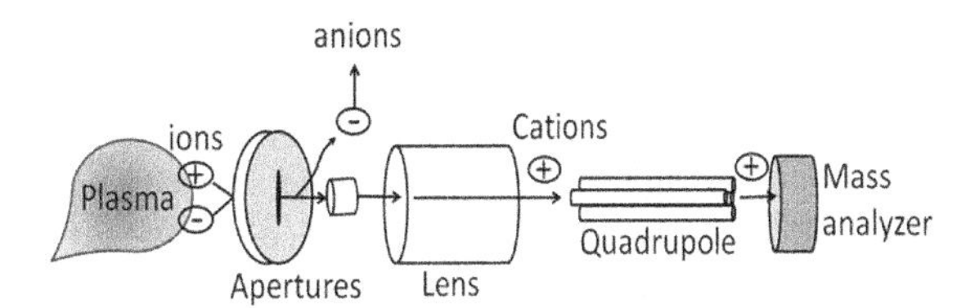

FIGURE 3 Components of inductivity coupled plasma (ICP) mass spectroscopy. Details are described in the text.

Specimens (solution—vaporized using a nebulizer, solids—sampled using laser ablation and gases—sampled directly) are introduced into an argon plasma consisting of electrons and positively charged argon ions. In the plasma, the material splits into individual atoms that lose electrons and become positively charged ions (anions are not detected by ICP-MS). The positive ion beam enters a quadrupole mass analyzer where the ions are separated according to their mass/charge (m/z) ratio. After passing the quadrupole, the ions hit a special mass detector that detects ions at the higher and lower range. ICP-MS, because of its excellent elemental sensitivity and specificity, is an ideal detection technique for the characterization of metal-based engineered nanoparticles. It allows simultaneous quantitative and confirmatory analysis of almost 100 metals in various matrices. In the single-particle analysis mode, ICP-MS may decipher more information (dissolved nanoparticles, presence of impurities, decay with time, etc.) in a relatively short time.

Selected references for detailed information on the application of ICP-MS in characterizing metal nanoparticles include the following: Telgmann et al. (2014), Tadjiki et al. (2013), Crayton (2012), Daniela et al. (2012), Mitrano et al. (2012), Gschwind et al. (2011), Helfrich and Bettmer (2011), Jimenez et al. (2010), and Allabashi et al. (2009).

3.4 Matrix-Assisted Laser Desorption Ionization TOF-MS

Matrix-assisted laser desorption ionization (MALDI) TOF-MS (Figure 4) is a powerful and highly sensitive soft-ionization (using laser beam) technique that is suitable for complex molecules, such as functionalized nanoparticles and proteins (Kim et al., 2014a,b). The laser beam usually desorbs and ionizes the particles that remain intact. Some equipment (MALDI-TOF/TOF-MS) may consist of two TOF chambers: TOF-1 for flight of desorbed ions and TOF-2 for ionization of each ion (1–5) and then flight of the secondary ions (i–iv). The TOF-2 allows characterization of each desorbed ion. Sample analysis is achieved in three stages (Figure 5):

The MALDI stage: Each specimen is mixed with an appropriate matrix and spotted onto a MALDI disc (Harrison, 1997; Sunner, 1993; Ehring et al., 1992). The matrix absorbs the ultraviolet nitrogen laser light (337 nm) and converts it to heat energy, resulting in desorption of the ionization of the particle. Ionization occurs by either addition ($[M+H]^+$, $[M+Na]^+$, or $[M+2H]^{2+}$, etc.) or removal ($[M-H]^-$) of protons. The three most commonly used matrices are 3,5-dimethoxy-4-hydroxycinnamic acid (sinapinic acid), α-cyano-4-hydroxycinnamic acid (CHCAα), and 2,5-dihydroxybenzoic acid (DHB). A solution of the selected molecules is made in a mixture of highly purified water, acetonitrile, and trifluoroacetic acid (50:50:0.1).

The time-of-flight stage: The ionized particles enter a TOF chamber where they are accelerated. The rate of acceleration decreases as the size (in terms of the m/z ratio) increases; thus, smallest particles reach the detector first, followed by the next in size. In TOF/TOF equipment, the particles leave the TOF-1 and enter TOF-2, where they are further fragmented into smaller-sized ions that travel to the detector according to their size (Spengler et al., 1991, 1992).

Detection of the ions: In TOF, the potential difference remains constant; thus, ions with smaller m/z value (lighter ions) move faster through the TOF space until they reach the detector. Consequently, the time of ion flight differs according to the m/z ratio of the ion. A limited list of references describing applications of MALDI-TOF-MS in different areas of nanoparticle characterization includes the following: Kim et al. (2014a,b),

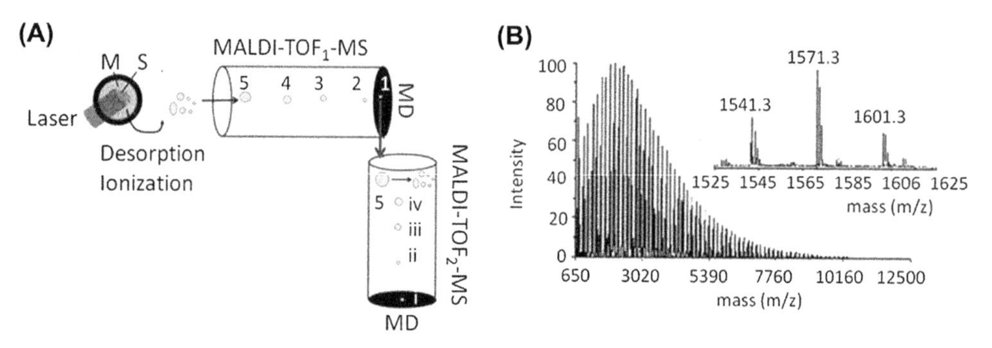

FIGURE 4 **Components of matrix-assisted laser desorption (MALDI) time-of-flight (TOF) mass spectrometry (MS).** (A) Samples are mixed with a matrix and spotted onto a MALDI disc that is then placed into the MALDI chamber. A laser beam desorbs and ionizes the specimen that enters into the TOF chamber. Ions get separated according to their m/z ratio and detected by a mass analyzer. In TOF/TOF equipment, individual ions, after detection, enter into another TOF, in which it is fragmented and the secondary ions are analyzed. M, matrix; S, sample; 1–5, primary ions in TOF_1; i–iv, secondary ions from TOF_2. (B) Mass spectrum of polymeric nanoparticles. Each peak represents the individual ions. *Reprinted from Bootza et al. (2005) with permission.*

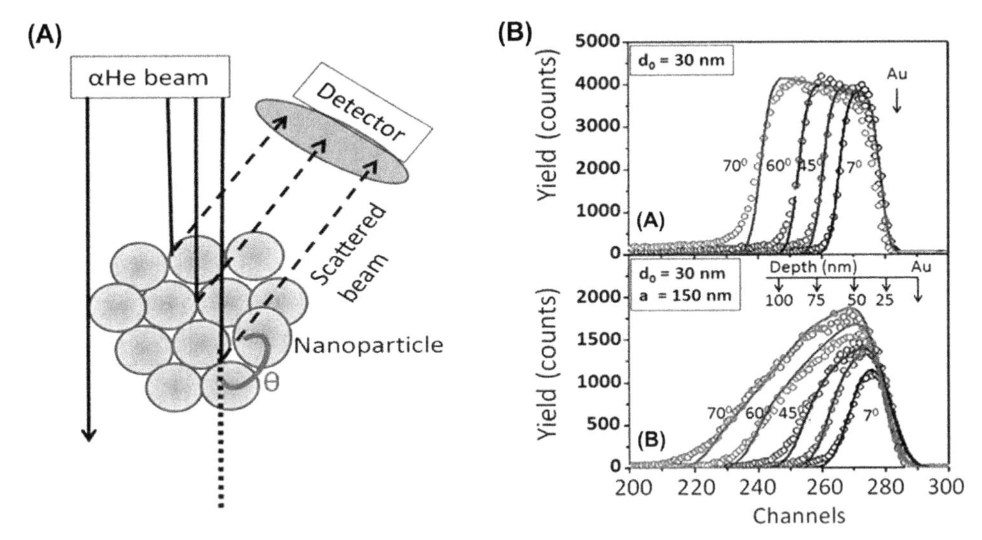

FIGURE 5 **Rutherford backscattering spectroscopy (RBS).** (A) Principle of RBS. Nanoparticles are bombarded with high-energy αHe beams that are either reflected by the surface atom or penetrate into inner layers and then are reflected by the inner atoms. The reflected atoms are screened and the reflection angle is measured. The procedure provides atomic configuration at different depths. (B) 1 MeV He^{+} RBS spectra of (a) a planar Au layer with thickness d_0 and (b) a tetragonal array of Au nano-bricks with height d_0 and width "a," for different sample tilt angles (Zolnai, 2013). The solid lines are the simulation with best RBS fit of the parameters d_0 and "a," respectively. The surface edge and depth scale for Au is indicated. At a 7° tilt angle, both samples exhibited inverted parabolic curves. The RBC spectral characteristics are highly dependent on the nanoparticle shape. *Reprinted from Zolnai (2013) with permission.*

Ramalinga et al. (2011), Guan et al. (2007), Alves et al. (2003), Bauer et al. (2003), Khitrov and Strouse (2003), and Dopke et al. (1998).

3.5 Rutherford Backscattering Spectrometry

Rutherford backscattering spectrometry (RBS) allows quantitative compositional analysis without the use of reference standards. For analysis, high-energy He^{2+}ions are directed onto the sample and the yield of the backscattered He^{2+} ions at a given angle is measured (Figure 5). Because the backscattering cross section for each element is known, it is possible to obtain a quantitative compositional depth profile from the RBS spectrum (Zolnai, 2013; Huttel et al., 2004; Stupik et al., 1992; Czanderna and Hercules, 1991; Feldman and Mayer, 1986; http://cnx.org/content/m43546/latest/?collection=col10699/latest). RBS has the following advantages: quantitative and nondestructive surface analysis, no need for reference standards, measures the composition and thickness of nanoparticles, measures mass and depth of the target sample, can determine the ratio of one element to another, and can determine the lattice location of impurities in single crystals.

3.6 Secondary-Ion Mass Spectrometry

Recently, sufficient interest has developed in devising multilayered intelligent devices whose function depends upon accurate interaction among the layers. To achieve this, a methodology that characterizes the layers of surfaces is needed. A mass-spectral technique, called the TOF secondary-ion mass spectrometry (SIMS), provides excellent surface sensitivity on the molecular level combined with the possibility of depth-profiling and imaging (Kim et al., 2014a,b; Muramoto et al., 2011; Benninghoven, 1994). In SIMS, nanoparticles are bombarded with primary ions of certain energy, depending upon the surface depth to be probed (Figure 6). This results in the release of secondary ions from the source. The secondary ions are separated in a TOF chamber for detection.

By incrementally increasing the ion energy, multiple layers can be probed and characterized. SIMS will ionize the surface nanoparticles as well as the functionalized surface groups, thus providing comprehensive information regarding the surface and the inner layers.

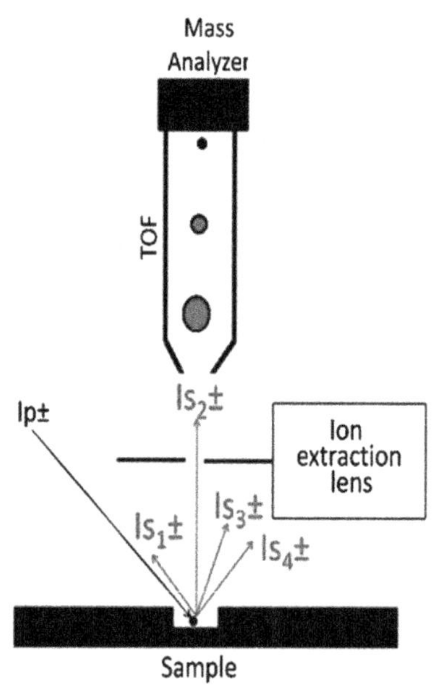

FIGURE 6 **Principle of secondary-ion (SI) mass spectrometry (SIMS).** The SI spectrometer consists of an ion source, a sample chamber, an ion extraction lens, and a TOF-MS. A high energy mixture of positive and negative primary ions (Ip±) strikes and fragments a sample into secondary ions (Is±). The ions go through an ion extraction lens that separates them according to their m/z before they enter the TOF-MS. Multilayer information can be obtained by increasing the ion energy so that the ions can penetrate the layers to a determined depth.

4. SURFACE ANALYSIS

4.1 Auger Electron Spectroscopy

Auger electron spectroscopy (AES) uses a focused electron beam to create Auger electrons near the surface of a solid sample (Ong and Lucas, 1998; Shimizu, 1983; Gallon and Matthew, 1972). The principle of the Auger electron is shown in Figure 7. When a high-energy electron beam collides with the 1s electron of the K orbital, it transfers some of its energy to the 1s electron and then exits the atom with lower energy (Figure 7(A)). The 1s electron is ejected from the K orbital if the energy absorbed is greater than the 1s electron's binding energy, resulting in the formation of an "electron hole" (discussed in Chapter 3) (Figure 7(B)). An electron from the L orbital (2s) moves to the 1s orbital and binds the 1s "electron hole" (Figure 7(C)), resulting in the release of energy equal to the energy difference between the two orbitals. If the released energy is greater than the L orbital's energy, another electron, called the Auger electron, will be ejected, resulting in the formation of two "electron holes" in L orbitals (Figure 7(D)). The Auger electron, characteristic of an element, is detected.

AES can provide elemental maps (except H and He) and depth profile (Figure 8). It is usually nondestructive, except when depth profiling requires surface scraping (http://www.rockymountainlabs.com/auger-electron-spectroscopy-aes.htm).

Some of the studies that have used AES in characterizing nanoparticles are Baer et al. (2012, 2013), Epicier et al. (2010), Bera et al. (2012), Martinez et al. (2011), and Vdovenkova et al. (2000). As shown in Figure 8, AES provides information regarding a particle by characterizing Auger electrons in atoms at different depths.

4.2 Atomic Force Microscopy

Atomic force microscopy (AFM; Binnig, 1987) is a type of scanning probe microscopy in which

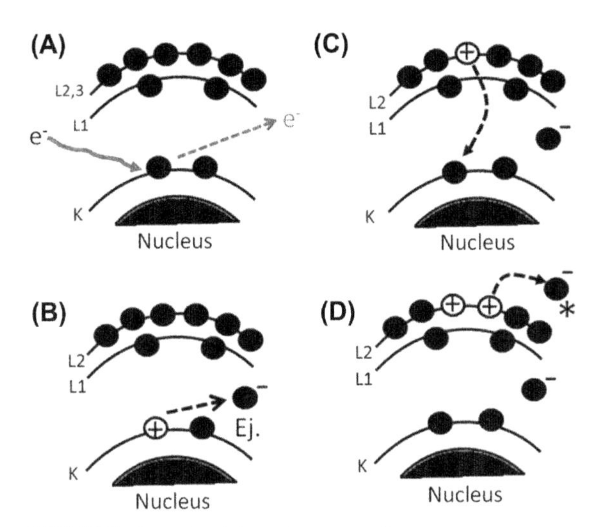

FIGURE 7 **Principle of Auger electron spectroscopy (AES).** In AES, an atom is bombarded with a high-energy electron having frequency matching the vibration of the inner K orbit (A). This will not affect other orbitals because of wavelength blocking (electrons of each orbital absorb photons of a certain wavelength characteristic of the orbital). An interaction between the incident electron (e^-) and the K-electron results in the transfer of energy from the incident electron to the K-electron (e^-); thus, the energy of the K-electron increases and the energy of the incident electron decreases. The low-energy incident electron then leaves the atom. If the energy transfer is sufficient to split the K-electron from its orbital and "electron hole," then the electron jumps into the conduction band, leaving the positive "hole" empty (B). The positive "hole" attracts an electron from a higher orbital, resulting in a new "hole" in another orbit (C). As an electron moves from a higher to lower orbital, it releases energy that ejects another electron into the conduction band (D). The second electron is called the Auger electron, which interacts with the surrounding atoms before exiting. The actual energy of the Auger electron is inversely related to the depth through which they interact. The concentration of electrons adsorbed by the surface and near-surface elements (C_{ads}) is measured as described by Vdovenkova et al. (2000) and Ong and Lucas (1998).

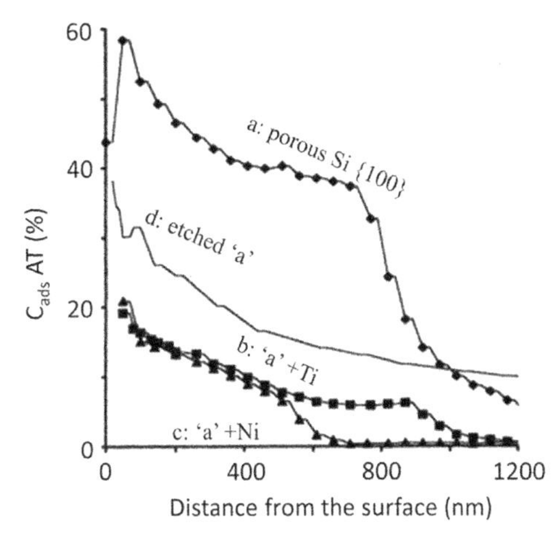

FIGURE 8 **Application of AES.** C_{ads} values for surface and near-surface (carbon + oxygen) elements were plotted against their distance from the surface. The different types of materials used were (a) porous silica {100} nanoparticle; (b) a + Ti (size of b > a); (c) a + Ni (size of c > b); and (d) etched "a" that reduces C and O atoms. The surface and near-surface layers contain about 40% of total elements. Thus, C_{ads} is highest at the surface and decreases gradually in a region near the surface; thereafter, it falls dramatically. Etching reduces the upper layers, thus decreasing the C and O atoms and causing an ensuing decrease in C_{abs} (d plot). Deposition of Ni or Ti decreased C_{abs} possibly due to an increase in the particle size (c > b). The AES can be used to determine a particle's depth as well as its chemical composition. *Reprinted from Baer et al. (2012, 2013) with permission.*

a probe systematically rides across the surface of a sample in a raster pattern (Figure 9). The vertical position is recorded as a spring attached to the probe that rises and falls in response to peaks and valleys on the surface. These deflections produce a topographic map of the sample. AFM is capable of almost nanometer resolution along the vertical axis. The relationship between interaction forces F and the tip vertical displacement ΔZ is as follows: $F = -K\Delta Z$, where K denotes the spring constant of the cantilever (Delvallee et al., 2013). AFM is a powerful tool to measure the nanoparticle surface with a great degree of accuracy (Figure 10).

A problem associated with AFM is that the tip size and geometry lead to a broadening of the lateral dimensions of the nanoparticles imaged, resulting in possible overestimation of their size. Guidelines for sample preparation, measurement, and analysis of results related to the use of AFM are provided by the Materials Measurement Science Division of the National

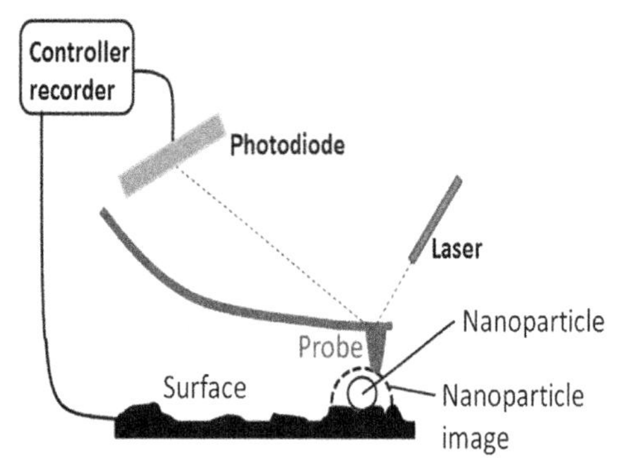

FIGURE 9 **Principle of atomic force microscopy (AFM).** Each surface, even though it appears perfectly flat, consists of peaks and valleys when viewed under an electron microscope. AFM consists of a probe that, while traveling near to a surface, experiences repulsive forces according to Hooke's law. The repulsion between the surface and the tip of the probe deflects the probe, which changes the reflection of a laser beam shining on the top surface of the probe onto an array of photodiodes. The variation of the laser beam is a measure of the applied forces. AFM software generates a three-dimensional map of the surface.

Institute of Standards and Technology (ASTM E2859). The method has been used to characterize different forms of nanoparticles (Klapetek et al., 2003, 2011; Gelinas and Vidal, 2010; Hoo et al., 2008; Fritzen-Garcia et al., 2009; Tello et al., 2003; Lee et al., 1998).

4.3 Brunauer–Emmett–Teller Surface Area Determination

The Brunauer–Emmett–Teller (BET) method (Brunauer et al., 1938) determines the specific surface area of a powder by physical adsorption of a gas on the solid's surface. Physical adsorption results from relatively weak forces (van der Waals forces) between the adsorbate gas molecules and the adsorbent surface area of the test powder. The determination is usually carried out at the temperature of liquid nitrogen. The amount of gas adsorbed can be measured

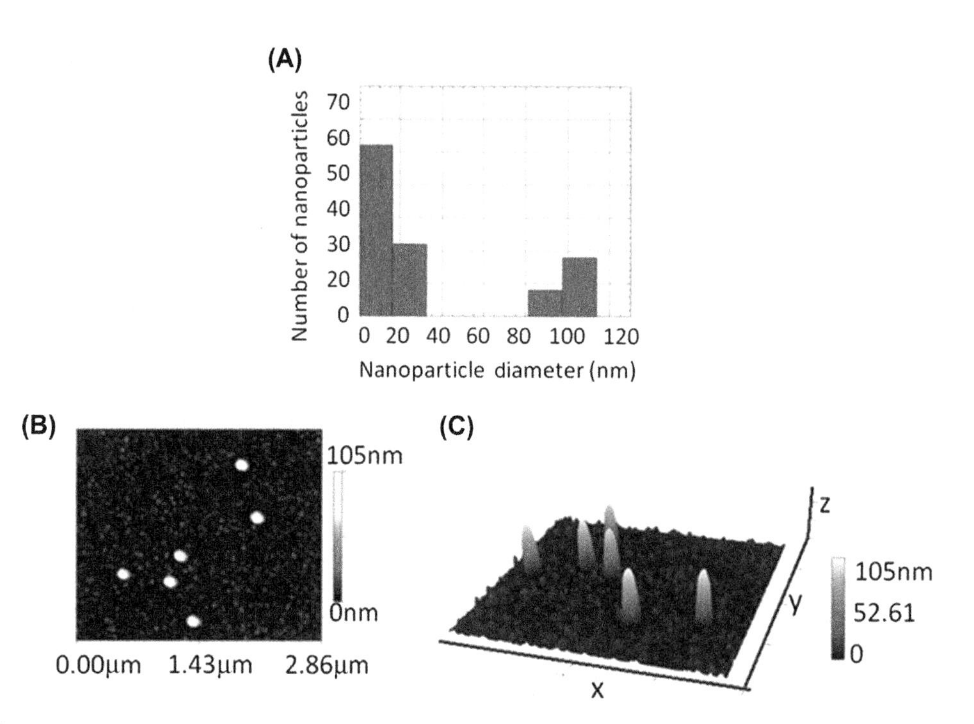

FIGURE 10 **Application of AFM in determining the surface properties of 20- and 100-nm polystyrene nanoparticles.** (A) The size distribution of nanoparticles. The 20- and 100-nm particles were used in this study. (B) AFM view of a mixture of 20- and 100-nm nanoparticles. (C) A 3D reconstruction of the AFM data; the 20-nm particles are shown in the background. *Reprinted from Hoo et al. (2008) with permission.*

by a volumetric or continuous flow procedure according to Eqn (1):

$$\frac{P}{V(P_0 - P)} = \frac{1}{V_m c} + \frac{(c-1)P}{V_m c P_0} \quad (1)$$

In this equation, P is the partial vapor pressure of adsorbate gas in equilibrium with the surface at 77.4 K (b.p. of liquid nitrogen), P_0 is the saturated pressure of adsorbate gas, V_a is the volume of gas adsorbed, V_m is the volume of gas adsorbed to produce an apparent monolayer on the sample surface, and C is a dimensionless constant related to the enthalpy of adsorption of the adsorbate gas on the powder sample. The $P/V(P_0 - P)$ versus P/P_0 plots give a straight line of slope $(c-1)/V_m C$ and an intercept of $1/V_m C$. Therefore, $V_m = (1/\text{slope} + \text{intercept})$. V_m represents the volume of gas adsorbed on the surface. Using V_m, the surface area can be calculated using the following equation:

$$A = \frac{PV_m}{RT} \cdot N_0 \sigma = \frac{N_0 \sigma V_m}{22400} \quad (2)$$

Here, N_0 is Avogadro's number and σ is the area of the molecule. The method is described in detail elsewhere (http://www.freestd.us/soft/258183.htm).

4.4 Chemical Force Microscopy

Chemical force microscopy (CFM) is a modification of AFM in which the tip is chemically modified (specific chemical groups are attached to the tip so that they can interact with specific chemical groups of the nanoparticles) to make them sensitive to specific chemical interactions (Figure 11; Frisbie et al., 1994; Noy et al., 1997). By using chemically functionalized tips, CFM can quantify forces between different molecular groups, probe surface free energies on a nanometer scale, determine pKa values of the surface acid/base groups locally, and map the spatial distribution of specific functional groups and their ionization state.

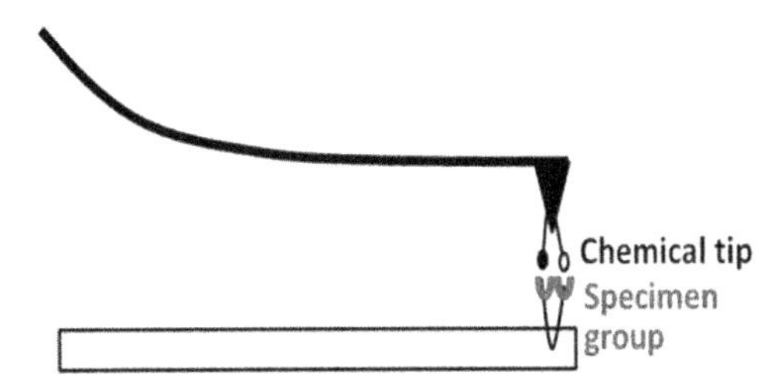

FIGURE 11 Chemical force microscopy (CFM). CFM is similar to the AFM, except that the tip is coupled to a reactive chemical group that repels or attracts the chemical moieties of the surface. This allows formation of the chemical profile of a surface.

4.5 Low-Energy Electron Diffraction

Low-energy electron diffraction (LEED) is an excellent method to study surface reconstruction (defined as the atoms near the surface that may be disturbed from their equilibrium positions in the bulk). The surface reconstruction changes the relative positions of the surface- and near-surface atoms. LEED involves bombardment of the surface with low-energy electrons (approx. 10–200 eV, the electron energy is gradually increased or decreased) and observation of diffracted electrons as spots on a *retarding-field analyzer* to detect the electrons. The relative position of the spots on the screen shows the surface crystallographic structure. The diffracted spots will move as the energy of the incident electrons changes; the intensity of the spots as a function of incident electron energy characterizes surface reconstructions (Figure 12). LEED has been used for surface characterization of nanoparticles (Gulde et al., 2012; Kamimura and Dobash, 2012; Gavaza et al., 2007; de la Figuera et al., 2006; Bauer, 1994; van Hove and Tong, 1979).

The retarding field analyzer consists of a series of concentric hemispherical grids. The first grid is usually grounded and the remaining grids have voltage applied to them. When a sample is struck by a beam of electrons, some electrons are diffracted by the sample toward the grids. An electron can pass through a grid or

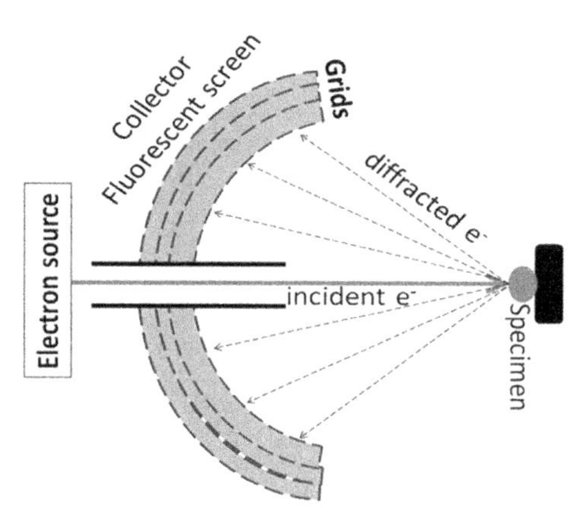

FIGURE 12 Low-energy electron diffraction (LEED). Electron diffraction is attributed to the wave-like nature of a beam of electrons when passing near a matter. A beam of electrons undergo reflection and/or diffraction. When reflected by the surface atoms, the reflected beam has the same energy and wavelength as the incident beam. When diffracted from the inner atoms, the diffracted electrons have lower energy and higher wavelength than the incident beam. A fluorescent-screen electron collector detects the reflected and diffracted electrons. The method can reveal the surface and inner atoms as well as the gases adsorbed onto it.

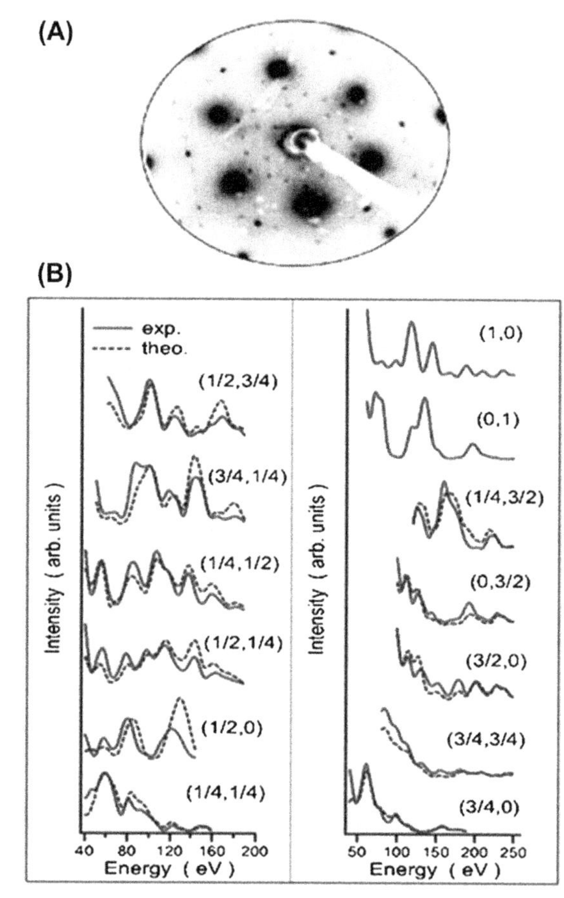

FIGURE 13 Determination of electron energy by LEED. (A) LEED (80 K with the incident electron energy of 115 eV) pattern of silicone on Ag {111}. The small black spots are LEED spots from the 4 × 4 structure and the large spots are the integer-order spots (review article by Braun and Held, 2005). The small white spots are associated with contamination. (B) Comparison between the experimental (red (gray in print versions)) and theoretical (blue (dark gray in print versions)) intensity values was made and plotted against eV. Data interpretation required extensive knowledge of the mathematical principles of LEED. *Reprinted from Kawahara et al. (2014) with permission.*

be reflected back, depending upon its energy. The detector is usually a phosphor screen; thus, light will be emitted wherever the electrons strike the screen (Figure 13).

4.6 Low-Energy Ion-Scattering Spectroscopy

In ion scattering (IS) spectroscopy, a beam of low energy (0.1–10 keV) ions is scattered from atoms at the surface of the nanoparticles (Figure 14(A)). When the incident ion (M_1, $KE = E_0$) strikes an atom (shown in red (gray in print versions)), some kinetic energy (hν) is lost from the scattered ion (M_1, $KE = E_s$), and the quantity of lost energy depends on the relative masses of the atom and ion. Low-energy IS spectroscopy (LEIS) characterizes the atomic layer resolution of the surface. The LEIS spectrum typically contains one peak for each element (separate peaks for the isotopes) in the sample. Their separation in energy depends on the relative atomic masses of the incident and scattering atoms. The method is extremely surface-sensitive. The E_s/E_0 ratio of the scattered

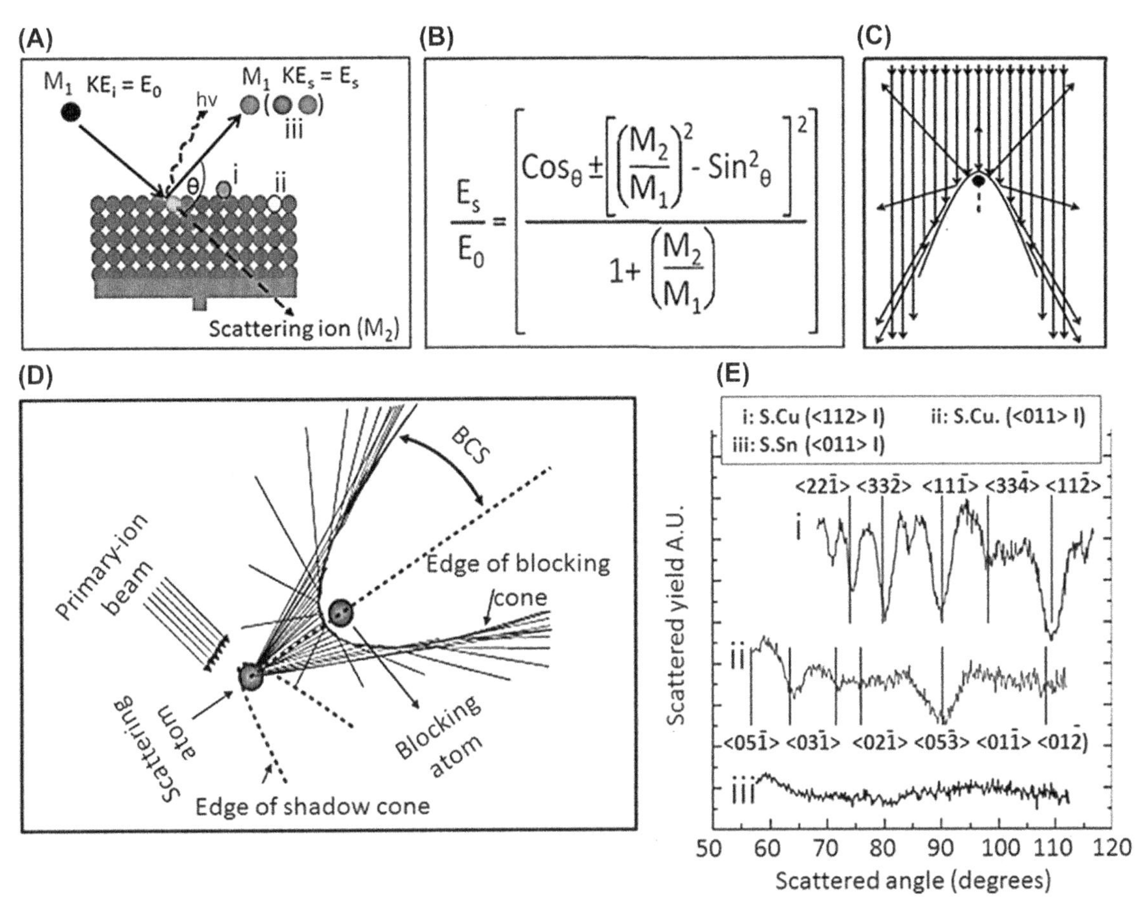

FIGURE 14 **Principle and application of low-energy ion scattering (LEIS).** (A) When a positively charged ion strikes a particle, it is scattered by the nucleus of the surface atom. The scattered ion has lower energy than the incident ion. The released energy is either transferred to another atom or released as a photon (hν). In LEIS, the surface atom are not ejected from the particle. (B) The ratio of the scattered particle energy (E_s) to the incident particle energy (E_0) is related to (i) mass of the incident and the scattering atoms and (ii) the critical angle. (C) Mechanism of shadowing. When a flux of positively charged ion beams approaches an atom, the nucleus repulses and bends the incident beams that interact with the nonscattered beams. Thus, an inverse parabolic area (the shadowing area) without ions develops. (D) Blocking: If the ions penetrate and strike the inner-layer's electron, the scattered ions approach the shadowing area of the surface atoms. The secondary scattered ions will not penetrate the shadowing area. The surface atom, therefore, blocks the ions scattered by the inner atom. (E) LEIS scan of Cu(i), Cu(ii) and Sn-deposited Cu(iii). Cu{112} exhibited a signal block at regular intervals, indicating deeper penetration of the incidence ions. Cu{011} exhibited poor blocking, suggesting poor penetration of the incident ions. Cu-Sn{011} completely lacked blocking, suggesting that Sn deposition occurred onto the crystal surface. *Reprinted from Walker et al. (2005) with permission.*

and incident ion energies and the angle through which the ions are scattered (θ) are used to characterize the ions (Figure 14(B)). For details, please review the following website: http://xpssimplified.com/ion_scattering_spectroscopy.php and the publications: Baer et al. (2012, 2013), Bonatto et al. (2012), and Shoji et al. (1989). Figure 14(C) shows that shadowing and blocking are key aspects of IS. A beam of primary positively charged ions, upon approaching the nucleus, are repulsed and create an inverse parabolic area, known as shadowing area, where no ions will be found.

If another atom, known as the blocking atom, is in the path of scattered beams, it will create another inverse parabolic blocking area that will reflect the secondary ions (Figure 14(D)). It is important to remember that (1) the surface signals do not show any blocking because of the absence of a blocking atom, (2) the surface atoms may block signals from inner atoms and (3) blocking will occur when the blocking atom is in the path of the secondary ions. Figure 14(E) shows the blocking curves for the surface of Cu (112 incidence) (top, (i)), Cu (011, incidence) (middle, (ii)), and Sn (011 incidence) bottom, (iii). The Cu (112) signal exhibits periodical blockings (sharp dip in scattering), Cu (011) exhibited minor blocking, while Sn (011) exhibited no blocking. Walker et al. (2005) put these and other data together and proposed the crystal structure for Cu (100).

4.7 Small-Angle X-ray Scattering and Small-Angle Neutron Scattering

The principles of small-angle X-ray scattering (SAXS) and small-angle neutron scattering (SANS) are that a collision between an incoming wave (X-ray for SAXS and neutrons for SANS) and a surface particle results in scattering of the waves in all directions. As the reflected waves interfere with one another, there will be constructive interference along certain angles, causing peaks to appear. This process is described by Bragg's law for simple scattering: $n\lambda = 2d \sin(\theta)$, in which "n" is an integer, λ is the wavelength, d is the vector representing the displacement between reflection sites, and θ is the angle between the reflected ray and the plane formed by the material's surface (Kirschbrown, 2007; Glatter and Kratky, 1982; Lifshin, 1999; Guinier et al., 1995). X-ray interacts with the surface particle's electron clouds; thus, inhomogeneity in the electron density will affect the scattering pattern. In addition, scattering falls off with the incident beam angle. Therefore, at small angles ($<10^\circ$), the inhomogeneity in the electron clouds can be observed, which will provide information about the size and shape of macromolecules in the sample (Kirschbrown, 2007). Neutrons interact with the nucleus; therefore, their scattering is independent of the incident angle. However, neutron scattering is different for each isotope. The composition of a low-angle scattering apparatus is shown in Figure 15(A), while the results of an experimental study are shown in Figure 15(B) (left: SAXS, right: SANS).

In general, wide-angle X-ray scattering (WAXS) is used to determine a crystal structure on the atomic length scale, while SAXS or SANS provide fast measurement of nanoparticle size distributions (Moglianetti et al., 2014; Agbabiaka et al., 2012, 2013; Disch et al., 2012; Prabhu and Reipa, 2012; Zheng et al., 2012; de la Venta et al., 2009; Masona and Lin, 2007; Pozzo and Walker, 2007; Nagao et al., 2004; Chu and Liu, 2000).

4.8 Ultraviolet-Visible Light Absorption Spectroscopy

As discussed in Chapter 3, electrons, depending upon their orbital energy, exhibit natural vibrational frequency. When electrons encounter a light wave of a frequency matching their vibrational frequencies, those electrons will absorb the energy of the light wave and acquire a vibrational motion. The vibrating electron interacts

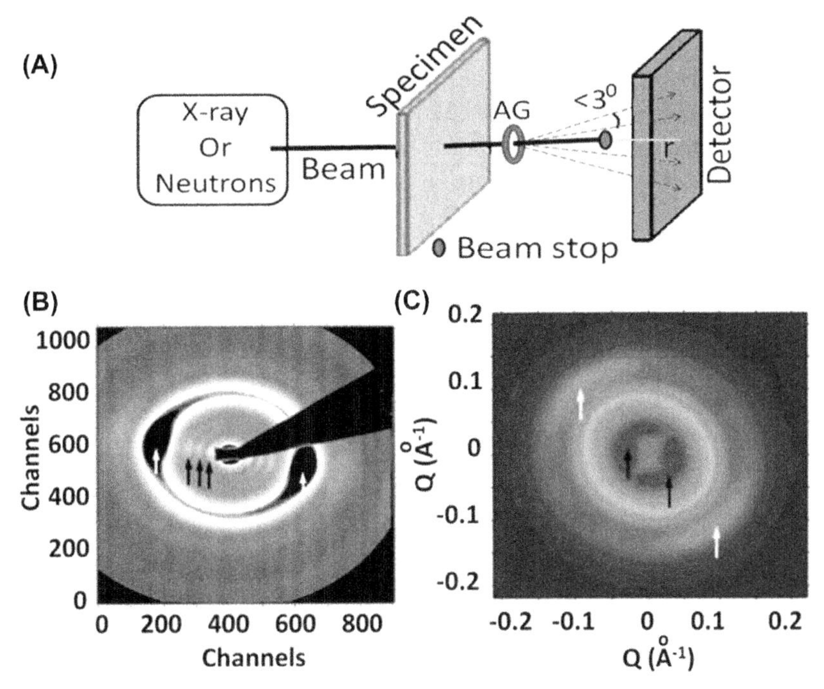

FIGURE 15 **Principle and application of small-angle X-ray scattering (SAXS) or neutron scattering (SANS).** (A) X-ray photon rays or neutron rays, upon encountering a specimen, penetrate them and then get scattered. In SAXS or SANS, photons or neutrons exhibiting less than 3° scattering are detected. (B) Two dimensional-detector images of a SAXS (left side) and a SANS (right side) patterns of tripalmitin nanodispersions. The first interference peaks due to the particle stacking are marked by black arrows. The reflections are marked by white arrows. In the X-ray image, the broad shadow of the beam stop for the primary beam is visible. *Reprinted from Bunjes and Unruh (2007) with permission.*

with the neighboring electrons and converts the vibration into heat energy. Absorption spectroscopy, therefore, refers to the measurement of the light absorption by a material as a function of the wavelength. The visible region of the spectrum is in the wavelength range of 380 nm (violet) to 740 nm (red), and the near-ultraviolet (UV) region extends to wavelengths of about 200 nm. The UV-visible spectrophotometer analyzes over the wavelength range of 200–900 nm. Absorption increases as the matter's electrons encounter light of a wavelength matching the electrons' vibrational frequency.

Because the electronic energy states vary from one atom or molecule to another, the radiation absorbed by one substance will usually differ from that absorbed by another substance. For example, an aqueous solution of bulk gold and bulk silver exhibited maximum absorption at 580 and 430 nm, respectively. However, 20-nm gold and silver nanoparticles exhibited maximum absorption at 540 nm (another strong absorbance occurs at 450 nm) and 390 nm, respectively. Gold nanoparticles absorb near the wavelengths of 520 and 450 nm, which appears to be due to the s-p (conduction band) and d-s (interband) transitions of electrons, respectively. As the size of gold nanoparticles increased from 10 to 100 nm, their absorption maxima increased from 400 to >560 nm with broadening of the peak. Because the wavelength of a light wave is inversely related to its energy, an increase in the nanoparticle size absorbs radiation of lower energy. This unique optical property of nanoparticles may be due to the quantum confinement of electrons (Hao et al., 2004). The UV-visible absorption curve, therefore, provides important information regarding the nanoparticles' electronic properties. Although the traditional UV-visible absorption technology (Figure 16) yields excellent reproducible results for bulk particles, it provides variable results for nanoparticles.

Gesquiere et al. (2008) have shown that nanoparticle absorption varied considerably (492–498 nm to 502–515 nm) in different samples prepared at different times, possibly due to differential size distribution or aggregation

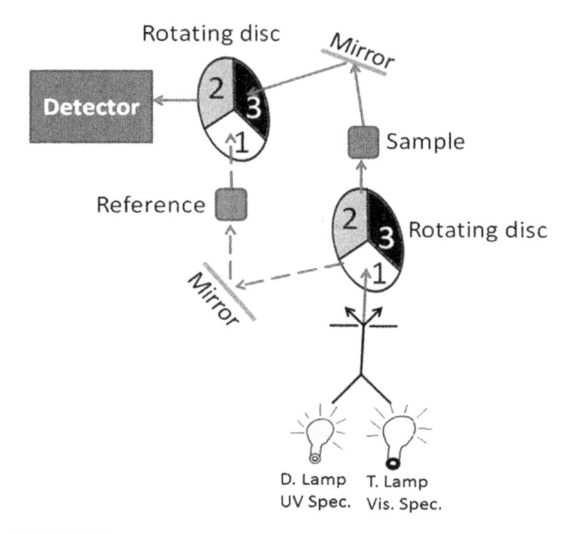

FIGURE 16 UV-visible spectrophotometer. The spectrometer has a light source covering the entire visible and the near ultraviolet region from 200 to 900 nm (a combination of a deuterium lamp for the UV region of the spectrum and tungsten or halogen lamp for the visible region). The light beam is then sent through a diffraction grating to achieve a monochromatic beam that falls onto a rotating disc consisting of three segments: an opaque black section (no light output), a transparent section (light transmitted), and a mirrored section (light reflected). The transmitted light passing through the sample cell is reflected by a mirror, hits the mirrored section of a second rotating disc, and then is collected by the detector. The reflected light passes through the reference cell, hits the transparent section of a second rotating disc, and then is collected by the detector. The absorption or transmittance is measured to characterize the specimen.

of nanoparticles (Traiphol et al., 2006; Schwartz, 2003). This has been attributed to a change in conjugation length, particle conformations that allow for pi-stacking inter-chain contacts (Hu et al., 2000; Huser et al., 2000), and/or interplane interactions (Kim et al., 2006a,b). Tomaszewska et al. (2013) used a preparation of monodispersed metal nanoparticles of average particle sizes 10 ± 5 nm, 55 ± 9 nm, and 80 ± 20 nm; the authors showed that the optical properties of nanoparticles are strongly dependent upon their size: a decrease in size below 100 nm increases light scattering, increases absorption, and red-shifts the absorption spectrum (Freud, 2012; Jain et al., 2006; Link and El-Sayed, 1999; Bohren and Huffmann, 1983; van de Hulst, 1981), Therefore, bulk UV-spectroscopy may not be suitable for quantitatively examining nanoparticles.

The currently used methods, however, only yield an average value associated with bulk UV-visible spectroscopy, which may result in absorption variability if the samples are not homogenous. This can be resolved by using modified procedures called single-particle absorption spectroscopy, fluorescence-voltage single-molecule spectroscopy, and fluorescence-voltage time-resolved single-molecule spectroscopy, which allow measurement of a single particle at a time (Gesquiere et al., 2004a, 2005). By measuring single nanoparticles, the complications presented by the "averaging" can be avoided (Ambrose et al., 1999; Moerner and Orrit, 1999; Xie and Trautman, 1998; Nie and Zare, 1997). With fluorescence techniques, conducting polymer molecules embedded in functioning devices can be studied (Gesquiere et al., 2004b; Lee et al., 2005).

4.9 Scanning Electron Microscopy

SEM is widely used to investigate the microstructure and chemistry of a range of materials. The main components of the SEM include a source of electrons, electromagnetic lenses to focus electrons, electron detectors, sample chambers, computers, and displays to view the images (Figure 17). Electrons, produced at the top of the column, are accelerated downwards where they passed through a combination of lenses and apertures to produce a fine beam of electrons. The electron beam hits the surface of the sample mounted on a movable stage under vacuum. The sample surface is scanned by moving the electron-beam coils. This beam scanning enables information about a defined area of the sample. The interaction of the electron beam with the sample generates a number of signals, which can then be detected by appropriate detectors.

The advantages of SEM include the detailed three-dimensional (3D) topographical imaging and the versatile information obtained from different detectors. The microscope is easy to operate and associated software is user-friendly. The disadvantages of SEM are its size and cost. SEM is expensive to operate. The preparation of samples can result in artifacts. A critical disadvantage is that SEM is limited to solid, inorganic samples small enough to fit inside a vacuum chamber that can handle moderate vacuum pressure.

The following manuscripts describe the application of SEM to decipher the surface properties of nanoparticles: de Siqueira et al. (2014), Polyakov et al. (2014), Brown and Hondow (2013), Ponce et al. (2012), Hatziantoniou et al. (2007), Masotti et al. (2007), and Puchalski et al. (2007).

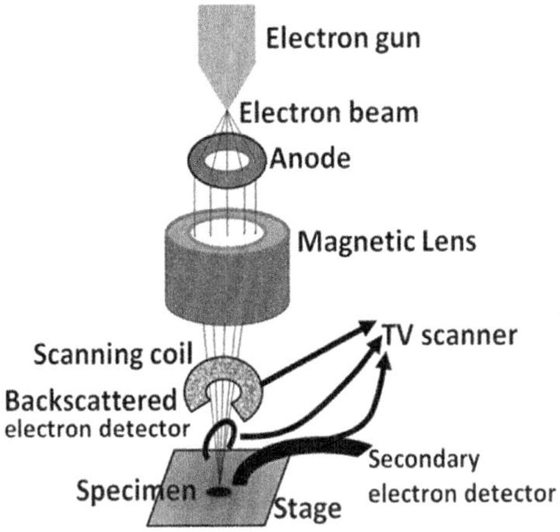

FIGURE 17 **Components of scanning electron microscopy (SEM).** SEM consists of an electron source that fires a beam of electrons at the object under examination. The beam goes through a couple of lenses made of magnets capable of bending the path of electrons. By doing so, the lenses focus and control the electron beam, ensuring that the electrons focus onto the specimen. The sample chamber is sturdy and insulated from vibration. The sample chambers also manipulate the specimen, placing it at different angles and moving it in all directions. SEM contains various types of detectors capable of (1) producing the most detailed images of an object's surface and (2) revealing the composition of a substance. The key point is that SEMs require a vacuum to operate. Without a vacuum, the scattered electron beam would distort the surface of the specimen.

4.10 Energy-Dispersive X-ray Spectroscopy

SEM uses a focused beam of high-energy electrons to generate a variety of signals at the surface of solid specimens, including nanoparticles. The incoming electrons also eject electrons from the specimen, resulting in the formation of a positively charged "electron hole" in the occupied bands. The "holes" attract electrons from higher states. The movement of electrons from a higher state to the "hole" results in the emission of X-rays. The energy characteristic of X-rays is unique to the atoms in nanoparticles. Energy-dispersive X-ray spectroscopy (EDX) detects X-rays emitted from the sample to characterize the elemental composition of the analyzed volume (Epicier et al., 2010; von Bohlen et al., 2010; Patri et al., 2009; Herzing et al., 2008).

5. PHYSIOCHEMICAL PROPERTIES

5.1 Atom Probe Tomography

Atom probe tomography (APT) offers both 3D imaging and chemical composition measurements at the atomic scale (around 0.1–0.3 nm resolution in depth and 0.3–0.5 nm laterally). In this technique, the sample is adsorbed onto a sharp tip that is applied with high DC voltage (5–20 kV), which induces a very high electrostatic field at the tip surface (Devaraj et al., 2014; http://www.cameca.com/instruments-for-research/atom-probe.aspx). Under laser, a few atoms are evaporated from the surface and projected onto a position-sensitive detector (PSD) with high detection efficiency (Figure 18). APT has been extensively used for characterization of nanometer-sized particles (selected references: Chen et al., 2013; Dawahre et al., 2013; De Luca et al., 2013; Diercks et al., 2013; Du et al., 2013a,b).

5.2 Electron Probe Microanalysis

A beam of accelerated electrons is focused on the surface of a specimen, resulting in generation of X-rays characteristic of the specimen. The rays are detected at particular wavelengths and their intensities are measured to determine concentrations. All elements (except H, He, and Li) can be detected because each element emits a specific set of X-rays. This analytical technique has a high spatial resolution and sensitivity, and individual analysis time is reasonably short, requiring only a minute or two in most cases.

As shown in Figure 19, multiple detectors can be assembled on the platform for multiple analyses. Additionally, the electron microprobe can function like SEM and obtain highly magnified secondary- and backscattered-electron images of a sample. The energy dispersive spectroscopy (EDS) system allows quantitative analysis by simply counting the X-rays received in the channels that correspond to a peak of interest. It is also used for analysis of the chemical composition of metals, alloys, ceramics, and glasses. The composition of individual particles or grains and chemical changes on a small scale can be assessed. A key disadvantage of EDS is that corrections must be made for overlapping peaks and the background noise is much higher, which limits its sensitivity (Chung et al., 2008; http://probelab.geo.umn.edu/electron_microprobe.html).

5.3 Electrospray Differential Mobility Analysis

Electrospray differential mobility analysis (ES-DMA) analyzes the effective particle size of aerosolized nanoparticles by measuring their electrical mobility under ambient conditions. The method is based on the principle that the mobility velocity (V) of a charged particle is proportional to the electric field (E) according to the equation $V = ZE$ where Z is the proportionality constant (Guha et al., 2012; Flagan, 2008). Z can be defined as C_cD_m/D_m, where D_m is the mobility equivalent *spherical* diameter of a particle and C_c is the Cunningham slip factor (Kim et al., 2005), which corrects for the nonslip boundary condition. For particles <50 nm C_c is inversely proportional to their diameter, while for larger particles, C_c increases linearly with increasing particle diameter (D_m). Thus, for very small particles (<50 nm), the differential mobility analysis separates particles based on differences in the particle area, whereas for larger particles (>100 nm) separation is based on differences in diameter (Figure 20).

The benefits of ES-DMA include the following:

1. ES-DMA provides a simple means to determine the nanoparticle size, and it has shown excellent agreement with more conventional imaging techniques such as transmission electron microscopy (TEM) and SEM (Grassian, 2008; Wacaser et al., 2007; Pease et al., 2007, 2008). ES-DMA does not have any issues with bubbles in the colloid, shows excellent accuracy (<10%) in the particle size distribution, and resolves peaks less than 1 nm apart (Allmaier et al., 2008; Kinney and Pui, 1991). Dynamic light scattering (DLS) and other indirect methods can have difficulty with nanoparticle size

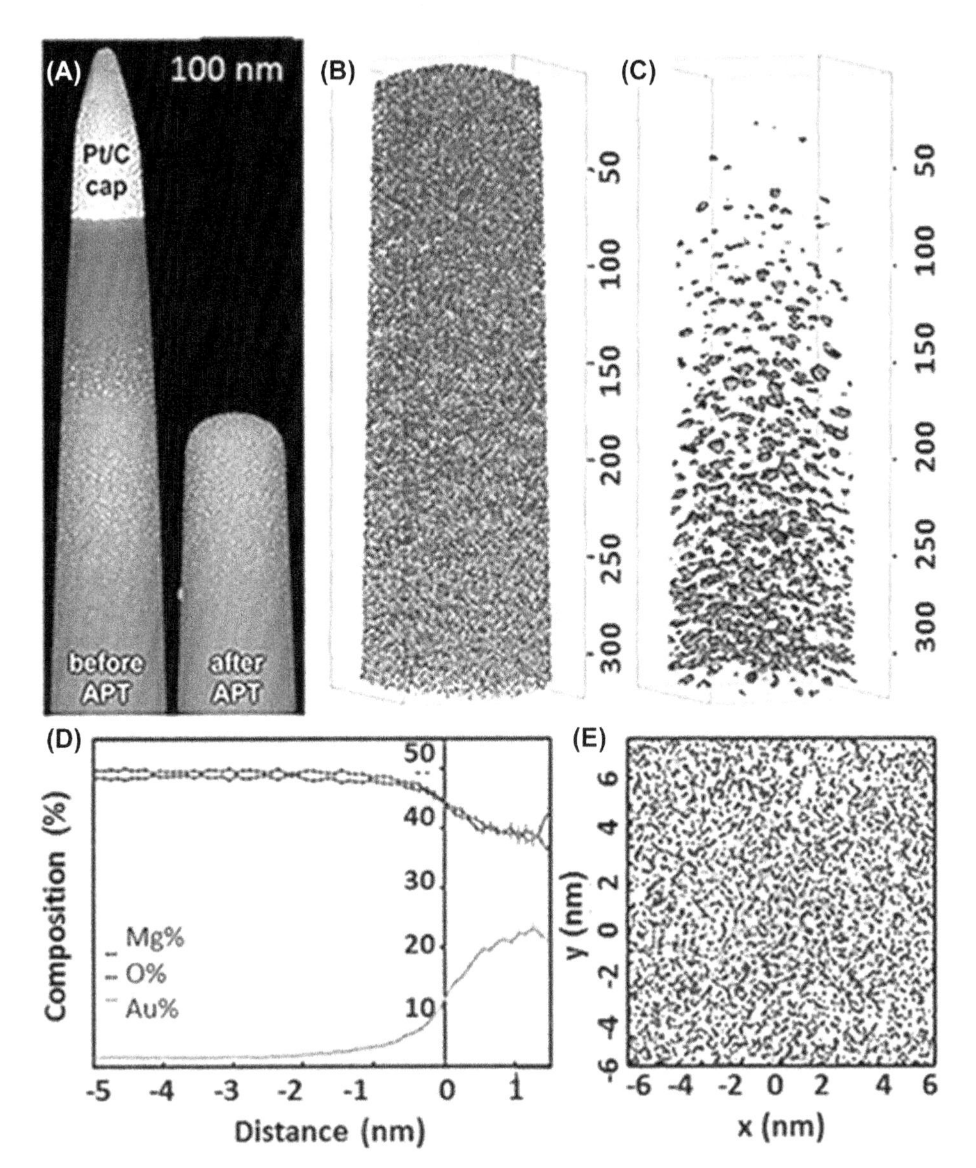

FIGURE 18 **Atomic probe tomography (APT) in conjunction with scanning tunneling electron microscopy (STM) of Au nanoparticles deposited over MgO.** (A) STM images of a sample before and after APT analysis. (B) Ionic view of the APT reconstruction where cyan points correspond to O, magenta to Mg, and yellow to Au. (C) Visualization of Au-rich regions. (D) Proximity histogram plotted across the Au nanoparticles. (E) The distribution of Au (yellow (light gray in the printed versions)) and MgO (blue (gray in the printed versions)) ions. This suggests that APT provides three-dimensional characterization of materials providing spatially resolved mass spectra at subnanometer spatial resolution and part-per-million level mass sensitivities. *Reprinted from Daveraj et al. (2014) with permission.*

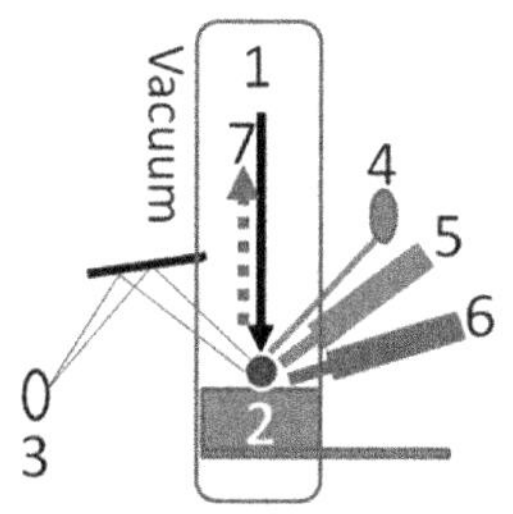

FIGURE 19 Structure of an electron probe microanalyzer.

accuracy (Elzey and Grassian, 2010a,b). The ability to detect small changes to particle size over time has made this method valuable in studying the kinetics of colloidal solutions (Johnson et al., 2008; Pease et al., 2008).

2. ES-DMP, with a single scan, can determine dispersed and aggregated nanoparticle colloids. The proportion of the aggregation state (dimers, trimers, and tetramers) can be identified rapidly from the multimode size distributions (Elzey and Grassian, 2010a,b).
3. ES-DMP measures the diameters of individual nanoparticles or macromolecules; thus, it is capable of measuring (1) subtle irregularities in colloidal solutions and (2) small changes in the nanoparticle size distribution due to mechanisms, such as dissolution or aggregation.
4. ES-DMP exhibits excellent sensitivity and precision for studying both small and large complexes (Kapellios et al., 2011).
5. The procedure requires less than 20 μL of solution containing approximately 2 ng material. DLS and the light scattering methods require a much larger sample.
6. A large number of particles are sized in a single run. This provides a representative size distribution—a task that is difficult to achieve with other techniques, such as TEM or SEM.
7. No size calibration is required for quantitative size analysis. Because ES-DMP does not rely on light scatter to size the particles, the technique is independent of the optical properties of the nanoparticle and solvent.

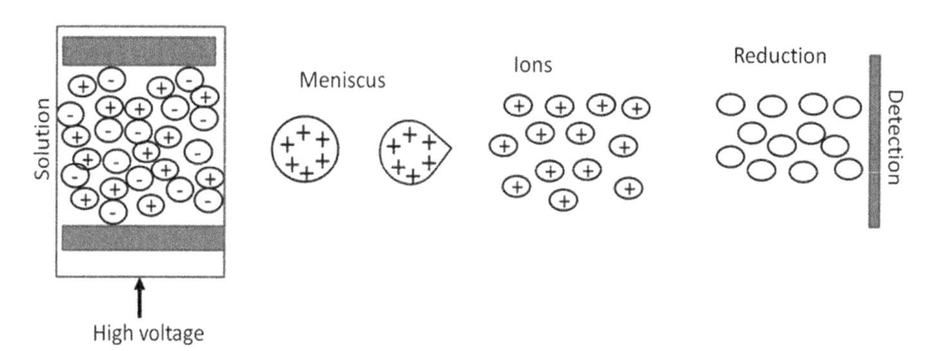

FIGURE 20 **Electrospray differential mobility analysis (ES-DMA).** The analyte dissolved in a volatile buffer solution is injected into a fused silica capillary under pressure and then electrosprayed to produce multiply charged droplets containing the analyte. The droplets are then mixed with air and the nebulized to achieve charged ions. A negative bias is applied to get positive ions (if needed, negative ions can be collected by applying a positive bias). The ions are reduced and then detected as described previously (Guha et al., 2012). Size distribution can be obtained using condensation particle counter or an electrometer.

5.4 Nuclear Magnetic Resonance

Nuclear magnetic resonance (NMR) spectroscopy is a highly versatile methodology used to study the structure of and interaction between molecules, the kinetics or dynamics of molecules, and the composition of biological and/or synthetic composites. The principle of NMR spectroscopy is based on nuclear magnetism. Because the nucleus is a positively charged particle in a magnetic field, it acts as a tiny magnet and tends to align as shown in Figure 21(A). As an example, a proton (^1H), when placed under a magnetic field, exhibits two possible alignments: either with the field direction (low-energy spin, $+1/2$) or opposed to it (high-energy spin, $-1/2$). Energy is required to change the more stable alignment ($+1/2$) to the less stable ($-1/2$) alignment.

In magnetic nanoparticles, magnetization of protons depends upon the electron cloud around it. The negatively charged electrons, in a magnetic field, spin in the direction of the applied field, while the nucleus aligns against the direction of the applied field. As the strength of the negative field around the nucleus increases, a relatively great strength of the magnetic field is required for magnetization of nucleus. The hypothetical compound B (Figure 21(B)) has a less dense electron cloud (relatively unshielded) than compound C (Figure 21(C)), which is highly shielded. Thus, the nuclear spin of compound B flips (from $+1/2$ to $-1/2$ state) at a relatively weak applied field than compound A. The shielding effect generates a structure-specific response in magnetic resonance imaging (MRI).

The typical MRI equipment consists of a sample chamber sandwiched between two magnets and connected to a radio wave (Figure 22). The sample is placed into the chamber and the radio wave is turned on. The magnetic field is increased gradually. At a critical magnetic field, the specimen's nuclear spin flips, resulting in generation of the electrical current that is traced. A spectrum is achieved that is detailed enough to serve as a useful "fingerprint" for a chemical.

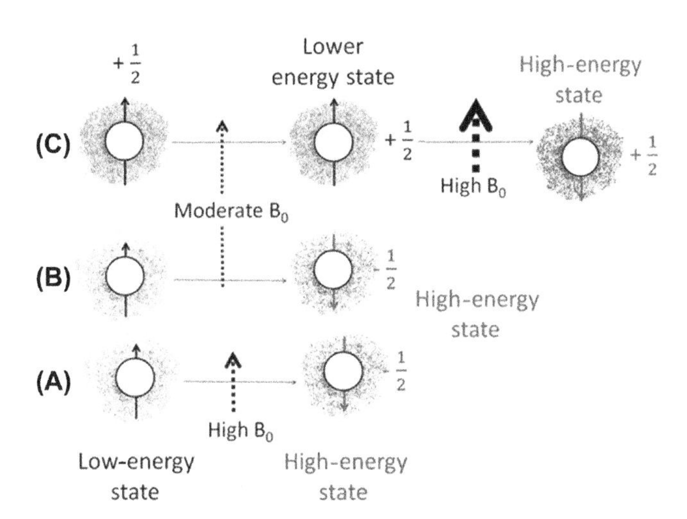

FIGURE 21 **The principle of NMR spectroscopy.** In the presence of a magnetic field, the nucleus acts as a tiny magnets and tends to be aligned as shown in (A). When a proton (^1H) is placed under a magnetic field, it exhibits two possible alignments: low-energy spin ($+1/2$) or high-energy spin ($-1/2$). The electron cloud also responds to the magnetic field and aligns against the field and protects the nucleus from the effects of the external field (B). Therefore, the density of electron cloud around the nucleus is inversely related to a nucleus's magnetization in a magnetic field. A relatively high field is needed to invert nuclear spin from $+1/2$ to $-1/2$ (C).

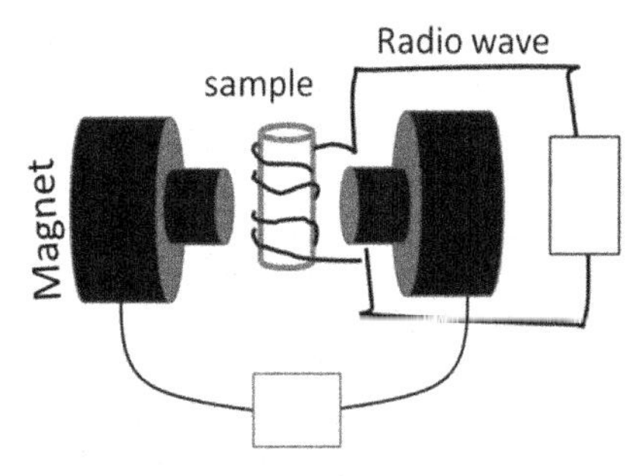

FIGURE 22 **Components of a magnetic resonance imaging (MRI).** Details are described in the text.

For chemicals that lack hydrogen atoms, the photon MRI will not generate a signal. A ^{13}C NMR can be useful for such chemicals. NMR has been extensively used to characterize nanoparticles (Cheng et al., 2012; Shao et al., 2010, 2012; Issa et al., 2011; Mayer, 2005).

5.5 Nuclear Reaction Analysis

Nuclear reaction analysis (NRA) provides a sensitive means to measure low-Z elements (e.g., carbon, nitrogen, oxygen, and boron) in nanoparticles. In NRA, a specimen is bombarded with a beam of heavy ions that induces a nuclear reaction with nuclei in the nanoparticle atoms (EAG, 2007). The products of the nuclear reaction are measured. NRA provides a simple way of measuring the depth profiles by measuring the nuclear reaction yields as a function of the beam kinetic energy (Figure 23(A)). Because NRA is based on nuclear reactions, it is element specific. The most commonly used beam in NRA is ^3He. The method is nondestructive and provides elemental composition and/or structure—namely, the depth profiles from the surface region. Figure 23(B) shows back scattering (detected by an RBS detector) and penetration of the incident ion in the specimen (ΔX region). When particles strike the specimen surface, they transfer energy to the specimen atom and initiate a nuclear reaction, resulting in the release of γ- and X-rays that can be captured by particle-induced X-ray emission (PIXE) and particle-induced gamma ray emission (PIGE) detectors, respectively. Taken together, these results provide key information regarding a nanoparticle's surface.

Some of the publications describing the application of NRA are Kokkoris (2012), Bielesch et al. (2013), Gurbich (2013), Gu et al. (2009), Ene et al. (2006), Zink et al. (1998), Endisch et al. (1992), Sie (1992), Schulte et al. (1993), Trocellier and Engelmann (1986), Walls (1989), Moller et al. (1978), and Barnes et al. (1977).

5.6 Raman Spectroscopy

Raman spectroscopy, widely used for studying the structure and properties of nanoparticles, is based on the inelastic light scattering by molecules (Raman (1928) and Raman and Krishnan (1928), winner of the Nobel Prize in Physics in 1930). The incoming photons interact transiently with atoms and then scatter into the surroundings in all directions (detected by an RBS detector). During the brief interaction with molecules, photons lose energy to the interacting atoms that emit γ- and X-rays, imaged by the respective detectors. Because the laser beam can be considered an oscillating electromagnetic wave with frequency v_0, its interaction with the specimen induces an electric dipole moment characterized by the equation $P = \alpha E$, where E is an electric field and α is the induced molecular polarity. The Raman effect causes a transient and periodical deformity in the molecules; thus, molecules start vibrating with the characteristic frequency "υ_m," which is a measurable index. The amplitude of vibration is called a nuclear displacement. The laser excites molecules and transforms them into oscillating dipoles (Figure 24(A)).

The oscillating dipoles emit light at three different frequencies (Figure 24(A)): Raman-1, -2, and -3:

Raman-1: A molecule absorbs a photon with frequency "υ_0." The excited molecule returns

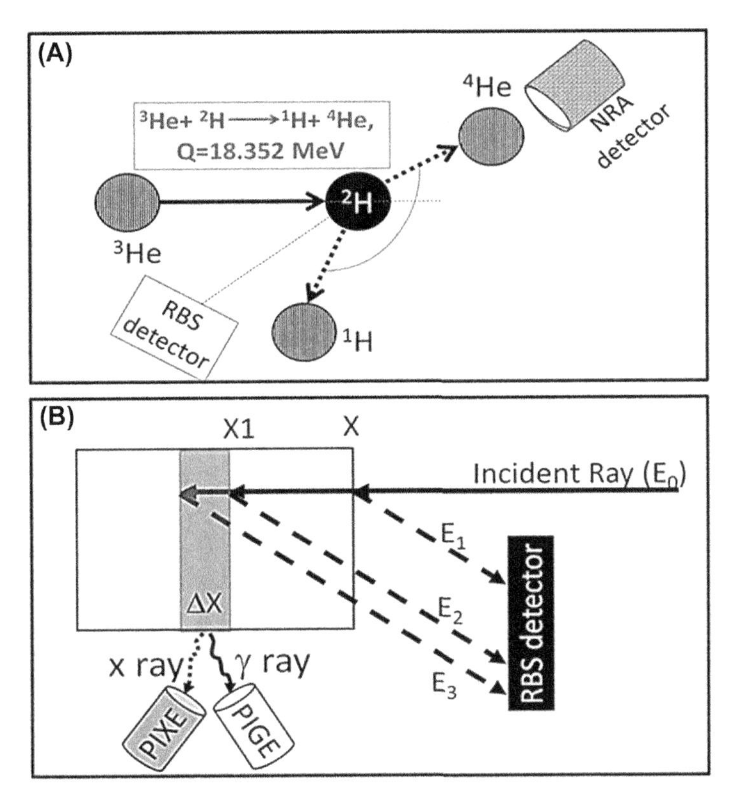

FIGURE 23 Principle of the nuclear reaction assay (NRA). (A) When low Z atoms are bombarded with ^{3}He, it is converted into ^{2}H, ^{1}H, and ^{4}H. In addition, some energy is released as X- and or γ-rays. ^{3}He is detected by an NRA detector, while backscattering can be detected by the RBS detector. (B) Release of X- and γ-rays during nuclear reaction. The ^{3}He reaction ($n + {}^3He \rightarrow {}^3H + {}^1H$) generates 0.764 MeV of energy that is released as γ- and X-rays. As ^{3}He (energy E_0) strikes the nanoparticle surface, it is reflected from the surface (E_1) or reflected after penetration (E_2 and E_3). The deeper the ^{3}He penetrates, the more energy it loses; thus, the exiting particle will have lower energy. Reflection, γ- and X-rays can be measured using RBS, PIXE, and PIGE detectors, respectively.

back to the same vibrational state and emits light with the same frequency (v_0) as the excitation source. This type of interaction is called an elastic *Rayleigh scattering* (Figure 24(B)).

Raman-2: A photon (with frequency υ_0) is absorbed by a molecule at the basal vibrational state. Part of the photon's energy is transferred to the Raman-active mode with frequency υ_m, and the resulting frequency of the scattered light is reduced to $\upsilon_0 - \upsilon_m$. This Raman frequency is called the *Stokes* frequency. In *Stokes* scattering, v_0 is higher than v_s.

Raman-3: A photon with frequency υ_0 is absorbed by a Raman-active molecule already in the excited vibrational state. The excessive energy of the excited Raman active mode is released, the molecule returns to the basic vibrational state, and the resulting frequency of scattered light goes up to $v_0 + v_m$. This Raman frequency is called the *anti-stokes* frequency. In *anti-stokes* scattering, v_0 is lower than v_s.

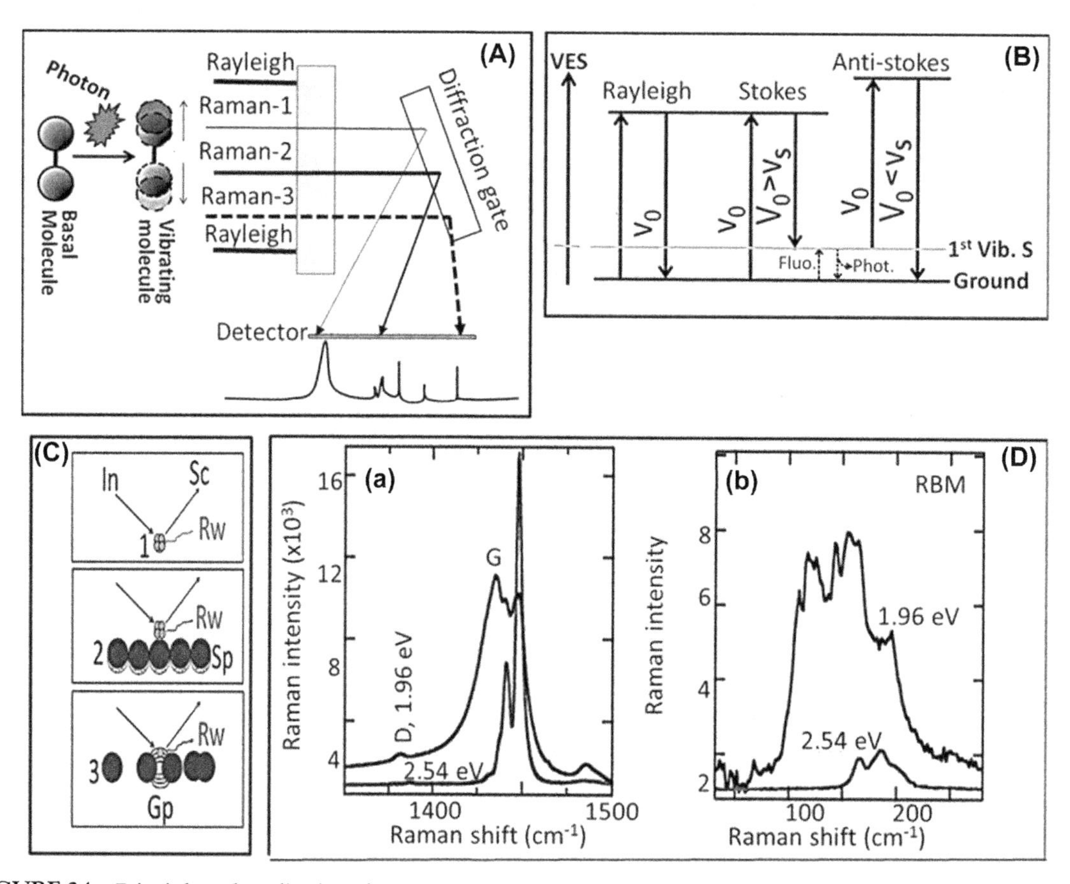

FIGURE 24 **Principle and application of Raman spectrometry.** (A) When a molecule absorbs high-intensity laser photons, an oscillating polarization in the molecules is formed at the virtual energy state or VES. The oscillating molecule emits photons of different wavelengths (Raman scattering) that can be measured by a detector. (B) Description of Rayleigh, Stokes, and anti-Stokes scattering (details are described in the text). Abbreviations: VES, virtual energy state that is a transitional energy state, 1st Vib; S, 1st vibrational energy state; V_0, vibrational energy of the incident laser photons; V_s, vibrational energy of scattered photons; Fluo., Fluorescence; Phot., Photons. (C) Principle of surface enhanced Raman spectroscopy (top). In traditional Raman spectroscopy, the inelastic scattered photons (Rw) are in low abundance compared to the Rayleigh scattered photons (Sc). However, when the specimen is placed on plasmonic metal nanoparticles (Sp) (middle) or placed within a gap-plasmon (Gp) (bottom), the Raman vibrational energy couples with the metal's plasmon, resulting in amplification of the Raman scattering. (D) Comparison of the Raman spectra of purified SWNTs acquired with 2.54 and 1.96 eV excitations for (1) G band (tangenital vibration related to the graphene), (2) D band (related to the disorder-induced vibrations or defects) and (3) radial breathing mode (RBM) band modes ((a) for 1 and 2, and (b) for 3). The G/D ratio deciphers the purity of the carbon nanotubes (>0.5 G/G ration indicated presence of large defects). *Reprinted from Dillon et al. (2004) with permission.*

A key disadvantage of traditional Raman spectroscopy is its low sensitivity, which has prevented its application in chemical biological systems. Developments of surface enhanced Raman spectroscopy (SERS) by Fleischmann et al. (1974) improved the method's sensitivity 10^6 folds. The single-atom SERS (Nie and Emory, 1997; Kneipp et al., 1995, 1997, 1998) has allowed measurement of the signal in a single atom (Figure 24(D)). These improvements have allowed widespread application of Raman spectroscopy, as follows.

- Characterization of optical and electronic properties of nanoparticles (Aleksandra et al., 2014; Ferrari and Basko, 2013; Maher, 2012; Soler and Qu, 2012; Alvarado et al., 2009)
- Identification of defects in carbon nanotubes and fullerenes (Tachibana, 2013; Dresselhaus et al., 2010; Costa et al., 2008; Hennrich et al., 2005)
- Scanning of nanoparticles in biological, food, and environmental samples (Zheng and He, 2014; Giovannozzi et al., 2014; Nima et al., 2014; Mustafa, 2013; Hargreaves et al., 2008)
- Identification of normal and cancerous cells (González-Solís et al., 2009, 2011, 2013; Shangyuan et al., 2011; Rodriguez-López, 2009; Pichardo-Molina et al., 2006; Shafer-Peltier et al., 2002)
- As a minimally invasive screening tool for acute renal transplantation rejection (Chung et al., 2008, 2009; Tu and Chang, 2012; Han et al., 2009; Brown et al., 2009a,b)
- Detection of various biomolecules (McGeehan et al., 2011; Carey, 1983; Lord and Yu, 1970)
- Applications in multiplexed Raman imaging, which is capable of detecting multiple targets simultaneously (Li et al., 2014; Zavaleta et al., 2014)
- Tissue engineering (Perlaki et al., 2014)
- Photothermal effect of metal nanoparticles (Bialkowski, 1996)

5.7 Scanning Tunneling Microscopy

Scanning tunneling microscopy (STM, for which Gerd Binnig and Heinrich Rohrer received the Nobel Prize in 1986 for developing) does not use a conventional microscope because it does not magnify the sample image. It consists of a fine electrically conductive tip (one atom at the end of the tip) that traverses just above the surface of an electrically conductive nanoparticle (Figure 25(A); Binnig and Rohrer, 1986; Bonnell and Huey, 2001; Hu et al., 2009). When a voltage is applied to the tip, electrons from the nanoparticles jump (called a "tunnel") across the space between the tip and sample. The flow of

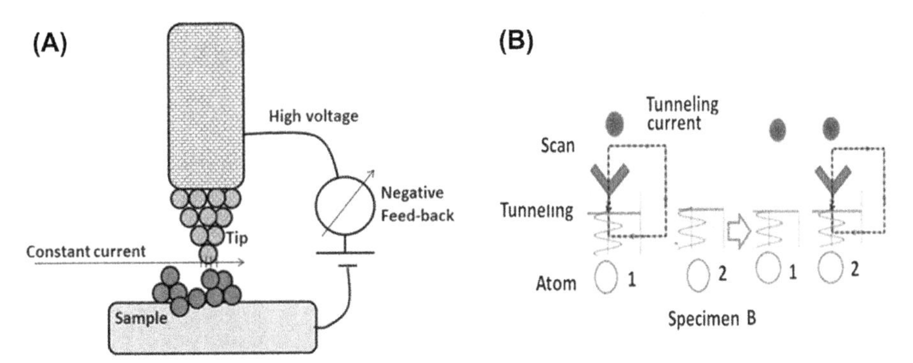

FIGURE 25 **Scanning tunneling microscope (STM).** (A) The STM consists of a probe that can move up/down and sideways, a sample chamber, an electrical circuit to apply potential to the tip and an ampere meter to measure the current developed by the tunneling electrons. The probe tip consists of one or two atom(s). When the tip is placed within a picometer (10^{-12} m) on the surface of a nanoparticle and a small positive potential is applied to the tip, the electron can tunnel or jump to the tip. Tunneling of the electron generates a current that is measured using an ammeter. (B) Repeatedly sliding the tip sideways a small distance, then raising or lowering the tip to maintain a constant current, builds up a one-dimensional profile of the surface height. Repeating the shifts and scans, over and over again, allows the creation of a two-dimensional profile of the surface.

electrons generates a current that can be detected (Swartzentruber, 1996). As the tip moves along the surface, its position is adjusted to make sure the tip's distance from the surface remains constant. This keeps the current constant. The movement of the tip is traced and recorded. The detailed features are graphically displayed on a computer screen.

The sensitivity of STM approaches that of atoms and molecules. The method has been applied in determining the electronic properties of semiconductors (Figure 25(B)) and the nanoparticle's surface characteristics (Moglianetti et al., 2014; Datar et al., 2012; Shi et al., 2011; Xu and Chen, 2009; Chaing, 1997; Avouris, 1996; Martel et al., 1996; Dai et al., 1996; Crommie et al., 1993, 1995; Shen et al., 1995).

5.8 Scanning Transmission Electron Microscopy

In scanning transmission electron microscopy (STEM), a highly focused electron probe is scanned across the material and various types of scattering are collected as a function of position. The transmitted electrons at a high scattering angle form high-resolution images, revealing the materials' atomic number (Z-) and chemical composition. The X-rays generated by the samples can be collected using an electron-disruptive X-ray spectroscopy detector. Electrons' energy loss is detected using a Gatan image filter to map the compositional and electronic properties of materials.

5.9 Surface Plasmon Resonance

Surface plasmon resonance (SPR)—in addition to becoming an important tool in biosensing technology, biochemistry, biology, and medical sciences—also enhances the surface sensitivity of several spectroscopic measurements, including fluorescence and Raman scattering. SPR has also found some applications in the detection of molecular adsorption and analysis of DNA, proteins, and metabolites. The aim of this section is to describe the principles and applications of SPR.

Plasmon resonance is an interfacial phenomenon in which light passing through a prism containing a media of higher refractive index (n1) to a media of lower refractive index (n2) is partly reflected and, depending upon the incidence angle, partly refracted (Figure 26(A)). Above a

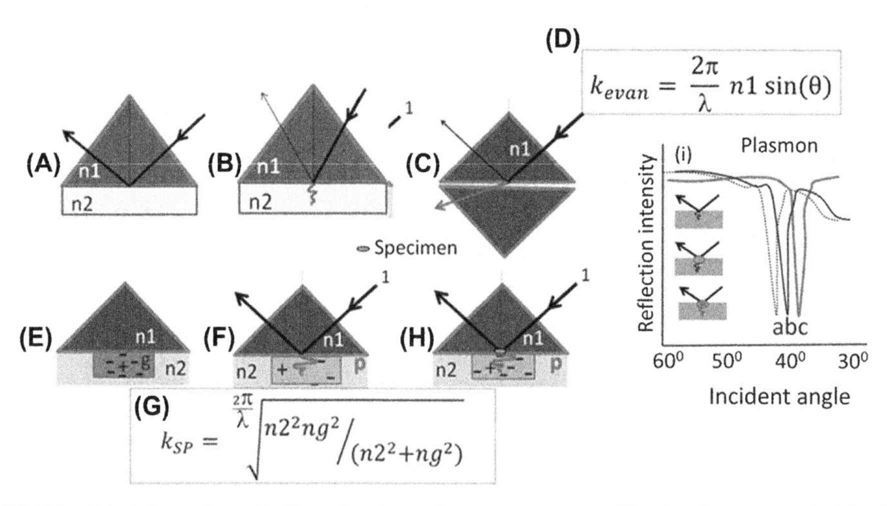

FIGURE 26 **Principle and application of surface plasmon resonance.** The details are described in the text.

certain critical angle (θ_c) of incidence, the incident light is completely reflected and does not cross into another medium. This stage is *total internal reflection* (TIR). However, when the incident light angle is below the critical angle, then part of the light wave is refracted into the medium with a lower refractive index (Figure 26(B)). However, the refracted beam remains at the surface in the form of an electromagnetic field, penetrating a short distance into the medium of a lower refractive index, creating an exponentially attenuating *evanescent wave*. If another prism is placed close to (picometer distance) the n1-prism, light refracts and passes through the second prism (Figure 26(C)). The wave function of the evanescent wave (k_{evan}) can be explained by the equation shown in Figure 26(G).

When a layer of gold nanoparticles is placed between the two media n1 and n2 Figure 26(E), and the incident light is monochromatic/p-polarized, the intensity of the reflected light is reduced at a specific incident angle, producing a sharp drop in reflectance (Figure 26(F), broken line). This process is called *surface plasmon resonance*, possibly due to the resonance energy transfer between evanescent wave and surface plasmon. The resonance conditions are influenced by the specimens adsorbed onto the gold nanoparticles (Figure 26(H)); therefore, in the presence of a specimen, the minimal reflectance may occur at a different angle (Figure 26(I), solid black line). If another particle binds to the specimen (line receptor-protein binding), there is a further change in the angle at which minimal absorbance occurs (Figure 26(I), solid red (gray in print versions) line). A linear relationship exists between resonance energy and mass concentration of molecules such as proteins, sugars, and DNA. The SPR, expressed in resonance units, is therefore a measure of mass concentration at the sensor chip surface. This means that the analyte and ligand association and dissociation can be observed, and ultimately rate constants as well as equilibrium constants can be calculated using SPR.

Metal nanoparticles, especially gold and silver nanoparticles, have great potential for chemical and biological sensor applications, due to their plasmonic properties, unique spectral response to the local environment of the nanoparticle surface, and ease of monitoring the light signal due to their strong scattering or absorption (Kumar, 2012; Lee and El-Sayed, 2006; Hutter and Fendler, 2004; Haes et al., 2004a,b; Haes and Van Duyne, 2002a,b, 2003a,b). An excitation of the surface by polarized monochromatic light results in a strong scattering of light and appearance of surface plasmon, the frequency and intensity of which are characteristic of the type of material (Hutter and Fendler, 2004).

For more information regarding SPR, please review the following manuscripts:

- *SPR and surface properties*: Haynes et al. (2003), Haynes and Van Duyne (2003a), Schatz and Van Duyne (2002), Kahl et al. (1998), and Freeman et al. (1995).
- *Optical filters*: Haynes and Van Duyne (2003b), and Dirix et al. (1999).
- *Plasmonic devices*: Maier et al. (2001, 2003), Andersen and Rowlen (2002), and Shelby et al. (2001).
- *Sensors*: Shashkov et al. (2010), Amanda et al. (2004), Haes et al. (2004a,b), Aizpurua et al. (2003), Hirsch et al. (2003), Bailey et al. (2003), Hase and Van Duyne (2002a,b, 2003a,b), Riboh et al. (2003), Fritzsche and Taton (2003), Obare et al. (2002), Nath and Chilkoti (2002), and Mucic et al. (1998). Aslan et al. (2004) has developed a nanogold-plasmon-resonance-based glucose sensor based on the aggregation and disassociation of 20-nm gold particles. The plasmon absorption was shifted due to interaction of nanoparticle and glucose. Hu et al. (2004) described a dielectric film (SiO_2)-based novel ultrahigh-resolution surface plasmon resonance biosensor having 0.1 pg/mm^2 sensitivity. This sensor can detect the interaction of small molecules in low concentrations. Chau et al. (2006) developed a

fiber-optic based localized surface plasmon resonance (LSPR) for sensing Ni^{2+} ion and label-free detection of streptavidin and staphylococcal enterotoxin B at the picomolar level using self-assembled gold nanoparticles. SPR has also been used for detection of explosives, such as trinitrotoluene (TNT) at the parts per trillion (ppt) level (Kawaguchi et al., 2007, 2008). Sensitivity was improved to ppt levels by LSPR using gold nanoparticles and further signal amplification.

- *Other applications of LSPR sensors* are detection of volatile organic compounds (Chen and Lu, 2009; Cheng et al., 2007), hydrogen peroxide (Filippo et al., 2009), and pathogens such as *Salmonella* (Fua et al., 2009).

5.10 X-ray or Ultraviolet Photoelectron Spectroscopy

In photoelectron spectroscopy, a specimen's surface is bombarded with either X-rays (XPS; for expulsion of K electrons) or UV photons (UPS; for valance electron). If electrons absorb sufficient energy, they are expelled from atoms in the near-surface region (photoelectrons) of the sample. The photoelectrons emitted by photons are unique because they are ejected from the valence or the inner orbits (such as K orbit) and they have energies characteristic of the atom they came from. Thus, XPS may allow elemental and chemical determinations of the nanoparticles. Compared to X-rays, UV photons have lower energies and probe the electrons in the outermost valence levels of the surface atoms. Thus, UPS yields information regarding the surface electronic structure. Many excellent reviews articles have described applications of XPS and UPS in characterization of nanoparticles. Selected manuscripts include the following:

- *Chemical properties*: Studies have evaluated the bonding environments of surface layers via XPS and valence states via UPS (Techane et al., 2011; Tonti and Zanoni, 2009; Turner and Al Jobory, 1962). Ultrahigh vacuum facilitates sample cleanliness. A high-resolution electron energy analyzer minimizes the spread of photoelectrons, allowing accurate determination of energies and enhanced peak separation for accurate chemical identifications (Kumiko, 2009; Jablonski, 2000; Millard and Bartholomew, 1977; Kobayashi, 1976; Carlson, 1975).
- *Electron binding energy and elemental characterization of the surface*: As described earlier, electrons remain bound to a positively charged "hole" in occupied orbitals. The electron–hole binding energy increases as the distance from the nucleus decreases; thus, electrons in the K-shell possess the highest binding energy, while the valence electrons possess the lowest binding energy. Photoelectrons are emitted when a molecule (M) is ionized by a beam of photons, in which the molecule loses an electron: $M + h\nu \rightarrow M^{+}(E_{int}) + e^{-}$, a reaction also known as photoionization reaction. This reaction occurs in three steps.
  - The molecule absorbs photons and activates valence or K-shell electrons.
  - For absorption, the photon's energy must match the electron's binding energy. The excited electron may dissociate from the "hole," leave the orbit, and travel to the surface of the molecule. If the excited electrons collide with another particle, they will lose energy.
  - The excited electron that escapes the surface of the molecule into the vacuum will be detected using appropriate detectors.

There is a linear relationship between the photoelectron count and the electron binding energy (photoelectron spectrum) coupled to the respective orbital energy (Figure 27). The energy spectrum of a photoelectron is unique for each atom. Therefore, XPS or UPS can characterize

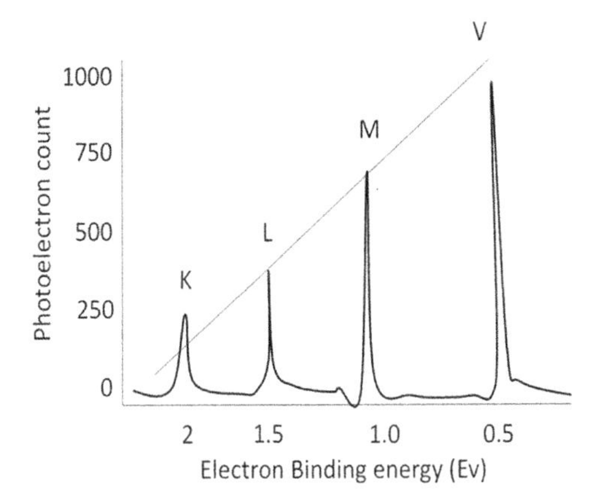

FIGURE 27 The relationship between the photoelectron count and the photoelectron energy spectrum that is coupled to the respective orbital energy.

the elemental character of a particle's surface (Brown et al., 2013; Xiong et al., 2013; Tonti and Zanoni, 2009; Grenha et al., 2008; Kopnov et al., 2008, 2009; Tsunekawa et al., 2000; Jablonski, 2000; Liu and Zhang, 1999).

XPS is a highly sensitive depth profiling and surface analysis method that can probe the top 10 nm of a nanofilm. When combined with sputtering sources, such as a single-atom sputtering probe (Nobuta and Ogawa, 2009; Hollander et al., 2007; Szakal et al., 2004) and fullerene C_{60}^{+} (Gilbert et al., 2013) to remove material without damaging underlying material, depth-profiling XPS enable high-resolution chemical analysis of polymer nanofilms. Cluster ion C_{60}^{+} sputtering is much less damaging than single-atom (Postawa et al., 2004) because most of the damaged material is removed from the surface, minimizing its interference with the proper analysis of the exposed surface (Mahoney, 2010).

UPS is analogous to XPS, except that the excitation source is a helium discharge source-1 (21.1 eV) or source-2 (44.8 eV). The photon energy is tuned to either 21.22 or 44.8 eV. Because of low photon energy, only the low binding-energy electrons (such as valence electrons) may be excited. UPS is more surface sensitive than XPS; thus, it is very sensitive to surface contamination.

5.11 X-ray Diffraction

X-ray diffraction (XRD) is commonly used to study the atomic structure of a nanoparticle (Taglieri et al., 2013; Dorofeev et al., 2012; Mahadevan et al., 2007; Petkov et al., 2007; Montejano et al., 2004; Shevchenko et al., 2002; Shevchenko and Madison, 2002).

The following terms are commonly used in X-ray methodologies:

- *Crystal*: A region of matter within which the atoms are arranged in a 3D transnationally periodic pattern (Buerger, 1956)
- *Lattice*: Crystal structure in an array of points in space
- *Unit cell*: The functional pattern of minimum number of atoms or molecules, which represents the full character of the crystal
- *Lattice plane*: Parallel equidistant planes passing through the lattice points.

XRD is based on constructive (*in* phase) interference of monochromatic X-rays and electrons in a crystalline (or powder) sample. When a beam of X-ray photons collides with electrons of an atom, some photons undergo elastic (no energy transfer)/inelastic (energy transfer) scattering (Figure 28(A)) according to Bragg's law, $n\lambda = 2d \sin \theta$, where d is an inner-plane distance in a given set of lattice and θ is certain degree of incidence angle (Figure 28(B)). The elastically scattered X-ray photons carry information about the electron distribution in nanoparticles; therefore, elastic electrons are measured in diffraction experiments. The scattered waves from different atoms can interfere with each other, and the resultant intensity depends upon whether the waves are in phase (which are additive) or out of phase (which are subtractive). Measuring the diffraction pattern therefore allows us to deduce the distribution of atoms in a material (Jenkins, 2000).

A simple variation of X-ray diffraction is powder X-ray diffraction, in which a powdery

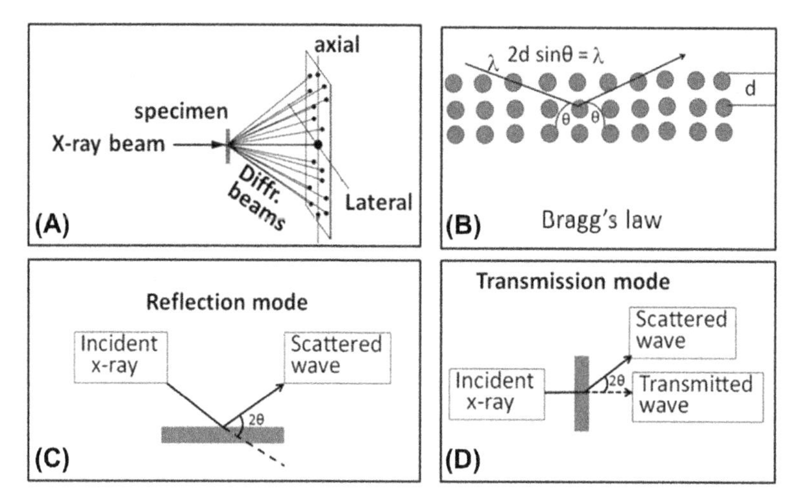

FIGURE 28 **Principle and application of X-ray diffraction.** (A) The scattering of X-rays by crystals. The intensity of the scattered rays differs due to interference effects. (B) Bragg's law. A crystal (each circle represents individual atom) is defined as a set of discrete parallel planes separated by a constant parameter "d." The incident X-ray photons would be reflected. If the phase shift is a multiple of 2π ($n\lambda = 2d \sin \theta n$) where n is an integer, λ is the wavelength of incident-ray photons, d is the spacing between the planes in the atomic lattice, and θ is the angle between the incident ray and the scattering planes, the interference among the scattered rays would be constructive and Bragg's peak will be generated. (C and D) Collection of data in reflection and transmission modes, respectively.

sample consisting of fine grains of single crystalline material is studied. Powder diffraction data can be collected using either transmission (Figure 28(C)) or reflection (Figure 28(D)) geometry. Because the particles in the powder sample are randomly oriented, these two methods will yield the same data. In general, an X-ray diffraction method is used to determine the following:

1. Orientation of a single crystal; the crystal structure of an unknown sample; and the size, shape, and internal stress of small crystalline regions
2. Structural properties such as lattice parameters, lattice strain, grain size, epitaxy, phase composition, preferred orientation, order-disorder transformation, thermal expansion, and thickness of thin films and multilayers
3. Atomic arrangement and electron configuration in crystals
4. Size and shape of nanoparticles.

Although many approaches have been described for the determination of size using XRD (Dorofeev et al., 2012; Gonçalves et al., 2012; Calvin et al., 2005; Gubicza et al., 2000), Sherrer's equation, $D = k\lambda/\beta \cos \theta$ (Langford and Wilson, 1978), is most commonly used for size determination. In this equation, D is the diameter in nm, k is the particle symmetry (0.9 for cubic symmetry (Qazi et al., 2009)), λ is the X-ray wavelength (1.542 Å), θ is Bragg's angle at which the peak is observed, and β is the full width of diffraction line at half of the maximum intensity (FWHM). Studies have shown that, in addition to diameter, other factors responsible for peak broadening are microstrain (β_ε) due to imperfection (dislocations, vacancies, interstitials, etc.) and instrument parameters, which are determined by measuring FWHM of a standard provided with the instrument and then using the equation $\beta_D = \beta_{observed} - \beta_{standard}$ (Gonçalves et al., 2012; Wajea et al., 2010).

6. MAGNETIC PROPERTIES

Magnetic nanoparticles are characterized by measuring magnetic susceptibility (χ), magnetic moment (μ), and hysteresis. χ is defined as the ratio of the magnetization of a material to the magnetic field strength. The μ of a magnet is a quantity that determines the torque (tendency of a force to rotate an object about an axis) it will experience in an external magnetic field. The magnetic moment of an electron is due to its rotation spin and angular momentum. Magnetic hysteresis defines a ferromagnetic particle's behavior in a magnetic field: it depends not only on the applied field strength but also on the previous history. The particle may retain magnetism even when the field is removed.

6.1 Magnetic Susceptibility and Magnetic Moment

Experimental approach: Magnetic susceptibility can be mass based (χ_g) or molar (χ_M or $\chi_g \times$ molecular weight). χ_M values for paramagnetic and diamagnetic samples are measured using the following methods: Gouy balance, Evans balance, and NMR methods. Gouy balance consists of a fixed magnet with separate N and S poles (Figure 29(A-6)), with a place for the specimen in the center. The specimen is placed in a glass tube (Figure 29(A-1)) within a glass isolator (Figure 29(A-2)). A thermometer is placed near the sample (Figure 29(A-3)). The sample is tied to a nylon thread (Figure 29(A-4)) that connects the sample to a balance to measure mass. The Evans balance is similar to the Gouy balance, except that it measures force, not mass (Figure 29(B)). In the Evans balance, the sample is placed between two poles (Figure 29(B-6)), but into the field of one magnet that experiences a force which would deflect the magnet. Then, a magnetic field is generated at the second magnet that, by negative feedback, restores the magnets to the original position. The current required to do this is proportional to the force exerted on the first magnet. In an NMR method, the specimen is dissolved in a solvent and placed in a glass NMR tube. Then, a capillary filled with the solvent alone is suspended into the specimen tube. In NMR analysis, the difference between the solvent and the specimen peak is measured (chemical shift).

Theory: Except for electrons in highest occupied molecular orbital (HOMO), lower orbitals (such as $2p_z$ or $2p_y$ orbitals) contain electron pairs in which each electron spins in an opposite direction, resulting in no net magnetic moment (overall spin quantum number, $S = 0$). In HOMO, unpaired electrons with uncompensated spin, high anisotropy, and net magnetic

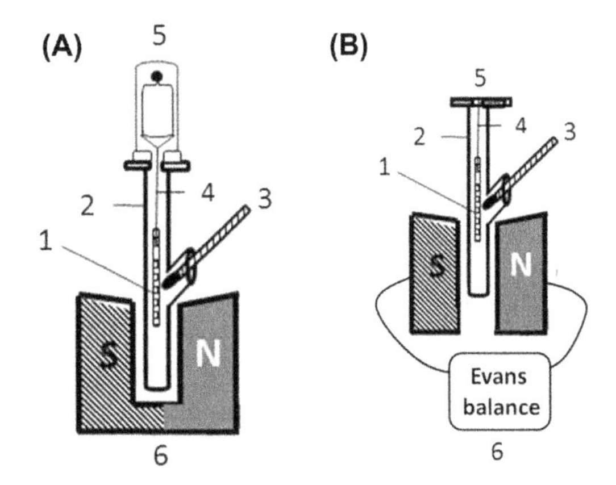

FIGURE 29 Sketches of Gouy (A) and Evans (B) balances. (A) In Gouy's balance (Sunderson, 1968), the specimen is packed in a capillary to a predetermined length (1) and hung in the center of a glass tube (2) via nylon thread (3). A thermometer is inserted close to the capillary (4). The thread is connected to a balance for measuring the specimen weight. A thermometer is placed to measure the sample temperature (5). The glass tube is centered in a magnetic field from fixed magnets (6). The sample is weighed with magnets "off" and then again with magnets "on." The weight difference is measured and used for calculation of magnetic susceptibility and moment as described in the text. (B) The Evans balance is similar to the Gouy balance, except for the following: The magnets are not fixed; they move forward and backward as force changes. Thus, the Evans balance measures force, not mass.

moment exist. Paired and unpaired electrons both revolve around the nucleus, exhibiting angular momentum and a net magnetic momentum. The electrons of each orbital exhibit a spin quantum number denoted as m_s, which is constituted by either upward ($m_s = +1/2$) or downward ($m_s = -1/2$) arrows. Thus, $2p_z$ orbitals contain $+1/2$ and $-1/2$ electrons. The particles containing all paired electrons are called diamagnetic particles, and they do not exhibit net magnetic moment (see Chapter 2). However, when a particle, in addition to "paired electrons" also contains unpaired electrons ($+1/2$ in the absence of a field), the paired electrons cause diamagnetic effects, while the unpaired electrons exhibit high anisotropy and cause paramagnetic effects ($S > 0$).

Diamagnetic particles, when placed between the poles of a strong magnet, exert a repulsion on the applied field (tending to reduce strength of the applied force). Because of repulsion, the diamagnetic particles experience weight loss (in the Gouy balance; Saunderson, 1968) or repel the magnetic poles (in Evans balance; Evans, 1974). Their mass magnetic susceptibility (χ_g) will be sum of the diamagnetic susceptibility (χ_α). Because χ_α represents repulsive forces, it has a negative value; thus, the diamagnetic susceptibility has a negative value. Unlike diamagnetic particles, when a paramagnetic particle is placed between the poles of a strong magnet, it will experience an attraction for the field and weigh more with a Gouy balance or show greater attraction of the magnetic poles with the Evans balance. The χ_M value of a paramagnetic particle is Σ (contribution of unpaired electrons or paramagnetic effect or χ_A that has positive values) + (contribution of the paired electrons or χ_α that has negative values). Because $\chi_M = \chi_A + (-\chi_\alpha)$, $\chi_A > \chi_M$ for paramagnetic particles.

Calculations: Mass (Gouy method) or force (Evans method) values are used to calculate mass magnetic susceptibility (χ_g), which is converted in molar magnetic susceptibility (χ_M) and paramagnetic contribution to the magnetic susceptibility (χ_A) using Eqns (3–5) shown below.

$$\chi_g = \frac{CL(\Delta R - \Delta R_0)}{m \times 10^9} \tag{3}$$

$$\chi_M = \mathrm{M} \times \chi_g \tag{4}$$

$$\chi_A = \chi_M - \sum \chi_\alpha \text{ (since } \chi_\alpha \text{ is negative, for paramagnetic particles } \chi_A > \chi_M) \tag{5}$$

In this equation, χ_M is molar magnetic susceptibility, χ_g is mass magnetic susceptibility, M is molecular weight, L is the sample length in centimeters, m is the sample mass in grams, and C is the balance calibration constant (different for each balance; printed on the back of the instrument). ΔR is the difference in mass (or force) of samples collected with and without magnetic field. ΔR_0 is the difference in mass (or force) of blank glass tubes collected with and without magnetic field. The effective magnetic moment (μ_{Eff}) is calculated using the equation below (T is the temperature in K):

$$\mu_{(Eff)} = 2.828\sqrt{\chi_A T} \tag{6}$$

The NMR method for the determination of magnetic susceptibility and moment was first described by Evans in 1959 and then improved by Loliger and Scheffold (1972), Grant (1995), and Piguet (1997). Evans (1959) devised Eqns (7) and (8) for calculation of mass magnetic susceptibility (χ_g) and molar magnetic susceptibility (χ_M), respectively,

$$\chi_g = \frac{\delta V}{\nu_0 S_f m} + \chi_0 + \chi_0 \frac{(d_0 - d_s)}{m} \tag{7}$$

$$\chi_M^p = \frac{\delta \nu^p M^p}{\nu_0 S_f m} + \chi_0 M^p + \chi_0 M^p \frac{\left(d_0 - d_s^0\right)}{M^p} - \chi_m^{dia} \text{ (for bulk particles)} \tag{8}$$

However, later studies Grant (1995) and Piguet (1997) have proposed that, in smaller dilute solutions, $(d_0 - d_s^0)/M^p$ can be neglected and μ_{Eff} can be calculated using Eqn (9):

$$\mu_{Eff} = 2.828\sqrt{\frac{TM^p}{S_f m^p}\left(\frac{\delta v^p}{v_0} + S_f m^p \chi_0 - \frac{S_f m^p}{M^p}\chi_M^{dia}\right)} \quad \text{(for bulk particles)} \tag{9}$$

Grant (1995) has shown that Eqn (9) is suitable only for bulk particles. He simplified Eqns (8) and (9) into Eqns (10) and (11), respectively, which are more suitable for magnetic nanoparticles:

$$\chi_M^p = \frac{\delta v^p M^p}{v_0 S_f m} - \chi_m^{dia} \text{ (for nanoparticles)} \tag{10}$$

$$\mu_{Eff} = 0.138\left(\sqrt{\frac{\delta v}{c}}\right)^{1/2} \quad \text{(for nanoparticles)} \tag{11}$$

In these equations, χ_M^p is the paramagnetic contribution toward χ_M, δv is the shift in NMR frequency from the value found in pure solvent (>0 for paramagnetism and <0 for diamagnetism), m is the concentration of the solute, S_f is the shape factor for the magnet (Field et al., 1988), χ_0 is the susceptibility of pure solvent, d_0 and d_s are densities of pure solvent and solution, respectively, χ_M^{dia} and χ_M^p are diamagnetic and paramagnetic susceptibility, respectively, M^p is the molar mass of dissolved paramagnetic compound, and d_s^p and d_s^{dia} are the density of solution containing paramagnetic and diamagnetic compounds, respectively.

6.2 Magnetic Hysteresis

One of the key properties of ferromagnetic materials is hysteresis—that is, as force field (H) is increased, magnetic flux (B) also increases nonlinearly from points 0 to 1, the saturation point (amount of H required to generate maximum flux density, B). When H is reduced to 0, a change in B takes a new path to points 2 (*remnant flux force*) and the specimen still remains partially magnetized. Then, the current must be reversed so that H increases in the opposite direction to demagnetize the specimen. The amount of field required to demagnetize the specimen is called the *coercive force* (Figure 30).

If the reversed current increases further, the particle saturates in the opposite direction (point 4). Then, if the current is reversed and increased, B increases, taking the path of 5, 6, and 1. This magnetization journey from 0 to 1, 2, 3, 4, 5, 6 and back to 1 is called the hysteresis loop. The hysteric loop is commonly studied using a "loop tracer," as shown in the block diagram (Figure 31). The tracers consist of four basic parts (Kulik et al., 1993):

- *Specimen coil*: nanoparticles or bulk particles.
- *Excitation system*: High current power amplifier, Helmholtz coil pair, control of current strength and direction.

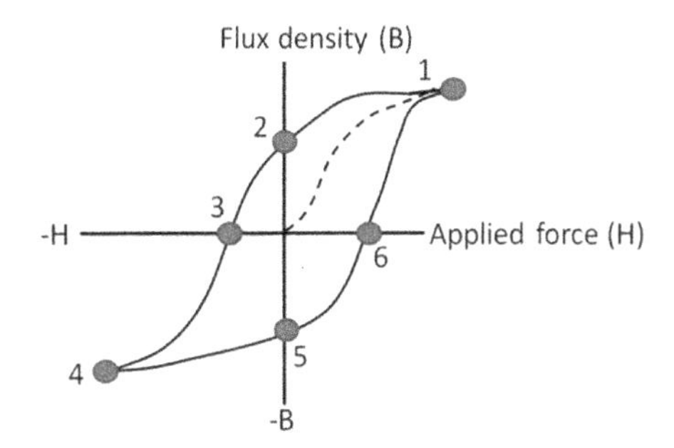

FIGURE 30 Description of hysteresis loop for ferromagnets. The point 0 represents lack of external magnetic force. As force field (H) is increases, magnetic flux (B) also increases nonlinearly from point 0 to point 1, the saturation point. Then, the field is reduced to 0. The flux, instead of returning to 0, goes to point 2 (the remnant flux force) and the specimen still remains partially magnetized. To demagnetize the specimen, the current must be reversed so that H increases in opposite direction. The amount of field required to demagnetize the specimen is called coercive force (point 3). A further increase in the field strength results in a second saturation point opposite the first one (point 4). At point 4, if the force is reversed and then increased, the flux will also increase nonlinearly but via a different path, passing through points 5, 6, and 1.

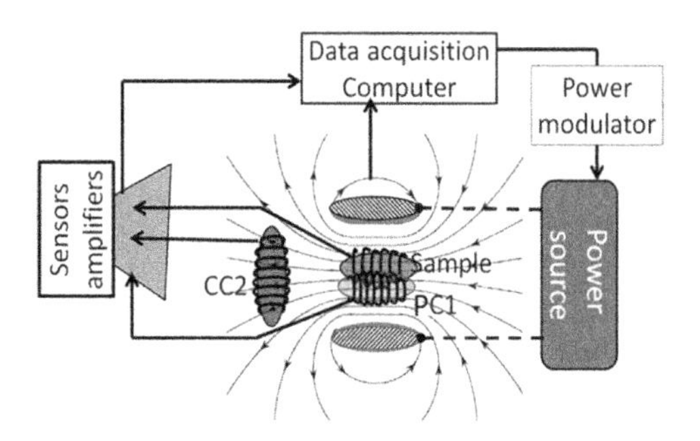

FIGURE 31 **Block diagram for a typical hysteresis loop tracer.** The tracer consists of four basic parts: the specimen pick-up coil (PC1), the excitation system consisting of high current power amplifier, Helmholtz coil pair, control of current strength and direction, the pick-up coil (PC), and compensation coil (CC) including the sensors and sense-amplifier and the control system including a computer and the software for data analysis.

- *Pickup coil (PC) and compensation coil (CC)*: Compensation coil and pickup coil, sensors, and sense-amplifier. The pickup and compensation coils, in the absence of a sample, create a net output voltage by changing the magnetic field. The flux change in the sample placed inside the pickup coil is analyzed by the software.
- *Control system*: Computer and software for data analysis.

From the hysteresis loop, a number of magnetic properties of a material can be determined. *Retentivity* is a material's ability to retain a certain amount of residual magnetic field when the magnetizing force is removed after achieving saturation (the value of B at point 1 on the hysteresis curve). *Residual magnetism or residual flux* is the magnetic flux density that remains in a material when the magnetizing force is zero. *Coercive force* is the amount of reverse magnetic field that must be applied to a magnetic material to make the magnetic flux return to zero. *Permeability* is the ease with which a magnetic flux is established in the component. *Reluctance* is the opposition that a ferromagnetic material shows to the establishment of a magnetic field. *Squareness factor* is a measure of how square the loop is and is a dimensionless quantity between 0 and 1; a squareness factor of 1 therefore corresponds to a perfectly square loop. *Hysteresis loop width* is an increase in the width of the hysteresis loop, which is associated with a decrease in permeability and an increase in coactivity, reluctance, and residual magnetism. Nanoparticles usually have narrower widths than the corresponding bulk particles.

7. THERMODYNAMIC CHARACTERIZATION

Accurate thermodynamic measurements are essential for understanding the fundamental properties of nanomaterials, providing direct insight into chemical reactions. The indices required for full characterization of the thermodynamic profile are enthalpy (H), entropy (S), heat capacity change (Cp), and free energy (G). It is possible to measure these reactions using the following methods.

7.1 Thermal Gravimetric Analysis

Thermal gravimetric analysis (TGA) determines endothermic and exothermic weight loss upon heating, cooling, etc. in nanoparticles. TGAs uses heat to force reactions and physical changes in materials. Thermogravimetric curves characterize specific compounds due to the unique sequence from physicochemical reactions occurring over specific temperature ranges (Moharram et al., 2014).

7.2 Differential Thermal Analysis

Differential thermal analysis (DTA) is based on the principle that the substance, upon heating, undergoes reactions and phase changes that involve absorption or emission of heat. The sample to be analyzed is placed with a control (known properties) and an inert material, and the temperature and phase changes are recorded. Identification of a substance is accomplished by comparing DTA curves obtained from the unknown with those of known elements. The DTA technique is widely used for identifying minerals and mineral mixtures (Ichiyanagi and Kimishima, 2002).

7.3 Differential Scanning Colorimeter

In a differential scanning colorimeter (DSC), heat released from a chemical process, either a chemical reaction or a conformational alteration, is recorded in test and control samples placed in separate chambers of a colorimeter. Then, the heat of reaction or $\Delta_r H$, defined as the change in enthalpy associated with a chemical reaction, is determined. Negative $\Delta_r H$ indicates an exothermic reaction and a positive $\Delta_r H$ indicates an endothermic reaction (Raymond, 2005).

7.4 Nanocalorimetry

Nanocalorimetery allows measurement of the heat capacity of very small samples, down to nanograms of material. The sample is placed into the nanocalorimeter via vapor deposition. The following studies have used nanocalorimetry in characterizing nanoparticles: Lai et al. (1996), Jiang et al. (2006), and Yi and LaVan (2013).

8. CONCLUSIONS

Nanoparticles, because of their unique physicochemical properties, are of great interest to the scientific, industrial, and medical communities. Gradually, nanoparticles are being incorporated in daily activities of our lives. Bulk materials (>500 nm) usually have constant physical properties regardless of their size or shape. However, the physicochemical and electronic properties of nanoparticles (less than 100 nm in at least one dimension) exhibit strong dependence on size. Thus, in particles less than 100 nm, the properties change drastically with minor changes in size because, at the nanoscale, the percentage of atoms at the surface becomes significant. For bulk materials, the percentage of atoms at the surface is insignificant in relation to the number of atoms in the bulk of the material.

The interesting and sometimes unexpected properties of nanoparticles are therefore largely due to the large surface area of the material, which dominates the contributions made by the small core of the material. It is well established that the biological activity of nanoparticles depends upon their physicochemical properties and size. Therefore, an accurate determination of a nanoparticle's size and properties is a prerequisite for controlling its biological properties. This chapter elucidates some of the commonly used methods for nanoparticle characterization.

References

Agbabiaka, A., Wiltfong, M., Park, C., 2012. Small angle X-ray scattering technique for the particle size distribution of nonporous nanoparticles. Int. J. Theor. Appl. Nanotechnol. 1, 1929–1948.

Agbabiaka, A., Wiltfong, W., Park, C., 2013. Small angle X-ray scattering technique for the particle size distribution of nonporous nanoparticle. J. Nanopart. 2013. ID 640436.

Aizpurua, J., Hanarp, P., Sutherland, D.S., Kall, M., Bryant, G.W., Garcia de Abajo, F.J., 2003. Optical properties of gold nanorings. Phys. Rev. Lett. 90, 057401.

Aleksandra, J., Kamilla, M., Katarzyna, M., 2014. An impact of the ring substitution in nicorandil on its adsorption on silver nanoparticles. Surface-enhanced Raman spectroscopy studies. Spectrochim. Acta A Mol. Biomol. Spectrosc. 129, 624–631.

Allabashi, R., Stach, W., de la Escosura-Muñiz, A., Liste-Calleja, L., Merkoçi, A., 2009. ICP-MS: a powerful technique for quantitative determination of gold nanoparticles without previous dissolving. J. Nanopart. Res. 11, 2003–2011.

Allmaier, G., Laschober, C., Szymanski, W.W., 2008. Nano ES GEMMA and PDMA, new tools for the analysis of nanobioparticles-protein complexes, lipoparticles, and viruses. J. Am. Soc. Mass Spectrom. 19, 1062–1068.

Alvarado, T., Norihiko, H., Satoshi, K., 2009. Tip-enhanced Raman spectroscopy for nanoscale strain characterization. Anal. Bioanal. Chem. 394, 1775–1785.

Alves, S., Kalberer, M., Zenobi, R., 2003. Direct detection of particles formed by laser ablation of matrices during matrix-assisted laser desorption/ionization. Rapid Commun. Mass Spectrom. 17, 2034–2038.

Amanda, J., Haes, A.J., Van Duyne, R.P., 2004. A unified view of propagating and localized surface plasmon resonance biosensors. Anal. Bioanal. Chem. 379, 920–930.

Ambrose, W.P., Goodwin, P.M., Jett, J.H., Orden, A.V., Werner, J.H., Keller, R.A., 1999. Single molecule fluorescence spectroscopy at ambient temperature. Chem. Rev. 99, 2929–2956.

Andersen, P.C., Rowlen, K.L., 2002. Brilliant optical properties of nanometric noble metal spheres, rods, and aperture arrays. Appl. Spectrosc. 56, 124A–135A.

Aslan, K., Lakowicz, J.R., Geddes, C.D., 2004. Nanogold-plasmon-resonance-based glucose sensing. Anal. Biochem. 330, 145–155.

Avouris, P., 1996. Manipulation of matter at the atomic and molecular levels. Acc. Chem. Res. 28, 95–101.

Baer, D.R., Engelhard, M.H., Johnson, G.E., Laskin, J., Lai, J., Mueller, K., Munusamy, P., Thevuthasan, S., Wang, H., Washton, N., Elder, A., Baisch, B.L., Karakoti, A., Kuchibhatla, S.V.N.T., Moon, D.W., 2013. Surface characterization of nanomaterials and nanoparticles: important needs and challenging opportunities. J. Vac. Sci. Technol. A 31, 050820.

Baer, D.R., Gaspar, D.J., Nachimuthu, P., Techane, S.D., Castner, D.G., 2012. Application of surface chemical analysis tools for characterization of nanoparticles. J. Nanopart. Res. 14, 1106.

Bailey, R.C., Nam, J.-M., Mirkin, C.A., Hupp, J.T., 2003. Real-time multicolor DNA detection with chemoresponsive diffraction gratings and nanoparticle probes. J. Am. Chem. Soc. 125, 13541–13547.

Barnes, C.A., Overley, J.C., Switkowski, Z.E., Tombrello, T.A., 1977. Measurement of hydrogen depth distribution by resonant nuclear reactions. App. Phys. Lett. 31, 239–241.

Bauer, E., 1994. Low energy electron microscopy. Rep. Prog. Phys. 57, 895–898.

Bauer, F., Sauerland, V., Ernst, H., Glasel, H.A., Naumov, S., Mehnert, R., 2003. Preparation of scratch- and abrasion-resistant polymeric nanocomposites by monomer grafting onto nanoparticles, 4-Application of MALDI-TOF mass spectrometry to the characterization of surface modified nanoparticles. Macromol. Chem. Phys. 204, 375–383.

Benninghoven, A., 1994. Chemical analysis of inorganic and organic surfaces and thin films by static time-of-flight secondary ion mass spectrometry (TOF-SIMS). Angew. Chem. Int. Ed. (Engl.) 33, 1023–1043.

Bera, S., Dhara, S., Velmurugan, S., Tyagi, A.K., 2012. Analysis on binding energy and Auger parameter for estimating size and stoichiometry of ZnO nanorods. J. Spectrosc. 2012, 371092.

Beyer, G.L., 1987. Particle counter for rapid determination of size distributions. J. Colloid Interface Sci. 118, 137–147.

Bialkowski, S.E., 1996. Photothermal spectroscopy methods for chemical analysis. In: Winefordner, J.D. (Ed.), Chemical Analysis: A Series of Monographs on Analytical Chemistry and Its Applications, vol. 134. John Wiley & Sons, Inc.

Bielesch, S., Oberkofler, M., Becker, H-W., Maier, H., Rogalla, D., Schwarz-Selinger, T., Linsmeier, C., 2013. Experimental resolution of deuterium and hydrogen depth profiling with the nuclear reactions $D(^{3}He,p)\alpha$ and $p(^{15}N,\alpha,\gamma)^{12}C$. Proceedings of the 19th International Workshop on Inelastic Ion-Surface Collisions (IISC-19), Frauenchiemsee, Germany, 16–21 September 2012. Nucl. Instrum. Methods Phys. Res. Sect. B 317, 121–125.

Binnig, G., Rohrer, H., 1986. Scanning tunneling microscopy. IBM J. Res. Dev. 30, 4.

Binnig, G., 1987. Atomic force microscopy. Phys. Scr. 19, 53–54.

von Bohlen, A., Brücher, A., Holland, B., Wagner, R., Hergenröder, R., 2010. X-ray standing waves and scanning electron microscopy – energy dispersive X-ray emission spectroscopy study of gold nanoparticles. Spectrochim. Acta Part B 65, 409–414.

Bohren, C.F., Huffmann, D.R., 1983. Absorption and Scattering of Light by Small Particles. Wiley-Interscience, New York.

Bonatto, F., Rovani, S., Kaufmann, I.R., Soares, G.V., Baumvol, I.J.R., Krug, C., 2012. Complementary low energy ion scattering and X-ray photoelectron spectroscopy characterization of polystyrene submitted to N_2/H_2 glow discharge. Nucl. Instrum. Methods Phys. Res. Sect. B 273, 189–191.

Bonnell, D.A., Huey, B.D., 2001. Basic principles of scanning probe microscopy. In: Bonnell, D.A. (Ed.), Scanning Probe Microscopy and Spectroscopy: Theory, Techniques, and Applications, second ed. Wiley-VCH, New York.

Bootza, A., Russb, T., Goresc, F., Karasb, M., Kreutera, J., 2005. Molecular weights of poly(butyl cyanoacrylate) nanoparticles determined by mass spectrometry and size exclusion chromatography. Eur. J. Pharm. Biopharm. 60, 391–399.

Bott, S.E., Hart, W.H., 1991. Particle Size Distribution II, vol. 472. American Chemical Society, Washington, DC.

Braun, W., Held, G., 2005. The exclusive use of integer-order spots for LEED-IV structure analysis of adsorption systems: p(2×2)-O on Ni{111}. Surface Science, 594, 203–211.

Brown, A., Hondow, N., 2013. Electron microscopy of nanoparticles in cells. Front. Nanosci. 5, 95–120.

Brown, K.L., Palyvoda, O.Y., Thakur, J.S., Nehlsen-Cannarella, S.L., Fagoaga, O.R., Gruber, S.A., 2009a. Differentiation of alloreactive versus CD3/CD28-stimulated T-lymphocytes using Raman spectroscopy: a greater specificity for noninvasive acute renal allograft rejection detection. Cytometry A 75, 917–923.

Brown, K.L., Palyvoda, O.Y., Thakur, J.S., Nehlsen-Cannarella, S.L., Fagoaga, O.R., Gruber, S.A., 2009b. Raman spectroscopic differentiation of activated versus non-activated T lymphocytes: an in vitro study of an acute allograft rejection model. J. Immunol. Methods 340, 48–54.

Brown, M.A., Jordan, I., Redondo, A.B., Kleibert, A., Wörner, H.J., van Bokhoven, J.A., 2013. *In situ* photoelectron spectroscopy at the liquid/nanoparticle interface. Surf. Sci. 610, 1–6.

Brunauer, S., Emmett, P.H., Teller, E., 1938. Adsorption of gases in multimolecular layer. J. Am. Chem. Soc. 60, 309–319.

Budiman, A.H., Dewi, E.L., Punvanto, W.W., Dalimi, R., Hwang, B.J., 2010. Atomic distribution of PtCo/C nanoparticles as investigated by X-ray absorption spectroscopy. Indones. J. Mater. Sci. 35–39. Special Edition on Materials/or Energy and Device.

Buerger, M.J., 1956. Partial Fourier synthesis and their application to the solution of certain crystal structure. PNAS 42, 776–781.

Bunjes, H., Unruh, T., 2007. Characterization of lipid nanoparticles by differential scanning calorimetry, X-ray and neutron scattering. Advanced Drug Delivery Reviews, 59, 379–402.

Calvin, S., Luo, S.X., Caragianis-Broadbridge, C., McGuinness, J.K., Anderson, E., Lehman, A., Wee, K.H., Morrison, S.A., Kurihara, L.K., 2005. Comparison of extended X-ray absorption fine structure and Scherrer analysis of X-ray diffraction as methods for determining mean sizes of polydisperse nanoparticles. Appl. Phys. Lett. 87, 233102.

Carey, P., 1983. Raman spectroscopy for the analysis of biomolecules. Trends Anal. Chem. 2, 275–277.

Carlson, T.A., 1975. Photoelectron spectroscopy. Annu. Rev. Phys. Chem. 26, 211–233.

Chaing, S., 1997. Force and tunneling microscopy. Chem. Rev. 97, 1015–1230.

Chau, K., Lin, Y.F., Cheng, S.F., Lin, Y.J., 2006. Fiber-optic, chemical and biochemical probes based on surface. localized, resonance. Sens. Actuators B 113, 100–105.

Chen, C.C., Zhu, C., White, E.R., Chiu, C.Y., Scott, M.C., Regan, B.C., Marks, L.D., Huang, Y., Miao, J.W., 2013. Three-dimensional imaging of dislocations in a nanoparticle at atomic resolution. Nature 496, 74–77.

Chen, Y.Q., Lu, C.J., 2009. Surface modification on silver nanoparticles for enhancing vapor selectivity of localized surface plasmon resonance sensors. Sens. Actuators B Chem. 135, 492–498.

Cheng, C.S., Chen, Y.Q., Lu, C.J., 2007. Organic vapour sensing using localized surface plasmon resonance spectrum of metallic nanoparticles self assemble monolayer. Talanta 3, 358–365.

Cheng, H.N., Asakura, T., English, A., 2012. NMR Spectroscopy of Polymers—innovative strategies for complex macromolecules. ACS Symposium Series 1077.

Chu, B., Liu, T., 2000. Characterization of nanoparticles by scattering techniques. J. Nanopart. Res. 2, 29–41.

Chung, K.-H., Hsiao, L.-Y., Lin, Y.-S., Duh, J.-G., 2008. Morphology and electrochemical behavior of Ag–Cu nanoparticle-doped amalgams. Acta Biomater. 4, 717–724.

Chung, Y.G., Tu, Q., Cao, D., Harada, S., Eisen, H.J., Chang, C., 2009. Raman spectroscopy detects cardiac allograft rejection with molecular specificity. Clin. Transl. Sci. 2, 206–210.

Costa, S., Borowiak-Palen, E., Kruszyńska, M., Bachmatiuk, A., Kaleńczuk, R.J., 2008. Characterization of carbon nanotubes by Raman spectroscopy. Mater. Sci. Pol. 26, 433–441.

Crayton, S., 2012. CP-MS Analysis of Lanthanide-Doped Nanoparticles: A Quantitative and Multiplexing Approach to Investigate Biodistribution, Blood Clearance, and Targeting. AAI3508988. http://repository.upenn.edu/dissertations/AAI3508988.

Crommie, M.F., Lutz, C.P., Eigler, D.M., Heller, E.J., 1995. Waves on a metal surface and quantum corrals. Surf. Rev. Lett. 2, 127–137.

Crommie, M.F., Lutz, C.P., Eigler, D.M., 1993. Confinement of electrons to quantum corrals on a metal surface. Science 262, 218–220.

Czanderna, Hercules, D.M., 1991. Ion Spectroscopies for Surface Analysis. Plenum Press, New York.

Dai, H., Hafner, J.H., Rinzler, A.G., Colbert, D.T., Smalley, R.E., 1996. Nanotubes as nanoprobes in scanning probe microscopy. Nature 384, 147–151.

Daniela, D., Charlotte, G., Heike, T., Ulrich, P., Janina, K., Norbert, J., 2012. Quantitative imaging of gold and silver nanoparticles in single eukaryotic cells by laser ablation ICP-MS. Anal. Chem. 84, 9684–9688.

Datar, S., Kumar, P.M., Sastry, M., Dharmadhikri, C.V., 2012. Scanning tunneling microscopy/spectroscopy of titanium dioxide nanoparticulate film on Au(111) surface. Int. J. Mass Spectrom. 315, 22–30.

Dawahre, N., Shen, G., Renfrow, S.N., Kim, S.M., Kung, P., 2013. Atom probe tomography of AlInN/GaN HEMT structures. J. Vac. Sci. Technol. B 31.

De Luca, A., Portavoce, A., Texier, M., Burle, N., Pichaud, B., 2013. Dynamic segregation of metallic impurities at SiO_2/Si interfaces, 18th Microscopy of Semiconducting Materials Conference (MSM XVIII). J. Phys. Conf. Ser. 471, 012029.

Delvallee, A., Feltin, N., Ducourtieux, S., Trabelsi, M., Hochepied, J.-F., 2013. Comparison of nanoparticle diameter measurements by atomic force microscopy and scanning electron microscopy. In: 16th International Congress of Metrology 06007. http://dx.doi.org/10.1051/metrology/201306007.

Devaraj, A., Colby, R., Vurpillot, F., Thevuthasan, S.J., 2014. Understanding atom probe tomography of oxide-supported metal nanoparticles by correlation with atomic-resolution electron microscopy and field evaporation simulation. Phys. Chem. Lett. 5, 1361–1367.

Diercks, D.R., Gorman, B.P., Kirchhofer, R., Sanford, N., Bertness, K., Brubaker, M., 2013. Atom probe tomography evaporation behavior of C-axis GaN nanowires: crystallographic, stoichiometric, and detection efficiency aspects. J. Appl. Phys. 114, 184903.

Dillon, A.C., Yudasaka, M., Dresselhaus, M.S., 2004. Employing Raman Spectroscopy to Qualitatively Evaluate the Purity of Carbon Single-Wall Nanotube Materials. Nanotechnology 4, 691–703.

Dirix, Y., Bastiaansen, C., Caseri, W., Smith, P., 1999. Oriented pearl-necklace arrays of metallic nanoparticles in polymers: a new route to polarization dependent color filters. Adv. Mater. 11, 223–227.

Disch, S., Wetterskog, E., Hermann, R.P., Wiedenmann, A., Vainio, U., Salazar-Alvarez, G., Bergström, L., Brückel, Th., 2012. Quantitative spatial magnetization distribution in iron oxide nanocubes and nanospheres by polarized small-angle neutron scattering. New J. Phys. 14, 013025.

Dopke, N.C., Treichel, P.M., Vestling, M.M., 1998. Matrix-assisted laser desorption/ionization time-of-flight mass spectrometry (MALDI-TOF MS) of rhenium(III) halides: a characterization tool for metal atom clusters. Inorg. Chem. 37, 1272–1277.

Dorofeev, G.A., Streletskii, A.N., Povstugar, I.V., Protasov, A.V., Elsukov, E.P., 2012. Determination of nanoparticle sizes by X-ray diffraction. Colloid J. 74, 675–685.

Dresselhaus, M.S., Jorio, A., Souza Filho, A.G., Saito, R., 2010. Defect characterization in graphene and carbon nanotubes using Raman spectroscopy. Philos. Trans. A 368, 5355–5377.

Du, S.C., Burgess, T., Loi, S.T., Gault, B., Gao, Q., Bao, P.T., Li, L., Cui, X.Y., Yeoh, W.K., Tan, H.H., Jagadish, C., Ringer, S.P., Zheng, R.K., 2013a. Full tip imaging in atom probe tomography. Ultramicroscopy 124, 96–101.

Du, S.C., Burgess, T., Gault, B., Gao, Q., Bao, P.T., Li, L., Cui, X.Y., Kong Yeoh, W., Liu, H.W., Yao, L., Ceguerra, A.V., Tan, H.H., Jagadish, C., Ringer, S.P., Zheng, R.K., 2013b. Quantitative dopant distributions in GaAs nanowires using atom probe tomography. Ultramicroscopy 132, 186–192.

EAG, 2007. Nuclear Reactor Analysis (ARA) for Li, B, C, N, O and F. AN-381. Evans Analytical Group.

Ehring, H., Karas, M., Hillenkamp, F., 1992. Role of photoionization and photochemistry in ionization processes of organic-molecules and relevance for matrix-assisted laser desorption ionization mass-spectrometry. Org. Mass Spectrom. 27, 472–480.

Elzey, S., Grassian, V.H., 2010a. Agglomeration, isolation and dissolution of commercially manufactured silver nanoparticles in aqueous environments. J. Nanopart. Res. 12, 1945–1958.

Elzey, S., Grassian, V.H., 2010b. Nanoparticle dissolution from the particle perspective: insights from particle sizing measurements. Langmuir 26, 12505–12508.

Endisch, D., Rauch, F., Götzelmann, A., Reiter, G., Stamm, M., 1992. Application of the ^{15}N nuclear reaction technique for hydrogen analysis in polymer thin films. Nucl. Instrum. Methods Phys. Res. Sect. B 62, 513–520.

Ene, A., Popescu, I.V., Badica, T., 2006. Determination of carbon in steels using particle-induced gamma ray spectrometry. J. Optoelectron. Adv. Mater. 8, 222–224.

Epicier, T., Sato, K., Tournus, F., TKonno, T., 2010. Chemical composition dispersion in bi-metallic nanoparticles: semi-automated analysis using HAADF-STEM. Anal. Bioanal. Chem. 396, 983–1002.

Evans, D.F., 1959. The Determination of the Paramagnetic Susceptibility of Substances in Solution by Nuclear Magnetic Resonance. Proc. Chem. Soc. 2003–2005.

Evans, D.F., 1974. A new type of magnetic balance. J. Phys. E Sci. Instrum. 7, 247. http://dx.doi.org/10.1088/0022-3735/7/4/007.

Feldman, L.C., Mayer, J.W., 1986. Fundamentals of Surface and Thin Film Analysis. North Holland-Elsevier, New York.

Ferrari, A.C., Basko, D.M., 2013. Raman spectroscopy as a versatile tool for studying the properties of graphene. Nat. Nanotechnol. 8, 235–246.

Field, L.D., Baker, M.V., Hambley, T.W., 1988. Paramagnetic – diamagnetic equilibria in solutions of bis(dialkylphosphino)ethane complexes of iron. Inorg. Chem. 27, 2872–2876.

de la Figuera, J., Puerta, J.M., Cerda, J.I., El Gabaly, F., McCarty, K.F., 2006. Determining the structure of Ru(0001) from low-energy electron diffraction of a single terrace. Surf. Sci. 600, L105–L109.

Filippo, E., Serra, A., Manno, D., 2009. Poly(vinyl alcohol) capped silver nanoparticles as localized surface plasmon resonance-based hydrogen peroxide sensor. Sens. Actuators B 138, 625–630.

Flagan, R.C., 2008. Differential mobility analysis of arisols: a tutorial. KONA Powder Part. J. 26, 254–268.

Fleischmann, M., Hendra, P.J., McQuillan, A.J., 1974. Raman-spectra of pyridine adsorbed at a silver electrode. Chem. Phys. Lett. 26, 163–166.

Freeman, R.G., Grabar, K.C., Allison, K.J., Bright, R.M., Davis, J.A., Guthrie, A.P., Hommer, M.B., Jackson, M.A., Smith, P.C., Walter, D.G., Natan, M.J., 1995. Self-assembled metal colloid monolayers: an approach to SERS substrates. Science 267, 1629–1632.

Freud, P.J., 2012. Nnoparticle Sizing – Dynamic Light Scattering Analysis in the Frequency Spectrum Mode. Application Note SL-AN-14 Rev C., Microtrac, Inc. Particle Size Measuring Instrumentation. http://www.microtrac.net/MTWP/wp-content/uploads/2012/10/Microtrac-Application-Notes-Nanoparticle-Sizing-Dynamic-Light-Scattering-Analysis-in-Frequency-Spectrum-Mode.pdf.

Frisbie, C.D., Rozsnyai, L.F., Noy, A., Wrighton, Lieber, C.M., 1994. Functional group imaging by chemical force microscopy. Science 265, 2071–2074.

Fritzen-Garcia, M.B., Zanetti-Ramos, B.G., de Oliveira, C.S., Soldi, V., Pasa, A.A., Creczynski-Pasa, T.B., 2009. Atomic force microscopy imaging of polyurethane nanoparticles onto different solid substrates. Mater. Sci. Eng. C 29, 405–409.

Fritzsche, W., Taton, T.A., 2003. Metal nanoparticles as labels for heterogeneous, chip-based DNA detection. Nanotechnology 14, R63–R73.

Fua, J., Park, B., Zhaoa, Y., 2009. Limitation of a localized surface plasmon resonance sensor for *Salmonella* detection. Sens. Actuators B 141, 276–283.

Gallon, T.E., Matthew, J.A.D., 1972. Auger electron spectroscopy and its application to surface studies. Rev. Phys. Technol. 3, 31.

Gavaza, G.M., Yu, Z.X., Tsang, Chan, C.H., Tong, S.Y., Van Hove, M.A., 2007. Theory of low-energy electron diffraction for detailed structural determination of nanomaterials: ordered structures. Phys. Rev. B 75, 014114.

Gelinas, V., Vidal, D., 2010. Determination of particle shape distribution of clay using an automated AFM image analysis method. Powder Technol. 203, 254.

Gesquiere, A.J., Park, S.J., Barbara, P.F., 2005. Hole-induced quenching of triplet and singlet excitons in conjugated polymers. J. Am. Chem. Soc. 127, 9556–9560.

Gesquiere, A.J., Park, S.J., Barbara, P.F., 2004a. F-V/SMS: a new technique for studying the structure and dynamics of single molecules and nanoparticles. J. Phys. Chem. B 108, 10301–10308.

Gesquiere, A.J., Park, S.J., Barbara, P.F., 2004b. Photochemistry and kinetics of single organic nanoparticles in the presence of charge carriers. Eur. Polym. J. 40, 1013–1018.

Gesquiere, A.J., Tenery, D., Hu, Z., 2008. Single-particle spectroscopy on conducting polymer-fullerene composite materials for application in organic photovoltaic devices. Spectroscopy 32–44. http://www.spectroscopyonline.com/spectroscopy/article/articleDetail.jsp?id=507431.

Giddings, J.C., 1993. Field-flow fractionation: analysis of macromolecular, colloidal and particulate materials. Science 260, 1456–1465.

Gilbert, J.B., Rubner, M.F., Cohen, R.E., 2013. Depth-profiling X-ray photoelectron spectroscopy (XPS) analysis of interlayer diffusion in polyelectrolyte multilayers. PNAS 110, 6651–6656.

Giovannozzi, A.M., Rolle, F., Sega, M., Abete, M.C., Marchis, D., Rossie, A.M., 2014. Rapid and sensitive detection of melamine in milk with gold nanoparticles by surface enhanced Raman scattering. Food Chem. 159, 250–256.

Glatter, O., Kratky, O., 1982. Small Angle X-Ray Scattering. Academic Press, 111 Fifth Avenue, New York, NY, 10003.

Gonçalves, N.S., Carvalho, J.A., Lima, Z.M., Sasaki, J.M., 2012. Size–strain study of NiO nanoparticles by X-ray powder diffraction line broadening. Mater. Lett. 72, 36–38.

González-Solís, J.L., Aguiñaga-Serrano, B.I., Martínez-Espinosa, J.C., Oceguera-Villanueva, A., 2011. Stage determination of breast cancer biopsy using Raman spectroscopy and multivariate analysis. AIP Conf. Proc. 1364, 33–40.

González-Solís, J.L., Luévano-Colmenero, G.H., Vargas-Mancilla, J., 2013. Surface enhanced Raman spectroscopy in breast cancer cells. Laser Ther. 22, 37–42.

González-Solís, J.L., Rodriguez-López, J., Martínez-Espinosa, J.C., Frausto-Reyes, C., Jave-Suárez, L.F., Aguilar-Lemarroy, A.C., Vargas-Rodríguez, H., Martínez-Cano, E., 2009. Detection of cervical cancer analyzing blood samples with Raman spectroscopy and multivariate analysis. AIP Conf. Proc. 1126, 91–95.

Grant, D.H., 1995. Paramagnetic susceptibility by NMR the "Solvent correction" reexamined. J. Chem. Educ. 72, 39–40.

Grassian, V.H., 2008. When size really matters: size-dependent properties and surface chemistry of metal and metal oxide. Nanoparticles in Gas and liquid phase environments. J. Phys. Chem. C 112, 18303–18313.

Grenha, A., Seijo, B., Serra, C., Remunan-Lopez, C., 2008. Surface characterization of lipid/chitosan nanoparticles assemblies, using X-ray photoelectron spectroscopy and time-of-flight secondary ion mass spectrometry. J. Nanosci. Nanotechnol. 8, 358–365.

Gschwind, S., Flamigni, L., Koch, J., Borovinskaya, O., Groh, S., Niemax, K., Günther, D., 2011. Capabilities of inductively coupled plasma mass spectroscopy for the detection of nanoparticles carried by monodisperse microdroplets. J. Anal. At. Spectrom. 26, 1166–1174.

Gu, Y.-J., Cheng, J., Lin, C.-C., Lam, Y.W., Cheng, S.H., Wong, W.-T., 2009. Nuclear penetration of surface functionalized gold nanoparticles. Toxicol. Appl. Pharmacol. 237, 196–204.

Guan, B., Lu, W., Fang, J., Cole, R.B., 2007. Characterization of synthesized titanium oxide nanoclusters by MALDI-TOF mass spectrometry. J. Am. Soc. Mass Spectrom. 18, 517–524.

Gubicza, J., Szepvolgyi, J., Mohai, I., Zsoldos, L., Ungar, T., 2000. Particle size distribution and dislocation density determined by high resolution X-ray diffraction in nanocrystalline silicon nitride powders. Mater. Sci. Eng. A 280, 263–269.

Guha, S., Li, M., Tarlov, M.J., Zachariah, M.R., 2012. Electrospray–differential mobility analysis of bionanoparticles. Trends Biotechnol. 30, 291–300.

Guinier, A., Fournet, G., Walker, C., Yudowitch, K., 1995. Small Angle Scattering of X-Rays. John Wiley and Sons, Inc.

Gulde, M., Schweda, S., Storeck, G., Maiti, M., Yu, H.K., Wodtke, A.M., Schäfer, S., Ropers, C., 2012. Ultrafast low-energy electron diffraction in transmission resolves polymer/graphene superstructure dynamics. Science 345, 200–204.

Gurbich, A., 2013. Nuclear Cross Sections Analysis and R-matrix Tools – Minischool.

Haes, A.J., Van Duyne, R.P., 2002a. A nanoscale optical biosensor: sensitivity and selectivity of an approach based on the localized surface plasmon resonance spectroscopy of triangular silver nanoparticles. J. Am. Chem. Soc. 124, 10596–10604.

Haes, A.J., Van Duyne, R.P., 2002b. A highly sensitive and selective surface-enhanced nanobiosensor. Mater. Res. Soc. Symp. Proc. 723, O3.1.1–O3.1.6.

Haes, A.J., Van Duyne, R.P., 2003a. Nanosensors enable portable detectors for environmental and medical applications. Laser Focus World 39, 153–156.

Haes, A.J., Van Duyne, R.P., 2003b. Nanoscale optical biosensors based on localized surface plasmon resonance spectroscopy. In: Proceedings of SPIE. The International Society for Optical Engineering, vol. 5221, pp. 47–58.

Haes, A.J., Zou, S., Schatz, G.C., Van Duyne, R.P., 2004a. A nanoscale optical biosensor: the short range distance dependence of the localized surface plasmon resonance of silver and gold nanoparticles. J. Phys. Chem. B 108, 6961–6968.

Haes, A.J., Zou, S., Schatz, G.C., Van Duyne, R.P., 2004b. A nanoscale optical biosensor: the long range distance dependence of the localized surface plasmon resonance of noble metal nanoparticles. J. Phys. Chem. B 10, 109–116.

Han, X.X., Huang, G.G., Zhao, B., Ozaki, Y., 2009. Label-free highly sensitive detection of proteins in aqueous solutions using surface-enhanced Raman scattering. Anal. Chem. 81, 3329–3333.

Hao, E., Li, S., Bailey, R.C., Zou, S., Schatz, G.C., Hupp, J.T., 2004. Optical properties of metal nanoshells. J. Phys. Chem. B 108, 1224–1229.

Hargreaves, M.D., Page, K., Munshi, T., Tomsett, R., Lynch, G., Edwards, H.G.M., 2008. Analysis of seized drugs using portable Raman spectroscopy in an airport environment—a proof of principle study. J. Raman Spectrosc. 39, 873–880.

Harrison, A.G., 1997. The gas-phase basicities and proton affinities of amino acids and peptides. Mass Spectrom. Rev. 16, 201–217.

Hatziantoniou, S., Deli, G., Nikas, Y., Demetzos, C., Papaioannou, G.T., 2007. Scanning electron microscopy study on nanoemulsions and solid lipid nanoparticles containing high amounts of ceramides. Micron 38, 819–823.

Haynes, C.L., McFarland, A.D., Zhao, L., Van Duyne, R.P., Schatz, G.C., Gunnarsson, L., Prikulis, J., Kasemo, B., Kall, M., 2003. Nanoparticle optics: the importance of the radiative dipole coupling in two-dimensional nanoparticle arrays. J. Phys. Chem. B 107, 7337–7342.

Haynes, C.L., Van Duyne, R.P., 2003a. Plasmon-sampled surface-enhanced Raman excitation spectroscopy. J. Phys. Chem. B 107, 7426–7433.

Haynes, C.L., Van Duyne, R.P., 2003b. Plasmon-sampled surface-enhanced Raman excitation spectroscopy. Nano Lett. 3, 939–943.

Helfrich, A., Bettmer, J., 2011. Analysis of gold nanoparticles using ICP-MS-based hyphenated and complementary ESI-MS techniques. Int. J. Mass Spectrom. 307, 92–98.

Hennrich, F., Krupke, R., Lebedkin, S., Arnold, K., Fischer, R., Resasco, D.E., Kappes, M.M., 2005. Raman spectroscopy of individual single-walled carbon nanotubes from various sources. J. Phys. Chem. B 109, 10567–10573.

Herzing, A.A., Watanabe, M., Edwards, J.K., Conte, M., Tang, Z.-R., Hutchings, G.J., Kiely, C.J., 2008. Energy dispersive X-ray spectroscopy of bimetallic nanoparticles in an aberration corrected scanning transmission electron microscope. Faraday Discuss. 138, 337–351.

Hirsch, L.R., Jackson, J.B., Lee, A., Halas, N.J., West, J.L., 2003. A whole blood immunoassay using gold nanoshells. Anal. Chem. 75, 2377–2381.

Hollander, A., Haupt, M., Oehr, C., 2007. On depth profiling of polymers by argon ion sputtering. Plasma Processes Polym. 4, 773–776.

Hono, K., Murayama, M., 1997. Nanoscale microstructural analyses by atom probe field ion microscopy. High Temp. Mater. Processes 17, 69–86.

Hoo, C.M., Starostin, N., West, P., Mecartney, M.L., 2008. A comparison of atomic force microscopy (AFM) and dynamic light scattering (DLS) method to characterize nanoparticle size distributions. J. Nanopart. Res. 10, 89–96.

van Hove, M.A., Tong, S.Y., 1979. Surface Crystallography by LEED. Springer, Berlin.

Hu, D.H., Yu, J., Wong, K., Bagchi, B., Rossky, P.J., Barbara, P.F., 2000. Collapse of stiff conjugated polymers with chemical defects into ordered, cylindrical conformations. Nature 405, 1030–1033.

Hu, W.P., Chen, S.J., Huang, K.T., Hsu, J.H., Chen, W.Y., Chang, G.L., Lai, K.A., 2004. A novel ultrahigh-resolution surface plasmon resonance biosensor with an Au nanocluster-embedded dielectric film. Biosens. Bioelectron. 19, 1465–1471.

Hu, Y., Wunsch, B.H., Sahni, S., Stellacci, F., 2009. Statistical analysis of scanning tunneling microscopy images of 'Striped' mixed monolayer protected gold nanoparticles. J. Scanning Probe Microsc. 4, 24–35.

van de Hulst, H.C., 1981. Light Scattering by Small Particles. Dover, New York.

Huser, T., Yan, M., Rothberg, L.J., 2000. Single chain spectroscopy of conformational dependence of conjugated polymer photophysics. Proc. Natl. Acad. Sci. U.S.A. 97, 11187–11191.

Huttel, Y., HGomez, H., Clavero, C., Cebollada, A., Armelles, G., Navarro, E., Ciria, M., Benito, L., Arnaudas, J.I., Kellock, A.J., 2004. Cobalt nanoparticles deposited and embedded in AlN: magnetic, magneto-optical, and morphological properties. J. Appl. Phys. 96, 1666.

Hutter, E., Fendler, J.H., 2004. Exploitation of localized surface plasmone resonance. Adv. Mater. 16, 1685–1706.

Ichiyanagi, Y., Kimishima, Y., 2002. Structural, magnetic and thermal characterizations of Fe_2O_3 nanoparticle systems. J. Therm. Anal. Calorim. 69, 919–923.

Issa, B., Qadri, S., Obaidat, I.M., Bowtell, R.W., Haik, Y., 2011. PEG coating reduces NMR relaxivity of Mn(0.5)Zn(0.5) Gd(0.02)Fe(1.98)O_4 hyperthermia nanoparticles. J. Magn. Reson. Imaging 34, 1192–1198.

Jablonski, A., 2000. Quantitative surface analysis by X-ray photoelectron spectroscopy. Pol. J. Chem. 74, 1533–1566.

Jain, P.K., Lee, K.S., El-Sayed, I.H., El-Sayed, M.A., 2006. Calculated absorption and scattering properties of gold nanoparticles of different size, shape, and composition: applications in biological imaging and biomedicine. J. Phys. Chem. B 110, 7238–7248.

Jenkins, R., 2000. X-ray techniques: overview. Encycl. Anal. Chem. http://www.spectroscopynow.com/userfiles/sepspec/file/specNOW/Enc_Anal_Chem_6801.pdf.

Jiang, H., Moon, K., Dong, H., Hua, F., Wong, C.P., 2006. Size-dependent melting properties of tin nanoparticles. Chem. Phys. Lett. 429, 492–496.

Jimenez, M.S., Gómez, M.T., Bolea, E., Laborda, F., Castillo, J., 2010. An approach to the natural and engineered nanoparticles analysis in the environment by inductively coupled plasma mass spectrometry. Atmos. Environ. 44, 5035–5052.

Johnson, A.C., Greenwood, P., Hagstrom, M., Abbas, Z., Wall, S., 2008. Aggregation of nanosized colloidal silica in the presence of various alkali cations investigated by the electrospray technique. Langmuir 2, 12798–12806.

Kahl, M., Voges, E., Kostrewa, S., Viets, C., Hill, W., 1998. Periodically structured metallic substrates for SERS. Sens. Actuators B 51, 285–291.

Kalbasi, R.J., Mosaddegh, N., 2012. Application of atomic absorption for determination of metal nanoparticles. In: Farrukh, M.A. (Ed.), Organic-Inorganic Nanocomposites. Atomic Absorption Spectroscopy. InTech Publisher.

Kamimura, O., Dobash, T., 2012. Low-energy electron diffractive imaging for three-dimensional light-element materials. Hitachi Rev. 61 (6), 269.

Kapellios, S., Karamanou, M.F., Sardis, M., Aivaliotis, A., Economou, S., Pergantis, A., 2011. Using nanoelectrospray ion mobility spectrometry (GEMMA) to determine the size and relative molecular mass of proteins and protein assemblies: a comparison with MALLS and QELSE. Anal. Bioanal. Chem. 399, 2411–2433.

Kawaguchi, T., Shankaran, D.R., Kim, S.J., Gobi, K.V., Matsumoto, K., Toko, K., Miura, N., 2007. Fabrication of a novel immunosensor using functionalized self-assembled monolayer for trace level detection of TNT by surface plasmon resonance. Talanta 72, 554–560.

Kawaguchi, T., Shankaran, D.R., Kim, S.J., Matsumoto, K., Toko, K., Miura, N., 2008. Surface plasmon resonance immunosensor using Au nanoparticle for detection of TNT. Sensors and Actuators B: Chemical 133, 467–472.

Kawahara, K., Shirasawa, T., Arafune, R., Lin, C.-L., Takahashi, T., Kawai, M., Takagi, N., 2014. Determination of atomic positions in silicene on Ag(111) by low-energy electron diffraction. Surf. Sci. 623, 25–28.

Khitrov, G.A., Strouse, G.F., 2003. ZnS nanomaterial characterization by MALDI-TOF mass spectrometry. J. Am. Chem. Soc. 125, 10465–10469.

Kim, B.H., Chang, H., Hackett, M.J., Park, J., Seo, P., Hyeon, T., 2014a. Size characterization of ultrasmall silver nanoparticles using MALDI-TOF mass spectrometry. Bull. Korean Chem. Soc. 35, 961–964.

Kim, H., So, W.W., Moon, S.J., 2006a. Effect of thermal annealing on the performance of P3HT/PCBM polymer photovoltaic cells. J. Korean Phys. Soc. 48, 441–445.

Kim, J.H., Mulholland, G.W., Kukuck, S.R., Pui, D.Y.H., 2005. Slip correction measurements of certified PSL nanoparticles using a nanometer differential mobility analyzer (nano-DMA) for Knudsen number from 0.5 to 83. J. Res. Natl. Inst. Stand. Technol. 110, 31–54.

Kim, Y., Cook, S., Tuladhar, S.M., Choulis, S.A., Nelson, J., Durrant, J.R., Bradley, D.D.C., Giles, M., McCulloch, I., Ha, C.S., Ree, M., 2006b. A strong regioregularity effect in self-organizing conjugated polymer films and high-efficiency polythiophene:fullerene solar cells. Nat. Mater. 5, 197–203.

Kim, Y.-P., Shon, S.K., Shin, S.K., Lee, T.G., 2014b. Probing nanoparticles and nanoparticle-conjugated biomolecules using time-of-flight secondary ion mass spectrometry. Mass Spectrom. Rev. http://dx.doi.org/10.1002/mas.21437. Early pub.

Kinney, P.D., Pui, D.Y.H., 1991. Use of the electrostatic classification method to size 0.1 um SRM particles – a feasibility study. J. Res. Natl. Inst. Stand. Technol. 96, 147–176.

Kirschbrown, J., 2007. Small-Angle X-Ray Scattering: A Concise Review. http://www.unc.edu/~justink/Justin_Kirschbrown_SAXS_A_Concise_Review.pdf.

Klapetek, P., Ohlídal, I., Franta, D., Montaigne-Ramil, A., Bonanni, A., Stifter, D., Sitter, H., 2003. Atomic force microscopy characterization of ZnTe epitaxial films. Acta Phys. Slovaca 53, 223.

Klapetek, P., Valtr, M., Necas, D., Salyk, O., Dzik, P., 2011. Atomic force microscopy analysis of nanoparticles in non-ideal conditions. Nanoscale Res. Lett. 6, 514.

Kneipp, K., Kneipp, H., Manoharan, R., Itzkan, I., Dasari, R.R., Feld, M.S., 1998. Surface-enhanced Raman scattering (SERS)—a new tool for single molecule detection and identification. Bioimaging 6, 104–110.

Kneipp, K., Wang, Y., Dasari, R.R., Feld, M.S., 1995. Approach to single-molecule detection using surface-enhanced resonance Raman scattering (SERRS)—a study of rhodamine 6G on colloidal silver. Appl. Spectrosc. 49, 780–784.

Kneipp, K., Wang, Y., Kneipp, H., Perelman, L.T., Itzkan, I., Dasari, R.R., Feld, M.S., 1997. Single-molecule detection using surface-enhanced Raman scattering (SERS). Phys. Rev. Lett. 7, 1667–1670.

Kobayashi, T., 1976. Photoelectron spectroscopy and its application to chemistry. J. Sci. Synth. Org. Chem. Jpn. 34, 17–29.

Kokkoris, M., 2012. Nuclear reaction analysis (NRA) and particle-induced gamma-ray emission (PIGE). Charact. Mater. 1–18.

Kopnov, F., Tenne, R., Späth, B., Jägermann, W., Cohen, H., Feldman, Y., Zak, A., Moshkovich, A., Rapoport, L., 2008. X-ray photoelectron spectroscopy and tribology studies of annealed fullerene-like WS_2 nanoparticles. NATO Sci. Peace Secur. Ser. B Phys. Biophys. 51–59.

Kopnov, F., Tenne, R., Späth, B., Jägermann, W., Cohen, H., Feldman, Y., Zak, A., Moshkovich, A., Rapoport, L., Nobuta, T., Ogawa, T., 2009. Depth profile XPS analysis of polymeric materials by c-60 (+) ion sputtering. J. Mater. Sci. 44, 1800–1812.

Kumar, G.V.P., 2012. Plasmonic nano-architectures for surface enhanced Raman scattering: a review. J. Nanophotonics 6, 064503.

Kulik, T., Savage, H.T., Hernando, A., 1993. A high-performance hysteresis loop tracer. J. Appl. Phys. 73, 6855–6857.

Kumiko, T., 2009. X-ray photoelectron spectroscopy (XPS, ESCA). Sen I Gakkaishi 65, 171–174.

Lai, S.L., Guo, J.Y., Petrova, V., Ramanath, G., Allen, L.H., 1996. Size-dependent melting properties of small tin particles: nanocalorimetric measurements. Phys. Rev. Lett. 77, 99–102.

Langford, J.I., Wilson, A.J.C., 1978. Scherrer after sixty years: a survey and some new results in the determination of crystallite size. J. Appl. Cryst. 11, 102–113.

Lee, G.W.J., Groves, M.J., 1981. A pragmatic approach to the particle sizing of submicrometre emulsions using a laser nephelometer. Powder Technol. 28, 49–54.

Lee, I., Chan, K.Y., Phillips, D.L., 1998. Atomic force microscopy of platinum nanoparticles prepared on highly oriented pyrolytic graphite. Ultramicroscopy 75, 69.

Lee, K.-S., El-Sayed, M.A., 2006. Gold and silver nanoparticles in sensing and imaging: sensitivity of plasmon response to size, shape, and metal composition. J. Phys. Chem. B 110, 19220–19225.

Lee, S., Rao, S.P., Moon, M.H., Giddings, J.C., 1996. Determination of mean diameter and particle size distribution of acrylate latex using flow field-flow fractionation, photon correlation spectroscopy, and electron microscopy. Anal. Chem. 68, 1545–1549.

Lee, Y.J., Park, S.J., Gesquiere, A.J., Barbara, P.F., 2005. Single molecule spectroscopy on organic light-emitting diodes. Appl. Phys. Lett. 87, 051906.

Li, M., Lu, J., Qi, J., Zhao, F., Zeng, J., Yu, J.C., Shih, W.-C., 2014. Stamping surface-enhanced Raman spectroscopy for label-free, multiplexed, molecular sensing and imaging. J. Biomed. Opt. 19, 050501.

Lifshin, E., 1999. X-Ray Characterization of Materials. Wiley-VCH.

Link, S., El-Sayed, M.A., 1999. Spectral properties and relaxation dynamics of surface plasmon electronic oscillations in gold and silver nanodots and nanorods. Phys. Chem. B 103, 8410–8426.

Liu, F.M., Zhang, L.D., 1999. X-ray photoelectron spectroscopy of GaSb nanoparticles embedded in SiO_2 matrices by radio-frequency magnetron co-sputtering. Semicond. Sci. Technol. 14, 710.

Loliger, J., Scheffold, R., 1972. Paramagnetic moment measurements by NMR. A micro technique. J. Chem. Educ. 49, 646–647.

Lord, R.C., Yu, N.-T., 1970. Laser-excited Raman spectroscopy of biomolecules. I. Native lysozyme and its constituent amino acids. J. Mol. Biol. 50, 509–524.

Mahadevan, S., Gnanaprakash, G., Philip, J., Rao, B.P.C., Jayakumar, T., 2007. X-ray diffraction-based characterization of magnetite nanoparticles in presence of goethite and correlation with magnetic properties. Phys. E 39, 20–25.

Maher, R.C., 2012. SERS hot spots. In: Kumar, C. (Ed.), Raman Spectroscopy for Nanomaterials Characterization. Springer Materials.

Mahoney, C.M., 2010. Cluster secondary ion mass spectrometry of polymers and related materials. Mass Spectrom. Rev. 29, 247–293.

Maier, S.A., Brongersma, M.L., Kik, P.G., Meltzer, S., Requicha, A.A.G., Atwater, H.A., 2001. Plasmonics—a route to nanoscale optical devices. Adv. Mater. 13, 1501–1505.

Maier, S.A., Kik, P.G., Atwater, H.A., Meltzer, S., Harel, E., Koel, B.E., Requicha, A.A.G., 2003. Local detection of electromagnetic energy transport below the diffraction limit in metal nanoparticle plasmon waveguides. Nat. Mater. 2, 229–232.

Martel, R., Avouris, Ph., Lyo, I.-W., 1996. Molecularly adsorbed oxygen species on Si(111)-(7 × 7): STM-induced dissociative attachment studies. Science 272, 385–388.

Martinez, E., Borowik, L., You, L., Chevalier, N., Guedj, C., Auvert, G., Roussey, A., Jousseaume, V., Boujamaa, R., Gros-Jean, M., Barbé, J.C., Feuillet, G., Bertin, F., Chabli, A., 2011. Nanoscale Chemical Characterization by Auger Electron Spectroscopy. The 2011 Conference, National Institute of Standards and Technology (NIST). http://www.nist.gov/pml/div683/conference/upload/tu-02.pdf.

Masona, T.G., Lin, M.Y., 2007. Time-resolved small angle neutron scattering measurements of asphaltene nanoparticle aggregation kinetics in incompatible crude oil mixtures. J. Chem. Phys. 119, 565–571.

Masotti, A., Marino, F., Ortaggi, G., Palocci, C., 2007. Fluorescence and scanning electron microscopy of chitosan/DNA nanoparticles for biological applications. In: Méndez-Vilas, A., Díaz, J. (Eds.), Modern Research and Educational Topics in Microscopy. Formatex.

Mayer, C., 2005. NMR studies of nanoparticles. Annu. Rep. NMR Spectrosc. 55, 205–258.

McGeehan, J.E., Bourgeois, D., Royant, A., Carpentier, P., 2011. Raman-assisted crystallography of biomolecules at the synchrotron: instrumentation, methods and applications. Biochim. Biophys. Acta 1814, 750–759.

Millard, M.M., Bartholomew, J.C., 1977. Surface studies of mammalian-cell grown in culture by X-ray photoelectron-spectroscopy. Anal. Chem. 49, 1290–1296.

Miller, M.K., 2000. The development of atom probe field-ion microscopy. Mater. Charact. 44, 11–27.

Mitrano, D., Ranville, J., Thomas, R., Neubauer, K., 2012. Field-flow fractionation coupled with ICP-MS for the analysis of engineered nanoparticles in environmental samples. TCGC Europe 25, 652–665.

Moerner, W.E., Orrit, M., 1999. Illuminating single molecules in condensed matter. Science 283, 1670–1676.

Moglianetti, M., Ong, Q.K., Reguera, J., Harkness, K.M., Mi, M.M., Radulescu, A., Kohlbrecher, J., Jud, C., Svergun, D.I., Stellacci, F., 2014. Scanning tunneling microscopy and small angle neutron scattering study of mixed monolayer protected gold nanoparticles in organic solvents. Chem. Sci. 5, 1232–1240.

Moharram, A.F., Mansour, S.A., Hussein, M.A., Rashad, M., 2014. Direct precipitation and characterization of ZnO nanoparticles. J. Nanomater. 2014, 716210.

Moller, W., Hufschmidt, M., Pfeiffer, Th., 1978. Diffusion studies by means of nuclear reaction depth profiling. Nucl. Instrum. Methods 149, 73–76.

Montejano, J.M., Rodríguez, J.L., Gutierrez-Wing, C., Miki, M., Jose-Yacaman, M., 2004. Crystallography and shape of nanoparticles and clusters. In: Nalwa, H.S. (Ed.), Encyclopaedia of Nanoscience and Nanotechnology, pp. 1–44.

Montesinos, E., Esteve, I., Guerrero, R., May 1983. Comparison between direct methods for determination of microbial cell volume: electron microscopy and electronic particle sizing. Appl. Environ. Microbiol. 45, 1651–1658.

Mucic, R.C., Storhoff, J.J., Mirkin, C.A., Letsinger, R.L., 1998. DNA-directed synthesis of binary nanoparticle network materials. J. Am. Chem. Soc. 120, 12674–12675.

Muramoto, S., Graham, D.J., Wagner, M.S., Lee, T.G., Moon, D.W., Castner, D.G., 2011. ToF-SIMS analysis of adsorbed proteins: principal component analysis of the primary ion species effect on the protein fragmentation patterns. J. Phys. Chem. C 115, 24247–24255.

Mustafa, C., 2013. Surface-enhanced Raman scattering: an emerging label-free detection and identification technique for proteins. Appl. Spectrosc. 67, 355–364.

Nagao, O., Harada, G., Sugawara, T., Sasaki, A., Ito, Y., 2004. Small-angle X-ray scattering method to determine the size distribution of gold nanoparticles chemisorbed by thiol ligands. Jpn. J. Appl. Phys. 43, 7742.

Nath, N., Chilkoti, A., 2002. Immobilized gold nanoparticle sensor for label-free optical detection of biomolecular interactions. Proc. SPIE Int. Soc. Opt. Eng. 4626, 441–448.

Nie, S., Zare, R., 1997. Optical detection of single molecules. Annu. Rev. Biophys. Biomol. Struct. 26, 567–596.

Nie, S., Emory, S.R., 1997. Probing single molecules and single nanoparticles by surface-enhanced Raman scattering. Science 275, 1102–1116.

Nima, Z.A., Biswas, A., Bayer, I.S., Hardcastle, F.D., Perry, D., Ghosh, A., Dervishi, E., Biris, A.S., 2014. Applications of surface-enhanced Raman scattering in advanced bio-medical technologies and diagnostics. Drug Metab. Rev. 46, 155–175.

Nobuta, T., Ogawa, T., 2009. Depth profile XPS analysis of polymeric materials by c-60 (+) ion sputtering. J. Mater. Sci. 44, 1800–1812.

Noy, A., Vezenov, D.V., Lieber, C.M., 1997. Chemical force microscopy. Annu. Rev. Mater. Sci. 27, 381–421.

Obare, S.O., Hollowell, R.E., Murphy, C.J., 2002. Sensing strategy for lithium ion based on gold nanoparticles. Langmuir 18, 10407–10410.

Ong, J.L., Lucas, L.C., 1998. Auger electron spectroscopy and its use for the characterization of titanium and hydroxyapatite surfaces. Biomaterials 19, 455–464.

Patri, A., Umbreit, T., Zheng, J., Nagashima, K., Goering, P., Francke-Carroll, S., Gordon, E., Weaver, J., Miller, T., Sadrieh, N., McNeil, S., Stratmeyer, M., 2009. Energy dispersive X-ray analysis of titanium dioxide nanoparticle distribution after intravenous and subcutaneous injection in mice. J. Appl. Toxicol. 29, 662–672.

Pease, L.F., Elliott, J.T., Tsai, D.H., Zachariah, M.R., Tarlov, M.J., 2008. Determination of protein aggregation with differential mobility analysis: application to IgG antibody. Biotechnol. Bioeng. 101, 1214–1222.

Pease, L.F., Tsai, D.H., Zangmeister, R.A., Zachariah, M.R., Tarlov, M.J., 2007. Quantifying the surface coverage of conjugate molecules on functionalized nanoparticles. J. Phys. Chem. C 111, 17155–17157.

Perlaki, C.M., Liu, Q., Lim, M., 2014. Raman spectroscopy based techniques in tissue engineering—an overview. Appl. Spectrosc. Rev. 49, 513–532.

Petkov, V., Ohta, T., Hou, Y., Ren, Y., 2007. Atomic-scale structure of nanocrystals by high-energy X-ray diffraction and atomic pair distribution function analysis: study of Fe_xPd_{100-x} (x = 0, 26, 28, 48) nanoparticles. J. Phys. Chem. C 111, 714–720.

Pichardo-Molina, J.L., Frausto-Reyes, C., Barbosa-García, O., Huerta-Franco, R., González-Trujillo, J.L., Ramírez-Alvarado, C.A., Gutiérrez-Juárez, G., Medina-Gutiérrez, C., 2006. Raman spectroscopy and multivariate analysis of serum simples from breast cancer patients. Laser Med. Sci. 10103, 432–438.

Piguet, C., 1997. Paramagnetic susceptibility by NMR: the "Solvent Correction" removed for large paramagnetic molecules. J. Chem. Educ. 74, 815–816.

Polyakov, B., Vlassov, S., Dorogin, L.M., Butikova, J., Antsov, M., Oras, S., Lohmus, R., Ilmar Kink, I., 2014. Manipulation of nanoparticles of different shapes inside a scanning electron microscope. Beilstein J. Nanotechnol. 5, 133–140.

Ponce, A., Mejía-Rosales, S., Jose-Yacaman, M., 2012. Scanning transmission electron microscopy methods for the analysis of nanoparticles. Methods Mol. Biol. 906, 453–471.

Postawa, Z., Czerwinsk, B., Szewczyk, M., Smiley, E.J., Winograd, N., Garrison, B.J., 2004. Microscopic insights into the sputtering of Ag{111} induced by C60 and Ga bombardment. J. Phys. Chem. B 108, 7831–7838.

Pozzo, D.C., Walker, 2007. Small-angle neutron scattering of silica nanoparticles template in PEO–P-PO–LMPEO cubic crystals. Colloids Surf. A 294, 117–129.

Prabhu, V.M., Reipa, V., 2012. In situ electrochemical small-angle neutron scattering (eSANS) for quantitative structure and redox properties of nanoparticles. J. Phys. Chem. Lett. 5, 646–650.

Puchalski, M., Dabrowski, P., Olejniczak, W., Krukowski, P., Kowalczyk, P., Polanski, K., 2007. The study of silver nanoparticles by scanning electron microscopy, energy dispersive X-ray analysis and scanning tunnelling microscopy. Mater. Sci. Pol. 25, 473–478.

Qazi, S.J.S., Renniea, A.R., Cockcroft, J.K., Vickers, M., 2009. Use of wide-angle X-ray diffraction to measure shape and size of dispersed colloidal particles. J. Colloid Interface Sci. 338, 105–110.

Ramalinga, U., Clogston, J.D., Patri, A.K., Simpson, J.T., 2011. Characterization of nanoparticles by matrix assisted laser desorption ionization time-of-flight mass spectrometry. Methods Mol. Biol. 697, 53–61.

Raman, C.V., Krishnan, K.S., 1928. A new type of secondary radiation. Nature 121, 501–502.

Raman, C.V., 1928. A new radiation. Indian J. Phys. 02, 387.

Raymond, C., 2005. Physical Chemistry for the Biosciences. University Science, New York.

Riboh, J.C., Haes, A.J., McFarland, A.D., Yonzon, C.R., Van Duyne, R.P., 2003. A nanoscale optical biosensor: real time immunoassay and nanoparticle adhesion. J. Phys. Chem. B 107, 1772–1780.

Rodriguez-López, J., 2009. Caracterización de Muestras de Sangre de Pacientes con Cáncer Cérvicouterino Usando Espectroscopia Raman y el Análisis Multivariado en la Plataforma MATLAB. Tesis de Licenciatura en Ing. Mecatrónica, Centro Universitario de los Lagos, Universidad de Guadalajara.

Rosse, P., Loizeau, J.L., 2003. Use of single particle counters for the determination of the number and size distribution of colloids in natural surface waters. Colloids Surf. A 217, 109–120.

Saunderson, A., 1968. A permanent magnet Gouy balance. Phys. Educ. 3, 272. http://dx.doi.org/10.1088/0031-9120/3/5/007.

Scaccia, S., Goszczynska, B., 2004. Sequential determination of platinum, ruthenium, and molybdenum in carbon-supported Pt, PtRu, and PtMo catalysts by atomic absorption spectrometry. Talanta 63, 791–796.

Schatz, G.C., Van Duyne, R.P. (Eds.), 2002. Electromagnetic Mechanism of Surface-Enhanced Spectroscopy. Wiley, New York, pp. 759–774.

Schulte, W.H., Ebbing, H., Becker, H.W., Berheide, M., Buschmann, M., Angulo, C., Rolfs, C., Amsel, G., Trimaille, I., Battistig, G., Mitchell, G.E., Schweitzer, J.S., 1993. High resolution depth profiling in near-surface regions of solids by narrow nuclear reaction resonances below 0.5 MeV with low energy spread proton beams. Vacuum 44, 185–190.

Schwartz, B.J., 2003. Conjugated polymers as molecular materials. How chain conformation and film morphology influence energy transfer and interchain interactions. Annu. Rev. Phys. Chem. 54, 141–172.

Shafer-Peltier, K.E., Haka, A.S., Fitzmaurice, M., Crowe, J., Myles, J., Dasari, R.R., Feld, M.S., 2002. Raman microespectroscopic model of human breast tissue: implications for breast cancer diagnosis in vivo. J. Raman Spectrosc. 33, 552–563.

Shangyuan, F., Rong, Ch., Jianji, P., Yanan, W., Yongzeng, L., Jiesi, Ch., Haishan, Z., 2011. Gastric cancer detection based on blood plasma surface-enhanced Raman spectroscopy excited by polarized laser light. Biosens. Bioelectron. 26, 3167–3174.

Shao, H., Min, C., Issadore, D., Liong, M., Yoon, T.-J., Weissleder, R., 2012. Magnetic nanoparticles and microNMR for diagnostic applications. Theranostics 2, 55–65.

Shao, H., Yoon, T.-T., Liong, M., Weissleder, R., Lee, H., 2010. Magnetic nanoparticles for biomedical NMR-based diagnostics. Beilstein J. Nanotechnol. 1, 142–154.

Shashkov, E.V., Galanzha, E.I., Zharov, V.P., 2010. Photothermal and photoacoustic Raman cytometry in vitro and in vivo. Opt. Express 18, 6929–6944.

Shelby, R.A., Smith, D.R., Schultz, S., 2001. Experimental verification of a negative index of refraction. Science 292, 77–78.

Shen, T.C., Wang, C., Abeln, G.C., Tucker, J.R., Lyding, J.W., Avouris, Ph, Walkup, R.E., 1995. Atomic-scale desorption through electronic and vibrational excitation mechanisms. Science 268, 1590–1592.

Shevchenko, V.Y., Madison, A.E., Smolin, Y.I., 2002. Specific features of diffraction by nanoparticles. Glass Phys. Chem. 28, 326–333.

Shevchenko, V.Y., Madison, A.E., 2002. Structure of Nanoparticles: I. Generalized crystallography of nanoparticles and magic numbers. Glass Phys. Chem. 28, 40–43.

Shi, L., Fleming, C.J., Riechers, S.L., Yin, N.-N., Luo, J., Lam, K.S., Liu, G.-Y., 2011. High-resolution imaging of dendrimers used in drug delivery via scanning probe microscopy. J. Drug Deliv. 2011, 254095.

Shimizu, R., 1983. Quantitative analysis by Auger electron spectroscopy. Jpn. J. Appl. Phys. 22, 1631.

Shoji, F., Kashihara, K., Oura, K., Hanawa, T., 1989. Surface hydrogen detection by low energy $^{4}He^{+}$ ion scattering spectroscopy. Surf. Sci. 220, L719–L725.

Sie, S.H., 1992. Rutherford backscattering spectrometry and nuclear reaction analysis. In: Springer Series in Surface Sciences, vol. 23, pp. 203–219.

de Siqueira, A.F., Cabrera, F.C., Pagamisse, A., Job, A.E., 2014. Segmentation of scanning electron microscopy images from natural rubber samples with gold nanoparticles using starlet wavelets. Microsc. Res. Tech. 77, 71–78.

Soler, M.A.G., Qu, F., 2012. Raman spectroscopy of iron oxide nanoparticles. In: Kumar, C. (Ed.), Raman Spectroscopy for Nanomaterials Characterization. Springer Methods, pp. 379–416.

Spengler, B., Kirsch, D., Kaufmann, R., Jaeger, E., 1992. Peptide sequencing by matrix-assisted laser-desorption mass-spectrometry. Rapid Commun. Mass Spectrom. 6, 105–108.

Spengler, B., Kirsch, D., Kaufmann, R., 1991. Metastable decay of peptides and proteins in matrix-assisted laser-desorption mass-spectrometry. Rapid Commun. Mass Spectrom. 5, 198–202.

Stupik, P.D., Donovan, M.M., Barron, A.R., Jervis, T.R., Nastasi, M., 1992. Thin Solid Films 207, 138.

Sunner, J., 1993. Ionization in liquid secondary-ion mass-spectrometry (Lsims). Org. Mass Spectrom. 28, 805–823.

Swartzentruber, B.S., 1996. Direct measurement of surface diffusion using atom-tracking scanning tunneling microscopy. Phys. Rev. Lett. 76, 459–462.

Szakal, C., Sun, S., Wucher, A., Winograd, N., 2004. C-60 molecular depth profiling of a model polymer. Appl. Surf. Sci. 231, 183–185.

Tachibana, M., 2013. Characterization of laser-induced defects and modification. In: Suzuki, S. (Ed.), Carbon Nanotubes by Raman Spectroscopy, Physical and Chemical Properties of Carbon Nanotubes. InTech, ISBN 978-953-51-1002-6. http://dx.doi.org/10.5772/52091. Available from: http://www.intechopen.com/books/physical-and-chemical-properties-of-carbon-nanotubes/characterization-of-laser-induced-defects-and-modification-in-carbon-nanotubes-by-raman-spectroscopy.

Tadjiki, S., Moldenhauer, E., Pfaffe, T., Analytics, P., Sannac, S., 2013. Characterization of metal nanoparticles using centrifugal FFF-ICP-MS. Agilent ICP-MS J. 53, 1–3.

Taglieri, G., Mondelli, G., Daniele, V., Pusceddu, E., 2013. A. synthesis and X-ray diffraction analyses of calcium hydroxide nanoparticles in aqueous suspension. Adv. Mater. Phys. Chem. 3, 30387.

Techane, S.D., Gamble, L.J., David, G., 2011. Castner X-ray photoelectron spectroscopy characterization of gold nanoparticles functionalized with amine-terminated alkanethiols. Biointerphases 6, 98.

Telgmann, L., Metcalfe, C.D., Hintelmann, H., 2014. Rapid size characterization of silver nanoparticles by single particle ICP-MS and isotope dilution. J. Anal. At. Spectrom. 29, 1265–1272.

Tello, M., Paulo, A.S., Rodríguez, T., Blanco, M., García, R., 2003. Imaging cobalt nanoparticles by amplitude modulation atomic force microscopy: comparison between low and high amplitude solutions. Ultramicroscopy 97, 171–175.

Tomaszewska, E., Soliwoda, K., Kadziola, K., Tkacz-Szczesna, B., Celichowski, G., Cichomski, M., Szmaja, W., Grobelny, J., 2013. Detection limits of DLS and UV-Vis spectroscopy in characterization of polydisperse nanoparticles colloids. J. Nanomater. 313081.

Tonti, D., Zanoni, R., 2009. Measurement methods electronic and chemical properties: X-ray photoelectron spectroscopy. Encycl. Electrochem. Power Sources 673–695.

Traiphol, R., Sanguansat, P., Srikhirin, T., Kerdcharoen, T., Osotchan, T., 2006. Spectroscopic study of photophysical change in collapsed coils of conjugated polymers: effects of solvent and temperature. Macromolecules 39, 1165–1172.

Trocellier, P., Engelmann, Ch., 1986. Hydrogen depth profile measurement using resonant nuclear reaction: an overview. J. Radioanal. Nucl. Chem. 100, 117–127.

Tsunekawa, S., Fukuda, T., Kasuya, A., 2000. X-ray photoelectron spectroscopy of monodisperse CeO_{2-x} nanoparticles. Surf. Sci. 457, L437–L440.

Tu, Q., Chang, C., 2012. Diagnostic applications of Raman spectroscopy. Nanomedicine 8, 545–558.

Turner, D.W., Al Jobory, M.I., 1962. Determination of ionization potentials by photoelectron energy measurement. J. Chem. Phys. 37, 3007.

Vdovenkova, T., Strikha, V., Tsyganova, A., 2000. Silicon nanoparticles characterization by Auger electron spectroscopy. Surf. Sci. 454–456, 952–956.

de la Venta, J., Bouzas, V., Pucci, A., Laguna-Marco, M.A., Haskel, D., te Velthuis, S.G.E., Hoffmann, A., Lal, J., Bleuel, M., Ruggeri, G., de Julián Fernández, C., García, M.A., 2009. X-ray magnetic circular dichroism and small angle neutron scattering studies of thiol capped gold nanoparticles. J. Nanosci. Nanotechnol. 9, 6434–6438.

Wacaser, B.A., Deppert, K., Böttger, M., Bi, Z., Adolph, D., Dick, K.A., Karlsson, L.S., Karlsson, M.A., 2007. Electrospraying of colloidal nanoparticles for seeding of nanostructure growth. Nanotechnology 18, 5304–5309.

Wajea, S.B., Hashima, M., Yusoff, W.D.W., Abbas, Z., 2010. X-ray diffraction studies on crystallite size evolution of $CoFe_2O_4$ nanoparticles prepared using mechanical alloying and sintering. Appl. Surf. Sci. 256, 3122–3127.

Walker, M., Brown, M.G., Draxler, M., Dowsett, M.G., McConville, C.F., Noakes, T.C.Q., Bailey, P., 2005. Structural analysis of the Cu(100)–$p(3\sqrt{2} \times \sqrt{2})$R45°–Sn surface using low and medium energy ion scattering spectroscopies. Surf. Sci. 594, 203–211.

Walls, J.M., 1989. Methods of Surface Analysis: Techniques and Applications. Cambridge University Press.

Wyatt, P.J., 1993. Light scattering and the absolute characterization of macromolecules. Anal. Chim. Acta 272, 1–40.

Wyatt Technology, 1998. ASTRA for Particles. Wyatt Technology Corp, Santa Barbara.

Xie, X.S., Trautman, J.K., 1998. Optical studies of single molecule at room temperature. Annu. Rev. Phys. Chem. 49, 441–480.

Xiong, W., Hickstein, D.D., Schnitzenbaumer, K.J., Ellis, J.E., Palm, B.B., Keister, K.E., Ding, C., Miaja-Avila, L., Dukovic, G., Jimenez, J.D., Murnane, M.M., Kapteyn, H.C., 2013. Photoelectron spectroscopy of CdSe nanocrystals in the gas phase: a direct measure of the evanescent electron wave function of quantum dots. Nano Lett. 13, 2924–2930.

Xu, L.-P., Chen, S., 2009. Scanning tunneling spectroscopy of gold nanoparticles: influences of volatile organic vapors and particle core dimensions. Chem. Phys. Lett. 468, 222–226.

Yi, F., LaVan, D.A., 2013. Electrospray-assisted nanocalorimetry measurements. Thermochim. Acta 569, 1–7.

Zavaleta, C.L., Smith, B.R., Walton, I., Doering, W., Davis, G., Shojaei, B., Natan, M.J., Gambhir, S.S., 2014. Multiplexed imaging of surface enhanced Raman scattering nanotags in living mice using noninvasive Raman spectroscopy. PNAS 106, 13511–13516.

Zheng, J., He, L., 2014. Surface-enhanced Raman spectroscopy for the chemical analysis of food. Compr. Rev. Food Sci. Food Saf. 13, 317–328.

Zheng, Y., Huang, S., Wang, L., 2012. Distribution analysis of nanoparticle size by small angle X-ray scattering. Electron probe micro-analysis (EPMA). Int. J. Theor. Appl. Nanotechnol. 1, 124–133.

Zink, F.F., Kerle, T., Klein, J., 1998. Near-surface composition profiles in isotopic polymer blends determined via high-resolution nuclear reaction analysis. Macromolecules 31, 417–421.

Zolnai, Z., 2013. Shape, size, and atomic composition analysis of nanostructures in 3D by Rutherford backscattering spectrometry. Appl. Surf. Sci. 281, 17–23.

CHAPTER

5

Principles of Nanotoxicology

OUTLINE

Engineered Nanoparticles
http://dx.doi.org/10.1016/B978-0-12-801406-6.00005-4

1. HISTORICAL PERSPECTIVES

In 1789, Lavoisier published *Traité Élémentaire de Chimie* (the English translation *Elements of Chemistry* followed in 1790), in which he explained the composition of air and water, and devised the term "oxygen" to explain the theory of combustion beyond the Greek's phlogiston theory that a fire-like element called phlogiston is contained within combustible bodies. Lavoisier proposed a caloric theory, a self-repellent fluid called caloric that flows from hotter bodies to colder bodies (Lavoisier, 1789). Gibbs, who introduced the idea of internal energy of a system in terms of entropy and combined the first and second laws of thermodynamics, further improved the concept of chemical energy (Gibbs, 1873). These and other discoveries led to the birth of synthetic chemistry, resulting in the synthesis of new chemicals. In 1874, Zeidler synthesized a chemical called dichloro-diphenyl-trichloroethane (DDT); later, following the discovery of its insecticidal properties by Paul H. Muller in 1949, DDT became a household name as a chemical that eradicated diseases such as typhus and malaria. To date, there are more than 61 million chemicals registered with Chemical Abstracts. How many of them have harmful effects? Probably all! This is because, as Paracelsus (1493–1541) suggested "All substances are poisons: there is none which is not a poison. The right dose differentiates a poison…"

Interestingly, the science of toxicology itself is thousands of years old. The ancient Vedic writings mention *ama*—poisons produced by the body that, if not removed, cause serious diseases. Early humans used plants and animals (e.g., frogs, containing curare) with toxic characteristics to poison laced arrows. The Egyptians were aware of many toxic substances, such as lead, opium, and hemlock, as early as 1500 BC. The toxins were used in political poisonings by the early Greeks and Romans. However, the Industrial Revolution of the eighteenth century is credited with the birth of modern toxicology. During the Industrial Revolution, many thousands of new chemicals were produced and used in consumer goods (plastics, pharmaceuticals, cosmetics, synthetics, etc.), pulp and paper, metallurgy, textiles, agriculture (pesticides and fertilizers), and construction. Many of the unused chemicals or industrial byproducts littered and contaminated the environment. Then came the fossil-fuel revolution of the twentieth century that released many of the modern pollutants.

In 1959, Richard Feyman (http://www.its.caltech.edu/~feynman/plenty.html) delivered a classic talk, "There's Plenty of Room at the Bottom," at the annual meeting of the American Physical Society at the California Institute of Technology. He introduced the concept of nanotechnology, especially of nanosized particles, and its future prospects. Since then, the field has developed exponentially, resulting in the synthesis of novel nanosystems for medicine, cosmetics, environmental clean-up, electronics, etc. Nanoparticles range from 1 to 100 nm in at least one dimension and possess unique physicochemical properties not found in bulk particles. Modern industries are taking advantage of the unique properties of nanoparticles and making products for our daily lives, including medicine, cosmetics, and food products.

In general, the adverse effects of nanoparticles have been assessed by extrapolating the data collected for bulk particles. Extensive commercial applications in the absence of appropriate regulation and toxicity information may pose serious risk to the public. Therefore, this chapter discusses the principles of classic toxicology (designed for bulk particles) and its relevance to nanotoxicology (for particles ranging from 1 to 100 nm)—a first step in understanding nanoparticle toxicity and the health risk it may pose to society.

2. CLASSIC TOXICOLOGY

2.1 Definitions of Key Terms Used in Toxicology

- *Toxicology*: Study of the adverse effects a chemical and/or physical agent causes in living organisms. In general, toxicology deals with the following:
  - Identification of poisons, poisoning, and treatments
  - Testing and detection of potentially toxic substances used in workplaces, at home, in agriculture (insecticides, herbicides, and fertilizers), in cosmetics, as food additives, and as drugs
  - Toxic waste in the air, water, and soil (e.g., chlorofluorocarbons, acid rain gases, dioxins, radioactive chemicals)
- *Poison* and *toxicant*: Toxicant and poison are used interchangeably and can be defined as substances that, when absorbed in sufficient quantities, cause damage, illness, or death to organisms. Regulatory agencies label highly toxic substances as poison, with less toxic substances referred to as harmful irritants or not labeled at all.
- *Venom and Toxin*: Living organisms can produce both *venom* and *toxins*, but there is a difference between organisms producing venoms and those producing toxins. Venom-producing animals inject venom into their prey when hunting or as a self-defense mechanism. Toxins are poisons produced by both plants and animals, and they are harmful when ingested or touched.
- *Toxicosis* and *poisoning*: Refer to the adverse effects of exposure to a poison, toxin, or venom.
- *Lethal dose* (*LD*): The dose (mass/kg) of an agent that is sufficient to cause death.
- LD_{50}: LD_{50} is the quantity of an agent sufficient to kill 50% of the experimental animals.

2.2 Examples of Poisons

1. Inorganic substances such as heavy metals (mercury, arsenic, cadmium, etc.), salts (oxides of metals, sulfates of metals, etc.), ammonia, nitrites, etc.
2. Organic substances such as pesticides, herbicides, preservatives, antibiotics, biotoxins, artificial colors, benzene, xylene, organic heavy metals, carcinogenic compounds, etc.
3. Biotoxins are chemicals synthesized by living organisms such as bacteria, fungus, plants, and animals. Animal toxins are mostly small peptides. Plant, bacterial, and fungal toxins have diverse structures.
   a. Environmental toxins are polychlorinated biphenyls, pesticides, mold, and other fungal toxins, phthalates, volatile organic compounds, dioxins, asbestos, heavy metals such as arsenic, mercury, lead,

aluminum, cadmium, chloroform, benzene and other organic solvents, chlorine, etc.

4. Household toxins: Houses, especially kitchens, basements, and garages, are the major source of toxins. A few examples are listed below:
   a. Triclosan: an antibacterial agent found in many liquid soaps and in some deodorants, toothpastes, cosmetics, kitchenware, and children's toys
   b. Phthalates: multiple uses and linked to endocrine, reproductive, and developmental problems
   c. Bisphenol A: used in epoxy resins and polycarbonate plastics for food containers and baby products
   d. Carbon monoxide: formed by incomplete combustion of fuel and decreases the delivery of oxygen to cells
   e. Perfluorinated chemicals: used to make stain repellents and nonstick surfaces
   f. Radon: odorless gas that forms as uranium in rocks and when soil breaks down
   g. Lead, mercury, arsenic, etc.: heavy metals that can build up in tissues
   h. Pesticides and herbicides: linked to problems with the nervous system and possibly a risk factor for cancer, developmental challenges, and reproductive problems
5. Industrial toxins are released into the environment and increase potential for human exposure. Some of the common toxins are listed below:
   a. Paper mills: dioxin, volatile organic chemicals, nitrous oxide, sulfur oxide, etc.
   b. Agricultural waste: manure estrogens, fertilizers and pesticides, release of methane gas, etc.
   c. Coal burning: carbon dioxide, tar, sulfur dioxide, mercury, etc.
   d. Electric transformers: polychlorinated biphenyls (PCBs)
   e. Flame retardants: tris(1,3-dichloro-2-propyl) phosphate.
   f. Organic solvents: benzene, toluene, chloroform, etc.

2.3 Acute or Chronic Exposure

2.3.1 *Acute Exposure*

Acute exposure is a single exposure of short period to a toxic substance, resulting in severe biological harm or death. The severity of immediate effects depends on the dose of the poison. Large acute exposures can result in death.

2.3.1.1 MILD SYMPTOMS

Mild symptoms include the following: behavior changes (e.g., restlessness, crankiness), diarrhea, dizziness, drowsiness, fatigue, headache, loss of appetite, skin or eye irritation, nausea, cough, soreness in the joints, and thirst.

2.3.1.2 MODERATE SYMPTOMS

Moderate poisoning may include the following: blurred vision, confusion and disorientation, difficulty in breathing, drooling, excessive tearing, fever, hypotension, muscle twitching, paleness, persistent cough, rapid heart rate, seizures, diarrhea, nausea, stomach cramps, sweating, thirst, trembling, and weakness.

2.3.1.3 SEVERE POISONING SYMPTOMS

These symptoms can be life threatening and result in permanent brain damage, disability, or death. Major symptoms include cardiopulmonary arrest, convulsions, uncontrolled bleeding or blood clotting, high fever, inability to breathe, loss of consciousness, muscle twitching (uncontrolled and severe), rapid heart rate with low blood pressure, respiratory distress requiring mechanical respiration, and seizures that do not respond to treatment.

2.3.2 *Chronic Exposure*

Chronic exposure is multiple exposures occurring over extended periods or over a significant fraction of an animal's or human's lifetime. Individuals suffering from chronic poisoning become ill gradually. This differs from acute poisoning in which the affected individuals experience severe symptoms within minutes or hours of ingesting a toxic substance. Chronic poisoning most often occurs with substances that can accumulate in the body over time, such as mercury, DDT, and several types of lead compounds.

2.3.2.1 CHRONIC FATIGUE SYNDROME

Chronic fatigue syndrome is one of the most common symptoms of chronic poisoning. The exposed individual experiences debilitating fatigue and a general feeling of being perpetually unwell. They may develop insomnia and suffer from muscle pain.

2.3.2.2 NEUROLOGICAL PROBLEMS

Neurological problems occur if the toxic substance accumulates within the brain. The symptoms include sudden and unexpected mood swings, irritability, and episodes of excessive anger punctuated by extreme depression.

2.3.2.3 ORGAN DAMAGE, ESPECIALLY LIVER AND KIDNEY DAMAGE

Organ damage, especially liver and kidney damage, develops when the accumulated toxins damage the tissues and cells. Liver and kidney damage may result in jaundice, fatigue, vomiting or diarrhea, unexplained weight loss, changes in urinary habits, and the amount and appearance of the urine.

2.3.2.4 BIRTH DEFECTS

Birth defects can occur if a pregnant woman is chronically exposed to a poison. Certain toxins may impair the development of the unborn child, resulting in a physical or neurological birth defect.

3. THE PRINCIPLES OF CLASSIC TOXICOLOGY AND NANOTOXICOLOGY

Toxicology, in general, is defined as a comprehensive and fundamental assumption that explains the basis for the adverse effects of poisons within the constraints of confounding factors. Because toxicology is not an absolute science like mathematics, the principles of toxicology cannot be absolute; they are subject to interpretations or disagreements. The goal of this section is to discuss the unity and diversity in the principles of bulk particle and nanoparticle toxicity.

3.1 Principles of Bulk Particle Toxicology

In general, there are five principles of bulk particle toxicology:

BP1 *The dose makes the poison.* All things are poison and nothing is without poison; only the dose separates poisons from nonpoisons (Paracelsus). The concept that all chemical agents are toxic at some dose is central to the inherent hazard of all chemicals. The dose–response curve is one of the pillars of toxicology. On the face, this appears to be a nonrefutable statement. However, if we look at the key part of this law, the dose, many discrepancies appear. Dose measured as mass per unit body weight (e.g., g/kg) may work fine for pure compounds, but when the administered material is a mixture, calculation of the dose–response relationship becomes a problem because of chemical interactions. Similarly, calculation of the dose in cases of inhalation is difficult because inhalation dose (mg/kg) = [Air concentration of agent (mg/ml) × volume of air inhaled per hour (ml/h) × duration of exposure (h)]/body weight (kg). In the case of essential minerals, an excess and a deficiency both cause an adverse response, giving an inverse U-shaped curve.

BP2 *The specificity of effects and the biological actions of chemicals are specific to each chemical* (Ambroise Paré). BP2 states that the biological actions of poisons are specific to each chemical, according to Ambroise Paré, a sixteenth-century French surgeon. He recognized that toxic agents may have different effects depending on their inherent nature. The toxicity of a chemical depends upon (1) structural determinants of their biological activity and (2) the biological niches in which chemicals interact. Very subtle changes in the chemical structure can make an enormous difference in biological effects.

BP3 *Humans are animals.* Humans and animals may respond similarly (to some degree) to a poisoning insult. Protection against the toxicity of chemicals today would be impossible without the ability to study the effects of toxic agents on laboratory animals.

BP4 *Risk = hazard × exposure.* A hazard is described as a potential source of harm or adverse health effect on a person. For example, a hazardous site may contain chemicals that have the potential to cause serious harm to humans. However, people living at a distance from the site will not be exposed to the toxins present in the hazardous site. Hazard is based on the characteristics of a chemical, while risk is a formal process of determining the potential of an agent to cause harm. If people visit a hazardous site, then there is possibility that they are at risk of exposure to the poisons.

BP5 *There are variations in the sensitivity of individuals.* The toxic effects of poisons exhibit significant inter- and intra-species differences that impose considerable limitations on the ability to extrapolate the toxicity of a compound from one species to another. For example, rats are considered to be one of the most sensitive species among experimental animals (Schwetz et al., 1973; Olson et al., 1980), although there is greater than 300-fold variability in LD_{50} values for tetrachlorodibenzo-*p*-dioxin among different strains of rats (Pohjanvirta and Tuomisto, 1987). The LD_{50} dose (the dose that kills 50% of animals within a predetermined time in experimental studies) for aflatoxin ranged from 0.3 to 17 mg/kg (rats < hamster < mouse < chick < monkey < sheep < guinea pig < dog < rainbow trout < pig < cat < duckling < rabbit) (Smith and Hamilton, 1970; Hanigen and Laishes, 1984; Newberne and Butler, 1969). Mature animals of a given species were more resistant than young ones (Patterson, 1973). These observations suggest that guinea pigs are more sensitive than rats to develop acute aflatoxin toxicity. Species differences in toxicity of poisons may be due to differences in their absorption, distribution, metabolism and excretion (ADME), pharmacokinetics, or pharmacodynamics (Rozman, 1988). This aspect will be discussed in detail in Chapter 6.

Taken together, these observations suggest that although the principles listed above may form general foundation of toxicology, disagreement may arise because of many unknown and confounding factors. The last principle—individual variations due to species, sex, and age—may enhance the uncertainty factors in risk determinations.

3.2 Principles of Nanotoxicology

In many commercial products such as cosmetics and sunscreen, titanium oxide nanoparticles are simply listed as titanium oxide because the Environmental Protection Agency (EPA) does not consider nanoparticles to be "new" substances just because of their size range.

According to the EPA, a particle must have a distinct molecular identity that is not shared with any other chemical on the Toxic Substances Control Act's existing Chemical Substance Inventory before they are considered to be new. As discussed in Chapters 3 and 4, the crystal structure, surface characteristics, physicochemical properties, and toxicity of nanoparticles are distinct from the corresponding bulk particles. There is substantial evidence in support of a hypothesis that the principles of classic toxicology do not govern nanoparticle toxicity. As described below, principles 2–5B overlap between nanoparticle and bulk particle toxicities, while principles NP1 to NP4 are unique for nanoparticles (NPs).

NP1 Dose in terms of size, particle number, surface activity, aggregation and, to some extent, mass determines nanoparticle toxicity.
NP2 Surface functionalization modifies nanoparticle toxicity.
NP3 The elemental component of nanoparticles modifies nanoparticle toxicity.
NP4 Differential protein adsorption results in higher accumulation of nanoparticles into the brain.

NP1: Dose in terms of size, particle number, surface activity, aggregation and, to some extent, mass determines nanoparticle toxicity. In classic bulk-particle toxicology, dose invariably relates to mass (g/kg bogy weight). The relationship between dose (in terms of mass) and response (some indices of toxicity, such as lethal dose) is a critical part of classic toxicology, especially in risk assessments. With nanoparticles less than 100 nm in at least one dimension, the toxicological effect does not correlate with the dose in terms of mass. This is because, in addition to mass, nanoparticle size also plays an important role in determining their adverse effects.

As shown in Figure 1, at a constant dose, size (Figure 1(A): 20 nm, Figure 1(B): 80 nm, and Figure 1(C): 113 nm) determined the adverse effects of nanoparticles: the smaller the particle, the larger the adverse effects (Figure 1(D): reactive oxygen species (ROS) or free radicals and Figure 1(E): lactate dehydrogenase (LDH) accumulation in extracellular fluid that indicates cell necrosis). Therefore, a mass-based dose–response curve may not provide a meaningful conclusion from in vitro and in vivo experiments and may not be suitable for conducting public health risk assessments. There has been considerable discussion on defining the most appropriate metric for assessing a nanoparticle's dose (Elsaesser and Howard, 2012; Elsaesser et al., 2010; Oberdörster, 2010; Song et al., 2009; Wittmaack, 2007). These and other studies have suggested that the particle number (particle count/g) or specific surface area (mm^2/g) may be a more discriminatory dose metric than mass. Earlier studies have shown (Figure 2) that for nanoparticles, dose in terms of surface area of the particles administered rather than dose in terms of mass is a more appropriate metric for toxicity studies. Ultrafine titanium dioxide appears to be less bioactive than ultrafine carbon black when a mass-based dose was used, but more bioactive than ultrafine carbon black when the dose in terms of surface area of particles was used (Sager and Castranova, 2013; Sager et al., 2008; Duffin et al., 2007; Stoeger et al., 2006). Thus, dose in terms of mass may not be critical in determining a nanoparticle's toxicity.

NP2: Surface functionalization modifies nanoparticle toxicity (Figure 3). This principle reflects the fact that the surface to volume ratio and the surface reactivity of nonfictionalized nanoparticles increase exponentially as the diameter decreases. Nonfunctionalized nanoparticles is more reactive than the surface of corresponding bulk particles. Thus, the reaction with biological structures increases exponentially with decreased diameter if the same amount

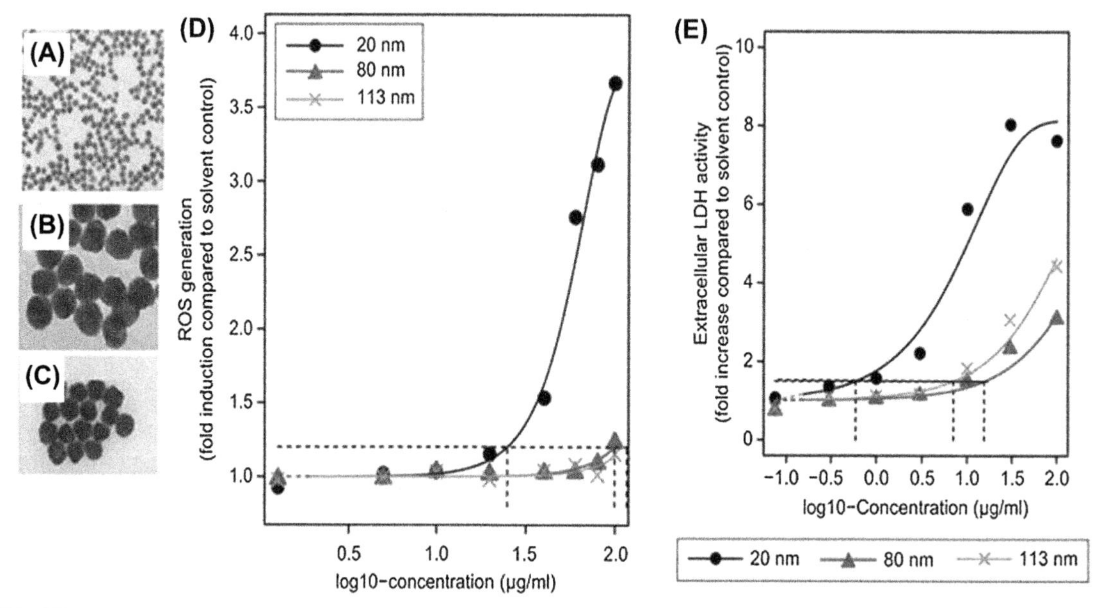

FIGURE 1 **Effects of size on nanoparticle toxicity.** (A–C) Transmission electron microscopy of silver nanoparticles, 20, 80, and 113 nm, respectively. The particles were well dispersed in the administered solution. (D) Effects of nanoparticle size on reactive oxygen species (ROS) accumulation in cell cultures in vitro. ROS accumulation is related inversely to the particle size. (E) Effects of particle size on extracellular lactate dehydrogenase (LDH) activity in cell cultures. *Reprinted from Park et al. (2011) with permission.*

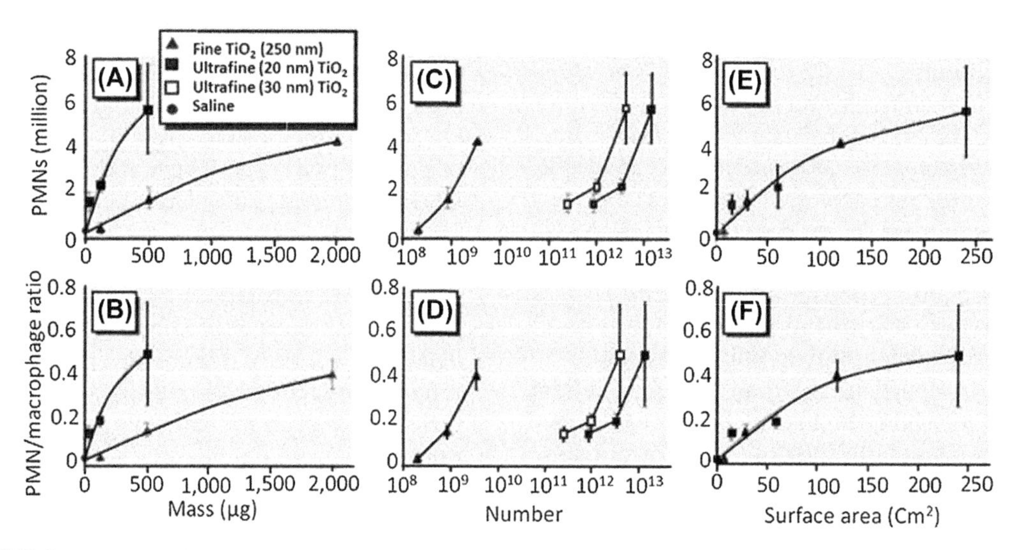

FIGURE 2 **Nanoparticle dose metric.** Inflammatory cell response (as PMN enumeration (A, C, and E) and PMN/macrophage ratio (B, D, and F) in lung lavage 24 h after intratracheal instillation of fine (~250 nm) and ultrafine (20–30 nm) TiO_2 nanoparticles (dose expressed by particle mass (A and B), number (C and D), and surface area (E and F)). The particle mass dose–response (A and B) and the particle number dose–response (C and D) relationships are several-fold apart for fine and ultrafine TiO_2, whereas the surface area plot (E and F) shows a good fit for the combined particle sizes. *Reprinted from Oberdörster et al. (2007) with permission.*

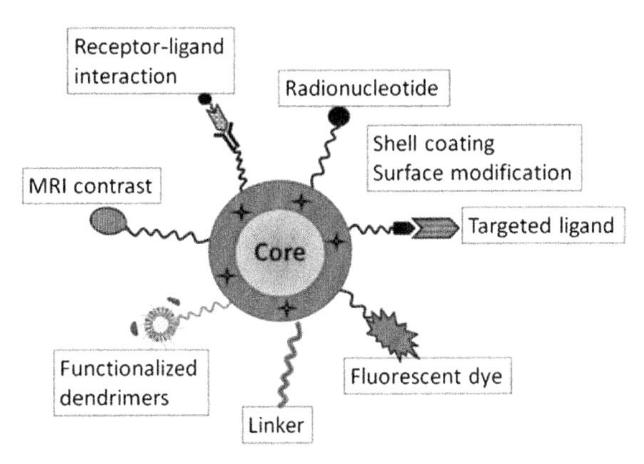

FIGURE 3 **Nanoparticle functionalization.** Pristine nanoparticles are highly reactive and poorly bioavailable; thus, they are not suitable for medicinal use. Coating the surface with various hydrophilic polymers such as polyethylene glycol (PEG) or certain polyamines such as poly-L-lysine improves the nanoparticle bioavailability. Nanoparticles can be functionalized into multifunctional particles that can be used for molecular imaging, drug delivery, and therapy (drugs embedded in the coating, attached to a linker, or loaded into a dendrimer). Optionally, nanoparticles can be functionalized for individualized diagnosis and treatments (nanoparticles coupled to the MRI contrast agent, fluorescent dye for near-infrared imaging, radionucleotides, target ligands, etc.). Actively functionalized nanoparticles can simultaneously provide imaging and treatment of cancer and other diseases.

is administered. Capping the surface bonds with different functionalization groups modifies the particles' reactivity (Figure 3).
NP3: The elemental component of nanoparticles modifies their toxicity. This principle states that, in addition to the changes in dimensions (i.e., going from bulk toward nano), the properties of the material and its composition, including impurities, will also influence its toxicity.
NP4: Differential protein adsorption results in higher accumulation of nanoparticles in the brain. Delivery of drugs across the blood–brain barrier (BBB) is still a major challenge. Either small hydrophobic particles or particles that can bind to a transporter or receptor (for bulk delivery) can accumulate in the brain. Engineered nanoparticles allow a novel means to deliver drugs into the brain using the principle of differential protein adsorption, in which nanoparticles are functionalized with a targeting group such as a cancer-selective peptide, which, upon reaching the desired site, binds with it and delivers the load (Figure 3). In this case, nonspecific distribution is minimal. Apolipoprotein E is the targeting moiety for the delivery of particles into the endothelial cells of the BBB (Wang et al., 2013; Micaelis et al., 2006). The key features of differential protein adsorption have been described in detail by Muller and Keck (2004).

In conclusion, the nanoparticle-induced toxic effects are dependent on particle size, shape, and surface. Thus, these factors need to be carefully considered when developing risk-characterization strategies.

4. NANOPARTICLE EXPOSURE

During the evolutionary process, as humans acquired consciousness (subjectivity, self-awareness, sentience, and the ability to perceive the relationship between oneself and one's environment), they have tried to modulate their surroundings through technology (application of knowledge to industrial or commercial objectives). Each new application is associated with new risks that the society has to deal with every day. For example, production of new chemicals has significantly improved the quality of life (by improving agriculture, pharmaceuticals, food products, etc.) of humans, but it has also resulted in a contaminated environment and ensuing adverse effects on human health.

Nanoparticles are the latest addition to the list of new chemicals. As society became more interested in reaping the potential benefits of nanotechnology, nanoparticle production is increasing and nanoparticle-based products are being integrated in our daily lives. With the continued explosion of the industrial applications of nanomaterials, it is anticipated that nanoparticle-based products will prevail in near-future marketplaces. This will increase the environmental release of nanoparticles throughout their lifecycle, including manufacture, consumer use, and disposal, thereby exposing workers and consumers. This chapter addresses the key issues related to nanoparticle exposure. An overview of this chapter is shown in Figure 4. The following axioms will help facilitate the discussion in this chapter:

- Engineered and/or nonengineered nanoparticles are already present in the environment and have always been, since beginning of the universe.
- Humans, animals, plants, and the environment are already in contact with nanoparticles.
- Nanoparticles can negotiate physiological barriers much easier than bulk particles because of their unique size and physicochemical properties.

Engineered nanoparticles are present in every aspect of our lives, such as household products,

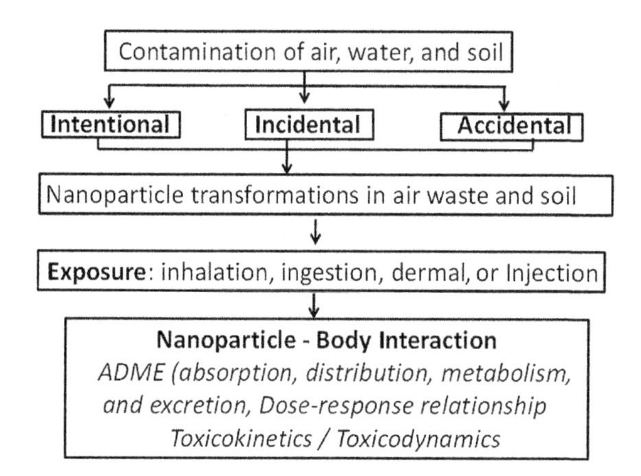

FIGURE 4 **An overview of the chapter discussion.** Nanoparticles enter the environment via intentional, accidental, and incidental modes. Humans and animals are exposed to nanoparticles via dermal contact, inhalation, ingestion, and injection (for medicinal use). The aim of this chapter is to discuss the nano-bio interaction that determines bioavailability and biodistribution of nanoparticles.

food items, antibiotics and other drugs, and automotive exhaust (Phoungthong et al., 2013; Myung and Park, 2012; Ribeiro et al., 2010; Giechaskiel et al., 2005, 2007, 2008a,b; Friends of the Earth, 2007; Vaaraslahti et al., 2005; Dockery et al., 1993; Pope et al., 1992; Ostro, 1984). Nanoparticles reach the environment via the following human activities:

- *Intentional activity*, such as dumping of nanoparticles in municipal waste or use of nanoparticle-based medicines, pharmaceutical cosmetics, silver utensils, and environmental cleanup products
- *Incidental activity*, such as nanoparticle washout from cosmetic use, excretion of nanoparticles from exposed organisms, and release from other sources
- *Accidental activity*, such as procedural/mechanical malfunction of large manufacturing facilities, leaks from storage devices

Nanoparticles readily transform in organisms and the environment. The transformed nanoparticles may have different properties that alter their transport, fate, and toxicity. Terrestrial organisms are exposed to unaltered and/or modified nanoparticles via inhalation, ingestion, dermal exposure, or injection (of medicinal nanoparticles). The following sections discuss the routes of exposure in detail.

4.1 Routes of Exposure

The overall pathways by which terrestrial organisms are exposed to nanoparticles are shown in Figure 5. Inhalation and skin contact are the most likely routes of exposure for people working in nanoparticle manufacturing and processing facilities, while exposure via ingestion may occur from contaminated food and water, cosmetics, or drug-delivery devices (Card et al., 2008; Lam et al., 2006). Brief descriptions of each exposure mode are provided in the following sections.

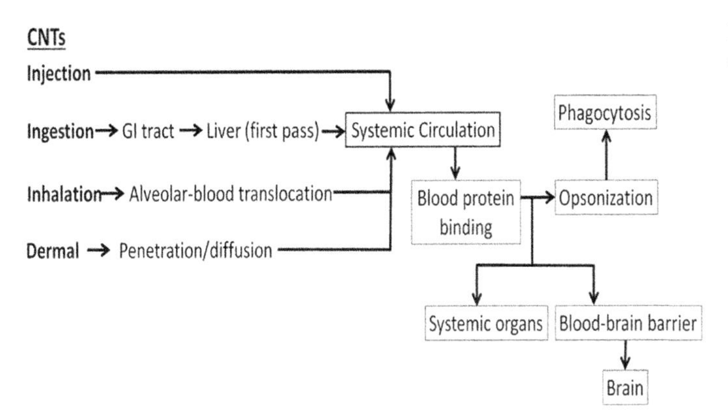

FIGURE 5 Different routes of nanoparticle exposure and associated biochemical pathways.

4.1.1 Skin Contact—Dermal Exposure

4.1.1.1 SKIN ANATOMY

The human skin has a surface area of approximately 1.4–2 m^2, the largest organ in human body. The skin consists of the epidermis, dermis, and hypodermis layers. The epidermis is the outermost layer (0.4–1.5 mm thick), providing the first defense against toxins or pathogens. The dermis (0.4–3 mm thick) contains nerves, blood vessels, hair follicles, muscles attached to hair follicles, and proteins. The innermost layer, the hypodermis, insulates the body and cushions tissues beneath the skin. Toxins may penetrate the skin either via the intercellular route through the lipid channels between the corneocytes or via the appendage route through hair follicles and sweat glands (Figure 6).

4.1.1.2 NANOPARTICLE DERMAL PENETRATION

In healthy individuals, 30-nm nanoparticles may penetrate stratum corneum (SC) through the lipid intercellular route or aqueous pores, while bulk particles may penetrate through the transfollicular route (Baroli, 2010). Titanium dioxide and beryllium nanoparticles also penetrate the SC by translocation via hair follicles (Lademann et al., 1999; Tan et al., 1996; Alvarez-Roman et al., 2004), but do not cross the skin (Tinkle et al., 2003; Baroli et al., 2007). Similarly, other nanoparticles ranging in diameter from 7 to 20 nm accumulate in follicular openings but do not penetrate the skin (Lauer et al., 1996; Toll et al., 2004; Meidan et al., 2005; Vogt et al., 2006). Murray et al. (2009) have shown that, in dermal exposure of mice to unpurified single walled carbon nanotubes (SWCNT), inflammation was localized around the hair follicles, indicating some SWCNT may be able to penetrate the SC, which may be associated with the development of skin toxicity. Some of the signs of skin toxicity were elevated myeloperoxidase activity, elevated dermal cell numbers, and skin

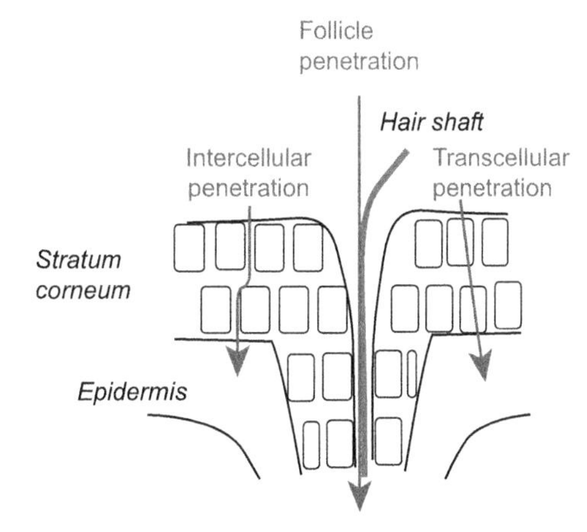

FIGURE 6 Dermal anatomy and possible pathways for nanoparticles' dermal penetration.

thickening resulting from the accumulation of polymorphonuclear leukocytes and mast cells.

The factors that modulate dermal penetration of toxins are their size and lipophilicity, overall surface area, adherence to the skin, residue transfer, frequency and duration of exposure, skin thickness, and condition of the skin (dry/wet condition, damaged skin, or exposed to chemicals that enhance penetration) (Christopher et al., 2012). Pristine small lipophilic nanoparticles penetrate the skin more easily than pristine large or hydrophilic nanoparticles. Depending on the defects, carbon nanotubes (CNTs) may acquire hydrophilic sites that may suppress their dermal penetration. The maximum particle flux across the skin decreases with the fourth power of the particle radius (Moghimi et al., 1996).

Engineered dermal-penetrating nanoparticles, including CNTs and quantum dots, allow percutaneous delivery of drugs (Rolland et al., 1993: polymeric microsphere; de Jalon et al., 2001: PLGA microparticles; Wissing and Muller, 2002: solid-lipid nanoparticles; Chu et al., 2008: quantum dots; Rouse et al., 2007: peptide nanoparticles; Zhang and Monteiro-Riviere, 2008: PEG-functionalized CNTs), but the nanoparticles may also increase the risk of dermal toxicity, although it is a highly polarizing issue. Nanoparticles have been shown to be essentially nontoxic (Tan et al., 1996: titanium dioxide nanoparticles; Bestak and Halliday, 1996: titanium dioxide; Lademann et al., 1999: titanium dioxide; Pflucker et al., 2001; Alvarez-Roman et al., 2004: polymeric nanoparticles; Sadrieh et al., 2010: titanium dioxide), as well as highly toxic (Oberdörster et al., 2005: ultrafine particles; Sayes et al., 2004: fullerenes; Halford, 2006: fullerenes; Duty et al., 2005: phthalates) to the skin.

The research, therefore, justifies the views of the cosmetic and the pharmaceutical industries, who claim the nanoparticles they use are nontoxic, as well as the views of health activists, who claim them to be highly toxic. Because different studies have used different nanoparticles and experimental designs, their results may not be comparable. Although metal nanoparticles are commonly used in cosmetics, a few patents have been approved for the applications of CNTs and fullerenes in hair coloring and other facial applications (Huang et al., 2007; Lens, 2011).

4.1.2 Exposure via Inhalation

4.1.2.1 NANOPARTICLE INHALATION

Inhalation is the most likely route of occupational exposure to nanoparticles, especially hydrophobic nanoparticles and CNTs, in people working in manufacturing and handling facilities (Hoet et al., 2004; Oberdörster et al., 2005a,b). The respiratory system extends from the conductive airways in the head and thorax to the alveoli, where respiratory gas exchange takes place between the alveoli and the capillary blood (Figure 7). When nanoparticle-contaminated air is inhaled, the particles are deposited onto the respiratory tract via impaction, gravitational sedimentation, and Brownian diffusion (Byron, 1986; Courrier et al., 2002). The distribution of nanoparticles in different regions of lungs depends upon their size because the size influences the particles' motion and ultimately the probability of their impaction, sedimentation, and diffusion. Accumulation of nanoparticles in the lungs depends on deposition, retention, and clearance (Mercer et al., 2010, 2011; Courrier et al., 2002).

4.1.2.2 DEPOSITION OF NANOPARTICLES IN THE RESPIRATORY TRACT

Deposition is the spontaneous adherence of particles to the surface. Total deposition (the sum of depositions in regions 2–4; Figure 7, plot 1) is highest for nanoparticles ranging from 1 to 10 nm. As the size increases above 10 nm, a rapid decrease in deposition occurs, with 100-nm particles exhibiting the smallest deposition. A further increase in particle size from 100 to 1000 nm resulted in a rapid increase of deposition (Figure 7, line 1). Smaller nanoparticles

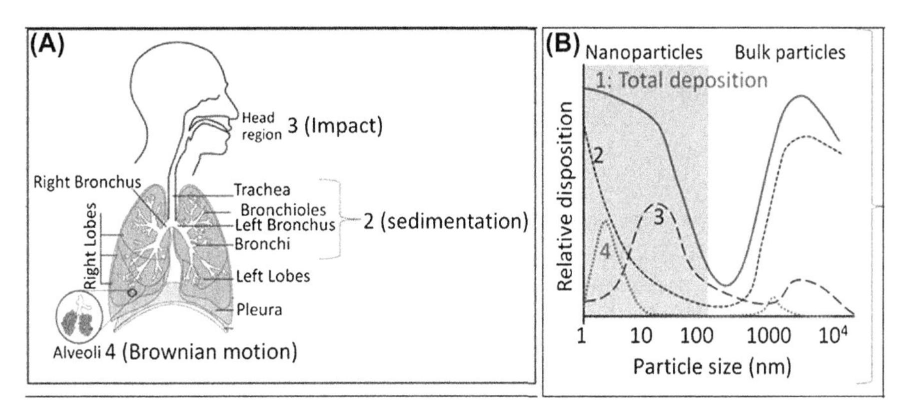

FIGURE 7 **Anatomy of the respiratory system (A) and kinetics of a nanoparticle's movement in different regions of the respiratory tract (B).** The distribution of nanoparticles in different regions of the tract depends upon their size; the size influences the particles' motion and ultimately the probability of their impaction (A3), sedimentation (A2), and diffusion (A4). Net accumulation in lungs depends upon the particles' deposition, retention, and clearance. (B) Total deposition (B1), sedimentation (B2), impaction (B3), and diffusion (B4).

(1–2 nm) exhibit the highest deposition in the tracheobronchial region. An increase in particle size for up to 100 nm rapidly decreases their deposition, with the lowest values reported for 100-nm particles. Thereafter, a further increase in deposition occurred in the tracheobronchial region (Figure 7, line 2). Nanoparticles ranging from 1 to 10 nm may reach the alveolar region, possibly due to Brownian diffusion. Larger nanoparticles may not reach the alveolus (Figure 7, line 4). The deposition in the head region was the highest for bulk particles (greater than 500 nm). Nanoparticles ranging from 10 to 100 nm exhibited much lower deposition.

4.1.2.3 NANOPARTICLE RETENTION

Retention is the accumulation of nanoparticles in specific regions where they can either remain unaltered for a considerable period or undergo some modification (Moller et al., 2008). The retention is either *direct* (example: one-dimensional (1D) nanoparticles) when the deposited nanoparticles penetrate the epithelial cells and induce localized inflammation or *indirect* when they are trapped by the alveolar macrophages (Alv. Mac.) or polymorphonuclear(PMN) lymphocytes (Figure 8). One well-recognized mechanism of retention is phagocytosis, or the engulfment of particles that deposit in the alveolar region or the particle-laden PMN leukocytes within hours after deposition. Most of the particles in the alveoli are engulfed within a 24-h

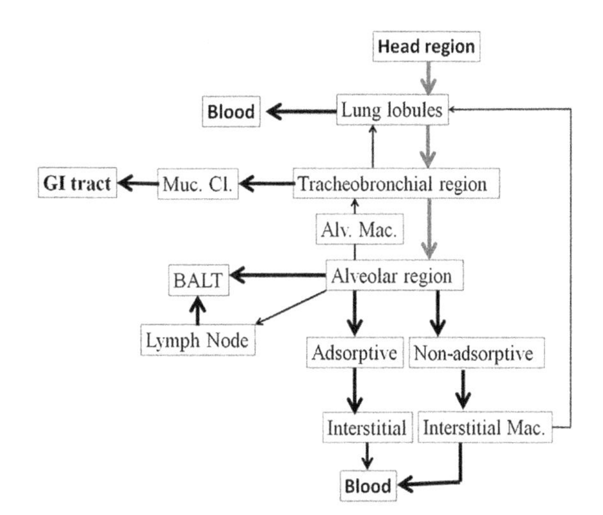

FIGURE 8 A flow diagram showing biodistribution of nanoparticles as they move from the head region into the alveoli. Nanoparticles, depending upon their size, are absorbed directly into the blood, absorbed by the GI tract, or deposited into the lung tissue, where they cause local inflammation.

period after a single bout of particle deposition (Schlesinger and Lippmann, 1972, 1978; Albert et al., 1969; Lippmann and Albert, 1969; Davies, 1952).

4.1.2.4 NANOPARTICLE CLEARANCE

The respiratory system has several mechanisms that can clear nanoparticles. The key mechanisms are mucociliary and cough clearance (McCC), macrophage engulfment and translocation, free particle clearance, and retention and clearance of 1D nanoparticles such as CNTs. The significance of each is described in the following sections.

4.1.2.4.1 MUCOCILIARY AND COUGH CLEARANCE McCC is important for keeping the airways cleansed of bacteria, viruses, and toxic particles including nanoparticles (Regnis et al., 1994). In McCC, particles deposited in the conducting airways are transported cephalad (toward the head or anterior section) on a layer of mucus that overlays a serous fluid layer. The mucus is propelled by ciliary action (Wanner et al., 1996; Mortensen et al., 1994). When the particles and mucus reach the larynx, they are swallowed or coughed out. Wagner and Foster (1996) have shown that blood supply from the bronchial artery is required for mucociliary clearance of insoluble particles. A reduction in blood supply significantly attenuated particle clearance. Because of the potential for airway injury from retained toxic materials, it is important to extend these studies and evaluate the mechanisms by which airway perfusion modulates particle clearance.

4.1.2.4.2 MACROPHAGE ENGULFMENT AND TRANSLOCATION (KNOWLES AND BOUCHER, 2002; WANNER ET AL., 1996) When particles are deposited in the alveolar spaces, they are ingested by alveolar macrophages. Then, the particle-loaded macrophages translocate to the base of the mucociliary escalator (it is composed of the mucus-producing goblet cells and the ciliated epithelium, which are continually beating, pushing mucus up and out into the throat). Once onto the base of the escalator, the particle-loaded macrophages are transported out of the lung by mucociliary clearance. Many infections and diseases, such as cystic fibrosis, interfere with the ciliary clearance, resulting in accumulation of particles in lungs (Camner et al., 1973, 1983; Chmiel and Davis, 2003).

4.1.2.4.3 FREE-PARTICLE CLEARANCE Often, particles move randomly into the interstitium or the lymphatic drainage system and leave alveoli. Unfortunately, the process takes several weeks to several months.

4.1.2.5 RETENTION AND CLEARANCE OF 1D NANOPARTICLES, SUCH AS CNTs

CNTs, being 1D nanoparticles, present a unique problem in their retention and clearance. As shown in Figure 9, long-multi walled carbon nanotubes(MWCNTs) may penetrate in alveolar epithelial cells (A), subpleural lymphatic vessel (B), the visceral pleura (C), and subpleural tissue (D) and induce local inflammation.

Long and short CNTs that are more dispersed and less agglomerated exhibit relatively great air-to-blood transport in lungs. In addition to the length and diameter, dispersion and rigidness also play an important role in determining MWCNTs fate in lungs (Nagai et al., 2011; Kostarelos, 2007, 2009; Donaldson et al., 2006, 2010). MWCNTs of approximately 50 nm in diameter were rigid and highly dispersed, thus showing mesothelium cell membrane piercing and ensuing cytotoxicity and inflammation. MWCNTs having a diameter of approximately 150 nm that were too thick or 2–20 nm that formed aggregates exhibited relatively low toxicity. This suggests the possibility of a bell-shaped diameter CNT penetration curve.

Figure 10 shows the overall fate of CNTs in the respiratory tract. Long and short SWCNTs and short MWCNTs aggregate and then are cleared via macrophage-mediated phagocytosis.

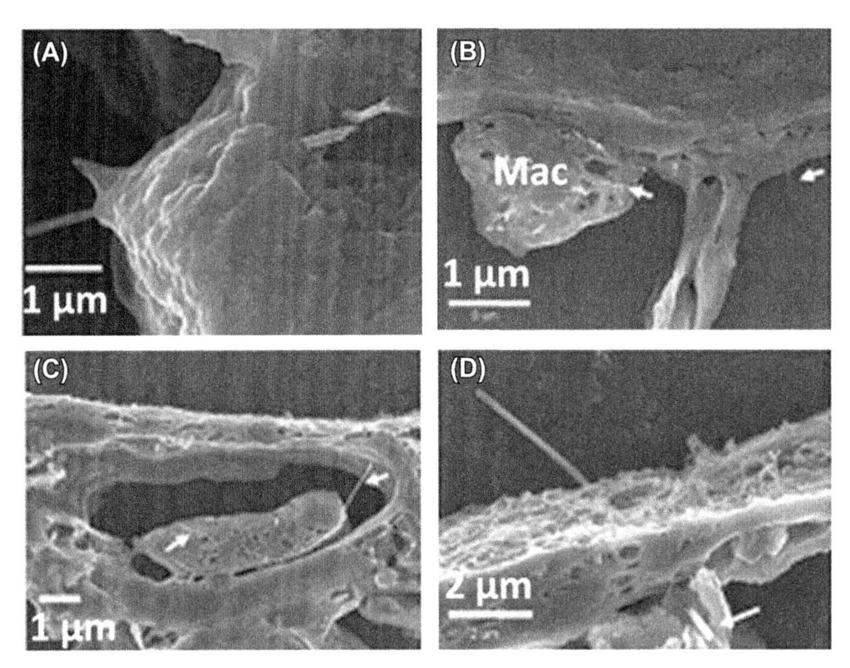

FIGURE 9 Penetration of multiwall carbon nanotubes (MWCNTs) into the lung tissues. Mac, macrophages. Penetration of alveolar epithelial cells (A), subpleural lymphatic vessel that contains a mononuclear inflammatory cell (B), the visceral pleura (C), and the subpleural tissue through the visceral pleura into the pleural space (D). *Reprinted from Mercer et al. (2010) with permission.*

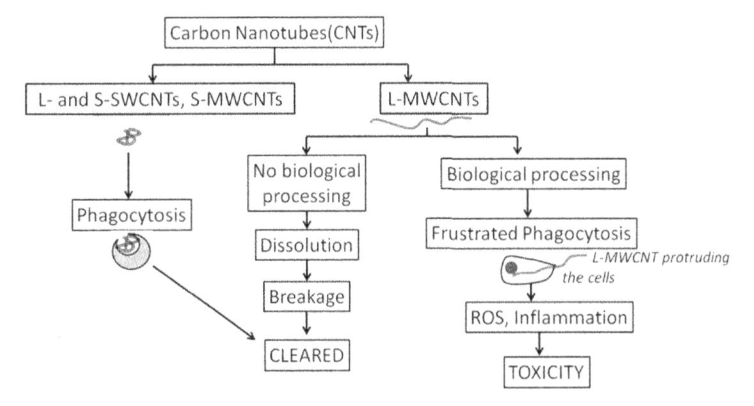

FIGURE 10 **Overall fate of CNTs in the respiratory tract.** Long (L) and short (S) SWCNTs and S-MWCNTs aggregated and then cleared via macrophage-mediated phagocytosis. L-MWCNTs were rigid and induced frustrated phagocytosis (macrophages were not able to completely engulf L-MWCNTs; also see Figure 9), resulting in induction of oxidative stress and inflammation.

Long MWCNTs induce frustrated phagocytosis (macrophages were not able to completely engulf long MWCNTs), resulting in the induction of oxidative stress and inflammation (Figures 9 and 10).

However, considerable differences in the fate of SWCNTs and MWCNTs in lungs have been reported (Mercer et al., 2008, 2011):

1. Poor dispersion of pristine SWCNTs allows approximately 80% of the applied SWCNTs to aggregate, while relatively better dispersion of pristine MWCNTs allows approximately 20% aggregation.
2. Approximately 10% of SWCNTs accumulated in macrophages, of which the interstitial fluid contained 90% of the applied dose. Approximately 70% of MWCNTs accumulated in macrophages, of which the interstitial fluid contained only 8% of the applied dose. This suggests that SWCNTs may not be recognized by the macrophages in lungs.
3. Rigid MWCNTs of approximately 50 nm in diameter pierced cells, while comparable SWCNTs did not.
4. SWCNTs induced a high fibrotic response (rate of connective tissue increases by 0.11 μm/μg) compared to MWCNTs (rate of connective tissue increases by 0.013 μm/μg). This may be because only 8% of MWCNT but 80% of SWCNT were present in the interstitial fluid.

These observations, taken together, suggest that the fate of nanoparticles in the respiratory tract is complex and depends on many factors including, but not limited to, their chemical properties, functionalization, length, diameter, dispersion, and rigidity.

4.1.3 *Exposure via the Gastrointestinal Tract*

Contaminated food and water, cosmetics, and drug-delivery devices and particles cleared from the lungs via the mucus pathway are the major sources of nanoparticle exposure via the gastrointestinal (GI) tract (Bergin and Witzmann, 2013). Smaller dispersed bare nanoparticles or functionalized nanoparticles can cross the GI barrier and reach the liver cells (Bockmann et al., 2000). Esch et al. (2014) have shown that the GI tract is a selective barrier preventing larger and aggregated nanoparticles, but allowing smaller nanoparticles to cross into systemic circulation. Larger nanoparticles interact with cell membranes and aggregate into clusters; thus, they are prevented much more effectively by the GI tract barrier. In addition to size, chemical properties also determine the nanoparticles' uptake via the GI tract Desai et al., (1996). Ensign et al. (2012) have also shown the polyethylene glycol (PEG)-coupled but not pristine nanoparticles are taken up by the GI tract and transferred into the systemic circulation. In systemic circulation, nanoparticles, depending on their physicochemical properties, interact with blood proteins (Ge et al., 2011).

A key difference between GI tract exposure and inhalation or dermal exposure is that nanoparticles entering blood from the GI tract, but not from lungs or skin, undergo first-pass metabolism (Lalka et al., 1993). Blood vessels from the GI tract first enter liver (where nanoparticles are modified and/or removed via the lymphatic circulation) and then to the systemic circulation, while blood vessels from the lungs or skin directly go to the systemic circulation (and to other organs/cells) before reaching the liver. Therefore, the therapeutic efficacy and toxicity of nanoparticles exposed via the GI tract and via inhalation or the dermal route may be different.

4.2 Interaction of Nanoparticles with Blood and Intracellular Components

Blood consists of the following: (1) many different types of cells, including red blood cells (RBCs) and innate inflammation cells, such as white blood cells, PMN, macrophages, etc.; (2) diverse groups of proteins and glycoproteins; and (3) lipids, carbohydrates, and electrolytes.

As shown in Figure 11, nanoparticles rapidly interact with many of the blood components, especially plasma proteins, binding to the cell membranes and/or entering the cells. It is an axiom that an interface formed between two different phases, such as nanoparticles and blood, has a higher standard free energy (surface tension) than the individual components. As a result, a nanoparticle's surface is thermodynamically stabilized by adsorption of proteins and other blood components. Adsorption of proteins on a nanoparticle surface often changes in its functions with important toxicological consequences.

4.2.1 *Protein Adsorption on a Nanoparticle's Surface*

Because of high surface reactivity, the surface of nanomaterials would be covered by protein corona upon their entrance to a biological medium (Singh, 2014; Walkey and Chan, 2012; Walkey et al., 2012; Monopoli et al., 2011; Simberg et al., 2009; Vroman et al., 1980). Adsorption of proteins at the nanoparticle's surface may involve noncovalent bonds such as hydrogen bonds, solvation forces, and Van der Waals interactions. However, binding of proteins to a nanoparticle's surface is a multifactorial process depending on the characteristics of the nanoparticles, the interacting proteins, and the medium. The details of nanoparticle–protein interactions are described in Section 5.

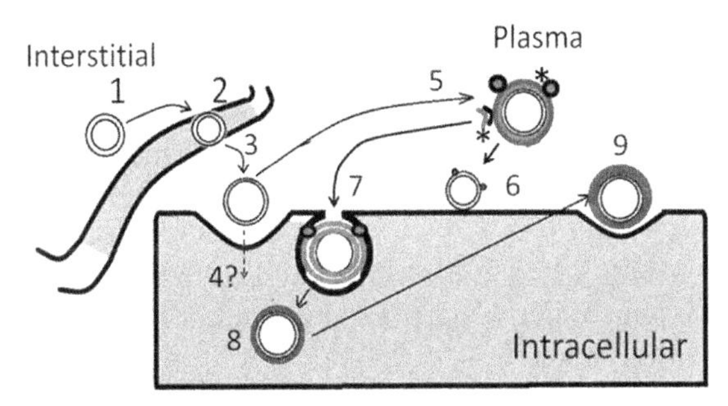

FIGURE 11 **Nanoparticle-blood interaction.** Nanoparticles, from interstitial fluid (1) cross the alveolar membrane (2) and enter the pulmonary blood supply (3). In blood, they either penetrate the cell (4; not established, less likely) or rapidly interact with plasma proteins and form a protein corona that may contain opsinins, such as IgG and complement ligand C3 (5, *—opsonins). The nanoparticle–corona complex binds the cell membrane (6) and/or enters the cell via receptor-mediated endocytosis (7). The internalized nanoparticles may have different corona composition than those in blood (8) and may exit the cells via exocytosis (9).

4.2.2 *Accumulation of Nanoparticles in Red Blood Cells*

In general, three mechanisms have been proposed for the uptake of nanoparticles by cells, as described in the following sections.

4.2.2.1 DIFFUSION THROUGH THE PLASMA MEMBRANE, INCLUDING DISRUPTION

Direct transport of nanoparticles can occur across the plasma membrane either via simple diffusion across the cell membrane or via disruption of the membrane, possibly by the nanoparticle itself. However, certain variables could affect the diffusion of nanoparticles across the cell membrane, such as the characteristics of the lipid bilayer; the nanoparticle's size, charge, hydrophobicity, composition, shape, and physical space limitations; and the properties of solid–liquid and nano-bio interfaces (Nel et al., 2009; Verma et al., 2008).

4.2.2.2 FLUID-PHASE OR RECEPTOR-MEDIATED ENDOCYTOSIS (MUKHERJEE ET AL., 1997)

Fluid-phase endocytosis occurs during the formation of vesicles at a time when the particles are suspended in the vicinity of a forming vesicle (but not bound to the vesicle wall). Particles are captured as the vesicle grows. In a receptor-mediated mechanism, nanoparticles bind to a component of the cell surface, which in turn internalizes with a vesicle, carrying the cargo into the cell.

4.2.2.3 ION CHANNEL AND TRANSPORTER PROTEINS

Numerous ion channels and transporter proteins reside in the plasma membrane and function to mediate the translocation of specific substances in or out of the cell. These aqueous pores can permit the rapid flux of molecules across the plasma membrane.

Although the above mechanisms are valid for a large variety of cells, they may not be viable for RBCs because the RBC membranes contain neither receptors nor ion channels. Rothen Rutishauser et al. (2006) have shown that 20-nm but not 200-nm gold nanoparticles (cationic, anionic, and uncharged) are internalized by the RBCs. They showed that RBCs accumulate nanoparticles of various materials, but the uptake mechanisms are neither endocytosis nor any receptor-based. Further research is needed to elucidate the uptake mechanisms.

5. NANO-BIO INTERFACE

Nanoparticles, because of their unique properties, are finding applications in the medical, pharmaceutical, cosmetic, and environmental industries. This increases the likelihood of engineered nanomaterials encountering humans, animals, and the environment. In an organism, nanoparticles interface with proteins, cell membranes, cell organelles, and DNA, determined by the interface free energy (surface tension) as well as the dynamics of interactions. An understanding of the nano-bio interaction is essential for designing biocompatible nanoparticles. The aim of this section is to elucidate the key aspects of the nano-bio interface.

5.1 Components of the Nano-Bio Interface

There are three basic components of the nano-bio interface.

1. *The nanoparticle surface*, which is characterized by (1) size, shape, surface area, surface free energy, functional groups, and ligands; (2) valence and conduction states; (3) crystallinity; and (4) hydrophobicity and hydrophilicity
2. *The solid–liquid interface* and the changes that occur when the particle interacts with components in the surrounding medium, as characterized by surface hydration and dehydration, release of free energy and surface reconstruction, ion adsorption and charge neutralization, zeta potential, isoelectric points, steric hindrance, electrostatic and steric interactions, aggregation and dispersion, and hydrophobic and hydrophilic interaction.
3. *The solid–liquid interface's contact zone with biological substrates*: This includes membrane interaction, receptor-ligand binding, membrane wrapping, biomolecule interaction and ensuing functional changes, free energy transfer, oxidation energy, and particulate damage (Bahrami et al., 2014; Nel et al., 2009; Velegol, 2007; Oberdörster et al., 2005; Sigmund et al., 2005; Gilbert et al., 2004; Vertegel et al., 2004)

The forces that participate in establishing nano-bio interface are the following:

1. Weak interactive forces, such as Van der Waals, electrostatic, solvation (affinity for a particular solvent), solvophobic (lacking affinity for particular solvent), and depletion forces (an entropic force that arises between large colloid particles that are suspended in a dilute solution of small force nonadsorbing depletants)
2. The fluid's ionic strength, which may determine the particle's hydrophobic interaction
3. Internal and external properties of nanoparticles (Baca et al., 2007)

5.2 Adsorption of Proteins on a Nanoparticle's Surface: Corona Formation

In organisms, nanoparticles rapidly interact with plasma or intracellular proteins and form protein-corona. For simple uncharged homopolymers, protein adsorption is largely driven by Van der Waals and hydrophobic interactions (Fleer et al., 1993). However, for complex nanoparticles, the driving force is generally more complex, including electrostatic interactions and occurrence of hydrophobic domains in the proteins. The protein–nanoparticle interaction results in either "soft" binding, in which large conformational changes occur in the adsorbed protein, and "hard" binding, in which the proteins remain intact (Singh, 2014; Aggarwal et al., 2009; Lundqvist et al., 2008; Malmstein, 1998; Norde, 1986; Haynes and Norde, 1994); a soft corona is due to binding of proteins onto the hard corona (Lundqvist et al., 2011). The nanoparticle–protein interaction is determined by the nanoparticle's (1) adsorption capacity (protein binding affinity, protein concentrations, and nanoparticle-protein equilibrium constant $K_A = k_{on}/k_{off}$), (2) solvation, and (3) surface charge.

5.2.1 *Nanoparticles' Protein Adsorption Capacity*

At equilibrium, the order of plasma protein binding to polymer nanoparticles was high-density lipoprotein (HDL), human serum albumin (HSA), and fibrinogen. (Dell'Orco et al., 2010), while the order for SWCNT was fibrinogen, immunoglobulin, transferrin, and HSA (Ge et al., 2011). For more information, review the articles by Jiang et al. (2009) and Rezwan et al. (2004).

The longevity of the nanoparticle–protein interaction depends on the association (k_{on}) and dissociation (k_{off}) rates for each protein (Figure 12(A)). Initially, abundant proteins such as albumin may occupy the nanoparticle surface (Figure 12(B2)); subsequently, they are replaced by other proteins, such as HDL, which have higher binding affinity for the surface (Figure 12(B3)) (Vroman, 1962). Dell'Orco et al. (2010) have

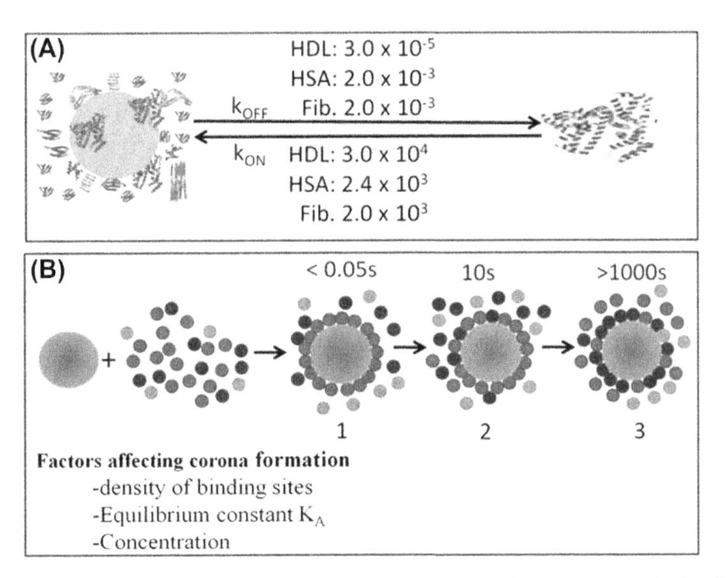

FIGURE 12 **Mechanism of nanoparticle-protein interaction.** (A) The equilibrium constant (K_A) is the ratio of the proteins' sorption (k_{on})/their desorption (k_{off}). (B) The nanoparticle–protein interaction is a complex process determined by (B1) affinity of proteins for nanoparticles, (B2) differential protein-binding sites on the nanoparticle surface and (B3) concentrations of different proteins. When nanoparticles encounter proteins, the high-abundance proteins, irrespective of their affinity, bind them rapidly (<0.05 s, blue particles—dark gray in print versions). Then, gradually, the high-affinity proteins (red particles—light gray in print versions) begin replacing the low-affinity proteins. Eventually, within 90 min, equilibrium is established. The final protein distribution in the corona depends on the distribution of different binding sites on the nanoparticle surface.

determined the k_{on} and k_{off} values for binding of HSA (concentration 40 mg/ml), HDL (concentration 3 mg/ml), and fibrinogen (concentration 3 mg/ml) to a synthetic *N*-iso-propylacrylamide/*N*-tert-butylacrylamide (NIPAM/BAM) copolymer nanoparticles of 70 nm diameter. The initial corona consisting of HSA was formed at 0.02–0.05 s after the nanoparticles were added to the protein solution (Figure 12(B)). Subsequently, HDL (the high-affinity protein) replaced HSA and became the prominent protein in the corona (Dell'Orco et al., 2010). The nanoparticle–protein interactions modulate the properties of both components (Karmali and Simberg, 2011; Lynch and Dawson, 2008; Lundqvist et al., 2008).

5.2.2 *Protein Solvency*

Solvency is defined as the potential of a solvent to dissolve solutes such as proteins. This is one of the key factors in determining the adsorption of proteins on nanoparticles. A decrease in solvency is associated with an increase in adsorption. The protein structure and stability may change drastically with a decrease in solvency. Studies have also shown that the maximum absorption of the proteins by nanoparticles occur at their isoelectric point (Asanov et al., 1997; Carter and Ho, 1994). As the protein charge decreases (protein charges are minimal at the isoelectric point), its solvency also decreases, with the lowest value occurring at the isoelectric point.

5.2.3 *Nanoparticle Surface Charge*

Cationic (positively charged) nanoparticles adsorb proteins with isoelectric points (pI) less than 5.5 (albumin), while the anionic (negatively charged) particles enhance the adsorption of proteins with pI greater than 5.5, such as immunoglobulin G (Moghimi et al., 2001). Gessner et al. (2002, 2003) showed that an increase in surface charge of cationic nanoparticles also increased protein adsorption. Bradley et al. (2011) reported the binding of complement (C1q) to anionic particles, which induces its opsonization.

5.3 Interaction of Nanoparticles with Cell Membranes

The cell membrane is a lipid-based membrane that envelops the cytoplasm and creates a semipermeable barrier. Membranes preserve the local chemical composition and play an active role in the interaction with foreign macromolecules, including nanoparticles. As shown in Figure 13, cell membranes are highly heterogeneous structures consisting of different units such as phospholipids, in which the outer and inner surfaces are negatively charged; the middle region is hydrophobic. Intrinsic proteins, such

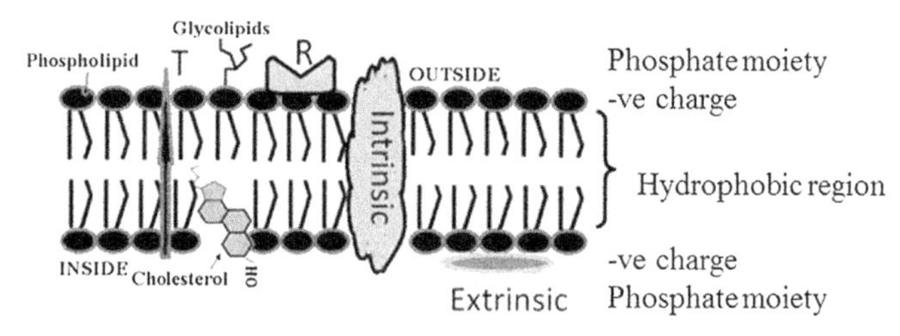

FIGURE 13 **Structure of cell membranes.** The cell membrane is a dynamic and fluid structure consisting of phospholipid bilayers, in which the negatively charged phosphate groups face the extra- and intracellular fluids, while the lipid group makes a highly hydrophobic inner environment. The membrane contains intrinsic (embedded into the membrane) and extrinsic (adsorbed onto the membrane surface) proteins (receptors, enzymes, and transporters), cholesterol and glycolipids. This unique structure of the membrane allows selective uptake of particles.

as receptors (R), ion-channels, transporters (T), and structural proteins, and extrinsic proteins are adsorbed onto the membrane.

Cell membranes exhibit certain unique characteristics and functions:

- Cell membranes are fluid mosaics, in which the phospholipids can move freely along the surface, while the embedded proteins have some flexibility.
- Specific transporters regulate transport in and out of the cell or subcellular domain.
- The membrane receptors bind selective signaling molecules and generate response, such as the release of second messengers and neurotransmitters. The receptor-regulated ion channels allow ions to migrate into and/or out of cells. Neurons are specialized cells that also participate in signal transductions.
- The membranes anchor cytoskeletal filaments or components of the extracellular matrix that allow the cells to maintain their shape.
- The membranes help to compartmentalize subcellular domains and provide a stable site for the binding and catalysis of enzymes.
- Specialized junctions known as gap junctions facilitate intercellular communications.

Nanoparticles, depending upon their composition, size, protein corona, and surface functionalization, interact with cell membranes by different mechanisms (Figure 14), such as adsorption onto the cell membrane (Figure 14(1)), penetrating into the membrane without exiting it (Figure 14(2)), binding to the membrane that initiates spontaneous wrapping for endocytosis (Figure14(3)), endocytosis of the nanoparticle inside the cells (Figure 14(4 and 5)), exiting the cells via exocytosis (Figure 14(6 and 7)), and direct diffusion of nanoparticles specially functionalized to penetrate the cell membrane (Figure 14(8 and 9)). The following sections elucidate the possible mechanisms underlying the nanoparticle-bio (nano-bio) interactions.

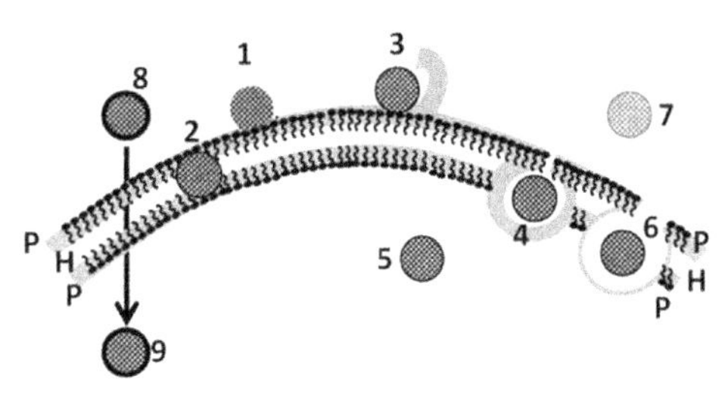

FIGURE 14 Nanoparticle–cell membrane interaction. Nanoparticles interact with cell membranes by (1) binding to the surface, (2) penetrating into the membrane without exiting it, (3) binding to the membrane that initiates spontaneous wrapping for endocytosis, (4 and 5) of nanoparticles inside the cells, (6 and 7) nanoparticles can exit the cells via exocytosis, and (8 and 9) direct diffusion of nanoparticles specially functionalized to penetrate the cell membrane.

5.3.1 Adsorption of Nanoparticles on Cell Membranes

The adsorption of proteins onto cell membranes is a key problem in the biomedical application of nanoparticles. Proteins adsorb from an aqueous solution onto the nanoparticle surfaces, forming a corona around the nanoparticles. The functional properties of the nanoparticles may depend upon the corona's composition. Therefore, an understanding of the driving forces and the kinetics of the formation of a corona is of fundamental importance for medicinal application of nanoparticles.

Lesniak et al. (2013) have shown that pristine nanoparticles are highly reactive due to their extremely high Gibbs free energy. Upon encountering plasma, culture media, or cytoplasm, proteins are rapidly adsorbed onto the surface, resulting in reduced Gibbs free energy and particle stabilization. The adsorption capacity of nanoparticles with a protein corona is much lower than the capacity of pristine particles. The number of nanoparticles adsorbed at a given time, t, is shown in Eqn (1). In this equation, c_o is the concentration, $N_{m,max}$ is the maximum number of nanoparticle adsorption possible, and k_{on} and k_{off} are rate constants for adsorption and desorption, respectively.

$$\frac{dN_m}{dt} = k_{om}c_o(N_{m,max} - N_m) - k_{mo}N_m \quad (1)$$

This equation determines adsorption in the absence of uptake (experiments are performed at 4 °C, which blocks uptake). To accommodate uptake, Lesniak et al. (2013) designed Eqn (2) for the calculation of adsorption and uptake. In this equation, $J(c_o)$ is the uptake rate at a given concentration:

$$J(c_o) = N_{m,max}k_{m1}/(1 + (k_{mo} + k_{m1})/k_{om}c_o) \quad (2)$$

Adsorption appears to be the first step of internalization. Mahmoudi et al. (2014) and Peetla et al. (2014) have addressed the question of whether all adsorbed nanoparticles undergo internalization. Mahmoudi et al. (2014) have also shown that (1) nanoparticles <20 nm (in at least one dimension) are adsorbed on the cell membrane but not internalized into the cells, (2) nanoparticles of about 40 nm are adsorbed as well as internalized, and (3) nanoparticles >50 nm are neither adsorbed nor absorbed, possibly due to their aggregation into larger particles.

5.3.2 Penetration of Nanoparticles in to the Cell Membrane Bilayer

Gold nanoparticles smaller than the membrane's bilayer diameter (approximately 7 nm) or cationic nanoparticles less than 20 nm penetrate the membrane (Lin et al., 2010), but they do not leave the membrane. They remain trapped in the membrane (Figure 15). If nanoparticles exit the membrane, they generate a severely disruptive pore, which in turn induces defective areas (areas on the bilayer surface that are devoid of lipid head groups) and altered bilayer surface texture (Figure 15(B)). The "pore" region of the bilayer is substantially hydrated and is water permeable (Figure 15(D)). As shown in Figure 15(E), the lower leaflet of the bilayer protruded downward to accommodate the penetration. The figures indicate the highly disruptive nature of cationic AuNPs.

The consequences of a nanoparticle's membrane penetration include the following:

1. Accumulation of hydrophobic nanoparticles within cell membranes increased their hydrophobicity and ensuing dehydration.
2. Some nanoparticles cause membrane condensation that inhibits endocytosis.
3. The nanoparticle-induced pores alter the cell's normal homeostasis, which may induce necrosis.

5.3.3 How Do Carbon Nanotubes cross the Cell Membrane and Enter the Cells?

In general, CNTs translocate across the cell membrane by two basic mechanisms: energy-dependent internalization and passive diffusion (Kostarelos et al., 2007; Kam and Dai, 2005; Kam et al., 2004). Pristine CNTs, being highly hydrophobic, interact with the lipid bilayer and penetrate the cell membrane, but they do not disperse into the cytoplasm. Lacerda et al. (2012) reported that approximately 30–50% of functionalized MWCNTs internalized via passive diffusion and the rest via the energy-dependent processes. They also identified three new translocation pathways: membrane wrapping, direct translocation, and vascular translocation. MWCNTs' length and diameter are two key factors determining their translocation across the cell membrane. MWCNTs less than 100 nm may be transported via clathrin and caveolae vesicles, while aggregated nanotubes or those longer than 400 nm will not be transported but will be taken up by the macrophages. Dispersed long nanotubes with diameters around 50 nm cross the cell membrane and enter the cells. Internalized SWCNTs may escape phagocytosis and accumulate in the cytoplasm, followed by their accumulation in lysosomes (Al-Jamal et al., 2011). The acidic environment of lysosomes may release the entrapped drugs by dissolving the acid-liable bonds of functionalized CNTs (Russier et al., 2011; Zhao et al., 2011; Kagan et al., 2010).

5.3.4 *Mechanisms of Endocytosis and Exocytosis*

Endocytosis and exocytosis are key bulk transport processes for the movement of nanoparticles and macroparticles, such as proteins or polysaccharides, in or out of the cell (Oh et al., 2014). Endocytosis moves materials into the cell, while exocytosis exports materials out of the cells via secretory vesicles. The next sections describe the relevance of these processes in nanoparticle transport.

5.3.4.1 ENDOCYTOSIS

As described above, positively charged nanoparticles enter the cells by forming pores. However, pore formation may cause serious deformity to the surrounding lipids. Endocytosis is a physiologically relevant energy-using process by which cells take up particles by engulfing them. Endocytosis occurs in the following stages:

- *Wrapping up*: A nanoparticle's adsorption activates a receptor-mediated process involving an invagination of the cell membrane that grows around the NP and surrounds it in a pinched-off vesicle (Dahan et al., 2001; Parak, 2011; Yang et al., 2005).

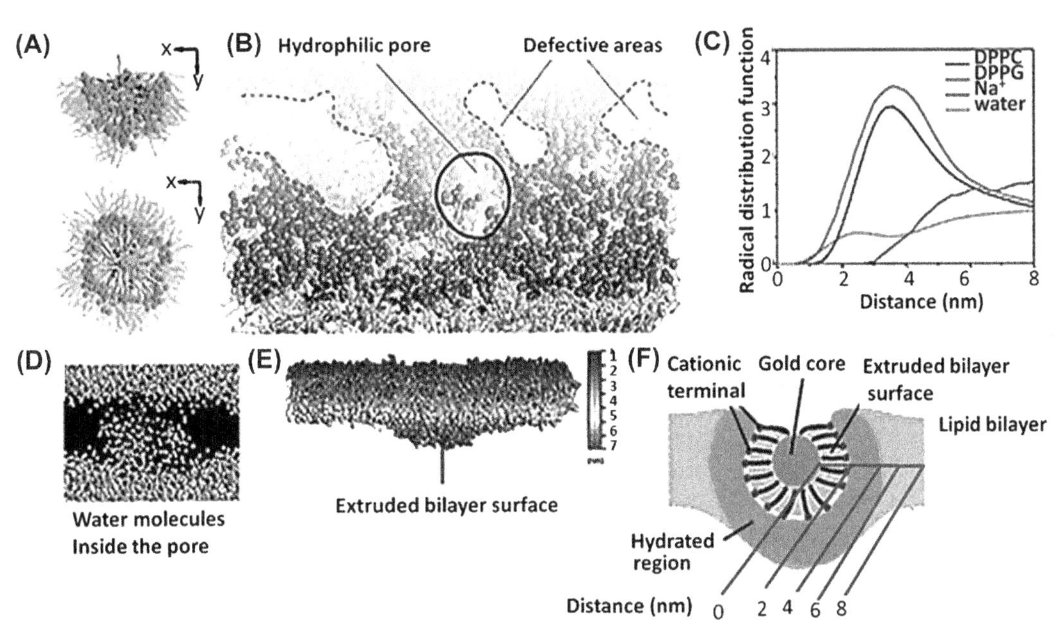

FIGURE 15 **Structural changes in cell membranes exposed to the cationic gold nanoparticles (AuNPs).** (A) When cationic gold nanoparticles encounter the lipid head groups, the lipids form a shell to enwrap the nanoparticles. This disrupts the membrane integrity. (B) Nanoparticles either adsorb onto the surface or penetrate into the bilayer. Upon penetration, a typical pore and defective areas are generated in the bilayer. (C) Radial distribution functions of gold atoms with respect to the lipid particles (DPPC, DPPG), water, and Na^+ (described in (F)). Very few water molecules can be found within 7 Å of the gold core, suggesting that a hydrophobic environment was maintained inside the alkyl thiol monolayer adsorbed onto the gold layer. (D) The penetration of the nanoparticles makes the bilayer water permeable. (E) The lower leaflet of the bilayer extrudes to accommodate the penetrated particle. (F) This sketch shows possible interaction of nanoparticles within the membrane environment. As nanoparticles penetrate the membrane, the lower layer extrudes and a hydrophobic region develops coveting the alkyl thiol monolayer (about 7 Å). Then a hydrated region is formed followed by the lipid monolayer. The nanoparticle-induced abnormalities impair membrane functions, including their permeability and free radical production. *Reproduced from Lin et al. (2010) with permission.*

- *Early endosomes*: The newly formed vesicles coming from the cell surface fuse and acquire a tubule-vascular characteristic having mildly acid pH.
- *Late endosomes*: Late endosomes often contain many membrane vesicles and have proteins characteristic of lysosomes, such as lysosomal membrane glycoproteins and acid hydrolases. The vesicles become more acidic, having a pH value of around 5.5 (Stoorvogel et al., 1991).
- *Lysosomes*: In the final step, the vesicles become strongly acidic (approx. pH of 4.8) and appear as large vacuoles having electron-dense material. Lysosomes may deplete the nanoparticles of their corona (Gruenberg and Maxfield, 1995). However, pristine nanoparticles may acquire a new corona from the cytoplasm.

5.3.4.2 EXOCYTOSIS

Exocytosis is an energy-consuming process that expels secretory vesicles containing nanoparticles (or other chemicals) out of the cell membranes into the extracellular space. Generally, these membrane-bound vesicles contain soluble proteins, membrane proteins, and lipids to be secreted to the extracellular environment. Studies have shown that nanoparticles may also be expelled from the cells via exocytosis (Sakhtianchi et al., 2013; Chung et al., 2012).

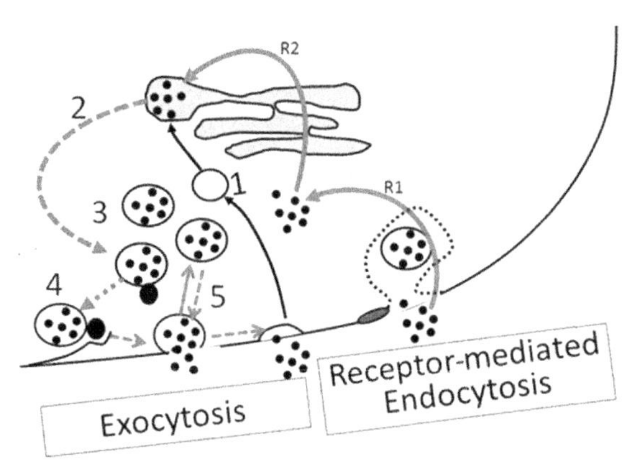

FIGURE 16 **Recycling of Golgi vesicles.** Nanoparticles enter the cells via receptor-mediated endocytosis (R1). From the cytoplasm, they translocate into the Golgi apparatus (R2, (1)). The nanoparticles accumulated in the Golgi apparatus are filled into empty vesicles (1). The refilled vesicles (2) travel across the cytoplasm (3) and are tethered or docked onto the cell membrane (4). Transient vesicle fusion to the cell membrane releases the vesicle contents into the extracellular space (5). The empty vesicles are recycled back to the Golgi apparatus (1).

The steps of exocytosis (Figure 16) are as follows: The vesicles that transport particles from Golgi apparatus to the cell membrane are recycled to transport new particles (step 1; Nichols et al., 2001). The vesicles containing nanoparticles are tethered or docked to the cell membrane (steps 2–4). Transient vesicle fusion to the cell membrane releases the vesicle contents into the extracellular space (step 5).

6. BIOAVAILABILITY AND BIODISTRIBUTION OF NANOPARTICLES

6.1 Bioavailability

6.1.1 General Principles

Bioavailability is the fraction of an administered dose of unchanged toxins that reaches the systemic circulation (for toxicokinetics) and/or the target site (toxicodynamics). One of the key factors determining the bioavailability of nanoparticles is the natural barrier it encounters when administered in an organism. These barriers can be physical, biochemical, or physiological—designed by nature to keep foreign material out and only allow small molecules with specific characteristics to cross. Figure 17 shows an overview of the barriers that nanoparticles may encounter when administered in an organism.

The key physical barriers are cell membranes, the BBB, the blood–placenta barrier, the small intestine, nasal passages, skin, and the mouth mucosa. The key biochemical barriers are blood

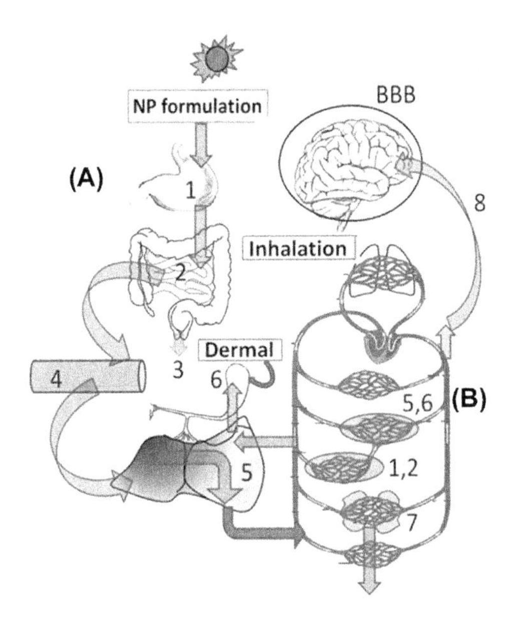

FIGURE 17 **Description of bodily barriers that nanoparticles encounter.** Upon ingestion, nanoparticles pass through stomach ((A1 and B1), pH < 4, highly acidic), duodenum (pH > 8, alkaline) and small intestine (high surface area and absorption capacity) and large intestine containing bacterial colonies (A2 and B2). Part of the unabsorbed particles (hydrophobic or large hydrophilic) are excreted via feces (A3). Nanoparticles absorbed from the GI tract first enter liver for first-pass metabolism (A5 and B5). Blood supply for GI tract, liver and lungs is shown in (B). Unmodified and modified nanoparticles either enter the blood supply or enter the bile duct (A6). Nanoparticles in the systemic circulation travel through different organs and tissues including the brain (B8). In blood, nanoparticles form a protein corona that allows their engulfment in macrophages. Smaller hydrophilic nanoparticles are excreted via kidney (B7). Nanoparticles entering the body via the dermal route either enter the capillaries (smaller hydrophilic particles) directly or are enter the bile circulation (larger particles). Nanoparticles entering the respiratory tract are deposited or reach the alveolar region where that are absorbed by the capillaries.

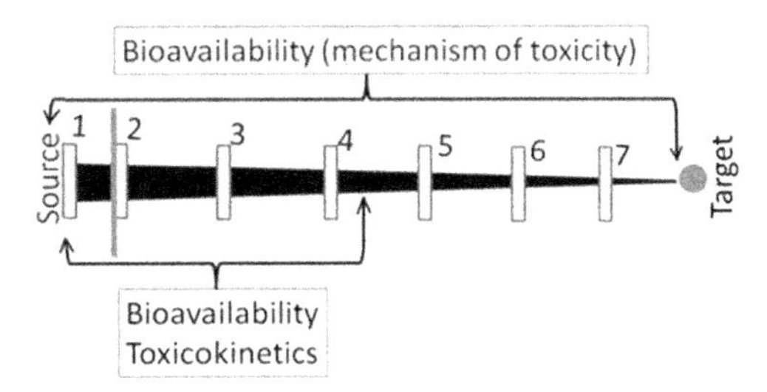

FIGURE 18 **Biological barriers reduce nanoparticles' bioavailability.** A sketch describing the decrease in nanoparticle concentrations as they cross barriers. As nanoparticles move through the physical and biochemical barriers, a progressive reduction in their concentration occurs. (1) In the environment, food item, or commercial products. (2) Exposure of organism via skin, inhalation, ingestion, or injection (in this case the particles enter systemic blood directly). (3) Barriers present in blood such as protein binding, accumulation in cells, or enzymes. (4) Nanoparticle removal via kidney or liver. (5) Membrane barriers, such as cell membranes or the blood–brain barrier. (6) Metabolism and biodegradation. (7) A small fraction reaches the target site.

and tissue enzymes that can degrade foreign substances. The physiological barriers include excretion via the kidneys (small ionic compounds) or bile route (large hydrophilic and all hydrophobic particles). In drug delivery, these barriers present a major challenge; however, for toxicity, these barriers play a protective role. Extensive research is being conducted to overcome natural barriers to improve the delivery of drugs. Nanoparticles are unique because they can be engineered to transport drugs across biological barriers and deliver them to the target site. This approach increases the bioavailability of drug-loaded nanoparticles, but it may enhance their toxicity.

As shown in Figure 18, the nanoparticle concentration decreases as it negotiates various barriers. Only a small fraction of the administered nonfunctionalized nanoparticle may reach the target site. The following sections provide a brief description of each barrier's characteristics.

6.1.2 Physical Barriers

The key physical barriers are the blood–tissue membrane (Figure 19(A)), skin (Figure 19(B)), intestinal mucosa (Figure 19(C)), and the BBB (Figure 19(D)).

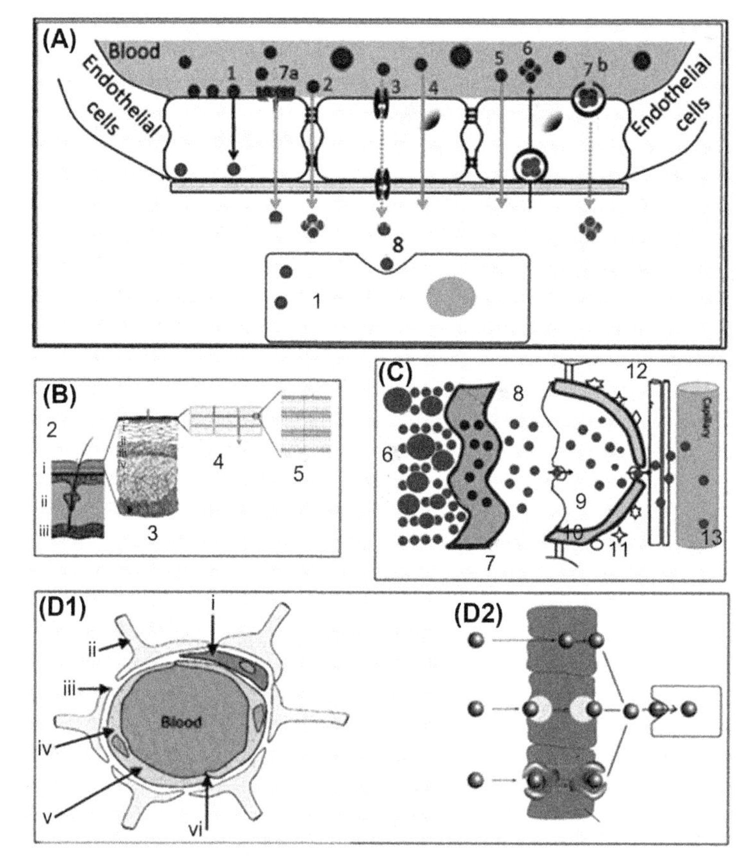

FIGURE 19 **Biophysical barriers.** (A) *The cell membranes and the intermembrane gap junctions*: Cell membranes are the key biophysical barrier between biological fluids (blood, extracellular fluid, cerebrospinal fluid) and the intracellular environment. Biological membranes are selectively permeable membranes that maintain homeostasis of the intracellular environment mostly by regulating permeation of molecules into and from the cells. In general, particles enter cells via two pathways:

- *Transcellular pathways* involving penetration of the particles in and out of the cells via diffusion (A1 and A5), ion channels (A3), bulk (A7a) or receptor-mediated (A7b) endocytosis and exocytosis (A6).
- *Paracellular pathways* involve permeation through the gap junctions between two cells (A2). The junction proteins occludines, claudines, connexins, and cadherins determine the permeation properties. The transported particles can be ingested by another cell via the transcellular mechanisms (Ai).

(B) *Dermal barrier*: Skin, the most important biophysical barrier, consists of three layers: epidermis, dermis, and hypodermis (B2). Epidermis consists of five diverse types of layers, stratum basale (B3i), stratum spinosum (B3ii), stratum granulosum (B3iii), stratum licidum (B3iv), and stratum corneum (B3v). The stratum basale consists of corneocytes (B4) with lipid bilayer membrane (B5). The rigidity of the surface may be associated with its impermeability, not absolute though because smaller hydrophilic particles may cross the dermal barrier or, at least, penetrate the top two layers (think of transdermal patches that deliver drugs such as nicotine, scopolamine, or birth-control drugs). (C) *Gastrointestinal (GI) barriers*: The GI barrier consists of a highly organized outer (C7) and inner (C10) mucosal surfaces lining onto a single-layered intestinal epithelial membrane (C9) that prevents the entry of microbes, different size hydrophobic particles, and large hydrophilic particles (C6) into the lamina propria. The aqueous area between the outer mucosal layer and endothelial cells prevents the uptake of hydrophobic particles. The inner mucosal layer contains Immunoglobulin A and antimicrobial peptides (AMPs). There are five types of intestinal endothelial cells: enterocytes, mucus-producing goblet cells, hormone-producing enteroendocrine cells, AMP-producing Paneth cells and M cells that sample antigens from the intestinal lumen in order to present them to nearby immune cells. When a heterogeneous mixture of nanoparticles (C6) encounters the outer mucosa, only the smaller particles penetrate the endothelial and enter capillary (C13) via the basement membrane (C12). (D) *The blood–brain barrier*: The overall structure of the blood–brain barrier is shown in (D1). The endothelial cells surrounding the brain form a restrictive tight junction (D1i) that prevents indiscriminate entry of particles from blood into the cerebrospinal fluid (D1v). The barrier also contains astrocytes (D1i) and nerve endings (D1ii). The transport mechanisms across the barrier are shown in (D2) and described in (A).

6.1.2.1 THE BLOOD–TISSUE BARRIER

The blood–tissue barrier modulates and restrains the permeability of molecules from extracellular fluid into the cells. The endothelial cells demonstrate closely associated membranes having selective permeability. *Occludines* regulate membrane occlusion, *claudines* regulate tightness of the cell–cell junction, *connexins* determine the gap junctions, and *cadherins* determine the adherence. A functional interaction among these proteins determines the permeability of nanoparticles in and out of cells. Figure 19(A) shows different processes involved in the uptake of molecules or nanoparticles from extracellular fluid into the cells. Uptake can occur via two basic pathways:

1. *Paracellular pathways* allow small nanoparticles to enter the lumen via gaps between two cells (average value of 4 nm). The gap junctions have a hydrophilic environment; thus, only hydrophilic particles pass the gap. A tight junction, such as the one found in the BBB, will not allow paracellular transport.
2. *Transcellular* pathways (Figure 19(A)) correspond to the passage of molecules through cells (Cheung and Brace, 2008). Nanoparticles in blood interact with the capillary endothelial cells and translocate into the extracellular fluid. From there, nanoparticles interact with the neighboring cells. As the nanoparticles negotiate different barriers, their concentrations decrease, resulting in a small number of particles entering the neighboring compartment. The different pathways involved are shown in Figure 19(A).

6.1.2.2 DERMAL BARRIER (FIGURE 19(B))

The skin is the largest organ in body, with a surface area ranging from 1.5 to 4 m^2, thickness ranging from 2 to 3 mm, and follicular density ranging from 600 to 800/cm^2. The skin consists of three major layers: the epidermis, dermis, and hypodermis. The epidermis (0.5–1.5 nm) consists of the stratum basale, stratum spinosum, stratum granulosum, stratum licidum, and SC; the first two layers contain dead cells. The epidermis also contains corneocytes, which have simple double lipid-layered membranes exhibiting hydrophilic gap junction. The dermis contains collagen, elastic tissues, reticular fibers, hair follicles, secretory glands, and blood vessels. The hypodermis contains large blood vessels and neurons.

The cell membranes are mostly lipophilic, while the cell junction pores are hydrophilic. Thus, lipophilic substances are transported mostly via the passive transcellular route, while hydrophilic substances are transported via the paracellular route. Campbell et al. (2012) have shown that polystyrene nanoparticles applied in aqueous suspension could infiltrate only the stratum layers in the final stages of desquamation and the infiltration was independent of the nanoparticle size.

6.1.2.3 INTESTINAL MUCOSA BARRIER (FIGURE 19(C))

The intestinal mucosa restricts nanoparticles to either the paracellular or the transcellular route in a size-dependent manner. Mucosal uptake is most effective within the nanometer ranges and impeded when the size is over 1 μm. The main barrier to the paracellular diffusion of molecules across the intestinal epithelium is the region of the tight junction or zonula occludens. As shown in Figure 19(C), the mucosal barrier consists of a mucus layer, water layer, epithelial cells, basement membrane, and capillaries that restrict the penetration of nanoparticles from the intestine to the blood capillaries.

6.1.2.4 THE BLOOD–BRAIN BARRIER

The BBB, consisting of endothelial cells joined by virtually impermeable tight junctions, protects the central nervous system (CNS) and maintains its homeostasis (Figure 19(D)). Due to the BBB, many potential drugs for the treatment of CNS diseases do not reach the brain in sufficient concentrations. Small lipophilic particles can penetrate the barrier via simple diffusion.

Some studies have demonstrated that engineered polymeric nanoparticles (attached to proteins that bind the surface receptors (for endocytosis) or nutrient transporters) can overcome the BBB (Youns et al., 2011; Petkar et al., 2011). The engineered nanoparticles can deliver proteins and other macromolecules across the BBB into the brain. Thus, they hold great promise for noninvasive therapy of CNS diseases (Holmes, 2013; Re et al., 2012). However, they also enhance CNS toxicity (Sharma et al., 2013).

6.1.3 *Biochemical Barriers*

The enzymes present in blood (such as esterases and amidases) and those released from the pancreas (lipase, protease, and amylase) into the intestinal lumen may play a role in the metabolism of surface ligands, which may defunctionalize the engineered nanoparticles. Defunctionalization may convert an active nanoparticle into more toxic and less bioavailable nonfunctionalized nanoparticles (Liu et al., 2010). Another biochemical barrier is the pH differences in different organs or organelles. The pH values of blood, lysosome, stomach, and duodenum are 7.4 (slightly basic), 5 (acidic), <4 (highly acidic), and 8 (basic), respectively. These pH differences may denature many proteins, depending upon their isoelectric point, in the stomach or duodenum; they also may alter a nanoparticle's water solubility (acidic particles will be insoluble in the stomach (nonionized $-COOH$ group) but soluble in the duodenum (ionized $-COO^-$ group). Basic particles will be insoluble in the duodenum (nonionized $-NH_2$ group) but soluble in the stomach (ionized $-NH_3^+$ group).

6.1.4 *Physiological Barriers*

Physiological barriers include, but are not limited to, the removal of nanoparticles via excretion (small water-soluble nanoparticles) or bile secretion (lipophilic nanoparticles or large water-soluble nanoparticles). Although the roles of physiological barriers in the bioavailability of bulk particles are well established, their roles in the bioavailability of engineered nanoparticles are not fully known.

Praetorius and Mandal (2007) have suggested that nanoparticle modifications, in addition to allowing for greater and more accurate tumor targeting, are also aiding in the crossing of biophysical barriers, such as the BBB, by reducing peripheral effects and increasing the relative amounts of drugs reaching the brain. Lee et al. (2010) have shown that magnetic iron oxide nanoparticles have great potential for overcoming the tumor stromal barrier, thus providing an enhanced therapeutic effect of nanoparticle drugs on pancreatic cancers.

Taken together, these observations suggest that bioavailability determines the percentage of the applied dose that will appear in the blood and the target site. For bulk particles, bioavailability is intrinsically related to the physicochemical properties and the biological barriers of the organisms. Nanoparticles, however, can be engineered to circumvent the barriers and yield artificially high bioavailability.

6.1.5 *Bioavailability of Bulk Particles*

Bioavailability of a toxin is a function of the toxin's physicochemical properties (water solubility, molecular weight, stability, electronic properties, etc.), the mode of exposure (intravenous, dermal, inhalation, and oral), and its fate (absorption, distribution, metabolism and excretion (ADME)) in an organism (Figure 20, discussed in Chapter 6). Intravenous, dermal, and inhalation exposures release toxins directly into the systemic blood and then to the organs and tissues, including the target site, liver (metabolism, transformation, and activation), and kidney (excretion). However, oral exposure does not release toxins directly into the systemic circulation, but releases toxins into the GI tract from where it is absorbed and transported into the

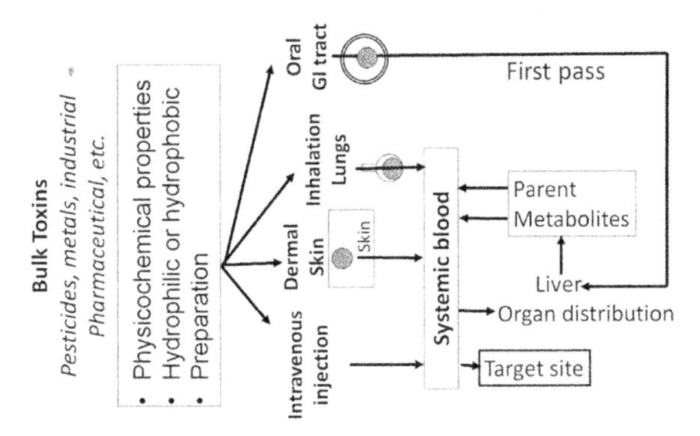

FIGURE 20 **Bioavailability of bulk particles.** Bioavailability of bulk particles is determined by the particles's physicochemical properties and the dose (mass based), the mode of exposure (intravenous, dermal, inhalation and oral, and its fate (nonspecific interactions, metabolism, and excretion). Intravenous, dermal, and inhalation exposures release the particles directly into the systemic blood followed by different organs and tissues, including the target site, liver, and kidney. Oral exposure first releases the particles into the GI tract and then to the liver for first-pass metabolism. In liver, the particles are metabolized, and the metabolite and the remaining native particles are released into the systemic blood. Thus, an organism exposed to a toxin via oral intake may receive much lower percentage of the administered dose than the organism exposed to the toxin intravenously, dermal application, or inhalation.

liver via hepatic vein. In liver, toxins are metabolized, and the metabolite and the remaining native toxin are released into the systemic blood. Because of these differences, toxin exposure via intravenous, dermal, and inhalation routs yields several folds higher (30% to 90% of the administered dose) bioavailability than exposure via the oral route (<1% to 10% of the administered dose).

6.1.5.1 FACTORS THAT INFLUENCE THE BIOAVAILABILITY OF BULK MATERIALS

Hepatic metabolism: The liver rapidly metabolizes drugs and toxins by either activating or inactivating (by glucuronidation and/or sulfation) them. The conjugated toxins, being hydrophilic, are excreted. The amount of activated and parent toxins that returns to the systemic circulation is decreased.

Solubility: Highly hydrophilic compounds are poorly absorbed because of their inability to cross the lipid-rich cell membranes. Paradoxically, particles that are extremely hydrophobic also poorly absorbed because they are insoluble in aqueous cytoplasm. For absorption, particles must exhibit a hydrophilic-lipophilic balance so that they can cross the membrane and then remain soluble in the cytoplasm.

Nature of the bulk particle formulations: Drug absorption may be altered by factors unrelated to the chemistry of the drug. For example, particle size, salt form, crystal polymorphism, enteric coatings, and the presence of excipients (e.g., binders and dispersing agents) can influence the ease of dissolution and, therefore, alter the rate of absorption.

Chemical instability: Some compounds may be unstable in acidic pH of the gastric contents, while others may be destroyed in the GI tract by degradation enzymes.

Hydrophilicity and lipophilicity: It is well established that the outer surface of cell membrane has a negative charge. Thus, negatively charged hydrophilic compounds, due to electrostatic repulsion, do not interact with the cell membrane and do not enter the cells. The positively charged hydrophilic drugs bind to the cell membranes and translocate into the cells via either simple diffusion or receptor-mediated-endocytosis. The lipophilic compounds dissolve in the hydrophobic site of the cell membrane, thus entering the cells. This theory, however, has the following drawbacks:

- The cell membrane is not absolutely lipophilic. It is a mosaic of hydrophilic and lipophilic molecules. The outer and inner surfaces are phospholipids (negatively charged), while the inner membrane is lipophilic. Thus, highly hydrophilic and highly lipophilic compounds may not cross the cell membrane.

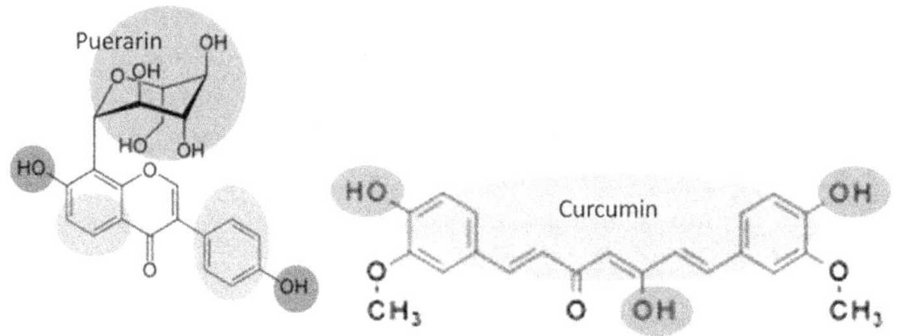

FIGURE 21 Hydrophilic (red—dark gray in print versions) and lipophilic (blue—light gray in print versions) surfaces of puerarin and curcumin.

- Molecules are not absolutely hydrophilic and lipophilic. They have both hydrophilic and lipophilic sites. As shown in Figure 21, puerarin (highly hydrophilic) and curcumin (highly lipophilic) both have hydrophilic (red—dark gray in print versions) and lipophilic (blue—light gray in print versions) groups. The hydrophilic-lipophilic balance determines a molecule's water solubility (Figure 21).
- The cytoplasm is hydrophilic; thus, lipophilic drugs may cross the cell membrane but, once inside the cell, may not translocate to the target site. The positively charged drugs may interact with the outer cell membrane, but they may not enter the membrane's lipophilic environment. An observation that puerarin, a highly hydrophilic isoflavone, and curcumin, a turmeric compound, both exhibit poor bioavailability in vitro stresses the drawback in this proposal.

6.1.6 *Lipinski's Rule-of -Five for Bulk Particles*

In 1997, Lipinski devised his "Rule of Five" (actually, multiples of five), which predicts that poor absorption or permeation is more likely when the following conditions exist:

1. H-bond donors >5 (sum of NH and OH) [5 × 1]
2. H-bond acceptors >10 (N and O) [5 × 2]
3. Molecular weight >500 [5 × 10]
4. Log P > 5 [5 × 1]

Lipinski's rule recognizes the need to balance the forces that enable a molecule to be at once both sufficiently hydrophilic to dissolve adequately in aqueous media and sufficiently lipophilic to penetrate more hydrophobic environments. Lipinski's indices for different drugs are listed in Table 1 (Lipinski et al., 1997). Four compounds were identified for poor permeation, although their log P value did not meet the criteria.

The Rule of Five for drug likeness has played an important role in routine drug discovery practice, especially in screening large sets of databases (with the exception of natural chemicals and enzyme substrates). However, people have started associating this simple guideline with screen absorption or permeation of chemicals with "drug likeness." In some cases, the Rule of Five was used to screen for biological or pharmaceutical properties of chemicals—a concept outside the scope of the rule. An earlier Food and Drug Administration study of 1200 drugs showed that approximately 20% of the drugs had excellent bioavailability, but they failed in at least one of the Rule of Five parameters. Therefore, the rule must be used only for prediction of a chemical's absorption and not for its biological activity.

6.1.7 *Bioavailability of Engineered Nanoparticles*

It is well established that the physicochemical properties of nanoparticles (10–100 nm) are different than those of atoms (≤1 nm) or bulk

TABLE 1 Examples of the Lipinski Rule

Name	Log P	OH + NH^+	MW	N + O^-
Acyclovir	−0.09	4		8
Alprazolam	4.74	0		4
Aspirin	1.70	1		4
Atenolol	0.92	4		5
Azithromycin[a]	0.14	5	749	14
AZT	−4.38	2		9
Benzyl penicillin	1.82	2		6
Caffeine	0.20	0		6
Candoxatril	3.03	2		8
Captopril	0.64	1		4
Carbamazepine	3.53	2		3
Chloramphenicol	1.23	3		7
Cimetidine	0.82	3		6
Clonidine	3.47	2		3
Cyclosporine[a]	−0.32	5	1202	23
Desipramine	3.64	3		5
Dexamethasone	1.85	0		3
Diazepam	3.36	2		3
Diclofenac	3.99	0		6
Diltiazem[a]	2.67	7	543	12
Doxorubicin	−1.33	2		7
Enalapril-maleate[a]	1.64	5	733	14
Erythromycin	−0.14	8		9
Famotidine	−0.18	1		5
Felodipine	3.22	2		4
Fluorouracil	−0.63	2		5

OH + NH^+, total H bond donor; MW, molecular weight; N + O^-, total H bond acceptor. Molecular weight >500 is shown.
[a] *Poor absorption.*

materials (>500 nm). Bulk materials have constant physicochemical properties regardless of size, but nanoparticles exhibit size-dependent properties, such as quantum confinement in semiconductor particles, surface plasmon resonance in some metal particles, and superparamagnetism in magnetic materials. The properties of materials change as their size approaches the nanoscale and as the percentage of atoms at the surface of a material becomes significant. For bulk materials larger than 1 μm, the percentage of atoms at the surface is minuscule relative to the total number of atoms of the material (Figure 22).

6.1.7.1 SIZE DEPENDENT BIOAVAILABILITY

In bulk particles, the proportion of core atoms is several-fold higher than surface molecules; thus, the core determines the surface properties that remain independent of the size. However, in nanoparticles, surface atoms are higher than the core atoms; thus, the surface atoms determine the nanoparticles' physicochemical properties. A change in size changes the surface/core atomic ratio, which alters the intermolecular interaction and ensuing changes in chemical, optical, and magnetic properties. This may explain why the color of a gold nanoparticle, depending on size, changes from red to blue (Figure 23). In addition, as the particles become smaller than 100 nm, their reactivity with blood proteins increases exponentially, which results in a size-dependent decrease in nanoparticle's bioavailability.

6.1.7.2 DOES LIPINSKI'S THEORY APPLY TO NANOPARTICLES?

In general, Lipinski's rule selects hydrophilic chemicals with some lipophilic properties. Unfortunately, pristine nanoparticles (not attached to a functional group) are either highly hydrophilic (e.g., amino acid or protein based) or highly lipophilic (carbon nanotubes, fullerenes,

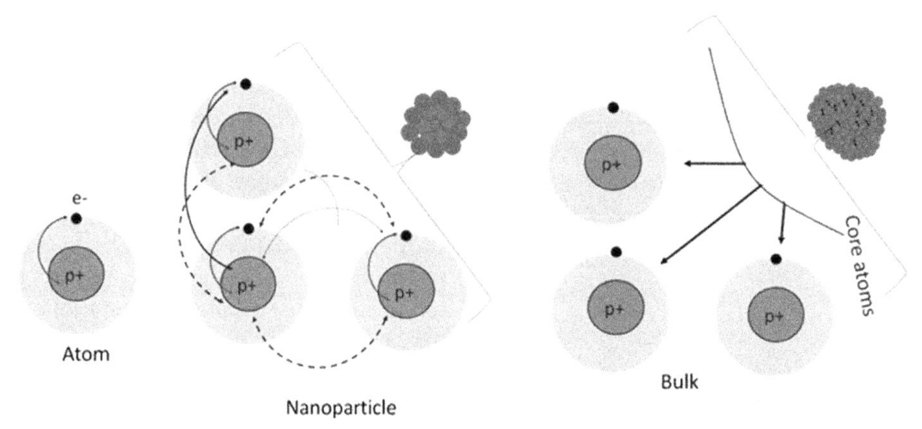

FIGURE 22 **Electronic structures of an atom, a nanoparticle, and a bulk particle.** In an atom, the net negative charge of electrons balances the net positive charge of protons. In a hypothetical cluster consisting of three atoms, the surface valence electrons are the key determinant of the cluster's properties. In fact, in nanoparticles smaller than 100 nm, surface atoms determine the particles' properties. In bulk particles, however, surface particles account for $<0.1\%$ of total atoms, thus the core atoms determine the particle's properties. The fundamental difference between nanoparticles and bulk particles is that surface atoms determine the properties of nanoparticles, while the core atoms determine the properties of bulk particles. Solid lines: attraction between electrons and protons, dotted lines: repulsion between electrons or between protons.

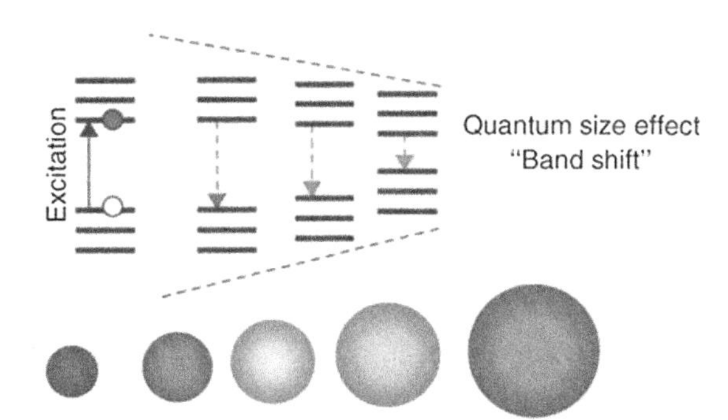

FIGURE 23 **Effects of size on electronic and optical properties of nanoparticles.** As the size decreases, the band-gap energy between the valence and conduction bands increases, resulting in blue shift of the optical properties. As the nanoparticle size increases, the energy of the emitted light decreases, resulting in different color light. Thus, it is possible to design a test in which free QD emits green light but the receptor-bound QD emits red light. The optical properties described above are limited to metal nanoparticles, not polymer nanoparticles. However, polymer nanoparticles can be coupled to the QDs for desired effects.

and polymer dendrimers). Therefore, pristine nanoparticles may not pass Lipinski's rules. However, the chemical and physical properties of nanoparticles can be engineered to achieve the desired solubility, selectivity, size, and function, as follows:

1. A nanoparticle functionalized with an aquaporin-receptor antibody is internalized into the cells via receptor internalization. This may cause artificial accumulation of nanoparticles in aquaporin-positive cells.
2. A nanoparticle coupled to transferrin receptor will bind to the brain endothelial cells and get internalized into the brain. This may cause artificial accumulation of nanoparticles into the brain.
3. A nanoparticle coupled to caspase-3 substrate will translocate into the pre-apoptotic cells.

These nanoparticles are designed to select a specific cellular protein for desired effects. Similarly, for imaging cancer cells, cancer-cell selective functional groups are attached to a particular size of nanoparticle that selectively binds cancer cells. Accumulation of nanoparticles in different tissues is determined by the size of nanoparticles. Smaller (<30 nm) particles tend to accumulate in tumor tissue much more than in normal tissues, possibly because of the tumor's newly formed abnormal blood vessels. A tumor's blood supply is usually abnormal in form and architecture, poorly aligned with wide fenestrations, and lacks a smooth muscle layer. In addition, tumors lack effective lymphatic drainage.

Therefore, unlike the bulk particles, nanoparticles can be customized to fit a desired criterion, independent of the Rule of Five. Therefore, the functionalized nanoparticle exhibits high absorption without passing Lipinski's rules.

6.1.8 *Opsonization and Phagocytosis of Nanoparticles*

One of the major barriers for the bioavailability of nanoparticles is their uptake by the reticuloendothelial system (RES) and their subsequent removal from systemic circulation (Owens and Peppas, 2006). The macrophage lineage in the liver and spleen identifies nanoparticles as a foreign body and removes them. The function of opsonins is to react with nanoparticles and make them more susceptible to ingestion by phagocytes (Winkelstein, 1973). Nanoparticles are opsonized when a specific antibody, alone or in concert with complement C3 (activated via the classic pathway of C1, C4, and C2), bind phagocytic cells expressing CR1. Fc receptors activate their phagocytosis (Aderem and Underhill, 1999; Carroll, 1998; Epstein et al., 1996; Sengelov, 1995; Tenner et al., 1995).

6.2 Biodistribution

Biodistribution is a method of tracking where and how the compounds of interest travel in an organism. This provides important information regarding a particle's bioavailability at the target site. In blood, nanoparticles interact with (1) plasma proteins, which result in the formation of a heterogeneous protein layer, called the protein corona, onto the nanoparticle surface; and (2) the blood-cell membranes, and they either remain adsorbed or penetrate into the cells via endocytosis. In some cases, nanoparticles can exit the cells via exocytosis. Pristine or modified nanoparticles are distributed in different organs and tissue, especially in the kidney and liver, which respectively excrete and modify the particles. Because of in vivo modifications, nanoparticles present in blood may be structurally and functionally different (different types of corona and/or metabolized products) than those entering the blood. The goal of this section is to discuss the fate of nanoparticles (and bulk particles for comparison) present in blood and different tissues.

6.2.1 *Biodistribution of Bulk Particles*

Biodistribution means the reversible transfer of chemicals from one location to another within the body. Once a chemical has entered the vascular system, it is distributed throughout the body fluids according to the physiochemical properties of the drug and its ability to penetrate the barriers. Biodistribution is characterized by their volume of distribution (V_D), defined as the volume that the total amount of administered chemical would have to occupy to provide the same concentration as it currently holds in the blood plasma. V_D, discussed in detail in Chapter 6, is defined as $V_D =$ total amount of a chemical in body/chemical's concentration in plasma. The average fluid volumes in different compartments of human body are as follows: extracellular fluid: 13–15 L, plasma: 3–4 L, interstitial fluids: 10–13 L, intracellular fluids: 25–28 L, and total body water: 40–46 L. The bulk particles may follow one of four patterns of distribution.

1. The drug may remain largely within the vascular system. Plasma substitutes, such as

dextran or the drugs strongly bound to plasma protein, may remain in plasma. A value of V_D in the region of 3–5 L (in an adult) would indicate that the drug remains mostly in plasma.

2. Some low-molecular-weight water-soluble compounds, such as ethanol and a few sulfonamides, are uniformly distributed throughout the body's water. These compounds would be expected to produce a V_D value of 30–50 L, corresponding to the total body water volume.
3. A few chemicals are concentrated specifically in one or more tissues that may or may not be the site of action. Some of the examples are iodine, which concentrates in the thyroid gland; chloroquine, which accumulates in the liver; tetracycline, which is almost irreversibly bound to bone and developing teeth; and highly lipid-soluble compounds, which distribute into fat tissue. Drugs such as chloroquine and nortriptyline, exhibiting distribution pattern 3, would exhibit very large V_D values due to the drug concentration effect.
4. If a drug is nonuniformly distributed in the body then, its V_D variations are largely determined by the ability of the drug to pass through membranes and their lipid/water solubility. The highest concentrations are often present in the kidney, liver, and intestine, usually reflecting the amount of excreted drugs. Total blood flow is greatest to brain, kidneys, liver, and muscle, with highest perfusion rates to the brain, kidney, liver, and heart. It would be expected that the total drug concentration would rise most rapidly in these organs.

V_D varies with an individual's height and weight; accumulation of fat (for lipid-soluble drugs), such as for obese patients; accumulation of fluids (for water-soluble drugs), such as ascites, edema, or pleural effusion; and changes in the proportion of each body compartment with age.

6.2.2 Biodistribution of Nanoparticles

Aggarwal et al. (2009) and Choi et al. (2009) have proposed that the biodistribution of nanoparticles in organisms depends upon their functionalization (Figure 24). Pristine nanoparticles, being highly reactive, rapidly bind to different blood proteins, including opsonins, which accelerate nanoparticles' opsonization and phagocytosis in the liver and spleen. PEGylation (polyethylene glycol (PEG) functionalization) suppresses phagocytosis of nanoparticles (Chaudhari et al., 2011), thus enhancing their systemic circulation and suppressing their accumulation in the liver and spleen. Engineered nanoparticles functionalized with ApoE selectively target the BBB and accumulate in the brain. Unlike nanoparticles, the biodistribution of bulk particles in an organism is not affected by its size or shape.

6.2.2.1 NANOPARTICLE BIODISTRIBUTION IN HEALTHY SUBJECTS

As discussed in earlier chapters, a nanoparticle's surface is more reactive than a corresponding bulk particle's surface (Brown et al., 2001; Hagens et al., 2007), possibly because the surface atoms of nanoparticles (smaller than 100 nm) dominate the core atoms; thus, they determine a nanoparticle's unique properties. Compared to nanoparticles, surface atoms in bulk particles play a minuscule role in determining their properties. When pristine nanoparticles come in contact with blood, they rapidly adsorb proteins and form a protein corona in a size-dependent manner, playing a major role in nanoparticles' interaction with the biological system (Foster and Hirst, 2005), including their distribution in organisms (De Jong et al., 2008; Hillyer and Albrecht, 2001).

A study conducted in our laboratory (Singh et al., 2013) and those from Manohar et al. (2011), Lasagna-Reeves et al. (2010), and De Jong et al. (2008) have shown differential distribution of 15–250 nm gold nanoparticles in

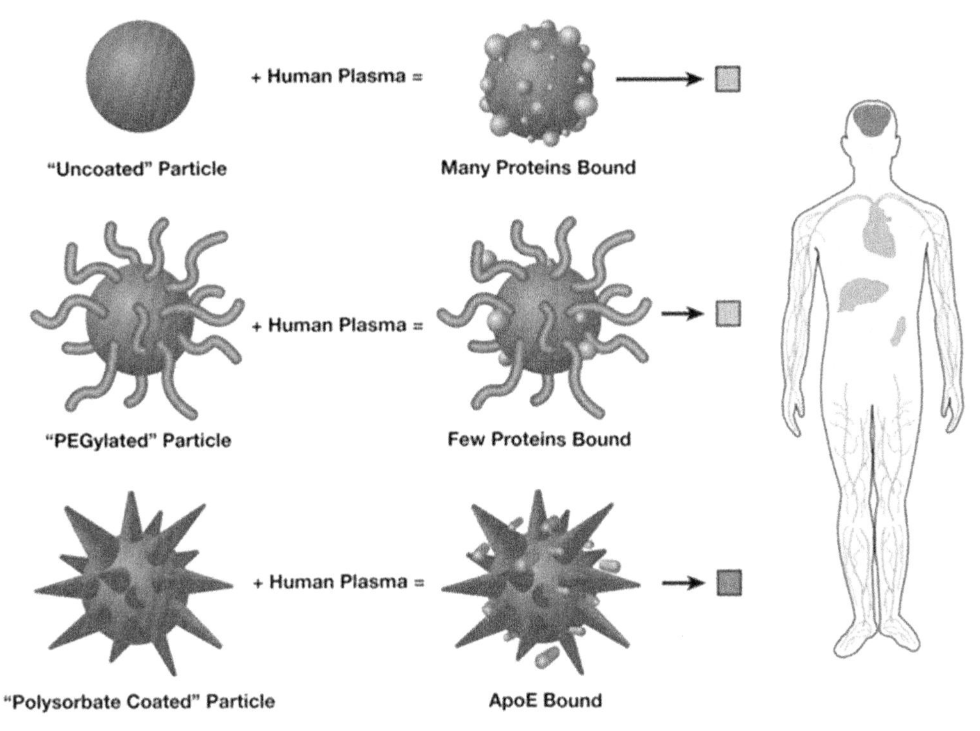

FIGURE 24 **Biodistribution of nanoparticles with varying coatings and bound proteins.** Uncoated (pristine) particles bind to different plasma proteins that may induce their uptake by the reticuloendothelial system (RES) containing phagocytic cells in liver and spleen. The PEG-functionalized particles bind very few proteins, avoid uptake by the RES, and are longer circulating in the blood. Polysorbate-coated particles that specifically bind ApoE, selectively target the brain across the blood–brain barrier. *Reprinted from Aggarwal et al. (2009) with permission.*

different tissues collected from nanoparticle-exposed rats. Larger nanoparticles (70–250 nm) exhibited greater accumulation in the liver and spleen, while smaller nanoparticles accumulate in all organs including the brain (Figures 25 and 26). Similar biodistribution has also been reported for silica. (Nabeshi et al., 2010, 2011; Kumar et al., 2012; He et al., 2008), magnetic nanoparticles (Lee et al., 2010; Zhu et al., 2009; Gamarra et al., 2007; Cole et al., 2011; Jain et al., 2008; Kwon et al., 2008; Lacava et al., 2002), and carbon nanotubes (de la Zerda et al., 2008; Singh et al., 2006; Saba, 1970). Hou et al. (2010), using real-time in vivo imaging in nanoparticle-injected mice (Figure 27), showed that the near-infrared fluorescence (NIRF) signal in the liver, spleen, and kidney gradually intensified after injection of nanoparticles, reaching maximum levels at 6 h; then, the NIRF signal gradually decreased, revealing a gradual clearance of nanoparticles.

6.2.2.2 NANOPARTICLE PERMEATION ACROSS THE BBB AND ACCUMULATION IN THE BRAIN

The brain and spinal cord both are anatomically separated from the systemic circulation by a relatively impermeable barrier consisting of endothelial cells joined by tight junctions. The brain also receives various perivascular cells, such as smooth muscle cells, pericytes, microglial cells, and astrocytes (Begley, 1996, 2004; Bernacki et al., 2008), which also regulate the blood–brain

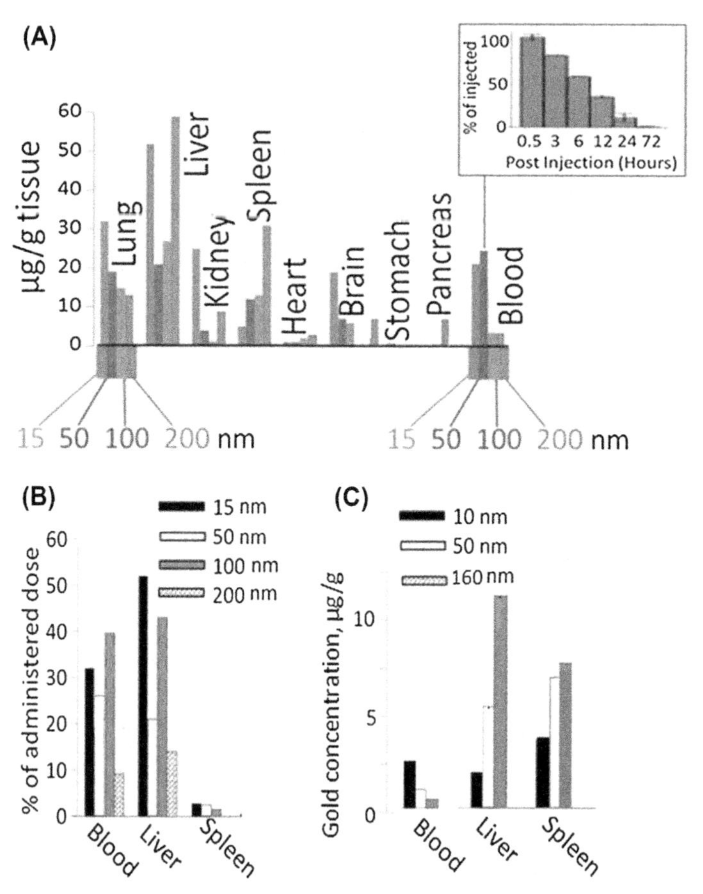

FIGURE 25 **Size-related biodistribution of nanoparticles.** (A) Size-related distribution of gold nanoparticles in rats injected intravenously 1 mg/ml dose of the gold nanoparticle *(data from Singh et al., 2013)*. Gold concentrations in different tissues at 24 h after nanoparticle injection is shown. The 15 nm nanoparticle exhibited greater accumulation than larger nanoparticles in liver, lungs, kidney, brain and stomach. The 15 and 200 nm nanoparticle exhibited comparable accumulation that was significantly greater than accumulation of 50 or 100 nm nanoparticles in liver. The 200 nm nanoparticles were most abundant in spleen. The nanoparticle distribution exhibited the following patterns: Liver: 200, 15 >> 100 > 50 nm, Spleen: 200 > 100 > 50 > 15 nm, Lungs, heart, and brain: 15 > 50 > 100 > 200 nm, Blood: 15, 50 >> 100, 200 nm. *Inset*: In another study, 50 nm gold nanoparticle was administered by intravenous injection and blood gold concentrations were determined at 0.5, 3, 6, 12, 24 and 72 h after injection. At 24 h, approximately 5% of the administered dose remained in blood. Results from the author's laboratory. (B and C) Size-related distribution of gold nanoparticles in rats *(data from De Jong et al. (2008) (B) and Terentyuk et al. (2009) (C) reprinted with permission)*. Distribution of gold in blood liver and spleen samples in rats (A, 15–200 nm gold nanoparticles) and rabbit (B, 15 and 50 nm gold, and 160 nm gold-SiO_2 nanoparticles) intravenously injected gold nanoparticles. There are some discrepancies in the accumulation of nanoparticles in liver and spleen samples. Singh et al. (2013) and de Jong et al. (2008) reported that 15 and 200 nm or 100 nm gold nanoparticles accumulated in liver, but only 200 nm nanoparticles accumulated in spleen. Terentyuk et al. (2009) showed that only 160 nm silica nanoparticles accumulated in both liver and spleen samples.

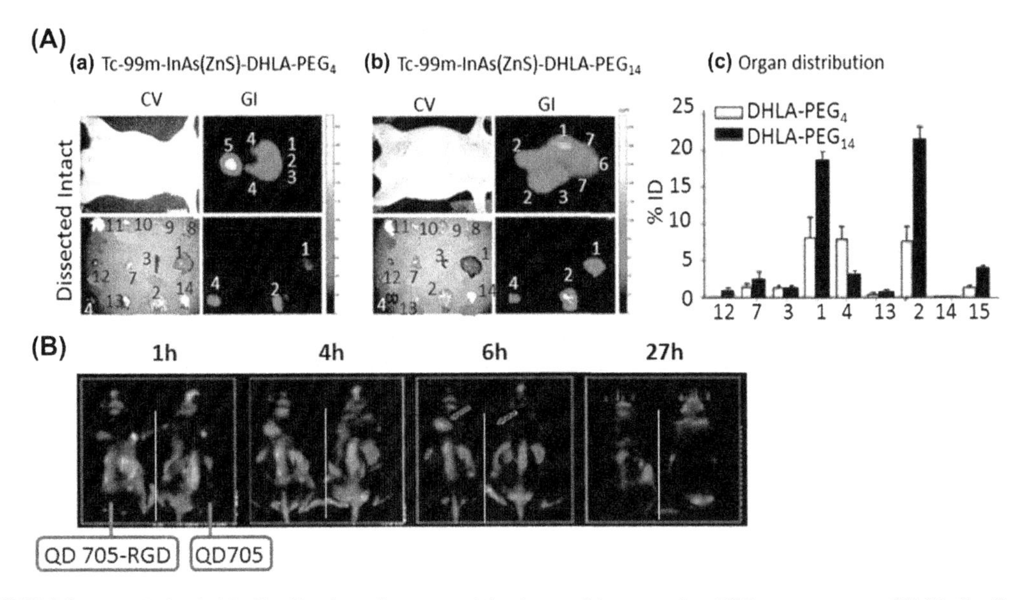

FIGURE 26 **Whole-body biodistribution of nanoparticles imaged in rats using NIR-spectroscopy.** (A) Biodistribution of ^{99m}Tc-conjugated near infrared (NIR) QDs (^{99m}Tc-InAs(ZnS)-DHLA) coated with PEG_4 (QD-P_4) or with PEG_{14} (QD-P_{14}) in CD-1 mice. (a and b) The color video (CV) and Anger camera gamma-ray images (GI) of QD-P_4 5.6 nm HD (a) and QD-P_{14} 8.7 nm HD (b), 4 h post-intravenous injection in intact animals immediately after sacrifice (top row). Bottom rows show images of dissected organs. *Abbreviations*: (1) liver; (2) intestine; (3) spleen; (4) kidney; (5) bladder; (6 and 12) heart; (7) lungs; (8) bone; (9) muscle; (10) adipose; (11) skin; (13) stomach; (14) brain; (15) feces. (c) Quantitative organ distribution (% Initial Dose (ID)) of QD-P_4 (white bar) QD-P_{14} (black bar). Each point represents the mean ± SD. QD-P_4 (5.6 nm HD) appeared in the bladder (47.0 ± 5.7% ID), kidney (7.9 ± 3.8% ID), liver (8.1 ± 5.6% ID), and intestine (7.5 ± 4.2% ID), likely representing QDs in the process of being excreted. Larger QDs (QD-P_{14}, 8.7 nm HD) exhibited relatively high uptake in the liver (18.6 ± 2.6%) and intestine (21.4 ± 3.2%) and almost no signal in the kidneys and bladder. *(Reprinted from Choi et al. (2009) with permission.)* (B) In vivo near infrared (NIR) fluorescence imaging of U87MG tumor-bearing mice (tumor is pointed by white arrows) injected with 200 pmol of QD705-RGD (tumor-selective probe) and QD705 (passive probe), respectively. The mice autofluorescence is color-coded green (dark gray in print versions) while the unmixed QD signal is color coded red (light gray in print versions). Prominent uptake in the liver, bone marrow, and lymph nodes is visible. Distribution of the two probes appears comparable. However, the QD 705-RGD exhibited greater accumulation than QD 705 in tumor cells. *Reprinted from Cai et al. (2006) with permission.*

	1 h	4 h	6 h	27 h
QD 705-RGD	3.08 ± 1.42	3.39 ± 1.13	4.42 ± 1.88	2.09 ± 1.17
QD 705	1.12 ± 0.14	0.78 ± 0.25	0.84 ± 0.21	0.67 ± 0.46

permeability (Wolburg and Lippoldt, 2002; Haseloff et al., 2005; Abbott et al., 2010). The BBB is not a statistic membrane, but it contains an asymmetric arrangement of membrane-bound transport systems with functional differences between the apical and basolateral sides (Farell and Pardridge, 1999; Vorbrodt, 1999). The BBB prevents passage of endogenic compounds, medicinal drugs, and potentially toxic xenobiotics (Hawkins et al., 2006; Alavijeh et al., 2005; Borges-Walmsley et al., 2003). The transporters at the membrane site are (1) the

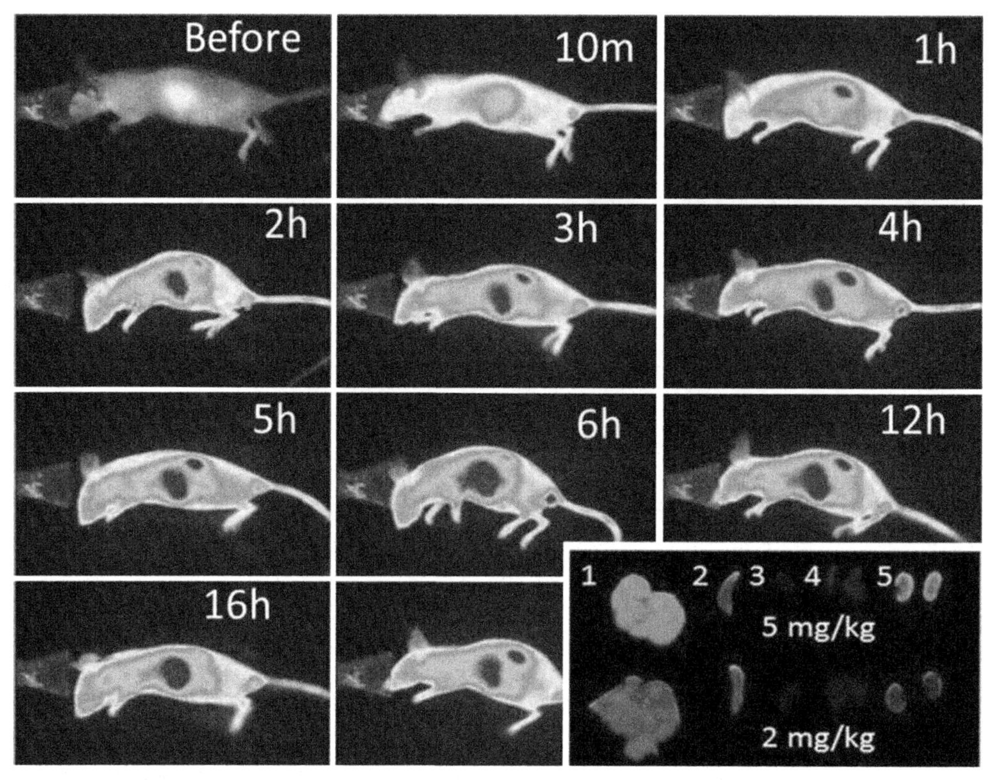

FIGURE 27 **Near-infrared fluorescence (NIRF) imaging of nanoparticle biodistribution in mouse administered the IRDy800CW-MNPs at a dose of 5 mg/kg body weight.** The images are displayed in a pseudo-color mode. The self-fluorescence background was assessed by imaging each mouse prior to the administration of the nanoparticles. Then, IRDy800CW-MNPs was administered (doses of 2 or 5 mg/kg) by intravascular injection and the animal was imaged at different time points. The NIRF signal in the liver region and kidneys gradually intensified after injection, reaching maximum levels at 6 h, then the NIRF signal gradually decreased. This suggests that the dynamic process of biodistribution of nanoparticles in the mouse could be monitored by the IRDy800CW nanoparticles. The last frame shows NIRF images of different excised organs ((1) liver; (2) spleen; (3) heart; (4) lungs; and (5) kidney). *Reprinted from Hou et al. (2010) with permission.*

ATP-binding cassette subfamilies B1, C, and G; (2) the MRP4/ABCC4 transporters that possess affinity for nucleoside analogs; and (3) the family of solute carrier transporters for the organic anion efflux transport system (consisting of organic anion-transporting polypeptides).

Particles in the systemic blood enter the brain mostly via perivascular (through the junctions) and transvascular (through cell penetration) mechanisms (Figure 28), although minor pathways that bypass the BBB, such as traveling through nerve terminals or diffusing through the neuronal or glial cytoplasmic membrane, are also available (Perez-Martınezc et al., 2012). In general, permeation of a nanoparticle across the BBB is directly related to its hydrophobicity but inversely related to its size, length, and molecular weight (Felgenhauer, 1980). Nanoparticles smaller than 5 nm are effectively excreted by the kidney, while particles greater than 5 nm do not cross the BBB; thus, pristine nanoparticles poorly accumulate in the brain.

Nanoparticles functionalized with ligands, which bind to the blood–brain transport receptors such as transferrin receptors, cross the BBB without damaging the tight junction.

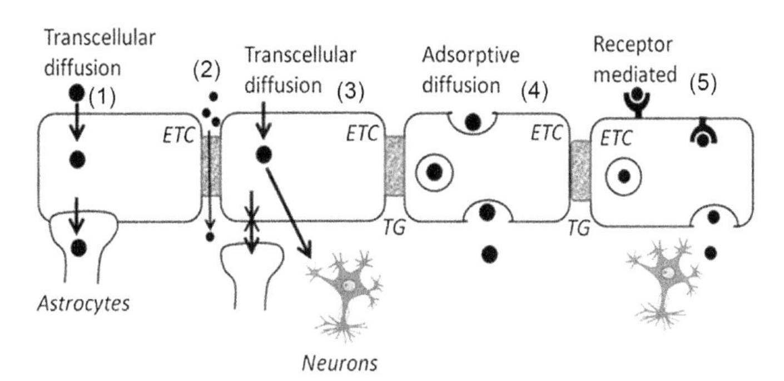

FIGURE 28 **Possible routes of nanoparticles' permeation across the blood–brain barrier.** *Trans-vascular diffusion into the astrocytes* (1) for small hydrophobic particles ($\leq$5 nm) that can diffuse across the endothelial cell and into the adjoining astrocyte. *Para-vascular transport* or *movement between cells* (2) is not important in the barrier's tight junction. *Trans-cellular diffusion when astrocytes and the endothelial cells are decoupled* (3) for small hydrophobic nanoparticles that can cross the barrier and then diffuse into the cerebrospinal fluid and to the nearest astrocyte or neurons. *Adsorptive endocytosis* (4) for cationic nanoparticles that are adsorbed onto the endothelial surface, thus initiating a budding process that encircles the particles and initiate their endocytosis. *Receptor mediated endocytosis* (5) is the process of opsonization that has been discussed earlier.

Nonfunctionalized (pristine) hydrophobic nanoparticles, such as metal nanoparticles or CNTs, have been shown to aggregate into variable-sized aggregated particles; thus, they may not cross the BBB. In comparison, functionalized CNTs remain dispersed and can negotiate the barrier. Although inconclusive, certain studies have shown that hydrophobic nanoparticles may induce oxidative stress, leading to the generation of free radicals that could disrupt the BBB and allow their entry into the brain (Shvedova et al., 2012; Sharma et al., 2007; Sriram et al., 2007).

Studies have shown that (1) nanoparticles deposited in the lungs may translocate into the circulatory system and eventually reach the brain (Nemmar et al., 2001, 2002) and (2) nanoparticles inhaled through the nose can translocate directly to the brain via the olfactory neurons (Elder et al., 2006; Oberdörster et al., 2004). Accumulation of CNT-filled macrophages in the alveoli may increase the permeability of the epithelial cell barrier and particle translocation to the alveolar interstitium, thus increasing the probability of CNTs reaching the brain (Ryman-Rasmussen et al., 2009; Mercer et al., 2010). However, more research is needed to establish the possible translocation of nanoparticles from lungs into the brain.

6.2.2.3 BLOOD-TUMOR PERMEABILITY

Tumors such as the astrocytoma also possess a relatively weak blood-tumor barrier (BTB), which restricts transport of large particles (Mosskin et al., 1989; Shivers et al., 1984). Three possible routes of transport across the BTB are (1) through the intercellular cleft, (2) via the pinocytotic vesicles, and (3) across the cell membrane and cytoplasm of the endothelial cells (Weihe et al., 1977). Sarin et al. (2009) have suggested that nanoparticles less than 12 nm may permeate from blood into the tumor.

Nanoparticles also exhibit an enhanced permeability and retention (EPR) effect by which, possibly due to the combination of enhanced blood supply, elevated osmotic pressure, and leaky nature of the vasculature (Liu et al., 2007; Cai et al., 2006; Tong et al., 2004; Boucher and Jain, 1992), certain sizes of nanoparticles tend to accumulate in tumor cells much more than they do in normal tissues (Matsumura and Maeda, 1986). The EPR effect is important for nanoparticle delivery to cancer tissue (Maeda, 2012).

Lin et al. (2011) have shown that the intravenously administered PEGylated nanoparticles (spherical or hexagonal) accumulate in tumor site via the EPR effect (Figure 29). The relative accumulation ratio ($ROI_{g\ of\ tumor}/ROI_{g\ of\ organ}$, where ROI is defined as the region of interest in the image) of hexagonal nanoparticles (HNPs) was higher than that of spherical nanoparticles (SNPs), possibly due to the inability of macrophages to ingest HNPs (discussed later).

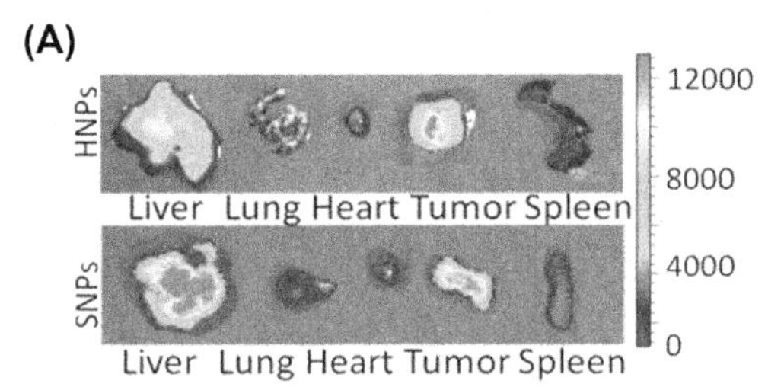

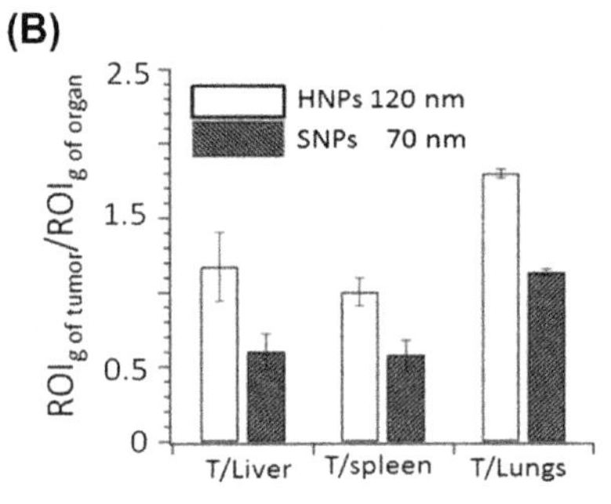

FIGURE 29 Relative accumulation of nanoparticles in liver, spleen, lungs, and tumor cells. (A) Images of liver, lungs, heart, tumor, and spleen collected from tumor bearing (inoculated with HeLa cells that are human cervical adenocarcinoma cells) nude mice at 48 h after nanoparticle (spherical (SNPs, 70 nm) or hexagonal (HNPs, 120 nm, exhibiting equal volume)) injection. SNPs accumulation pattern was liver > tumor > spleen > lungs and heart; while HNPs' accumulation pattern was tumor >> liver > spleen (no accumulation in lungs and heart). The fluorescence images were acquired with an exposure time of 1 s and used the Cy5.5 filter channel. (B) The relative accumulation ratio ($ROI_{g\ of\ tumor}/ROI_{g\ of\ organ}$) showed HNPs exhibiting greater accumulation in tumor cells than in other organs (ratio > 1). The $ROI_{g\ of\ tumor}/ROI_{g\ of\ liver\ or\ spleen}$ ratio for SNPs were approximately 0.5, suggesting greater accumulation in liver and spleen than in tumor cells. Possible mechanism for differences between SNPs and HNPs is shown in Figure 31. *Reprinted from Lin et al. (2011) by permission.*

6.2.2.4 PHAGOCYTOSIS OF NANOPARTICLES

As discussed, nanoparticles preferentially distribute in liver and spleen, which are closely related to the RES, containing abundant phagocytic cells that may remove nanoparticle from blood (Ferrucci and Stark, 1990). Approximately 75% of the administered magnetic nanoparticles were rapidly sequestered into RES macrophages in the liver and spleen (Chouly et al., 1996). However, as shown in Figure 30, sequestration of nanoparticles by macrophages depends on the nanoparticles' size and shape, protein corona, and PEG density on the nanoparticle (Walkey et al., 2012).

1. *Effects of size on nanoparticle phagocytosis*: Earlier studies have shown that particle parameters such as size, shape, surface chemistry, and mechanical properties influence phagocytosis (Champion et al., 2008; Champion and Mitragotri, 2006; Ahsan et al., 2002; Beningo and Wang, 2002). Immunogloblin G opsonized particles (acting via Fc receptors) with diameters between 1.7 and 3 μm were phagocytosed more readily than both smaller and larger particles (Rudt and Muller, 1993; Tabata and Ikada, 1990; Kawaguchi et al., 1986; Schroeder and Kinden, 1983), while the nonopsonized particles (acting via scavenger receptors) exhibited slightly different size requirements (Champion et al., 2008). An increase in size increases the phagocytosis of hydrophobic particles, but decreases the phagocytosis of hydrophilic particles (Chen et al., 1997; Simon and Schmidschonbein, 1988).
2. *Effects of shape on nanoparticle phagocytosis*: Shape is important in a nanoparticle's phagocytosis. Morphology did not have a significant effect on the phagocytosis of relatively small nanoparticles, but it had considerable influence on the phagocytosis of relatively large nanoparticles (Lin et al., 2011). Spherical nanoparticles (SNPs) with

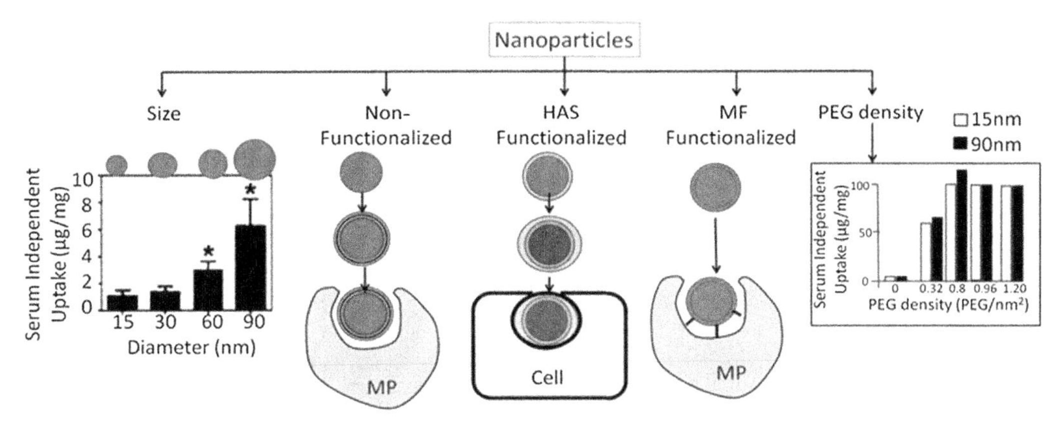

FIGURE 30 **Size-, shape-, and surface-dependent phagocytosis of nanoparticles.** Phagocytosis is related to the nanoparticle size. Phagocytic cells efficiently recognize particles greater than 50 nm and the extent of phagocytosis increases as the nanoparticle size increases. Non-functionalized nanoparticles rapidly interact with plasma proteins and form protein corona that may include opsonins such as IgG or complement proteins. Opsonins facilitate phagocytosis of nanoparticles. Human serum albumin (HAS) coated nanoparticles can enter non-phagocytic cells via receptor-independent endocytosis. Nanoparticles functionalized with macrophage-selective ligands (MF) are rapidly engulfed by macrophages. Phagocytosis exhibits an inverse relationship with the density of polyethylene glycol (PEG) on nanoparticle surface. Thus, PEG-coated nanoparticles exhibit greater half-life than nonfunctionalized particles in vivo.

70 nm diameter and hexagonal nanoparticles (HNPs) with 70 nm for each axis exhibited comparable phagocytosis, and macrophages engulfed both nanoparticles equally efficiently. However, SNPs and HNPs had significant phagocytosis (SNPs, but not HNPs, were engulfed by macrophages) when the nanoparticle size was 120 nm (Figure 31). The distribution half-life ($t_{1/2}\ \alpha$) of HNPs and SNPs was 2.867 and 0.615 h, respectively, which is indicative of the longer circulation of HNPs in the bloodstream. It is proposed that, for a particle to avoid uptake by macrophages, it must be small enough to be engulfed by macrophages and/or a hydrophilic surface, which reduces opsonization reactions and subsequent clearance. Nonspherical (e.g., HNPs) particle shape, measured by tangent angles (Figure 31) at the point of initial contact, also dictates whether or not macrophages initiate phagocytosis (Lin et al., 2011).

3. *Effects of protein corona on nanoparticle uptake in nonphagocytic cells*: When nanoparticles encounter plasma or culture media, proteins are adsorbed onto the nanoparticle surface and form a protein corona almost instantly. The protein corona, not the bare nanoparticle surface, interacts with cellular structures and initiates the nanoparticle's internalization in nonphagocytic cells (e.g., the protein corona containing albumin, caevolae, and related proteins) or their sequestration into the macrophages (e.g., the corona containing immunoglobulin G or opsinogens, such as complement proteins; Figure 32). Dutta et al. (2007) showed that albumin adsorbed on the surface of SWCNTs reduced their phagocytosis by inducing an anti-inflammatory pathway in RAW macrophages. The presence of apolipoproteins on the surface assists in the transport of nanoparticles across the BBB (Kreuter et al., 2002; Kreuter, 2013). While it is generally accepted that protein binding

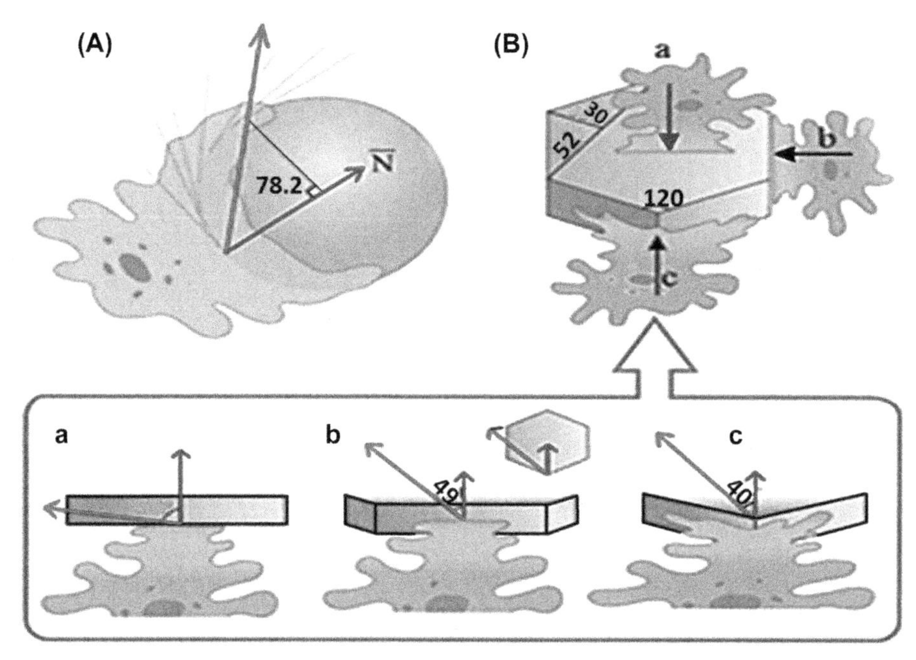

FIGURE 31 **Effect of morphology on nanoparticle phagocytosis.** On an equal volume basis (70 nm spherical nanoparticles (SNP) and 120 nm hexagonal nanoparticles(HNP) have equal volume), SNPs (A) exhibit greater phagocytosis than HNPs (B). For HNPs, the angle of approach is important in the particle's phagocytosis. This explains the observations of Figure 29. SNPs exhibits lower bioavailability because they exhibit greater phagocytosis. *Reprinted from Lin et al. (2011) by permission.*

on the nanoparticle surface facilitates its uptake, there have been some contradictory reports in the literature. Uptake of FePt nanoparticles by HeLa cells was suppressed in the presence of the protein corona (Jiang et al., 2009). Silica nanoparticles dispersed in serum-free media were taken up more efficiently by lung epithelial cells as compared to ones in the presence of 10% serum. Further research is needed to fully characterize nanoparticle uptake in cells.

4. *Effects of PEG density on nanoparticles' phagocytosis*: Intravenously injected nanoparticles $\geq$70 nm rapidly bind plasma proteins, including immunoglobulin G and opsinogens, which facilitates their capture by macrophages. This greatly decreases nanoparticles' blood persistence and strongly limits the efficacy of the drug load because the phagocytized nanoparticles accumulate in RES organs. Studies have shown that the PEG- or polyoxyethylene-coated nanoparticles are not captured by RES macrophages; thus, their plasma lifetime is enhanced. This is associated with a decrease in the particles' zeta potential (Gref et al., 1999, 2000) (Figure 33).

6.2.3 Comparison of Bulk Particles and Nanoparticles

As shown in Figure 34(A), bulk particles' distribution in the body depends on the particles' chemical and ionic properties and the organism's weight, body fat, and fluid variations (Figure 34(left)). On average, bulk particles non-specifically distributed 25%, 10%, 50% and 15%, respectively, in fat, plasma(dissolved in), tissue

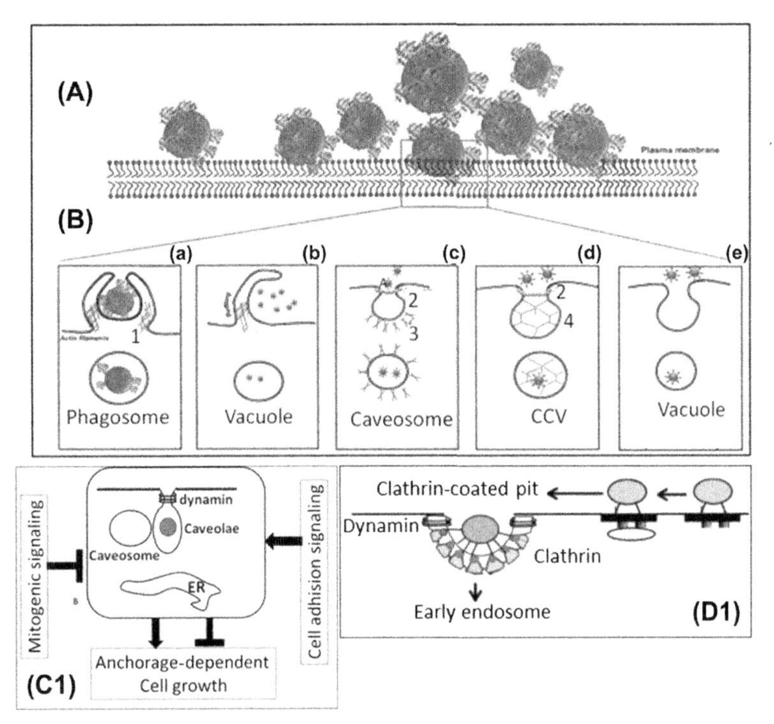

FIGURE 32 **Effects of protein corona on the nanoparticle uptake in non-phagocytic cells.** Nanoparticle-corona interacts with cell membranes (A) and initiates several endocytosis pathways (B) described below.

- *Phagocytosis* (a) is engulfment of nanoparticles by early-immune cells such as macrophages.
- *Macropinocytosis* (b) is a form of bulk uptake of fluid and solid cargo into cytoplasmic vacuole-type endocytic vesicles. This process is growth factor-induced, but dynamin-independent (dynamin is a large GTPase that participates in the budding and scission nascent vesicles from parent membranes). The nanoparticle corona may contain certain proteins that initiate micropinocytosis.
- *Caveolin-dependent endocytosis* (c): Caveolins (CAV-1, -2, -3) are specialized 50–70 nm flask-shaped invaginations of the plasma membrane. Internalization via caveolae occurs upon cell stimulation. Internalized nanoparticles bound to the cell-receptor are delivered to the early endosome. *(Reprinted from Saptarshi et al. (2013) with permission.)* (c1) RNAi-mediated inhibition of mitogenic signaling (left) stimulates, while RNAi-mediated inhibition of cell adhesion signaling (right) inhibits caveolae-mediated endocytosis. In addition, activation of caveolae-mediated endocytosis may be required to allow mitogenic signaling and suppress integrin signaling (depicted by dashed suppressing arrow). *(Reprinted from Lucas (2005) with permission.)*
- *Clathrin-mediated endocytosis* (d): Clathrin is a network of proteins that forms clathrin-coated pits on the cell membrane in response to the attachment of ligands (such as TGFβ) to their receptors. The pit becomes the inside surface of the clathrin-coated vesicle (CCV) during endocytosis. (d1) Nanoparticles binding to the surface induce formation of the clathrin-coated pits that engulf the nanoparticles in clathrin-coated vesicles, the early endosomes.
- Caveolin and clathrin independent endocytosis: Cells also possess many clathrin- and caveolin-independent endocytosis such as those using glycosylphosphatidylinositol (GPI)-anchored proteins (Mayor and Riezman, 2004; Sabharanjak et al., 2002).

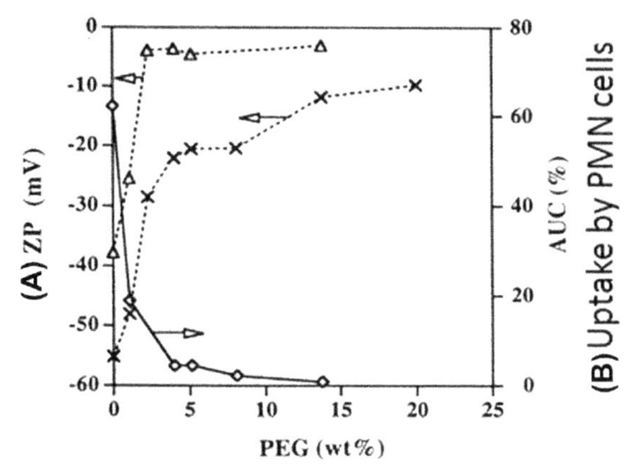

FIGURE 33 Influence of the PEG on (A) zeta potential (ZP) measured in PBS with glucose (△) and in NaCl 10^{-3} M solution (×) and (B) opsonization by PMN cells (◇) in terms of AUC of nanospheres prepared from diblock PEG-PLA or PEG-PCL copolymers, or from blends with PEG-PLA:PLA. An increase in PEG (weight%) exponentially decreased opsonization of nanoparticles by PMN cells, but increased their ZP (glucose caused greater increase than NaCl). Comparable results were observed in human monocytes strain, THP-1 (Gref et al., 1999).

and plasma proteins (bound to). Biodistribution of nanoparticles depends upon their functionalization. Nanoparticles functionalized with site-specific antibodies may accumulate into the specific site and exhibit poor nonspecific binding. A bare (nonfunctionalized) nanoparticle may exhibit nonspecific binding comparable to the bulk particles. In general biodistribution of nanoparticles in the body may depend on (1) the nanoparticle's chemical and ionic properties, size and shape, surface functionalization, and aggregation; (2) the organism's height and weight, fluid variation, disease, etc.; (3) formation of a protein corona; and (4) possible defunctionalization of nanoparticles (Figure 34(right)).

7. THE TARGET TISSUE/ORGAN

Target tissue/organ is defined as the site on which a poison exerts its action or an organ that accounts for the primary toxicity of a poison. For example, organophosphate insecticides distribute throughout the body, but selective inhibition of acetylcholinesterase in the nervous system accounts for their toxicity, especially seizures and respiratory failure. In the following sections, the interaction between the target organ and bulk particles (Section 7.1) and nanoparticles (Section 7.2) are discussed.

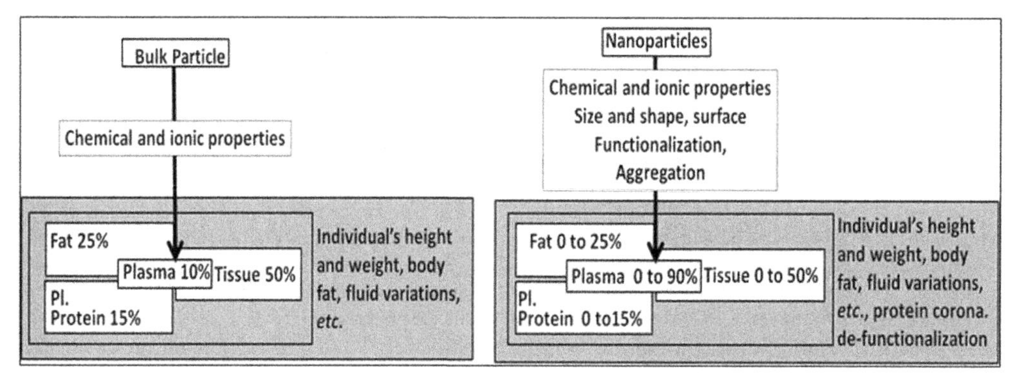

FIGURE 34 **Comparative biodistribution of bulk particles and nanoparticles.** Biodistribution of a bulk particle, in a given species and mode of exposure, depends upon its chemical and ionic properties. In general, bulk particle's distributions in plasma, plasma proteins, tissues, and fat are 50%, 25%, 15% and 25%, respectively (*caveat*: In fat, hydrophobic compounds may exhibit greater distribution than hydrophilic compounds). Biodistribution of nanoparticles, however, can be highly variable because of their size, shape, and surface functionalization. For example, PEGylated nanoparticles exhibit greater residence time in blood and lower accumulation in liver or spleen, transferrin-coupled nanoparticles may selectively accumulate in the brain and pristine nanoparticles may accumulate exclusively in liver and spleen.

7.1 Bulk Particles and Target-Mediated Toxicity

In general, the distribution of bulk particles in the body depends on their water solubility. The key target sites are blood/cardiovascular systems (cardiovascular toxicity), skin and oral systems (dermal toxicity), genetic and reproductive organs (reproductive toxicity), liver (hepatotoxicity), immune system (immunotoxicity), kidney (nephrotoxicity), nervous system (neurotoxicity), and the respiratory tract (respiratory toxicity).

- *Blood and cardiovascular toxicity* results from poisons that directly interact with cells in blood, bone marrow, and the heart. Some examples of toxins causing blood and cardiovascular toxicity are blood pressure medicines, chloramphenicol, carbon monoxide (inhibition of hemoglobin–oxygen binding), benzene (bone marrow cells), and cholesterol accumulation (arteriosclerosis).
- *Dermal toxicity* results from direct contact or internal distribution of toxins to the skin, resulting in disruption of the skin's barrier properties and ensuing accumulation of toxins in other tissues, as well as skin diseases such as mild irritation, corrosivity, hypersensitivity, and skin cancer. Some of the examples of toxins causing dermal toxicity are strong acids and bases, plant poisons such as poison ivy, arsenic ingestion, and ultraviolet exposure.
- *Eye toxicity* results from direct contact or internal distribution of toxins to the cornea and conjunctiva, resulting in conjunctivitis and corneal erosion, respectively. Key toxins include strong acids and alkalis, corticosteroids, and alcohols.
- *Hepatotoxicity* occurs due to the accumulation of toxins in the liver, bile duct, and gallbladder. The liver is particularly susceptible to poisons due to a large blood supply and its role in metabolism. The direct effects of hepatotoxicity are as follows:
  - Chemical hepatitis, characterized by lipid accumulation in the liver
  - Necrosis (death) of the hepatic cells
  - Cirrhosis, characterized by chronic fibrosis, often due to alcoholism
  - Intrahepatic cholestasis, characterized by backup of bile salt in the liver
  - Hepatic cancer (liver cancer)
  - Hypersensitivity

  Indirect effects include abnormal metabolism of toxins, which may result in their activation.
- *Immunotoxicity*, the toxicity of the immune system, occurs when poisons interact with the immune cells and either activate or suppress it. Immunotoxicity may take several forms: hypersensitivity (allergy and autoimmunity), immunodeficiency, and uncontrolled proliferation (leukemia and lymphoma).
- *Nephrotoxicity* occurs due to the adverse effects of poisons on the kidney, which is highly susceptible to toxicants for two reasons. A high volume of blood flows through it and it filtrates large amounts of toxins, which can concentrate in the kidney tubules. Nephrotoxicity can also result in systemic toxicity, such as a decreased ability to remove waste, electrolyte imbalance, and hormonal disorders.
- *Neurotoxicity* represents toxicant damage to cells of the central nervous system (brain and spinal cord) and the peripheral nervous system (nerves outside the CNS). The primary types of neurotoxicity are neuropathy, axonopathy, demyelination, and abnormal neurotransmission. Some examples are organophosphate and carbamate insecticides that inhibit acetylcholinesterase in neuromuscular junctions, heavy metals such as lead and mercury that cause neuropathy, and calcium-channel blockers that impair neurotransmission.

- *Reproductive abnormalities and developmental toxicity* occur when poisons damage either the reproductive system or impair normal development of infants and children (https://www.osha.gov/SLTC/reproductive hazards/index.html).
- *Respiratory toxicity* relates to effects on the upper respiratory system (nose, pharynx, larynx, and trachea) and the lower respiratory system (bronchi, bronchioles, and lung alveoli). The primary types of respiratory toxicity are pulmonary irritation, asthma/bronchitis, reactive airway disease, emphysema, allergic alveolitis, fibrotic lung disease, pneumoconiosis, and lung cancer.

A key point is that bulk particles do not accumulate exclusively in the target organs, but their accumulation in the target organ results in primary toxicity. At a certain dose, a toxin's nonspecific accumulation may also cause adverse effects (Figure 35), thus complicating the diagnosis.

7.2 Nanoparticles and Target Toxicity

Nanoparticles used as a pharmaceutical or drug transporter are functionalized with PEG (to prevent protein binding and phagocytosis), a target protein for brain uptake (e.g., transferrin, which acts via its binding site onto the BBB), and

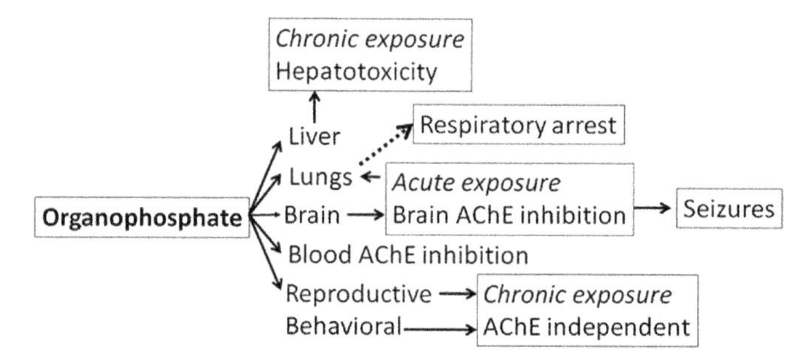

FIGURE 35 **Relationship between biodistribution of bulk particle (organophosphate insecticide) and site-directed toxicity.** After an acute exposure, organophosphates may distribute in all peripheral tissues and the brain, but their accumulation in the brain may be causally related to their toxic effects. Chronic exposure may cause some nonspecific effects such as hepatotoxicity and behavioral toxicity.

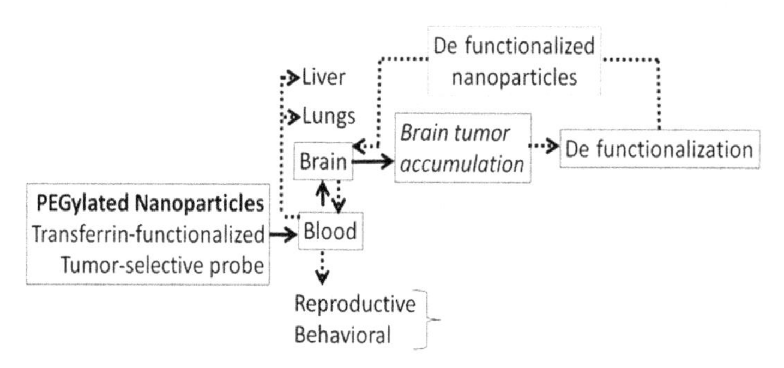

FIGURE 36 **Relationship between the biodistribution of nanoparticles (PEGylated and transferrin functionalization) and site-directed toxicity.** Unlike bulk particles, actively functionalized nanoparticles may selectively distribute into the brain and then accumulate tumor cells. However, a PEGylated nanoparticle without transferrin may not enter the brain and remain in blood. There are some evidence that nanoparticles may be defunctionalized in brain and the defunctionalized particles return to blood via exocytosis.

a disease-selective probe for selective targeting of the diseased tissues/organs (Figure 36). In tumors, the drugs are released at the tumor site due to the pH difference between normal tissue (pH 7.4) and the tumor (pH < 6.0) environment. Functionalized nanoparticles, therefore, may exhibit highly selective biodistribution from systemic blood into the brain and the tumor cells. The possibilities are limitless because nanoparticles can be engineered to accumulate in any specific organ or tissues, as long as an appropriate probe is available. Because the anticancer drug is released selectively at the tumor site, the nonspecific toxicity is significantly reduced. However, studies have shown that nanoparticles may be defunctionalized in the body, resulting in nonspecific accumulation (Yang et al., 2009; Azizian et al., 2012).

Because defunctionalized nanoparticles may be relatively more hydrophobic and stable, their accumulation in tissues may pose significant health hazards. Earlier studies have shown defunctionalization of PEG-SWNT in the liver but not in the spleen under similar conditions (Yang et al., 2009). Nunes et al. (2012) showed that the brain microglia effectively engulf and defunctionalize amino-functionalized CNTs in vivo. This initial observation of CNT defunctionalization within the brain necessitates further investigations to establish their pharmacological and toxicological consequences. If defunctionalization includes removal of the PEG corona, the resulting nanoparticles may exhibit significant neurotoxicity. This aspect of nanomedicine has been largely ignored and needs further research so that nanoparticle safety can be accurately assessed.

8. CONCLUSIONS

The overall toxicity of bulk particles and nanoparticles is determined by their interaction with plasma and cell components (membranes, proteins, fat, and carbohydrates), bioavailability, and biodistribution. Humans and animals are exposed to poisons via dermal, inhalation, and ingestion modes. Dose in terms of mass (mg/kg) characterizes bulk particles' interaction with cell components and their bioavailability and biodistribution. However, dose in terms of size, reactivity, surface functionalization, and surface reactivity characterizes nanoparticles' interaction with cells and their bioavailability and biodistribution. Upon reaching blood, nanoparticles rapidly adsorb plasma protein and form a protein corona that eventually determines the biological properties of nanoparticles. If the protein corona contains immunoglobulin G and certain complement factors, nanoparticles are opsonized and undergo phagocytosis in liver and spleen. However, if the nanoparticles are functionalized with PEG and lack opsonization proteins, they accumulate in blood with higher half-lives.

References

Abbott, N.J., Patabendige, A.A., Dolman, D.E., Yusof, S.R., Begley, D.J., 2010. Structure and function of the blood–brain barrier. Neurobiol. Dis. 37, 13–25.

Aderem, A., Underhill, D.M., 1999. Mechanisms of phagocytosis in macrophages. Annu. Rev. Immunol. 17, 593–623.

Aggarwal, P., Hall, J.B., McLeland, C.B., Dobrovolskaia, M.A., McNeil, S.E., 2009. Nanoparticle interaction with plasma proteins as it relates to particle biodistribution, biocompatibility and therapeutic efficacy. Adv. Drug Deliv. Rev. 61, 428–437.

Ahsan, F.L., Rivas, I.P., Khan, M.A., Suarez, A.I.T., 2002. Targeting to macrophages: role of physicochemical properties of particulate carriers—liposomes and microspheres—on the phagocytosis by macrophages. J. Control. Rel. 79, 29–40.

Alavijeh, M.S., Chishty, M., Qaiser, M.Z., Palmer, A.M., 2005. Drug metabolism and pharmacokinetics, the blood–brain barrier, and central nervous system drug discovery. NeuroRx 2, 554–571.

Albert, R.E., Lippmann, M., Briscoe, W., 1969. The characteristics of bronchial clearance in humans and the effect of cigarette smoking. Arch. Environ. Health 18, 738–755.

Al-Jamal, K.T., Nerl, H., Müller, K.H., Ali-Boucetta, H., Li, S., Haynes, P.D., Jinschek, J.R., Prato, M., Bianco, A., Kostarelos, K., Porter, A.E., 2011. Cellular uptake mechanisms of functionalised multi-walled carbon nanotubes by 3D electron tomography imaging. Nanoscale 3, 2627–2635.

Alvarez-Román, R., Naik, A., Kalia, Y.N., Guy, R.H., Fessi, H., 2004. Enhancement of topical delivery from biodegradable nanoparticles. Pharm. Res. 21, 1818–1825.

Asanov, A.N., DeLucas, L.J., Oldham, P.B., Wilson, W.W., 1997. Interfacial aggregation of bovine serum albumin related to crystallization conditions studied by total internal reflection fluorescence. J. Colloid Interf. Sci. 196, 62–73.

Azizian, G., Riyahi-Alam, N., Haghgoo, S., Moghimi, H.R., Zohdiaghdam, R., Rafiei, B., Gorji, E., 2012. Synthesis route and three different core-shell impacts on magnetic characterization of gadolinium oxide-based nanoparticles as new contrast agents for molecular magnetic resonance imaging. Nanoscale Res. Lett. 7, 549.

Baca, H.K., Carnes, E., Singh, S., Ashley, C., Lopez, D., Brinker, C.J., 2007. Cell-directed assembly of bio/nano interfaces: a new scheme for cell immobilization. Acc. Chem. Res. 40, 836–845.

Bahrami, A., Raatz, M., Agudo-Canalejoa, J., Michel, R., Curtis, E.M., Hall, C.K., Gradzielski, M., Lipowskya, R., Weikl, T.R., 2014. Wrapping of nanoparticles by membranes. Adv. Colloid Interf. Sci. 208, 214–224.

Baroli, B., Ennas, M.G., Loffredo, F., Isola, M., Pinna, R., López-Quintela, M.A., 2007. Penetration of metallic nanoparticles in human full-thickness skin. J. Invest. Dermatol. 127, 1701–1712.

Baroli, B., 2010. Penetration of nanoparticles and nanomaterials in the skin: fiction or reality? J. Pharm. Sci. 99, 21–50.

Begley, D.J., 1996. The blood–brain barrier: principles for targeting peptides and drugs to the central nervous system. J. Pharm. Pharmacol. 4, 136–146.

Begley, D.J., 2004. Delivery of therapeutic agents to the central nervous system: the problems and the possibilities. Pharmacol. Ther. 104, 29–45.

Beningo, K.A., Wang, Y.L., 2002. Fc-receptor-mediated phagocytosis is regulated by mechanical properties of the target. J. Cell Sci. 115, 849–856.

Bergin, I.L., Witzmann, F.A., 2013. Nanoparticle toxicity by the gastrointestinal route: evidence and knowledge gaps. Int. J. Biomed. Nanosci. Nanotech. 3, 163–210.

Bernacki, J., Dobrowolska, A., Nierwińska, K., Malecki, A., 2008. Physiology and pharmacological role of the blood–brain barrier. Pharmacol. Rep. 60, 600–622.

Bestak, R., Halliday, G.M., 1996. Sunscreens protect from UV-promoted squamous cell carcinoma in mice chronically irradiated with doses of UV radiation insufficient to cause edema. Photochem. Photobiol. 64, 188–193.

Bockmann, J., Lahl, H., Eckert, T.H., Unterhalt, B., 2000. Titan-Blutspiegel vor und nach Belastungsversuchen mit Titandioxid. Pharmazie 55, 140–143.

Borges-Walmsley, M.I., McKeegan, K.S., Walmsley, A.R., 2003. Structure and function of efflux pumps that confer resistance to drugs. Biochem. J. 376, 313–338.

Boucher, Y., Jain, R.K., 1992. Microvascular pressure is the principal driving force for interstitial hypertension in solid tumors: implications for vascular collapse. Cancer Res. 52, 5110–5114.

Bradley, D.T., Zipfel, P.F., Hughes, A.E., 2011. Complement in age-related macular degeneration: a focus on function. Eye (Lond.) 25, 683–693.

Brown, D.M., Wilson, M.R., MacNee, W., Stone, V., Donaldson, K., 2001. Size-dependent proinflammatory effects of ultrafine polystyrene particles: a role for surface area and oxidative stress in the enhanced activity of ultrafines. Toxicol. Appl. Pharmacol. 175, 191–199.

Byron, P.R., 1986. Prediction of drug residence times in regions of the human respiratory tract following aerosol inhalation. J. Pharm. Sci. 75, 433–438.

Cai, W., Shin, D.-W., Chen, K., Gheysens, O., Cao, Q., Wang, S.X., Gambhir, S.S., Chen, X., 2006. Peptide-labeled near-infrared quantum dots for imaging tumor vasculature in living subjects. Nano Lett. 6, 669–676.

Camner, P., Jarstrand, C., Philipson, K., 1973. Tracheobronchial clearance in patients with influenza. Am. Rev. Resp. Dis. 108, 131–135.

Camner, P., Mossberg, B., Afzelius, B.A., 1983. Measurements of tracheobronchial clearance in patients with immotile-cilia syndrome and its value in differential diagnosis. Eur. J. Resp. Dis. Suppl. 127, 57–63.

Campbell, C.S., Contreras-Rojas, L.R., Delgado-Charro, M.B., Guy, R.H., 2012. Objective assessment of nanoparticle disposition in mammalian skin after topical exposure. J. Control. Rel. 162, 201–207.

Card, J.W., Zeldin, D.C., Bonner, J.C., Nestman, E.R., 2008. Pulmonary applications and toxicity of engineered nanoparticles. Am. J. Physiol. 95, L400–L411.

Carroll, M.C., 1998. The role of compliment and compliment receptors in induction and regulation of immunity. Annu. Immunol. 16, 545–568.

Carter, D.C., Ho, J.X., 1994. Structure of serum albumin. Adv. Protein Chem. 45, 153–203.

Champion, J.A., Mitragotri, S., 2006. Role of target geometry in phagocytosis. PNAS U.S.A. 103, 4930–4934.

Champion, J.A., Walker, A., Mitragotri, S., August 2008. Role of particle size in phagocytosis of polymeric microspheres. Pharm. Res. 25, 1815–1821.

Chaudhari, K.R., Ukawala, M., Manjappa, A.S., Kumar, A., Mundada, P.K., Mishra, A.K., Mathur, R., Monkkönen, J., Murthy, R.S., 2011. Opsonization, biodistribution, cellular uptake and apoptosis study of PEGylated PBCA nanoparticle as potential drug delivery carrier. Pharm. Res. 29, 53–68.

Chen, J., Carey, K., Godowski, P.J., 1997. Identification of Gas6 as a ligand for Mer, a neural cell adhesion molecule related receptor tyrosine kinase implicated in cellular transformation. Oncogene. 14, 2033–2039.

Cheung, C.Y., Brace, R.A., 2008. Unidirectional transport across cultured ovine amniotic epithelial cell monolayer. Reprod. Sci. 15, 1054–1058.

Chmiel, J.F., Davis, P.B., 2003. Why do the lungs of patients with cystic fibrosis become infected and why can't they clear the infection? Resp. Res. 4, 8.

Choi, H.S., Ipe, B.I., Misra, P., Lee, J.H., Bawendi, M.G., Frangioni, J.V., 2009. Tissue- and organ-selective biodistribution of NIR fluorescent quantum dots. Nano Lett. 9, 2354–2359.

Chouly, C., Pouliquen, D., Lucet, I., Jeune, J.J., Jallet, P., 1996. Development of superparamagnetic nanoparticles for MRI: effect of particle size, charge and surface nature on biodistribution. J. Microencapsul. 13, 245255.

Christopher, S.J., Campbell, L., Contreras-Rojas, R., Delgado-Charro, M.B., Guy, R.H., 2012. Objective assessment of nanoparticle disposition in mammalian skin after topical exposure. J. Control. Rel. 162, 201–207.

Chu, M., Wu, Q., Wang, J., Hou, S., Miao, Y., Peng, J., Sun, Y., 2008. In vitro and in vivo transdermal delivery capacity of quantum dots through mouse skin. Nanotech 2007, 18.

Chung, H.-E., Park, D.-H., Choy, J.-H., Choi, S.-J., 2012. Intracellular trafficking pathway of layered double hydroxide nanoparticles in human cells: size-dependent cellular delivery. Appl. Clay Sci. 65–66, 24–30.

Cole, A.J., David, A.E., Wang, J., Galban, C.J., Yang, V.C., 2011. Magnetic brain tumor targeting and biodistribution of long-circulating PEG-modified, crosslinked starch-coated iron oxide nanoparticles. Biomaterials 32, 6291–6301.

Courrier, H.M., Butz, N., Vandamme, T.F., 2002. Pulmonary drug delivery systems: recent developments and prospects. Crit. Rev. Ther. Drug Carr. Syst. 19, 425–498.

Dahan, M., Laurence, Pinaud, T.F., Chemla, D.S., Alivisatos, A.P., Sauer, M., et al., 2001. Time-gated biological imaging by use of colloidal quantum dots. Opt. Lett. 26, 825–827.

Davies, C.N., 1952. Dust sampling and lung disease. Br. J. Ind. Med. 9, 120–126.

De Jong, W.H., Hagens, W.I., Krystek, P., Burger, M.C., Sips, A.J.A.M., Geertsma, R.E., 2008. Particle size-dependent organ distribution of gold nanoparticles after intravenous administration. Biomaterials. 29, 1912–1919.

de la Zerda, A., Zavaleta, C., Keren, S., Vaithiligam, S., Bodapati, S., Liu, Z., Levi, J., Smith, B., Ma, T.-J., Oralkan, O., Cheng, Z., Chen, X., Dai, H., Khuri-Yakub, B.T., Gambhir, S.S., 2008. Carbon nanotubes as photoacoustic molecular imaging agents in living mice. Nat. Nanotech. 3, 557–562.

Dell'Orco, D., Lundqvist, M., Oslakovic, C., Cedervall, T., Linse, S., 2010. Modeling the time evolution of the nanoparticle-protein corona in a body fluid. PLoS One 5, e10949.

Desai, M.P., Labhasetwar, V., Amidon, G.L., Levy, R.J., 1996. Gastrointestinal uptake of biodegradable microparticles: effect of particle size. Pharm. Res. 13, 1838–1845.

Dockery, D., Pope, C., Wu, X., 1993. An association between air pollution and mortality in six US cities. N. Eng. J. Med. 329, 1753–1759.

Donaldson, K., Aitken, R., Tran, L., Stone, V., Duffin, R., Forrest, G., Alexander, A., 2006. Carbon nanotubes: a review of their properties in relation to pulmonary toxicology and workplace safety. Toxicol. Sci. 92, 5–22.

Duffin, R., Tran, L., Brown, D., Stone, V., Donaldson, K., 2007. Proinflammogenic effects of low-toxicity and metal nanoparticles in vivo and in vitro: highlighting the role of particle surface area and surface reactivity. Inhal. Toxicol. 19, 849–856.

Dutta, D., Sundaram, S., Teeguarden, J., Riley, B., Fifield, L., Jacobs, J., Addleman, S., Kaysen, G., Moudgil, B., Weber, T., 2007. Adsorbed proteins influence the biological activity and molecular targeting of nanomaterials. Toxicol. Sci. 100, 303–315.

Duty, S., Ackerman, R., Calafat, A., Hauser, R., 2005. Personal care use predicts urinary concentration of some phthalate monoesters. Environ. Health Persp. 113, 1530–1535.

Elder, A., Gelein, R., Silva, V., Feikert, T., Opanashuk, L., Carter, J., Potter, R., Maynard, A., Ito, Y., Finkelstein, J., Oberdörster, G., 2006. Translocation of inhaled ultrafine manganese oxide particles to the central nervous system. Environ. Health Persp. 114, 1172–1178.

Elsaesser, A., Howard, C.V., 2012. Toxicology of nanoparticles. Adv. Drug Deliv. Rev. 64, 129–137.

Elsaesser, A., Taylor, A., de Yanes, G.S., McKerr, G., Kim, E.M., O'Hare, E., Howard, C.V., 2010. Quantification of nanoparticle uptake by cells using microscopical and analytical techniques. Nanomed. (Lond.) 5, 1447–1457.

Ensign, L.M., Cone, R., Hanes, J., 2012. Oral drug delivery with polymeric nanoparticles: the gastrointestinal mucosa barriers. Adv. Drug Deliv. Rev. 64, 557–565.

Epstein, J., Eichbaum, Q., Sheriff, S., Ezekowitz, R.A., 1996. The collection in innate immunity. Curr. Opin. Immunol. 8, 29–35.

Esch, M.B., Mahler, G.J., Stokol, T., Shuler, M.L., 2014. Body-on-a-chip simulation with gastrointestinal tract and liver tissues suggests that ingested nanoparticles have the potential to cause liver injury. Lab Chip 14, 3081–3092.

Farell, C.L., Pardridge, W.M., 1999. Blood–brain-barrier glucose transporter is asymmetrically distributed on brain capillary endothelial luminal and abluminal membranes: an electronic microscopic immunogold study. PNAS U.S.A. 88, 5779–5783.

Felgenhauer, K., 1980. Protein filtration and secretion at human body fluid barriers. Pflugers Arch. 384, 9–17.

Ferrucci, J.T., Stark, D.D., 1990. Iron oxide-enhanced MR imaging of the liver and spleen: review of the first 5 years. Am. J. Roentgenol. 155, 943950.

Fleer, G.J., Cohen Stuart, M.A., Scheutjens, J.M.H.M., Cosgrove, T., Vincent, B., 1993. Polymers at Interfaces. Chapman & Hall, London.

Foster, N., Hirst, B.H., 2005. Exploiting receptor biology for oral vaccination with biodegradable particulates. Adv. Drug Deliv. Rev. 57, 431–450.

Friends of the Earth, 2007. Nanotechnology & Sunscreens: A Consumer Guide for Avoiding Nano-Sunscreens. http://www.foe.org/nano_sunscreens_guide/Nano_Sunscreens.pdf.

Gamarra, L.F., Pontuschka, W.M., Amaro, E., Costa-Filho, A.J., Brito, G.E.S., Vieira, E.D., Carneiro, S.M., Escriba, D.M., Falleiros, A.M.F., Salvador, V.L., 2007. Kinetics of elimination and distribution in blood and liver of biocompatible ferrofluids based on Fe_3O_4 nanoparticles: an EPR and XRF study. Mat Sci. Eng. C-Bio. S. 28, 519–525, 10.1016.

Ge, C., Du, J., Zhao, L., Wang, L., Liu, Y., Li, D., Yang, Y., Zhou, R., Zhao, Y., Chai, Z., Chen, C., 2011. Binding of blood proteins to carbon nanotubes reduces cytotoxicity. PNAS U.S.A. 11, 16968–16973.

Gessner, A., Lieske, A., Paulke, B.R., Muller, R.H., 2002. Influence of surface charge density on protein adsorption on polymeric nanoparticles: analysis by two-dimensional electrophoresis. Eur. J. Pharm. Biopharm. 54, 165–170.

Gessner, A., Lieske, A., Paulke, B.R., Muller, R.H., 2003. Functional groups on polystyrene model nanoparticles: influence on protein adsorption. J. Biomed. Mater. Res. A 65, 319–326.

Gibbs, J.W., 1873. A method of geometrical representation of the thermodynamic properties of substances by means of surfaces. Trans. Conn. Aca. Arts Sci. 2, 382–404.

Giechaskiel, B., Ntziachristos, L., Samaras, Z., Scheer, V., Casati, R., Vogt, R., 2005. Formation potential of vehicle exhaust nucleation mode particles on-road and in the laboratory. Atmos. Environ. 39, 3191–3198.

Giechaskiel, B., Munoz-Bueno, R., Rubino, L., Manfredi, U., Dilara, P., Santi, G.D., Andersson, J., 2007. Particle Size and Number Emissions before, during and after Regeneration Events of a Euro 4 DPF Equipped Light-Duty Diesel Vehicle. SAE. Paper No. 2007-01-1944.

Giechaskiel, B., Alessandrini, S., Forni, F., Martinez-Lozano, P., Lesueur, D., Carriero, M., Martini, G., 2008a. Particle Measurement Programme (PMP): Heavy-Duty (HD) Inter-Laboratory Exercise. Exploratory Work at JRC (October 2007–February 2008), EU Report 23426.

Giechaskiel, B., Martinez-Lozano, P., Alessandrini, S., Forni, F., Montigny, F., Fumagalli, I., Carriero, M., Martini, G., 2008b. Particle Measurement Programme (PMP): Heavy Duty (HD) Inter-Laboratory Exercise. Validation Exercise Tests at JRC (February 2008), EU Report 23496.

Gilbert, B., Huang, F., Zhang, H., Waychunas, G.A., Banfield, J.F., 2004. Nanoparticles: strained and stiff. Science 305, 651–654.

Gref, R., Miralles, G., ESdith, D., 1999. Polyoxyethylene-coated nanospheres: effect of coating on zeta potential and phagocytosis. Polym. Int. 48, 251256.

Gref, R., Lück, M., Quellec, P., Marchand, M., Dellacherie, E., Harnisch, S., Blunk, T., Müller, R.H., 2000. 'Stealth' corona-core nanoparticles surface modified by polyethylene glycol (PEG): influences of the corona (PEG chain length and surface density) and of the core composition on phagocytic uptake and plasma protein adsorption. Colloids Surf. B 18, 301–313.

Gruenberg, J., Maxfield, F.R., 1995. Membrane transport in the endocytic pathway. Curr. Opin. Cell Biol. 7, 552–563.

Hagens, W.I., Oomen, A.G., de Jong, W.H., Cassee, F.R., Sips, A.J.A.M., 2007. What do we (need to) know about the kinetic properties of nanoparticles in the body? Regul. Tox. Pharm. 49, 217–229.

Halford, B., 2006. Fullerene for the face: cosmetics containing C60 nanoparticles are entering the market, even if their safety is unclear. Chem. Eng. News 84, 47.

Hanigen, H.M., Laishes, B.A., 1984. Toxicity of aflatoxin B1 in rat and mouse hepatocytes in vivo and in vitro. Toxicology 30, 185–193.

Haseloff, R.F., Blasig, I.E., Bauer, H.C., Bauer, H., 2005. In search of the astrocytic factor(s) modulating blood–brain barrier functions in brain capillary endothelial cells in vitro. Cell Mol. Neurobiol. 25, 25–39.

Hawkins, R.A., O'Kane, R.L., Simpson, I.A., 2006. Structure of the blood–brain barrier and its role in the transport of amino acids. J. Nutr. 136, 218S–226S.

Haynes, C.A., Norde, W., 1994. Globular proteins at solid/liquid interfaces. Colloids Surf. B 2, 517–566.

He, X., Nie, H., Wang, K., Tan, W., Wu, X., Zhang, P., 2008. In vivo study of biodistribution and urinary excretion of surface-modified silica nanoparticles. Anal. Chem. 80, 9597–9603.

Hillyer, J.F., Albrecht, R.M., 2001. Gastrointestinal persorption and tissue distribution of differently sized colloidal gold nanoparticles. J. Pharm. Sci. 90, 1927–1936.

Hoet, P.H., Bruske-Hohlfeld, I., Salata, O.V., 2004. Nanoparticles—known and unknown health risks. J. Nanobiotechnol. 2, 12.

Holmes, D., 2013. The next big things are tiny. Lancet Neurol. 12, 31–32.

Hou, Y., Liu, Y., Chen, Z., Gu, N., Wang, J., 2010. Manufacture of IRDye800CW-coupled Fe_3O_4 nanoparticles and their applications in cell labeling and in vivo imaging. J. Nanobiotech. 8, 25.

Huang, X., Kobos, R., Xu, G., 2007. Hair Coloring and Cosmetic Compositions Comprising Carbon Nanotubes. US 7276088 B2.

Jain, T.K., Reddy, M.K., Morales, M.A., LesliePelecky, D.L., Labhasetwar, V., 2008. Biodistribution, clearance, and biocompatibility of iron oxide magnetic nanoparticles in rats. Mol. Pharmaceut. 5, 316–327.

de Jalon, E.G., Blanco-Prieto, M.J., Ygartua, P., Santoyo, S., 2001. Topical application of acyclovir-loaded microparticles: quantification of the drug in porcine skin layers. J. Control. Rel. 75, 191–197.

Jiang, X., Weise, S., Hafner, M., Rocker, C., Zhang, F., Parak, W.J., Nienhaus, G.U., 2009. Quantitative analysis of the protein corona on FePt nanoparticles formed by transferrin binding. J. R. Soc. Interf. 11, S5–S13.

Kagan, V.E., Konduru, N.V., Feng, W., Allen, B.L., Conroy, J., Volkov, Y., Vlasova II, Belikova, N.A., Yanamala, N., Kapralov, A., Tyurina, Y.Y., Shi, J., Kisin, E.R., Murray, A.R., Franks, J., Stolz, D., Gou, P., Klein-Seetharaman, J., Fadeel, B., Star, A., Shvedova, A.A., 2010. Carbon nanotubes degraded by neutrophil myeloperoxidase induce less pulmonary inflammation. Nat. Nanotechnol. 5, 354–359.

Kam, N.W.S., Dai, H., 2005. Carbon nanotubes as intracellular protein transporters: generality and biological functionality. J. Am. Chem. Soc. 127, 6021–6026.

Kam, N.W.S., Jessop, T.C., Wender, P.A., Dai, H., 2004. Nanotube molecular transporters: internalization of carbon nanotube-protein conjugates into mammalian cells. J. Am. Chem. Soc. 126, 6850–6851.

Karmali, P.P., Simberg, D., 2011. Interactions of nanoparticles with plasma proteins: implication on clearance and toxicity of drug delivery systems. Exp. Opin. Drug Deliv. 8, 343–357.

Kawaguchi, H., Koiwai, N., Ohtsuka, Y., Miyamoto, M., Sasakawa, S., 1986. Phagocytosis of latex-particles by leukocytes. 1. Dependence of phagocytosis on the size and surface-potential of particles. Biomaterials 7, 61–66.

Knowles, M.R., Boucher, R.C., 2002. Mucus clearance as a primary innate defense mechanism for mammalian airways. J. Clin. Invest. 109, 571–577.

Kostarelos, K., Lacerda, l, Pastorin, G., Wu, W., Wieckowski, S., Luangsivilay, J., Godefroy, S., Pantarotto, S.D., Briand, J.P., Muller, S., Prato, M., Bianco, A., 2007. Cellular uptake of functionalized carbon nanotubes is independent of functional group and cell type. Nat. Nanotechnol. 2, 108–113.

Kostarelos, K., Bianco, A., Prato, M., 2009. Promises, facts and challenges for carbon nanotubes in imaging and therapeutics. Nat. Nanotechnol. 4, 627–633.

Kreuter, D.S., Valery, P., Peter, R., Klaus, C., Claudia, K.-B., Renad, A., 2002. Apolipoprotein-mediated transport of nanoparticle-bound drugs across the blood–brain barrier. J. Drug Target 10, 317–325.

Kreuter, J., 2013. Mechanism of polymeric nanoparticle-based drug transport across the blood–brain barrier (BBB). J. Microencapsul. 30, 49–54.

Kumar, D., Meenan, B.J., Mutreja, I., D'Sa, R., Dixon, D., 2012. Controlling the size and size distribution of gold nanoparticles: a design of experiment study. Int. J. Nanosci. 11, 1250023 (7 pages).

Kwon, J.T., Hwang, S.K., Jin, H., Kim, D.S., Minai-Tehrani, A., Yoon, H.J., Choi, M., Yoon, T.J., Han, D.Y., Kang, Y.W., 2008. Body distribution of inhaled fluorescent magnetic nanoparticles in the mice. J. Occup. Health 50, 16.

Lacava, L.M., Lacava, Z.G.M., Azevedo, R.B., Chaves, S.B., Garcia, V.A.P., Silva, O., Pelegrini, F., Buske, N., Gansau, C., Da Silva, M.F., Morais, P.C., 2002. Use of magnetic resonance to study biodistribution of dextran-coated magnetic fluid intravenously administered in mice. J. Magn. Magn. Mater. 252, 367369.

Lacerda, L., Russier, J., Pastorinb, G., Herreroc, M.A., Venturelli, E., Dumortier, H., Al-Jamal, K.T., Prato, M., Kostarelos, K., Bianco, A., 2012. Translocation mechanisms of chemically functionalized carbon nanotubes across plasma membranes. Biomaterials 33, 3334–3343.

Lademann, J., Weigmann, H.J., Rickmeier, C.H., Barthelmes, H., Schaefer, H., Mueller, G., Sterry, W., 1999. Penetration of titanium dioxide microparticles in a sunscreen formulation into the horny layer and the follicular orifice. Skin Pharmacol. Appl. Skin Physiol. 2, 247–256.

Lalka, D., Griffith, R.K., Cronenberger, C.L., 1993. The hepatic first-pass metabolism of problematic drugs. J. Clin. Pharmacol. 33, 657–669.

Lam, C.W., James, J.T., McCluskey, R., Arepalli, S., Hunter, R.L., 2006. A review of carbon nanotube toxicity and assessment of potential occupational and environmental health risks. Crit. Rev. Toxicol. 36, 189–217.

Lasagna-Reeves, C., Gonzalez-Romero, D., Barria, M.A., Olmedo, I., Clos, A., Sadagopa, V.M., Urayama, R.A., Vergara, L., Kogan, M.J., Soto, C., 2010. Bioaccumulation and toxicity of gold nanoparticles after repeated administration in mice. Biochem. Biophys. Res. Comm. 393, 649–655.

Lauer, A.C., Ramachandran, C., Lieb, L.M., Niemiec, S., Weiner, N.D., 1996. Targeted delivery to the pilosebaceous unit via liposomes. Adv. Drug Deliv. Rev. 18, 311–324.

Lavoisier, A., 1789. Traité Élémentaire de Chimie, présenté dans un ordre nouveau, et d'après des découvertes modernes, first ed. Cuchet, Paris. Libraire, (retrieved 15.04.15 via Gallica).

Lavoisier, A., 1790. Elements of Chemistry in New Systematic Order, Containing All Modern Discoveries, Illustrated with 13 Copperplates, Translated from the French by

Robert Kerr, first ed. William Creech, Edinburgh (retrieved 15.04.15.).

Lee, P.W., Hsu, S.H., Wang, J.J., Tsai, J.S., Lin, K.J., Wey, S.P., Chen, F.R., Lai, C.H., Yen, T.C., Sung, H.W., 2010. The characteristics, biodistribution, magnetic resonance imaging and biodegradability of superparamagnetic core-shell nanoparticles. Biomaterials 31, 1316–1324.

Lens, M., 2011. Recent progresses in application of fullerenes in cosmetics. Recent Pat. Biotechnol. 5, 67–73.

Lesniak, A., Salvati, A., Santos-Martinez, M.J., Radomski, M.W., Dawson, K.A., Aberg, C., 2013. Nanoparticle adhesion to the cell membrane and its effect on nanoparticle uptake efficiency. J. Am. Chem. Soc. 135, 1438–1444.

Lin, J., Zhang, H., Chen, Z., Zheng, Y., 2010. Penetration of lipid membranes by gold nanoparticles: insights into cellular uptake, cytotoxicity, and their relationship. ACS Nano 4, 5421–5429.

Lin, S.-Y., Hsu, W.-H., Lo, J.-M., Tsai, H.-C., Hsiue, G.-H., 2011. Novel geometry type of nanocarriers mitigated the phagocytosis for drug delivery. J. Control. Rel. 154, 84–92.

Lipinski, C.A., Lombardo, F., Dominy, B.W., Feeney, P.J., 1997. Experimental and computational approaches to estimate solubility and permeability in drug discovery and development settings. Adv. Drug Deliv. Rev. 23, 3–25.

Lippmann, M., Albert, R.E., 1969. The effect of particle size on the regional deposition of inhaled aerosols in the human respiratory tract. Am. Ind. Hyg. Assoc. J. 30, 257–275.

Liu, Z., Cai, W., He, L., Nakayama, N., Chen, K., Sun, X., Chen, X., Dai, H., 2007. In vivo biodistribution and highly efficient tumor targeting of carbon nanotubes in mice. Nat. Nanotechnol. 2, 47–52.

Liu, J.J.A., Anilkumar, P., Wang, X., Yang, S.-T., Luo, P.G., Wang, H., Lu, F., Meziani, M.J., Liu, Y., Korch, K., Sun, Y.-P., 2010. Cytotoxicity evaluations of fluorescent carbon nanotubes. Nano Life 01, 153.

Lucas, P., 2005. Secrets of caveolae- and lipid raft-mediated endocytosis revealed by mammalian viruses. Biochim. Biophys. Acta 1746, 295–304.

Lundqvist, M., Stigler, J., Elia, G., Lynch, I., Cedervall, T., Dawson, K.A., 2008. Nanoparticle size and surface properties determine the protein corona with possible implications for biological impacts. Proc. Natl Acad. Sci. U.S.A. 105, 14265–14270.

Lundqvist, M., Stigler, J., Cedervall, T., Berggard, T., Flanagan, M.B., Lynch, I., Elia, G., Dawson, K., 2011. The evolution of the protein corona around nanoparticles: a test study. ACS Nano. 5, 7503–7509.

Lynch, I., Dawson, K.A., 2008. Protein-nanoparticle interactions. Nano Today 3, 40–47.

Maeda, H., 2012. Macromolecular therapeutics in cancer treatment: the EPR effect and beyond. J. Control. Rel. 164, 138–144.

Mahmoud, M.A., O'Neil, D., El-Sayed, M.A., 2014. Hollow and solid metallic nanoparticles in sensing and in nanocatalysis. Chem. Mater. 26, 44–58.

Malmsten, M. (Ed.), 1998. Biopolymers at Interfaces. Marcel Dekker, New York.

Manohar, S., Ungureanu, C., Van Leeuwen, T.G., 2011. Gold nanorods as molecular contrast agents in photoacoustic imaging: the promises and the caveats. Cont. Media Mol. Imag. 6, 389–400.

Matsumura, Y., Maeda, H., 1986. A new concept for macromolecular therapeutics in cancer chemotherapy: mechanism of tumoritropic accumulation of proteins and the antitumor agent smancs. Cancer Res. 46, 6387–6392.

Mayor, S., Riezman, H., 2004. Sorting GPI-anchored proteins. Nat. Rev. Mol. Cell. Biol. 5, 110–120.

Meidan, V.M., Bonner, M.C., Michniak, B.B., 2005. Transfollicular drug delivery—is it a reality? Int. J. Pharm. 306, 1–14.

Mercer, R.R., Scabilloni, J.F., Wang, L., Kisin, E., Murray, A.R., Schwegler-Berry, D., S, A.A., Castranova, V., 2008. Alteration of deposition pattern and pulmonary response as a result of improved dispersion of aspirated single-walled carbon nanotubes in a mouse model. Am. J. Physiol. Lung Cell Mol. Physiol. 294, L87–L97.

Mercer, R.R., Hubbs, A.F., Scabilloni, J.F., Wang, L., Battelli, L.A., Schwegler-Berry, D., Castranova, V., Porter, D.W., 2010. Distribution and persistence of pleural penetrations by multi-walled carbon nanotubes. Part. Fibre Toxicol. 7, 28.

Mercer, R.R., Hubbs, A.F., Scabilloni, J.F., Wang, L., Battelli, L.A., Friend, S., Castranova, V., Porter, D.W., 2011. Pulmonary fibrotic response to aspiration of multi-walled carbon nanotubes. Part. Fibre Toxicol. 8, 21.

Michaelis, K., Hoffmann, M.M., Dreis, S., Herbert, E., Alyautdin, R.N., Kreuter, J., Langer, K., 2006. Covalent linkage of apolipoprotein e to albumin nanoparticles strongly enhances drug transport into the brain. J. Pharmacol. Exp. Ther. 317, 1246–1253.

Moghimi, H.R., Williams, A.C., Barry, B.W., 1996. A lamellar matrix model for stratum corneum intercellular lipids II. Effect of geometry of the stratum corneum on permeation of model drugs 5-fluorouracil and oestradiol. Int. J. Pharma. 131, 117–129.

Moghimi, S.M., Hunter, A.C., Murray, J.C., 2001. Long-circulating and target specific nanoparticles: theory to practice. Pharm. Rev. 53, 283–318.

Moller, W., Felten, K., Sommerer, K., Scheuch, G., Meyer, G., Meyer, P., Häussinger, K., Kreyling, W.G., 2008. Deposition, retention and translocation of ultrafine particles from the central airways and lung periphery. Am. J. Respir. Crit. Care Med. 177, 426–432.

Monopoli, M.P., Walczyk, D., Campbell, A., Elia, G., Lynch, I., Bombelli, F.B., Dawson, K.A., 2011. Physical-chemical aspects of protein corona: relevance to in vitro and in vivo biological impacts of nanoparticles. J. Am. Chem. Soc. 133, 2525–2534.

Mortensen, J., Lange, L., Nyboe, J., Groth, S., 1994. Lung mucociliary clearance. Eur. J. Nucl. Med. 21, 953–961.

Mosskin, M., Erickson, K., Hindmarsh, T., Vonholst, H., Collins, V.P., Bergstrom, M., Ericksson, L., Johnstrom, P., 1989. Positron emission tomography compared with magnetic-resonance imaging and computed tomography in supratentorial gliomas using multiple streotactic biopsies as reference. Acta Radiol. 30, 225–232.

Mukherjee, S., Ghosh, R.N., Maxfield, F.R., 1997. Acetoacetylated lipoproteins used to distinguish fibroblasts from macrophages in vitro by fluorescence microscopy. Arteriosclerosis 1 Endocytosis. Physiol. Rev. 77, 759–803.

Muller, R.H., Keck, C.M., 2004. Drug delivery to the brain – realization by novel drug carriers. J. Nanosci. Nanotechnol. 4, 471–483.

Muller, P.H., 1949. Dichlorodiphenyläthan und neuere Insektizide. Nobel lecture, delivered December 11, 1948. In: Boktryckeriet, P.A., Norstedt & Söner (Eds.), Les Prix Nobel en 1948. Kungl, Stockholm, pp. 122–123.

Murray, A.R., Kisin, E., Leonard, S.S., Young, S.H., Kommineni, C., Kagan, V.E., Castranova, V., Shvedova, A.A., 2009. Oxidative stress and inflammatory response in dermal toxicity of single-walled carbon nanotubes. Toxicology 257, 161–171.

Myung, C.L., Park, S., 2012. Exhaust nanoparticle emission from internal combustion engines: a review. Int. J. Automot. Technol. 13, 9–22.

Nabeshi, H., Yoshikawa, T., Matsuyama, K., Nakazato, Y., Arimori, A., Isobe, M., Tochigi, S., Kondoh, S., Hirai, T., Akase, T., Yamashita, T., Yamashita, K., Yoshida, T., Nagano, K., Abe, Y., Yoshioka, Y., Kamada, H., Imazawa, T., Itoh, N., Tsunoda, S., Tsutsumi, Y., 2010. Size-dependent cytotoxic effects of amorphous silica nanoparticles on Langerhans cells. Pharmazie 65, 199–201.

Nabeshi, H., Yoshikawa, T., Matsuyama, K., Nakazato, Y., Matsuo, K., Arimori, A., Isobe, M., Tochigi, S., Kondoh, S., Hirai, T., Akase, T., Yamashita, T., Yamashita, K., Yoshida, T., Nagano, K., Abe, Y., Yoshioka, Y., Kamada, H., Imazawa, T., Itoh, N., Nakagawa, S., Mayumi, S.T., Tsunoda, S., Tsutsumi, Y., 2011. Systemic distribution, nuclear entry and cytotoxicity of amorphous nanosilica following topical application. Biomaterials 32, 2713–2724.

Nagai, H., Okazaki, Y., Chew, S.H., Misawa, N., Yamashita, Y., Akatsuka, S., Ishihara, T., Yamashita, K., Yoshikawa, Y., Yasui, H., Jiang, L., Ohara, H., Takahashi, T., Ichihara, G., Kostarelos, K., Miyata, Y., Shinohara, H., Toyokuni, S., 2011. Diameter and rigidity of multiwalled carbon nanotubes are critical factors in mesothelial injury and carcinogenesis. PNAS 108, E1330–E1338.

Nel, A.E., Mädler, L., Velegol, D., Xia, T., Hoek, E.M., Somasundaran, P., Klaessig, F., Castranova, V., Thompson, M., 2009. Understanding biophysicochemical interactions at the nano-bio interface. Nat. Mater. 8, 543–557.

Nemmar, A., Vanbilloen, H., Hoylaerts, M.F., Hoet, P.H., Verbruggen, A., Nemery, B., 2001. Passage of Intratracheally Instilled Ultrafine Particles from the Lung into the Systemic Circulation in Hamster.

Nemmar, A., Hoet, P.H.M., Vanquickenborne, B., Dinsdale, D., Thomeer, M., et al., 2002. Passage of inhaled particles into the blood circulation in humans. Circulation 105, 411–414.

Newberne, P.M., Butler, W.H., 1969. Acute and chronic effects of aflatoxin on the liver of domestic and laboratory animals: a review. Cancer Res. 29, 236–250.

Nichols, B.J., Kenworthy, A.K., Polishchuk, R.S., Lodge, R., Roberts, T.H., Hirschberg, K., Phair, R.D., Lippincott-Schwartz, J., 2001. Rapid cycling of lipid raft markers between the cell surface and Golgi complex. J. Cell Biol. 153, 529–541.

Norde, W., 1986. Adsorption of proteins from solution at the solid–liquid interface. Adv. Colloid Interf. Sci. 25, 267–340.

Nunes, A., Bussy, C., Gherardini, L., Meneghetti, M., Herrero, M.A., Bianco, A., Prato, M., Pizzorusso, T., Al-Jamal, K.T., Kostarelos, K., 2012. In vivo degradation of functionalized carbon nanotubes after stereotactic administration in the brain cortex. Nanomed. (Lond.) 7, 1485–1494.

Oberdörster, G., Sharp, Z., Atudorei, V., Elder, A., Gelein, R., Kreyling, W., Cox, C., 2004. Translocation of inhaled ultrafine particles to the brain. Inhal. Toxicol. 16, 437–445.

Oberdörster, G., Oberdörster, E., Oberdörster, J., 2005a. Nanotoxicology: an emerging discipline evolving from studies of ultrafine particles. Environ. Health Persp. 113, 823–839.

Oberdörster, G., Maynard, A., Donaldson, K., Castranova, V., Fitzpatrick, J., Ausman, K., Carter, J., Karn, B., Kreyling, W., Lai, D., Olin, S., Monteiro-Riviere, N., Warheit, D., Yang, H., 2005b. Principles for characterizing the potential human health effects from exposure to nanomaterials: elements of a screening strategy. Part. Fibre Toxicol. 2, 8.

Oberdörster, G., Oberdörster, E., Oberdörster, J., 2007. Concepts of nanoparticle dose metric and response metric. Environ. Health Persp. 115, A290.

Oberdörster, G., 2010. Safety assessment for nanotechnology and nanomedicine: concepts of nanotoxicology. J. Intern. Med. 267, 89–105.

Oh, N., Park, J.-H., 2014. Endocytosis and exocytosis of nanoparticles in mammalian cells. Int. J. Nanomed. 9 (Suppl. 1), 51–63.

Olson, J.R., Holscher, M.A., Neal, R.A., 1980. Toxicity of 2,3,7,8-tetrachlorodibenzo-*p*-dioxin in the Golden Syrian hamster. Toxicol. Appl. Pharmacol. 55, 67–78.

Ostro, B., 1984. A research for a threshold in the relationship of air pollution to mortality: a reanalysis of London winters. Environ. Health Persp. 58, 397–399.

Owens III, D.E., Peppas, N.A., 2006. Opsonization, biodistribution, and pharmacokinetics of polymeric nanoparticles. Int. J. Pharmceut. 307, 93–102.

Parak, W.J., 2011. Complex colloidal assembly. Science 334, 1359–1360.

Park, M.V.D.Z., Neigh, A.M., Vermeulen, J.P., de la Fonteyne, L.J.J., Verharen, H.W., Briedé, J.J., van Loveren, H., de Jong, W.H., 2011. The effect of particle size on the cytotoxicity, inflammation, developmental toxicity and genotoxicity of silver nanoparticles. Biomaterials 32, 9810–9817.

Patterson, D.S.P., 1973. Metabolism as a factor in determining the toxic action of the aflatoxins in different animal species. Food Cosmatic Toxicol. 11, 287–294.

Peetla, C., Jin, S., Weimer, J., Elegbede, A., Labhasetwar, V., 2014. Biomechanics and thermodynamics of nanoparticle interactions with plasma and endosomal membrane lipids in cellular uptake and endosomal escape. Langmuir 30, 7522–7532.

Perez-Martínez, F.C., Carrión, B., Ceña, V., 2012. The use of nanoparticles for gene therapy in the nervous system. J. Alzheimers Dis. 31, 697–710.

Petkar, K.C., Chavhan, S., Agatonovik-Kustrin, S., Sawant, K., 2011. Nanostructured materials in drug and gene delivery: a review of the state of the art. Crit. Rev. Ther. Drug Carrier Syst. 28, 101–164.

Pflucker, F., Wendel, V., Hohenberg, H., Gärtner, E., Will, T., Pfeiffer, S., Wepf, R., Gers-Barlag, H., 2001. The human stratum corneum layer: an effective barrier against dermal uptake of different forms of topically applied micronised titanium dioxide. Skin Pharmacol. Appl. Skin Physiol. 14 (Suppl. 1), 92–97.

Phoungthong, K., Tekasakul, S., Tekasakul, P., Prateepchaikul, G., Jindapetch, N., Furuuchi, M., Hata, M., 2013. Emissions of particulate matter and associated polycyclic aromatic hydrocarbons from agricultural diesel engine fueled with degummed, deacidified mixed crude palm oil blends. J. Environ. Sci. (China) 25, 751–757.

Pohjanvirta, R., Tuomisto, J., 1987. Han/Wistar rats are exceptionally resistant to TCDD. Arch. Toxicol. 11, 344–347.

Pope, C., Schwartz, J., Ransom, M., 1992. Daily mortality and PM10 pollution in Utah valley. Arch. Environ. Health 47, 211–217.

Praetorius, N.P., Mandal, T.K., 2007. Engineered nanoparticles in cancer therapy. Recent Pat. Drug Deliv. Formulation 1, 37–51.

Re, F., Gregori, M., Masserini, M., 2012. Nanotechnology for neurodegenerative disorders. Nanomed. NBM 8 (Suppl. 1), S51–S58.

Regnis, J.A., Robinson, M., Bailey, D.L., Cook, P., Hooper, P., Chan, H.K., Gonda, I., Bautovich, G., Bye, P.T., 1994. Mucociliary clearance in patients with cystic fibrosis and in normal subjects. Am. J. Resp. Crit. Care Med. 150, 66–71.

Rezwan, K., Meier, L.P., Rezwan, M., Vörös, J., Textor, M., Gauckler, L.J., 2004. Bovine serum albumin adsorption onto colloidal Al_2O_3 particles: a new model based on zeta potential and UV–vis measurements. Langmuir 11, 10055–10061.

Ribeiro, J., Flores, D., Ward, C.R., Silva, L.F.O., 2010. Identification of nanominerals and nanoparticles in burning coal waste piles from Portugal. Sci. Total Environ. 408, 6032–6041.

Rolland, A., Wagner, N., Chatelus, A., Shroot, B., Schaefer, H., 1993. Site-specific drug delivery to pilosebaceous structures using polymeric microspheres. Pharm. Res. 10, 1738–1744.

Rothen-Rutishauser, B.M., Schürch, S., Haenni, B., Kapp, N., Gehr, P., 2006. Interaction of fine particles and nanoparticles with red blood cells visualized with advanced microscopic techniques. Environ. Sci. Technol. 40, 4353–4359.

Rouse, J.G., Yang, J., Ryman-Rasmussen, J.P., Barron, A.R., Monteiro-Riviere, N.A., 2007. Effects of mechanical flexion on the penetration of fullerene amino acid-derivatized peptide nanoparticles through skin. Nano Lett. 7, 155–160.

Rozman, K.K., 1988. Disposition of xenobiotics: species differences. Toxicol. Pathol. 16, 123–129.

Rudt, S., Muller, R.H., 1993. In vitro phagocytosis assay of nano-and microparticles by chemiluminescence III uptake of differently sized surface-modified particles, and its correlation to particle properties and in vivo distribution. Eur. J. Pharm. Sci. 1, 31–39.

Russier, J., Ménard-Moyon, C., Venturelli, E., Gravel, E., Marcolongo, G., Meneghetti, M., Doris, E., Bianco, A., 2011. Oxidative biodegradation of single- and multiwalled carbon nanotubes. Nanoscale 3, 893–896.

Ryman-Rasmussen, J.P., Cesta, M.F., Brody, A.R., Shipley-Phillips, J.K., Everitt, J.I., Tewksbury, E.W., Moss, O.R., Wong, B.A., Dodd, D.E., Andersen, M.E., Bonner, J.C., 2009. Inhaled carbon nanotubes reach the subpleural tissue in mice. Nat. Nanotech. 4, 747–751.

Saba, T.M., 1970. Physiology and physiopathology of the reticuloendothelial system. Arch. Int. Med. 126, 1031–1052.

Sabharanjak, S., Sharma, P., Parton, R.G., Mayor, S., 2002. GPI-anchored proteins are delivered to recycling endosomes via a distinct cdc42-regulated, clathrin-independent pinocytic pathway. Dev. Cell 2, 411–423.

Sadrieh, N., Wokovich, N.M., Gopee, N.V., Zheng, J., Haines, D., Parmiter, D., Siitonen, P.H., Cozart, C.R., Patri, A.K., McNeil, S.E., Howard, P.C., Doub, W.H., Buhse, L.F., 2010. Lack of significant dermal penetration of titanium dioxide from sunscreen formulations containing nano- and submicron-size TiO_2 particles. J. Toxicol. Sci. 115, 156–166.

Sager, T.M., Castranova, V., November 2013. Surface area of particle administered versus mass in determining the pulmonary toxicity of ultrafine and fine carbon black: comparison to ultrafine titanium dioxide. Int. J. Mol. Sci. 14 (11), 21266–21305.

Sager, T., Kommineni, C., Castranova, V., 2008. Pulmonary response to intratracheal instillation of ultrafine versus fine titanium dioxide: role of surface area. Part. Fibre Toxicol. 5, 17.

Sakhtianchi, R., Minchin, R.F., Lee, K.-B., Alkilany, A.M., Serpooshan, V., Mahmoudi, M., 2013. Exocytosis of nanoparticles from cells: role in cellular retention and toxicity. Adv. Colloid Interf. Sci. 201–202, 18–29.

Saptarshi, S.R., Duschl, A., Lopata, A.L., 2013. Interaction of nanoparticles with proteins: relation to bio-reactivity of the nanoparticle. J. Nanobiotechnol. 19 (11), 26.

Sarin, H., Kanevsky, A.S., Wu, H., Sousa, A.A., Wilson, C.M., Aronova, M.A., Griffiths, G.L., Leapman, R.D., Vo, H.Q., 2009. Physiologic upper limit of pore size in the blood–tumor barrier of malignant solid tumors. J. Transl. Med. 7, 51.

Sayes, C., Fortner, J., Guo, W., Lyon, D., Boyd, A., Ausman, K., Tao, Y., Sitharaman, B., Wilson, L., Hughes, J., West, J., Colvin, V., 2004. The differential cytotoxicity of water-soluble fullerenes. Nano Lett. 4, 1881–1887.

Schlesinger, R.B., Lippmann, M., 1972. Particle deposition in casts of the human upper tracheobronchial tree. Am. Ind. Hyg. Assoc. J. 33, 237–251.

Schlesinger, R.B., Lippmann, M., 1978. Selective particle deposition and bronchogenic carcinoma. Environ. Res. 15, 424–431.

Schroeder, F., Kinden, D.A., 1983. Measurement of phagocytosis using fluorescent latex beads. J. Biochem. Biophys. Methods 8, 15–27.

Schwetz, B.A., Norris, J.M., Sparschu, G.L., Rowe, U.K., Gehring, P.J., Emerson, J.L., Gerbig, C.G., 1973. Toxicology of chlorinated dibenzo-*p*-dioxins. Environ. Health Persp. 5, 87–99.

Sengelov, H., 1995. Compliment receptors in neutrophils. Crit. Rev. Immunol. 15, 107–131.

Sharma, A., Muresanu, D.F., Patnaik, R., Sharma, H.S., 2013. Size- and age-dependent neurotoxicity of engineered metal nanoparticles in rats. Mol. Neurobiol. 48, 386–396.

Sharma, N.C., Sahi, S.V., Nath, S., Parsons, J.G., Gardea-Torresdey, J.L., Pal, T., 2007. Synthesis of plant mediated gold nanoparticles and catalytic role of biomatrix-embedded nanomaterials. Environ. Sci. Technol. 41, 5137–5142.

Shivers, R.R., Edmonds, C.L., Del Maestro, R.F., 1984. Microvascular permeability in induced astrocytomas and peritumor neurophil of rat brain. Acta Neuropathol. (Berl.) 64, 192–202.

Shvedova, A.A., Pietroiusti, A., Fadeel, B., Kagan, V.E., 2012. Mechanisms of carbon nanotube-induced toxicity: focus on oxidative stress. Tox. Appl. Pharmacol. 261, 121–133.

Sigmund, W., Pyrgiotakis, G., Daga, A., 2005. Chemical Processing of Ceramics. CRC.

Simberg, D., Park, J.H., Karmali, P.P., Zhang, W.M., Merkulov, S., McCrae, K., Bhatia, S.N., Sailor, M., Ruoslahti, E., 2009. Differential proteomics analysis of the surface heterogeneity of dextran iron oxide nanoparticles and the implications for their in vivo clearance. Biomaterials 30, 3926–3933.

Simon, S.I., Schmidschonbein, G.W., 1988. Biophysical aspects of microsphere engulfment by human-neutrophils. Biophys. J. 53, 163–173.

Singh, R., Pantarotto, D., Lacerda, L., Pastorin, G., Klumpp, C., Prato, M., Bianco, A., Kostarelos, K., 2006. Tissue biodistribution and blood clearance rates of intravenously administered carbon nanotube radiotracers. PNAS U.S.A. 103, 3357–3362.

Singh, A.K., Jiang, Y., Gupta, S., Younus, M., Ramzan, M., 2013. Anti-inflammatory potency of nano-formulated puerarin and curcumin in rats subjected to the lipopolysaccharide-induced inflammation. J. Med. Food 16, 899–911.

Singh, A.K., 2014. Challenges in the medicinal applications of carbon nanotubes (CNTs): toxicity of the central nervous system and safety issues. J. Nanomed. Nanotech. 1, 002.

Smith, J.W., Hamilton, P.B., 1970. Aflatoxicosis in the broiler chicken. Poult. Sci. 49, 207–215.

Song, Y., Li, X., Du, X., 2009. Exposure to nanoparticles is related to pleural effusion, pulmonary fibrosis and granuloma. Eur. Respir. J. 34, 559–567.

Sriram, K., Porter, D., Turuoka, S., Endo, M., Jefferson, A., Wolfarth, W., et al., 2007. Neuroinflammatory responses following exposure to engineered nanomaterials. Toxicologist A1390.

Stoeger, T., Reinhard, C., Takenaka, S., Schroeppel, A., Karg, E., Ritter, B., 2006. Instillation of six different ultrafine carbon particles indicates surface area threshold dose for acute lung inflammation in mice. Environ. Health Persp. 114, 328–333.

Stoorvogel, W., Strous, G.J., Geuze, H.J., Oorschot, V., Schwartz, A.L., 1991. Late endosomes derive from early endosomes by maturation. Cell 65, 417–427.

Tabata, Y., Ikada, Y., 1990. Phagocytosis of polymer microspheres by macrophages. Adv. Polym. Sci. 94, 107–141.

Tan, M., Commens, C., Burneitt, L., Snitch, P.J., 1996. A pilot study on the percutaneous absorption of microfine titanium dioxide from sunscreens. Austr. J. Dermatol. 57, 185–187.

Tenner, A.J., Robinson, S.L., Ezekowitz, R.A., 1995. Mannose binding protein (MBP) enhances mononuclear phagocyte function via a receptor that contains the 126,000 Mr component of the C1q. Immunity 3, 485–493.

Terentyuk, G.S., Maslyakova, G.N., Suleymanova, L.V., Kogan, B.Y., Khlebtsov, B.N., Akchurin, G.G., Shantrocha, A.V., Maksimova, I.L., Khlebtsov, N.G., Tuchin, V.V., 2009. Circulation and distribution of gold nanoparticles and induced alterations of tissue morphology at intravenous particle delivery. J. Biophoton. 2, 292–302.

Tinkle, S.S., Antonini, J.M., Rich, B.A., Roberts, J.R., Salmen, R., DePree, K., Adkins, E.J., 2003. Skin as a route of exposure and sensitization in chronic beryllium disease. Environ. Health Persp. 111, 1202–1208.

Toll, R., Jacobi, U., Richter, H., Lademann, J., Schaefer, H., Blume-Peytavi, U., 2004. Penetration profile of microspheres in follicular targeting of terminal hair follicles. J. Invest. Dermatol. 123, 168–176.

Tong, R.T., Boucher, Y., Kozin, S.V., Winkler, F., Hicklin, D.J., Jain, R.K., 2004. Vascular normalization by vascular endothelial growth factor receptor 2 blockade induces a pressure gradient across the vasculature and improves drug penetration in tumors. Cancer Res. 64, 3731–3736.

Vaaraslahti, K., Keskinen, J., Giechaskiel, B., Murtonen, T., Solla, A., 2005. Effect of lubricant on the formation of heavy-duty diesel exhaust nanoparticles. Environ. Sci. Technol. 39, 8497–8504.

Velegol, D., 2007. Assembling colloidal devices by controlling interparticle forces. J. Nanophoton. 1, 012502.

Verma, A., Uzun, O., Hu, Y., Hu, Y., Han, H.S., Watson, N., Chen, S., Irvine, D.J., Stellacci, F., 2008. Surface-structure-regulated cell-membrane penetration by monolayer-protected nanoparticles. Nat. Mater. 7, 588–595.

Vertegel, A.A., Siegel, R.W., Dordick, J.S., 2004. Silica nanoparticle size influences the structure and enzymatic activity of adsorbed lysozyme. Langmuir 20, 6800–6807.

Vogt, A., Combadiere, B., Hadam, S., Stieler, K.M., Lademann, J., Schaefer, H., Autran, B., Sterry, W., Blume-Peytavi, U., 2006. 40 nm, but not 750 or 1,500 nm, nanoparticles enter epidermal CD1a+ cells after transcutaneous application on human skin. J. Invest. Dermatol. 126, 1316–1322.

Vorbrodt, A.W., 1999. Morphological evidence of the functional polarization of brain microvascular endothelium. In: Pardridge, W.M. (Ed.), The Blood–Brain Barrier: Cellular and Molecular Biology. Raven Press, New York, NY.

Vroman, L., Adams, A.L., Fischer, G.C., Munoz, P.C., 1980. Interaction of high molecular-weight kininogen, factor-XII, and fibrinogen in plasma at interfaces. Blood 55, 156–159.

Vroman, L., 1962. Effect of adsorbed proteins on the wettability of hydrophilic and hydrophobic solids. Nature 11, 476–477.

Wagner, E.M., Foster, W.M., 1996. Importance of airway blood flow on particle clearance from the lung. J. Appl. Physiol. 81, 1878–1883.

Walkey, C.D., Chan, W.C., 2012. Understanding and controlling the interaction of nanomaterials with proteins in a physiological environment. Chem. Soc. Rev. 41, 2780–2799.

Walkey, C.D., Olsen, J.B., Guo, H., Emili, A., Chan, W.C.W., 2012. Nanoparticle size and surface chemistry determine serum protein adsorption and macrophage uptake. J. Am. Chem. Soc. 134, 2139–2147.

Wang, D., El-Amouri, S.S., Dai, M., Kuan, C.Y., Hui, D.Y., Brady, R.O., Pan, D., 2013. Engineering a lysosomal enzyme with a derivative of receptor-binding domain of apoE enables delivery across the blood–brain barrier. Proc. Natl. Acad. Sci. U.S.A. 110, 2999–3004.

Wanner, A., Salathe, M., O'Riordan, T.G., 1996. Mucociliary clearance in the airways. Am. J. Respir. Crit. Care Med. 154, 1868–1902.

Weihe, E., Hartschuh, W., Metz, J., Bruhl, U., 1977. The use of ionic lanthanum as a diffusion tracer and as a marker of calcium binding sites. Cell Tissue Res. 178, 285–302.

Winkelstein, J.A., 1973. Opsonins: their function, identity, and clinical significance. J. Pediatrics 82, 747–753.

Wissing, S.A., Muller, R.H., 2002. Solid lipid nanoparticles as carrier for sunscreens: in vitro release and in vivo skin penetration. J. Control. Rel. 81, 225–233.

Wittmaack, K., 2007. In search of the most relevant parameter for quantifying lung inflammatory response to nanoparticle exposure: particle number, surface area, or what? Environ. Health Persp. 115, 187–194.

Wolburg, H., Lippoldt, A., 2002. Tight junctions of the blood–brain barrier: development, composition and regulation. Vascul. Pharmacol. 38, 323–337.

Yang, P.-H., Sun, X., Chiu, J.-F., Sun, H., He, Q.-Y., 2005. Transferrin-mediated gold nanoparticle cellular uptake. Bioconjug. Chem. 16, 494–496.

Yang, H., Liu, C., Yang, D., Zhang, H., Xi, Z., 2009. Comparative study of cytotoxicity, oxidative stress and genotoxicity induced by four typical nanomaterials: the role of particle size, shape and composition. J. Appl. Toxicol. 29, 69–78.

Youns, M., Hoheisel, J.D., Efferth, T., 2011. Therapeutic and diagnostic applications of nanoparticles. Curr. Drug Targets 12, 357–365.

Zeidler, O., 1874. Verbindungen von Chloral mit Brom- und Chlorbenzol. Berichte der deutschen chemischen Gesellschaft 7, 1180–1181.

Zhang, L.W., Monteiro-Riviere, N.A., 2008. Assessment of quantum dot penetration into intact, tape-stripped, abraded and flexed rat skin. Skin Pharm. Physiol. 21, 166–180.

Zhao, Y., Wei, J., Vajtai, R., Ajayan, P.M., Barrera, E.V., 2011. Iodine doped carbon nanotube cables exceeding specific electrical conductivity of metals. Sci. Rep. 1, 83.

Zhu, M.T., Feng, W.Y., Wang, Y., Wang, B., Wang, M., Ouyang, H., Zhao, Y.L., Chai, Z.F., 2009. Particokinetics and extrapulmonary translocation of intratracheally instilled ferric oxide nanoparticles in rats and the potential health risk assessment. Toxicol. Sci. 107, 342351.

CHAPTER

6

Nanoparticle Pharmacokinetics and Toxicokinetics

OUTLINE

Engineered Nanoparticles
http://dx.doi.org/10.1016/B978-0-12-801406-6.00006-6

1. INTRODUCTION

Pharmacokinetics (PK) and toxicokinetics (TK) characterize a nanoparticle's *a*bsorption, *d*istribution, *m*etabolism, and *e*limination (ADME). In general, TK has been defined as the application of PK principles to characterize the adverse effects of the particle or study of "the rate a chemical will enter the body and what happens to it once it is in the body" (http://en.wikipedia.org/wiki/Toxicokinetics). TK can also be viewed as PK at relatively high drug doses. While this characterization of TK may be valid for bulk particles, it is not valid for toxicants because many toxicants, such as nerve gases, biopoisons, venoms, etc., cause adverse effects at very low doses. In some cases such as allergy and cancer, there may not be a safe dose as is commonly found in the case of drugs. Assuming for the sake of argument that, for drugs and toxicants both, TK is an extension of PK at higher doses, we will not be able to characterize the beneficial or toxic effects of nanoparticles, especially for the engineered nanoparticles designed for drug delivery. This is because nanoparticles do not comply with the commonly used metric of dose: mass per unit body weight or volume such as g/kg, g/cm^2 skin surface area, or g/cm^3 inhaled. For nanoparticles, the dose in terms of particle surface area per unit weight ($\Sigma nm^2/kg$) or surface activity per unit body weight is more relevant than a mass-based dose metric.

Despite the key dose-metric-related difference between bulk and nanoparticles, a mass-based dose metric has been used in most studies characterizing nanoparticles. This has blurred the distinction between PK and TK. Therefore, a new term—*biokinetics*—has been devised to include both PK and TK. Biokinetics is also used in cases where the data cannot be clearly defined for one or the other. Because an understanding of the nanoparticles' biokinetics is essential to design effective and safe nanoproducts, this chapter provides a thorough discussion of the principles of ADME and integrated kinetics as they may relate to nanoparticles. The chapter includes three sections: (1) an introduction, (2) principles of ADME and integrated kinetics for bulk particles, and (3) nanoparticle biokinetics.

Terrestrial and atmospheric organisms are exposed to drugs and toxicants, including nanoparticles, via contaminated dermal, inhalation, and ingestion routes. In some cases, such as medicine delivery or drug abuse situations, chemicals are administered deliberately via injection or ingestion. Aquatic animals, especially fish, may also get exposed to toxins via the gills. Reversible or irreversible damage may occur if the organism receives an appropriate dose of a toxicant. The toxicant-induced adverse effects occur in three phases:

1. *The exposure phase*: In this phase, drugs, toxicants, or nanoparticles encounter and penetrate an organism and accumulate in blood, organs, and tissues (discussed in Chapter 5).
2. *The biokinetic phase*: There are two components of the biokinetic phase:
   a. *PK* that is relevant to exposure doses at which the medicinal and beneficial effects of a drug or nanoparticle outweigh the adverse effects
   b. *TK* that is relevant to the doses at which the adverse effects outweigh the beneficial effects (in case of the drugs) or the adverse effects begin to appear (in case of a toxicant)

Biokinetics is the key focus of this chapter. It encompasses the ADME of a drug, toxicant, or nanoparticle. PK defines the medicinal properties of the pharmaceutical drugs, while TK defines the adverse effects of the drugs or toxicants.

3. *The biodynamic phase*: This phase refers to the kinetics of the mechanisms underlying

beneficial (pharmacodynamics, or PD) and/or adverse (toxicodynamics, or TD) effects. In the case of drugs, the target site for beneficial effects may be different from that for their toxicity, and the dose causing toxicity may be higher than the dose causing pharmacological effects. In the case of toxicants, the dose causing toxicity may depend on physicochemical properties, formulation, and species. Biodynamics will be discussed in Chapter 7.

The biokinetics, including PK and TK, are characterized by the mode of exposure, the dose at the target site, ADME, the ability of a chemical or its metabolites to come into contact with the target site, and the amount and duration of storage of the substance (or its metabolites) in body tissues. Therefore, an understanding of the biokinetic parameters is important in designing toxicological experiments and accurate risk characterization for toxicants (Bessems and Geraets, 2013).

2. ABSORPTION, DISTRIBUTION, METABOLISM, AND ELIMINATION

The ADME involve the bioavailability of a toxicant, drug, or nanoparticle to organisms.

2.1 Absorption

Absorption is the transfer of a toxicant from its site of exposure (or administration) to the bloodstream and then to the target site. In general, chemicals or particles translocate from the extracellular fluid into the cells via diffusion through the gaps between cells (paracellular) or through the lipid bilayer (transcellular). Transcellular passage occurs via (1) simple diffusion through the cell membranes or through the pores, (2) carrier-mediated transport, (3) active transport, and (4) endocytosis or exocytosis (Chapter 5). The efficiency of absorption may depend on the route of administration. For intravenous delivery, the total dose of toxicants reaches the systemic circulation. Exposure by other routes may require the toxicant to cross many barriers; thus, bioavailability may be lower (Chapter 5).

In *simple diffusion*, particles move according to the toxicants' concentration difference between the two sides of the membrane: from higher to lower concentration. Nonpolar substances dissolve freely in membrane lipids; therefore, they penetrate the cell membrane freely. Fick's law for small, nonpolar molecules defines the rate of simple diffusion (dn/dt) that is directly proportional to the concentration gradient (dC/dx) across the membrane, where dC is the differential substrate concentrations inside (C_{in}) and outside (C_{out}) the cell and dx is the width of the cell membrane. The rate of diffusion is directly proportional to the concentration gradient, as defined by the following equation:

$$\frac{dn}{dt} = \mathrm{P} \times \mathrm{A} \times \left(\frac{dC}{dx}\right) \tag{1}$$

In Eqn (1), A is the membrane area and P is the permeability constant (the ease of entry of a molecule into the cell depending on the molecule's size and lipid solubility). Because A and P are constant, this equation describes a straight line where dn/dt is a function of dC/dx.

Although it occurs from high to low concentrations, *facilitated diffusion* requires carrier proteins to transport the chemicals or particles across the cell membrane. Often, more than one chemical is co-transported via facilitated diffusion. For example, glucose is transported along with sodium from the gastrointestinal (GI) tract membrane. The transporter proteins are target-specific and can be saturated at sufficiently higher concentrations. At low concentrations, facilitated diffusion follows the laws of simple diffusion as above. As the solute concentration increases beyond the transporter capacity, a maximum rate (V_{max}) is achieved and no further increase in the rate of diffusion occurs. This is where facilitated diffusion differs from simple diffusion, as shown in Eqn (2).

$$\frac{dn}{dt} = \frac{V_{max}}{1 + K(dC/dx)} \quad (2)$$

In this equation, K is the concentration gradient at which the rate of diffusion is 1/2 V_{max}. K and V_{max} depend on properties of the diffusing molecule, such as its permeability (P), as well as the surface area (A) of the cell.

Active transport is an energy (ATP)-requiring carrier-mediated form of transport that moves molecules/particles against the concentration gradient. The carrier binds the particles and then it transfers the particle to the other side of the membrane. Active diffusion is energy mediated. Ions such as K^+, Ca^{++}, and Na^+ are transferred via energy-coupled receptor-mediated transporters. The environmental toxin Pb^{2+} can replace Ca^{2+} for Ca-ion channels.

In *pinocytosis*, a particle coming in contact with the cell membranes induces an invagination, such as pseudopods that trap the particle and form vesicles in which the particle is present and taken into the cell. For example, insulin can enter the blood–brain barrier (BBB) by this process. The details are included in Chapter 5.

2.1.1 Absorption from the Gastrointestinal Tract

The GI tract—a system consisting of organs with considerable species diversity (Figure 1)—performs six key functions: ingestion, propulsion, mechanical digestion, chemical and

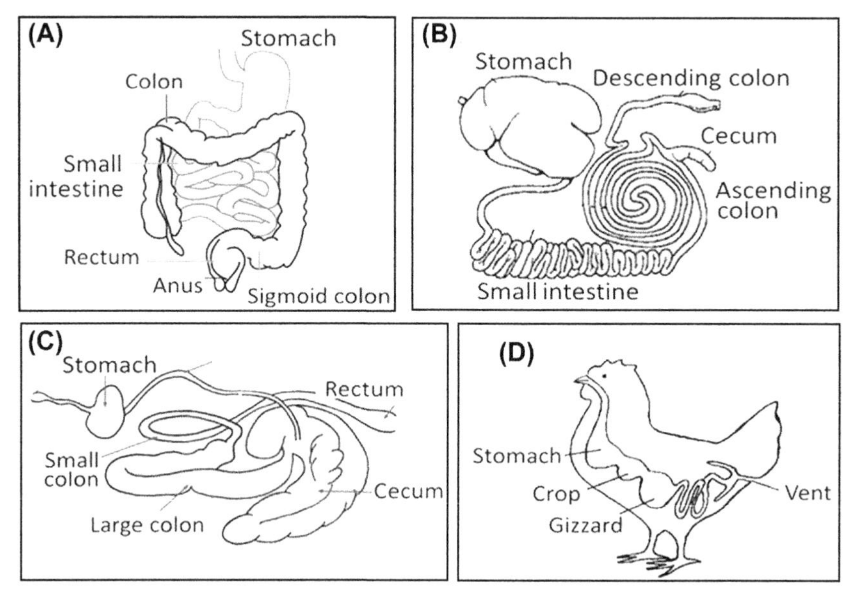

FIGURE 1 Anatomy of the GI tract in a human (A), cow (B), horse (C), and hen (D). Digestive tracts can be divided into two groups: ruminants and nonruminants (humans, pigs, most carnivores, and birds). Ruminants (cows or sheep) are usually herbivores; thus, they have a specialized esophagus to support a pregastric microbial fermentation process. From the esophagus, ingested substances enter the rumen, the omasum, and then the abomasum (the true stomach of ruminants). Hindgut fermentation (in the cecum and colon) also occurs in ruminants. Nonruminant humans, pigs, and the carnivores contain simple stomach and fermentation occurs in colon. The esophagus of many birds allows for the accumulation of food, called the crop. The crop leads via the esophagus to the proventriculus (site of gastric secretions) which is anterior to the ventriculus (muscular organ; commonly referred to as the gizzard). Horses have small stomach where all true digestion occurs by enzymatic digestion and bacterial fermentation occur in large cecum. These anatomical and physiological differences may cause differential ADME in different species.

enzymatic digestion, absorption, and excretion. The tract is responsible for extracting energy and nutrients and expelling the remaining waste. Nanoparticles or bulk particles present in food or water also go through the barriers as well as the physical and biochemical processes of the tract, resulting in significant transformations and alteration of their pharmacological/ toxicological properties. In biological systems, absorption of a chemical depends on two factors: (1) the intrinsic characteristics of the GI tract and (2) physicochemical properties of the chemical or particle.

2.1.1.1 INTRINSIC CHARACTERISTICS OF THE GI TRACT

The bioavailability and absorption of nutrients and contaminants in the GI tract are regulated by the following factors: blood and GI tract pH, gastric emptying, intestinal motility, mucosa and unique features, and blood flow to the GI tract (Candice et al., 2013; Marathe et al., 2000).

The blood and GI tract pH: The GI tract exhibits large variations in the pH, ranging from less than 4 in the stomach to greater than 7 in the small intestine (Figure 2). Thus, weakly acidic and basic

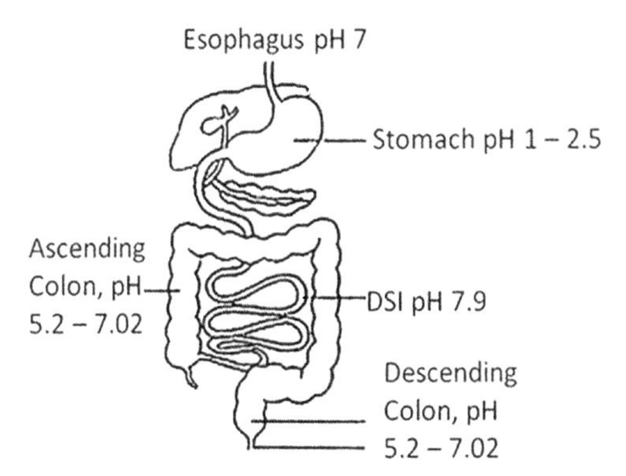

FIGURE 2 The differential pH of the GI tract in nonruminants. The stomach is highly acidic and the small intestine is weakly basic. The remaining components are weakly acidic to neutral. This will strongly influence the ionization state of hydrophilic chemicals (discussed in detail later).

molecules remain unionized and ionized, respectively, in the stomach, but remain ionized and unionized, respectively, in the intestine (Hogben et al., 1957; Hogben, 1959; Schanker, 1978). Shore et al. (1957) established a possible relationship between ionization and absorption. This aspect is discussed in detail in a later section.

Gastric emptying and absorption: Gastric emptying is essential for digestion and movement of food in the GI tract. Two opposing forces coordinate stomach emptying:

1. Intense peristaltic contractions in the antrum push the food down through the duodenum.
2. Varying degrees of resistance occur for chyme in stomach, which has been thoroughly mixed with stomach secretions before passage at the pylorus.

The rate of chime expulsion out of stomach depends upon the pressure generated by antrum against pylorus resistance.

The intestine has the following unique characteristics:

1. *Motility*: The GI tract consists of circular muscles that, upon contraction, squeeze the food in one direction. A synchronized contraction causes peristalsis that pushes the food downstream. A nonsynchronized muscle contraction has the effect of mixing the contents without moving them up or down. In most animals, the small intestine cycles through two states, each of which is associated with distinctive patterns of motility. Impairments in intestinal motility also impair adsorption of drugs and toxicants from the intestinal lumen.
2. *High absorptive surface area*: This is due to the villi, microvilli, and ion transporters (Craddock et al., 1998; Koepsell, 1998; Ito et al., 1999; Zhang et al., 1998). The duodenum and jejunum have the highest concentration of villi and microvilli, and thus the greatest surface areas (Magee and Dalley, 1986).

The enterocyte membrane contains pathways for influx, (ASBT, MCT1, and PEPT1), efflux (BCRP, MRP2, and MDR1), and metabolism (CYP, UGT, GST, PSE, etc.) of drugs and toxicants (Figure 3).

3. The tract cells express the chemical-metabolizing and conjugation enzymes including pro-oxidative cytochrome P450 3A–CYP3A4 (Thummel et al., 1996; Paine et al., 1996, 1997; Ilett et al., 1990; Dubey and Singh, 1988). Therefore, similar to the liver, the GI tract may also inactivate or metabolize toxicants. The enterocytes may metabolize toxicants immediately after absorption and before the drug can pass into the portal venous blood.
4. Interaction of chemicals with the efflux transporters (Figure 4) at the apical and basolateral membranes often results in poor absorption and low oral bioavailability as the particles are readily transported back into the intestinal lumen and excreted out of the body (Wacher et al., 2001; Chen et al., 2009;

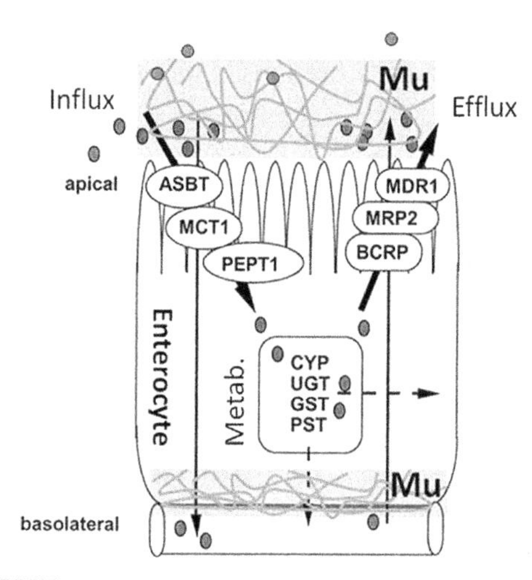

FIGURE 3 Schematic diagram of the enterocyte of the intestine showing absorptive and efflux transporters at the apical and basolateral membranes, and enzymes for intracellular metabolism; and mucus (Mu) polymeric composition (enterocyte model *reproduced from Pang, 2003*; mucus model *reproduced from Peppas et al., 1984*). The enterocyte contains pathways for influx, efflux, and metabolism of toxicants or drugs. Intestinal mucus gel-barrier consisting of a dilute entangled network of macromolecular chains composed of glycoproteins. *Abbreviations*: ASBT, apical sodium-dependent bile acid transporter; PEPT1, peptide transporter 1; MDR1, multidrug resistance gene product 1; MRP2, multidrug resistance-associated protein 2; BCRP, breast cancer-resistant protein; Mct1, monocarboxylic acid transporter 1; UGT, UDP-glucuronosyltransferase; PST, sulfotransferase; GST, glutathione *S*-transferase; P450, CYP, cytochrome P450.

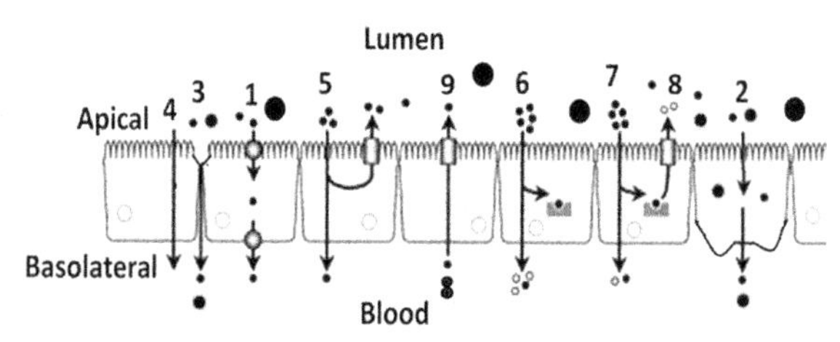

FIGURE 4 Nanoparticle translocation across the gastrointestinal tract. *Reprinted from Chan et al. (2004) with permission.*

1. *Endocytosis*: It allows translocation of nanoparticles less than 50–100 nm in diameter.
2. *M-cell uptake*: M cells are special epithelial cells associated with Peyer's patches and lymphoid follicles that actively take up particulate matter from the intestinal contents. These cells are the portal of entry for large nanoparticles (nanoparticles 20–100 nm and small microparticles >100–500 nm).
3. *Per-sorption*: Passage of nanoparticles through gaps at the villous tip following loss of enterocyte(s) to the gut lumen. Uptake via per-sorption may be size-independent.
4. *Putative para-cellular uptake*: Very small (<5 nm) nanoparticles may pass through the intercellular junction, especially when the junctions are compromised.
5. Absorption via the paracellular pores restricted to smaller particles.
6. Apical efflux transporters actively transport the compounds back into the intestinal lumen, thus restricting their absorption into the blood.
7. Intracellular enzymes metabolize the internalized compounds. The metabolite and the parent compounds enter blood.
8. The metabolites may exit the cells via apical receptors into the lumen.
9. Apical efflux transporters facilitate the intestinal clearance of the metabolized particles from the blood into the lumen.

Suzuki and Sugiyama, 2000; Lin et al., 1999). The major efflux transporters of the ABC family include MDR1 (P-glycoprotein (P-gp), ABCB1), BCRP (ABCG2), and MRP2 (ABCC2). The efflux transporters, in addition to the intestine, are also localized to barrier tissues of the body, such as the liver, kidney, BBB, and placenta. Coordination between the efflux transporters and intracellular metabolizing enzymes forms an effective barrier against the absorption of toxicants.

5. The mucus layer is the first physical barrier to transport of solutes through the intestinal wall (Peppas et al., 1984). Mucus is a highly viscous product secreted by the goblet cells of the small intestine (Forstner et al., 1973).
6. The flow of blood (A and B) and bile (C) to the intestine is much greater than the flow to the stomach; thus, absorption from the intestine is favored over that from the stomach. The GI tract receives blood supply from the following vessels (Figure 5(A)):
   a. The *coeliac artery* divides into the *lineal artery* (for spleen), the *left gastric artery*, and the *common hepatic artery*, which also becomes the right gastric artery.
   b. The *superior mesenteric artery* supplies the whole small intestine and extends branches up to the middle third of the transverse colon.

The portal veins are blood vessels located in the region between two capillary systems (Figure 5(B) and (C-4–6)). Portal veins transport nutrients and toxins absorbed from the intestines and stomach to the liver. Thus, blood is processed in the liver prior to it returning to the heart via the hepatic veins and is pumped into the whole body. The GI tract is also connected to gall bladder (C-4) and liver (C-2) via the bile duct (C-3). Hydrophobic or larger particles can enter the liver via the bile duct (Heckly et al., 1960; May and Whaler, 1958; Dack and Wood, 1927).

These observations suggest that intestinal absorption is a complex process with multiple barriers.

2.1.1.2 PHYSICOCHEMICAL PROPERTIES AND GI ABSORPTION

The physicochemical properties of bulk and nanoparticles are described in detail in Chapter 2. Therefore, the focus of this section is the pH differences in the intestine and its effects on the ionization properties of chemicals. As shown in Figure 6, the pKa value for either the acidic or the basic molecules indicates the amount of unionized molecules that is available for absorption in the stomach and intestine. The Henderson–Hasselbach equation (Figure 6(i)) determines the relative amount of ionized and

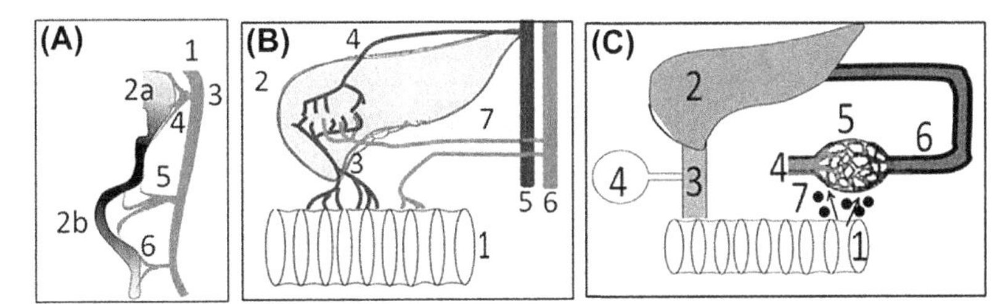

FIGURE 5 (A) and (B): Blood supply to the intestine and liver. (C): Bile duct connects intestine, gall bladder, and liver. A1: aorta, A2a: stomach, A2b: intestine, A3 the coeliac artery, A4: the left gastric artery, A5: the common hepatic artery, which somewhat later becomes the right gastric artery and A6: the superior mesenteric artery supplies the whole small intestine and extends branches up to the middle third of the transverse colon. B1: intestine, B2: liver, B3: portal vein, B4: hepatic vein, B5: vena cava, B6: aorta, and B7: hepatic artery. C1: Intestine, C2: liver, C3: bile duct, C4: gall bladder, C4 to C5: capillary formation, C7: particles absorbed from the intestine enter the capillaries.

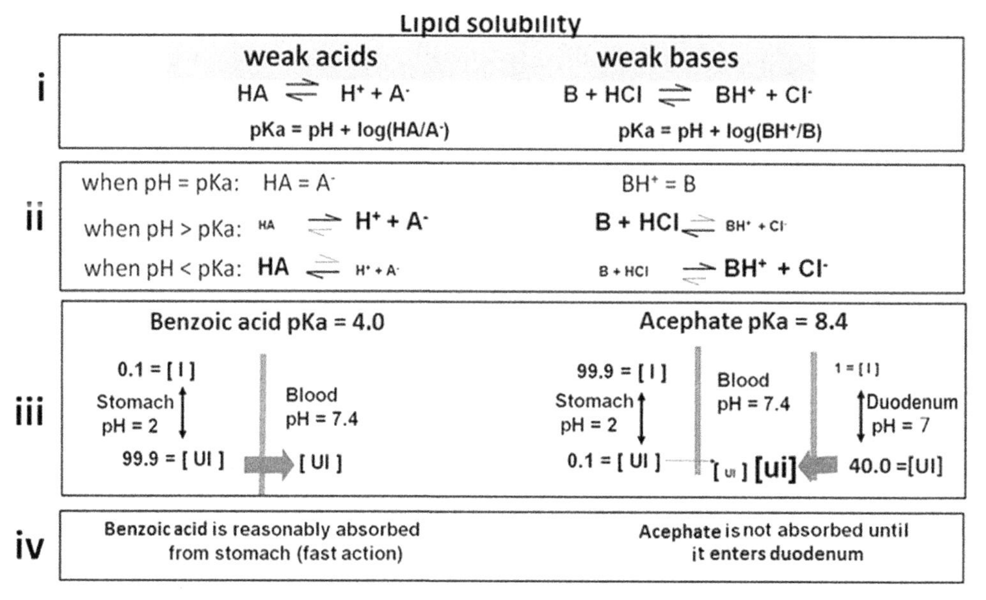

FIGURE 6 Brodie's theory for absorption of acidic and basic chemicals in stomach and intestine. i: Ionization schemes and the Henderson-Hasselbach equation for acids and bases. pKa = pH at which [unionized molecule] = [ionized molecules]. ii: Ionization of acids and bases at pH greater than or smaller than the pKa. Acids' ionization increases, while bases' ionization decreases as the pH increases above their pKa. iii: Acidic toxins (pKa higher than the stomach's pH, but lower than the intestine's pH) remain unionized (thus are absorbed) in stomach but are ionized (remain unabsorbed) in intestine. iv: Basic toxins (pKa higher than the stomach's and the intestine's pH) are ionized (thus remain unabsorbed) in stomach but are unionized (are absorbed) in intestine.

unionized molecules in a solution at a particular pH.

In *Brodie's theory of pH partition*, the possible distribution of unionized and ionized molecules between the blood and the GI tract is shown in Eqn (3) (Brodie et al., 1960; Shore et al., 1957):

$$D = [\text{toxicant}_{\text{blood}}]/[\text{toxicant}_{\text{intestine}}]. \quad (3)$$

Equation (3) is modified to accommodate the ionized and unionized toxins (Eqn (4)), in which U is unionized, I is ionized, $_B$ is blood, and $_G$ is the GI tract.

$$D = \frac{[U]_B + [I]_B}{[U]_G + [I]_G} \quad (4)$$

For an acid with a pKa value of 4 in the stomach (pH of 2), the $[U]/[I]$ will be $10^{\text{pKa}-\text{pH}}$ or $10^{4-2} = 100$ or $[I] = 0.01 \times [U]$. In blood, the $[U]/[I]$ is $10^{4-7} = 0.001$ or $[I] = 1000 \times [U]$. The calculated value of D = 1000 suggests that acids will remain unionized in the stomach, thus translocating in blood where it will be ionized (Eqn (5)):

$$D = \frac{[U]_B}{[U]_G} \cdot \frac{1000 + 1}{0.01 + 1} = 1000 \quad (5)$$

For an acid with a pKa value of 4 in the intestine (pH of 7), the $[U]/[I]$ will be $10^{4-7} = 0.001$ or $[I] = 1000 \times [U]$. In blood, the $[U]/[I]$ is $10^{4-6} = 0.01$ or $[I] = 100 \times [U]$. The calculated value of $D = 10$ suggests that acid absorption in the intestine will be 10-fold lower than that in stomach (Eqn (6)). However, the acid molecules will ionize in blood.

$$D = \frac{[U]_B}{[U]_G} \cdot \frac{1000 + 1}{100 + 1} = 10 \quad (6)$$

The transport of particles across the intestinal epithelium is a complex process; thus, the pH--partition hypothesis provides an oversimplified view of the process. It does not consider the solubility of the chemicals. To address this issue, Dressman et al. (1985) developed the *absorption potential theory*, which takes into account not only the partition coefficient but also the solubility and dose. Using a dimensionless analysis approach, Eqn (7) was designed for determining the absorption potential:

$$AP = \log\left(\frac{PF_{un}}{D_0}\right) \text{ where } D_0 = \frac{C_0}{S} = \frac{D/V_0}{S} \quad (7)$$

In this equation, AP is the absorption potential (the fraction of dose absorbed), P is the partition coefficient, and F_{un} is the fraction in the unionized form at pH 6.5 (the intestinal pH). F_{un} brings in the pH-partition hypothesis in this equation. D_0 in Eqn (7) is a dimensionless dose number and is defined as the ratio of dose concentration to solubility, (C_0/S), where S is the physiological solubility, D is the dose, and V_0 is the volume of water generally set to be 250 mL (Dressman et al., 1985).

Lipinski et al. (1997) *improved the pH partition theory* with the *Rule of Five*. As discussed in Chapter 5, this rule predicts poor oral availability when there are more than 5 H-bond donors, 10 H-bond acceptors, the log$P$ value is greater than 5, and the molecular weight exceeds 500. Lipinski suggested the significance of hydrophilic-lipophilic balance in absorption of toxicants. For highly lipophilic toxicants, absorption may be rate-limited by the inability of toxicants traversing the hydrophilic cytoplasm. For highly hydrophilic or polar toxicants, membrane lipophilicity is the limiting factor. However, when molecules possess both hydrophilic and lipophilic qualities, they permeate the hydrophilic cytoplasm and lipophilic membrane both. In this case, the blood perfusion rate becomes the overall rate-limiting step for absorption.

The pH partition theory explains the overall influence of GI tract pH and a chemical's pKa on the extent of drug transfer or drug absorption. The absorption of a molecule (molecular weight greater than 100) that is primarily transported across the membrane by passive diffusion is governed by three important factors:

1. The dissociation constant (pKa) of the molecule (*assumption*: the larger the fraction of unionized drug, the faster is its absorption)
2. The lipid solubility of the unionized molecule (partition coefficient, $K_{o/w}$; *assumption*: the greater the lipophilicity of the nonionized molecules, the better is the absorption)
3. pH at the absorption site (*assumption*: the GI tract is a simple lipid-bilayer barrier to the transport of molecules across the GI tract)

2.1.1.3 GASTROINTESTINAL ABSORPTION OF NANOPARTICLES: A BRIEF REVIEW

As described in Chapter 2, pristine nanoparticles are highly reactive and either form large aggregates that are not absorbed or disrupt the cell membranes and cause cytotoxicity. Similar to bulk particles, pristine nanoparticles may also encounter the GI tract barrier including the elimination via phagocytosis and the lymphatic system. However, unlike the bulk particles, nanoparticles can be functionalized with different ligands such as polyethylene glycols (PEGs), receptors, transporters, or cell-specific ligands that can assist the particles in bypassing the intestinal barriers (Figure 7). Some of the approaches commonly used to improve the GI absorption of nanoparticles are described here.

Nanoparticles PEGylation enhances their absorption: Nanoparticles, in order to be absorbed in the GI tract (structure is shown in Figure 7) and circulate systemically, must withstand pH changes ranging from 3 (highly acidic) in the stomach to >7 (basic) in the intestine, physical churning, the mucus and the water layer

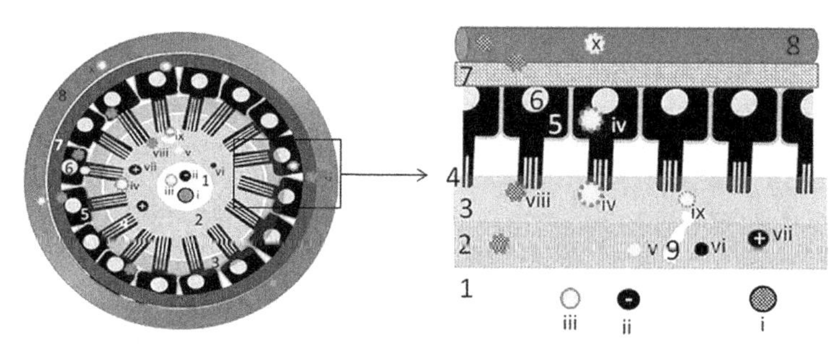

FIGURE 7 Sketch of intestine's cross-sectional view. 1: Lumen, 2: mucus, 3: water layer, 4: micro-villa, 5: enterocyte, 6: enterocyte nucleus, 7: basement membrane, 8: capillaries, i: large (>100 nm) hydrophobic nanoparticles, ii: negatively charged particles remain in lumen, iii: PEG_{12} coupled large nanoparticles, iv: hydrophilic nanoparticles conjugated with PEG_4 adsorbed onto the villa, v: retention of small hydrophilic nanoparticles in mucus, vi: retention of small hydrophobic nanoparticles in mucus, vii: positively charged nanoparticles retained by the mucus, viii: PEG_4-coupled hydrophobic nanoparticles adhere to the villa, ix and 9: PEG_4-coupled nanoparticles negotiate the mucus barrier by disrupting its properties, x: PEG_4-coupled hydrophilic nanoparticles are absorbed by the enterocytes and reach blood capillaries, xi: PEG_4-coupled hydrophobic nanoparticles are absorbed by the enterocytes and reach blood capillaries.

(Figure 7(1 and 2)), and intestinal enzymes that can metabolize the functional groups. Mucus, a neutral to negatively charged heterogeneous anionic gel consisting of glycoprotein and lipid polymers suspended in water, protects the GI tract lining from harmful substances in the diet, while allowing absorption of nutrients. The water layer allows the particles with hydrophilic surfaces to bind the villi and enter enterocytes (Figure 7).

For absorption, a substance, including nanoparticles, must diffuse through the unstirred mucus and travel through the aqueous layer to contact the epithelial cells (Figure 7). Negatively charged nanoparticles (Figure 7(ii)), hydrophilic nanoparticles coupled to PEG_{12} (Figure 7(iii)), or large hydrophobic nanoparticles (Figure 7(i)) may remain in the lumen. Hydrophobic and hydrophilic particles both adhere to the mucus because hydrophobic particles form multiple low-affinity hydrophobic interactions (Figure 7), while hydrophilic bare nanoparticles form weak hydrogen bonds and electrostatic bonds with mucus (Figure 7). The binding of nanoparticles to mucus may delay or prevent their contact with the enterocyte membrane and reduce their absorption. Certain nanoparticles can disrupt the mucus layer and reach the site of absorption (Figure 7(9 and ix)).

Hillyer and Albrecht (2001) and Schleh et al. (2012) have shown that bare (without functionalization) Au nanoparticles, ranging from 2 to 150 nm in diameter, were not absorbed into the systemic blood and that nanoparticles were eliminated via the lymphatic system or the feces. Smith et al. (2013) showed that PEG_4-conjugated Au nanoparticles were effectively absorbed by the GI tract and accumulated in blood, while PEG_{12}- or PEG_{24}-conjugated Au nanoparticles were not. Thus, Au nanoparticles coupled to smaller PEG were absorbed in the GI tract and circulated systemically (Figure 7). Yuan et al. (2013) have also shown that PEG conjugation of solid-lipid nanoparticles (SNL) promoted the cellular uptake of nanoparticles in villi as well as enhanced their permeation across the epithelium due to the PEGylated nanoparticles' penetration of mucus (Figure 7).

Mucoadhesion—a novel means to increase nanoparticle absorption: As discussed earlier, intestinal mucus may retain (via mucoadhesion) and then eliminate certain nanoparticles that do not diffuse through it. However, mucoadhesion is being investigated as a novel means of drug delivery (Sarparanta et al., 2012; He et al., 2012; Dunnhaupt et al., 2011). Studies have shown that mucoadhesive drug delivery systems prolong the residence time of the dosage, resulting in an increase of their absorption and better therapeutic performance. Studies have shown that chitosan, a nontoxic and biocompatible polymer, has found use in a number of drug delivery applications, including a mucoadhesion drug delivery system (Mazzaferro et al., 2012; He et al., 2012; Petit et al., 2012). The limited solubility of chitosan can be improved using chitosan derivatives such as trimethyl chitosan chloride, which has been shown to considerably increase the permeation of neutral and cationic peptide analogs across Caco-2 intestinal epithelia (Thanou et al., 2001, 2000). Chitosan and its derivatives act by reversibly interacting with components of the tight junctions, leading to widening of the paracellular routes. The chitosan derivatives do not damage the cell membranes or alter the viability of intestinal epithelial cells.

2.1.2 *Dermal Absorption*

Dermal absorption allows the passage of compounds across the three skin layers—the epidermis (Figure 8(E)), dermis (Figure 8(D)), and hypodermis (Figure 8(H)), as described in Chapter 5. The dermis and hypodermis, but not epidermis, contain an interconnected network of vessels characterized by regular structures on all levels. Two networks of blood vessels are present in the skin: a superficial network at the interface between the papillary and reticular dermis and a lower network located on the border between the dermis and hypodermis. Vertical vessels connect both networks. The skin's blood vessel system supplies the cells and tissue with nutrients and oxygen, regulates the blood pressure, regulates the body temperature, and provides a barrier against absorption. Dermal absorption can be divided into three steps (Schaefer and Schalla, 1980):

- *Penetration* is the entry of a substance via the skin layers. The compound or particle may accumulate in a particular layer of stratum corneum (Figure 8(2)) or enter the blood supply (Figure 8(5)).
- *Permeation* is the movement of chemicals through one layer (Figure 8(3 and 4)) into another via hydrophilic channels (Figure 8(1)).
- *Resorption* is the uptake of a substance into the vascular system (lymph and/or blood vessel), which acts as the central compartment (Figure 8(5)). Dermal absorption is the total sum of the three processes.

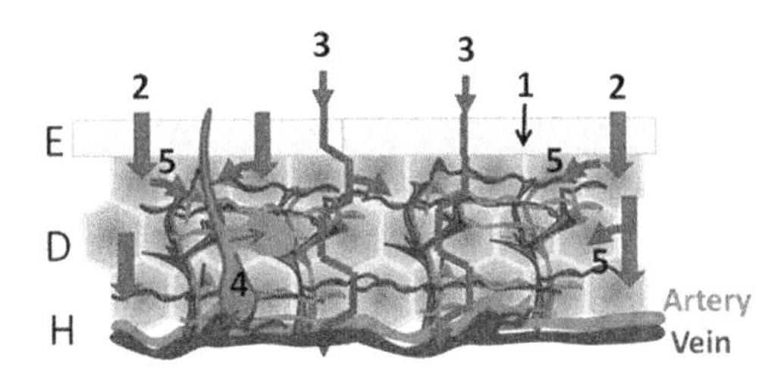

FIGURE 8 Mechanism of dermal transport. A sketch of skin cross-section showing epidermis (E), dermis (D), and hypodermis layers, main artery and vein, blood capillaries penetrating different layers of skin, and hydrophilic intercellular channels (8-1). Chemicals or particles are absorbed by the skin via penetration (8-2), permeation (8-3: intercellular routes and 8-4: hair shaft route), and resorption (8-5).

2.1.2.1 FACTORS AFFECTING DERMAL ABSORPTION

Factors affecting dermal absorption include the following:

The applied dose: Wester and Noonan (1980) have shown that total absorption increases, while the specific absorption (% of the initial dose) decreases with an increase in the applied dose.

Formulation: Chemicals penetrate the skin in two stages: first, they diffuse out of the vehicles and then they bind and enter the skin (Ostrenga et al., 1971). Formulations can modulate either diffusion or penetration of chemicals. Bonina et al. (1995) have shown that chemicals in a lipid formulation exhibited greater skin penetration than in a nonlipid matrix.

Physical state: Chemicals can penetrate skin in both solution and powder states, although chemicals in solution are more effectively absorbed than chemical powder (Higuchi, 1960; Roberts et al., 2002; Romonchuk and Bunge, 2003).

Size and molecular weight: Permeation decreases exponentially with molecular weight (Kasting, 1987; Potts and Guy, 1992; Magnusson et al., 2004). Compounds with molecular weight over 500 D poorly permeate through a normal human skin (Bos and Meinardi, 2000; Hostjnek, 1997).

Maximum flux (maximum flow rate per unit area): The flux (J) of a chemical across the cell membranes is negatively related to the concentration, as described by Fick's law of diffusion (Eqn (8)). The negative sign indicates the direction from high concentration to low concentration.

$$J\left(\frac{\text{mass}}{\text{cm}^2\cdot\text{s}}\right) = -D\frac{\delta C}{\delta x} \quad (8)$$

The maximum J values of a chemical's dermal permeation occur at the solubility limit of the chemical (Roberts et al., 2002). Measurement of flux is a better indicator of absorption than the chemical's physicochemical properties. In normal skin, dermal absorption is not larger than the maximum flux. However, this can be achieved if the skin is damaged or the solute is in some unusual supersaturated solutions (Davis and Hadgraft, 1991).

Ionization: The stratum corneum permeability coefficients for nonionized chemicals are manyfold greater than the coefficients for their ionized counterparts (Vecchia and Bunge, 2003). Thus, ionization inhibits dermal absorption.

Site of application: Dermal absorption by skin exhibits high-region-dependent variations. Earlier studies have shown the forearm skin to forehead skin absorption ratio for parathion (an organophosphate insecticide) and hydrocortisone to be 1:4 and 1:6–13, respectively (Guy and Maibach, 1984; Maibach et al., 1971; Fredriksson, 1961). This suggests that the forehead skin is more permeable than the forearm skin to both parathion and hydrocortisone.

Condition of skin: Although the outer stratum corneum is lipophilic, the intercellular channels and cytoplasm are hydrophilic. In dry skin, the intercellular pores are impermeable to chemicals. Wetting the pores increases their absorption into and through the skin. In addition, lipid solubility is required initially, followed by water solubility, to pass through the water-based gel portion of the skin (Stoughton, 1965; Cronin and Stoughton, 1962, 1963).

2.1.2.2 MODELING DERMAL ABSORPTION

Dermal absorption is characterized by the following indices (Berge, 2009):

- *Permeation coefficient* (K_P), which is a measure of the conductance of skin to a particular chemical from a particular vehicle (Eqn (9))
- *Time lag* (t_{lag}), which is the time during which a steady-state absorption is established (Eqn (10))
- *Diffusivity* (D), which is a measure of how quickly a substance diffuses through a medium. D_{sc} is how quickly a substance diffuses through the stratum corneum layer (Eqn (11)). As shown in Eqn (12), K_P, t_{lag}, and D_{sc} are interrelated indices, determining the total dermal uptake.

$$\text{Permeation coefficient } (K_P) = \frac{P_{sc/w} * D_{sc}}{h_{sc}} \text{ (cm/h)} \quad (9)$$

$$\text{Lag time } (t_{lag}) = \frac{h_{sc}^2}{6D_{sc}} = \frac{h_{sc} * P_{sc/w}}{6K_P} \text{ (h)} \quad (10)$$

$$\text{Diffusivity stratum cornetum } (D_{sc}) = \frac{h_{sc}^2}{6t_{lag}} = \frac{h_{sc} * K_P}{P_{sc/w}} \text{ (cm}^2\text{/h)} \quad (11)$$

$$\text{Dermal uptake (mg)} = K_P\ (\text{cm/h}) \times \text{Conc (mg/mL)} \times \text{Contact time (h)} \times \text{Contact area } (\text{cm}^2) \quad (12)$$

In these equations, $P_{sc/w}$ is the (stratum corneum/water) partition coefficient, D_{sc} is the diffusivity of the stratum corneum, and h_{sc} is the thickness of the stratum corneum.

2.1.2.3 DERMAL ABSORPTION OF NANOPARTICLES: A SHORT REVIEW

As discussed earlier, nanoparticles penetrate the skin via (1) simple diffusion through the epidermal membrane; (2) the hair shaft, hair follicle, sebaceous gland, and sweat gland pores; and (3) intercellular hydrophilic channels (Scheuplein, 1966). Small lipophilic or ionic nanoparticles (<5 nm) penetrate the skin passively (Barry, 2001; Laresea et al., 2009). Baroli et al. (2007) showed that metallic (iron) nanoparticles <10 nm in diameter can penetrate the skin through the stratum corneum lipidic matrix, reaching the deepest layers of the stratum corneum, but they did not permeate the skin. Larger nanoparticles (greater than 10 nm) penetrate the skin via the pores or channels. Particles with a diameter ranging between 7 and 200 nm have been shown to be present in the hair follicle infundibulum (Filon et al., 2009, 2007, 2004; Vogt et al., 2006; Meidan et al., 2005; Alvarez-Romàn et al., 2004; Schaefer et al., 1990) Elastic nanoparticles of 100–150 nm could penetrate the stratum corneum matrix through channel-like structures (Honeywell-Nguyen et al., 2004; van den Bergh et al., 1999a,b). Accordingly, dermal exposure may pose significant health risks to people (Donaldson et al., 2006; Geys et al., 2006, 2007; Lam et al., 2004; Magrez et al., 2006; Muller et al., 2005; Nel et al., 2006; Oberdorster et al., 2005; Rotoli et al., 2008; Shimada et al., 2006; Shvedova et al., 2005). However, it is difficult to compare the results of these studies due to differences in their experimental methodologies.

2.1.3 *Absorption from the Respiratory Tract*

Ingestion and inhalation are two of the most common routes of exposure to a drug, toxicant, or nanoparticle. However, ingestion, but not inhalation, subjects the toxicants, particles, or drugs to first-pass metabolism in the liver, resulting in their metabolism prior to reaching the systemic circulation. In addition, the lungs have only a small fraction of the efflux transporter activity of the gut and liver (Tronde et al., 2003; Chun-mei et al., 1995; Keith et al., 1987). In organisms orally exposed to chemicals, the systemic blood, before reaching the organs, may contain the parent as well as the transformed chemicals; while, in organisms exposed via inhalation, the systemic blood may contain mostly the parent, unaltered chemicals or particles. Some of the unique feature of absorption via the lungs include the following.

1. The lungs are far more permeable to macromolecules than any other portal of entry into the body (Patton et al., 2004, 1999; Patton, 1996; Schanker, 1978).
2. Proteins are more rapidly absorbed via lungs than via subcutaneously injection (Patton et al., 1999).
3. Small hydrophobic molecules are absorbed rapidly after inhalation (Hung et al., 1995).

In terms of toxicity, oral ingestion, because of the uncertainty associated with first-pass metabolism, presents a bigger challenge than inhalation (Lim and Shen, 2005). First pass in the liver and gut may make a particle inactive or more toxic and reduce its bioavailability by excretion of conjugated metabolites (discussed previously and in Chapter 5).

In lungs, chemical and particle absorption occur in the following stages: deposition of

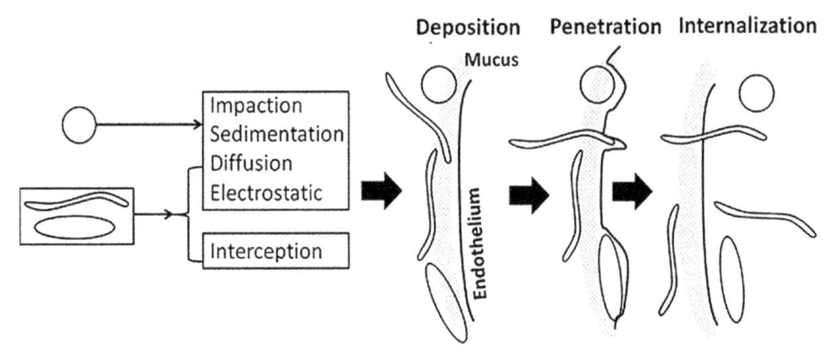

FIGURE 9 Particle adsorption onto and penetration into the endothelial cell lining. Particle absorption occurs in the following stages: deposition of particles at the cell surface, their penetration into the cell membrane, and then their internalization and dissolution into the blood capillaries.

particles at the cell surface, their cell membrane penetration, and then their dissolution into the blood capillaries (Figure 9). The deposition in the respiratory tract is dependent on the following factors:

1. The material's physical-chemical characteristics
2. The host factors, such as the anatomy of the respiratory tract (diameter of the airways, branching angle of airways and tidal volume) and health status, such as bronchoconstriction, hypersecretion of mucus, etc. (Fievez et al., 2009; Frampton et al., 2004; Oberdorster et al., 1994, 2002; Daigle et al., 2003; Borm, 2002).

The mechanisms that characterize deposition of spherical and nonspherical particles in the lungs are shown in Figure 9. The cell membrane penetration mechanisms are described in Chapter 5. The following section describes the deposition of spherical and nonspherical particles onto the respiratory track's endothelium.

2.1.3.1 ANATOMY OF AND AIRFLOW IN THE RESPIRATORY TRACT

The respiratory tract consists of an upper tract (nose, nasal cavity, mouth, pharynx, and larynx), middle track (lungs and includes the trachea, bronchi, and bronchioles) and lower tract (trachea, bronchi, bronchioles, alveoli, and muscles of respiration; Figure 10).

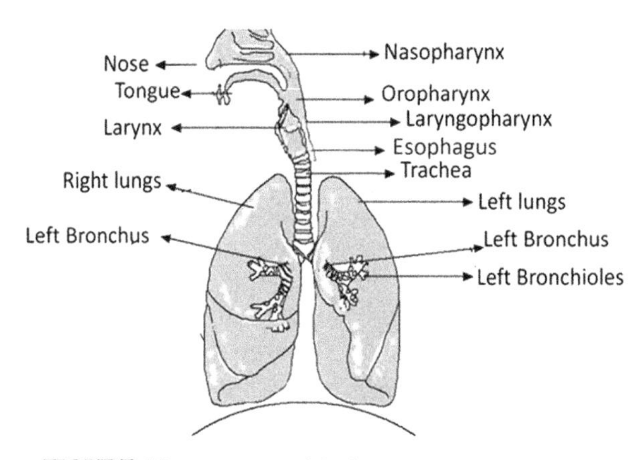

FIGURE 10 Anatomy of the human respiratory system.

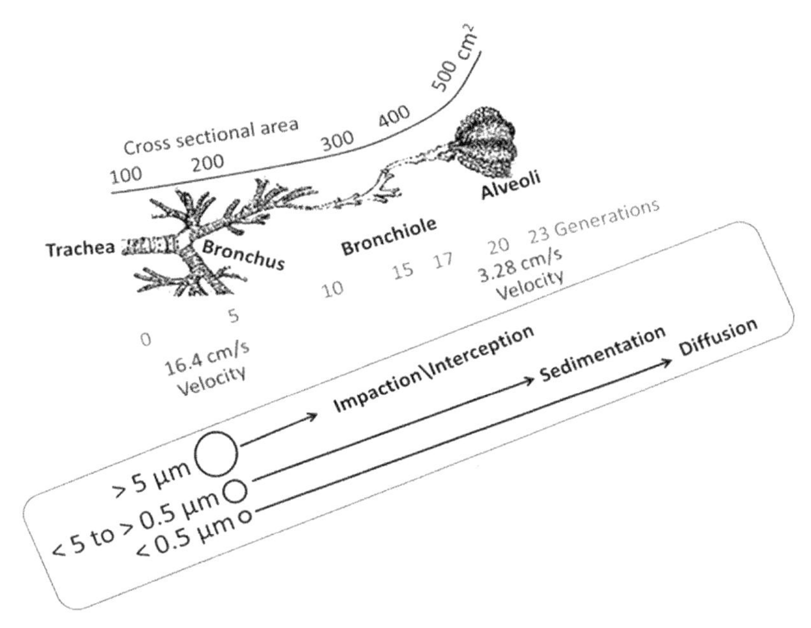

FIGURE 11 Microanatomy of the bronchial tree from trachea to alveoli including their cross-sectional area, generations and air velocity and particle distribution. Particles larger than 5 μm deposit due to impaction, particles smaller than 5 but larger than 0.5 μm deposit due to sedimentation and particles less than 0.5 μm deposit in alveoli via simple diffusion.

Figure 11 shows a cross-sectional area of the lower tract. The cross-sectional area increases as the distance from the trachea increases. The surface area of peripheral branches (generations 17–23) is about five-fold greater than the surface area of the central branches. During inspiration, as air flows from the nose (or mouth) through the conducting tracks into the alveolar sac, the rate of the air flow decreases and reaches to almost zero in the alveolar regions (Figure 11). Particles larger than 5 μm, 0.5–5 μm, and less than 0.5 μm undergo impaction, sedimentation, and diffusion, respectively. Nonspherical (fibrous and platelets) particles also undergo interception, in addition to the factors listed above.

2.1.3.2 MECHANISMS OF PARTICLE DEPOSITION

Most of the earlier models characterizing particle deposition were designed for impaction, sedimentation, or diffusion of spherical particles. The models for nonspherical particles are based on the concept of *aerodynamic diameter* (d_{ae}) based on the diameter of a unit-density (1 g/cm^3) sphere with exactly the same aerodynamic volume and properties as the particle of interest (Baron and Willeke, 1964; Hinds, 1999). In the following sections, the significance of each model is discussed.

2.1.3.2.1 MODELS DEVELOPED FOR SPHERICAL PARTICLES

1. *Impaction*: Abrupt changes in the direction of air flow and the ensuing inertia forces allow adsorption of particles onto the lung's surface, according to Stokes' (Stk) law (Eqn (13); Darquenne, 2012):

$$\text{Stk} = \left(\rho_p d_p^2 u\right) / (18 \mu d) \tag{13}$$

In this equation, the d_p, $\boldsymbol{\rho}_p$, u, $\boldsymbol{\mu}$, and d indices are the particle diameter, particle density, the mean velocity of the carrier gas, dynamic viscosity of the carrier gas, and a characteristic length equal to the diameter of the airway, respectively. The higher the Stk number, the more likely that particles will deposit by inertial impaction. Because u, μ, and d are usually constant, the d_p and ρ_p may determine a particle's impaction.

2. *Sedimentation*: Particles can sediment from the air when the air speed is lower than the settling velocity (V_s) of the particle (Eqn (14); Darquenne, 2012):

$$V_s = \frac{\rho_p d_p^2}{18\mu} g \qquad (14)$$

The equation suggests that sedimentation, similar to impaction, is dependent on air velocity as well as size and on the relative density of the particle.

3. *Diffusion—Brownian motion*: Collision of particles in the air or water with each other or with other molecules initiates random movement that allows the particles to move from the lumen to the epithelium. Diffusion becomes the primary force of deposition when the airflow velocity is low and the particles are light and small. In the alveolar region, the air speed is almost zero; thus, impaction and sedimentation are not possible.
4. *Electrically charged particles*: Particles can interact with the epithelium due to charge differences. In this mode, the deposition may differ from the expected deposition because the regional deposition will be based on characteristics such as size, shape, and density.

2.1.3.2.2 MODELS DEVELOPED FOR NONSPHERICAL PARTICLES The deposition of nonspherical particles at the bronchial airway, especially where it bifurcates, is determined by the particles' orientation (parallel or perpendicular) relative to the direction of the air stream. Studies have shown that nonspherical particles may rotate around their center of gravity (Myojo and Takaya, 2001; Zhang et al., 1996; Asgharian and Yu, 1988; Cai and Yu, 1988; Yeh and Schum, 1980). As the air flow direction changes, smaller particles are able to track the flow, but the larger particles, because of their inertia, keep the initial direction and hit the airway's epithelial walls (*interception*). Some rigid fibers may penetrate the cells and induce local inflammation. As the smaller particles move to the periphery, the flow decreases due to an increase in volume, becoming almost zero in the alveolar region. Thus, smaller particles begin to sediment or diffuse to the surface, resulting in further deposition.

As mentioned, d_{ae} (Eqn (15)) can be used to determine the deposition of nonspherical particles.

$$d_{ae} = \sqrt{\frac{1}{\chi} \cdot \frac{\rho_p}{\rho_0} \cdot \frac{C_c(d_{ev})}{C_c(d_{ae})}} \qquad (15)$$

In this equation, d_{ev} denotes the equivalent volume diameter (the diameter of a sphere for example, fullerene for carbon nanotubes) with exactly the same characteristics and volume as the particle of interest (e.g., carbon nanotubes), whereas χ, ρ_p, ρ_0, $C_c(d_{ev})$, and $C_c(d_{ae})$, respectively, represent the dynamic shape factor, the density (g/ cm^3) of the particle of interest, unit density (g/ cm^3), the Cunningham slip correction factor for a particle with diameter d_{ev}, and the correction factor for a particle with diameter d_{ae}. As shown in Figure 13, the dynamic shape factor (χ) is closely related to the aspect ratio (β). For spheres, both β and χ are 1. An increase and decrease in the aspect ratio increase the χ values. The dynamic shape index, χ, has two components: the χ_{II} that represents the nonspheres that travel parallel to the airflow and $\chi_{\perp}$ that represents the nonspheres that travel perpendicular to the

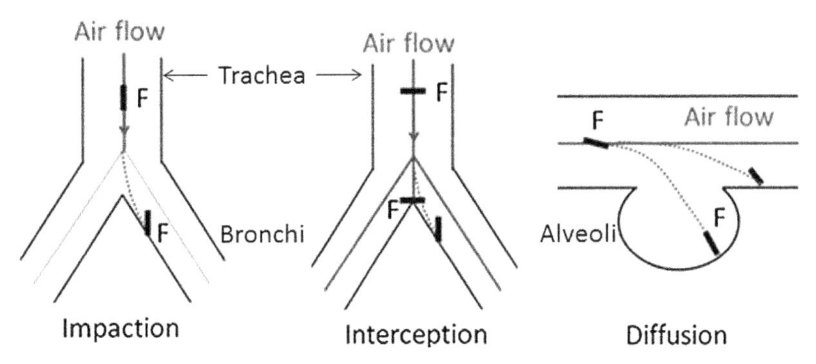

FIGURE 12 Particle deposition onto the endothelial mucosa. As shown in Figure 11, air velocity is high in trachea and decreases as air travels down the tract with essentially stagnant air in alveoli. Thus, nanoparticles may deposit via impaction, interception, sedimentation, and diffusion according to the air velocity and the nanoparticles shape, and their orientation within the air stream. *Impaction*: If a 1D particle (such as nanotube) is oriented parallel to the airflow, then, as the airflow direction changes in bronchi, the particles, because of their inertia, keep the initial path and deposit onto the airways epithelial walls. Some rigid fibers may penetrate the cells and induce local inflammation. *Interception*: If the particle orients perpendicular to the airflow, it is intercepted upon reaching close to the surface cells of the airway passages. This method of deposition is most important for fibers such as asbestos Sturm (2009). The fiber length determines where the particle will be intercepted or impacted. *Sedimentation and diffusion*: As the smaller particles move to the periphery, the flow decreases due to an increase in volume, becoming almost zero in alveolar region. Thus, smaller particles begin to sediment or diffuse to the surface, resulting in further deposition. Equations (16)–(18) shown are for deposition of particles in lungs. The χ_{Π} equation represents the nonspheres that travel parallel to the airflow, while $\chi_{\perp}$ represents the nonspheres that travel perpendicular to the air flow.

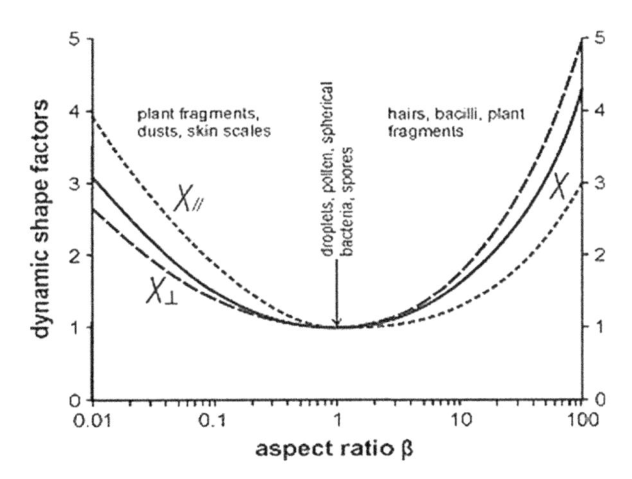

FIGURE 13 The particle's aspect ratio versus dynamic shape factor plots for oval discs (ratio < 1), spheres (ratio = 1) and nanotubes or fibers (ratio > 1). The χ_{Π} plot represents the nonspheres that travel parallel to the airflow, $\chi_{\perp}$ plot represents the nonspheres that travel perpendicular to the airflow as shown in Figure 12. The χ values decrease as the aspect ratio increases, with a minimal value of 1 occurring at aspect ratio 1 (sphere). An increase in the aspect ratio thereafter increases the χ values. *Reprinted from Sturm (2012, 2010a, 2010b) with permission.*

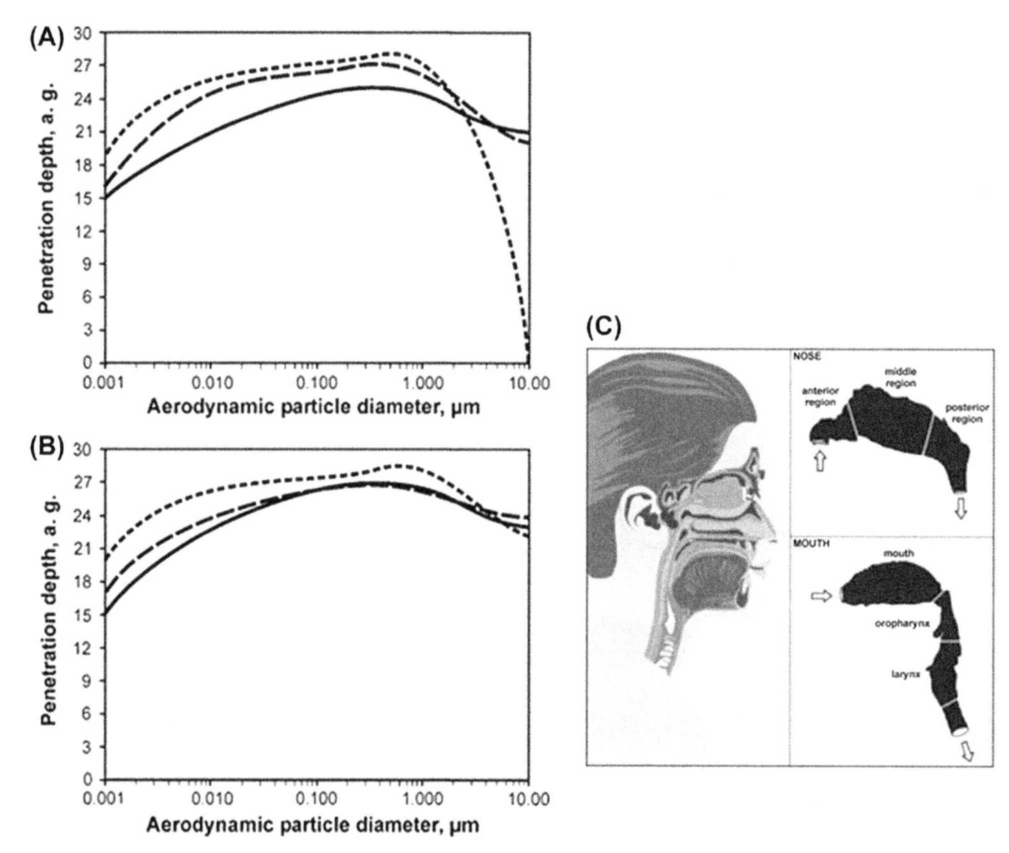

FIGURE 14 Penetration depth (deposition >0.01%) and its dependence on aerodynamic particle diameter (sitting breathing: solid line; light-work breathing: dashed line; heavy-work breathing: dotted line): (A) inhalation through the nose, (B) inhalation through the mouth (e.g., is airway generation), and (C) anatomy of nose and mouth. *Reprinted from Sturm (2012) with permission.*

airflow, as shown in Figure 12. Equations (16)–(18) show the relationship between χ, χ_{Π}, and $\chi_{\perp}$.

$$\frac{1}{\chi} = \frac{1}{3\chi_{\Pi}} + \frac{3}{2\chi_{\perp}} \tag{16}$$

$$\chi_{\Pi} = \frac{\frac{4}{3}\cdot\left(\beta^2 - 1\right)\cdot\beta^{\frac{1}{3}}}{\left\{\frac{(2\beta^2-1)}{\sqrt{\beta^2-1}}\cdot\ln\left[\beta + \sqrt{\beta^2 - 1]} - \beta\right]\right\}} \tag{17}$$

$$\chi_{\perp} = \frac{\frac{8}{3}\cdot\left(\beta^2 - 1\right)\cdot\beta^{\frac{1}{3}}}{\left\{\frac{(2\beta^2-1)}{\sqrt{\beta^2-1}}\cdot\ln\left[\beta + \sqrt{\beta^2 - 1]} - \beta\right\}} \tag{18}$$

Sturm (2012, 2010a,b) has shown that particles with $d_{ae} = 0.5$ may exhibit the highest penetration depth, up to the alveolar region, while particles with $d_{ae} = 0.001$ and $d_{ae} = 10$ are characterized by a rather limited ability to penetrate the lung (penetration limited to generations 15–18) (Figure 14). Inhalation from the nose and mouth may exhibit different levels of penetration because of their anatomical differences. The penetration depth of diverse size particles is modulated by the breathing mode: sitting, light-work, or heavy-work breathing. During nose breathing, light and heavy breathing increased the penetration depth of 0.001–1 µm particles, but decreased the penetration depth of 10-μm particles. The penetration depth was almost zero for a 10-μm particle under heavy activity. However, during mouth breathing, the penetration changes during sitting, light activity, and heavy activity were fairly uniform.

The differences between nose and mouth breathing may be related to their anatomical differences. The nasal cavity consists of several flow-splitting conches, with a 90° turn in the airflow stream in the posterior nasopharynx. Thus, most of the inhaled particulate mass is already filtered in this extrathoracic compartment (Figure 14(C)).

2.1.3.3 ABSORPTION OF NANOPARTICLES VIA THE RESPIRATORY SYSTEM

The respiratory system represents a unique target for efficient drug delivery and the potential toxicity of nanoparticles (Hickey, 2014). There is potential for exposure of the lungs to countless numbers of nanoparticles, such as carbon-based (e.g., nanotubes, nanowires, fullerenes), metal-based (e.g., gold, silver, quantum dots, metal oxides such as titanium dioxide and zinc oxide), polymeric dendrimers, and biological (e.g., chitosan) (Bharatwaj et al., 2014; Baek et al., 2011). Exposure may occur via inhalation and via any other exposure routes, resulting in systemic distribution because the lungs receive the entire cardiac output. The respiratory route is commonly used to deliver contrast agents for imaging and for drug delivery (Borm and Kreyling, 2004; Borm, 2002; Pandey et al., 2003; Sung et al., 2007). Studies have shown that the polysaccharide coating of nanoparticles may provide steric protection against the particles' protein binding and phagocytosis, thus prolonging their systemic residence time. Absorption may depend not only on their size but also on the properties of their coating layer, which may allow targeting to specific tissue, such as lymph nodes or brain tumors. For respiratory tract absorption, the nanoparticles, in addition to appropriate aerosolization, may also require an advanced carrier system for safe and efficient delivery to their target. These carriers could be either large porous nanoparticles with a low aerodynamic diameter or lectin-functionalized liposomes (Ehrhardt et al., 2002).

2.2 Distribution

Distribution of chemicals and particles in body is defined as the reversible accumulation of the particles in different organs and tissues. Once chemicals or particles enter the systemic blood, they distribute throughout the body and/or are eliminate from the body. Typically, distribution is influenced by regional blood flow (% of total cardiac output: brain, 14; bones and other tissues, 9; GI tract and liver, 27; heart muscles, 4; kidneys, 20; lungs, 100; skeletal muscle, 21 (can be up to 75% during exercise); skin, 5) and is much more rapid than elimination. In general, chemicals or particles are distributed in three compartments:

1. The *central compartment* includes the well-perfused organs, such as the heart, blood, liver, brain, and kidney, in which chemicals or particles equilibrate rapidly.
2. The *peripheral compartment* includes the less-perfused organs, such as adipose tissues and skeletal muscle. Equilibration occurs more slowly in this compartment. Although redistribution from one compartment to another is possible, it often alters the duration of effect at the target tissue.
3. The *special compartments* include the brain (due to the BBB), pericardial fluid, bronchial secretions, and middle ear fluid. The entry of chemicals is restricted in these sites because of tight junctions in the barrier or restricted blood flow. Figure 15 describes an overview of chemical's distribution.

2.2.1 Fate of Chemicals and Particles in Blood and Cells

Many chemicals and nanoparticles rapidly bind to plasma proteins and form a reversible or irreversible complex called a *corona*, which

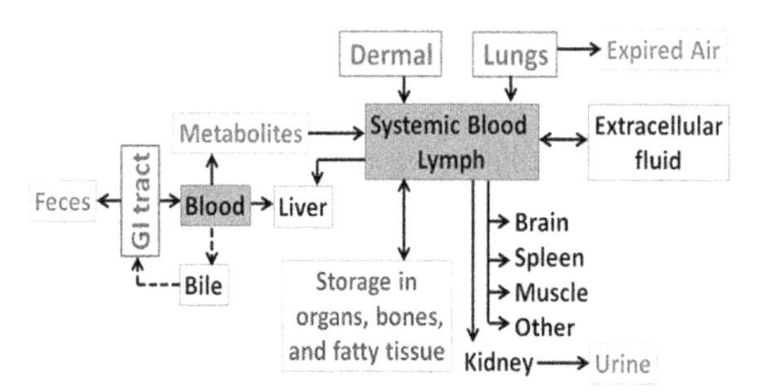

FIGURE 15 Flow diagram shows biodistribution and fate of nanoparticles administered via GI tract, lungs, and skin.

alters their surface properties. Nanoparticles functionalized with weak acidic and neutral ligands may bind to albumin and related proteins, in particular, while the particles functionalized with basic ligands tend to bind to alpha-1-acid glycoprotein and related proteins (Rowland et al., 2004; Rowland and Tozer, 1989). Some nanoparticles may also adhere to the blood cell's surface proteins. This suggests that a chemical or a particle in blood exists in two forms: protein-bound and free form. If the protein binding is reversible, then a chemical equilibrium will exist between the bound and unbound forms. The unbound fraction (*fu*) is calculated by Eqn (19) where $[D]$ is the free drug concentration and $[DP]$ is the concentration of drug protein complex. The percentage of protein binding can be determined using Eqn (20), where $\%PB$ is the percentage of bound drug. Protein-bound chemicals and particles are protected from the blood enzymes.

$$fu = \frac{[D]}{[D] + [DP]} \tag{19}$$

$$\%PB = 100 \times \frac{[DP] - [D]}{[DP]} \tag{20}$$

2.2.1.1 EFFECTS OF PROTEIN BINDING ON DRUG DISTRIBUTION

One of the basic rules of chemical distribution in body is that only the free (unbound) chemicals/particles can penetrate cell membranes. Protein binding therefore decreases the net transfer of drugs/particles across cell membranes. Binding of bulk particles to plasma proteins is generally reversible, while binding of nanoparticles to proteins can be irreversible or replaceable with other proteins. Because the protein concentration of extravascular fluids (e.g., cerebrospinal fluid, lymph and synovial fluid) is very low, at equilibrium, the total drug concentration in plasma is usually higher than that in extravascular fluid.

2.2.1.2 EFFECTS OF PROTEIN BINDING ON DRUG ELIMINATION

Protein binding decreases the rate of elimination of drugs/particles via excretion by renal glomerular filtration. The rate of renal excretion of protein-bound drugs is inversely related to their extent of plasma protein binding. However, protein binding may enhance drug elimination via hepatic or biliary circulation.

2.2.1.3 BINDING OF CHEMICALS AND NANOPARTICLES TO CELL MEMBRANES

Binding of chemicals and nanoparticles to tissue proteins may cause local accumulation at the binding site. For example, if a drug binds more extensively to the intracellular proteins than the extracellular sites, then the intracellular and extracellular concentrations of free drug may be equal, but the total intracellular drug concentration may be much greater than the total extracellular concentration (Figure 16).

2.2.2 *Apparent Volume of Distribution (Vd)*

The volume of distribution is a nonphysiological parameter describing a relationship between the total amount of a drug in the body (example: 100 mg) and the drug concentration in the plasma compartment (example: 10 mg/L) after a single intravenous injection. In general, V_D (unit: L) = (total chemical amount in the body (unit: mg)/ plasma concentration (unit: mg/L)). In case of the data shown above, $V_D = 100$ mg/10 mg/ L = 10 L. Some of the examples of V_D values

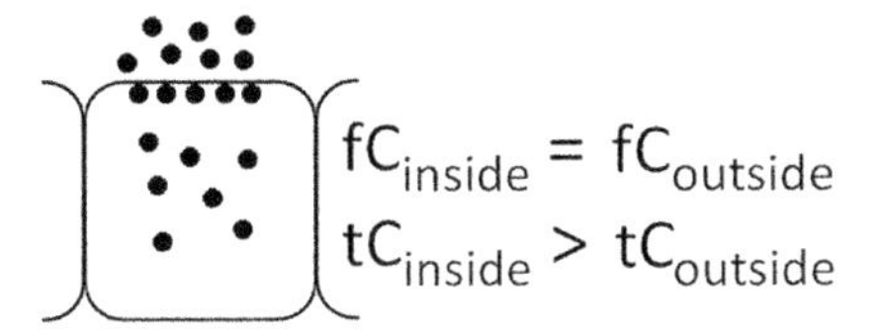

FIGURE 16 Extracellular-intracellular equilibration of nanoparticles. If chemicals do not bind to the cell membrane, extracellular fluid proteins, or intracellular fluid proteins, then free and total chemicals inside are equal to that outside the cells ($fC_{inside} = fC_{outside}$ and $tC_{inside} = tC_{outside}$). However, if the chemicals bind to the cell membrane, then the free chemicals concentration inside the cell is equal to that of outside the cells, while total concentration inside is greater than those outside ($fC_{inside} = fC_{outside}$ and $tC_{inside} > tC_{outside}$).

are: heparin, 5 L; warfarin, 10 L; phenytoin, 44 L; diazepam, 168 L; digoxin, 420 L; and chloroquine, 13,000 L.

How can V_D be 168, 420, or 13,000 L when total plasma volume in humans is around 40 L? This is because V_D is just a ratio of two numbers, a nonphysiological parameter. Figure 17 shows the V_D values for three compounds:

Compound 1 (Figure 17, left) exhibits 1% plasma bound and 1% cell absorption.

Compound 2 (Figure 17, center) exhibits 50% plasma bound and no cell absorption.

Compound 3 (Figure 17, right) exhibits 50% cell absorption and no plasma binding.

A total of 100 mg of each compound was added in a glass tube containing equal volumes of plasma and red blood cell suspension. The V_D values for total and free compounds are shown in Figure 17. As the free compound concentration decreases, V_D increases. This is because V_D is the volume of plasma needed to achieve 100 mg of the total body compound. As the plasma concentration decreases, more plasma is needed for 100 mg of the compound.

How does V_D relate to the accumulation of chemical and particles in body? It is an axiom that the distribution of a chemical or particles in the body depends on its physicochemical properties, the blood flow rates in various tissues, and plasma protein binding and tissue uptake. In general, extensive tissue uptake is characterized with a high volume of distribution, as is seen for basic lipophilic drugs that accumulate in fatty tissues. Lysosomal trapping and phospholipid

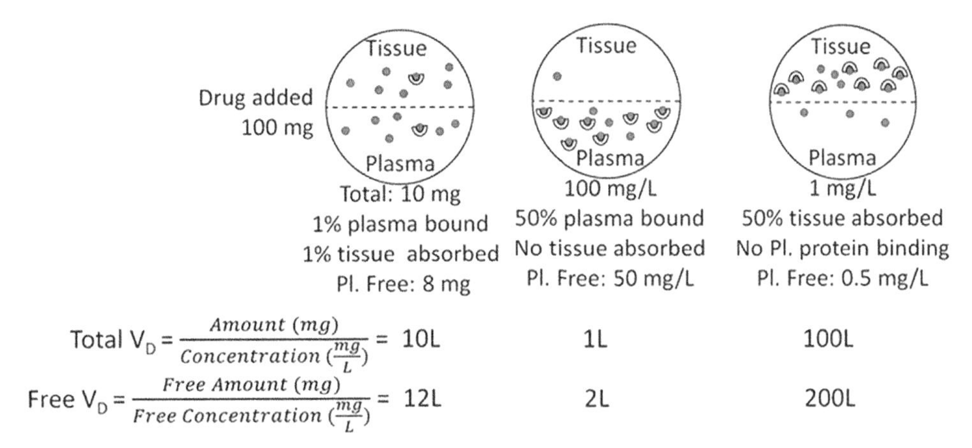

FIGURE 17 Distribution of chemicals between plasma and tissue, total volume of distribution (total V_D and free chemical volume of distribution free V_D). Left: When chemicals mostly remain in free form in blood and tissue both, they equally distribute between them. When only 1% of the chemicals remain in bound form, the total V_D is 10 L and free V_D is 12 L. Center: When 50% of the chemicals in plasma is protein-bound and no binding occurrs in tissues, the V_D is 1 L for total and 2 L for free chemical. Right: When 50% of the chemicals in tissue is bound and no binding occurrs in plasma, the total V_D is 100 L and free V_D 200 L.

binding are two key mechanisms of distribution of lipophilic drugs. Certain drugs, such as neuroleptics and antidepressants, may competitively inhibit their lysosomal uptake, thus reducing their distribution (Daniel, 2003).

Challenges in studying nanoparticle's biodistribution: The application of nanotechnology to the area of drug development is increasing rapidly because of their ability to overcome the biological barriers, effectively deliver hydrophobic drugs and biologics, and preferentially target sites of disease. Despite these potential advantages, only a few nanoparticle-based medicines have received approval for clinical use. This may be because of the numerous challenges and hurdles at different stages of development and characterization of nanodrugs. A review of the literature shows considerable variations in the volume of distribution and biodistribution of nanoparticles. Some examples are described below:

- Metal and polymer nanoparticles exhibited greater distribution in different organs than their bulk counterparts did. The biodistribution was dependent on the size, shape, and surface functionalization of the nanoparticles (Browning et al., 2013; Simpson et al., 2013; Tan et al., 2013; Yue et al., 2013; Liu et al., 2012; Khlebtsov and Dykman, 2011; Sonavane et al., 2008; Longmire et al., 2008; Balogh et al., 2007).
- Hirn et al. (2011) showed that the size-dependent increase in nanoparticle biodistribution was dependent on their volumetric specific surface area, which was related inversely to the diameter:

$$VSSA = \left(\frac{2}{d}\right) \cdot \left(\frac{1}{h} + \frac{1}{r}\right)$$

- Li et al. (2011) showed that bulk paclitaxel and SLN-coupled paclitaxel exhibited comparable V_D and accumulations in various tissues. Nanotization did not show any beneficial effects.
- Purvin et al. (2014) showed that after ingestion, a nano-AZT engineered to target specific sites accumulated in tissues via lymphatic circulation, thus avoiding blood circulation. The nano-AZT, but not bulk AZT, entered the brain. This presents a unique problem because PK calculations depend on blood drug concentrations. A bypass of the blood circulation will yield unusually high V_D values and a false impression that the compound is accumulating in tissues.
- Chu et al. (2013) studied the bioaccumulation in tumor-bearing rats of two-dimensional (2D) taxotere nanoparticles (dimensions: 200 nm length and 200 nm depth or 320 nm length and 80 nm depth) using poly-(lactide-co-glycolide). The V_D values were 8509, 257, and 474 mg/kg for bulk taxotere, $taxotere_{200\times200}$ and $taxotere_{80\times320}$, respectively. In addition, $taxotere_{80\times320}$ exhibited smaller accumulation than $taxotere_{200\times200}$ in the spleen, liver, and lungs, but greater accumulation in tumor cells and blood. The greater accumulation of $taxotere_{200\times200}$ in the spleen may be due to its ingestion by macrophages. Champion et al. (2008) have shown that nanoparticles are subject to phagocytosis in a size-dependent manner: larger particles are ingested more efficiently than smaller particles. Zhang et al. (2013, 2009) have shown that galactosylated trimethyl chitosan-cysteine nanoparticles selectively deposited siRNA to the activated macrophages.
- Nigavekar et al. (2004) showed that bioaccumulation of dendrimer nanoparticles depends upon their generations and net surface charge. Positively charged nanoparticles exhibit greater accumulation than negatively charged particles. PEGylation modulated the bioaccumulation of nanoparticles in a size-dependent manner (Kaminskas et al., 2008).

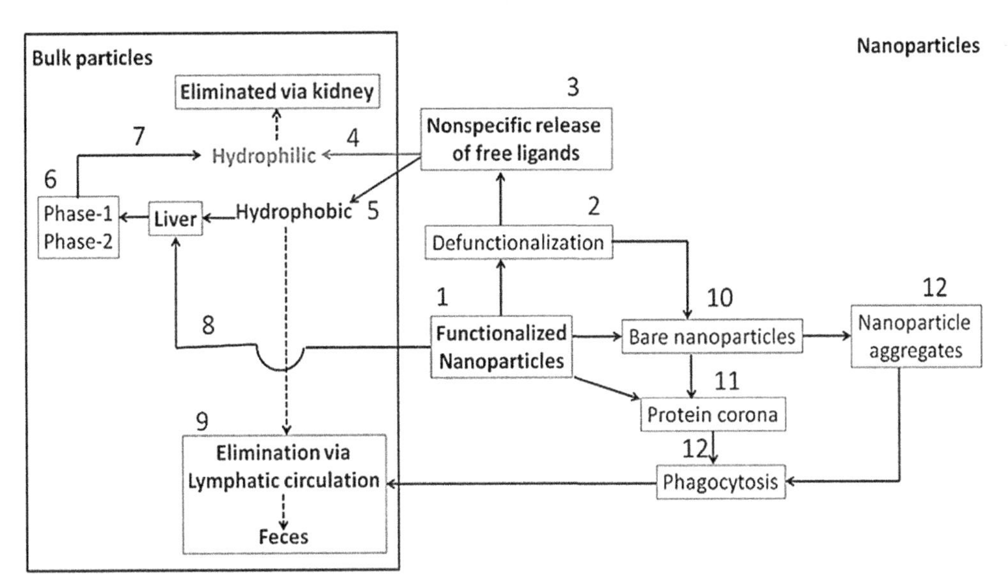

FIGURE 18 Metabolism and elimination pathways for bulk particles and nanoparticles. Functionalized nanoparticles (18-1) are either defunctionalized (18-2), resulting in the release of the functional groups (18-3) from nanoparticles (18-10). Hydrophilic functionalized groups (18-4) are eliminated via kidney, while hydrophobic groups (18-5) undergo liver metabolism (18-6) where they are converted into hydrophilic metabolite (18-7). Hydrophobic moieties larger than 5 nm are eliminated via the lymphatic circulation (18-9). Bare and functionalized nanoparticles both form protein corona (18-11) that induces their phagocytosis (18-12) and elimination via the lymphatic system. Bare nanoparticles also undergo aggregation and phagocytosis (18-13). Hydrophilic bulk particles are eliminated via kidney, while the hydrophobic bulk particles are metabolized in liver.

Therefore, the complexities of nanoparticles suggest that the safety and efficacy of nanoparticles can be influenced by minor variations in multiple parameters listed above. This necessitates a careful examination of nanoparticle biodistribution, targeting the intended sites, and potential modifications.

2.3 Metabolism

Metabolism is a set of the biochemical pathways that modify the physicochemical properties of drugs, toxicants, and nanoparticles. Metabolism occurs through specialized enzymatic systems that often convert lipophilic compounds into hydrophilic compounds for elimination via the kidney. The rate of metabolism determines the duration of a chemical or particle's residence in body. Although metabolic reactions often act to detoxify poisonous compounds, in some cases, the metabolized products are more toxic than the parent compounds (Figure 18).

2.3.1 *Metabolism of Bulk Particles (Drugs or Toxicants)*

As mentioned above, the general goal of metabolism is to transform lipophilic compounds into more polar, water-soluble products (Sipes and Gandolfi, 1991). The products of drug/toxin metabolism are usually less active than the parent compound. For example, many lipophilic anesthetic compounds are short acting because they are rapidly converted into a more water-soluble product that lacks biological activity. However, in some cases, such as mycotoxins and certain pesticides, the metabolites are more potent than the parent compounds. Another

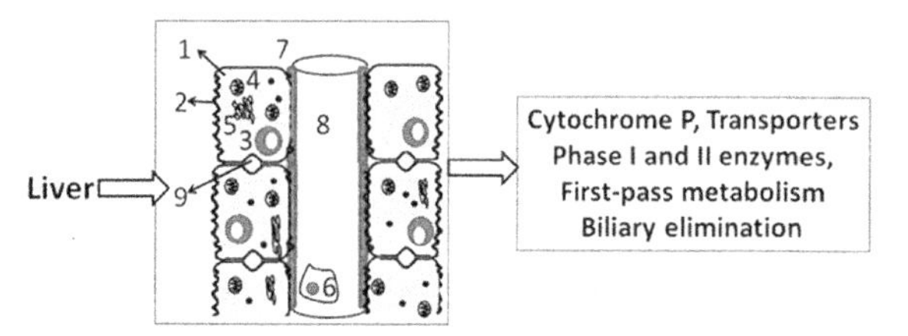

FIGURE 19 Microstructure of hepatocytes. Hepatocytes (19-1), microvilli (19-2), intracellular organelles such as nucleus (19-3), mitochondria 4, ER 5, etc., kupffer cells (19-6), endothelium (19-7), sinusoid lumen (19-8), and bile canaliculus (19-9).

example is cyclophosphamide, a prodrug, which is an inert compound; upon metabolism in the liver, it is converted into a highly active anticancer drug. Therefore, metabolism (biotransformation) plays a critical role in determining the pharmacological or toxicological properties of bulk particles.

Sites of drug metabolism: The liver is the primary site of metabolism. The functional units of the liver are hepatocytes, Kepffer cells, and the hepatic macrophages. Hepatocytes are rich in transporters, enzymes of phase I and phase II metabolism, and biliary circulation for excretion of larger and/or lipophilic particles (Figure 19).

Hepatocytes, constituting about 60–80% of the liver mass, have a polygonal plate-like shape that are in contact with neighboring cells and facilitate functions such as metabolism (including of carbohydrate, protein, fat, and the xenobiotics such as drugs, toxins, and nanoparticles), biliary excretion, and protein synthesis and storage. Bile leaves the liver along the bile duct and plays a role in digestion. The xenobiotic metabolism in the liver occurs in two phases (Figure 20):

- Phase I reactions introduce or create (via oxygenation and hydrolysis) a reactive functional group for phase II reactions. This step is also responsible for activation of pro-drugs and toxicants via the microsomal mixed function oxidase cascade requiring a NADPH, molecular oxygen, and Cyt P_{450} (a hemoprotein; the carbon monoxide derivative absorbs light at 450 nm).
- Phase II reactions conjugate a highly polar group to the phase I product via the group-selective transferases shown in Table 1.

Conjugation may occur directly on the parent compounds that contain appropriate structural motifs (–OH groups) or, as is usually the case,

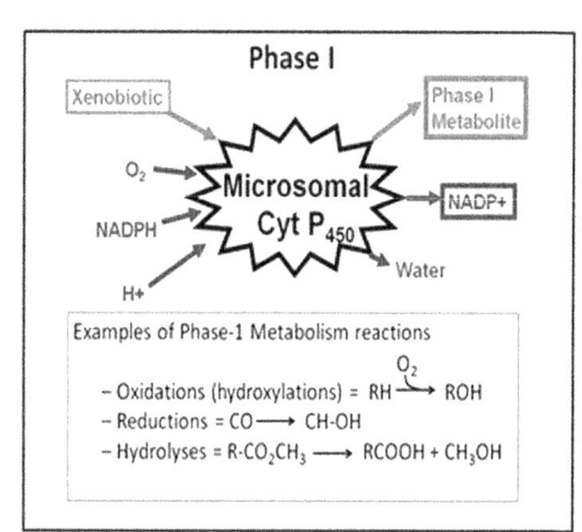

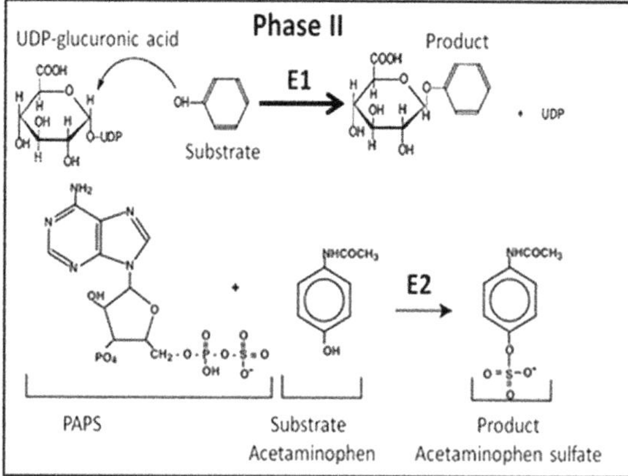

FIGURE 20 Basic mechanisms of phase-I and phase-II metabolism in liver. In phase-I metabolism, the microsomal cytochrome p450 enzymes add a reactive group to the molecule. In phase-II metabolism, a hydrophilic group (glucuronic acid or sulfate) is conjugated to the molecule that is excreted by the kidney.

TABLE 1 Examples of Key Phase 1 Reactions

Functional group	Enzyme	Co-substrate
Glucuronic acid	UDP-glucuronosyltransferases	UDP-glucuronide
Sulfate	Sulfotransferases	PAPS
Glutathione	Glutathione-*S*-transferases	Free glutathione
Acetate	*N*-acetyltransferases	Acetyl-CoA
Methyl	*N*-, *S*-, and *O*-methyl-transferases	SAM
Amino acids	Amino acid transferases	Free amino acids/ATP

PAPS, 3′-Phosphoadenosine-5′-phosphosulfate; SAM, *S*-adenosylmethionine.

on functional groups added or exposed by phase I oxidation. Conjugation reactions usually increase the molecular weight and water solubility of the compound, in addition to adding a negative charge to conjugates.

In addition to the liver, the GI tract epithelium is a key extrahepatic site for chemical metabolism. Chemicals absorbed via the intestine are subjected to first-pass metabolism in the liver. The lungs also participate in metabolism because of their enormous perfusion rate. The kidney, skin, and placenta can also carry out xenobiotic metabolizing reactions to some extent (Pannatier et al., 1978).

2.3.2 *Metabolism of Nanoparticles*

Unlike bulk particles, engineered nanoparticles are structurally complex, consisting of a core, protein corona, and primary surface ligands (e.g., PEG), which improve the particles' dispersion and provide reactive groups (NH_2, COOH, SH, etc.) for secondary functionalization (drugs, imaging molecules, or site-selective proteins). The nanoparticle core may consist of diverse chemical units (metals, carbon, polymers, etc.), but their primary surface ligands are usually PEG or poly-L-amino acids (Figure 21). This diversity in the core but similarity in functionalization complicates their metabolism in organisms. Usually, the metabolism of an engineered nanoparticle includes the processing all of its components, core, and functionalized groups (Figure 21).

2.3.2.1 METABOLISM AND BIOTRANSFORMATION OF PROTEIN CORONA AND SURFACE LIGANDS

Protein corona consists of proteins adsorbed from plasma and/or intracellular fluid. The dissociated and the adsorbed proteins may be hydrolyzed into the component amino acids that are either metabolized into the corresponding keno-acids or reincorporated into other proteins into the liver. Unlike the protein corona, the metabolism of surface ligands is not known. If nanoparticles are defunctionalized, the free functional ligands will be metabolized in liver (Liu et al., 2010a,b; Yang et al., 2009; Fu et al., 2001). If the engineered nanoparticles are programmed to release drugs and/or imaging molecules at a specific site such as a tumor, trauma, inflammation, or diseased sites, then the ligands may gradually translocate into the blood circulation and metabolize as they pass through the liver. Metabolism and biotransformation may alter the functional and medicinal properties of nanoparticles. Therefore, a nanoparticle engineered to shrink a tumor or treat a disease may lose its efficacy as they are metabolized in body. In addition, significant toxicity may occur if

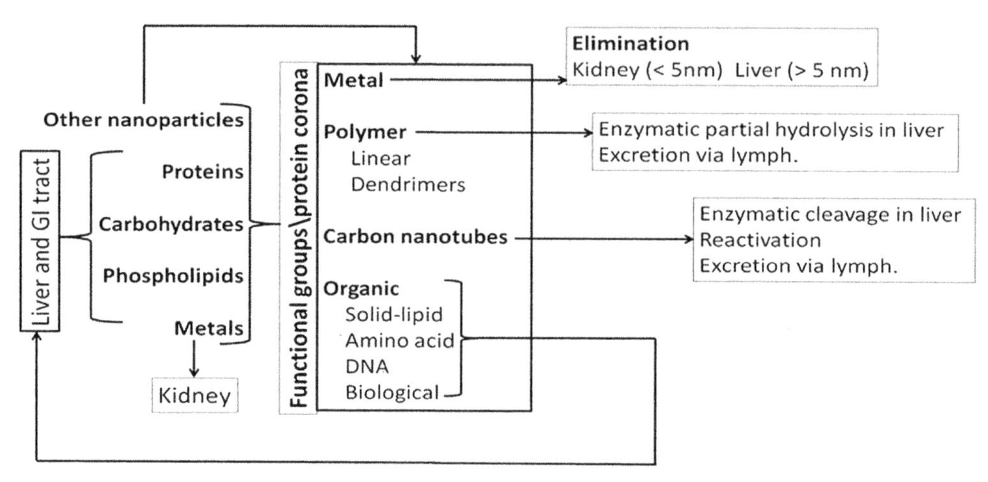

FIGURE 21 Fate of functionalized nanoparticles in body. Metal nanoparticles smaller than 5 nm are excreted via kidney, while larger ones are eliminated via liver. Polymers are partially degraded in liver and the metabolites excreted via kidney or liver. Carbon nanotubes are defunctionalized and the bare nanotubes are enzymatically hydrolyzed and excreted via the lymphatic circulation. Organic nanoparticles are metabolized via liver, GI tract and kidney.

defunctionalization results in the formation of bare nanoparticles.

Yang et al. (2012, 2009) have shown that removal of the functional groups is one of the major metabolic pathways in vitro and in vivo. Covalently functionalized CNTs appear to be more stable than noncovalently dispersed CNTs. However, the bare CNT skeleton remains unchanged and accumulates in different tissues, remaining intact for up to 3 months in vivo (Deng et al., 2007; Yang et al., 2008). Singh et al. (2006) and McDevitt et al. (2007a,b) have suggested that a minor change in functionalization may cause large change in the CNT's residence time and ensuing toxicity in vivo.

2.3.2.2 METABOLISM AND BIOTRANSFORMATION OF CARBON NANOTUBES

Carbon nanotubes, once considered to be biologically stable, are degraded by many plant and animal enzymes in vitro as well as in many intracellular organelles in vivo. The plant horseradish peroxidase (Kotcheyv et al., 2012, 2011; Russier et al., 2011; Zhao et al., 2011; Allen et al., 2009, 2008; Azevedo et al., 2003) and the animal myeloperoxidase and eosinophil peroxidase (Andon et al., 2013; Shvedova et al., 2012; Vlasova et al., 2011; Kagan et al., 2010) catalyze the biodegradation of carbon nanotubes in vitro. Elgrabli et al. (2008) first demonstrated in vivo degradation of carbon nanotubes (MWCNTs) in the lungs of rats installed intratracheally with the nanotubes. Significant diminution of MWCNT length was noticed, suggesting the cleavage of MWCNTs by rat lungs. Although the mechanism of MWCNT cleavage remains unknown, the authors have implicated that alveolar macrophages modify the MWCNTs upon phagocytosis. Shvedova et al. (2012) have shown that oxidized SWCNTs underwent biodegradation, as evidenced by the shortening of the CNTs, an increase in the ratio of the D to G bands in Raman spectra (indication of an increase in defects), and the Myeloperoxidase (MPO) playing a critical role in oxidatively biodegrading SWCNTs in vivo (Liu et al., 2012, 2010a,b; Wepasnick et al., 2011; Kagan et al., 2010) (Figure 22).

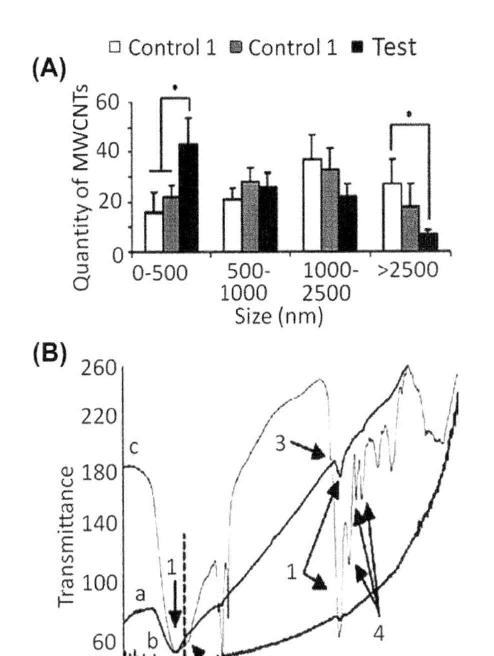

FIGURE 22 (A): Length distribution of peroxidase-digested MWCNTs in vivo and in vitro. *Control 1*: MWCNTs alone. *Control 2*: MWCNTs incubated for 15 days with suspension of lysed cells from broncheoalveolar lavage. Test: MWCNTs injected to rats after 15 days. *Significant from control ($P < 0.05$). Error bars represent standard deviation. (B): Infrared spectra of samples from (a) Control 1, (b) Control 2, and (c) MWCNTs injected to rats after 15 days. Possible peak assignment: 1. OH from H_2O, 2. OH from the alcohol group, 3. low quantity of carbonyl group, 4. nitrogen functional group. *Reprinted from Kotchey et al. (2013) with permission.*

2.3.2.3 SOLID-LIPID NANOPARTICLES

The administration route, distribution process, adsorption of biological material on the particle surface, and desorption determine the in vivo fate of the SLN (Jores et al., 2005; Mehnert and Mader, 2001). Because SLN are composed of physiological or physiologically related lipids, pathways for transportation and metabolism are present in the body may contribute to a large extent to the in vivo fate of the carrier. Probably the most important enzymes of SLN degradation are lipases, which are present in various organs and tissues (Olbrich et al., 2002; Muller et al., 1996). Lipases split the ester linkage and form partial glycerides or glycerol and free fatty acids. Most lipases require activation by an oil–water interface, which opens the catalytic center (lid opening). In vitro experiments indicate that SLNs show different degradation velocities by the lipolytic enzyme pancreatic lipase as a function of their lipid composition (Mukherjee et al., 2009). Surfactants can promote or almost completely inhibit SLN degradation (Muller et al., 1996).

The biological fate of SLNs depends on their initial physicochemical characteristics (e.g., composition, structure, dimensions, interfacial properties, physical state), as well as any changes in these properties as it passes through the different regions of the GI tract (McClements, 2013). In general, the fate of ingested nanoparticles is determined by three distinct phases: the digestive phase, the absorption phase, and the circulatory uptake. The digestive phase physically breaks the down the nanoparticles into a coarse emulsion due to antral contraction, retropulsation, and gastric emptying. The emulsion is hydrolyzed by the gastric lipase into more polar monoglycerides and fatty acids that enter the metabolic process. The majority of orally administered nanoparticles or metabolites enter the systemic circulation via direct absorption into the portal blood or, in case of extremely lipophilic drugs, via the lymphatic route, which avoids hepatic first-pass metabolism (Alex et al., 2011). Because of the extensive metabolism, oral delivery may exhibit low bioavailability. Studies have demonstrated that ligand-mediated targeting improves absorption of nanoparticles (Zhang and Wu, 2014; Zhang et al., 2014).

2.3.2.4 LINEAR POLYMERIC AND DENDRIMER NANOPARTICLES

Polymeric nanoparticles are one of the earliest nanoparticles used in drug delivery as well as for food and agricultural uses. However, the in vivo fate of dendrimers is not fully understood. Determination of the fate of nanoparticles in vivo can be challenging due to the diversity of their chemical composition, size, shape, surface chemistry, agglomeration, degradation, and interaction with proteins and other cellular components (Lim et al., 2014; Mu et al., 2014; Rao et al., 2014; Voigt et al., 2014; Wu et al., 2014; Zhao et al., 2014; Yang et al., 2013; Sarkar et al., 2012; Dallas et al., 2011; Shia et al., 2005; Mulders et al., 1997). Some of the key studies describing the biodegradation and metabolism of dendrimer nanoparticles are described below.

1. Studies have shown that the rate of biodegradation of pristine G4 polymer nanoparticles (Boltorn™ H40 based on 2,2-bis-(methylol)-propionic acid *bis*-MPA) with 64 terminal OH groups (Zagar and Zigon, 2002) was greater than the rates for either their lipophilic (Boltorn™ U3000) or hydrophilic (Boltorn™ W3000) derivatives (Reul et al., 2011; Domanska et al., 2009; Mezzenga et al., 2001; Domaska and Zołek-Tryznowska, 2009). Thus, functionalization reduced the hydrolysis of polymer dendrimers (Figure 23).
2. Feliu et al. (2012) conducted mass spectral analysis (using MALDI-TOF-MS) of polyamidoamine G1−G5 dendrimers containing cationic (−COOH) or neutral (−OH) groups as a function of pH, temperature, and time (Figure 24). They showed that the cationic dendrimers were cytotoxic and relatively resistant to hydrolysis. The neutral dendrimers, especially the G4-OH dendrimers, are relatively stable at acidic pH, but hydrolyze as the pH increases. The MS was characteristic of polymer degradation and almost all of the parent G4-OH dendrimers were hydrolyzed at physiological pH in about 14 days (Figure 24).

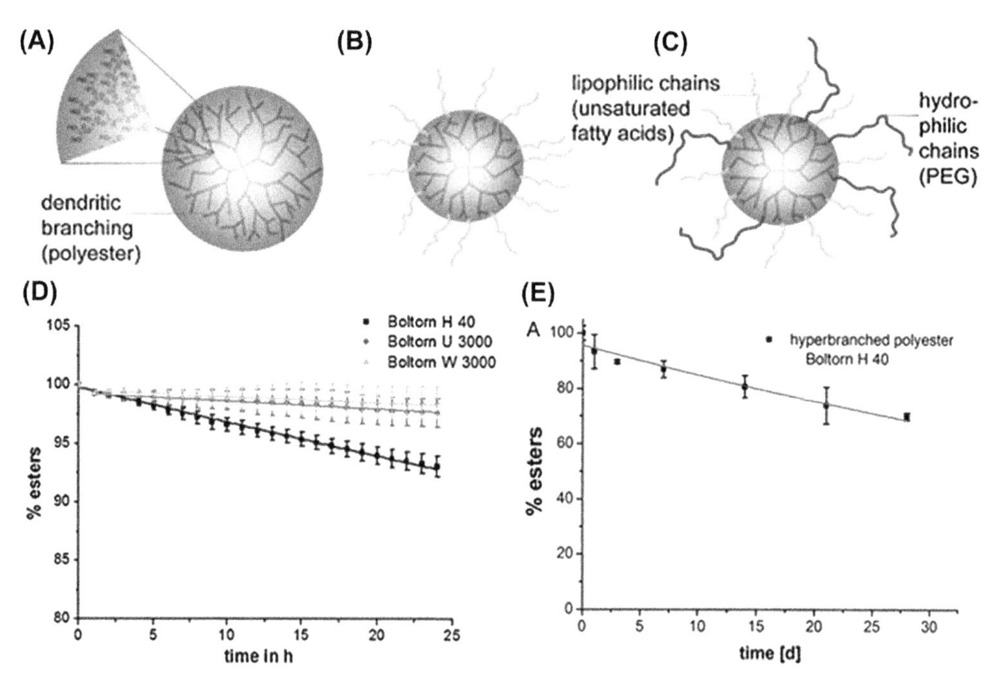

FIGURE 23 Schematic representation of the hyperbranched polyesters: (A) Boltorn H40, (B) Boltorn U3000 and (C) Boltorn W3000. (D): pH stat titration profiles of all three dendritic polymers (expressed in % esters). (E): Mass loss of the hyperbranched polyester at pH 7.4 and 37 °C during 4 weeks. *Reprinted from Reul et al. (2011) with permission.*

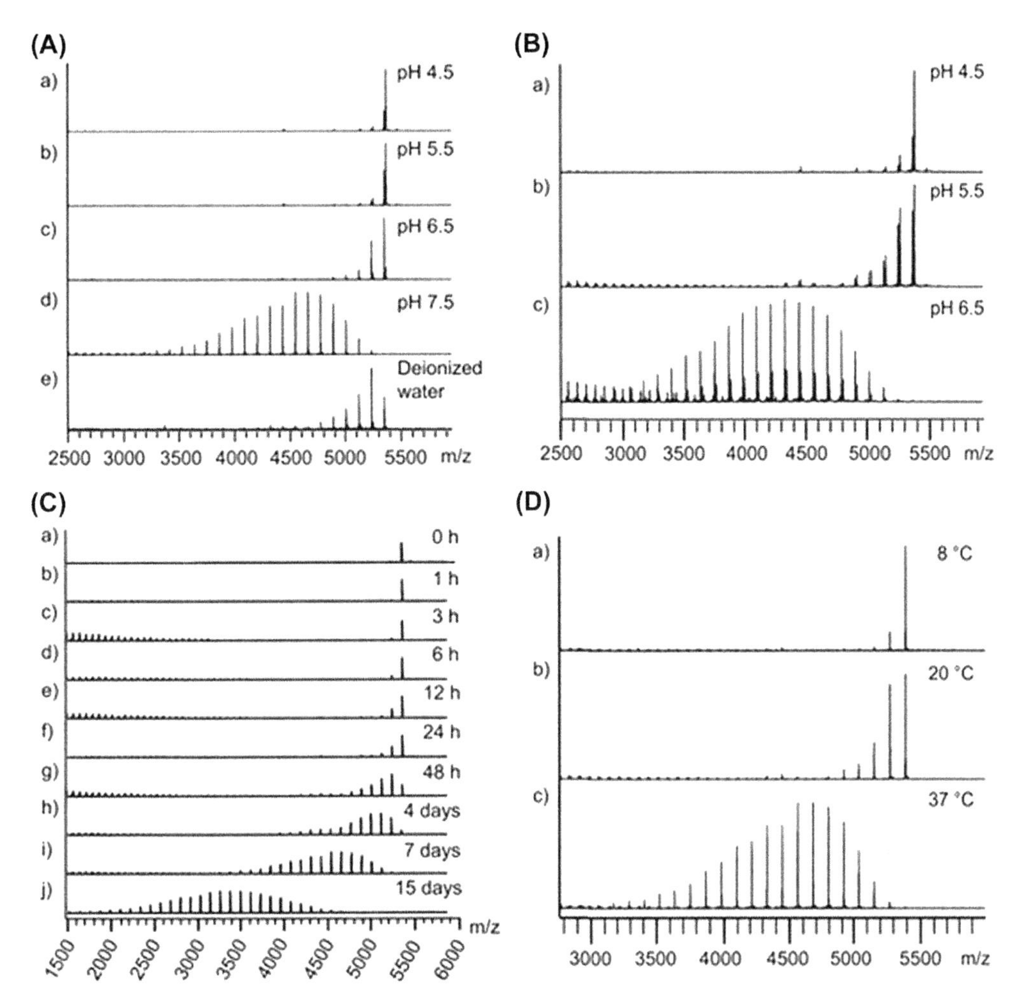

FIGURE 24 The stability of the G4–OH dendrimers at different pH, incubation time, and incubation temperature. (A) Effect of pH on the stability of the G4 hydroxyl (TMP-derivatized) dendrimer. MALDI-TOF MS spectra of the dendrimer after 7 days at 37 °C and different pH: (a) pH 4.5, (b) pH 5.5, (c) pH 6.5, (d) pH 7.5, and (e) deionized water. The particles were stable at acidic pH, but became unstable at neutral pH, including pure water. The spectrum is typical of a polymer. (B) MALDI-TOF MS spectra of the TMP-G4-OH dendrimer after 60 days at 37 °C at different pH: (a) pH 4.5, (b) pH 5.5, and (c) pH 6.5. At acidic pH, the dendrimers were stable for up to 60 days. (C) Assessment of time-dependent stability of the TMP-G4-OH dendrimer at pH 7.5 (37 °C) at different time points. MALDI-TOF MS spectra: (a) 0 h, (b) 1 h, (c) 3 h, (d) 6 h, (e) 12 h, (f) 24 h, (g) 48 h, (h) 4 days, (i) 7 days, and (j) 15 days. (D) The stability of the G4-OH at different temperatures. MALDI-TOF MS spectra of the TMP-G4-OH dendrimer were measured at pH 7.5 after 7 days at different temperatures: (a) 8 °C, (b) 20 °C, and (c) 37 °C. *Reprinted from Feliu et al. (2012) with permission.*

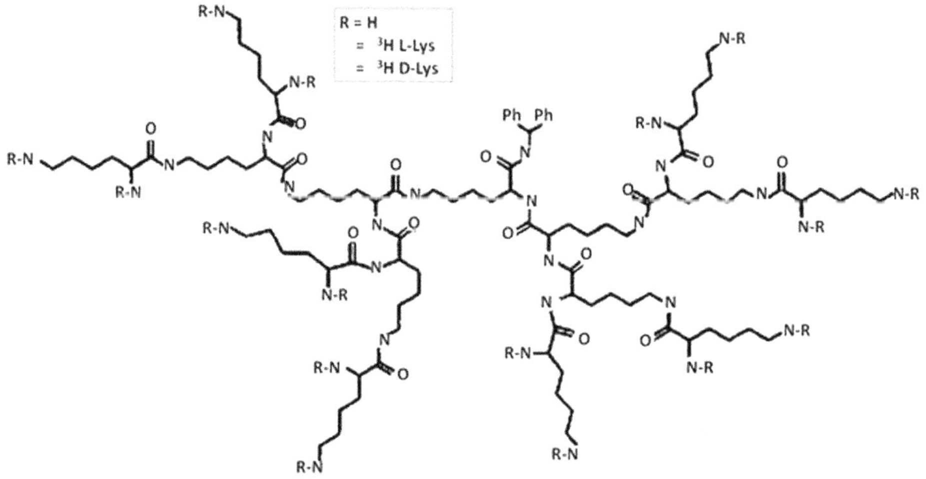

FIGURE 25 Structure of D- and L-Lys dendrimers.

3. Boyd et al. (2006) prepared ^3H-labeled poly-L-lysine dendrimers (Figure 25) with either third-generation (BHALys $[Lys]_4$ $[^3H\text{-}Lys]_8$ $[NH_2]_{16}$, called D3) or fourth-generation (BHALys $[Lys]_8$ $[^3H\text{-}Lys]_{16}$ $[NH_2]_{32}$, called D4) cores and with the surfaces left uncapped as L-lysine. They also prepared a third dendrimer consisting of the Lys core capped with a D-lysine outer layer (BHALys $[Lys]_4$ $[^3H\text{-}Lys]_8$ $[dLys]_{16}[NH_2]_{32}$, called D3-Lys). Groups of rats were administered bulk ^{3}HLys (Lys rats), $^3HD_3$ (D3 rats), ^3H D_4 (D_4 rats), and ^3H D_3-dLys (D_3-dLys rats) and ^3H l-Lys concentrations in plasma were determined (Figure 25).

As shown in Figure 26, In D_3-dLys rats, ^{3}HLys concentrations exponentially decreased from >10,000 ng/mL to <100 ng/mL. In Lys rats and $^3HD_3$ rats, ^{3}HLys concentrations decreased rapidly from >10,000 mg to about 1000 ng/mL and then began increasing again, peaking at about 10–12 days after injection. The concentrations remained elevated thereafter. The latter increase is due to the incorporation of Lys into proteins. Because dLys is not incorporated in proteins, the dLys levels continue to fall. As shown in Figure 26, less than 10% of the dose was excreted; thus a considerable amount of the administered dose was metabolized. These observations suggest that free Lys is released from Lys-dendrimers.

2.3.2.5 MAGNETIC NANOPARTICLES

Mejias et al. (2013) have studied biotransformation of magnetic nanoparticles by determining their out-of-phase alternate current (AC) magnetic susceptibility ($\chi''(T)$) profile and then correlating this with the particles' degradation. The $\chi''(T)$ profile is useful for detecting superparamagnetic material in tissue samples since the magnetic activity of the indigenous iron-containing species do not contribute to $\chi''(T)$ (Gutierrez et al., 2011, 2012; Lopez et al., 2007). The principle of this method is as follows: in intact magnetic nanoparticles, $\chi''(T)$ shows a single peak in both out-of-phase and in-phase components, indicative of blocking of particle magnetic moments. As the particles are metabolized or aggregated, the $\chi''(T)$ peak location on the temperature axis changes due to variations in dipolar interactions generated by the distinct dipole moments of metabolites or aggregated particles (Gutierrez et al., 2011). Mejias et al. (2013) reported that the relative height of the $\chi''(T)$ peak at different time points tended to decrease in the

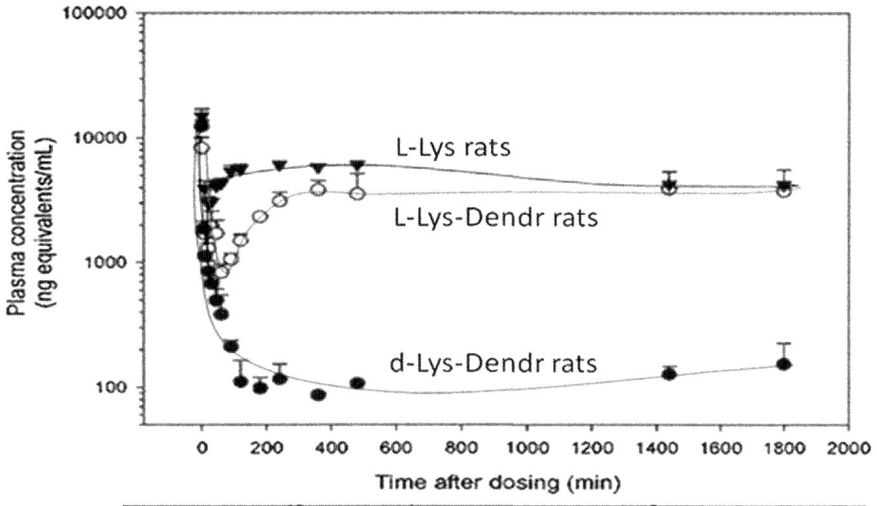

recovery (% of injected dose)		
dendrimer	in urine	in feces
L-Lys-Dendr rats	7.5 ± 1.3	below LOQ (LOQ = 1.6%)
d-Lys-Dendr rats	3.9 ± 0.9	below LOQ (LOQ = 1.2%)

FIGURE 26 Time course of changes in ^3H labeled D- and L-Lys dendrimers in rats. D-Lys concentration decreased rapidly from 10,000 ng/mL to 100 ng/mL in about 3 h and then the values remained consistently lower. L-Lys and L-Lys-dendrimer concentrations decreased exponentially similar to the d-Lys dendrimers, but in about an hour, the radioactivity began to increase and peaked to about 8000 nm/mL. The values remained elevated thereafter. The later increase in radioactivity is due to the incorporation of radioactivity into the proteins. The table shows percentage recovery of L-Lys-dendrimer (Dendr.) and d-Lys-dendrimer in urine.

spleen, liver, and lungs in rats at different time intervals after DMSA-MNP administration. At 3 months postparticle administration, approximately half the nanoparticle quantity was observed. The greatest reduction occurred in the lungs (Figure 27(A-i)). The scaled susceptibility (adjusted for the tissue weight) showed out-of-phase peaks in all three samples. These observations indicate that, with time, the nanoparticles acquired dipole moments either due to aggregation (Figure 27(A-ii)) or due to metabolism.

2.3.2.6 METAL NANOPARTICLES

Gold nanoparticles, considered the most inert nanoparticles, resist biodegradation in cells (Alkilany and Murphy, 2010) but phagocytize and are retained in the liver and the spleen (Chen et al., 2009). Iron oxide nanoparticles and quantum dots, however, can be gradually degraded in the acidic environment of endosomes or lysosomes, thereby releasing ferric ions (Arbab et al., 2005; Mahendra et al., 2008; Mancini et al., 2008). The degradation of quantum dots releases heavy metal ions, such as Cd^{2+}, that are highly cytotoxic. Often, the semiconductors or metal nanoparticles are coated with polymers, lipids, or surfactants to preserve their physicochemical properties (Nam et al., 2013). Because these coatings undergo oxidative biodegradation, the properties of nanoparticles may change with time. These observations indicate considerable diversity in the mechanisms of

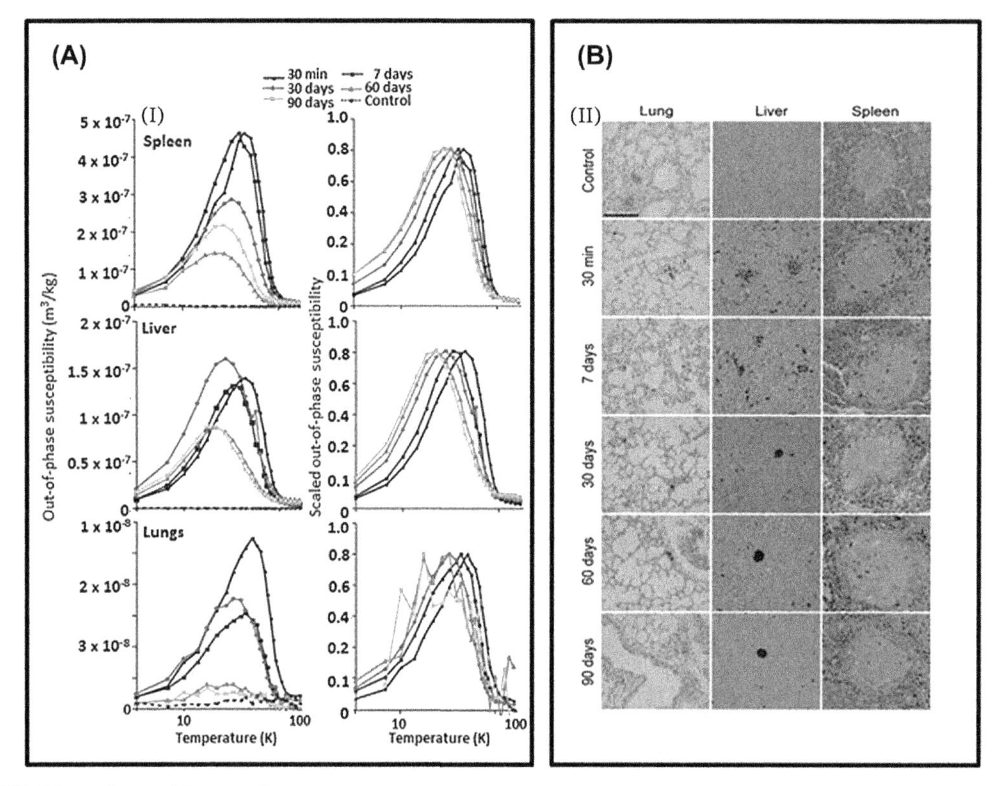

FIGURE 27 Effects of biotransformation on magnetic activity (measured as (A) out-of-phase alternate current (AC) magnetic susceptibility ($\chi''(T)$) profile and (B) scaled susceptibility) and distribution of super-paramagnetic nanoparticles in different tissues in vivo. (C): The $\chi''(T)$ values for super-paramagnetic nanoparticles decreased gradually (Ai to Av) after their administration in experimental animals. However, the pattern of change was different in different tissues. The out-of-phase signal (indicated by an increase in peak temperature (K) (vii)) indicated loss of magnetism. (C): Histological analysis. All three tissue samples exhibited aggregation of nanoparticles, but their distribution and extent of aggregation were different. This suggests that nanoparticle aggregation and/or interaction with the plasma or tissues modulated their magnetism. Modified from Mejias et al. (2013) with permission.

degradation of different nanoparticles and their surface ligands. Because nanoparticle safety for humans and animals is a major concern, an understanding of the fate of the nanoparticle is essential for the responsible development and use of nanomaterials in food and agriculture.

2.4 Excretion

Toxicants, drugs, and nanoparticles are excreted via the kidneys (small hydrophilic) and/or liver-biliary system (large hydrophilic or hydrophobic). In the kidney, the glomerular filtration allows hydrophilic particles of <5 nm diameter to pass into the urine. The hydrophobic particles are mostly eliminated via the biliary route into the feces. Other minor routes are milk (lactating females), lungs (volatile agents, such as general anesthetics), and the enterohepatic shunt.

2.4.1 Excretion via the Kidney

Each kidney contains approximately 1 million nephrons, the structural and functional units of

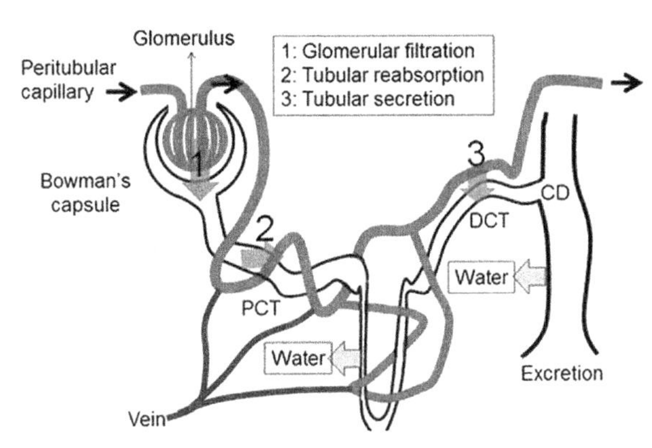

FIGURE 28 Microstructure of the glomerular apparatus. PCT, proximal convoluted tubules; DCT, distal convoluted tubules; CD, collecting duct.

the kidneys that carry out processes that form urine (Figure 28). Nephrons consist of the glomerulus corpuscle (composed of capillaries) surrounded by a glomerular capsule (Bowman's capsule) and a renal tubule. The renal tubule begins at the glomerular capsule as the proximal convoluted tubule (PCT), continues through the loop of Henle, and turns into a distal convoluted tubule (DCT) before emptying into a collecting duct. The collecting ducts collect filtrate from many nephrons and empty into a minor calyx.

Renal Glomerular Filtration: Glomerular filtration is a passive, nonselective process in which hydrostatic pressure (approximately 10 mm Hg) forces fluids and small (<5 nm) hydrophilic particles through the glomerular membrane (Deen et al., 2001). Glomeruli restrict the passage of the hydrophobic or nanoparticles. The glomerular filtration rate (GFR) is the volume of filtrate formed each minute by all the glomeruli of the kidneys combined. The normal adult GFR is 120–125 mL/min. Changes in the GFR affect the rate of elimination of chemicals, which are primarily eliminated by filtration (e.g., digoxin, kanamycin). In addition to size, the filtration of molecules also depends on their charge. The filtration-size threshold for globular proteins has been accepted to be less than 5 nm in hydrodynamic diameter (HD). For example, inulin (HD: 3 nm) achieves 100% renal filtration with a blood half-life of only 9 min (Prescott et al., 1991).

Renal Tubular Reabsorption: The intertubular pressure allows filtration of all nonprotein bound hydrophilic components of plasma (nutrients: glucose, amino acids, and essential elements; particles <5 nm) in PCT. This involves near total reabsorption of organic nutrients and the hormonally regulated reabsorption of water and ions. Nonresorbed polar molecules remain in the renal filtrate and are excreted via urine. Urine pH plays an important role in the excretion of polar chemicals. For weak acids, urine alkalization favors the ionized form and promotes excretion. Different areas of the tubules have different absorptive capabilities.

- The PCT is most active and selective in reabsorption.
- The descending limb of the loop of Henle is permeable to water, while the ascending limb is impermeable to water but permeable to electrolytes.
- The DCT and collecting duct have Na^+ and water permeability regulated by the hormones aldosterone, antidiuretic hormone, and atrial natriuretic peptide.

Renal Tubular Secretion: Tubular secretion disposes unwanted solutes and solutes that were reabsorbed, rids the body of excess K^+, and controls blood pH. Tubular secretion is most active in the PCT, but it also occurs in the DCT and collecting ducts. The kidney can actively transport some filtered drugs (e.g., dicloxacillin) against a concentration gradient, even if the drugs are protein-bound, but dissociates rapidly. A drug called probenecid competitively inhibits the tubular secretion of the penicillin, and it may be used clinically to prolong the duration of effect of the antibiotics.

2.4.2 Biliary Excretion

Compared to the excretion of chemicals via kidney, little is known about hepatic

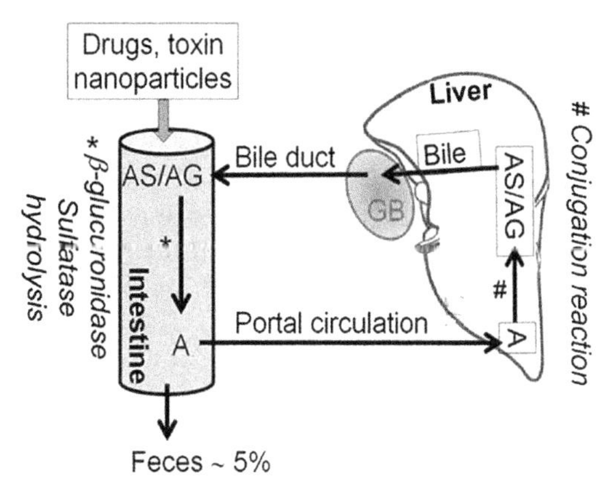

FIGURE 29 Enterohepatic clearance. In liver, the hydrophobic particles (A) converted into hydrophilic particles via glucuronide (AG) or sulfate (AS) conjugation. Larger conjugated particles that cannot be excreted via kidney are eliminated via gallbladder (GB) into the intestine where the particles are deconjugated and transported into the liver via portal circulation. Approximately 5% of the conjugate is excreted via feces.

chemical/nanoparticle elimination. Many chemicals and metabolites are passed into the small intestine via bile and may undergo enterohepatic cycling, as shown in Figure 29. Hepatocytes conjugate chemicals mainly to glucuronide or sulfate (discussed earlier). These conjugates travel with bile acids to the gallbladder then to the duodenum, where the conjugates are hydrolyzed into free drugs that travel back to the liver via portal circulation (Figure 29). About 5% of the administrated dose is excreted in feces. The liver secretes about 0.25–1 L of bile each day. Anions, cations, and nonionized molecules containing both polar and lipophilic groups are excreted into the bile, provided that the molecular weight is greater than about 300 Da. Lower molecular-weight compounds are absorbed before being excreted from the bile duct. Conjugates, glucuronides (drug metabolites), are often of sufficient molecular weight for biliary excretion. This leads to biliary recycling, as described above.

3. BIOKINETICS

3.1 Introduction

PK and TK both describe the rate at which a chemical will enter the body and what happens to it once it is in the body (Gibaldi, 1984; Gibaldi and Perrier, 1982). In other words, TK is concerned with the variation in toxicant concentration with time as a result of ADME. TK can be used for establishing relationships between exposures in toxicology experiments in animals and the corresponding exposures in humans, as well as for determining environmental risks in order to determine the potential effects of toxicants to the environment and its inhabitants. Although TK and PK are based on similar principles, they have different focus and differ in the doses (TK >> PK) that are analyzed. To evaluate the toxicity of a drug, the higher doses push both absorption and elimination processes to the limit, and sometimes past it, because, in toxicity studies, higher drug concentrations are required to induce undesirable side effects. Therefore, the TK parameters may differ in magnitude from the PK parameters of the same drug. In both TK and PK, the time course of pharmacological or toxicological action depends on the drug/toxicant dose, route of administration, and ADME as described above. The concentration–effect (dose–response) relationship may also play a key role. A hypothetical time–concentration curve for a toxin administered via intravenous or oral routes is shown in Figure 30. The toxic dose is usually several-fold greater than the pharmacological dose.

3.2 Principles of Biokinetic Models

Mathematical models of ADME describe the PK and TK properties of a drug, toxicant, or nanoparticle in organisms. These models quantitatively describe the temporal change in the concentrations of chemicals of nanoparticles and/or their metabolites in different organs/tissues of

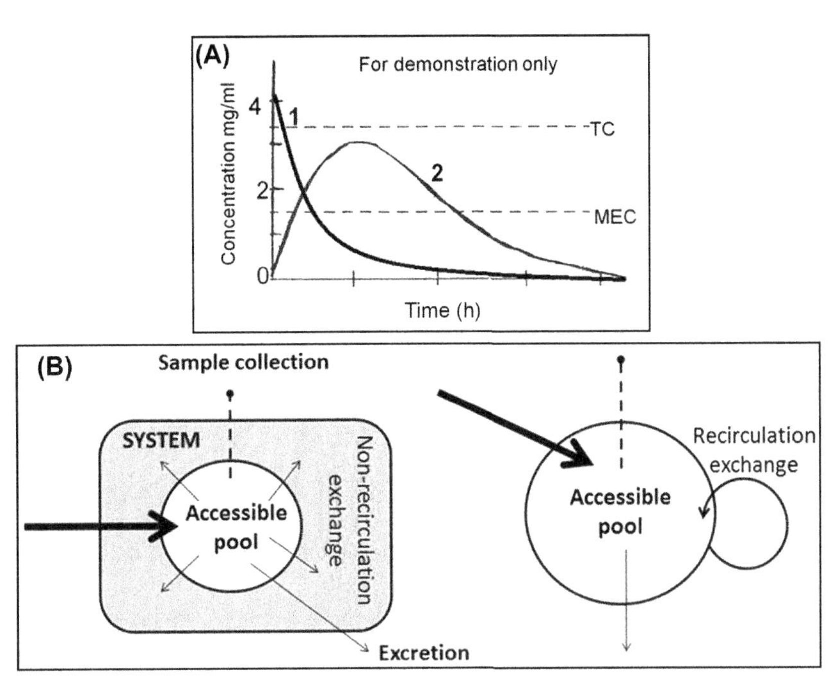

FIGURE 30 (A): The time course of changes in chemical concentration in plasma from experimental animals administered a single intravenous (plot 1) and oral (plot 2) dose. The interrupted horizontal lines show the minimum effective concentration (MEC) and toxic concentration (TC). A therapeutic effect can be expected only when the plasma level is above the MEC and below the TC. (B): The basic principles of pharmacokinetics/toxicokinetics.

exposed organisms. Three types of kinetic models are commonly used: the classic compartmental models (Figure 30(B1)), the physiologic-based compartmental models, and the noncompartmental models (Figure 30(B2)). A detailed description is provided in the following publications: Gabrielsson and Weiner (2012), Jacquez (1996), and DiStefano (1982).

1. In *compartmental models*, an organism is considered to be a set of compartments (usually one to three) characterized empirically. The compartmental model assumes that the body contains one access pool, such as plasma, in which a chemical can be introduced or sampled for analysis. Parts of the body, such as organs and tissues, are not accessible for test input and/or data collection.
2. *Physiologically based* models are based on interrelationships among critical biological, physicochemical, and biochemical determinants. Each organ and tissue can be considered a compartment.
3. *Non-compartmental models*: The difference between noncompartmental and compartmental models is the way in which the nonaccessible portion of the system is described (Gillespie, 1991). In a noncompartmental model, the administered chemicals leave the accessible pool via a reversible-exchange mechanism in which the chemicals in the accessible pool and nonaccessible pool may establish an equilibrium.

3.2.1 *Classic Pharmacokinetics Models*

The classic models include one compartment (chemicals distributed rapidly in the entire body), two compartments (a central and a highly

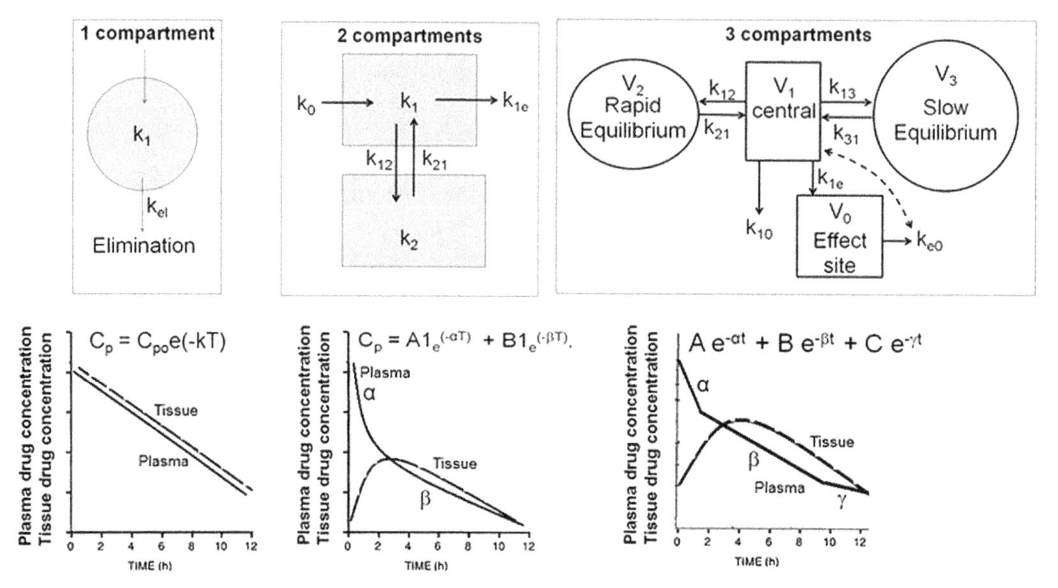

FIGURE 31 Description of one, two, and three compartment pharmacokinetic models. In one compartment, the tissues, plasma, and interstitial fluid are all lumped as one compartment in which a chemical or particle can distribute freely. The plasma concentrations reflect the tissue concentration. In a two-compartment model, plasma and tissues may be considered separate compartments giving a biphasic curve reflecting the distribution phase (α) and the excretion phase (β). In a three-compartment model, the plasma, highly perfused tissues and poorly perfused tissues are considered separate components. This adds a new component γ that represents the secretion phase. Plasma concentration may not reflect the tissue concentrations in two- and three-compartment models.

perfused peripheral pool), or three compartments (a central, a highly perfused, and a poorly perfused pool) models. Chemicals or particles in the central compartment alone can be metabolized and eliminated. In a three-compartment model, the tissues are separated as fast-equilibrium and slow-equilibrium compartments and a third compartment representing the effect site (Figure 31). The three models yield distinct plasma and tissue time-course curves (Figure 31, bottom).

3.2.1.1 ONE-COMPARTMENT LINEAR MODEL

This linear model is based on the following assumptions:

1. Chemicals are evenly distributed throughout the body.
2. Chemicals are eliminated from the body in a first-order fashion. That is, the rate of elimination is proportional to the amount of drug in the body. The first-order elimination rate constant, k_{el} (unit: reciprocal time usually 1/h), is defined by the equation $k_{el} = (\ln C_{po} - \ln C_{pt})/t$, where C_{po} is the peak plasma toxicant concentration and C_{pt} is the plasma toxicant concentration at time t. In the one-compartment model, the plasma concentration follows the equation $C_p = C_{po}^{e(-kT)}$.

This model is appropriate for chemicals that rapidly and readily distribute between the plasma and other body tissues. Because the body is conceptualized as a single compartment, the one-compartment model has only one volume term, Vd, usually expressed in liters. The one-compartment model yields a straight line when using a log scale on the y-axis. The Vd can be defined by the equation $Vd = dose/C_{po}$.

3.2.1.2 TWO-COMPARTMENT MODEL

For a one-compartment model, there must be a rapid equilibrium between chemical or particle concentrations in systemic blood and in different tissues; thus, the blood concentration represents the tissue concentration (Figure 31). This assumption is valid mostly for the lipophilic chemicals. In a two-compartment model, a chemical, immediately after administration, rapidly distributes in the central compartment (usually blood). From there, chemicals or particles are distributed into the second compartment or eliminated (Figure 31). The plasma time-course curve exhibits two phases: the α-phase, which represents distribution, and the β-phase, which represents elimination. A1 and B1 are the intercepts for α and β curves. C_p can be calculated using the equation shown in Figure 31. The microconstants can be calculated using the following equations:

$$k_{21} = \frac{(\alpha \cdot B) + (\beta \cdot A)}{A + B}$$

$$k_{10} = \frac{\alpha \cdot \beta}{k_{21}}$$

$$k_{12} = \alpha + \beta - k_{10} - k_{21}$$

3.2.1.3 THREE-COMPARTMENT MODEL

In a three-compartment model, the peripheral compartment is divided into rapid equilibrating (V2) and slow equilibrating (V3) compartments (Figure 31). The assumption is that the V2 compartment is rich in blood vessels, which allow rapid equilibrium in chemical distribution, and that V3 corresponds to the vessel poor compartment rich in fat cells. The toxin distribution in the three-component model is characterized by the equation: $C(t) = A\,e^{-\alpha t} + B\,e^{-\beta t} + C\,e^{-\gamma t}$, where α (distribution), β (elimination), and γ (tissue secretion) represent the slope of the three linear phases, while A, B, and C represent the respective intercepts.

3.2.2 Physiologically Based Biokinetics

The classic kinetics involve fitting parametric, nonmechanistic functions (volume of distribution, area under the curve (AUC), and intrinsic clearance rates) with constants or exponential coefficients (Mager and Jusko, 2001; Cheung and Levy, 1989; Hazelett and Preusch, 1988). The classical models, although characterizing the fate of drugs and toxicants, do not reveal information about the chemical kinetics in different tissues, biotransformation, mechanisms of action, and interspecies scaling. In addition, the models do not satisfy the conservation laws to ensure a balance in the chemical species. The Physiologically based toxicokinetics (PBTK)/physiologically based pharmacokinetics (PBPK) models circumvent the disadvantages of the classic PK (Dickschen et al., 2012; Krishnan and Andersen, 2007; Haddad and Krishnan, 1998), as shown below.

- *Chemical distribution* in an individual organ is determined using organ-selective mathematical models.Figure 32(A) shows the input and output pathways for different organs in body, while Figure 32(B) shows differences between exposure (1) via ingestion when chemicals first go to liver and then to systemic blood (first pass) and (2) exposure via inhalation, in which case chemicals first go to the systemic circulation and then distribute to different organs including liver. Figure 33 shows the generic equations for distribution in different organs (top) and in blood (bottom).
- *The fate of drugs* in cells is determined by selected biochemical reactions. Metabolism is scaled according to chemical principles that have to be studied independently, for example, in cell cultures.
- *Quantitative biotransport*, such as the mass transfer between blood and interstitial fluid, can be determined using appropriate equations for each organ (Elliott et al., 1980). Mass transfer is a function of the size of mass surface area. In general, the specific mass

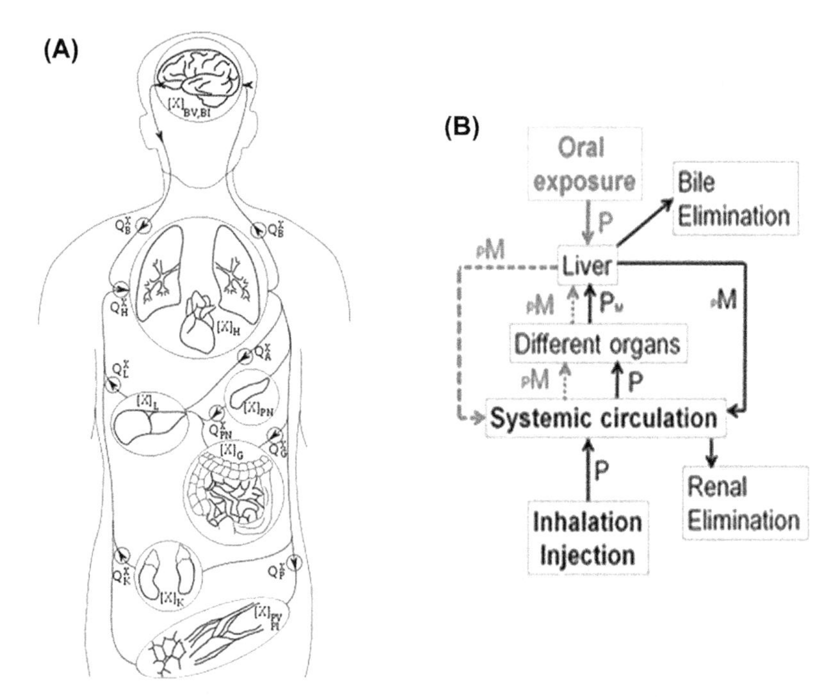

FIGURE 32 Physiologically based biokinetics. (A): The key organs commonly used in determining the PBPK or PBTK. *Abbreviations*: X, concentration; BV, brain vessels; Bl, blood; Q, blood flow; H, heart; L, liver; PN, pancreas; G, GI tract; K, kidney; P, parent chemical; M, metabolite. (http://en.wikipedia.org/wiki/File:WholeBody_wiki.svg.) (B): Overall PBPK or PBTK approach. P, parent compound; M, metabolite. The size of P and M represents relative distribution of the parent compounds and their metabolites.

$$\underbrace{V_K \frac{dC_x}{dt}}_{\text{Accumulation in an organ}} = \overbrace{Q_K C_B}^{\text{Rate of inflow* with blood}} - \underbrace{Q_k C_k}_{\text{Rate of outflow with blood}} - \overbrace{k_K C_B}^{\text{Clearance by kidney}} - \underbrace{\left(\frac{V_{max,K}\, C_K}{K_{m,K} + C_K}\right) V_{KT}}_{\text{Rate of metabolism}}$$

*: for liver a blood flow correction is made since liver receives blood from both arterial and portal circulation.

$$\underbrace{V_B \frac{dC_B}{dt}}_{\text{Accumulation in blood}} = \underbrace{(Q_H C_H + Q_{Li} C_{Li} - Q_M C_M + Q_K C_K + Q_{Le} C_{Ke})}_{\text{Rate of inflow from tissue compartment}} - \underbrace{Q_B C_B}_{\text{Rate of outflow to tissue compartment}} - \overbrace{\left(\frac{V_{max,B}\, C_B}{K_{m,B} + C_B}\right) V_B}^{\text{Rate of metabolism}} + \underbrace{Mglt)}_{\text{Rate of injection}}$$

FIGURE 33 Basic PBPK and PBTK equations. In these equations V is compartment size (mL or g), C is concentration in mass per unit volume (mg/mL), *t* is time, Q is blood flow to the organ, P is parent, M is metabolite, k is clearance in mL/min, v_{max} is enzyme activity at saturation (μg/g·min), K_m is Michaelis constant (μg/mL), subscript $_K$ and $_{KT}$ are kidney (including blood) and kidney tissue, M is the dose of chemical (μg), g(t) is injection function (min^{-1}), and subscripts have tissue identification.

transfer rate of the blood–organ interface may be constant. This can be scaled independently to account for the differences in organ surfaces/mass transfers.

- *Biodistribution* in organs is determined by physiological and anatomical parameters, such as blood perfusion rate and age. Physiological and biodistribution parameters can be determined experimentally.
- *Population pharmacokinetic variations* are determined by distinct phenotypes, most prominently by anatomical and physiological differences, which are strong functions of body mass.
- *Interspecies pharmacokinetics* can be scaled based on well-known anatomical relationships, such as the changes in organ weight as a function of total body weight or blood perfusion rates as a function of different cardiac output.

The PBTK/PBPK provides a mechanistic basis of interaction between a toxicant and an organism. It enables the species, dose, and route extrapolations of toxic occurrences; in addition, it allows the assessment of an interaction-based mixture risk assessment. Therefore, PBTK facilitates the evaluation of the magnitude and relevance of chemical interactions in assessing the risks of low-level human exposures to complex chemical mixtures.

3.2.3 Factor Affecting Biokinetics of Bulk Particles

The chemical, physicochemical, and formulation characteristics of the chemical and the physiological properties of the organism modulate the chemical's PK and TK. The key factors are listed below (Clewell et al., 2008, 2007, 2004, 2002, 2001, 1999; Shargel and Andrew, 1999; Brahmankar and Jaiswal 1995; Swarbrick and Boylan, 1988).

Chemical factors include the following:

a. An ester or a salt of a weak acid is absorbed more efficiently than the free acids.
b. Amorphous forms dissolve more efficiently than crystals of a chemical.
c. Solubility, permeability, ionization, and lipophilicity are the major physicochemical factors affecting the rate and extent of absorption.
d. Formulations modify the disintegration time and dissolution of the drugs.

The organism's factors: The organism's physiological properties, such as membrane permeability, GI pH, gastric emptying, small intestinal transit time, bile salt, and absorption mechanism, also modulate a chemical's absorption and clearance.

4. NANOPARTICLE BIOKINETICS

It is well established that nanoparticles (1–100 nm in at least one dimension) possess unique features not found in the corresponding bulk particles (>100 nm). As the nanoparticle size decreases, its surface-to-volume ratio, chemical reactivity, electronic band gap, transparency, ultraviolet-filtering efficiency, dispersion properties, excretion via kidneys, and toxic effects increase (<5 nm), while its phagocytosis and excretion via feces decrease (Sokohara and Ishida, 1998). The unique physicochemical and electronic properties of nanoparticles may be responsible for their unique biological activity and toxicity, which may be different from their bulk-sized counterparts (Oberdorster et al., 2005). In addition, the biological properties of nanoparticles of different sizes may be different from each other. Thus, a 50-nm gold nanoparticle, a 5-nm gold nanoparticle, and a bulk gold ionic solution may have different chemical and biological characteristics. In addition to the size, the shape, surface properties, surface charge, and aggregation state may also determine the biological properties of nanoparticles (Figure 34). Gold nanoparticles coupled to an anticancer drug via

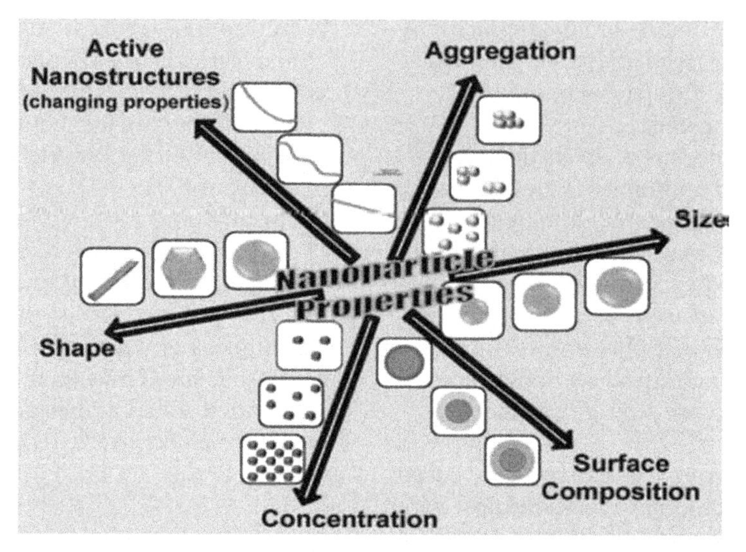

FIGURE 34 The unique characteristics of nanoparticles resulting in interference with PBPK and PBTK calculations. The key factors are changing properties of nanoparticles in vivo, size, shape, aggregation, surface properties, and concentrations. *Reprinted from Tinke et al. (2006) with permission.*

a cancer-specific linker, but not bare gold nanoparticles, may release the drug at the cancer site, while passively functionalized gold nanoparticle, may release the drug nonspecifically as it disintegrates.

Taken together, these observations suggest that different engineered nanoparticles made up of the same element may exhibit differing PK and TK properties in vivo. This presents a major challenge for extrapolation of data to assess the risk posed by different nanoparticles in humans and animals. Therefore, to understand the differences between the PK and TK of nanoparticles, it is important to understand the changes of nanoparticles in different media, the role of formation of protein corona in nanoparticle–organism interactions, and the size-, shape-, and surface-related ADME.

4.1 Do Bulk Particles' Pharmacokinetic Criteria Apply to Nanoparticles?

Although only a small fraction of the total dose of a chemical or nanoparticle reach the target site, the absorbed fraction may diffuse more readily through the biological barriers; thus, the blood level may be somewhat in equilibrium and related to the target tissue levels. Therefore, the standard approach of determining the drug or toxin PK using the blood or plasma concentrations is acceptable for bulk particles. However, the standard PK approaches may not be applicable to nanoparticles because of their unique properties not in common with the bulk particles. One of the key technological problems commonly encountered in interpreting nanoparticles is differences in the formulations of bulk chemicals and those of the nanoparticles. In case of bulk particles, the blood concentration of active compounds (drugs or toxins) is directly measured, which allows direct fitting of the data to one-, two-, or three-compartment equations (classic PK) or to the PBPK and/or PBTK models. However, the fate of nanoparticles in blood and tissue samples is much more complex, possibly due to their size, shape, and surface functionalization being dependent on unique physiochemical and biological properties. The PK of the drugs loaded

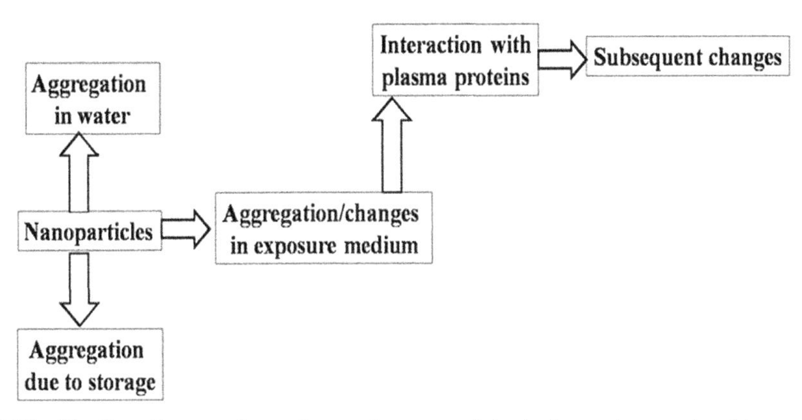

FIGURE 35 The flow diagram shows changes in nanoparticles in the environment and in an organisms.

in nanoparticles depends not only on the properties of the drugs, but also on the properties of the nanoparticles and the rate of drug release. Immediately after an intravenous administration of a nanoparticle-drug formulation (prior to the release of the drugs), the nanoparticle PK may be important. However, as the drugs are released from the formulation, the drug PK becomes relevant.

A key interfering factor in the determination of nanoparticle biokinetics is the particle aggregation. As-manufactured (bare) nanoparticles are highly reactive and rapidly aggregate into differing size particles with uncontrolled properties (Figure 35). Their aggregation and properties are highly dependent on their size. Engineered nanoparticles, although they disperse in aqueous media, undergo drastic changes in protein-containing culture/incubation media or when administered into an organism. They rapidly interact with the media or blood proteins and form a protein corona. The characteristics, composition, and properties of the corona depend on the nanoparticle size, shape, and surface charge and exposure time (Zhang et al., 2011; Casals et al., 2010; Monopoli et al., 2012). Lundqvist et al. (2008) have shown that, even for identical material types, the size and surface charge of the nanoparticle may entirely change the nature of the biologically active proteins in the corona (Figure 36). The different-sized nanoparticles or nanoparticles at different sites in an organism may be surrounded by protein coronas of a different nature that may modulate their interaction with cell membrane proteins, entry in cells, and compatibility with physiological systems (Mu et al., 2014; Walczyk et al., 2010; Lynch et al., 2009) (Figure 37).

Figure 38 shows the fate of bare or passively functionalized nanoparticles in the body. As the particles enter the blood (systemic or portal), they form a protein corona. The presence of component factors in the corona may, in a size- and shape-dependent manner, initiate phagocytosis and the ensuing elimination via the lymphatic system. The particles that escape phagocytosis are (1) internalized by the blood cells, (2) hydrolyzed by the blood enzymes, resulting in the release of the loaded chemicals and accumulation of the empty nanoparticles; and (3) transported into the liver, where they may be excreted prior to or after releasing the drugs. However, the fate of engineered nanoparticles (containing ligands to suppress complement binding, attenuate liver uptake, selectively transported to the target site and on-demand drug-release system) may be much simpler because functionalization prevents phagocytosis, rapid transport to the target site, and release of drug as needed (Figure 38).

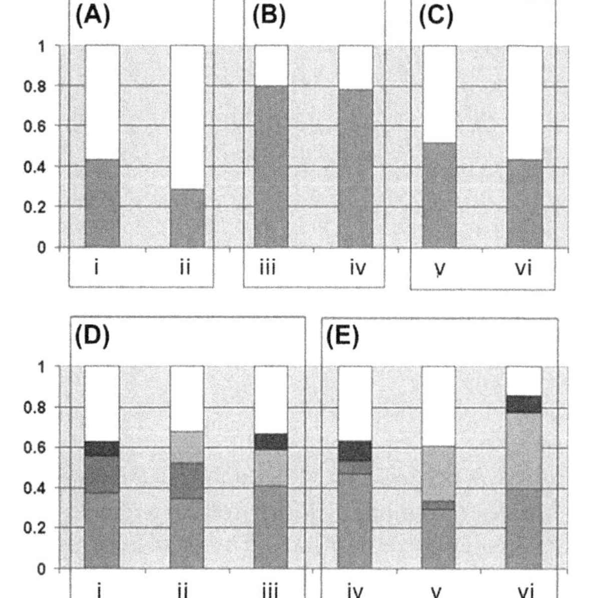

FIGURE 36 Composition of the protein corona in different nanoparticles. (i) 100 nm amine-modified, (ii) 50 nm amine-modified, (iii) 100 nm plain, (iv) 50-nm plain, (v) 100 nm carboxyl-modified and (vi) 50 nm carboxyl-modified. (A–C): Protein composition in corona of different sized nanoparticles having similar surface properties. *Green* (gray in print versions): Proteins common on both 50 and 100 nm particles and *white*: Differential protein distribution, found on one size but not the other. (D) and (E): Comparison of the corona for particles of the same size but different surface charges: fraction of proteins found on all three particles (*green*), fraction of proteins found on the amine-modified and plain particles (*blue* (dark gray in print versions)), fraction of proteins found on the plain and carboxyl-modified particles (*yellow* (light gray in print versions)), fraction of proteins found on the amine- and carboxyl-modified particles (*red* (black in print versions)), and fraction of proteins found on just one specific particle surface (*white*). *Reprinted from Lundqvist et al. (2008) with permission.*

These observations suggest a number of differences between nanoparticles and bulk particles that may be relevant to their PK and TK:

1. *Bulk particles*, depending upon the mode of exposure, enter the blood and then equilibrate in different tissues. Pharmacokinetic indices are calculated from the blood particle concentrations. Size, surface characteristics, and surface engineering are not relevant to bulk particles, but these factors are critical for *nanoparticles*. A minor change in a nanoparticle's size of surface properties may cause drastic alterations in its physiochemical properties and the corona composition.
2. The lipophilic properties and hydrophilic-lipophilic balance determine the ADME of bulk properties. However, the ADME of nanoparticles is determined by complex multi-faceted mechanisms (Figure 39), as described in an excellent review article by Moghimi et al. (2012).
3. As discussed in Chapter 3, the core atoms determine the electronic and physicochemical properties of bulk particles, while their surface atoms determine the properties of nanoparticles (diameter 1–100 nm). Because the surface activity increases exponentially with a decrease in nanoparticle size, a minor alteration in the surface may cause major changes in a particle's reactivity and its interaction with the biological system. In general, similarly functionalized but different-size nanoparticles of the same element may exhibit different physicochemical properties. In addition, differently functionalized nanoparticles of same size and elemental composition may also exhibit different physicochemical properties.
4. Immediately after administration, the physicochemical characteristics of the nanoparticles, not the loaded drug, determine the kinetics of a nanoparticle–drug complex (nanodrug) in an organism. If the nanodrug releases the loaded drug in blood, then the kinetics of the released drug and the empty particles will depend on the drug's and the particles' properties, respectively. If the nanodrug is engineered to release the loaded drug at a specific site such as the tumor, then the time course of change in drug concentration in plasma may not be of use in traditional PK calculations. New models

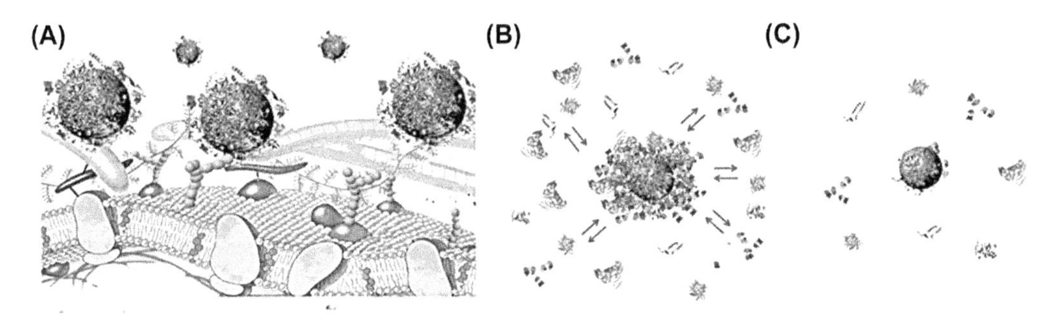

FIGURE 37 Kinetics of protein corona formation. (A) The exchange and interaction of proteins at the bio–nano interface-cellular level. (B) Drawing of the structure of nanoparticle–protein complexes in plasma: the "core" nanoparticle is surrounded by the protein corona composed of an outer weakly interacting layer of protein rapidly exchanging with a collection of free proteins. (C) A "hard" slowly exchanging corona of proteins. Diagram is not to scale in representing the proportions of the different objects. *Reprinted from Walczyk et al., 2010 with permission.*

need to be developed to determine the kinetics of nanoparticle–drug complexes in vivo. Although many studies have described procedures to determine the kinetics of nanoparticles (some examples are shown later) in vivo, their conclusions may be limited to the specifics of the particles used in the study. Extrapolation of their results to assess the kinetics of other, differently functionalized nanoparticles may not be possible.

5. *Concentration metric*: For bulk particles, the concentration calculation is simple:

$$1 \text{ mole} = \text{molecular weight in grams, } 1 \text{ mole of C} = 12 \text{ g}$$

Thus, bulk NaCl crystals contain 58.44 MW molecules that, when dissolved in 1 L water, will yield a concentration of 58.4 g/1000 mL or 0.0584 g/mL. If we weigh 58.44 g of NaCl, it will always be 1 mol. This may not be the case for nanoparticles, as shown in Figure 40. If particle density is kept constant—that is, we take 1000 particles/cm^3 of 500-, 50-, and 5-nm nanoparticles, then the mass density of 500-, 50-, and 5-nm NPs will be 10^5 μg/cm^3, 10 μg/cm^3, and 1 μg/cm^3, respectively. However, if particle mass is kept constant to 1 μg/cm^3 of 500-, 50-, and 5-nm NPs, then the number density of 500-, 50-, and 5-nm NPs will be 10 μg/cm^3, 10^4 μg/cm^3, and 10^7 μg/cm^3, respectively.

This suggests that size is an important issue when determining the nanoparticle dose in animal experiments. If an investigator is studying the toxicity of atmospheric nanoparticles, then size control may not be possible. It will be important to determine the distribution of particle size.

6. *Effects of a nanoparticle's size, shape, topology, and surface functionalization on its biokinetics*: The PK properties of nanoparticles depend on how well they negotiate their passage through a membrane pore (e.g., kidney glomerulus pore: 5–10 nm, interendothelial junction in healthy tissue: 10 nm, tumor: 40–80 nm, and cancerous tissue: 100–500 nm). Thus, particles can penetrate the tumors and the cancer cells much easier than they can penetrate the normal healthy cells. In addition to the size, a nanoparticle's shape, topology, volume, molecular conformation in solution, surface functionalization, flexibility, branching, and anisotropy also play important roles in their transport across cell membranes (Nangia and Sureshkumar, 2012) (Figure 41(A)). The flexible, loosely coiled, polymer nanoparticles

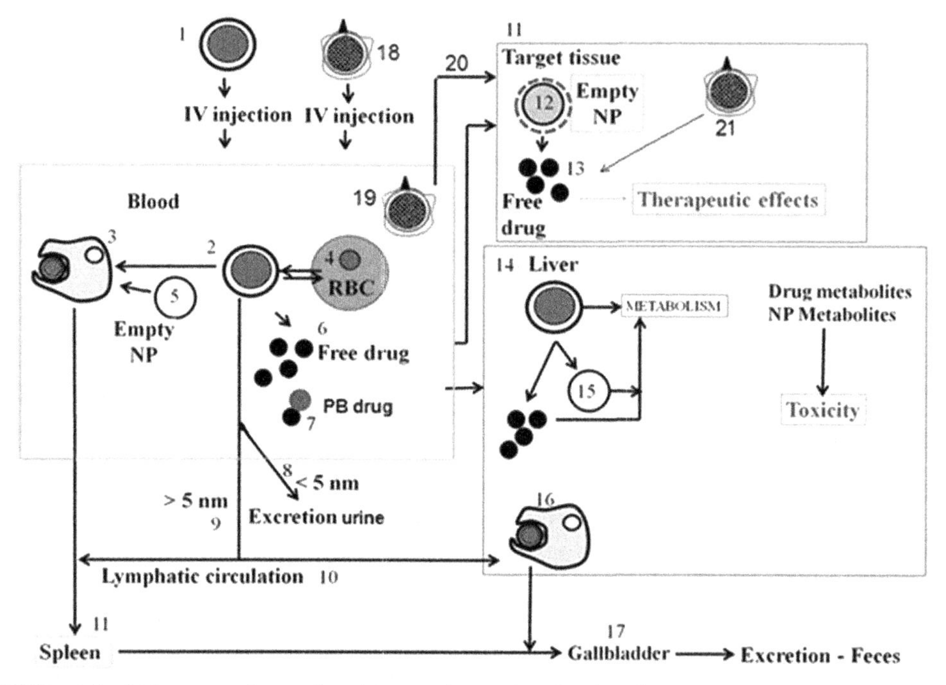

FIGURE 38 Fate of nanoparticles in body. 1: Intravenous injection of bare nanoparticles. A corona is formed if the matrix contains proteins. 2: Nanoparticles in blood. The corona composition changes because the blood proteins may replace the matrix proteins. 3: The protein corona may be recognized by the microphages, resulting in the particle's phagocytosis. 4: The nanoparticles may accumulate in blood cells. 5 and 6: Nanoparticles may disintegrate in blood and release drugs, resulting in the accumulation of empty particles. 7: The released drugs may bind proteins. 8: Hydrophilic nanoparticles (bare or engineered) less than 5 nm is excreted via kidney. 9: Hydrophilic nanoparticles larger than 5 nm or hydrophobic nanoparticles enter the lymphatic circulation (10) and transported to liver or spleen (11). 12: Engineered nanoparticles are transported to the target site where they release the drugs as needed (13). 14: From blood circulation, nanoparticles are transported into the liver where they may undergo defunctionalization or depletion (15) or phagocytosis (16). The phagocytized nanoparticles enter the gallbladder (17). 18: Engineered nanoparticles enter blood (19) and then transported (20) to the target site (21) where they release the load.

could readily deform to pass through a pore (Figure 41(B)). Many hyperbranched polymers and dendrimers, and poorly solvated polymers, may adopt more rigid conformations and have difficulty passing through a pore. Rigid, elongated polymers can, depending on their diameter, pass the pore. The detailed information can be found in the following manuscripts: Kettler et al. (2014), Moghimi et al. (2011, 2012), Nangia and Sureshkumar (2012), Chou et al. (2011) and Yang and Ma (2010).

4.2 Selected Examples of Nanoparticle Biokinetics

Biokinetic models are developed using three key strategies:

1. *Classic or compartmental* (one, two, or three): As described earlier, this method requires

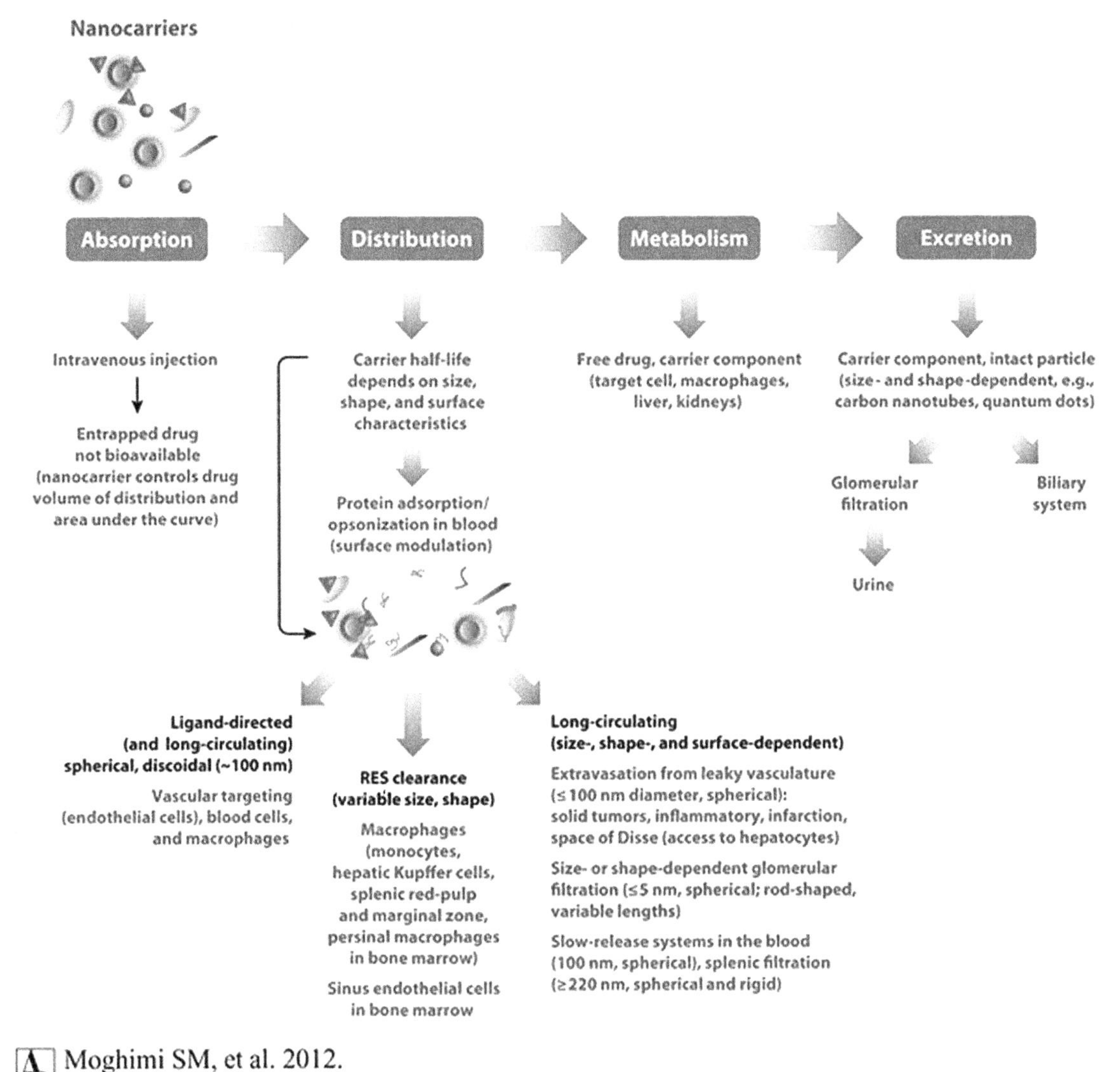

FIGURE 39 The factors affecting nanoparticle pharmacokinetics.

linear and nonlinear curve fitting of experimental data. The concentration–time profile is regarded as an expression of exponents. Compartmental models are useful for most of the applications.

2. *PBPK* or *PBTK*: One key disadvantage of the compartmental method is that curve fitting does not take into account the physiological and/or biochemical parameters. PBPK and PBTK models rely a priori on the anatomical, physiological, and biochemical processes. They usually have more than three components corresponding to predefined organs or tissues (Figure 32) characterized by a

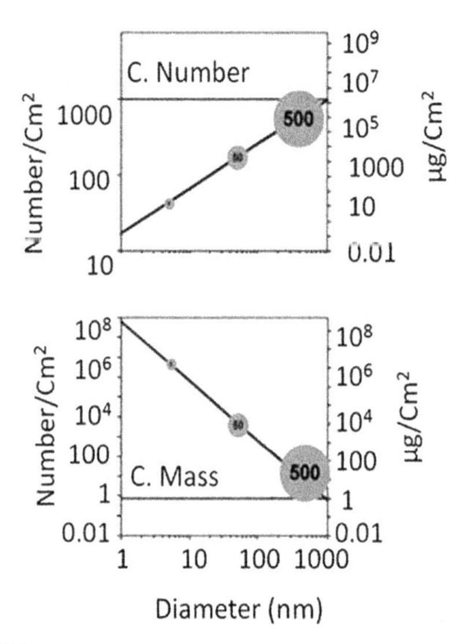

FIGURE 40 The dose metric for nanoparticles. Top: At the constant particle number (C. number), mass per unit area increases as the particle size increases. Bottom: At the constant mass per unit area (C. Mass), the inverse relation between particle size and particle number (number unit area decreases as the particle size increases).

set of differential equations for concentration or quantity as well as parameters representing blood flows, pulmonary ventilation rate, organ volumes, enzyme activities, etc., as available in scientific publications. By entering needed data, the software can yield meaningful kinetic values that can be used to decipher mechanisms or set up risk assessment.

3. *Noncompartmental calculations*: These models do not require curve fitting of the data, but they rely on simple algebraic equations that are limited to linear PK. The concentration−time profile is regarded as statistical distribution. Calculations are simple and rapid, but the usefulness is limited to clinical PK, bioavailability, and bioequivalence studies.

In this section, an example of each pharmacokinetic model is described. This includes a brief description of the methodology, results, and applicability.

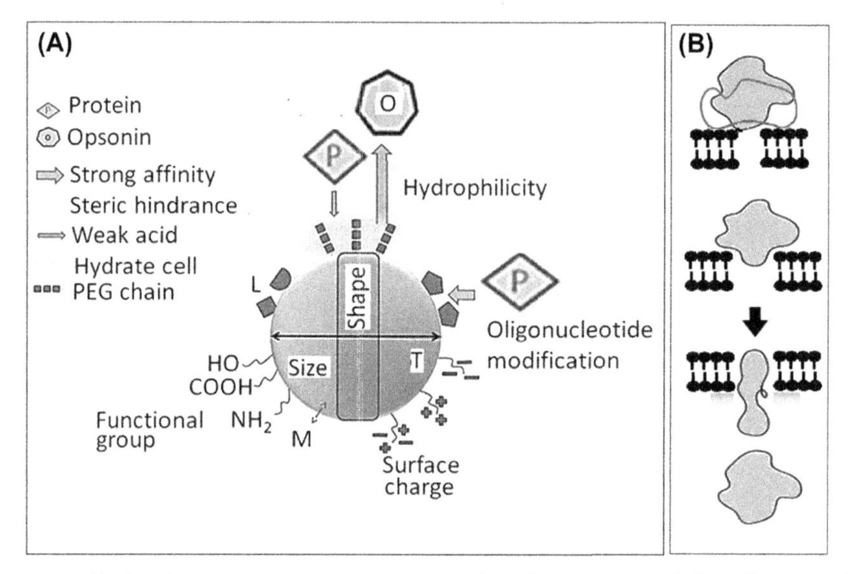

FIGURE 41 Factors affecting the nanoparticle's transport across the cell membrane. (A): Size, shape, topology (T) and aggregation state (Chou et al., 2011) are the key determinants of the interaction of nanoparticles with cell membranes. Other factors include protein corona, opsonins, ligands (L), and functionalizations described by Chou et al. (2011). PEG (polyethylene glycol) improves the particles' movement across the cell membrane. (B): Nanoparticle flexibility plays an important role in their transport across the cell membrane.

4.2.1 *Compartmental Models*

4.2.1.1 EXAMPLE-1: NANO-ADVANTAGE IN ENHANCED DRUG DELIVERY WITH BIODEGRADABLE NANOPARTICLES: CONTRIBUTION OF REDUCED CLEARANCE (KADAM ET AL., 2012)

The objective of this study was to develop compartmental PK models for bulk clozapine or polymer nanoparticle-encapsulated clozapine after oral, intraduodenal, and intravenous administration in experimental animals. The goal was to understand the influence of altered drug clearance on enhanced drug exposure with nanoparticle formulations (Table 2).

Figure 42 shows the typical curves and PK values for bulk and nanotized clozapine using two compartmental models. The results are summarized as follows:

- The drug–nanoparticle complex showed lower apparent plasma clearance, lower volume of distribution, longer circulation half-life, higher absorption rate constant (ka), and higher fraction absorbed than the corresponding plain drugs in animals subjected to intravenous or oral drug administration. A reduction in the apparent drug clearance from plasma enhances the drug's circulation half-life and potential cumulative drug delivery to the target tissues.
- After intravenous administration, neutral clozapine nanoparticles showed higher apparent clearance and lower circulation half-life compared with positively charged nanoparticles. This suggests higher delivery of positively charged nanoparticles than neutral and negatively charged nanoparticles.
- After oral administration, negatively charged nanoparticles had higher apparent clearance and lower circulation half-life than positively

TABLE 2 Characteristics of the Drug-Loaded Nanoparticles, Animals Used, Doses, and the Data Reference Used by Kadam et al. (2012)

Drug name	NPs	Particle size	Particle charge	Route	Species	Dose	References
		nm				mg/kg	
Camptothecin IT-101	1	78	N.A.	i.v.	F SDR	1	Schluep et al. (2006)
9-Nitrocamptothecin	2	207 ± 26	N.A.	i.v.	MWR	2	Dadashzadeh et al. (2008)
Epirubicin	3	208	(−ve)	i.v.	MWR	10	Li et al. (2010)
Vinpocetine	4	70.3 ± 7.8	(−ve)	Oral	MWR	10	Luo et al. (2006)
Clozapine	5	233 ± 13	(+ve)	Du, i.v.	MWR	20	Manjunath and Venkateswarlu (2005)
Clozapine	6	163 ± 0.7	Neutral	Du, i.v.	MWR	20	El-Shabouri (2002)
Cyclosporine (+ve)	7	148 ± 29	(+ve)	Oral	BD	7.5	El-Shabouri (2002)
Cyclosporine (ive)	8	104 ± 18	(−ve)	Oral	BD	7.5	El-Shabouri (2002)

Abbreviations: N.A., information not available; +ve, positively charged; −ve, negatively charged nanoparticleds; Du, duodenal exposure; i.v., intravenous exposure; F, female; SD, Sprague Dawley; W, wistar; R, rats; D, dog; and b, beagle.
Nanoparticles: 1: Polymer conjugate of camptothecin and β-cyclodextrin-based polymer. 2: Nanoparticles made with PLGA (50:50) and polyvinyl alcohol (PVA). (Yin et al., 2006) 3: Self-assembled system made with carboxymethyl curdlan coupled with cholesterol chitosan. 4: Solid-lipid nanoparticles made with glycerol monostearate, polysorbate 80, and soya lecithin. 5: Solid-lipid nanoparticles made with triglyceride, phosphatidylcholine, poloxamer 188, and tripalmitin along with stearylamine. 6: Solid-lipid nanoparticles made with triglyceride, phosphatidylcholine, poloxamer 188, and tripalmitin. 7: Nanoparticles made with lecithin, poloxamer 188, and chitosan. 8: Nanoparticles made with lecithin, poloxamer 188, and sodium glycocholate.

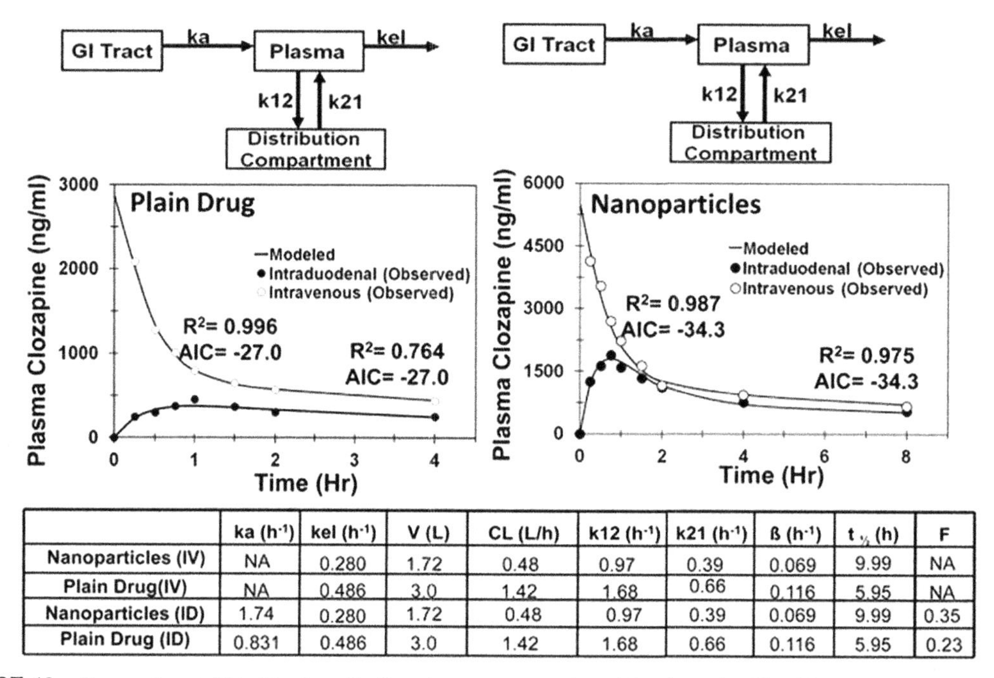

	ka (h^{-1})	kel (h^{-1})	V (L)	CL (L/h)	k12 (h^{-1})	k21 (h^{-1})	ß (h^{-1})	$t_{1/2}$ (h)	F
Nanoparticles (IV)	NA	0.280	1.72	0.48	0.97	0.39	0.069	9.99	NA
Plain Drug(IV)	NA	0.486	3.0	1.42	1.68	0.66	0.116	5.95	NA
Nanoparticles (ID)	1.74	0.280	1.72	0.48	0.97	0.39	0.069	9.99	0.35
Plain Drug (ID)	0.831	0.486	3.0	1.42	1.68	0.66	0.116	5.95	0.23

FIGURE 42 Compartmental bio-kinetics of bulk and nano-preparation of the clozapine. Top: Two-compartment model for bulk- and nano-formulations (ka, absorption rate constant from GI tract to the plasma; k_{el}, elimination rate constant from the plasma compartment; k_{12}, rate constant for transfer of drug from the plasma compartment to the distribution compartment; k_{21}, rate constant for transfer of drug from the distribution compartment to the plasma compartment). Middle: Model-predicted (line) and observed (data points) concentrations of clozapine in plasma after intraduodenal and intravenous administration of the two drug forms in male Wistar rats. The data fit a two-compartment model for both formulations. Bottom: A table showing the values for different kinetic parameters.

charged clozapine nanoparticles. They suggested that, in addition to reduced clearance in the lumen of the GI tract, liver, and circulation, reduced clearance from tissues at the site of administration or absorption may also contribute to enhanced drug exposure with positively charged nanoparticles.

The effects of a nanoparticle's surface properties, including surface charge, hydrophobicity, and functional groups, on PK are controversial. Levchenko et al. (2002) showed that negatively charged liposomes (200 nm) have higher plasma clearance than the neutral liposomes, while Yamamoto et al. (2001) and Roser et al. (1998) reported that surface charge has no effect on the plasma clearance of polymer and albumin nanoparticles (500–600 nm). Piskin et al. (1994) and Alexis et al. (2008) hypothesized that positively charged nanoparticles or have high nonspecific internalization and short blood circulation half-life, indicating lower availability; this is in contrast to other studies showing that, after intravenous administration, positively charged clozapine nanoparticles have longer apparent circulation half-life and lower apparent clearance than neutral charged nanoparticles (Xu et al. (2009) and Chonn et al. (1992)). In Addition, these investigators also showed that cationic nanoparticles have 9- and 31-fold longer circulation half-life than untreated and anionic polymeric nanoparticles after intravenous administration in rats.

4.2.1.2 EXAMPLE-2: EFFECTS OF SURFACE CHARGE ON A NANOPARTICLE'S PHARMACOKINETICS

For effective medicinal properties, a nanoparticle must overcome biological barriers, release the drug at the target site, and resist rapid clearance. Nanoparticle size is one of the key determinants of a nanoparticle's biodistribution. Another important but little studied factor determining the biodistribution and PK of nanoparticles is surface charge. A study by Arvizo et al. (2011) has demonstrated the possible role of the surface charge of gold nanoparticles in modulating the PK and the tumor uptake of the neutral (TEGOH), positively charged (TTMA), negatively charged (TCOOH), and zwitterion (TZwit) nanoparticles (Figure 43(A)). Because nanoparticle PK is dependent on the route of injection, mice were administered each NP via intravenous (iv) or intraperitoneal (ip) injection (Figure 43(B)). As shown in Figure 43(C), the *zeta* potential for the NPs exhibited the following pattern: TEGOH (neutral) < TZwit (+/−) << TTMA (positive) < TCOOH (negative). From this data, one would expect that TTMA and TCOOH would exhibit greater stability in vivo.

In *iv*-injected mice, TEGOH, TZwit, and TCOOH exhibited higher C_o (concentration immediately after injection) than TTMA, but TCOOH exhibited a rapid decline (Figure 43(B)) and significantly higher $t_{1/2}$ values (Figure 43(C)). TTMA exhibited the highest $t_{1/2}$ values, which may explain the lowest plasma TTMA concentrations (Figure 43(C)). In *ip*-injected mice, TEGOH and TZwit exhibited the highest plasma concentration, although TZwit levels were lower than TEGOH levels (reflected by the AUC values (Figure 43(B) and (C))). Neutral and zwitterionic particles provide high systemic exposure and low clearance when administered through *iv* administration. In addition, they are rapidly absorbed in the circulation after *ip* administration. Negative particles provide moderate systemic exposure, but positive particles clear rapidly. After *ip* administration, positive and negative particles both are absorbed poorly in the circulation, indicating the inability of these particles to cross the

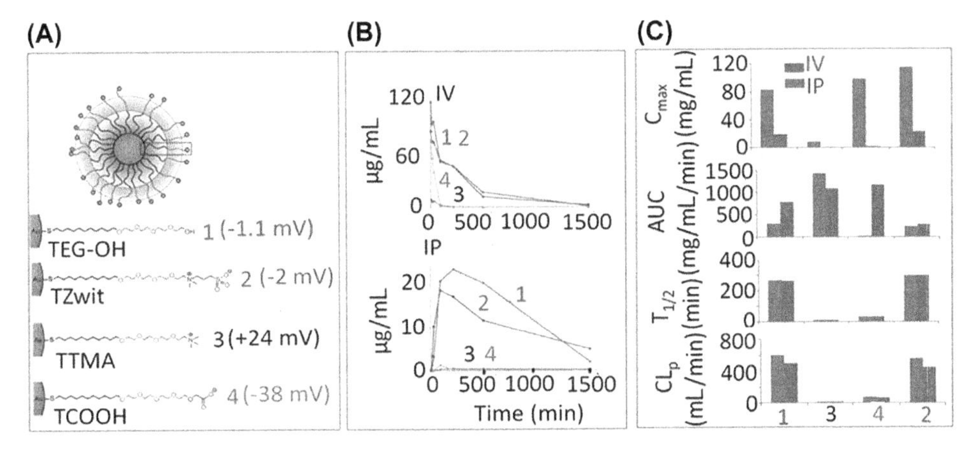

FIGURE 43 (A): Structure of engineered gold nanoparticles functionalized with −OH (1, TEG-OH, neutral particle), $-NH_3^+$ (3, T_{TMA}, positively charged), −COOH (4, TCOOH, negative charge), and $-NH_3^+(CH_2)SO_3^-$ (2, T_{Zwit}, zwitterion). (B): Plasma Au concentrations in rats exposed to Au nanoparticles via intravenous (iv) and intra-peritoneal (IP) injection. (C): Pharmacokinetic analysis of Au nanoparticles in rats administered different nanoparticles.

peritoneal barrier. These observations suggest that surface charge is a key determinant of a nanoparticle's bioavailability.

4.2.2 Physiologically Based Models

4.2.2.1 EXAMPLE-3: A PHYSIOLOGICALLY BASED PHARMACOKINETIC MODEL FOR IONIC SILVER AND SILVER NANOPARTICLES (BACHLER ET AL., 2013)

Biokinetic modeling provides insight into the relationships between an external dose and nanoparticle distribution in internal organs and blood, and their elimination from the body, which are increasingly used in a variety of risk assessments (Christensen et al., 2010). Bachler et al. (2013) developed different models for Ag^+ and NanoAg, as shown in Figure 44. In general, ionic silver (Ag^+) is an atom of silver (Ag^0) missing an electron that gives it a positive charge, while nanoAg is a colloid of Ag^0. Once in an organism, Ag^+ rapidly binds Cl^- and forms AgCl, resulting in their rapid excretion and lower toxicity, while the Ag nanoparticles (nanoAg), having distinct physicochemical and electronic properties, behave differently than Ag^+. Therefore, the same PBPK/PBTK model could not describe Ag^+ and nanoAg. The dual approach allowed Bachler et al. (2013) to evaluate the biokinetics of Ag^+ and nanoAg in different tissues following five exposure scenarios (dietary intake, use of three separate consumer products, and occupational exposure) in rats (experimental) and in humans (predicted by the model). The key difference between the two models is the enhanced role of mononuclear phagocyte system (MPS) in elimination of nanoAg. This is because the MPS, mainly the liver and the spleen, rapidly clear metallic and polymeric nanoparticles from body (Ferguson et al., 2013; Brandenberger et al., 2010; Christensen et al., 2010; Decuzzi et al., 2010; Owens and Peppas, 2006). The models developed predicted ADME in rats and humans.

Figure 45 shows the model-predicted organ Ag levels in rats and their comparison with the experimentally determined values for humans. The model-predicted values generally agreed with the experimentally determined values. The rate of nanoAg uptake in the liver (key MPS organ) increased with an increase in the blood Ag level. The liver uptake in low-, middle-, and high-dose groups was 0%, 25%, and 100% of the original rate, respectively (original dated from Lankveld et al., 2010). This suggests that the nanoAg uptake in the MPS compartments is negligible below a blood silver concentration of 180 ng/g. The Ag^+ is not taken up at any Ag concentration. The model also suggests that Ag^+ is converted into silver nanoparticles in the intestine (Figure 45(C-b)). The models also reasonably predicted the values (not confirmed by the experimental values) for the time course of the

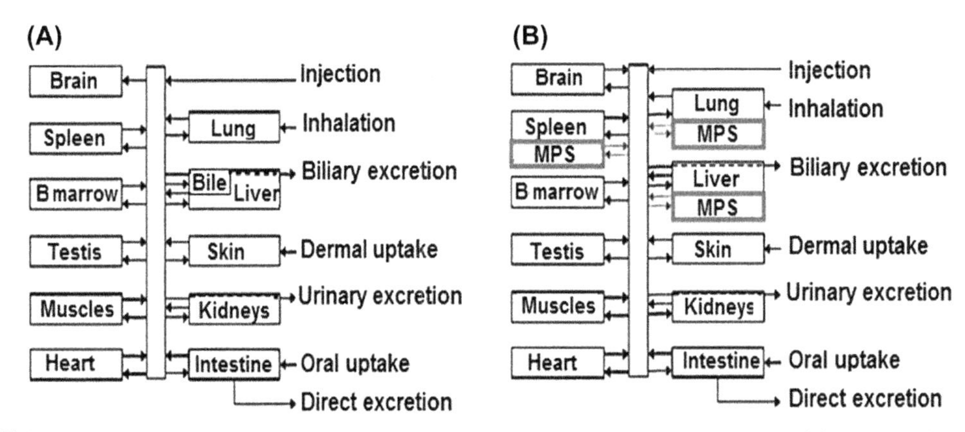

FIGURE 44 Schematic diagram of PBPK models for (A) ionic silver and (B) nano-silver. The models were used to predict the silver concentrations in rats and in humans. Abbreviations: MPS, mononuclear phagocyte system; PBPK, physiologically based pharmacokinetic. *Reprinted from Bachler et al. (2013) with permission.*

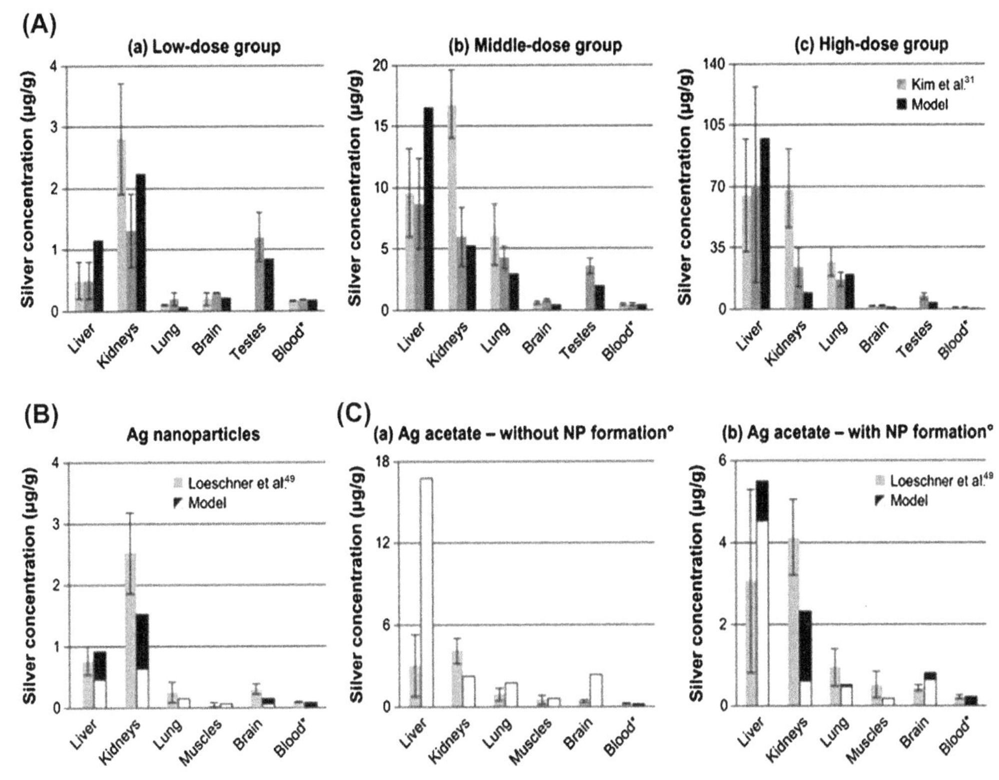

FIGURE 45 The experimental and model-simulated toxicokinetic data for 60 nm or 14 nm nanoparticles in rats after oral exposure for 28 days. (A) 60 nm *particles*: 30 (a), 300 (b), and 1000 (c) mg/kg/day dose, (B) 14 nm *particles*: 12.6 mg/kg/day and (C) 9 mg/kg/day silver acetate without (a) and with (b) consideration of the formation of nanoparticles in the GI tract. *Light gray*: female rats; dark gray: male rats; *black*: nanosilver; *white*: ionic silver. Error bars represent standard deviations; n = 5 to n = 10. In general, the predicted and experimental results were comparable. However, some differences (especially in kidney) were noted. *Abbreviations*: PBPK, physiologically based pharmacokinetic; NP, nanoparticle; Ag, silver; GI, gastrointestinal. *Reprinted from Bachler et al. (2013) with permission. They used data from Kim et al. (2008) and Loeschner et al. (2011) for model development.*

biodistribution of (A) Ag^+ and (B) nanoAg in the human body (organs with predicted concentrations higher than 2 ng/g are shown). A key difference between Ag^+ and nanoAg is the nanoAg's accumulation in the bone marrow. This shows that bone marrow may clear large amounts of nanoAg from the blood circulation. The predicted observation for bone marrow is supported by the observations that silver and other metal nanoparticles selectively induce necrosis and inflammation in bone marrow (Dobrzyńska et al., 2014; Syama et al., 2014; Iannitti et al., 2010). Taken together, this study demonstrates that PBPK models can provide important mechanistic information regarding the nanoparticles in vivo. The uptake of nanoAg by the MPS compartments is negligible below a threshold concentration of 180 ng silver/gram blood. Most surface coatings do not influence nanoAg biodistribution, which might be attributed to the presence of opsonization proteins in the corona (Figure 46).

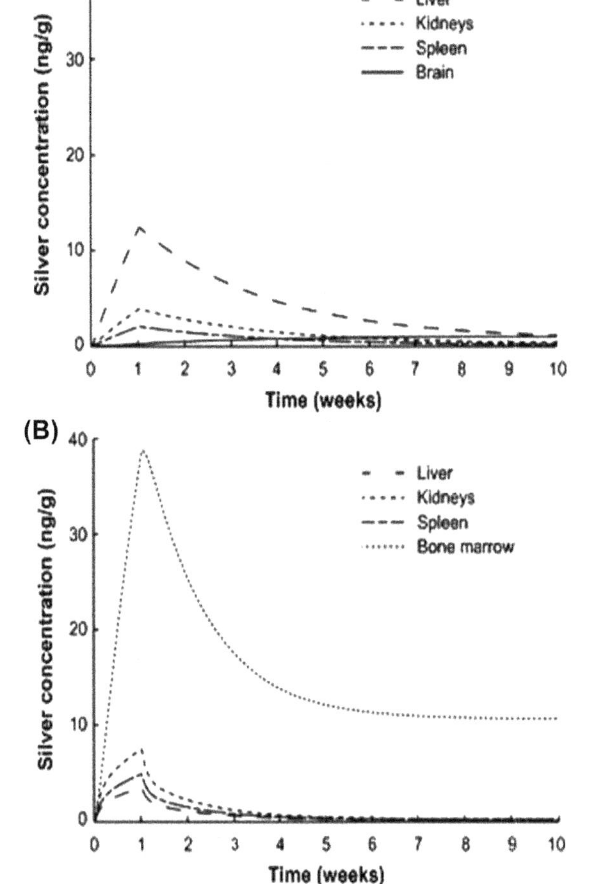

FIGURE 46 Comparison of the biodistribution of silver ion and nanosolver predicted by the PBPK model for humans. (A) Predicted release of ionic silver. (B) Predicted release of silver nanoparticles. *Abbreviation*: PBPK, physiologically based pharmacokinetic.

4.2.3 Noncompartmental Model

4.2.3.1 EXAMPLE 4: PLASMA, TUMOR, AND TISSUE PHARMACOKINETICS OF DOCETAXEL DELIVERED VIA NANOPARTICLES OF DIFFERENT SIZES AND SHAPES IN MICE BEARING SKOV-3 HUMAN OVARIAN CARCINOMA XENOGRAFT

Chu et al. (2013) have studied possible interdependent effects of size and shape on the biodistribution of unprocessed docetaxel (Doc) and its nanopreparations using 2D particle replication in nonwetting templates (PRINT) nanoparticles with dimensions of 200 × 200 nm (PRINT-Doc-200 × 200) or 80 × 320 nm (PRINT-Doc-80 × 300). PRINT is a top-down fabrication technique that produces size- and shape-specific particles (Jeong et al., 2010; Rolland et al., 2005). Using unprocessed Doc and both nanoparticles, PRINT-Doc-80 × 300 and PRINT-Doc-200 × 200, Chu et al. (2013) demonstrated that nano-formulation improved plasma PK and tumor delivery of Doc compared to the unprocessed Doc. Differences in clearance for the two PRINT particles suggest that shape plays a role in reducing clearance by the MPS and enhancing tumor delivery. The researchers were able to determine the drug PK for different tissues, including tumor cells. The overall noncompartmental procedure is described in the legend of Figure 47.

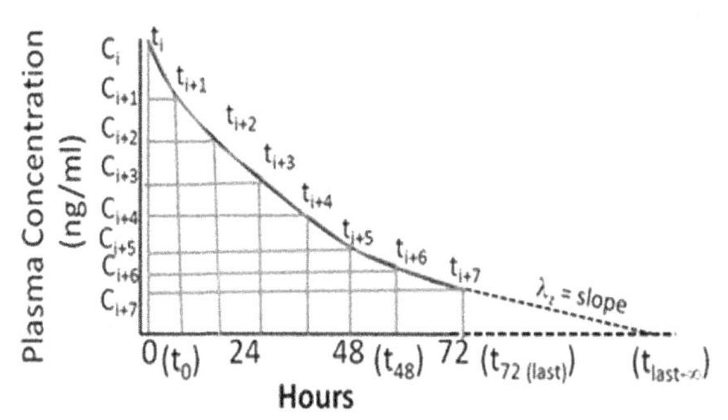

FIGURE 47 Principle of noncompartmental pharmacokinetic analysis. In non-compartmental analysis, the trapezoidal rule that approximates the area under a curve (AUC) is used. This requires dividing the area into a number of strips of equal width. Then, the area of each strip is approximated by the area of the trapezium formed. The sum of these approximations gives the final numerical result of the area under the curve. The general equations for AUC and mean residence time (MRT) is shown below.

$$\mathrm{AUC} = \int_0^\infty Cdt = \frac{C_i - \left(C_{i-(i+1)}\right)\cdot\left(t_{(i+1)} - t_i\right.}{\left(Ln\, C_i - Ln\, C_{(i+1)}\right.} + \frac{Cp_{last}}{\lambda_z}$$

$$\mathrm{MRT} = \int_0^\infty (tC\cdot dt)/\mathrm{AUC}$$

In these equations, C is concentration, lower case I represents the area of a strip and *t* is the time.

Female C.B.-17 SCID mice, aged 6–8 weeks and 14–18 g in body weight, were injected with cancer cells (5.0×10^6 cells in 200 μL 1 × PBS) subcutaneously (SC) into the right flank of each mouse. Tumor volume was calculated using the formula: tumor volume (mm^3) = $(w^2 \times l)/2$, where w = width and l = length in millimeters of the tumor. Forty-two days after injection, the animals were injected with the drug, PRINT-Doc-200 × 200, or PRINT-Doc-80 × 320 (10 mg/kg Doc dose) via the tail vein. Mice were sacrificed at different time intervals after the injection; blood, tissues and the tumor samples, were collected and analyzed for the drug using mass spectrometry. Pharmacokinetic indices were calculated using a noncompartmental software WinNonlin Professional Edition version 5.2.1 (Pharsight Corp, Cary, NC). The AUC was calculated using the linear up/log down rule. Volume of distribution (Vd) and clearance (CL) were calculated using standard equations. The maximum concentration (C_{max}), time of C_{max} (T_{max}), last measured concentration (C_{last}), and time of C_{last} (T_{last}) were determined by visual inspection of the concentration versus time curve data.

The PK values are shown in Table 3. The docetaxel plasma C_{max} values were approximately 20-fold higher for both PRINT particles compared to nonfunctionalized Doc. The volume of distribution (Vd) of nonfunctionalized Doc was about 18-fold higher than those for PRINT-Doc-80 × 320 and 33-fold higher than those for PRINT-Doc-200 × 200. The prolonged duration of Doc in plasma, when dosed with PRINT formulations, subsequently led to an increased tumor exposure of Doc from 0 to 168 h (about 53% higher for PRINT-Doc-80 × 320 and about 76% higher for PRINT-Doc-200 × 200 particles). Thus, the use of particles with a smaller feature size may be preferred to decrease clearance by organs of the MPS. These observations suggest that the noncompartmental method provides rapid calculation of PK indices that can be used in the clinical setting for establishing doses or determining the clearance of drugs, toxicants, and nanoparticles.

TABLE 3 Pharmacokinetic Parameters for Taxotere, PRINT-Doc-200 × 200 and PRINT-Doc-80 × 320

			Formulation		
Specimen	**Parameter**	**Units**	**Taxotere (Doc)**	**PRINT-Doc-200 × 200**	**PRINT-Doc-80 × 320**
Plasma	AUC_{0-t}	ng/mL·h	5809 (0–72 h) 5358 (0–24 h)	138,359 (0–168 h) 134,429 (0–72 h) 136,942 (0–24 h)	136,419 (0–168 h) 133,462 (0–72 h) 132,041 (0–24 h)
	C_{max}	ng/mL	10,550 ± 1,431	75,167 ± 2843	52,233 ± 3656
	CL	mL/h/kg	1757	72	73
	V_d	mL/kg	8509	257	474
Tumor	AUC_{0-t}	ng/mL·h	224,481(0–168 h) 63,197 (0–24 h)	396,104 (0–168 h) 79,644 (0–24 h)	342,937 (0–168 h) 95,516 (0–24 h)
	C_{max}	ng/mL	3564 ± 949	4413 ± 323	4187 ± 23
	T_{max}	h	1	1	6

(*Continued*)

TABLE 3 Pharmacokinetic Parameters for Taxotere, PRINT-Doc-200 × 200 and PRINT-Doc-80 × 320—cont'd

Specimen	Parameter	Units	Formulation		
			Taxotere (Doc)	PRINT-Doc-200 × 200	PRINT-Doc-80 × 320
	C_{last}	ng/mL	568	2321	757
	T_{last}	H	168	168	168
Liver	AUC_{0-t}	ng/mL·h	10,902 (0–24 h)	264,273(0–168 h) 87,255 (0–24 h)	91,770 (0–168 h) 64,274 (0–24 h)
	C_{max}	ng/mL	15,167 ± 2,324	13,400 ± 964	10,920 + 2275
	C_{last}	ng/mL	19.4	682	16.3
	T_{last}	h	24	168	168
Spleen	AUC_{0-t}	ng/mL·h	13,298 (0–24 h)	2,258,411 (0–168 h) 712,579 (0–24 h)	470,351 (0–168 h) 324,978 (0–24 h)
	C_{max}	ng/mL	2947 ± 605	32,333 ± 5659	18,038 ± 6260
	C_{last}	ng/mL	59.6	4106.3	67.8
	T_{last}	h	24	168	168
Lung	AUC_{0-t}	ng/mL·h	13,575 (0–72 h)	62,791.9 (0–168 h) 50,887 (0–72 h)	44,377 (0–72 h)
	C_{max}	ng/mL	4070 ± 590	7280 ± 225	4967 ± 709
	C_{last}	ng/mL	21.3	43.9	46.3
	T_{last}	h	72	168	72

5. CONCLUSIONS

PK and TK characterize the ADME of drugs, toxicants, and nanoparticles in organisms. In both PK and TK, the distribution of an administered drug, toxicant, or nanoparticle is compared to the time course of observed effects. PK, because of its application in drug design and synthesis, has evolved into a highly interactive discipline. TK is relatively recent but is based on similar principles. In the beginning, the TK studies were carried out to demonstrate that the newly developed drugs do not cause toxicity at pharmacologically relevant doses (Figure 48).

To achieve this, ADME at toxic doses and the occurrence and time course of toxic events are measured. Unfortunately, different studies have used different dose levels and experimental

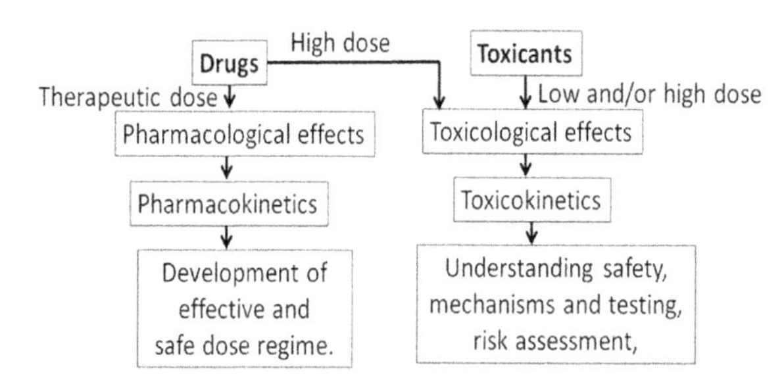

FIGURE 48 Relationship between pharmacokinetics and toxicokinetics of nanoparticles.

designs to establish solubility, stability, absorption, presystemic clearance, protein binding, and metabolism. These differences prevent direct comparison of their results. For bulk particles, the dose (in terms of mass/unit body weight) determined the effects (response). However, for nanoparticles, the dose (in terms of mass/unit body weight) may not correlate with the pharmacological or toxicological response. The dose in terms of size (diameter or specific area/unit body weight) or reactivity (activity/unit body weight) may be more important than the dose in terms of mass. In addition, the nanoparticle's size, shape, and surface properties also play a role in determining its biokinetics (PK and TK). Therefore, the Ag^+, 10-nm nanoAg, and 100-nm nanoAg may exhibit distinct PK and TK. Inside the body, nanoparticles adsorb plasma and/or intracellular proteins and form protein corona that significantly alter their PK. This chapter provided a detailed description of bulk and nanoparticle PK and TK, and their relevance to nanoparticle toxicity.

References

Alex, M.R.A., Chacko, A.J., Jose, S., Souto, E.B., 2011. Lopinavir loaded solid lipid nanoparticles (SLN) for intestinal lymphatic targeting. Eur. J. Pharm. Sci. 42, 11–18.

Alexis, F., Pridgen, E., Molnar, L.K., Farokhzad, O.C., 2008. Factors affecting the clearance and biodistribution of polymeric nanoparticles. Mol. Pharm. 5, 505–515.

Alkilany, A.M., Murphy, C.J., 2010. Toxicity and cellular uptake of gold nanoparticles: what we have learned so far? J. Nanopart. Res. 12, 2313–2333.

Allen, B.L., Kichambare, P.D., Gou, P., Vlasova, I.I., Kapralov, A.A., Konduru, N., Kagan, V.E., Star, A., 2008. Biodegradation of single-walled carbon nanotubes through enzymatic catalysis. Nano Lett. 8, 3899–3903.

Allen, B.L., Kotchey, G.P., Chen, Y., Yanamala, N.V.K., Klein-Seetharaman, J., Kagan, V.E., Star, A., 2009. Mechanistic investigations of horseradish peroxidase-catalyzed degradation of single-walled carbon nanotubes. J. Am. Chem. Soc. 131, 17194–17205.

Alvarez-Romàn, R., Naik, A., Kalia, Y.N., Guy, R.H., Fess, H., 2004. Skin penetration and distribution of polymeric nanoparticles. J. Controlled Release 99, 53–62.

Andon, F.T., Kapralov, A.A., Yanamala, N., Feng, W., Baygan, A., Chambers, B.J., Hultenby, K., Ye, F., Torak, M.S., Brandner, B.D., Fornara, A., Klein-Seetharaman, J.K., Kotchey, G.P., Star, A., Shvedova, A.A., Fadeel, B., Kagan, V.E., 2013. Biodegradation of single-walled carbon nanotubes by eosinophil peroxidase. Small 16, 2721–2729.

Arbab, S., Wilson, L.B., Ashari, P., Jordan, E.K., Lewis, B.K., Frank, J.A., 2005. A model of lysosomal metabolism of dextran coated superparamagnetic iron oxide (SPIO) nanoparticles: implications for cellular magnetic resonance imaging. NMR Biomed. 18, 383–389.

Arvizo, R.R., Miranda, O.R., Moyano, D.F., Walden, C.A., Giri, K., et al., 2011. Modulating pharmacokinetics, tumor uptake and biodistribution by engineered nanoparticles. PLoS One 6, e24374.

Asgharian, B., Yu, C.P., 1988. Deposition of inhaled fibrous particles in the human lung. J. Aerosol Med. 1, 37–50.

Azevedo, A.M., Martins, V.C., Prazeres, D.M.F., Vojinović, V., Cabral, J.M.S., Fonseca, L.P., 2003. Horseradish peroxidase: a valuable tool in biotechnology. Biotechnol. Annu. Rev. 9, 199–247.

Bachler, G., von Goetz, N., Hungerbühler, K., 2013. A physiologically based pharmacokinetic model for ionic silver and silver nanoparticles. Int. J. Nanomed. 8, 3365–3382.

Baek, M., Kim, M.K., Cho, H.J., Lee, J.A., Yu, J., Chung, H.E., Choi, S.J., 2011. Factors influencing the cytotoxicity of zinc oxide nanoparticles: particle size and surface charge. J. Phys. Conf. Ser. 304, 012044.

Balogh, L., Nigavekar, S.S., Nair, B.M., Lesniak, W., Zhang, C., Sung, L.Y., Kariapper, M.S., El-Jawahri, A., Llanes, M., Bolton, B., et al., 2007. Significant effect of size on the in vivo biodistribution of gold composite nanodevices in mouse tumor models. Nanomed. Nanotechnol. Biol. Med. 3, 281–296.

Baroli, B., Ennas, M.G., Loffredo, F., Isola, M., Pinna, R., Lopez-Quintela, M.A., 2007. Penetration of metallic nanoparticles in human full-thickness skin. J. Invest. Dermatol 127, 1701–1712.

Baron, P.A., Willeke, K., 1964. Aerosol Fundamentals, Aerosol Measurement: Principles, Techniques, and Applications.

Barry, B.W., 2001. Novel mechanisms and devices to enable successful transdermal drug delivery. Eur. J. Pharm. Sci. 14, 101–114.

Berge, T.W., 2009. A simple dermal absorption model: derivation and application. Chemosphere 75, 1440–1445.

Bessems, J.G.M., Geraets, L., 2013. Proper knowledge on toxicokinetics improves human hazard testing and subsequent health risk characterization: a case study. Regul. Toxicol. Pharmacol. 67, 325–334.

Bharatwaj, B., Dimovski, R., Conti, D.S., da Rocha, S.R.P., 2014. Polymeric nanocarriers for transport modulation across the pulmonary epithelium: dendrimers, polymeric nanoparticles, and their nanoblends. AAPS J. 16, 522–538.

Bonina, F.P., Montenegro, L., Scrofani, N., Esposito, E., Cortesi, R., Menegatti, E., Nastruzzi, C., 1995. Effects of phospholipid based formulations on in vitro and in vivo percutaneous absorption of methyl nicotinate. J. Controlled Release 34, 53–63.

Borm, P.J., 2002. Particle toxicology: from coal mining to nanotechnology. Inhal. Toxicol. 14, 311–324.

Borm, P.J., Kreyling, W., 2004. Toxicological hazards of inhaled nanoparticles–potential implications for drug delivery. J. Nanosci. Nanotechnol. 4, 521–531.

Bos, J.D., Meinardi, M.M., 2000. The 500 Dalton rule for the skin penetration of chemical compounds and drugs. Exp. Dermatol. 9, 165–169.

Boyd, B.J., Kaminskas, L.M., Karellas, P., Krippner, G.Y., Lessene, R., Porter, C.J.H., 2006. Cationic poly-L-lysine dendrimers: pharmacokinetics, biodistribution, and evidence for metabolism and bioresorption after intravenous administration to rats. Mol. Pharm. 3 (5), 614–627.

Brahmankar, D.M., Jaiswal, S.B., 1995. "Absorption of Drugs" Biopharmaceutics and Pharmacokinetics – A Treatise, First ed. Vallabh Prakashan, Delhi. pp. 5–75.

Brandenberger, C., Mühlfeld, C., Ali, Z., et al., 2010. Quantitative evaluation of cellular uptake and trafficking of plain and polyethylene glycol-coated gold nanoparticles. Small 6, 1669–1678.

Brodie, B.B., Kurz, H., Schanker, L.S., 1960. The importance of dissociation constant and lipid-solubility in influencing the passage of drugs into the cerebrospinal fluid. J. Pharmacol. Exp. Ther. 130, 20–25.

Browning, L.M., Huang, T., Xu, X.H.N., 2013. Real-time in vivo imaging of size-dependent transport and toxicity of gold nanoparticles in zebrafish embryos using single nanoparticle plasmonic spectroscopy. Interface Focus 3, 320120098.

van den Bergh, B.A., Bouwstra, J.A., Junginger, H.E., Wertz, P.W., 1999a. Elasticity of vesicles affects hairless mouse skin structure and permeability. J. Controlled Release 62, 367–379.

van den Bergh, B.A., Vroom, J., Gerritsen, H., Junginger, H.E., Bouwstra, J.A., 1999b. Interactions of elastic and rigid vesicles with human skin in vitro: electron microscopy and two-photon excitation microscopy. Biochim. Biophys. Acta 1461, 155–173.

Cai, F.S., Yu, C.P., 1988. Inertial and interceptional deposition of spherical particles and fibers in a bifurcating airway. J. Aerosol Sci. 19, 679–688.

Candice, A., Smith, C.A., Simpson, C.A., Kim, G., Carter, C.J., Feldheim, D.L., 2013. Gastrointestinal bioavailability of 2.0 nm diameter gold nanoparticles. ACS Nano 7, 3991–3996.

Casals, E., Pfaller, T., Duschl, A., Oostingh, G.J., Puntes, V., 2010. Time evolution of the nanoparticle protein corona. ACS Nano 4, 3623–3632.

Champion, J.A., Walker, A., Mitragotri, S., 2008. Role of particle size in phagocytosis of polymeric microspheres. Pharm. Res. 25, 1815–1821.

Chen, Y.S., Hung, Y.C., Liau, I., Huang, G.S., 2009. Assessment of the in vivo toxicity of gold nanoparticles. Nanoscale Res. Lett. 4, 858–864.

Cheung, W.K., Levy, G., 1989. Comparative pharmacokinetics of coumarin anticoagulants XLIX: nonlinear tissue distribution of S-warfarin in rats. J. Pharm. Sci. 78, 541–546.

Chonn, A., Semple, S.C., Cullis, P.R., 1992. Association of blood proteins with large unilamellar liposomes in vivo. Relation to circulation lifetimes. J. Biol. Chem. 267, 18759–18765.

Chou, L.Y.T., Ming, K., Chan, W.C.W., 2011. Strategies for the intracellular delivery of nanoparticles. Chem. Soc. Rev. 40, 233–245.

Christensen, F.M., Johnston, H.J., Stone, V., Aitken, R.J., Hankin, S., Peters, S., Aschberger, K., 2010. Nano-silver – feasibility and challenges for human health risk assessment based on open literature. Nanotoxicology 4, 284–295.

Chu, K.S., Hasan, W., Rawal, S., Walsh, M.D., Enlow, E.M., Luft, J.C., Bridges, A.S., Kuijer, J.L., Napier, M.E., Zamboni, W.C., DeSimone, J.M., 2013. Plasma, tumor and tissue pharmacokinetics of Docetaxel delivered via nanoparticles of different sizes and shapes in mice bearing SKOV-3 human ovarian carcinoma xenograft. Nanomed. Nanotechnol. Biol. Med. 9, 686–693.

Chun-mei, J.I., Cardosa, W.V., Gebremichael, A., Philpot, R.M., Buckpitt, A.R., Plopper, C.G., Pinkerton, K.E., 1995. Pulmonary cytochrome P-450 monooxygenase system and Clara cell differentiation in rats. Am. J. Physiol. 269, L393–L402.

Clewell, H.J., Tan, Y.M., Campbell, J.L., Andersen, M.E., 2008. Quantitative interpretation of human biomonitoring data. Toxicol. Appl. Pharmacol. 231, 122–133.

Clewell, H.J., Andersen, M.E., Barton, H.A., 2002. A consistent approach for the application of pharmacokinetic modeling in cancer and noncancer risk assessment. Environ. Health Perspect. 110, 85–93.

Clewell, H.J., Gearhart, J.M., Gentry, P.R., Covington, T.R., VanLandingham, C.B., Crump, K.S., Shipp, A.M., 1999. Evaluation of the uncertainty in an oral reference dose for methylmercury due to interindividual variability in pharmacokinetics. Risk Anal. 19, 547–558.

Clewell, H.J., Gentry, P.R., Covington, T.R., Sarangapani, R., Teeguarden, J.G., 2004. Evaluation of the potential impact of age- and gender-specific pharmacokinetic differences on tissue dosimetry. Toxicol. Sci. 79, 381–393.

Clewell, H.J., Gentry, P.R., Gearhart, J.M., Allen, B.C., Andersen, M.E., 2001. Comparison of cancer risk estimates for vinyl chloride using animal and human data with a PBPK model. Sci. Total Environ. 274, 37–66.

Clewell, H.J., Reddy, M.B., Lave, T., Andersen, M.E., 2007. Physiologically based pharmacokinetic modeling. In: Gad, S.C. (Ed.), Preclinical Development Handbook. Johnon Wiley and Sons, Hoboken, NJ.

Craddock, A.L., Loe, M.W., Daniel, R.W., Kirby, L.C., Walters, H.C., Wong, M.H., Dawson, P.A., 1998. Expression and transport properties of the human ileal and renal sodium-dependent bile acid transporter. Am. J. Physiol. 274, G157–G169.

Cronin, E., Stoughton, R.B., 1963. Nicotinic acid and ethyl nicotinate in excised human skin. Arch. Dermatol. 87, 445–449.

Cronin, E., Stoughton, R.B., 1962. Percutaneous absorption. Regional variations and the effect of hydration and epidermal stripping. Br. J. Dermatol. 74, 265–272.

Chan, M.S.L., Lowes, S., Hirst, B.H., 2004. The ABCs of drug transport in intestine and liver: efflux proteins limiting drug absorption and bioavailability. Eur. J. Pharm. Sci. 21, 25–51.

Dack, G.M., Wood, W.L., 1927. Impermeability of the small intestine of rabbits to botulinum toxin. J. Infect. Dis. 40, 585–587.

Dadashzadeh, S., Derakhshandeh, K., Shirazi, F.H., 2008. 9-Nitrocamptothecin polymeric nanoparticles: cytotoxicity and pharmacokinetic studies of lactone and total forms of drug in rats. Anticancer Drugs 19, 805–811.

Daigle, C.C., et al., 2003. Ultrafine particle deposition in humans during rest and exercise. Inhal. Toxicol. 15, 539–552.

Dallas, P., Sharma, V.K., Zboril, R., 2011. Silver polymeric nanocomposites as advanced antimicrobial agents: Classification, synthetic paths, applications, and perspectives. Adv. Colloid Interface Sci. 166, 119–135.

Daniel, W.A., 2003. Mechanisms of cellular distribution of psychotropic drugs. Significance for drug action and interactions. Prog. Neuropsychopharmacol. Biol. Psychiatry 27, 65–73.

Darquenne, C., 2012. Aerosol deposition in health and disease. J. Aerosol Med. Pulm. Drug Deliv. 25, 140–147.

Davis, A.F., Hadgraft, J., 1991. Effect of supersaturation on membrane transport: 1. Hydrocortisone acetate. Int. J. Pharm. 76, 1–8. J. Controlled Res. 62, 13–23.

Decuzzi, P., Godin, B., Tanaka, T., Lee, S.Y., Chiappini, C., Liu, X., Ferrari, M., 2010. Size and shape effects in the biodistribution of intravascularly injected particles. J Control Release 141, 320–327.

Deen, W.M., Lazzara, M.J., Myers, B.D., 2001. Structural determinants of glomerular permeability. Am. J. Physiol. Renal Physiol. 281, F579–F596.

Deng, X.Y., Jia, G., Wang, H.F., Sun, H.F., Wang, X., Yang, S.T., Wang, T.C., Liu, Y.F., 2007. Translocation and fate of multi-walled carbon nanotubes in vivo. Carbon 45, 1419–1424.

Dickschen, K., Willmann, S., Thelen, K., Lippert, J., Hempel, G., Eissing, T., 2012. Physiologically based pharmacokinetic modeling of tamoxifen and its metabolites in women of different CYP2D6 phenotypes provides new insight into the tamoxifen mass balance. Front. Pharmacol. 3, 92.

DiStefano III, J.J., 1982. Noncompartmental versus compartmental analysis: some bases for choice. Am. J. Physiol. 243, R1–R6.

Dobrzyńska, M.M., Gajowik, A., Radzikowska, J., Lankoff, A., Dusinska, M., Kruszewski, M., 2014. Genotoxicity of silver and titanium dioxide nanoparticles in bone marrow cells of rats in vivo. Toxicology 315, 86–91.

Domanska, U., Zołek-Tryznowska, Z., Pobudkowska, A., 2009. Separation of hexane/ethanol mixtures LLE of ternary systems (ionic liquid or hyperbranched polymer ethanol hexane) at T 298.15 K. J. Chem. Eng. Data 54, 972–976.

Domaska, U., Zołek-Tryznowska, Z., 2009. Mass-fraction activity coefficients at infinite dilution measurements for organic solutes and water in the hyperbranched polymer Boltorn W3000 using inveurse gas chromatography. J. Chem. Eng. Data 42, 363–370.

Donaldson, K., Aitken, R., Tran, L., Stone, V., Duffin, R., Forrest, G., Alexander, A., 2006. Carbon nanotubes: a review of their properties in relation to pulmonary toxicology and workplace safety. Toxicol. Sci. 92, 5–22.

Dressman, J.B., Amidon, G.L., Fleisher, D., 1985. Absorption potential: estimating the fraction absorbed for orally administered compounds. J. Pharm. Sci. 74, 588–589.

Dubey, R.K., Singh, J., 1988. Localization and characterization of drug-metabolizing enzymes along the villus-crypt surface of the rat small intestine. II. Conjugases. Biochem. Pharmacol. 37, 177–184.

Dunnhaupt, S., Barthelmes, J., Hombach, J., Sakloetsakun, D., Arkhipova, V., Bernkop-Schnurch, A., 2011. Distribution of thiolated mucoadhesive nanoparticles on intestinal mucosa. Int. J. Pharm. 408, 191–199.

Ehrhardt, C., Fiegel, J., Fuchs, S., Abu-Dahab, R., Schaefer, U.F., Hanes, J., Lehr, C.-M., 2002. Drug absorption by the respiratory mucosa: cell culture models and particulate drug carriers. J. Aerosol Med. 15, 131–139.

Elgrabli, D., Floriani, M., Abella-Gallart, S., Meunier, L., Gamez, C., Delalain, P., Rogerieux, F., Boczkowski, J., Lacroix, G., 2008. Biodistribution and clearance of instilled carbon nanotubes in rat lung. Part. Fibre Toxicol. 5, 20.

Elliott, R.L., Amidon, G.L., Lightfoot, E.N., 1980. A convective mass transfer model for determining intestinal wall permeabilities: laminal flow in a circular tube. J. Theor. Biol. 87, 757–771.

El-Shabouri, M.H., 2002. Positively charged nanoparticles for improving the oral bioavailability of cyclosporin-A. Int. J. Pharm. 249, 101–108.

Fellu, N., Walter, M.V., Montanez, M.I., Kunzmann, A., Hult, A., Nyström, A., Malkoch, M., Fadeel, B., 2012. Stability and biocompatibility of a library of polyester dendrimers in comparison to polyamidoamine dendrimers. Biomaterials 33, 1970–1981.

Ferguson, P.M., Feindel, K.W., Slocombe, A., MacKay, M., Wignall, T., Delahunt, B., Tilley, R.D., Hermans, I.F., 2013. Strongly magnetic iron nanoparticles improve the diagnosis of small tumours in the reticuloendothelial system by magnetic resonance imaging. PLoS One 8, e56572.

Fievez, V., Plapied, L., des Rieux, A., Pourcelle, V., Freichels, H., Wascotte, V., Vanderhaeghen, M.L., Jerôme, C., Vanderplasschen, A., Marchand-Brynaert, J., Schneider, Y.J., Préat, V., 2009. Targeting nanoparticles to M cells with non-peptidic ligands for oral vaccination. Eur. J. Pharm. Biopharm. 73, 6–24.

Filon, L.F., Adami, G., Venier, M., Maina, G., Renzi, N., 2007. In vitro percutaneous absorption of metal compounds. Toxicol. Lett. 170, 49–56.

Filon, L.F., Maina, G., Adami, G., Venier, M., Coceani, N., Bussani, R., Massiccio, M., Barbieri, P., Spinelli, P., 2004. In vitro percutaneous absorption of cobalt. Int. Arch. Environ. Health 77, 85–89.

Filon, L.F., D'Agostin, F., Crosera, M., Adami, G., Renzi, N., Bovenzi, M., Maina, G., 2009. Human skin penetration of silver nanoparticles through intact and damaged skin. Toxicology 255, 33–37.

Forstner, J., Taichman, N., Kalnins, V., Forstner, G., 1973. Intestinal goblet cell mucus: isolation and identification by immunofluorescence of a goblet cell glycoprotein. J. Cell Sci. 12, 585–602.

Frampton, M.W., Utell, M.J., Zareba, W., Oberdörster, G., Cox, C., Huang, L.S., Morrow, P.E., Lee, F.E., Chalupa, D., Frasier, L.M., Speers, D.M., Stewart, J., 2004. Effects of exposure to ultrafine carbon particles in healthy subjects and subjects with asthma. Res. Rep. Health Eff. Inst. 126, 1–47.

Fredriksson, T., 1961. Studies on the percutaneous absorption of parathion and paraoxon. II. Distribution of ^{32}P-labelled parathion within the skin. Acta Derm.-Venereol. 41, 344–352.

Fu, K., Huang, W., Lin, Y., Riddle, A., Carroll, D.L., Sun, Y.-P., 2001. Defunctionalization of functionalized carbon nanotubes. Nanoletters 1, 439–441.

Gabrielsson, J., Weiner, D., 2012. Non-compartmental analysis. Methods Mol. Biol. 929, 377–389.

Geys, J., Coenegrachts, L., Vercammen, J., Engelborghs, Y., Nemmar, A., Nemery, B., Hoet, P.H., 2006. In vitro study of the pulmonary translocation of nanoparticles: a preliminary study. Toxicol. Lett. 160, 218–226.

Geys, J., Nemery, B., Hoet, P.H., 2007. Optimisation of culture conditions to develop an in vitro pulmonary permeability model. Toxicol. In Vitro 21, 1215–1219.

Gibaldi, M., Perrier, D., 1982. Pharmacokinetics. Marcel Dekker, New York.

Gibaldi, M., 1984. Biopharmaceutics and Clinical Pharmacokinetics, third ed. Lea and Febiger, Philadelphia.

Gillespie, W.R., 1991. Noncompartmental versus compartmental modeling in clinical pharmacokinetics. Clin. Pharmacokinet. 20, 253–262.

Gutierrez, L., Mejías, R., Barber, D.F., Veintemillas-Verdaguer, S., Serna, C.J., Lázaro, F.J., Morales, M.P., 2011. AC magnetic susceptibility study of in vivo nanoparticle biodistribution. J. Phys. D: Appl. Phys. 44, 255002.

Gutierrez, L., Spasic, M.V., Muckenthaler, M.U., Lazaro, F.J., 2012. Quantitative magnetic analysis reveals ferritin-like iron as the most predominant iron-containing species in the murine Hfe-haemochromatosis. Biochim. Biophys. Acta 1822 (2012), 1147–1153.

Guy, R.H., Maibach, H.I., 1984. Correction factors for determining body exposure from forearm percutaneous absorption data. J. Appl. Toxicol. 4, 26–28.

Haddad, S., Krishnan, K., 1998. Physiological modeling of toxicokinetic interactions: implications for mixture risk assessment. Environ. Health Perspect. 106 (Suppl. 6), 1377–1384.

Hazelett, S.E., Preusch, P.C., 1988. Tissue distribution and warfarin sensitivity of vitamin K epoxide reductase. Biochem. Pharmacol. 37, 929–934.

He, C., Yin, L., Tang, C., Yin, C., 2012. Size-dependent absorption mechanism of polymeric nanoparticles for oral delivery of protein drugs. Biomaterials 33, 8569–8578.

Heckly, R.J., Hildebrand, G.J., Lamanna, C., 1960. On the size of the toxic particles passing the intestinal barrier in botulism. J. Exp. Med. 111, 745–759.

Hickey, A.J., 2014. Controlled delivery of inhaled therapeutic agents. J. Controlled Release 190, 182–188.

Higuchi, T.J., 1960. Physical chemical analysis of percutaneous absorption process from creams and ointments. J. Soc. Cosmet. Chem. 11, 85–87.

Hillyer, J.F., Albrecht, R.M., 2001. Gastrointestinal persorption and tissue distribution of differently sized colloidal gold nanoparticles. J. Pharm. Sci. 90, 1927–1936.

Hinds, W.C., 1999. Aerosol Technology: Properties, Behavior, and Measurement of Airborne Particles. John Wiley, New York, NY.

Hirn, S., Semmler-Behnke, M., Schleh, C., Wenk, A., Lipka, J., Schäffler, M., Takenaka, S., Möller, W., Schmid, G., Simon, U., Kreyling, W.G., 2011. Particle size-dependent and surface charge-dependent biodistribution of gold nanoparticles after intravenous administration. Eur. J. Pharm. Biopharm. 77, 407–416.

Hogben, C.A.M., 1959. The electrophysiology of the elasmobranch stomach. Science 129, 1224–1225.

Hogben, C.A.M., Schanker, L.S., Tocco, D.J., Brodie, B.B., 1957. Absorption of drugs from the stomach, II. The human. J. Pharmacol. Exp. Ther. 120, 540–545.

Honeywell-Nguyen, P.L., Gooris, G.S., Bouwstra, J.A., 2004. Quantitative assessment of the transport of elastic and rigid vesicle components and a model drug from these vesicle formulations into human skin in vivo. J. Invest. Dermatol. 123, 902–910.

Hostjnek, J.J., 1997. Modelling in vivo human skin absorption. Quant. Struct.-Act. Relat. 16, 473–479.

Hung, O.R., Whynot, S.C., Varvel, J.R., Shafer, S.L., Mezei, M., 1995. Pharmacokinetics of inhaled liposome-encapsulated fentanyl. Anesthesiology 83, 277–284.

Iannitti, Y., Capone, S., Gatti, A., Capitani, F., Cetta, F., Palmieri, B., 2010. Intracellular heavy metal nanoparticle storage: progressive accumulation within lymph nodes with transformation from chronic inflammation to malignancy. Int. J. Nanomed. 5, 955–960.

Ilett, K.F., Tee, L.B., Reeves, P.T., Minchin, R.F., 1990. Metabolism of drugs and other xenobiotics in the gut lumen and wall. Pharmacol. Ther. 46, 67–93.

Ito, K., Kusuhara, H., Sugiyama, Y., 1999. Effects of intestinal CYP3A4 and P-glycoprotein on oral drug absorption—-theoretical approach. Pharm. Res. (NY) 16, 225–231.

Jacquez, J.A., 1996. Compartmental Analysis in Biology and Medicine, third ed. BioMedware, Ann Arbor.

Jeong, W., Napier, M.E., DeSimone, J.M., 2010. Challenging nature's monopoly on the creation of well-defined nanoparticles. Nanomedicine (Lond) 5, 633–639.

Jores, K., Haberland, A., Wartewig, S., Mäder, K., Mehnert, W., 2005. Solid Lipid Nanoparticles (SLN) and oil-loaded SLN studied by Spectrofluorometry and Raman spectroscopy. Pharm. Res. 22, 1887–1897.

Kadam, R.S., Bourne, D.W.A., Kompella, U.B., 2012. Nano-advantage in enhanced drug delivery with biodegradable nanoparticles: contribution of reduced clearance. Drug Metab. Dispos. 40, 1380–1388.

Kagan, V.E., Konduru, N.V., Feng, W., Allen, B.L., Conroy, J., Volkov, Y., Vlasova, I.I., Belikova, N.A., Yanamala, N., Kapralov, A., Tyurina, Y.Y., Kisin, J.S., Murray, A.R., Franks, J., Stolz, D., Gou, P., Klein-Seetharaman, J., Fadeel, B., Star, A., Shvedova, A.A., 2010. Carbon nanotubes degraded by neutrophil myeloperoxidase induce less pulmonary inflammation. Nat. Nanotechnol. 5, 354–359.

Kaminskas, L.M., Boyd, B.J., Karellas, P., Krippner, G.Y., Lessene, R., Kelly, B., Porter, C.J.H., 2008. The impact of molecular weight and PEG chain length on the systemic pharmacokinetics of PEGylated poly l-lysine dendrimers. Mol. Pharm. 5, 449–463.

Kasting, J.F., 1987. Theoretical constraints on oxygen and carbon dioxide concentrations in the Precambrian atmosphere. Precambrian Res. 34, 205–229.

Keith, I.M., Olson, E.B., Wilson, N.M., Jefcoate, C.R., 1987. Immunological identification and effects of 3-methylcholanthrene and phenobarbital on rat pulmonary cyctochrome P-450. Cancer Res. 47, 1878–1882.

Kettler, K., Veltman, K., van de Meent, D., van Wezel, A., Hendriks, A.J., 2014. Cellular uptake of nanoparticles as determined by particle properties, experimental conditions, and cell type. Environ. Toxicol. Chem. 33, 481–492.

Khlebtsov, N., Dykman, L., 2011. Biodistribution and toxicity of engineered gold nanoparticles: a review of in vitro and in vivo studies. Chem. Soc. Rev. 40, 1647–1671.

Kim, Y.S., Kim, J.S., Cho, H.S., Rha, D.S., Kim, J.M., Park, J.D., Choi, B.S., Lim, R., Chang, H.K., Chung, Y.H., Kwon, I.H., Jeong, J., Han, B.S., Yu, I.J., 2008. Twenty-eight-day oral toxicity, genotoxicity, and gender-related tissue distribution of silver nanoparticles in Sprague-Dawley rats. Inhal. Toxicol. 20, 575–583.

Koepsell, H., 1998. Organic cation transporters in intestine, kidney, liver and brain. Annu. Rev. Physiol. 60, 243–266.

Kotchev, G.P., Allen, B.L., Vedala, H., Yanamala, N., Kapralov, A.A., Tyurina, Y.Y., Klein-Seetharaman, J., Kagan, V.E., Star, A., 2011. The enzymatic oxidation of graphene oxide. ACS Nano 5, 2098–2108.

Kotchey, G.P., Hasan, S.A., Kapralov, A.A., Ha, S.H., Kim, K., Shvedova, A.A., Kagan, V.E., Star, A., 2012. A natural vanishing act: the enzyme-catalyzed degradation of carbon nanomaterials. Acc. Chem. Res. 45, 1770–1781.

Kotchey, G.P., Zhag, Y., Kagan, V.E., 2013. Peroxidase-mediated biodegradation of carbon nanotubes in vitro and in vivo. Adv. Drug Deliv. Rev. 65, 1921–1932.

Krishnan, K., Andersen, M.E., 2007. Physiologically based pharmacokinetic modeling in toxicology. In: Hayes, A.W. (Ed.), Principles and Methods of Toxicology, fifth ed. Taylor & Francis, Boca.

Lam, C.W., James, J.T., McCluskey, R., Hunter, R.L., 2004. Pulmonary toxicity of single-wall carbon nanotubes in mice 7 and 90 days after intratracheal instillation. Toxicol. Sci. 77, 126–134.

Lankveld, D.P., Oomen, A.G., Krystek, P., Neigh, A., Troost-de Jong, A., Noorlander, C.W., Van Eijkeren, J.C., Geertsma, R.E., De Jong, W.H., 2010. The kinetics of the tissue distribution of silver nanoparticles of different sizes. Biomaterials 31, 8350–8361.

Laresea, F., D'Agostin, F., Croserab, M., Adamib, G., Renzic, N., Bovenzia, M., Mainad, G., 2009. Human skin penetration of silver nanoparticles through intact and damaged skin. Toxicology 255, 33–37.

Levchenko, T.S., Rammohan, R., Lukyanov, A.N., Whiteman, K.R., Torchilin, V.P., 2002. Liposome clearance in mice: the effect of a separate and combined presence of surface charge and polymer coating. Int. J. Pharm. 240, 95–102.

Li, B., Yang, X., Xia, L., Majeed, M.I., Tan, B., 2013. Hollow microporous organic capsules. Sci. Rep. 2013, 3.

Li, R., Eun, J.S., Lee, M.-K., 2011. Pharmacokinetics and biodistribution of paclitaxel loaded in pegylated solid lipid nanoparticles after intravenous administration. Arch. Pharm. Res. 34, 331–337.

Li, L., Gao, F.P., Tang, H.B., Bai, Y.G., Li, R.F., Li, X.M., Liu, L.R., Wang, Y.S., Zhang, Q.Q., 2010. Self-assembled nanoparticles of cholesterol-conjugated carboxymethyl curdlan as a novel carrier of epirubicin. Nanotechnology 21, 265601.

Lim, C.J., Shen, W.C., 2005. Comparison of monomeric and oligomeric transferrin as potential carrier in oral delivery of protein drugs. J. Controlled Release 106, 273–286.

Lim, E.K., Kang, B., Choi, Y., Jang, E., Han, S., Lee, K., Suh, J.S., Haam, S., Huh, Y.M., 2014. Gadolinium-based nanoparticles for highly efficient T1-weighted magnetic resonance imaging.

Lin, J.H., Chiba, M., Baillie, T.A., 1999. Is the role of the small intestine in first-pass metabolism overemphasized? Pharmacol Rev. 51, 135–158.

Lipinski, C.A., Lombardo, F., Dominy, B.W., Feeney, P.J., 1997. Experimental and computational approaches to estimate solubility and permeability in drug discovery and developmental settings. Adv. Drug Deliv. Rev. 46, 3–25.

Liu, J.-H., Anilkumar, P., Cao, L., Wang, X., Yang, S.T., Luo, P.G., Wanh, H., 2010a. Cytotoxicity evaluations of fluorescent carbon nanoparticles. Nano Life 01, 153.

Liu, X., Hurt, R.H., Kane, A.B., 2010b. Biodurability of single-walled carbon nanotubes depends on surface functionalization. Carbon N.Y. 48, 1961–1969.

Liu, J.-H., Yang, S.-T., Wang, H., Chang, Y., Cao, A., Liu, Y., 2012. Effect of size and dose on the biodistribution of graphene oxide in mice. Nanomedicine 7 (12), 1801–1812.

Loeschner, K., Hadrup, N., Qvortrup, K., et al., 2011. Distribution of silver in rats following 28 days of repeated oral exposure to silver nanoparticles or silver acetate. Part Fibre Toxicol. 8, 18.

Longmire, M., Choyke, P.L., Kobayashi, H., 2008. Clearance properties of nano-sized particles and molecules as imaging agents: considerations and caveats. Nanomedicine 3, 703–717.

Lopez, A., Gutierrez, L., Lazaro, F.J., 2007. The role of dipolar interaction in the quantitative determination of particulate magnetic carriers in biological tissues. Phys. Med. Biol. 52, 5043–5056.

Lundqvist, M., Stigler, J., Elia, G., Lynch, I., Cedervall, T., Dawson, K.A., 2008. Nanoparticle size and surface properties determine the protein corona with possible implications for biological impacts. Proc. Natl Acad. Sci. U.S.A. 105, 14265–14270.

Luo, Y., Chen, D., Ren, L., Zhao, X., Qin, J., 2006. Solid lipid nanoparticles for enhancing vinpocetine's oral bioavailability. J. Controlled Release 114, 53–59.

Lynch, I., Salvati, A., Dawson, K.A., 2009. What does the cell see? Fluorescence correlation spectroscopy is used as a quantitative method to understand the binding and exchange behavior of proteins on the surfaces of nanoparticles. Nat. Nanotechnol. 4, 546–547.

Magee, D.F., Dalley II, A.F., 1986. Digestion and the Structure and Function of the Gut. In: Karger Continuing Education Series, vol. 8. Karger, Basel.

Mager, D.E., Jusko, W.J., 2001. General pharmacokinetic model for drugs exhibiting target-mediated drug disposition. J. Pharmacokinet. Pharmacodyn. 28, 507–532.

Magnusson, B.M., Anissimov, Y.G., Cross, S.E., Roberts, M.S., 2004. Molecular size as the main determinant of solute maximum flux across the skin. J. Invest. Dermatol. 122, 993–999.

Magrez, A., Kasas, S., Salicio, V., Pasquier, N., Seo, J.W., Celio, M., Catsicas, S., Schwaller, B., Forro, L., 2006. Cellular toxicity of carbon-based nanomaterials. Nano Lett. 6, 1121–1125.

Mahendra, S., Zhu, H., Colvin, V.L., Alvarez, P.J., 2008. Quantum dot weathering results in microbial toxicity. Environ. Sci. Technol. 42, 9424–9430.

Maibach, H.I., Feldmann, R.J., Milby, T.H., Serat, W.F., 1971. Regional variation in percutaneous penetration in man. Arch. Environ. Health 23, 208–211.

Mancini, M.C., Kairdolf, B.A., Smith, A.M., Nie, S., 2008. Oxidative quenching and degradation of polymer-encapsulated quantum dots: new insights into the long-term fate and toxicity of nanocrystals in vivo. J. Am. Chem. Soc. 130, 10836–10837.

Manjunath, K., Venkateswarlu, V., 2005. Pharmacokinetics, tissue distribution and bioavailability of clozapine solid lipid nanoparticles after intravenous and intraduodenal administration. J. Controlled Release 107, 215–228.

Marathe, P.H., Wen, Y., Norton, J., Greene, D.S., Barbhaiya1, R.H., Wilding, I.R., 2000. Effect of altered gastric emptying and gastrointestinal motility on metformin absorption. Br. J. Clin. Pharmacol. 50, 325–332.

May, A.J., Whaler, B.C., 1958. The absorption of Clostridium botulinum type A toxin from the alimentary canal. Br. J. Exp. Pathol. 39, 307–316.

Mazzaferro, S., Bouchemal, K., Skanji, R., Gueutin, C., Chacun, H., Ponchel, G., 2012. Intestinal permeation enhancement of docetaxel encapsulated into methyl-beta-cyclodextrin/poly(isobutylcyanoacrylate) nanoparticles coated with thiolated chitosan. J. Controlled Release 162, 568–574.

McClements, D.J., 2013. Edible lipid nanoparticles: digestion, absorption, and potential toxicity. Prog. Lipid Res. 52, 409–423.

McDevitt, M.R., Chattopadhyay, D., Jaggi, J.S., Finn, R.D., Zanzonico, P.B., Villa, C., Rey, D., Rey, D., Mendenhall, J., Batt, C.A., Njardarson, J.T., Sheinberg, D.A., 2007a. PET imaging of soluble Yttrium-86-labeled carbon nanotubes in mice. PloS One 9, e907.

McDevitt, M.R., Chattopadhyay, D., Kappel, B.J., Jaggi, J.S., Schiffman, S.R., Antczak, C., Njardarson, J.T., Brentjens, R., Scheinberg, D.A., 2007b. Tumor tar-geting with antibody-functionalized, radiolabeled carbon nanotubes. J. Nucl. Med. 48, 1180–1189.

Mehnert, W., Mader, K., 2001. Solid lipid nanoparticles: production, characterization and applications. Adv. Drug Deliv. Rev. 47, 165–196.

Meidan, V.M., Bonner, M.C., Michniak, B.B., 2005. Transfollicular drug delivery—is it a reality? Int. J. Pharm. 306, 1–14.

Mejias, R., Gutiérrez, L., Salas, G., Pérez-Yagüe, S., Zotes, T.M., Lázaro, F.J., Morales, M.P., Barber, D.F., 2013. Long term biotransformation and toxicity of dimercaptosuccinic acid-coated magnetic nanoparticles support their use in biomedical applications. J. Controlled Release 171, 225–233.

Mezzenga, R., Pettersson, B., Manson, J.A., 2001. Thermodynamic evolution of unsaturated polyester-styrene-hyperbranched polymers. Polym. Bull. 46, 419–426.

Moghimi, S.M., Hunter, A.C., Andresen, T.L., 2011. Factors controlling nanoparticle pharmacokinetics: an integrated analysis and perspective. Annu. Rev. Pharmacol. Toxicol. 52, 481–503.

Moghimi, S.M., Hunter, A.C., Andresen, T.L., 2012. Factors controlling nanoparticle pharmacokinetics: an integrated analysis and perspective. Annu. Rev. Pharmacol. Toxicol. 52, 481–503.

Monopoli, M.P., Aberg, C., Salvati, A., Dawson, K.A., 2012. Biomolecular coronas provide the biological identity of nanosized materials. Nat. Nanotechnol. 7, 779–786.

Mu, Q., Jiang, G., Chen, L., Zhou, H., Fourches, D., Tropsha, A., Yan, B., 2014. Chemical basis of interactions between engineered nanoparticles and biological systems. Chem. Rev. 114, 7740–7781.

Mukherjee, S., Ray, S., Thakur, R.S., 2009. Solid lipid nanoparticles: a modern formulation approach in drug delivery system. Indian J. Pharm. Sci. 71, 349–358.

Mulders, S.J.E., Brouwer, A.J., Liskamp, R.M.J., 1997. Molecular diversity of novel amino acid based dendrimers. Tetrahedron Lett. 38, 3085–3088.

Muller, J., Huaux, F., Moreau, N., Misson, P., Heilier, J.F., Delos, M., Arras, M., Fonseca, A., Nagy, J.B., Lison, D., 2005. Respiratory toxicity of multi-wall carbon nanotubes. Toxicol. Appl. Pharmacol. 207, 221–231.

Muller, R.H., Rfihl, D., Runge, S.A., 1996. Biodegradation of solid lipid nanoparticles as a function of lipase incubation time. Int. J. Pharm. 144, 115–121.

Myojo, T., Takaya, M., 2001. Estimation of fibrous aerosol deposition in upper bronchi based on experimental data with model bifurcation. Ind. Health 39, 141–149.

Nam, J., Won, N., Bang, J., Jin, H., Park, J., Jung, S., Jung, S., Park, Y., Kim, Y.S., 2013. Surface engineering of inorganic nanoparticles for imaging and therapy. Adv. Drug Deliv. Rev. 65, 622–648.

Nangia, S., Sureshkumar, R., 2012. Effects of nanoparticle charge and shape anisotropy on translocation through cell membranes. Langmuir 21 (28), 17666–17671.

Nel, A., Xia, T., Mädler, L., Li, N., 2006. Toxic potential of materials at the nanolevel. Science 311, 622–627.

Nigavekar, S.S., Sung, L.Y., Llanes, M., El-Jawahri, A., Lawrence, T.S., Becker, C.W., Balogh, L., Khan, M.K., 2004. ^{3}H dendrimer nanoparticle organ/tumor distribution. Pharm. Res. 21, 476–483.

Oberdorster, G., Oberdorsters, E., Oberdorster, J., 2005. Nanotoxicology: an emerging discipline evolving from studies of ultrafine particles. Environ. Health Perspect. 113, 823–839.

Oberdorster, G., Sharp, Z., Atudorei, V., Elder, A., Gelein, R., Lunts, A., Kreyling, W., Cox, C., 2002. Extrapulmonary translocation of ultrafine carbon particle following whole-body inhalation exposure of rats. J. Toxicol. Environ. Health A 65, 1531–1543.

Oberdorster, G., Ferin, J., Lehnert, B.E., 1994. Correlation between particle size, in vivo particle persistence, and lung injury. Environ. Health Perspect. 102 (Suppl. 5), 173–179.

Olbrich, C., Kayser, O., Müller, R.H., 2002. Lipase degradation of Dynasan 114 and 116 Solid Lipid Nanoparticles (SLN)—effect of surfactants, storage time and crystallinity. Int. J. Pharm. 237, 119–128.

Ostrenga, J., Steinmetz, C., Poulsen, B., 1971. Significance of vehicle composition I: relationship between topical vehicle composition, skin penetrability and clinical efficacy. J. Pharm. Sci. 60, 1175–1183.

Owens, D.E., Peppas, N.A., 2006. Opsonization, biodistribution, and pharmacokinetics of polymeric nanoparticles. Int. J. Pharm. 307, 93–102.

Paine, M.F., Khalighi, M., Fisher, J.M., Shen, D.D., Kunze, K.L., Marsh, C.L., Perkins, J.D., Thummel, K.E., 1997. Characterization of interintestinal and intraintestinal variations in human CYP3A-dependent metabolism. J. Pharmacol. Exp. Ther. 283, 1552–1562.

Paine, M.F., Shen, D.D., Kunze, K.L., Perkins, J.D., Marsh, C.L., McVicar, J.P., Barr, D.M., Gilles, B.S., Thummel, K.E., 1996. First-pass metabolism of midazolam by the human intestine. Clin. Pharmacol. Ther. 60, 14–24.

Pandey, R., Sharma, A., Zahoor, A., Sharma, S., Khuller, G.K., Prasad, B., 2003. Poly (DL-lactide-co-glycolide) nanoparticle-based inhalable sustained drug delivery system for experimental tuberculosis. J. Antimicrob. Chemother. 52, 981–986.

Pang, K.S., 2003. Modeling of intestinal drug absorption: roles of transporters and metabolic enzymes. Drug Metab. Dispos. 31, 1507–1519.

Pannatier, A., Jenner, B., Testa, B., Etter, J.C., 1978. The skin as a drug-metabolising organ. Drug Metab. Rev. 8, 319–343.

Patton, J.S., Bukar, J., Nagarajan, S., 1999. Inhaled insulin. Adv. Drug Deliv. Rev. 35, 235–247.

Patton, J.S., Fishburn, C.S., Jeffry, G., Weers, J.G., 2004. The lungs as a portal of entry for systemic drug delivery. Proc. Am. Thorac. Soc. 1, 338–344.

Patton, J.S., 1996. Mechanisms of macromolecular absorption by the lungs. Adv. Drug Deliv. Rev. 19, 3–36.

Peppas, N.A., Hansen, P.J., Buri, P.A., 1984. A theory of molecular diffusion in the intestinal mucus. Int. J. Pharm. 20, 107–118.

Petit, B., Bouchemal, K., Vauthier, C., Djabourov, M., Ponchel, G., 2012. The counterbalanced effect of size and surface properties of chitosan-coated poly(isobutylcyanoacrylate) nanoparticles on mucoadhesion due to pluronic F68 addition. Pharm. Res. 29, 943–952.

Piskin, E., Tuncel, A., Denizli, A., Ayhan, H., 1994. Monosize microbeads based on polystyrene and their modified forms for some selected medical and biological applications. J. Biomater. Sci. Polym. Ed. 5, 451–471.

Potts, R.O., Guy, R.H., 1992. Predicting skin permeability. Pharm. Res. 9, 663–669.

Prescott, L.F., McAuslane, J.A., Freestone, S., 1991. The concentration-dependent disposition and kinetics of inulin. Eur. J. Clin. Pharmacol. 40, 619–624.

Purvin, S., Vuddanda, P.R., Singh, S.K., Jain, A., Singh, S., 2014. Pharmacokinetic and tissue distribution study of solid lipid nanoparticles of zidovudine in rats. J. Nanotechnol. 2014, 854018.

Rao, J.P., Gruenberg, P., Geckeler, K.E., 2014. Magnetic zero-valent metal polymer nanoparticles: current trends, scope, and perspectives. Prog. Polym. Sci. Available online July 14, 2014.

Reul, R., Renette, T., Bege, N., Kissel, T., 2011. Nanoparticles for paclitaxel delivery: a comparative study of different types of dendritic polyesters and their degradation behavior. Int. J. Pharm. 407, 190–196.

Roberts, M.S., Cross, S.E., Pellett, M.A., 2002. Skin transport. In: Walters, K.A. (Ed.), Dermatological and Transdermal Formulations, Drugs and the Pharmaceutical Sciences, vol. 119. Marcel Dekker, New York, pp. 89–195.

Rolland, J.P., Maynor, B.W., Euliss, L.E., Exner, A.E., Denison, G.M., DeSimone, J.M., 2005. Direct fabrication and harvesting of monodisperse, shape-specific nanobiomaterials. J. Am. Chem. Soc. 127, 10096–100100.

Romonchuk, W.J., Bunge, A.L., 2003. Absorption of 4-cyanophenol from powder and saturated aqueous solution into silicone rubber membranes and human skin. In: American Association of Pharmaceutical Scientists Annual Meeting and Exposition, Salt Lake City, UT, October 26–30, 2003.

Roser, M., Fischer, D., Kissel, T., 1998. Surface-modified biodegradable albumin nano- and microspheres. II: Effect of surface charges on in vitro phagocytosis and biodistribution in rats. Eur. J. Pharm. Biopharm. 46, 255–263.

Rotoli, B.M., Bussolati, O., Bianchi, M.G., Barilli, A., Balasubramanian, C., Bellucci, S., Bergamaschi, E., 2008. Non-functionalized multiwalled carbon nanotubes alter the paracellular permeability of human airway epithelial cells. Toxicol. Lett. 17, 95–102.

Rowland, M., Tozer, T.N., 1989. Clinical Pharmacokinetics: Concepts and Applications', second ed. Lea & Febiger, Philadelphia. 141.

Rowland, M., Balant, L., Peck, C., 2004. Physiologically based pharmacokinetics in Drug Development and Regulatory Science: a workshop report. AAPS PharmSci. 6, E6.

Russier, J., Menard-Moyon, C., Venturelli, E., Gravel, E., Marcolongo, G., Meneghetti, M., Doris, E., Bianco, A., 2011. Oxidative biodegradation of single- and multi-walled carbon nanotubes. Nanoscale 3, 893–896.

Sarkar, S., Guibal, E., Quignard, F., SenGupta, A.K., 2012. Polymer-supported metals and metal oxide nanoparticles: synthesis, characterization, and applications. J. Nanoparticle Res. 14, 715.

Sarparanta, M.P., Bimbo, L.M., Makila, E.M., Salonen, J.J., Laaksonen, P.H., Helariutta, A.M., Linder, M.B., Hirvonen, J.T., Laaksonen, T.J., Santos, H.A., Airaksinen, A.J., 2012. The mucoadhesive and gastroretentive properties of hydrophobin-coated porous silicon nanoparticle oral drug delivery systems. Biomaterials 33, 3353–3362.

Schaefer, H., Schalla, W., 1980. Kinetics of percutaneous absorption of steroids. In: Mauvais-Jarvis, P., Vickers, D.F.H., Wepierre, J. (Eds.), Percutaneous Absorption of Steroids. Adademic Press, New York, p. 54.

Schaefer, H., Watts, F., Brod, J., Illel, B., 1990. Follicular penetration. In: Scott, R.C., Guy, R.H., Hadgraft, J. (Eds.), Prediction of Percutaneous Penetration. Methods, Measurements, Modelling, vol. 163. IBC Technical Services, London, p. 732.

Schanker, L.S., 1978. Drug absorption from the lung. Biochem. Pharmacol. 27, 381–385.

Scheuplein, R.J., 1966. Mechanisms of percutaneous adsorption. I. Routes of penetration and the influence of solubility. J. Invest. Dermatol 45, 334–346.

Schleh, C., Semmler-Behnke, M., Lipka, J., Wenk, A., Hirn, S., Schäffler, M., Schmid, G., Simon, U., Kreyling, W.G., 2012. Size and surface charge of gold nanoparticles determine absorption across intestinal barriers and accumulation in secondary target organs after oral administration. Nanotoxicology 6, 36–46.

Schluep, T., Cheng, J., Khin, K.T., Davis, M.E., 2006. Pharmacokinetics and biodistribution of the camptothecin-polymer conjugate IT-101 in rats and tumor-bearing mice. Cancer Chemother. Pharmacol. 57, 654–662.

Shargel, L., Andrew, B.C., 1999. Fourth Edition "Physiologic Factors Related to Drug Absorption" Applied Biopharmaceutics and Pharmacokinetics. Prentice Hall International, INC., Stanford,, pp. 99–128.

Shia, X., Bányaib, I., Islama, M.T., Lesniaka, W., Davisa, D.Z., Jra, J.R.B., Balogha, L.P., 2005. Generational, skeletal and substitutional diversities in generation one poly(amidoamine) dendrimers. Polymer 46, 3022–3034.

Shimada, A., Kawamura, N., Okajima, M., Kaewamatawong, T., Inoue, H., Morita, T., 2006. Translocation pathway of the intratracheally instilled ultrafine particles from the lung into the blood circulation in the mouse. Toxicol. Pathol. 34, 949–957.

Shore, P.A., Brodie, B.B., Hogben, C.A.M., 1957. The gastric secretion of drugs: a pH partition hypothesis. J. Pharmacol. Exp. Ther. 119, 361–369.

Shvedova, A.A., Pietroiusti, A., Fadeel, B., Kagan, V.E., 2012. Mechanisms of carbon nanotube-induced toxicity: focus on oxidative stress. Toxicol Appl Pharm. 261, 121–133.

Shvedova, A.A., Kapralov, A.A., Feng, W.H., Kisin, E.R., Murray, A.R., Mercer, R.R., St Croix, C.M., Lang, M.A., Watkins, S.C., Konduru, N.V., Allen, B.L., Conroy, J., Kotchey, G.P., Mohamed, B.M., Meade, A.D., Volkov, Y., Star, A., Fadeel, B., Kagan, V.E., 2012. Impaired clearance and enhanced pulmonary inflammatory/fibrotic response to carbon nanotubes in myeloperoxidase-deficient mice. PLoS One 7, e30923.

Shvedova, A.A., Kisin, E.R., Mercer, R., Murray, A.R., Johnson, V.J., Potapovich, A.I., Tyurina, Y.Y., Gorelik, O., Arepalli, S., Schwegler Berry, D., 2005. Unusual inflammatory and fibrogenic pulmonary responses to single-walled carbon nanotubes in mice. Am. J. Physiol. Lung Cell Mol. Physiol. 289, L698–L708.

Simpson, C.A., Salleng, K.J., Cliffel, D.E., Eldheim, D.L., 2013. In vivo toxicity, biodistribution, and clearance of glutathione-coated gold nanoparticles. Nanomed. Nanotechnol. Biol. Med. 9, 257–263.

Singh, R., Pantarotto, D., Lacerda, L., Pastorin, G., Klumpp, C., Prato, M., Bianco, A., Kostarelos, K., 2006. Tissue biodistribution and blood clearance rates of intravenously administered carbon nanotube radiotracers. PNAS 103, 3357–3362.

Sipes, I.G., Gandolfi, A.J., 1991. Biotransformation of toxicants. In: Amdur, M.O., Doull, J., Klaassen, C.D. (Eds.), Toxicology, the Basic Science of Poisons, fourth ed. Pergamon Press, New York, pp. 88–126.

Sokohara, S., Ishida, M., 1998. Visible luminescence and surface properties of nanosized ZnO colloids prepared by hydrolyzing zinc acetate. J. Phys. Chem. B 102, 10169–10175.

Sonavane, G., Tomoda, K., Makino, K., 2008. Biodistribution of colloidal gold nanoparticles after intravenous administration: effect of particle size. Colloids Surf. B 66, 274–280.

Smith, C.A., Simpson, C.A., Kim, G., Carter, C.J., Feldheim, D.L., 2013. Gastrointestinal bioavailability of 2.0 nm diameter gold nanoparticles. ACS Nano. 7, 3991–3996.

Stoughton, R.B., 1965. Percutaneous absorption, influence of temperature and hydration. Arch. Environ. Health 11, 551–554.

Sturm, R., 2009. A theoretical approach to the deposition of cancer-Inducing asbestos fibers in the human respiratory tract. Open Lung Cancer J. 2, 1–11.

Sturm, R., 2010a. Deposition and cellular interaction of cancer-inducing particles in the human respiratory tract: theoretical approaches and experimental data. Thorac. Cancer 4, 141–152.

Sturm, R., 2010b. Theoretical approach to the hit probability of lung cancer sensitive epithelial cells by mineral fibers with various aspect ratios. Thorac. Cancer 3, 116–125.

Sturm, R., 2012. Modeling the deposition of bioaerosols with variable size and shape in the human respiratory tract – a review. J. Adv. Res. 3, 295–304.

Sung, J.C., Pulliam, B.L., Edwards, D.A., 2007. Nanoparticles for drug delivery to the lungs. Trends Biotechnol. 25, 563–570.

Suzuki, H., Sugiyama, Y., 2000. Role of metabolic enzymes and efflux transporters in the absorption of drugs from the small intestine. Eur. J. Pharm. Sci. 12, 3–12.

Swarbrick, J., Boylan, J.C., 1988. "Absorption" Encyclopedia of Pharmaceutical Technology, 1. Marcel Dekker, INC., New York, pp. 1–32.

Syama, S., Sreekanth, P.J., Varma, H.K., Mohanan, P.V., 2014. Zinc oxide nanoparticles induced oxidative stress in mouse bone marrow mesenchymal stem cells. Toxicol. Mech. Methods 1–10. Ahead of Print.

Tan, J., Shah, S., Thomas, A., Ou-Yang, H.D., Liu, Y., 2013. The influence of size, shape and vessel geometry on nanoparticle distribution. Microfluid. Nanofluidics 14, 77–87.

Thanou, M., Verhoef, J.C., Junginger, H.E., 2001. Oral drug absorption enhancement by chitosan and its derivatives. Adv. Drug Deliv. Rev. 52, 117–126.

Thanou, M.M., Kotze, A.F., Scharringhausen, T., Luessen, H.L., de Boer, A.G., Verhoef, J.C., Junginger, H.E., 2000. Effect of degree of quaternization of N-trimethyl chitosan chloride for enhanced transport of hydrophilic compounds across intestinal caco-2 cell monolayers. J. Controlled Release 64, 15–25.

Thummel, K.W., O'Shae, D., Paine, M.F., Shen, D.D., Kunze, K.L., Perkins, J.D., Wilkinson, G.R., 1996. Oral first pass elimination of midazolam involves both gastrointestinal and hepatic CYP3A-mediated metabolism. Clin. Pharmacol. Ther. 59, 491–502.

Tinke, A.P., Govoreanu, R., Vanhoutte, K., 2006. Particle size and shape characterization of nano-and submicron liquid dispersions. Am. Pharm. Rev. 9, 33–37.

Tronde, A., Norden, B., Marchner, H., Wendel, A.K., Lennernas, H., Bengtsson, U.H., 2003. Pulmonary absorption rate and bioavailability of drugs in vivo in rats: structure-absorption relationships and physicochemical profiling of inhaled drugs. J. Pharm. Sci. 92, 1216–1233.

Vecchia, B.E., Bunge, A.L., 2003. Skin absorption databases and predictive equations. In: Guy, R., Hadgraft, J. (Eds.), Transdermal Drug Delivery, Drugs and the Pharmaceutical Sciences, second ed., vol. 123. Marcel Dekker, New York, pp. 57–141.

Vlasova, I., Sokolov, A., Chekanov, A., Kostevich, V., Vasilyev, V., 2011. Myeloperoxidase-induced biodegradation of single-walled carbon nanotubes is mediated by hypochlorite. Russ. J. Bioorg. Chem. 37, 453–463.

Vogt, A., Combadiere, B., Hadam, S., Stieler, K.M., Lademann, J., Schaefer, H., 2006. 40 nm, but not 750 or 1500 nm, nanoparticles enter epidermal CD1a^{+} cells after transcutaneous application on human skin. J. Invest. Dermatol. 126, 1316–1322.

Voigt, J., Christensen, J., Shastri, V.P., 2014. Differential uptake of nanoparticles by endothelial cells through polyelectrolytes with affinity for caveolae. PNAS 111, 2942–2947.

Wacher, V.J., Salphati, L., Benet, L.Z., 2001. Active secretion and enterocytic drug metabolism barriers to drug absorption. Adv. Drug Deliv. Rev 46, 89–102.

Walczyk, D., Bombelli, F.B., Monopoli, M.P., Lynch, I., Dawson, K.A., 2010. What the cell "Sees" in bionanoscience. J. Am. Chem. Soc. 132, 5761–5768.

Wepasnick, K.A., Smith, B.A., Schrote, K.E., Wilson, H.K., Diegelmann, S.R., Fairbrother, D.H., 2011. Surface and structural characterization of multi-walled carbon nanotubes following different oxidative treatments. Carbon 49, 24–36.

Wester, R.C., Noonan, P.K., 1980. Relevance of animal models for percutaneous absorption. Int. J. Pharm. 7, 99–110.

Wu, X., Li, H., Xu, Y., Xu, B., Tong, B.H., Wang, L., 2014. Thin film fabricated from solution-dispersible porous hyperbranched conjugated polymer nanoparticles without surfactants. Nanoscale 6, 2375–2380.

Xu, F., Yuan, Y., Shan, X., Liu, C., Tao, X., Sheng, Y., Zhou, H., 2009. Long-circulation of hemoglobin-loaded polymeric nanoparticles as oxygen carriers with modulated surface charges. Int. J. Pharm. 377, 199–206.

Yamamoto, Y., Nagasaki, Y., Kato, Y., Sugiyama, Y., Kataoka, K., 2001. Long-circulating poly(ethylene glycol)-poly(D,L-lactide) block copolymer micelles with modulated surface charge. J. Controlled Release 77, 27–38.

Yang, K., Ma, Y.-Q., 2010. Computer simulation of the translocation of nanoparticles with different shapes across a lipid bilayer. Nat. Nanotechnol. 5, 579–583.

Yang, S.-T., Jianbin Luo, J., Zhou, Q., Wang, Q., 2012. Pharmacokinetics, metabolism and toxicity of carbon nanotubes for biomedical purposes. Theranostics 2, 271–282.

Yang, S.T., Wang, H., Meziani, M.J., Liu, Y., Wang, X., Sun, Y.P., 2009. Bio-defunctionalization of functionalized single-walled carbon nanotubes in mice. Biomacromolecules 10, 2009–2012.

Yang, S.-T., Wang, X., Jia, G., Gu, Y., Wang, T., Nie, H., Ge, C., Wang, H., Liu, Y., 2008. Long-term accumulation and low toxicity of single-walled carbon nanotubes in intravenously exposed mice. Toxicol. Lett. 181, 182–189.

Yang, X., Li, B., Majeed, I., Liang, L., Long, X., Tan, B., 2013. Magnetic microporous polymer nanoparticles. Polym. Chem. 4, 1425–1429.

Yeh, H.C., Schum, G.M., 1980. Models of the human lung airways and their application to inhaled particle deposition. Bull. Math. Biol. 142, 461–480.

Yin, Y., Chen, D., Qiao, M., Lu, Z., Hu, H., 2006. Preparation and evaluation of lectin-conjugated PLGA nanoparticles for oral delivery of thymopentin. J. Controlled Release 116, 337–345.

Yuan, H., Chen, C.Y., Chai, G.H., Du, Y.Z., Hu, F.Q., 2013. Improved transport and absorption through gastrointestinal tract by PEGylated solid lipid nanoparticles. Mol. Pharm. 10, 1865–1873.

Yue, J., Liu, S., Xie, Z., Xing, Y., Jing, X., 2013. Size-dependent biodistribution and antitumor efficacy of polymer micelle drug delivery systems. J. Mater. Chem B 1, 4273–4280.

Zagar, E., Zigon, M., 2002. Characterization of a commercial hyperbranched aliphatic polyester based on 2,2-bis(methylol)propionic acid. Macromolecules 35, 9913.

Zhang, G., Yang, Z., Lu, W., Zhang, R., Huang, Q., Tian, M., Li, L., Liang, D., Li, C., 2009. Influence of anchoring ligands and particle size on the colloidal stability and in vivo biodistribution of polyethylene glycol-coated gold nanoparticles in tumor-xenografted mice. Biomaterials 30, 1928–1936.

Zhang, H., Burnum, K.E., Luna, M.L., Petritis, B.O., Kim, J.S., Qian, W.J., Moore, R.J., Heredia-Langner, A., Webb-Robertson, B.J., Thrall, B.D., Camp 2nd, D.G., Smith, R.D., Pounds, J.G., Liu, T., 2011. Quantitative proteomics analysis of adsorbed plasma proteins classifies nanoparticles with different surface properties and size. Proteomics 11, 4569–4577.

Zhang, J., Tang, C., Yin, C., 2013. Galactosylated trimethyl chitosan-cysteine nanoparticles loaded with Map4k4 siRNA for targeting activated macrophages. Biomaterials 34, 3667–3677.

Zhang, L., Asgharian, B., Anjilvel, S., 1996. Inertial and interceptional deposition of fibers in a bifurcating airway. J. Aerosol Med. 9, 419–430.

Zhang, L., Brett, C.M., Giacomini, K.M., 1998. Role of organic cation transporters in drug absorption and elimination. Annu. Rev. Pharmacol. Toxicol. 38, 431–460.

Zhang, X., Wu, W., 2014. Ligand-mediated active targeting for enhanced oral absorption. Drug Discov. Today 19, 898–904.

Zhang, X., Qi, J., Lu, Y., He, W., Li, X., Wu, W., 2014. Biotinylated liposomes as potential carriers for the oral delivery of insulin. Nanomedicine 10, 167–176.

Zhao, Y., Allen, B.L., Star, A., 2011. Enzymatic degradation of multiwalled carbon nanotubes. J. Phys. Chem. A 115, 9536–9544.

Zhao, Y.C., Zhang, L.M., Wang, T., Han, B.H., 2014. Microporous organic polymers with acetal linkages: synthesis, characterization, and gas sorption properties. Polym. Chem. 2014 (5), 614–621.

CHAPTER

7

Mechanisms of Nanoparticle Toxicity

OUTLINE

Engineered Nanoparticles
http://dx.doi.org/10.1016/B978-0-12-801406-6.00007-8

1. INTRODUCTION

Nanoparticles are particles 1–100 nm in diameter at least in one dimension. In this size range, the physicochemical and electronic properties of nanoparticles are different from the atoms (or molecules) and the bulk particles. Their properties are not constant over their size and exhibit unique size dependency and quantum electronic properties. Nanoparticles, depending on the situation, may be classified as a new entity or a starting bulk material. If classified as a starting material, then the toxic effects reported for bulk particles may be applicable to the nanoparticles. However, if classified as a new entity, then the toxic effects reported for bulk particles may not be applicable to the nanoparticles.

There is compelling evidence in support of nanoparticles being a separate entity rather than starting materials. Nanoparticles exhibit novel physicochemical properties not present in the starting bulk materials, unique interactions with biological systems, and uncertainties about their fate in the environment. Because the toxic effects of a chemical or particles depend on their physicochemical properties and their interactions with biological systems, we can hypothesize that the biological and adverse effects of nanoparticles may be different from those of the starting materials. As described in Chapter 5, the mode of a nanoparticle's preparation affects its likely interaction with biological systems. Incorporation of nanoparticles in a composite material reduces their toxicity. However, if applications result in direct exposure (e.g., drug delivery systems, sunscreen, cosmetics, etc.), the toxicity and ensuing health risk will be much higher. It is also hypothesized that because of mechanical, physicochemical, and electronic differences between bulk particles and nanoparticles, the toxicological data for bulk toxins cannot be used to decipher a nanoparticle's toxicity potential. One example is metal ions (titanium, silver, and gold ions) that are chemically inert and essentially nontoxic. However, reduction of ionic metals into nanoparticles causes an exponential increase in their surface-to-volume ratio and surface reactivity; thus, the particles become more toxic. Unfortunately, the toxicity potential of metal nanoparticles is not fully established.

A review of the literature yielded contradictory information: nanoparticles have been shown to be highly toxic (Jia et al., 2005; Bottini et al., 2006; Magrez et al., 2006; Pulskamp et al., 2007), poorly toxic, as well as nontoxic (Fiorito et al., 2006; Flahaut et al., 2006; Wang et al., 2011; Kolosnjaj-Tabi et al., 2010). The differing and often contradictory results of different studies were possibly due to the heterogenic nature of nanoparticles, differences in the experimental design, and lack of common indices of toxicity measurements. As discussed in Chapters 3 and 6, the properties of nanoparticles change enormously with a minor change in their structure or diameter. Carbon nanotubes (CNTs) are unique because, depending upon their folding (n, m) pattern, they can acquire either metallic or semiconductor properties (Chapter 3). In addition, differences in the incubation media may also affect the functional properties of CNTs and other nanoparticles. Because of numerous compounding factors, studies using different types of nanoparticles and experimental designs may have yielded different results. Therefore, to understand the health risk potential of nanoparticles, it is important to understand their unique toxicity.

2. OVERVIEW OF TOXICOLOGY

As discussed in Chapter 1, there are three basic principles of toxicology:

- The dose determines toxicity (Paracelsus).
- Toxic effects are unique to the structure of each chemical (Ambriose Paré).
- Humans are animals.

Paracelsus, a physician from the early sixteenth century who is often called the father of toxicology, defined poisons as follows: "All things are poison, and nothing is without poison; only the dose permits something not to be poisonous." The Royal Society of Chemistry (RSC), to make Paracelsus's definition friendly for pharmacists, modified it by stating, "While there is no such thing as a safe chemical, it must be realized there is no chemical that cannot be used safely by limiting the dose or exposure. Poisons can be safely used and be of benefit to society when used appropriately." Both statements place the emphasis on dose in terms of mass/kg, but in a different perspective. The RSC's modified statement has allowed many poisonous substances, by controlling their dose or exposure, to be used as medicines. The dose (mass/kg)–response relationship has become a central theme in classic toxicology and in determining "safe" and "hazardous" levels for drugs, potential pollutants, and other substances to which humans or other organisms are exposed. The size, shape, or molecular topology may not determine the toxicity of bulk chemicals. As the research in the area of toxicology progressed, an understanding of the mechanisms underlying the toxic effects became clearer and new principles of toxicology emerged.

- Dose determines the response
- Risk = Hazard × Exposure
- Individual Sensitivity determines toxicity

The dose metric for bulk particle toxicity is mass per unit body-weight, area, or volume. Thus, mass is the commonly used dose metric to judge a nanoparticle's hazard. However, nanoparticles are seriously challenging the central role of the dose in terms of mass/kg in defining their toxicity. For nanoparticles, the dose in terms of size (surface area/kg) and shape, if not more, then at least as important as the dose in terms of mass/kg in characterizing their beneficial effects and toxicity (Oberdorster et al., 2007; Albanese et al., 2012; Peretz et al., 2012). Therefore, nanoparticles may threaten the concept of the dose (mass/kg)–response relationship in favor of either size (diameter [nm] or surface area/kg)–response or surface activity–response relationship to characterize their toxicity.

As described previously and in Chapters 1, 3, and 6, nanoparticles exhibit unique physical and chemical properties that are not present in their bulk counterparts. The unique properties of nanoparticles make them beneficial for commercial and medical uses (Atala, 2004; Cherukuri et al., 2004; Brannon-Peppas and Blanchette, 2004; Aoki et al., 2005; Cai et al., 2005; Kawasaki and Player, 2005; Farokhzad and Langer, 2006; Groneberg et al., 2006; Wagner et al., 2006). Unfortunately, the same unique properties may also be responsible for their unique toxicity (Gwinn and Vallyathan, 2006). Many inert bulk materials, such as amorphous silica, gold, and silver, may, at the nanoscale, become biologically active and causes adverse effects (El-Ansary and Al-Daihan, 2009; Donaldson et al., 2006).

During the last decade, many excellent reviews covering different aspects of nanoparticle toxicity and epidemiology have been published (Lam et al., 2004, 2006; Madl and Pinkerton, 2009; Aillon et al., 2009; Tsuda et al., 2009; Aschberger et al., 2010; Boczkowski and Hoet, 2008; Johnston et al., 2010; Shvedova and Kagan, 2010; Beg et al., 2011; Haniu et al., 2011; Nerl et al., 2011; Uo et al., 2011; van der Zande et al., 2011; Donaldson et al., 2006; Gulati and Gupta, 2012; Love et al., 2012; Shvedova et al., 2012, 2010, 2009, 2008a,b, 2005; Luyts et al., 2013; Yang et al., 2012; Zhao and Liu, 2012). Although the results are diverse, they bring up one common theme: the physicochemical properties and toxic effects of nanoparticles (<100 nm diameter) are different from those of the corresponding bulk particles. However, the correlations between the physicochemical properties and toxicity of nanoparticles are not fully understood, possibly because of a lack of communication between the solid-state physicists working on the electronic and mechanical properties of CNTs and biological scientists working on the medicinal properties and toxicity of nanoparticles.

3. FACTORS DETERMINING NANOPARTICLE TOXICITY

Unlike the toxicity of bulk particles, the toxic effects of nanoparticles are modulated by their size, shape, topology, surface functionalization and aggregation status. Toxicity of one-dimensional (1D) and two-dimensional (2D) nanoparticles also depends on the particles' aspect ratio (length/diameter). The nanoparticles' unique properties dictate their fundamental characteristics, including their ability to get into cells and cause toxicity. Listed in the following sections are some of the determining factors.

3.1 Physicochemical Properties

Chapters 4 and 6 have described the nanoparticles' physicochemical properties in detail. For particles less than 100 nm in diameter, a decrease in size increases the particle's surface area and reactivity, and thus their potency to cause adverse effects. CNTs are unique because, depending on their folding patterns, they may exhibit metallic or semiconducting properties. CNTs also exhibit different types of defects that increase molecular strain and reactivity.

3.2 Aggregation

In aqueous environments, pristine nanoparticles may tend to aggregate because of interparticle van der Wall interactions (Thess et al., 1996; Kwon et al., 1998; Vigolo et al., 2000). The aggregated nanoparticles are much larger than the primary size of the particles. In the case of CNTs, pristine (nonfunctionalized) large single-walled carbon nanotubes (SWCNTs) and small multi-walled carbon nanotubes (MWCNTs) aggregate, while long-MWCNTs of about 50 nm in diameter remain dispersed in aqueous environments. The properties and mobilities of aggregated and dispersed nanoparticles are considerably different in the host environment.

The aggregation behavior of nanoparticles and their adverse effects depend on their functionalization, hydrophobicity, length, and diameter (Tripathi et al., 2014; Fayol et al., 2012; Ntim et al., 2012). For 1D nanotubes, aggregation also showed a dependence on their length: the shorter MWCNTs were less prone to aggregation than the longer ones. Dispersed nanoparticles enter the cells via nonphagocytic processes, while aggregated particles do so via phagocytosis (Geyser et al., 2005). Therefore, aggregated nanoparticles mostly accumulate in cytosol and lysosomes (Monteiro-Riviera et al., 2005). Wick et al. (2007) showed that agglomerated SWCNTs were more toxic than SWCNT bundles, which were more toxic than dispersed CNTs, in the mesothelioma cell line MSTO-211H. Contrarily, Wang et al. (2011) have shown that dispersed MWCNTs elicited greater profibrogenic responses than aggregated CNTs in vitro (epithelial cells, macrophages, and fibroblast trophic cell unit) and in vivo in mouse. This suggests that aggregation compromises the relationship between the nanoparticle size and toxicity, thus compromising their risk assessment.

3.3 Concentration

As discussed, size is an important issue when determining the nanoparticle dose in animal experiments. The dose in terms of mass (e.g., g/kg) may not be appropriate for nanoparticles. Other metrics, such as surface area or surface activity per unit body weight, may be more relevant.

3.4 Functionalization

Pristine nanoparticles are either highly hydrophobic (metal, polymer, or solid-lipid nanoparticles) or highly hydrophilic (amino acid, PAMAM, and DNA dendrimers). The hydrophobic dendrimers agglomerate and exist as an entangled bundle with amplified defects in an aqueous environment (Kasaliwal et al., 2010). The hydrophilic dendrimers may not aggregate, but they lack bioavailability. Smaller CNTs, for their biological application, may remain dispersed in the host media. Chemical functionalization (1) prevents aggregation of hydrophobic nanoparticles and (2) improves bioavailability of hydrophilic nanoparticles. For example, adding an amino group, carboxylic group, or sulfonated-4-chlorophenyl group to pristine hydrophobic nanoparticles may improve their aqueous solubility (Foldvari and Bagonluri, 2008a,b; Chen et al., 2008; Ntim et al., 2012). Surface functionalization, in addition to dispersing the nanoparticles, also enables them to achieve specific functions, such as selective tissue targeting using antibodies, cell membrane penetration using specialized peptides, and tumor selective antibodies for site-directed drug release (Sun et al., 2002; Kuzmanya et al., 2004; Balasubramanian and Burghard, 2005; Lerner et al., 2013).

3.5 Length and Aspect Ratio (Length/Diameter)

The concept of the causal relationship between the aspect ratio and toxicity originated from a pioneering study by Stanton et al. (1981), who showed that thin (<0.5 μm) and long asbestos fibers (>8 μm) caused more mesothelioma than short fibers. Poland et al. (2008) and Murphy et al. (2011) first showed that long, but not short MWCNTs, produced inflammation and fibrogenic response in mice, suggesting a possible correlation between asbestos and CNT toxicities. The question is whether asbestos and CNTs cause length-dependent toxicity via identical mechanisms. Nagai and Toyokuni (2012) addressed this question by looking into their uptake in nonphagocytic cells. The authors reported the following differences:

- Asbestos fibers are actively endocytosed by nonphagocytic cells independent of their diameter. CNTs directly pierce through the cell and nuclear membranes and enter mesothelium cells. CNTs' internalization is directly related to the fibers' rigidity and inversely related to their diameters.
- Long CNT fibers, but not asbestos fibers, contain Fe, Co, or Ni impurities that have been shown to modulate the CNTs' adverse effects (Ge et al., 2007).
- Following the internalization, the asbestos fibers, but not CNTs, bind small GTPase, Rab5a, which is a marker of the early endosome and phagosome (Nagai and Toyokuni, 2012). Depending on the length, asbestos fibers could be internalized by a phagocytosis-like mechanism (>3 μm) or an endosome-based mechanism (<2 μm).

As described by Nagai et al. (2011) and shown in Figure 1, long CNTs (Figure 1(a)) may pierce the membrane and nucleus of a nonphagocytic cell (Figure 1(i); Vakarelski et al., 2007; Kostarelos et al., 2007; Lacerda et al., 2007) similar to the "frustrated phagocytosis" in phagocytotic cells.

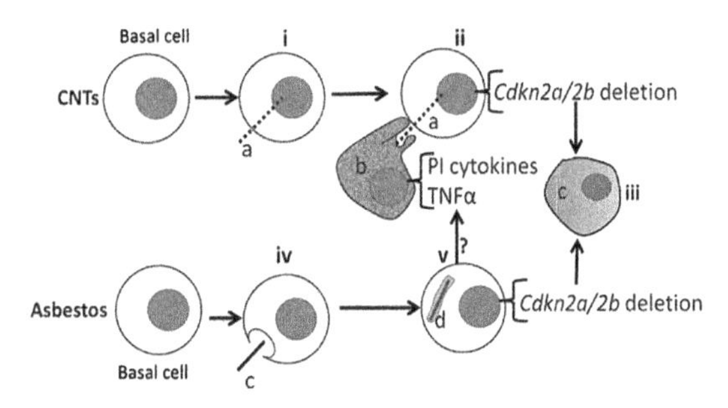

FIGURE 1 Mechanisms underlying toxicity of CNTs and asbestos. CNTs enter the basal cells via penetration and phagocytosis (i), while asbestos enters the cells via endocytosis (ii). The extruding CNTs attach to macrophages that release pro-inflammatory (PT) cytokines, including tumor necrosis factor alpha (TNFα). Intracellular CNTs or asbestos cause deletion of *Cdkn2a/2b* that activates carcinogenesis signaling. *Reprinted from Nagai et al. (2011).*

The damaged cells induce inflammation via interaction with phagocytes, resulting in secretion of pro-inflammatory cytokines and tumor necrosis factor-α (Figure 1(ii)) (Donaldson et al., 1989; Poland et al., 2008). If CNTs directly pierce the nucleus, they block expression of *Cdkn 2a/2b*, a tumor suppressor gene (Figure 1(iv)), resulting in transformation of normal cells into cancer cells (Figure 1(iii)) (Nagai et al., 2011). Asbestos (c) fibers enter the cells via active endocytosis, resulting in accumulation of the fibers in the cytoplasm where the fibers bind small GTPases, *Rab5a*. Intracellular asbestos fiber augments carcinogenic signaling and blocks expression of *Cdkn 2a/2*. Possible mechanisms for an increase in inflammation in asbestos-exposed cells are debatable.

Studies have shown that asbestos may cause lung cancer and mesothelioma (NIOSH, 2011) and a number of nonmalignant respiratory diseases such as asbestosis and thickening of pleura (ATS, 2004) in humans. Despite extensive effort to remove asbestos from housings, there has been an increase in deaths from asbestosis and/or cancer, possibly because chronic diseases may take years to develop (NIOSH, 2007). Development of cancer has also been reported in animals exposed to asbestos fibers (Campbell et al., 1979; Davis et al., 1991). Bernstin et al. (2013) showed that amphibole, a stable form of asbestos, was more potent than chrysotile, an unstable form of asbestos, in causing cancer. Parallel to the observations of Bernstin et al. (2013), earlier studies have also shown that long MWCNTs of 50 nm diameter, but not smaller (not rigid) or larger (too thick) diameter, caused lung cancer in experimental animals (Nagai et al., 2011). Thus, long MWCNTs may potentially cause asbestosis and lung cancer, comparable to asbestos. This may pose a serious public health concern to populations that are exposed to MWCNTs.

3.6 Dispersion Media

As discussed earlier, pristine hydrophobic nanoparticles aggregate in hydrophilic media. Aggregated nanoparticles lack the unique properties of dispersed particles. Thus, synthetic dispersants and surfactants, which bind to the nanoparticles noncovalently, are commonly used. Such noncovalent functionalization has great promise because the procedure minimizes alterations in the electronic and mechanical properties of nanoparticles. However, it is important to ensure that the dispersants alone or the nanoparticle–dispersant complex is nontoxic (Alpatova et al., 2009). Recently, Gasser et al. (2012) have shown that precoating MWCNTs with a pulmonary surfactant, curosurf, increased oxidative stress and inflammation in vitro (Figure 2).

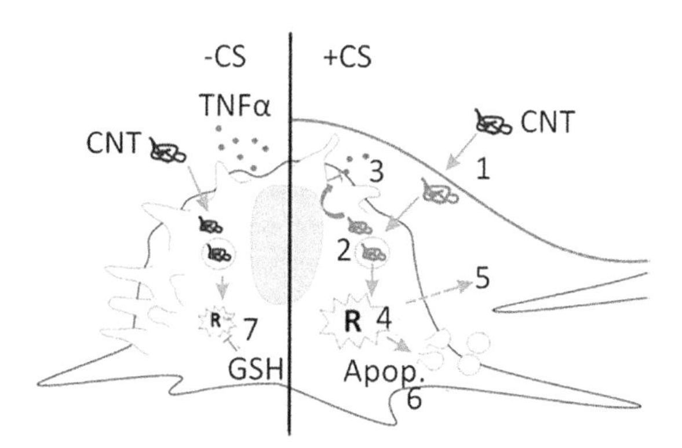

FIGURE 2 Possible effects of curosurf (CS)-coated (+CS) or uncoated (−CS) MWCNTs. CS, being a surfactant, binds to the MWCNTs and alters their surface characteristics (1). The coated MWCNTs accumulate in vesicles of monocyte-derived macrophages as well as in the cytoplasm (2). Curosurf directly decreased TNFα release (3), increased ROS production (5) and induced apoptosis (6) in macrophages. Uncoated MWCNTs, although depleted intracellular glutathione (7), did not cause macrophage apoptosis. This suggests that toxicity inherent to the dispersants must be considered when evaluating the toxicity of dispersed CNTs. *Reprinted from Gasser et al. (2012).*

4. MECHANISMS OF NANOPARTICLE TOXICITY

Nanoparticles, depending on the mode of exposure, may cause local and systemic toxicity,

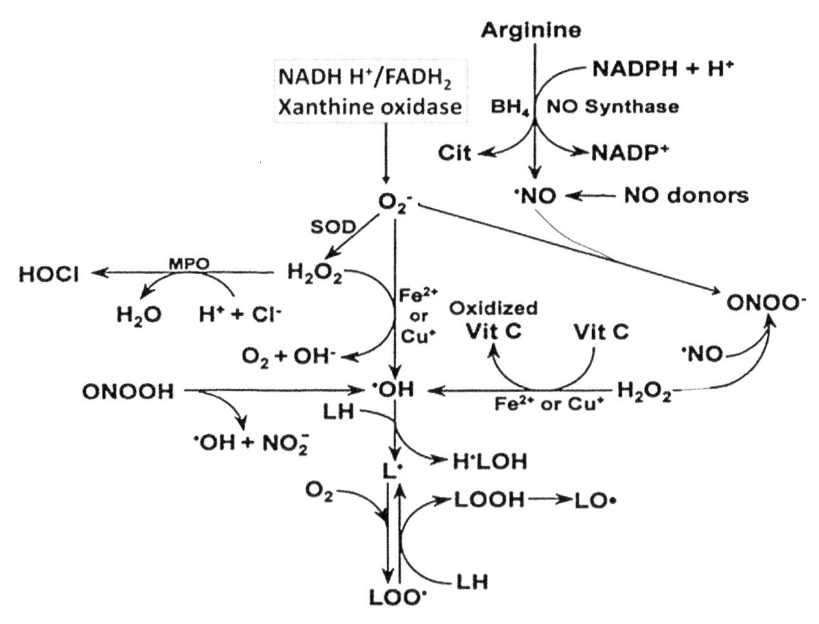

FIGURE 3 Production of oxygen and nitrogen free radicals and other reactive species in mammalian cells. AA, amino acid; Arg, L-arginine; BH_4, (6R)-5,6,7,8,-tetrahydro-L-biopterin; CH_2O, formaldehyde; Cit, L-citrulline; DQ, diquat; ETS, electron transport system; FAD, flavin adenine dinucleotide (oxidized); $FADH_2$, flavin adenine dinucleotide (reduced); Gly, glycine; H_2O_2, hydrogen peroxide; HOCl, hypochlorous acid; H˙LOH, hydroxy lipid radical; IR, ionizing radiation; L˙, lipid radical; LH, lipid (unsaturated fatty acid); LO˙, lipid alkoxyl radical; LOO˙, lipid peroxyl radical; LOOH, lipid hydroperoxide; MPO, myeloperoxidase; NAD^+, nicotinamide adenine dinucleotide (oxidized); NADH, nicotinamide adenine dinucleotide (reduced); $NADP^+$, nicotinamide adenine dinucleotide phosphate (oxidized); NADPH, nicotinamide adenine dinucleotide phosphate (reduced); ˙NO, nitric oxide; O_2^-, superoxide anion radical; ˙OH, hydroxyl radical; $ONOO^-$, peroxynitrite; P-450, cytochrome P-450; PDG, phosphate-dependent glutaminase; Sar, Sarcosine; SOD, superoxide dismutase; Vit C, vitamin C; Vit E, vitamin E (α-tocopherol). *Reprinter from Fang et al. (2002).*

possibly due to an increase in oxidative stress that induces inflammation, the immune response, fibrogenesis and tissue remodeling, blood clotting, vascular injury, neurotoxicity and genotoxicity (Gasser et al., 2012; Hsieh et al., 2012; Murphy et al., 2012; Shvedova et al., 2012, 2008a,b; Crouzier et al., 2010; Moller et al., 2010; Uttara et al., 2009; Pulskamp et al., 2007; Yang et al., 2007; Kagan et al., 2006).

4.1 Free Radicals and Oxidative Stress

An imbalance between the production and detoxification of reactive oxygen species (ROS) and reactive nitrogen species (RNS) that results in their accumulation is termed *oxidative stress.* ROS constitute a pool of oxidative species, such as superoxide anion ($O_2^{\cdot -}$), hydroxyl radical ($OH^{\cdot}$), hydrogen peroxide (H_2O_2), singlet oxygen ($^1O_2$), and hypochlorous acid (HOCl). Nitric oxide ($NO^{\cdot}$) and peroxynitrite ($ONOO^{\cdot}$) are key examples of RNS. Figure 3 shows an integrative map for the generation of free radicals on mammaliam cells (Fang and Zheng, 2002; Fang et al., 2002).

Oxygen plays a central role in the generation of reactive species (Freidvich, 1999). Superoxide ($O^{\cdot -}$) is generated from O_2 by multiple pathways (Evans and Halliwll, 2001; Gilbert, 2000):

1. Oxidation of (1) NADPH by NADPH oxidase, (2) xanthine or hypoxanthine by xanthine oxidase, and (3) NADH, NADPH, and

FADH$_2$ via the mitochondrial electron transport system
2. Autoxidation of monomines (e.g., dopamine, epinephrine, and norepinephrine), flavin, and hemoglobin in the presence of trace amounts of transition metals
3. One-electron-based reduction of O_2 by cytochrome P-450 or NO synthase

NO is formed from L-arginine, a reaction catalyzed by NO synthase (NOS) isoforms: constitutive neuronal NOS, inducible NOS, and endothelial NOS (Wu and Morris, 1998). NO reacts with $O_2^{\bullet -}$ or H_2O_2 to form $ONOO^-$, whose oxidant potential is greater than that of $O_2^{\bullet -}$ or H_2O_2 alone (Fang and Zheng, 2002; Fang et al., 2002). Hydrogen peroxide is produced by superoxide dismutase (SOD), while HOCl is generated from H_2O_2 and Cl^- by myeloperoxidase in activated phagocytes (Fang and Zheng, 2002; Fang et al., 2002). When free radicals such as $^{\bullet}OH$, $HOO^{\bullet}$, and $ONOO^-$ extract a hydrogen atom from an unsaturated fatty acid, a carbon-centered lipid radical ($L^{\bullet}$) is produced, which is further converted into a lipid peroxyl radical, $LOO^{\bullet}$ and $LO^{\bullet}$.

Mitochondrial respiration accounts for most of the superoxide generated in a cell. The free radicals may leak out of the mitochondria via leakage sites at complex I and at ubisemiquinone. The steady-state concentration of $O_2^{\bullet -}$ is kept low by activation and translation of the antioxidation transcription factor activator, Nrf2, which, in the absence of oxidative stress, remains trapped in the cytoplasm as an Nrf2-Keap1 complex. An increase in oxidative stress dissociates the Nrf2-Keap1 complex and Nrf2 translocases into the nucleus, where it induces expression of antioxidation proteins that neutralize the free radicals. When the generation of free radicals exceeds the antioxidation capacity, a net increase in the oxidative stress occurs in the cells. Oxidative stress induces various kinases, such as the Src kinase family, protein kinase C, mitogen-activated protein kinase, MAPK, and receptor tyrosine kinases; and transcriptional factors such as AP-1 and NF-kB (Hubbard, 2001; Hubbard et al., 2002, 2001; Palmer and Paulson, 1997; Sundaresan et al., 1995; Foncea et al., 2000; Cho et al., 1999). In the absence of oxidative stress, NFκB remains in the cytoplasm as NFκB—IκB complex. Oxidative stress dissociates the complex, resulting in translocation of free NFκB (p50-p65 dimer) into the nucleus, where they induce expression of prion-inflammatory genes (Figure 4).

The MAPK pathways include extracellular signal-regulated kinases (ERK $^1/_2$), c-Jun amino-terminal kinases (JNK), p38, and MAPK-activated protein kinase (MKK) (Roux and Blenis, 2004). Studies have shown that arsenate treatment potently activated both JNK and p38, but only moderately activated ERK and that activation of all three kinases was prevented by the free radical scavenger *N*-acetyl-L-cysteine, suggesting that an oxidative signal initiates the responses (Liu et al., 1996). These results suggest that JNK and p38 may share common upstream regulators distinct from those involved in ERK activation. Functionally, p38 and JNK, but not ERK, responds to an increase in oxidative stress and their activation is essential for expression of pro-inflammatory cytokines and evolution of inflammation. ERK is mostly associated with the cellular proliferation and differentiation (Herlaar and Brown, 1999a,b). These observations suggest that Nrf2, NFκB, and MAPK pathways play a critical role in the development of adverse effects in response to oxidative stress.

4.1.1 *Nanoparticles and Oxidative Stress*

4.1.1.1 MECHANISMS OF FREE RADICAL GENERATION

Nanoparticles have been shown to increase oxidative stress by stimulating the respiratory burst in phagocytic cells with increased oxygen consumption, resulting in production of free

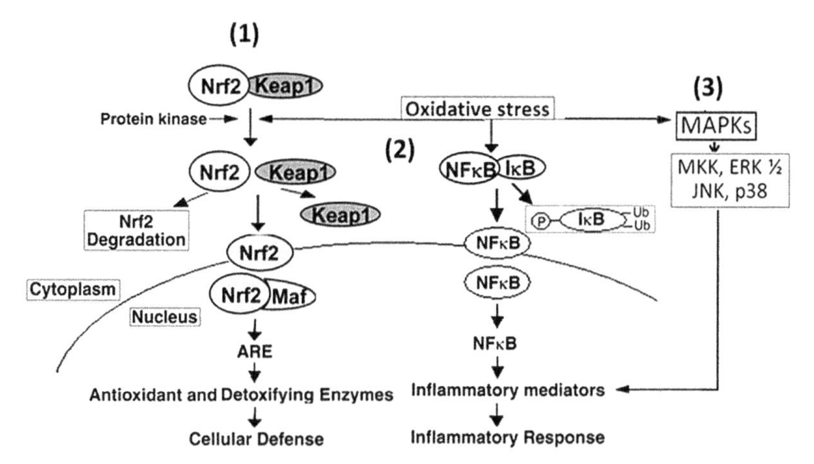

FIGURE 4 Oxidative stress and down-stream signaling. Oxidative stress, in general, induces three sets of signaling pathways. (1) *The Nrf2 pathways*: Nrf2 binds to and then induces the genes expressing antioxidation proteins, thus it maintains an oxidative homeostasis in cells. In cytoplasm Nrf2 exists as an Nrf2-Keap1 complex. An increase in oxidative stress dissociates the complex and Nrf2 translocates into the nucleus, interacts with Maf nuclear protein, and activates the genes that induces the synthesis of antioxidation enzymes, resulting in neutralization of free radicals. (2) *The NFκB pathways*: The pro-inflammatory NFκB consists of two subunits p50 and p65. In cytoplasm NFκB exists as an NFκB − IκB complex. An increase in oxidative stress dissociates the complex and NFκB translocates into the nucleus and activates the genes that induces pro-inflammatory cytokines and chemokines. After an appropriate response, a new NFκB consisting of p50-p50 is activated that induces activation of anti-inflammatory genes, resulting in resolution of inflammation (not shown in the figure). (3) *The MAPK pathways*: Oxidative stress also activates the MAPKs such as ERKs, JNKs, and p38 that, upon phosphorylation, modulate inflammatory activity.

radicals such as $O^{\bullet -}$, H_2O_2, and NO (Manke et al., 2013; Vallyathan et al., 1992; Blackford et al., 1994; Vallyathan and Shi, 1997; Castranova et al., 1991). Metal nanoparticles also induce oxidative stress in different mammalian cell lines in vitro (Huerta-Garcia et al., 2014; Montiel-Dávalos et al., 2012; Jaeger et al., 2012; Wan et al., 2012; Petkovic et al., 2011; Zhang et al., 2013; Liu et al., 2010; Xue et al., 2010; Bhattacharya et al., 2009; Valko et al., 2007; Gurr et al., 2005). Nanoparticles' reactive surface plays a key role in the generation of free radicals via the following mechanisms:

- Nanoparticle-mediated pore formation in cell membranes allows the influx of free radicals from the extracellular fluid into the cytoplasm.
- Mitochondrial damage allows the release of free radicals from mitochondria into the cytoplasm (Xia et al., 2006a,b; Sioutas et al., 2005).
- Reaction of the oxidant with the nanoparticle surface generates free radicals that remain bound to the surface. Fubini and Hubbard (2003) and Knaapen et al. (2004) have shown that surface bound $SiO^{\bullet}$ are responsible for the formation of ROS, such as $OH^{\bullet}$.
- Nanoparticles, such as CNTs, possess structural defects with altered electronic properties that can interact with molecular O_2 to generate ROS via Fenton-type reactions.
- Free radicals bound to the nanoparticle surface are released in an aqueous suspensions, generating H_2O_2, $OH^{\bullet}$, and $^1O_2$.

- Ar, Be, Co, and Ni nanoparticles promote the activation of an intercellular radical-generating system, such as the MAPK and NF-κB pathways (Smith et al., 2001).

4.1.1.2 OXIDATIVE STRESS PATHOPHYSIOLOGY

Nel et al. (2006), in a review article, have proposed that oxidative stress is central to the development of nanoparticle toxicity, as described below:

- *Nanoparticles generate ROS and RNS* that increase oxidative stress and ensuing protein, DNA, and membrane injury.
- *Oxidative stress* induces phase 2 enzymes, inflammation, mitochondrial damage, and activation of the reticulo-endothelial system.
- *Mitochondrial damage* may include damage of the inner membrane, pore opening, energy failure, cytotoxicity, and apoptosis.
- *Oxidative stress increases expression of pro-inflammatory cytokines* that induce inflammatory cell infiltration and systemic inflammation. Unregulated inflammation may lead to fibrosis, granulomas, and atherogenesis.
- *Nanoparticles may activate the reticuloendothelial system*, resulting in their sequestration and storage in liver, spleen, and lymph nodes. This process may further increase oxidative stress.

Jaeger et al. (2012) have shown that oxidative stress induced by TiO_2 nanoparticles resulted in peroxidation of lipids and an ensuing increase in malondialdehyde (MDA) concentration, oxidative damage of mitochondrial DNA, alterations in the mitochondrial membrane potential, and changes in the expression of antioxidant enzymes, such as SOD2.

Figure 5 shows an integrated pathway for possible effects of nanoparticle-induced oxidative stress in the development of toxicity. Oxidative stress directly affects platelet and vascular endothelial damage, DNA damage, membrane damage, pro-inflammatory NFκB signaling, pro-inflammatory cytokine expression, and mitochondrial damage, resulting in the development of blood-clotting abnormalities, diseases associated with DNA damage, neuropathy and apoptosis, cell survival defects, inflammation abnormalities, and abnormalities of the mitochondrial functions, respectively.

Taken together, these changes may be etiologically related to many diseases, such as cancer, atherosclerosis, stem cell abnormalities, etc. Table 1 lists the oxidative stress-related pathways targeted by different metal nanoparticles, while Table 2 lists the oxidative stress-mediated effects of metal nanoparticles. The studies listed in Tables 1 and 2 show that nanoparticles, although increased oxidative stress, caused differential modulation of the signaling pathways and adverse effects. Thus, oxidative stress may not be the sole mechanism of nanoparticle toxicity. This hypothesis is supported by earlier studies showing oxidative stress-independent induction of cytotoxicity of polystyrene and chitosan nanoparticles (Qi et al., 2005).

4.1.2 Carbon Nanotubes, Metal Impurities, and Oxidative Stress

As described in Chapter 2, different-sized CNTs are synthesized using iron nanoparticles as the catalyst. Although the completed products are washed thoroughly to deplete the iron catalysts, a considerable amount of iron remains attached to the CNTs (Murray et al., 2009; Pacurari et al., 2008; Kagan et al., 2006). Therefore, in the case of CNTs, whether the increase in oxidative stress is due to the CNTs' metallic contaminations, CNTs alone, or a combination of both is not fully established. Clichici et al. (2011) have shown that functionalized SWCNTs (dose: 270 mg/L, i.p. injection) increased concentrations of MDA, protein carbonyls, and antioxidant capacity in blood—an indication

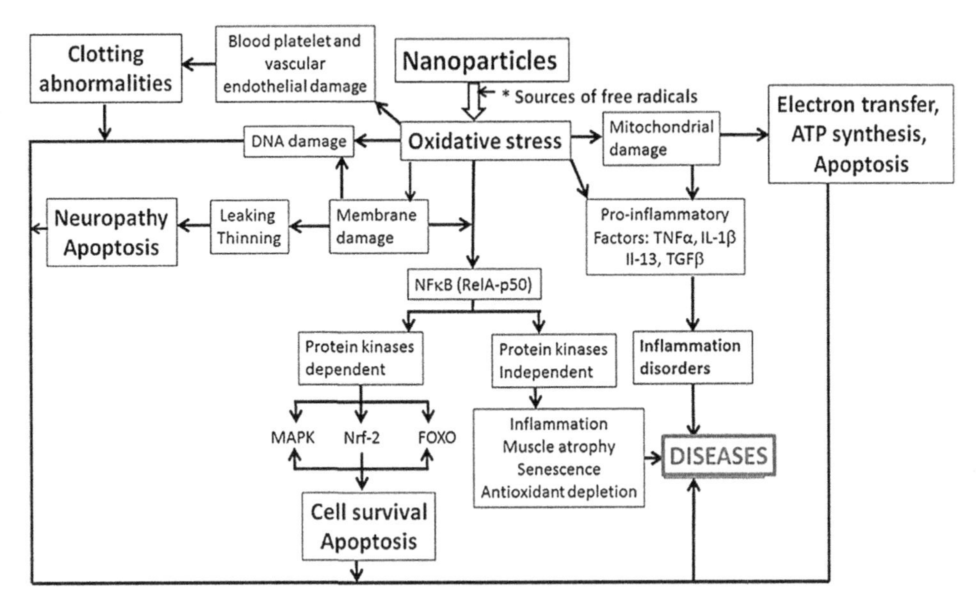

FIGURE 5 Overall mechanism of oxidative stress-mediated toxicity of nanoparticles. Nanoparticles generate free radicals as described in Figure 3. Oxidative stress causes toxicity either by damaging the membranes, blood components, DNA, and mitochondria or by deregulating cell signaling (NFκB, MAPK, pro-inflammatory, etc.). Taken together, these abnormalities cause clotting abnormalities, apoptosis, neuropathy, inflammation, etc. that result in induction of disease states.

TABLE 1 The Signaling Pathways Activated by Oxidative Stress

Pathways	Particle Type Size	Genes or Proteins	References
NFκB cAMP, PKA, JNK, ERK, cell cycle, C-myc	ZnO/Cds, TiO_2, Au, CdSe/ZnS and other metal nanoparticles	CRE, E2F, SMAD, NFκB, ILs, *p*-IκB, TLR4	Wu et al. (2010), Pujalte et al. (2011), Chen et al. (2011), Deng et al. (2011), Romoser et al. (2011), and Rallo et al. (2011)
HIF-1α	WCCo composit	SOX2, ID2, FOXO4	Busch et al. (2010)
NFκB Nrf-2 AP-1	uCP, SootL SootH Ptx90, PtxG	CSF2, CXCL1/5, IL6, MT $^1/_2$, NQ01, OGG1	Stoeger et al. (2009) and Ahamed et al. (2011a,b)
MAPK, ERK1/2 EGF-R, Akt	CdSe/ZnS QD Printex 90	TNFα, CXCL8, *p*-p38, ERK1/2, JNK1/2, Casp3	Lee et al. (2009), Unfried et al. (2008), and Ge et al. (2011)
MAPK Nrf2	Si porous 5–15 nm	SOD, HO-1	Eom (2009)
ERK, p38, Nrf2	CeO_2 15–45 nm	Ner2, HO-1, p38, JNK, SOD	Eom and Choi (2009)
Cell cycle	TiO_2, P25 125 nm	SOD2,	Ge et al. (2011)

TABLE 2 Oxidative Stress-Dependent Effects of Metal Nanoparticles

Nanoparticles	ROS-dependent Effect	References
Iron oxide	Necrosis, apoptosis, acute toxicity, endothelial permeability, and inflammation	Naqvi et al. (2010), Stone et al. (1998), Sohaebuddin et al. (2010), and Raghunathan et al. (2013)
Copper oxide	Oxidative damage, genotoxicity, cytotoxicity, nephrotoxicity	Ahamed et al. (2010), Fahmy and Cormier, 2009, Karlsson et al. (2009), and Lei et al. (2008)
Cerium oxide	Inflammation, apoptosis, p38-Nrf2 signaling, membrane damage	Kumar et al. (2011), Ma et al. (2011), and Eom and Choi (2009)
Zinc oxide	Mitochondrial dysfunction, apoptosis, necrosis, inflammation	Alarifi et al. (2013), Guo et al. (2013), Sharma et al. (2012), Ahamed et al. (2011a,b), Xia et al. (2008)
Nanosilica	Cytotoxicity, apoptosis, G1 phase arrest in vitro	Chen et al. (2013) and Ye et al. (2010)
Nickel oxide	Lipid peroxidation and apoptosis	Siddiqui et al. (2012) and Ahamed et al. (2011a,b)
Titanium dioxide	Apoptosis, cytotoxicity, genotoxicity	Yoo et al. (2012), Sharma, (2009)
Aluminum oxide	Mitochondria mediated oxidative stress and cytotoxicity in human mesenchymal stem cells	Ashrani et al. (2008, 2009, 2011)
Gold	Lipid peroxidation and autophagy in vitro in MRC-5 lung fibroblasts	Chairuangkitti et al. (2013), and Li et al., (2010)
Ag-NP	Mitochondrial damage and genotoxicity in human lung fibroblast cells (IMR-90) and human glioblastoma cells (U251)	Papageorgiou et al. (2007)
Co-Cr NP	Oxidative DNA damage, micronuclei induction, reduced cell viability in human dermal fibroblasts	Papageorgiou et al. (2007)

of a generalized increase in oxidative stress in rats.

Pulskamp et al. (2007) have shown that a commercial preparation of CNT containing metallic impurities increased ROS production in rat NR8383 macrophages and human A549 lung cells; in addition, metal stripping by acid wash abolished the increase in ROS. Metals such as Fe^{2+} can increase ROS biologically by inducing the pro-oxidation proteins and chemically via Fenton reaction: $LFe^{2+} + H_2O_2 \rightarrow LFe^{3+} + HO^{\cdot +} HO^- \rightarrow LFe = O^{2+} + H_2O$ (Fenton, 1894; Prousek, 2007). Taken together, these observations suggest that metal ions, not the CNTs, induced oxidation stress. As shown in Figure 6, Fe^{2+} impurities dissociate (Figure 6(ii)) when the CNTs, especially short CNTs, are internalized (Ge et al., 2007). Release of CNTs into the cytoplasm (Figure 6(iii)) and the nucleus (Figure 6(v)) induced expression of the gene encoding pro-oxidative enzymes (thioxanthinase, NADPH oxidase 4, nitric oxidase synthase 2, uncoupling protein 3, etc.) and downregulated the genes that express antioxidative enzymes (isocitrate dehydrogenase, SOD, glutathione peroxidase, succinate dehydrogenase, neutrophil cytosolic factor 1, etc.). Lam et al. (2004) observed that a dose of 0.5 mg/mouse of CNTs (sonicated suspension) containing 26% nickel and 5% yttrium killed about 50% of the animals within 7 days after dosing, while metal depleted CNTs were essentially nontoxic. The deaths

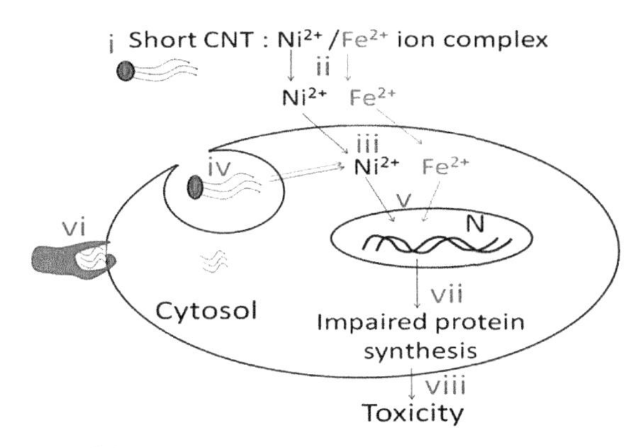

FIGURE 6 Effects of metal impurities on CNT toxicity. Partially purified short-CNTs may contain Ni^{2+} and Fe^{2+} impurities (i) that may dissociate from the CNTs in extracellular fluid (ii). Free metal ions may enter the cells (iii) and the nucleus (v), and induce expression of the genes expressing pro-oxidative enzymes (vii). The CNTs may enter the cells via endocytosis (iv). The aggregated CNTs may bind to the cell membrane and get engulfed by macrophages (vi). The metal-induced impairment in gene expression may initiate toxic reaction (viii).

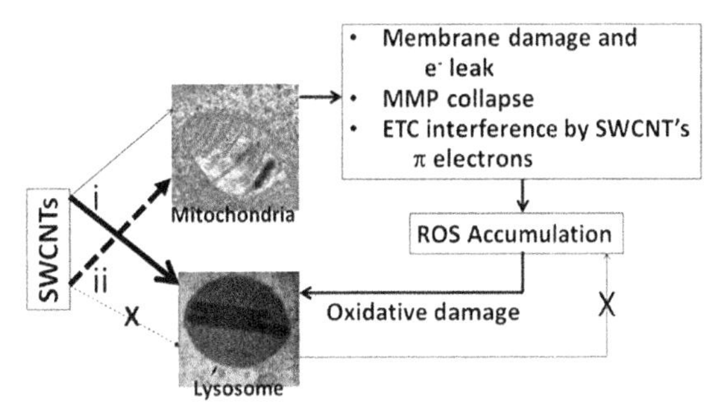

FIGURE 7 Distribution of SWCNTs between lysosome and mitochondria determines the induction of oxidative stress. Administration of SWCNTs in rodents resulted in their accumulation mostly in lysosomes and, to some extent, in mitochondria (i). If CNTs accumulation in lysosomes is blocked, then they exclusively accumulate in mitochondria. In mitochondria, CNTs caused (1) membrane damage and ensuing leak in electrons, (2) matrix metalloprotein (MMP) collapse, and (3) interference of the electron transport chain (ETC). This results in an increase of ROS production and accumulation, and further toxicity. *Adopted from Yang et al. (2007).*

were attributed to nickel that was released during ultrasonication. Thus, the presence of metals may be a compounding factor in the determination of CNT toxicities.

Yang et al (2007) have shown that purified SWCNTs (dose: 5–500 mg/kg in mice via gastric gavage), upon entering the cells, preferentially accumulated in lysosome; Figure 7(i)). Blockage of the autophagy of lysosomes decreased the quantity of SWCNTs in lysosomes, but proportionally increased that in mitochondria (Figure 7(ii)). Despite their higher accumulation in lysosomes, SWCNTs did not induce ROS production in lysosomes. However, an increase in SWCNT concentration in mitochondria caused a proportional increase in ROS production from mitochondria. This suggests that mitochondria may be the central organelles for ultrastructural damage of the cells. In mitochondria, electrons leaking from the electron transport chains can produce ROS. Under normal conditions, leaked electrons are eliminated by the mitochondrial protective enzyme systems. The π electrons of SWCNTs may block the electron transport, resulting in an increase of ROS that would damage mitochondrial membrane. The damaged mitochondrial membranes allow ROS to diffuse out and damage lysosomes. This mitochondrial production of ROS was not mediated by the presence of metal ions because purified SWCNTs were used. The differential effects of SWCNTs on lysosome and mitochondria present a serious challenge to the ability of nanomaterial to deliver drugs unless their adverse effects on mitochondria are established.

Kagan et al. (2006) reported the effects of Fe-depleted CNT and CNTs containing either traces of or excess Fe on induction of oxidative stress and NO in RAW 264.7 macrophages that contain phosphatidylserine-coupled toll-like receptors (TLR) 2 and 4 (PtdSer-TLR2 and PtdSer-TLR4). Upon ligand binding, the receptors activate NADPH-oxidase and NO synthase, respectively. However, CNTs with trace levels (<1%) of Fe, but not Fe-depleted CNTs or Fe-excess CNTs, interacted with the

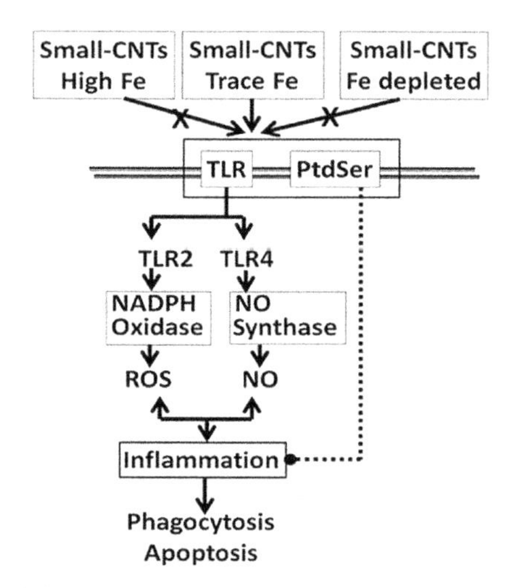

FIGURE 8 Possible roles of reactive oxygen species (ROS) and nitric oxide (NO) in induction of inflammation mediated by the toll-like receptors (TLRs) in microphages. Small CNTs containing trace levels of Fe, but not the Fe-depleted or Fe-excess CNTs, interacted with TLR-phosphatidyl serine (PtdSer) complex and induced NADPH oxidase (mediated by TLR2) and NO synthase (mediated by TLR4), resulting in an increase in generation of ROS and NO, respectively. Together they caused inflammation. *Data from Kagan et al. (2006) was used to construct the flow diagram.*

TLR-PtsSer receptor complex and activated NADPH oxidase and NO synthase (Figure 8). An increase in NADPH oxidase enhanced ROS and an activation of NO synthase enhanced production of NO and NO-derived RNS. For macrophages, the enhanced production of ROS and NO is required for effective phagocytosis of bacteria, but their uncontrolled production may result in their apoptotic death.

An increase in the ROS and RNS has also been shown to increase the pathogenesis of cancer, cardiovascular disease, atherosclerosis, hypertension, ischemia/reperfusion injury, diabetes mellitus, neurodegenerative diseases (Alzheimer disease and Parkinson disease), rheumatoid arthritis, and aging (Valko et al., 2005, 2006, 2007). This suggests that, at least for macrophages, a threshold level of iron may be required to activate ROS production. The observations of Kagan et al. (2006) provide a mechanistic basis for earlier observation that exposure to CNTs and infection mutually enhances inflammation and depresses bacterial clearance. Nanotubes may increase susceptibility to lung infection in exposed populations (Shvedova et al., 2008a).

CNTs generate ROS either by directly interacting with cell components (frustrated phagocytosis) or by indirectly impairing the mitochondrial function. The most likely mechanism for CNT-induced oxidative stress and lung toxicity involves mitochondrial dysfunction. However, as discussed above, an important controversy regarding the pro-oxidative effects of CNTs is whether the oxidative stress observed in response to the CNT is due to the nanotube alone or to the metal impurities, such as Fe, Co, and Ni, introduced within the CNT during their synthesis. As shown in Table 3, bare CNTs, purified (without metals) CNTs, and CNTs containing metal impurities yielded different results. CNTs (SWCNT or MWCNT) with metal impurities caused severe oxidative stress and ensuing abnormalities (inflammation, apoptosis, DNA damage, cytotoxicity, etc.) in cell cultures in vitro or in experimental animals in vivo. However, purified CNTs did not cause oxidative stress but activated pro-inflammatory signaling. Thus, metal-depleted CNTs may cause certain abnormalities not yet characterized.

4.1.3 *Is Oxidative Stress a Hierarchical Event Leading to the Development of a Nanoparticle's Toxicity?*

Despite the common belief that oxidative stress is the hierarchical event leading to the development of CNT toxicity (Castranova, 2004; Kamp et al., 1992; Brown et al., 2000), a few studies have shown the development of adverse effects without a significant increase in oxidative stress (Shvedova et al., 2005; Flahaut et al., 2006). Crouzier et al. (2010) have shown that mice exposed to DWCNTs of 1.3–3.2 nm diameter and 10-μm length exhibited thickening of alveolar wall, a reduction in alveolar space, an increase in alveolar macrophages, and an

TABLE 3 Oxidative Stress-Related Toxicity of Carbon Nanotubes

	CNT	
SWCNT with 30% iron by mass	Lipid peroxidation, activation of AP-1 and NF-κB, cytotoxicity, DNA damage and pro-inflammatory response in vitro and in vivo.	Murray et al. (2009), Pacurari et al. (2008), Sharma et al. (2007), and He et al. (2011)
Acid-treated MWCNTs with Co and Ni	Decreased cell viability, altered mitochondrial membrane potential in rat macrophages (NR8383) and human A549 lung cells	Pulskamp et al. (2007)
SWCNT	Reduced cell proliferation, activation of NF-κB in human keratinocytes	Manna et al. (2005)
Unpurified SWCNT (17.7% w/w iron)	Lipid peroxidation, acute inflammatory response, decreased respiratory function in adult C57BL/6 mice	Shvedova et al. (2008)
Raw MWCNT	Dose-dependent cytotoxicity in RAW 264.7 macrophages and A549 cells: Cell inflammation, membrane leakage, lipid peroxidation, and protein release	Chen et al. (2011b)
MWCNT	Increase in cell permeability, cell migration, and endothelial permeability in human microvascular endothelial cells (HMVEC)	Azad et al. (2012)
MWCNT	Activation of NF-κB, fibroblast-myofibroblast transformation, pro-fibrogenic cytokine, and growth factor induction in vitro (BEAS-2B, WI-38, and A549 cell lines)	He et al. (2011)

increase in signs of systemic inflammation without increasing oxidative stress or tumor necrosis factor (TNFα) expression. Similar to the observation of Crouzier et al. (2010), other studies have also shown an increase in macrophage migration and local increased pro-inflammatory cytokines in response to CNT (diameter <5 nm) exposure (Inoue et al., 2008; Mangum et al., 2006; Muller et al., 2005; Tsukahara and Haniu, 2011). Contrarily, Yamashita et al. (2010b) showed that CNT-induced inflammation was highest for MWCNTs (5–15 μm length) having 20–60 nm diameter, while the response was minimal for CNTs having <2 nm diameter. Thus, the use of thinner DWCNTs having 1.2–3.2 nm diameter should have resulted in poor inflammation reaction. Differential aggregation and/or free radical scavenger activity of the CNTs may account for the discrepancies between these studies. These observations, taken together, indicated that CNTs, in addition to inducing oxidative stress and downstream signaling, also initiate oxidative stress-independent toxicity pathways. Shvedova et al. (2012) and Coxon et al. (1996) have proposed both oxidative stress-dependent and -independent mechanisms for the development of CNT toxicity. They showed that CNTs caused an oxidative stress-dependent activation of mitochondrial Caspase and ensuing development of apoptosis and inflammation. However, CNTs also activated myeloperoxidase in an oxidative stress-independent mechanism in neutrophils, resulting in aneuploidy, interference with actin cytoskeleton, phagocytosis, and steric hindrance of ion channels. More research is needed to characterize the role of oxidative stress in the development of CNT toxicities.

The oxidative stress-based theory of toxicity is based on a global increase in oxidative stress characterized with a nonspecific increase in oxidation of lipids (order of oxidation: phosphatidylcholine > phosphatidylethanolamine >> phosphatidylinositol > phosphatidylserine [PS] > cardiolipin [CL]). An increase in peroxidation of cell membrane's polyunsaturated phospholipids is one of the major mechanisms of lung injury triggered by SWCNTs. Tyurina et al. (2011) and Samhan-Arias et al. (2012) showed that inhalation of SWCNTs selectively induced peroxidation of CL and PS, rather than nonspecific increase in lipid oxidation. Because CL and PS are selectively associated with the *Cyt c*-mediated mitochondrial apoptosis, the authors proposed that the SWCNT-induced phospholipid oxidation in lungs is associated with the *Cyt c*-mediated apoptosis as well as macrophage disposal of apoptotic cells. This suggests that the macrophage-induced and the mitochondria-induced mechanisms both participate in the development of CNT toxicity.

4.1.4 Free Radical Scavenging by CNTs

As described elsewhere in this manuscript, unpurified (containing metal impurities) and, to some extent, purified CNTs induced generation of ROS, inflammation, and fibrotic reactions in the lungs (Muller et al., 2005; Castranova, 2004; Fubini and Hubbard, 2003; Mossman and Churg, 1998). However, many studies have shown the absence of ROS generation in macrophages exposed to purified CNTs (Shvedova et al., 2005; Tsukahara and Haniu, 2011; Vittorio et al., 2009; Xu et al., 2009). Poor detection of ROS in response to purified CNTs may be due to their free radical scavenging activity (Martínez-Morlanes et al., 2012; Galano, 2008; Fenoglio et al., 2006; Mylvaganam and Zhang, 2004; Peng et al., 2003; Watts et al., 2003; Khabashesku et al., 2002; Holzinger et al., 2001). Interestingly, CNTs that scavenged free radicals retained their ability to induce inflammation, apoptosis, and fibrogenic activity (Fenoglio et al., 2006). Thus, MWCNT features, other than their ability to free radical generation, may be responsible for development of their toxicity. The free radical scavenging activity of CNTs also impairs the ability of phagocytotic cells (neutrophils, monocytes, macrophages, and eosinophil) to phagocyte and clear bacteria from the host. Phagocytotic cells, in response to an inflammatory stimulus such as bacterial infection, generate oxidative bursts strong enough to kill the bacteria (Rosen et al., 1995). The common free radicals released are $O_2^{\bullet -}$, $HO^{\bullet}$, and $NO_2^{\bullet}$ According to the following reaction (Squadron and Pryor, 1998),

$$NO^{\bullet} + O_2^{\bullet -} \rightarrow ONTO^{-} \text{(peroxynitrate)}$$
$$ONOO^{-} + H^{+} \rightarrow ONOOH$$
$$ONOOH \rightarrow HO^{\bullet} + NO_2^{\bullet}$$

$O_2^{\bullet -}$, $HO^{\bullet}$, and $NO_2^{\bullet}$, in addition to providing a highly toxic environment, also induce toxic signaling in the microbe. A study by Shvedova et al. (2008) has shown that a combined exposure of animals to SWCNTs and *Listeria monocytogenase* augmented inflammation, but attenuated NO generation and the clearance of the bacteria, possibly due to free radical scavenging activity of SWCNTs.

4.2 Inflammation

Inflammation is an organism's response to harmful stimuli including, but not limited to, pathogens, poisons, or irritants. Inflammation is induced by an oxidative stress-dependent activation of NFκB's pro-inflammatory dimer Rel A-p50 that induces expression and release of pro-inflammatory cytokines interleukin-1 (IL-1), IL-18, TNFα, γ-interferon, the granulocyte-macrophage colony-stimulating factor, etc. (Gilmore, 2006; Singh and Jiang, 2004; Collins et al., 1995). Figure 9 shows different pathways responsible for activation of NKκB RelA-p50 (Gilmore, 2006). After an appropriate response, inflammation is resolved by activation of NFκB's anti-inflammatory dimer p50-p50, which

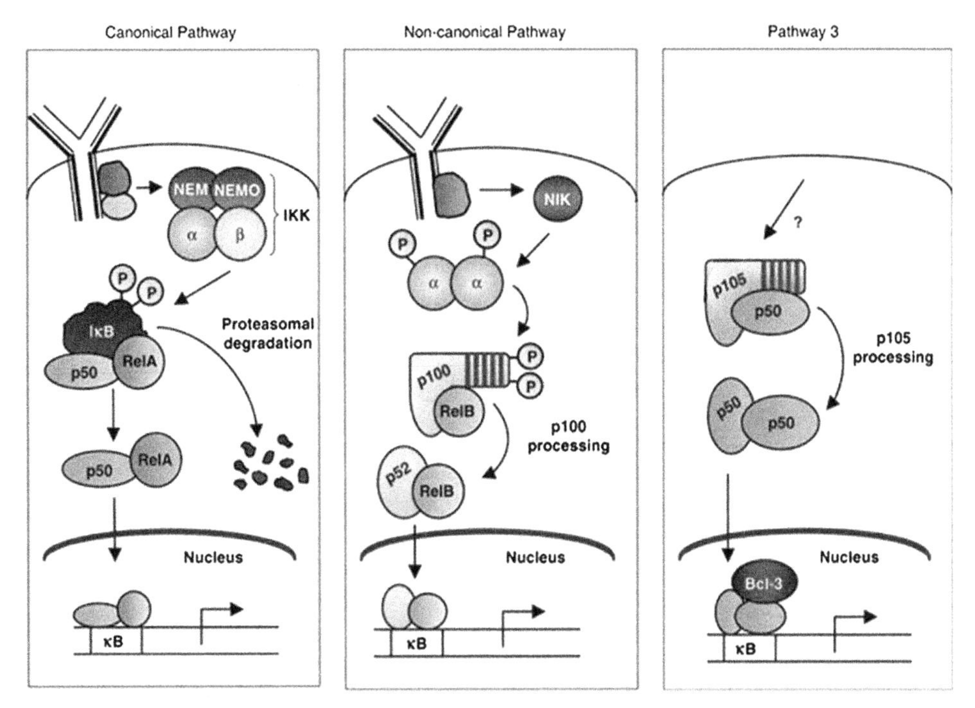

FIGURE 9 The NFκB-mediated pro-inflammatory signal transduction pathways. In the canonical pathway, NFκB dimer p50-RelA remain in the cytoplasm as p50-RelA-IκB complex. The binding of a ligand to a cell surface receptor activates the inhibitor of IκB kinase (IKK) complex, resulting in phosphorylation and degradation of IκB.This releases active NF-κB dimer p50-RelA that translocates into the nucleus to turn on target genes. NFκB also activates expression of IκB gene that, by forming a p50-RelA-IκB complex, sequester NF-κB back into the cytoplasm. The non-canonical pathways include p100/RelB-mediated and p50/p50-mediated pathways. Receptor binding activates the NFκB-inducing kinase that liberates the pro-inflammatory p52/RelB complex. Activation of p50 (or p52) homodimers, when activated, releases anti-inflammatory cytokines. *Reprinter from Gilmore (2006).*

suppresses the pro-inflammatory response and induces release of the anti-inflammatory cytokines, such as IL-4, IL-10, IL-13, IFNα, and the transforming growth factor β (TGF) beta (Singh and Jiang, 2004).

In addition to the NFκB signaling, studies have also proposed a possible role of inflammasomes (Schroder and Tschopp, 2010), consisting of Caspase 1, apoptosis-associated speck-like protein containing a Caspase recruitment domain (ASC), and nucleotide-binding oligomerization domain-like receptor protein (NALRP) (Yazdi et al., 2010; Yazdi and Zhou, 2011), in the release of IL-1β and possibly IL-1α (Figure 10). Nanoparticles, especially CNTs, induce inflammation by activating Nlrp3 inflammasome (Hamilton et al., 2012; Meunier et al., 2012; Palomaki et al., 2011). As shown in Figure 10, short CNTs mostly accumulate in the lysosome and induce release of cathepsin B, while long CNTs, via frustrated phagocytosis, activate NADPH oxidase in mitochondria. Although both CNTs result in an increase in ROS, short CNTs are less potent than long CNTs. ROS accumulation facilitates polymerization of NLRP3, ASC, and procaspase-1 into NLRP3 inflammasome and ensuing release of IL1β and IL1α. Long CNTs, via the inflammasome pathway, may deregulate inflammation induced by the pathogens.

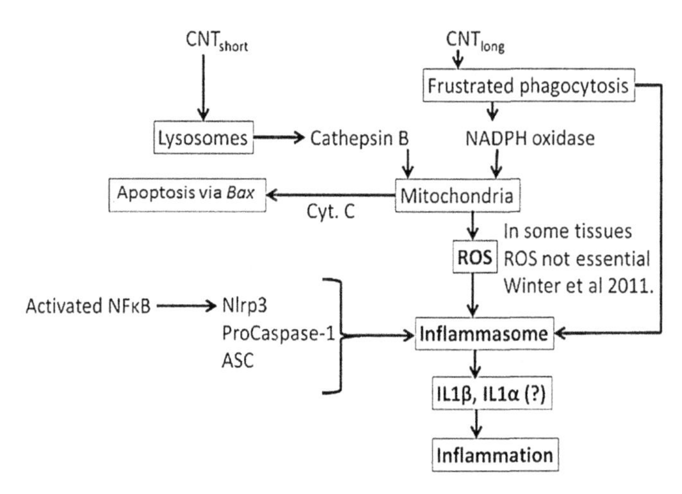

FIGURE 10 Activation of inflammasomes by CNTs. Short and long CNTs both activated ROS production by activating Cathepsin B and NADPH oxidase, respectively. Frustrated phagocytosis may be involved in NADPH oxidase activation. Activating Cathepsin B and NADPH oxidase, by altering mitochondrial membrane, induces ROS generation and activation of inflammasome that further induces (i) IL-1β release and ensuing inflammation and (ii) Bax mediated apoptosis (Zhou et al. (2011).

4.3 Genomics, Proteomics, Transcriptomics, and Metabolomics

The recent developments of high-throughput screening techniques that can analyze thousands of genes, proteins, transcripts, or metabolites simultaneously have resulted in the generation of large quantities of diverse sets of data that, on the face value, are difficult to interpret manually. The "omics" technologies provide a scientific means to integrate and standardize the multiple streams of data. At present, the omics areas include genomics, proteomics, transcriptomics, metabolomics, glycomics, lipidomics, endocrinomics, secretomics, and physiomics (Figure 11). Of these, genomics, proteomics, transcriptomics, and metabolomics are relevant to understanding nanoparticle toxicity. Insufficient information is available for other omics areas.

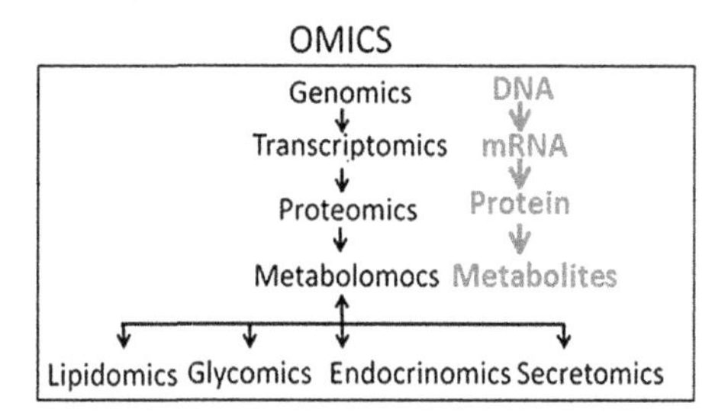

FIGURE 11 Different areas included in OMICS.

Genomics is the study of all aspects of the genome (gene structure, function, regulation of transcription, mutations, etc.) (Jonathan, 2009). *Proteomics* is the large-scale study of the proteome (Σ all bodily proteins), particularly their identification, structures, and functions (Anderson and Anderson, 1998). *Metabolomics* is the systematic study of the unique metabolic fingerprints (enzymes, metabolites, and mRNAs) that specific cellular processes leave behind (Bennett, 2005). *Transctiptomics* is the study of all RNA molecules, including mRNA, rRNA, tRNA, and other non-coding RNA produced in one or a population of cells. This section discusses the four omics approaches to decipher nanoparticle toxicity.

4.3.1 *Genomics*

Bulk particles have been shown to regulate approximately 5000 genes in living cells (Mu et al., 2014). Recently, Garcia et al. (2012) showed greater than 1000 downregulated and 1251 upregulated genes in fish (*Fundulus grandis*) that were collected from the deep-water horizon oil well area. The majority of the upregulated genes were aryl-hydrocarbon receptor (AHR) pathway genes. Unlike the bulk particles, nanoparticles usually affect much fewer genes (about 500), possibly because nanoparticles are either less active and/or more selective toward cells (Bouwmeester et al., 2011 (silver nanoparticles), Liu et al., 2011 (DMSA coated Fe_3O_4 magnetic nanoparticles), Mu et al., 2009 (carboxylated SWCNTs), Zhang et al., 2006 (silica coated QD), Cui et al., 2005 (SWCNTs)). Pristine SWCNTs, carboxylated SWNTs, and MWCNTs induced different sets of genes:

1. Pristine SWNTs or MWNTs affected a large number of genes related to the cell cycle, signaling transduction, apoptosis, oxidative stress, metabolism, transport, and immune responses in human cells (Chou et al., 2008: SWCNTs; Cui et al., 2005: SWCNTs; Ding et al., 2005: MWCNTs).
2. Carboxylated-SWCNTs inhibited DNA binding ID-genes, a family of helix-loop-helix (HLH) proteins that lack a basic DNA-binding domain and inhibit transcription through formation of nonfunctional dimers that are incapable of binding to DNA (Ellmeier et al., 1992) and bone morphogenetic protein (BMP) genes in HEK293 cells. PEG-modified SWNTs, however, affected fewer genes (38 genes at low doses [8 nm] and 12 genes at high doses [80 nm]) than pristine SWCNTs.

Metal nanoparticles, such as Fe_3O_4 nanoparticles, altered mostly inflammation-related genes such as MAPK, TLR, and JAK-STAT signaling pathway genes in a time-dependent manner. These observations are supported by the bioinformatics analyses that metal nanoparticles activated inflammatory and immune responses and inhibited biosynthesis and metabolism (Liu et al., 2011). Gold, silver, and copper nanoparticles affected significantly fewer genes (approximately 100 genes) than Fe_3O_4 or TiO_2 nanoparticles (500–700 genes) (Bouwmeester et al., 2011; Hanagata et al., 2011).

4.3.2 *Transcriptomics*

Genomics deals with the entire genome, which resides in the nucleus (eukaryotic cells) and contains the instructions needed to build and maintain cells. However, for the instructions to be carried out, DNAs must be transcribed into corresponding mRNAs, which are called *transcripts*. In general, humans, nematode (earthworms), and *Drosophila* have approximately 30,000, 20,000, and 14,000 protein-coding genes, respectively (Venter et al., 2001; Adams et al., 2000; The C. elegans Sequencing Consortium, 1998). Interestingly, the total mRNAs transcript numbers are greater than the total gene numbers (approximately 60,000 transcripts in humans), but all of the transcripts are not synthesized simultaneously. The differences in gene and transcript numbers may be because of alternative splicing, RNA editing, or alternative transcription initiation and termination sites. In addition to mRNA, cells also contain various other kinds of RNAs such as transfer RNAs, risosomal RNAs, etc., that participate in protein synthesis. In multicellular organisms, nearly every cell contains the same genome and thus the same genes. However, not every gene is transcribed in every cell, that is, different cells show different patterns of gene expressions that underlie the wide range of physical, biochemical, and developmental differences seen among various cells and tissues. These differences in transcriptomes may play a role in the difference between healthy and diseased cells. Thus, any changes in transcriptional activity may reflect or contribute to diseases.

Measurement of transcriptional activity provides a powerful approach to determine how an organism responds to a particular biotic/abiotic condition. Modern high-throughput methodologies can quantify the levels of mRNA transcripts in organisms under health and disease conditions (Chen et al., 2012). Effects on specific physiological or biochemical processes can be revealed by changes in the transcript levels of the genes. The transcriptome sequencing (RNA-seq) approach generates a vast amount of gene transcripts using massive parallel DNA sequencing technologies, bioinformatics, and sequence databases. The transcriptome profiling provides substantial and detailed information about an organism's toxicological responses. However, transcriptome profiling has certain limitations: it does not reveal changes due to effects at the translational, posttranslational, cell biological, or organismal levels.

Kong et al. (2013) have developed a unique database called NanoMiner, which contains experimental results from different nanoparticle-related gene expression microarray studies (Barrett et al., 2011; Parkinson et al., 2011). NanoMiner provides links to the original studies, access to the annotations of the data samples, and various visualization and statistical analysis options to aid nanoparticle research. This software is a unique tool for researchers working in toxicogenomics, especially to anticipate the outcome of the interaction of nanoparticles with biological systems and thus the future risk of using these materials. For example, Figure 12 shows a data-set in NanoMiner measured with Affymetrix, Agilent, Illumina, or Spotted Oligonucleotide platforms (Figure 12(A)). The data are composed of at least six human samples treated with nanoparticles (Figure 12(B): number of genes, Figure 12(C): number of samples, Table 4).

Simon et al. (2013) characterized the transcriptomic effects of four metal-based nanoparticles—Ag NPs, ZnO NPs, TiO_2 NPs, and CdTe/CdS quantum dots (QDs)—using eukaryotic green alga, *Chlamydomonas reinhardtii*. RNA-seq was used to characterize the effects of nanoparticles on transcription to identify distinct and similar toxicological effects with the ultimate goal of determining the modes of action induced by each nanoparticle. Their studies showed at least twofold changes in approximately 400 transcripts, although all four nanoparticles differently affected the transcript expressions: Au NP 86% upregulated (ur) and 77% downregulated

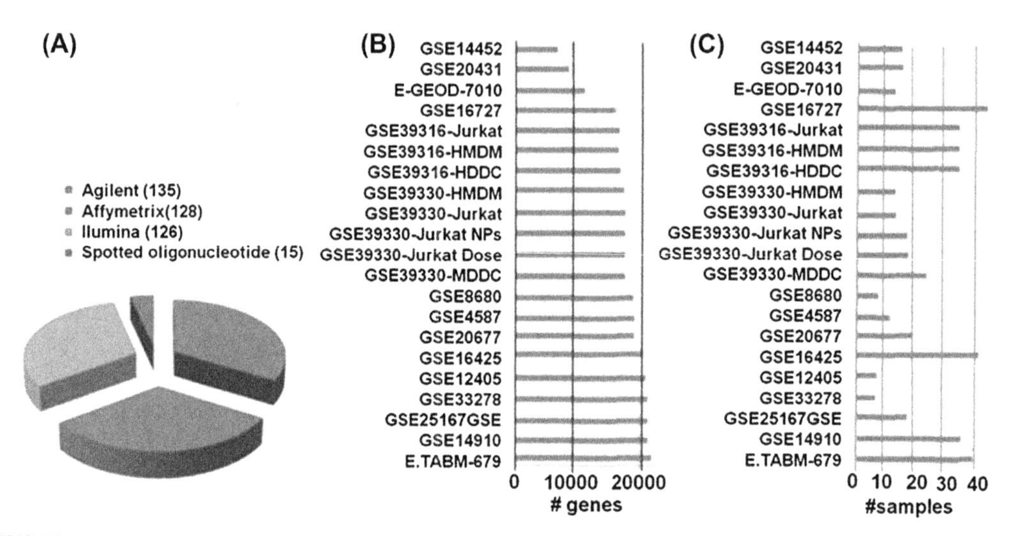

FIGURE 12 Data in NanoMiner. (A) The samples in NanoMiner measured with different platforms, (B) Number of genes measured in the datasets in NanoMiner, (C) Number of samples in each dataset. For gene abbreviations, refer to references shown in Table 4.

TABLE 4 Description of the Gene Symbols, Platforms Used, Cell Types, and Nanoparticles Used

Reference	Platform	Cell Type	Nanoparticle
Busch et al. (2010), GSE16727	Agelent-014850	HaCaT	WC ± Co
Fujita et al. (2009), GSE16425	Agelent-014850	HaCaT	TiO_2
Gras et al. (2009), GSE12405	Agelent-014850	Macrophages	2G-NN16
Hofer et al. (2008), GSE4567	Afymetrix U133 + 2	MDM	TiO_2
Huang et al. (2011), GSE7010	Afymetrix U133 + 2	Epithelial	PM
Karoly et al. (2007), GSE4567	Afymetrix U133 + 2	HPAEC	uPM
Kawata et al. (2009), GSE14452	Afymetrix target array	HepG2	AgNP
No reference, E-TABM-679	Illumina-12 v3.0	Epithelial	Carbon black
Kim et al. (2012), GSE20677	Afymetrix U133 + 2	Embr. Kidney	Au
Tuomela et al. (2013), GSE39330	Afymetrix U219	HDMD	ZnO
Tuomela et al. (2013), GSE39316	Illimina −12 v3.0	HMDM	ZnO
Moos et al. (2011), GSE14910	Agelent-014850	CaCo2	Multiple
Moos et al. (2011), GSE25167	Agelent-014850	HaCaT/SK Mel 28	ZnO TiO_2
Balakumaran et al. (2010), GSE20431	Spotted Oligonucl. CD/DTM	Bone merrow	Au
Hanagata et al. (2011), GSE33278	Agelent-014850	Lung epithelial	CuO

(dr), TiO_2 52% ur and 15% dr, QD 29% ur and 0% dr, and ZnO 72% ur and 50% dr (Simon et al., 2013). In addition, the changes occurring in Ag NP-exposed samples were opposite to the changes occurring in TiO_2 NP, QD, or ZnO NP exposed samples. ZnO NPs caused nonspecific global stress to the cells under environmentally relevant conditions, while TiO_2 NPs downregulated the transcripts regulating photosynthesis. Exposures to TiO_2 NP, ZnO NP, and QDs elevated the levels of transcripts encoding subunits of the proteasome, a phenomenon believed to underlie the development and progression of several major diseases, including Alzheimer disease, and used in chemotherapy against multiple myeloma in humans. Au NPs downregulated the transcripts that regulated protein synthesis, cell wall synthesis, and phospholipase. Taken together, this study suggests that transcriptome analysis provides critical information regarding gene expression and pathophysiological responses of organisms that genomic analysis will not. Further information regarding the effects of nanoparticles on transcriptomes can be found in the following publications (Boenigk et al., 2014; Fisichella et al., 2014; Hanagata et al., 2011; Lee et al., 2009b; Jin et al., 2008).

4.3.3 *Proteomics*

Protein synthesis is downstream of gene transcription and controlled by a number of regulatory factors. Because proteins directly regulate the organism's metabolism and physiology, an evaluation of proteomic expression may be more useful for evaluating nanoparticles than genomic and transcriptome analyses. In general, the global protein expression profile is examined using 2D gel electrophoresis combined with mass spectrometry (MS). This method consists of the following steps: extraction of total proteins from cell suspension, separation of proteins by 2D gel electrophoresis, cutting out each protein spot, digesting it in aqueous solution, and then analyzing the proteins via MS analysis. MS data are analyzed by searching databases to identify the proteins.

Although extensively used, the 2D electrophoresis method has many disadvantages, such as lack of reproducibility, recovery, and quantification. Another approach is the use of isobaric tags for a relative and absolute quantitation (iTRAQ) technique. In this method, the extracted proteins are labeled with iTRAQ, which are analyzed using liquid chromatography to separate isotopically labeled peptides and MS/MS analysis to identify and quantitate peptides (Casado-Vela et al., 2010). This method exhibits excellent sensitivity but lacks high-throughput.

The high-throughput protein microarray methods allow detection of multiple proteins and phosphorylated proteins in a single experiment. Yuan et al. (2011), using iTRAQ-coupled LC-MS/MS analyses, showed that SWNTs altered expression of the proteins involved in metabolic pathways, redox regulation, signaling pathways, cytoskeleton formation, and cell growth. Yuan et al. (2011), using 2D gel electrophoresis and matrix-assisted laser desorption-time of flight mass spectrometry (MALDI-TOF MS), showed that abundances of 45 proteins associated with metabolism, biosynthesis, stress response, and differentiation that were altered by MWCNTs in human monoplastic leukemia U937. Tsai et al. (2011), using phosphorylated protein antibodies arrays, revealed that Au NPs activated endoplasmic reticulum stress response in human chronic myelogenous leukemia K562 cells. Ge et al. (2011) showed that TiO_2 NPs significantly altered 46 proteins (involved in the stress response, metabolism, adhesion, cytoskeletal dynamics, cell growth, cell death, and signaling pathways) in human bronchial epithelial BEAS-2B cells (Ge et al., 2011). In general, nanoparticles perturb approximately 20 proteomic pathways, consistent with pathways from genomic data analyses. For further information, please review the following studies (Sund et al., 2014; Gomes et al., 2014; Lin et al., 2014; Mirzajani et al., 2011; Zhang et al., 2011, 2014).

1.3.1 Metabolomic Alterations

Metabolomics is the study of metabolism at the organism/population level. Metabolomics is the concept that the metabolic state is representative of the overall physiologic status of the organism. In metabolomic studies, global biochemical events are assessed assaying thousands of small molecules in cells, tissues, organs, or biological fluids, followed by the application of informatics techniques to define metabolomics signatures (Figure 13). Metabolomic studies can lead to the enhanced understanding of disease mechanisms as well as enhanced understanding of mechanisms for a nanoparticle's effect.

Since nanoparticles cause physiological changes in living systems and generate altered metabolic profiles (Kim et al., 2011; Feng et al., 2011; Lenz and Wilson, 2007), an investigation of global metabolomics profiles may provide insight into the physiological processes that are perturbed by nanoparticles. Nanoparticle-induced metabolomics alterations are studied in live animals or cells and the analytical methods used are shown in Figure 13. Lenz and Wilson (2007), using high-resolution nuclear magnetic resonance spectroscopy, demonstrated that Ag NPs (i) decreased the levels of glutathione (GSH), lactate, and taurine and (ii) increased the levels of most amino acids, choline analogs, and pyruvate. This suggests that GSH depletion may induce conversion of lactate and taurine to pyruvate, providing a biochemical mechanism for oxidative stress (Mu et al., 2014). Feng et al. (2011) have shown that nanoparticles decreased the levels of triglycerides, essential amino acids, and choline metabolites while increasing glycerophospholipids, tyrosine, phenylalanine, lysine, glycine, and glutamate,

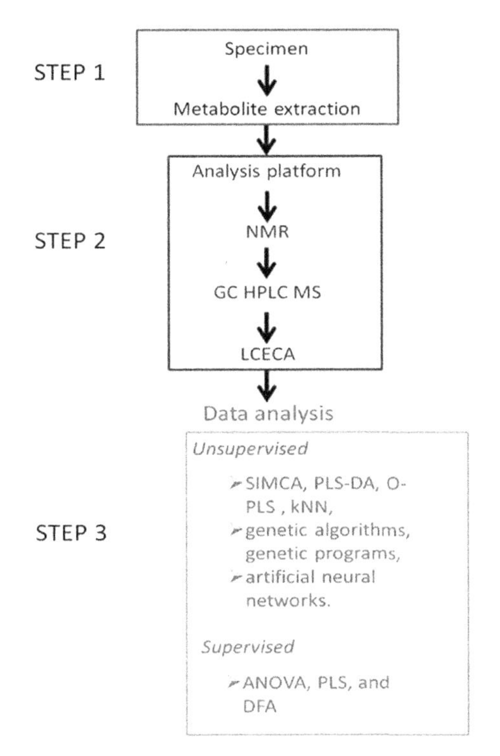

FIGURE 13 The procedure commonly used for studying metabolomics. GC HPLC MS: gas chromatography or high-performance liquid chromatography mass spectrometry, LCECA: Liquid chromatography–electrochemistry array, NMR: nuclear magnetic resonance, SIMCA: Soft Independent Modeling of Class Analogy, PLS: principal component analysis, OPLS: orthogonal PLS, kNN: k-nearest neighbors algorithm, DFA: Detrended fluctuation analysis.

possibly due to phagocytosis and cell membrane perturbation. Mu et al. (2014) have suggested that nanoparticles alter metabolism via direct and indirect mechanisms. In the direct mechanism, nanoparticles bind cell-membrane phospholipids, receptors, cytosol proteins, and DNA molecules, and then modulate the cells' microenvironment. Indirect effects of nanoparticles include activation of the signaling pathway via the nanoparticle-induced ROS that causes cytotoxicity. Direct interactions are much more specific and nanoparticle dependent compared to indirect interactions.

5. SYSTEM/ORGAN TOXICITY OF NANOPARTICLES

Mammals and other multicellular organisms consist of a complex network of biologically relevant entities, called systems. Depending upon the scale of the biological organization spans, biological systems can be classified as populations of organisms in an ecosystem, the organ and tissue scale in mammals and other animals (the circulatory system, the respiratory system, the nervous system, etc.). On the microscopic to nanoscopic scale, examples of biological systems are cells, organelles, macromolecular complexes, and regulatory pathways.

5.1 Dermal Toxicity

Dermal toxicity is the ability of a substance to cause local reaction and/or systemic poisoning in people or animals by contact with the skin. The absorption of toxic materials through the skin depends, to various degrees, on their chemical composition and solubility. A sound understanding of the relationship between dermal exposure and systemic exposure is important in assessing the effect that nanoparticle might have on people coming in contact dermally. Murray et al. (2009), using an epidermal cell line, showed that unpurified, but not purified CNTs, caused dermal toxicity associated with free radical generation and inflammation. They also showed a significant increase in oxidative stress and skin thickness possibly due to an increase in enumeration of dermal cells, polymorphonuclear leukocytes, and mast cells in SKH-1 mice topically exposed to unpurified SWCNT.

Shvedova et al. (2005), using human epidermal keratinocytes, showed that unrefined, but not refined, SWCNTs induced oxidative stress and ensuing dermal cytotoxicity characterized by accumulation of peroxidative products, reduction of total sulfhydryl, and a decrease in the vitamin E content. This suggests that metal impurities, but not pure CNTs, may be associated with the dermal toxicity of CNTs. Contrarily, Tian et al. (2006), using human dermis fibroblast cells, showed that the surface area was the best predictor of refined SWCNTs' toxicity and that refined SWCNTs were more toxic than their unrefined counterparts. Thus, SWCNTs' toxicity in dermal fibroblasts may be metal-independent. This was further confirmed by Monteiro-Riviere et al. (2005) who, using iron-free MWCNTs, showed that the CNTs were capable of both localizing within and initiating an irritation response in the epithelial cell in vitro. Taken together, the studies described above provide a conflicting view regarding the role of metal impurities in the adverse effects of CNTs on skin.

5.2 Respiratory System Toxicity

As discussed elsewhere, the distribution of CNTs in the lungs is determined by their length, diameter, defects, aggregation and agglomeration status, dissolution, purity, and surface characteristics. The proceeding paragraphs describe these factors in detail.

5.2.1 Diameter- and Length-Dependent Toxicity

Studies have shown that inhaled CNTs exhibited diameter-dependent dispersion in lungs (Asgharian and Price, 2007). CNTs greater than 100 nm deposited in all regions, CNTs less than 10 nm diameter deposited in the tracheobronchial region, CNTs 10–20 nm diameter deposited in the alveolar region, and CNTs greater than 20 nm deposited in the nasopharyngeal-laryngeal region. CNTs of 50 nm in diameter would pierce the cell membrane, while the thinner CNTs would aggregate and thicker CNTs (>100 nm) would not enter the cells (Nagai et al., 2011; Vakarelski et al., 2007; Lacerda et al., 2007). This suggests that the CNTs of 50-nm diameter may exhibit the greatest toxicity. The overall toxicity is further compounded by the length-dependent effects of CNTs. Studies have shown that short and long SWCNTs, short MWCNTs, and long MWCNTs (>10 μm) having less than 20 nm diameter may enter the cells and then aggregate into the cytoplasm. Long MWCNTs having 50–60 nm diameter may enter the cells and remain dispersed, while long MWCNTs having greater than 100 nm diameter may not enter the cells (Murphy et al., 2011). Aggregated CNTs would be recognized by the resident macrophages and undergo phagocytosis. Long MWCNTs will also be recognized by the macrophages but phagocytosis will be incomplete due to their length—a process known as frustrated phagocytosis (Murphy et al., 2011).

Frustrated phagocytosis is a strong pro-inflammatory stimulus resulting in an increase in oxidative stress and pro-inflammatory cytokines in lungs. Small CNTs enter the pleural space and, via the stoma, enter into the lymphatic capillary endothelium where microphages engulf and clear them (Figure 14(A) upper panel). Black spots can be formed transiently in parietal pleura, as shown in tissue histology (Figure 14(A): lower panel shows fiber accumulation at 1 day, which is completely cleared at 7 days after small CNT). Unlike small CNTs, large CNTs are retained at the stromal opening (Figure 14(B), upper panel). Macrophages present in the pleural space attack large CNTs and try to engulf them unsuccessfully, resulting in frustrated phagocytosis and ensuing toxicity. Black spots appear at day 1 and increased further at day 7 (Figure 14(B), lower panel). Murphy et al. (2011) also reported that long MWCNTs were more persistent and exhibited greater toxicity than short MWCNTs. After intratracheal administration in rats, short MWCNTs, but not long MWCNTs, appeared in the cranial mediastinal region, indicating that short MWCNTs efficiently exited stoma. The longer residence time may be responsible for the greater toxicity of long MWCNTs.

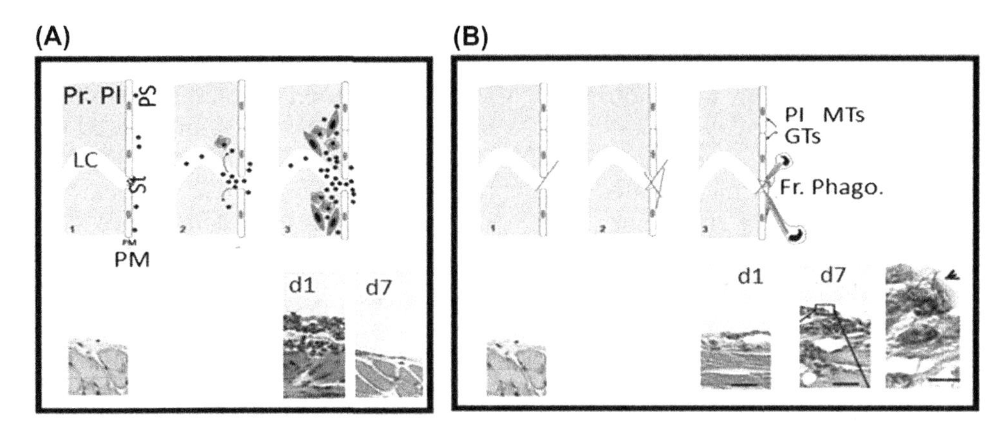

FIGURE 14 Processing of long and short CNTs in lungs. (A) Processing of small CNTs. (A1) Particles enter the pleural space (PS) and, via stoma (St), they enter into the lymphatic capillary (LC) endothelium (A2) where microphages engulf and clear the CNTs (A3), thus black spots (accumulation of fibers in parietal pleura (Pr Pl)) do not form or form transiently. The tissue histology (A lower panel) shows fiber accumulation at d1 after that cleared completely at d7 after small CNT exposure. (B) Processing of long CNTs. The long fibers accumulate at the stromal opening (B1 and B2). Macrophages accumulate in the region and try to engulf the particles but undergo frustrated phagocytosis, releasing pro-inflammatory, genotoxic and mitogenic mediators close to the pleural mesothelial cells (B3). The macrophage fiber complex yields the black color visible under a microscope. The tissue histology (B lower panel) shows fiber accumulation at d1 that grows larger at days 7 and 21 after CNT administration.

5.2.2 CNT Functionalization and Pulmonary Toxicity

Pristine CNTs may exhibit strong intertube van der Waal attraction that hinders their dissolution and dispersion in hydrophilic environments. Undispersed CNTs remain aggregated with physicochemical properties that are different from the dispersed particles. Compared to pristine MWCNTs, anionic-functionalized (COOH and PEG) CNTs decreased the production of pro-fibrogenic cytokines and growth factors, including IL-1β, TGF-β1, and PDGF-AA in BEAS-2B and THP-1 cells. Weakly cationic CNTs (NH_2 functionalized) were less effective than anionic CNTs in protecting BEAS-2B and THP-1 cells. In contrast to the weakly cationic CNTs, the strongly cationic polyethylenimine (PEI)-functionalized CNTs exhibited potent fibrotic potential in vivo. This suggests that functionalization may be a key modifier of CNT toxicity in vitro and in vivo.

Sohn et al. (2012) studied the effects of hyperbranched poly(ether-ketone) on the surface of both SWCNTs and MWCNTs. They observed that hyperfunctionalization decreased the SWCNT diameter from 40 to 60 nm for pristine CNTs and from 5 to 25 nm for functionalized CNTs and the particles' overall state of dispersion was homogeneous. In the case of MWCNTs, hyperbranched polyfunctionalization increased their diameters. Because the physicochemical properties of CNTs depend on their size, the functionalization-induced alteration in diameters of SWCNTs and MWCNTs may also alter their properties. Ali-Boucetta et al. (2013) functionalized long MWCNTs with tri(ethylene glycol) and found that functionalization dispersed MWCNT bundles into individual fibers with relatively short lengths. In addition, an asbestos-like pathogenicity observed for long MWCNTs was significantly attenuated by functionalization with tri(ethyl glycol). This may be because the functionalized dispersed MWCNTs were small enough to be phagocytized by the macrophages and cleared from the lungs. Lam et al. (2006) have shown that long CNTs induced time-dependent granulomas and interstitial inflammation in a dose-dependent manner in the mouse. In some mice, the lungs revealed peribronchial inflammation and necrosis extending into the alveolar septa. The lungs of mice treated with carbon black or quartz were either normal or exhibited mild inflammation. Thus, CNTs may be a serious occupational health hazard under chronic inhalation exposure situations.

5.3 Liver Toxicity

The liver is the main organ of bodily metabolism and detoxification of drugs and toxins; thus, it is susceptible to their toxicity. Jain et al. (2011) compared to the toxicities of the liver, spleen, kidney, and lungs by measuring the organ index (a ratio of an organ weight to body weight) in mice exposed to pristine or variably (0% to >6% carboxylation) functionalized MWCNTs. The organ index values for the liver were significantly higher than the values for spleen, kidney, or lungs, and an increase in the degree of functionalization decreased the liver index. This suggests that CNTs may selectively target the liver, possibly because they were trapped by the reticuloendothelial system and remained in the liver for a long time after intravenous injection (Deng et al., 2007, 2008). Ji et al. (2009) showed that pristine CNTs were more hepatotoxic than functionalized ones. Pristine-nonfunctionalized MWCNTs increased aspartate aminotransferase, alanine aminotransferase, total bilirubin, and creatinine levels—the key indicators of liver damage (Murakami et al., 2004)—while PEG-functionalized MWCNTs did not cause these changes (Ji et al., 2009). All of the above studies showed development of toxicity without any increase in ROS or reactive ROS metabolites. Thus, oxidative stress was not the key etiological factor in causing the CNT-induced hepatocyte toxicity.

Patlolla et al. (2011), using COOH-functionalized (2–7% COOH) MWCNTs (diameter 15–20 nm, length 15–20 μm), showed that the COOH-MWCNTs increased oxidative stress, decreased glutathione concentration, increased liver enzymes, and caused hepatotoxicity. Patolla et al. (2011) and Lewinski et al. (2008) proposed a key role for toxic metabolites of lipids, such as lipid peroxidases (LOOHs), in the development of COOH-MWCNT-mediated hepatotoxicity (Cejas et al., 2004). CNTs accumulated primarily in Kupffer cells and caused hepatocyte injury via inducing oxidative stress (Laskin and Laskin, 2001; Laskin and Pendino, 1995; Jaeschke et al., 2002). Administration of TNFα-antagonists and depletion of Kupffer cells protected hepatocytes against the drug-induced toxicity, suggesting a direct effect of inflammation on liver toxicity (Jaeschke et al., 2002). Exposure to COOH-MWCNTs induced ROS; enhanced the activities of serum aminotransferases, alkaline phosphatases, and concentrations of lipid hydro peroxide; and caused hepatotoxicity, which was attributed to the CNT-induced increase in oxidative stress (Bourdon et al., 2012). These observations further elucidate controversies in the causal relationship between oxidative stress and hepatotoxicity. More research is needed to resolve this issue.

5.4 Cardiovascular Toxicity

Cardiovascular toxicity includes damage to the heart due to toxin-induced electrophysiological abnormalities or/and muscle damage, as well as vascular atherosclerosis due to oxidative stress and inflammation. Taken together, these abnormalities may impair blood flow and circulation. Oxidative stress and inflammation may be the central pathogenic mechanisms underlying cardiovascular toxicity (Lefer and Granger, 2000; Ceriello and Motz, 2004; Kovacic and Thurn, 2005; Juni et al., 2013). The causal relationship between nanoparticles and cardiovascular disorders were first demonstrated by the studies investigating airborne particulate matter (Brook et al., 2004; Pope et al., 2004). In general, short-term exposure to elevated particle matters (PM) increased acute cardiovascular mortality, while prolonged exposure enhanced the risk of atherosclerosis. Sun et al. (2005) showed that PMs, either 10 μm or 2.5 μm diameter, develop advanced coronary atherosclerosis, although the severity of the effects were size dependent: PMs with smaller diameters were more toxic than those with larger diameters (Brook et al., 2004; Pope et al., 2004; Araujo et al., 2008). Although the mechanisms underlying the adverse effects of a PM are not yet known, oxidative stress and inflammation appear to be key causative factors.

Although the cardiovascular toxicity of CNTs is not fully established, the PM results provide strong evidence for a central role of oxidative stress and inflammation in a CNT's cardiovascular toxicity. The evidence for the proposed hypothesis stems from the following observations:

1. Li et al. (2007) showed that, in experimental animals, SWCNTs increased cardiovascular oxidative stress and mitochondrial dysfunction, which are recognized factors in causing cardiovascular diseases (Ballinger, 2005).
2. CNTs and other nanoparticles caused platelet aggregation and acute systemic prothrombiotic response following the induction of inflammatory activity (Radomski et al., 2005; Nemmar et al., 2004).

Another unique aspect of the cardiovascular toxicity of CNTs unrelated to either oxidative stress or inflammation is that CNTs interfere with the arterial baroreceptors that regulate blood pressure and cardiovascular variability (Nafz et al., 1986; Eckberg et al., 1991; Sleight and Eckberg, 1992; Pang, 2001). Thus, CNTs (SWCNTs and MWCNTs both) may directly modulate the regulation of autonomic cardiovascular control (Legramante et al., 2012, 2009; Stapleton et al., 2012).

There is sufficient but not compelling evidence for coupling between pulmonary exposure and cardiovascular toxicity of CNTs. As shown in Figure 15, pulmonary exposure to CNTs, in addition to causing pulmonary toxicity, also induces cardiovascular toxicity either directly via translocation of dissociated CNTs from lungs to blood (Ma-Hock et al., 2009; Pauluhn, 2010) or indirectly by inducing mitochondrial oxidative stress, inflammation, or platelet activation (Li et al., 2007; Legramante et al., 2012, 2009; Stapleton et al., 2012).

5.5 Central Nervous System Toxicity

The central nervous system (CNS) consists of the brain and spinal cord and the blood–brain barrier. The CNS, due to the blood–brain barrier's tight junctions around the capillaries, does not receive blood directly from the systemic circulation (Crone and Christensen, 1981; de Vries et al., 1997). Therefore, many excellent drugs, because of their inability to cross the blood–brain barrier, do not effectively treat brain disorders. Observations that nanoparticles, including CNTs, can pass through the blood–brain barrier without damaging the tight junction have proliferated development of diverse nanoparticle-based agents designed for treatment of different brain disorders (Beg et al., 2011; de Boer and Gaillard, 2007a,b). Because pristine CNTs are not viable options for drug delivery, a number of studies have functionalized CNTs with PEG and proteins, such as antitransferrin receptor antibodies, for their transport across the blood–brain barrier (Xiao and Gan, 2013; Lee et al., 2000; Chang et al., 2009; Ulbrich et al., 2009; Masserini, 2013). However, as described elsewhere in this text, functionalized CNTs can be defunctionalized (Fu et al., 2001), resulting in the accumulation of pristine nanotubes and ensuing toxicity in tissues. Because CNTs' defunctionalization in the brain is not well established, their effects on human health are of serious concern.

In general, CNS toxicity is evaluated using in vitro and in vivo methods, which have their

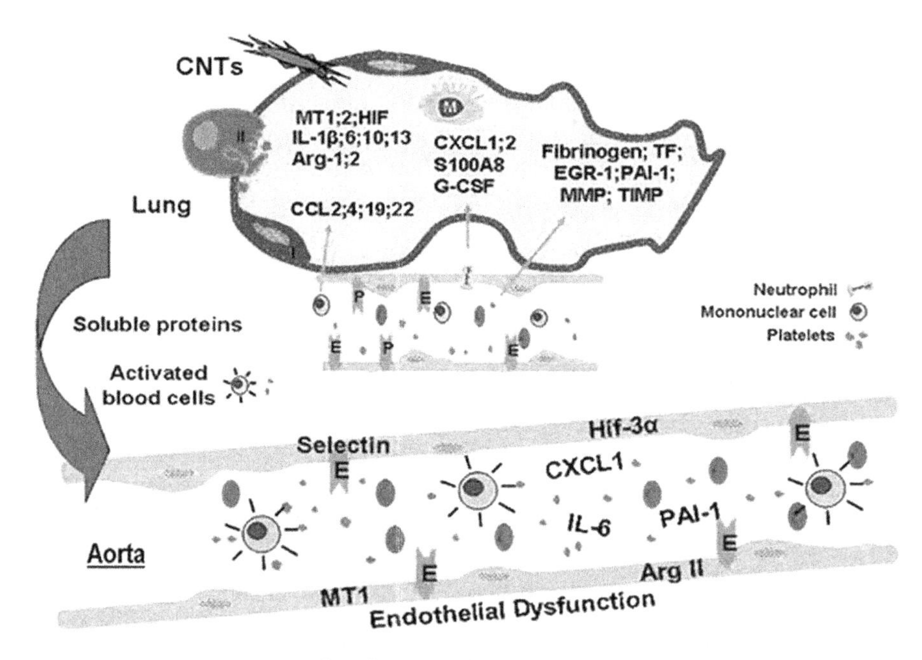

FIGURE 15 Possible cross-talk between lung and systemic circulation that may lead to the potential risk for cardiovascular adverse outcomes. Abbreviations include (I) alveolar type I cell; (II) alveolar type II cell; (M) alveolar macrophage; (MT) metallothionein; (HIF) hypoxia inducible factor; (IL) interleukin; (Arg) arginase; (G-CSF) granulocyte-colony stimulating factor; (TF) tissue factor; (EGR) early growth response; (PAI) plasminogen activator inhibitor; (MMP) matrix metalloproteinase; (TIMP) tissue inhibitor of metalloproteinase. *Reprinted from Erdely et al. (2009).*

own advantages and disadvantages. In in vitro models using organotypic or dissociated cultures, cell lines or stem cells are exposed to different doses (in terms of mass, size, or surface area) of the toxins and then the indices of toxicity are measured. The commonly measured indices are cell growth and proliferation, apoptosis, necrosis, neurotransmitter release or reuptake, inflammation, oxidative stress, frustrated phagocytosis, receptor and ion channel functions, signaling mechanisms, electrophysiology, and morphological changes. In in vivo models, experimental animals (wild-type or genetically modified) are exposed to different doses of the toxins and, at different time intervals, the indices of toxicity (behavior, blood and tissue chemistry, toxins' biodistribution and toxicokinetics, systemic and central inflammation, oxidative stress, brain neurotransmitter systems, electrophysiology, etc.) are measured. The proceeding paragraphs discuss the advantages and disadvantages of the two methods.

5.5.1 *In Vitro Toxicity*

The in vitro toxicity of CNTs has been studied using mostly nonneural models, such as dermal fibroblasts (Patlolla et al., 2010), bronchial epithelial cells (Lindberg et al., 2009), macrophages (Fiorito et al., 2006), and embryonic kidney (HEK293) cells (Reddy et al., 2010, 2012). A few studies have used Caco2 intestinal endothelial cell lines and Calu-3 airway epithelial cell lines to study possible effects of CNTs on tight junction and cytotoxicity. Both MWCNTs and SWCNTs decreased transepithelial resistance (TER) without causing cytotoxicity, even at the highest dose used (Banga et al., 2012; Lee and Geckeler, 2012). Clark et al. (2012) showed that the functionalized CNTs, although opening the tight junction, did not enter the Caco-2 cells. Coyuco et al. (2011) showed that COOH functionalized, but not PEG-functionalized, SWCNTs disrupted organization of the tight-junction protein, ZO-1. Contrary to these observations, Pichardo et al. (2012) and Jos et al. (2009) showed that the COOH-functionalized SWCNTs (1) significantly increased ROS levels and catalase, SOD and glutathione peroxidase activities; (2) decreased glutathione levels; and (3) increased LDH leak, indicating development of cytotoxicity in Caco-2 cells. Mahajan et al. (2010) and Xu et al. (2008) have evaluated permeability of transferrin (TR)-conjugated nanoparticles such as QDs across an in vitro model of the blood–brain barrier. The TR-coupled nanoparticles permeated across the blood–brain barrier via a receptor-mediated transport mechanism (Mahajan et al., 2012; Chang et al., 2009; He et al., 2011; Pang et al., 2011; Gan and Feng, 2010; Ulbrich et al., 2009). However, these studies have not looked at the possible toxicity of nanoparticles on the blood–-brain barrier; thus, it is not possible to predict their safety in vivo.

Belyanskaya et al. (2009) have compared sensitivities of dissociated spinal cord (SPC) cells (peripheral nervous system) and dorsal root ganglia (DRG) cell (CNS) to aggregated SWCNTs (SWCNT-a) and dispersed SWCNTs (SWCNT-d) in vitro. In SPC cells, the SWCNT-a was more potent than SWCNT-d; in DRG cells, only the SWCNT-d exhibited toxicity. SWCNT-a and SWCNT-d both impaired Na^+-channels in SPC cells, while SWCNT-d, but not SWCNT-a, suppressed Ca^{2+} conductance and Ca^{2+} resting potential in DRG cells. This may be because the SPC neurons contain mostly Na^+, while the DRG neurons contain Ca^{2+} channels. The adverse effects of SWCNTs on neurons may be either due to accumulation of SWCNTs in neurons or secondary to their effects on astrocytes. This study demonstrates differential effects of CNTs on cells of the central and peripheral nervous systems.

Zhang et al. (2011) studied the toxicities of pristine SWCNTs ($_{p}$SWCNTs, diameter 0.7–1.6 nm, length 0.2–3 μm, dose: 0.1–100 μg/ml, PC-12 cells) and PEG-functionalized nanotubes ($_{PEG}$SWCNTs, diameter 2.5–4.5 nm, length 0.1–1 μm, dose as above) in PC-12 cells in vitro. The indices measured were Raman and dark-field imaging, MTT (3-(4,5-dimethylthiazol-2-yl)-2,5-diphenyl tetrazolium bromide) live cell assay, XTT (2,3-bis[2-methoxy-4-nitro-5-sulfophenyl]-2H-tetrazolium-5-carboxanilide) assay for mitochondrial function, LDH assay for neuronal necrosis, ROS assay, GSH assay for the antioxidation capacity, and proteomic analysis for oxidative and antioxidative enzymes. In general, all of the tests listed above, except the proteomic analysis of oxidative stress, showed that $_{p}$SWCNTs and $_{PEG}$SWCNT both ($_{p}$SWCNTs > $_{PEG}$SWCNT) decreased cell viability, damaged mitochondria, increased cytoplasmic LDH, increased ROS, and decreased GSH concentration in PC-12 cells. The proteomic studies showed that $_{p}$SWCNTs and $_{PEG}$SWCNTs modulated different sets of genes: $_{p}$SWCNTs and $_{PEG}$SWCNT both downregulated expressions of antioxidative enzymes, while $_{p}$SWCNT, but not $_{PEG}$SWCNT, upregulated expressions of pro-oxidative enzymes. These observations present a controversial picture of the toxic effects of CNTs on cells having tight junctions. However, extensive variations in the nanoparticle groups, functionalization, and experimental conditions do not allow direct comparison of different results.

5.5.2 In Vivo Toxicity

CNTs, in addition to acting as a drug transporter (Tran et al., 2009), can also integrate with neural cells and form synapses, thus allowing the possible use of CNTs in neuroelectrical interfacing in vivo (Liopo et al., 2006; Gabay et al., 2007; Ricci, 2008; Cui et al., 2010, 2012). This may allow treatment of many diseases that are currently untreatable. However, the safety of such devices is of serious concern because only a few studies have been conducted regarding the biocompatibility of CNT within the brain in vivo. As described above in the in vitro toxicity section, despite large variations in types of CNTs, dose regimens, cell models, and duration of interaction, the studies reflect a common theme that chemical functionalization of CNTs improved their aqueous dispersibility and biocompatibility, and reduced their toxicity. Unfortunately, under in vivo situations, functionalized CNTs may undergo defunctionalization (dissociation of functional groups), resulting in accumulation of pristine CNTs that complicates the safety of functionalized CNTs in vivo. The proceeding sections describe some of the important issues associated with CNTs' in vivo neurotoxicity.

5.5.3 *How Do CNTs Cross the Blood–Brain Barrier In Vivo?*

Particles in systemic blood enter the brain mostly via crossing the blood–brain barrier (Figure 16), although minor pathways that bypass the blood–brain barrier, such as traveling through nerve terminals or diffusing through the neuronal or glial cytoplasmic membrane, are also available (Perez-Martınezc et al., 2012). In addition to the endothelial cell tight junction, an astrocyte–endothelial interaction also regulated the blood–brain permeability (Wolburg et al., 1994; Haseloff et al., 2005; Abbott et al., 2006). In general, permeation of a toxin across the blood–brain barrier is directly related to its hydrophobicity, but it is inversely related to the size, diameter, length, and molecular weight (Felgenhauer, 1980). CNTs and other nanoparticles, functionalized with ligands that bind to the blood–brain transport receptors such as transferrin receptors, cross the blood–brain barrier without damaging the tight junction.

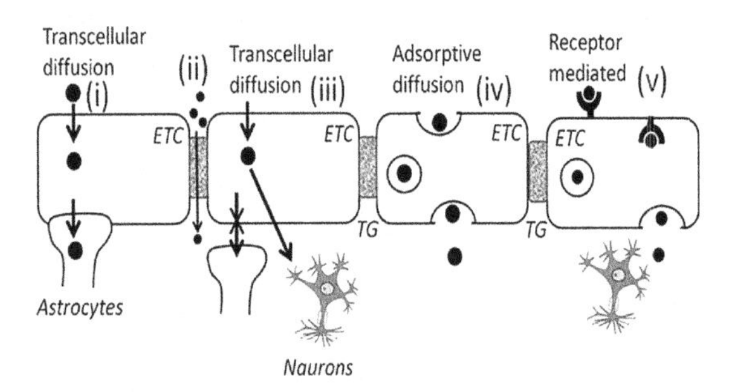

FIGURE 16 Transport of nanoparticles across the blood–brain barrier. (i) Transcellular diffusion from the extracellular fluid into the astrocytes that penetrate the barrier-particles are internalized through the cell according to their concentration gradient, (ii) paracellular diffusion through the intercellular gaps, (iii) transcellular diffusion from extracellular fluid into the neurons that do not contact the barrier, (iv) adsorptive diffusion – particles such as hydrophobic or functionalized particles, are adsorbed into the lipid bilayer and thus, cross the barrier, and (v) receptor/transporter mediated – particles bind to a transporter that internalize the particles.

Pristine CNTs, being highly hydrophobic, have been shown to aggregate into different-sized aggregated particles; thus, they may not cross the blood–brain barrier. In comparison, functionalized CNTs remain dispersed and can negotiate the barrier. Although inconclusive, certain studies have shown that CNTs may induce oxidative stress, leading to the generation of free radicals that could disrupt the blood–brain barrier and allow their entry into the brain (Sriram et al., 2009, 2007). Studies have shown that (1) nanoparticles deposited in the lungs may translocate into the circulatory system and eventually reach the brain (Nemmar et al., 2001, 2002) and (2) nanoparticles inhaled through the nose can translocate directly into the brain via the olfactory neurons (Elder et al., 2006; Hunter and Undem, 1999). Ryman-Rasmussen et al. (2009), Porter et al. (2009), and Mercer et al. (2010) have shown that accumulation of CNT-filled macrophages in the alveoli increased permeability of the epithelial cell barrier and particle translocation to the alveolar interstitium, thus increasing the probability of CNTs reaching the brain. However, more research is needed to establish the possible translocation of CNTs from lungs into the brain.

5.5.4 Do CNTs Defunctionalize into the Brain?

As discussed above, a key problem associated with the long-term application of functionalized CNTs is their defunctionalization (Fu et al., 2001; Yang et al., 2009; Azizian et al., 2012). Because defunctionalized CNTs may be relatively more hydrophobic and stable, their accumulation in tissues may pose significant health hazards. Earlier studies have shown the defunctionalization of PEG-SWNT in the liver but not in the spleen under similar conditions (Yang et al., 2009). Whether or not the brain-incorporated CNTs undergo defunctionalization is not yet known. Recently, Nunes et al. (2012) have shown that the brain microglia effectively engulf and defunctionalize the amino-functionalized CNTs in vivo. This initial observation of CNT defunctionalization within the brain necessitates further investigations to establish their pharmacological and toxicological consequences.

5.5.5 What Are Possible Mechanisms of CNT Toxicity in the Brain In Vivo?

As discussed in the in vitro toxicity section, CNTs may cause toxicity via development of oxidative stress-dependent and -independent inflammation. Similar to the in vitro studies, in vivo studies have also shown variable results. Intranasal instillation of ultrafine carbon black in vivo induced the expression of pro-inflammatory cytokines and chemokines in the olfactory bulb (Shwe et al., 2005; Tin-Tin-Win-Shwe et al., 2006, 2007). Unique to CNTs is the relationship between their aspect ratio (diameter to length) and potential adverse health effects in the lungs and the brain in vivo (Bonner, 2011). Animals exposed to MWCNTs developed systemic inflammation, progressive fibrosis, and granulomas (Kobayashi et al., 2012; Ryman-Rasmussen et al., 2009) in a length-dependent manner. MWCNTs with greater than 20 μm length generated frustrated phagocytosis and ensuing polymorphonuclear leukocytes and granulomas (Donaldson et al., 2010). There is some evidence for development of carcinogenesis via frustrated phagocytosis in animals exposed to MWCNTs (Fraczek et al., 2008; Takagi et al., 2008; 2012). SWCNTs and MWCNTs both may also enter the brain via the resident microglial cells which, for long CNTs, may result in frustrated phagocytosis and ensuing inflammatory activity (Konduru et al., 2009).

5.6 Reproductive and Developmental Toxicity

Nanoparticles may cause reproduction and developmental toxicity directly by either entering germ cells or fetuses or indirectly by dysregulating the endocrine system or interfering with proliferation and differentiation of the germ cells (Campagnolo et al., 2012). There is some (but not compelling) experimental evidence showing that nanoparticles, such as TiO_2 particles, cause reproductive and developmental toxicities in zebra fish and abalone embryos (Hussain et al., 2011; Zhu et al., 2011). Arndt et al. (2014) have shown that carbon nanomaterials (fullerenes (C_{60}), SWCNTs, and MWCNTs with neutral, positive, and negative functional groups) impaired reproduction of daphnia across generations. A total of 50 ppm of SWCNT, SWCNT-$CONH_2$, C_{60}-OH, and 5 ppm C_{60}-malonate suppressed reproduction of F_1, while only 50 ppm of C_{60} and SWCNT-$CONH_2$ suppressed F_2 reproduction. Possible mechanisms proposed are ingestion of nanoparticles in the digestive tract, inhibition of protein activity (Zuo et al., 2013), and disruption of membrane transport activities in the cell (Park et al., 2003), which may impede normal daphnid reproduction and growth. AuNPs, because of their greater biocompatibility, are less toxic than other nanoparticles in several in vivo studies on the zebrafish (Bar-Ilan et al., 2009; Ashrani et al., 2008). Martins et al. (2007) have proposed that ecotoxicity assays with *Daphnia magna* and *Danio rerio* can be used as alert systems in water contaminated with toxic substances.

Unlike invertebrates, limited data are available to define the developmental or reproductive toxicity of nanoparticles in mammals. With respect to in vivo studies, Takeda et al. (2009) demonstrated that prenatal subcutaneous exposure of mice to TiO_2 NPs damaged genital and cranial nerve systems in offspring. Shimizu et al. (2009) have shown that prenatal TiO_2 exposure in pregnant mice altered the expression of genes associated with brain development, cell death, response to oxidative stress, and mitochondrial activity in the brain. Yamashita et al. (2010a) reported that silica and TiO_2 NPs with diameters of 70 and 35 nm, respectively, could cause pregnancy complications when injected intravenously into pregnant mice that had smaller uteri and smaller fetuses than untreated controls. The nanoparticles were found in the placenta, fetal liver, and fetal brain. Qi et al. (2014), in an extensive study, have shown that the oxidized-MWCNTs administered to the mice at the gestational age of 14 days crossed the placental barrier and caused placental tissue damage. The specific changes were following:

1. The number of blood vessels in the placenta tissue decreased.
2. Estradiol secretion and ensuing excitability of uterine contraction increases, which may compromise pregnancy.
3. Fetal growth was restricted and abortion increased in the exposure groups.

The limited in vivo and in vitro studies provide significant, but not conclusive, evidence for the reproductive toxicity of nanoparticles. Further studies will clarify the mechanisms underlying these toxicity results.

6. CONCLUSIONS

The range of applications for nanomaterials continues to grow at a rapid rate. Nanoparticles and CNTs are present in toothpastes, beauty products, sunscreens, coatings, drug delivery systems, sensors, building materials, and textiles. As the applications grow, so will the potential for nanoparticle exposure in humans and animals. Thus, an understanding of the interactions between nanoparticles and target cells may be necessary to decipher their local and systemic effects. In general, the most commonly studied nanoparticles are metal or metal oxides (including Cu, CuO, Zn, and Ag) as well as carbon-based nanoparticles. There is sufficient but not compelling evidence in support of the adverse effects of nanoparticles on neuronal cells as well as their interactions with cell membranes, including the blood–brain barrier. Nanoparticles may cause local and systemic effects in models of cell cultures in vitro and in experimental animals in vivo. The systemic toxicity includes skin, respiratory system, liver, brain, and reproductive toxicity. The key mechanisms include oxidative stress and inflammation.

The previous studies describing functionality/nanotoxicity of nanoparticles should be the basis for conducting future research toward understanding the mechanism of action, and their systemic effects using animal and in vitro models. Transgenic animal models may provide further assessment of functionality and toxicity. Some of the existing problems that need to be addressed are nanoparticle agglomeration and aggregation within both biotic and abiotic environments; nanoparticle dispersion in air with different sizes, materials, and morphologies; the kinetics of protein corona formation; and the possibility of nanoparticle defunctionalization.

In addition, the current knowledge on engineered nanoparticles and their interactions with the biological membranes are extremely limited and require more research. In the CNS, nanoparticles' interaction with microglial cells, a type of macrophage found in the brain, will provide key information regarding their stability in the brain. Biological interactions linked to particle size, surface energy, composition, and aggregation will be the focal points of some future studies.

References

Abbott, N.J., Rönnbäck, L., Hansson, E., 2006. Astrocyte-endothelial interactions at the blood–brain barrier. Nat. Rev. Neurosci. 7, 41–53.

Adams, M.D., Celniker, S.E., Holt, R.A., et al., 2000. The genome sequence of *Drosophila melanogaster*. Science 287, 2185–2195.

Ahamed, M., Siddiqui, M.A., Ahmad, J., Musarrat, J., AlKhedhairy, A.A., AlSalhi, M.S., Alrokayan, S.A., 2011a. Oxidative stress mediated apoptosis induced by nickel ferrite nanoparticles in cultured A549 cells. Toxicology 283, 101–108.

Ahamed, M., Akhtar, M.J., Raja, M., et al., 2011b. ZnO nanorod-induced apoptosis in human alveolar adenocarcinoma cells via p53, survivin and bax/bcl-2 pathways: role of oxidative stress. Nanomedicine 7, 904–913.

Ahamed, M., Siddiqui, M.A., Akhtar, M.J., Ahmad, I., Pant, A.B., Alhadlaq, H.A., 2010. Genotoxic potential of copper oxide nanoparticles in human lung epithelial cells. Biochem. Biophys. Res. Commun. 396, 578–583.

Aillon, K.L., Xie, Y., El-Gendy, N., Berkland, C.J., Forrest, M.L., 2009. Effects of nanomaterial physicochemical properties on in vivo toxicity. Adv. Drug Deliv. Rev. 61, 457–466.

Alarifi, S., Ali, D., Alkahtani, S., et al., 2013. Induction of oxidative stress, DNA damage, and apoptosis in a malignant human skin melanoma cell line after exposure to zinc oxide nanoparticles. Int. J. Nanomed. 8, 983–993.

Albanese, A., Tang, P.S., Chan, W.C.W., 2012. The effect of nanoparticle size, shape, and surface chemistry on biological systems. Ann. Rev. Biomed. Eng. 14, 1–16.

Ali-Boucetta, H., Nunes, A., Sainz, R., Herrero, M.A., Tian, B., Prato, M., Bianco, A., Kostarelos, K., 2013. Asbestos-like pathogenicity of long carbon nanotubes alleviated by chemical functionalization. Ange. Chem. 52, 2274–2278.

Alpatova, A.L., Shan, W., Babica, P., Upham, B.L., Rogensues, A.R., Masten, S.J., Drown, E., Mohanty, A.K., Alocilja, E.C., Tarabara, V.V., January 2010. Single-walled carbon nanotubes dispersed in aqueous media via non-covalent functionalization: effect of dispersant on the stability. Water Res. 44 (2), 505–520.

Anderson, N.L., Anderson, N.G., 1998. Proteome and proteomics: new technologies, new concepts, and new words. Electrophoresis 19, 1853–1861.

Aoki, N., Yokoyama, A., Nodasaka, Y., Akasaka, T., Uo, M., Sato, Y., Tohji, K., Watari, F., 2005. Cell culture on a carbon nanotube scaffold. J. Biomed. Nanotechnol. 1, 402–405.

Araujo, J.A., Barajas, B., Kleinman, M., Wang, X., Bennett, B.J., Gong, K.W., Navab, M., Harkema, J.Sioutas, Lusis, A.J., Ne, A.E., 2008. Ambient particulate pollutants in the ultrafine range promote early atherosclerosis and systemic oxidative stress. Circ. Res. 102, 589–596.

Arndt, D.A., Chen, J., Moua, M., Rebecca, D.K., 2014. Multigeneration impacts on Daphnia magna of carbon nanomaterials with differing core structures and functionalizations. Environ. Toxicol. Chem. 33, 541–547.

Aschberger, K., Johnston, H.J., Stone, V., Aitken, R.J., Hankin, S.M., Peters, S.A., Tran, C.L., Christensen, F.M., 2010. Review of carbon nanotubes toxicity and exposure–appraisal of human health risk assessment based on open literature. Crit. Rev. Toxicol. 40, 759–790.

Asgharian, B., Price, O.T., 2007. Deposition of ultrafine (nano) particles in the human lung. Inhal. Toxicol. 19, 1045–1054.

Asharani, P.V., Lianwu, Y., Gong, Z., Valiyaveettil, S., 2011. Comparison of the toxicity of silver, gold and platinum nanoparticles indeveloping zebrafish embryos. Nanotoxicology 5, 43–54.

Asharani, P.V., Mun, G.L.K., Hande, M.P., Valiyaveettil, S., 2009. Cytotoxicity and genotoxicity of silver nanoparticles in human cells. ACS Nano 3, 279–290.

Ashrani, P.V., Wu, W.L., Gong, Z., Valiyaveettil, S., 2008. Toxicity of silver nanoparticles in zebrafish models. Nanotechnology 19, 255102.

Atala, A., 2004. Tissue engineering and regenerative medicine: concepts for clinical application. Rejuvenation Res. 7, 15–31.

ATS (American Thoracic Society), 2004. Diagnosis and initial management of nonmalignant diseases related to asbestos. Am. J. Respir. Crit. Care Med. 170, 691–715.

Azad, N., Iyer, A.K., Wang, L., Liu, Y., Lu, Y., Rojanasakul, Y., 2012. Reactive oxygen species-mediated p38 MAPK regulates carbon nanotube-induced fibrogenic and angiogenic responses. Nanotoxicology 7, 157–168.

Azizian, J., Zomorodbakhsh, S., Shameli, A., Entezari, M., 2012. Enviromentaly friendly functionalization of carboxylated shortend multi-wall nanotubes with sunset yellow dye. Orient. J. Chem. 28, 115–121.

Balakumaran, A., Pawelczyk, E., Ren, J., Sworder, B., Chaudhry, A., et al., 2010. Superparamagnetic iron oxide nanoparticles labeling of bone marrow stromal (mesenchymal) cells does not affect their "stemness". PLoS One 5, e11462.

Balasubramanian, K., Burghard, M., 2005. Chemically functionalized carbon nanotubes. Small 1, 180–192.

Ballinger, S.W., 2005. Mitochondrial dysfunction in cardiovascular disease. Free Radic. Biol. Med. 38, 1278–1295.

Banga, A., Witzmann, F.A., Petrache, H.I., Blazer-Yost, B.L., 2012. Functional effects of nanoparticle exposure on Calu-3 airway epithelial cells. Cell Physiol. Biochem. 29, 197–212.

Bar-Ilan, O., Albrecht, R.M., Fako, V.E., Furgeson, D.Y., 2009. Toxicity assessments of multisized gold and silver nanoparticles in zebrafish embryos. Small 5, 1897–1910.

Barrett, T., Troup, D.B., Wilhite, S.E., Ledoux, P., Evangelista, C., et al., 2011. NCBI GEO: archive for functional genomics data sets—10 years on. Nucleic Acids Res. 39, D1005—D1010.

Beg, S., Rizwan, M., Sheikh, A.M., Hasnain, M.S., Anwer, K., Kohli, K., 2011. Advancement in carbon nanotubes: basics, biomedical applications and toxicity. J. Pharm. Pharmacol. 63, 141—163.

Belyanskaya, L., Weigel, S., Hirsch, C., Tobler, U., Krug, H.F., Wick, P., 2009. Effects of carbon nanotubes on primary neurons and glial cells. Nanotoxicology 30, 702—711.

Bennett, D., 2005. Growing pains for metabolomics. Scientist 19, 25—28.

Bernstein, D., Dunnigan, J., Hesterberg, T., Brown, R., Velasco, J.A., Barrera, R., Hoskins, J., Gibbs, A., 2013. Health risk of chrysotile revisited. Crit. Rev. Toxicol. 43, 154—183.

Bhattacharya, K., Davoren, M., Boertz, J., Schins, R.P., Hoffmann, E., Dopp, E., 2009. Titanium dioxide nanoparticles induce oxidative stress and DNA-adduct formation but not DNA-breakage in human lung cells. Part. Fibre Toxicol. 21, 6—17.

Blackford, J.A., Antonini, J.M., Castranova, V., Dey, R.D., 1994. Intratracheal instillation of silica up-regulates inducible nitric oxide synthase gene expression and increases nitric oxide production in alveolar macrophages and neutrophils. Am. J. Respir. Cell Mol. Biol. 11, 426—431.

Boczkowski, J., Hoet, P., 2008. What's new in nanotoxicology? Implications for public health from a brief review of the 2008 literature. Nanotoxicology 4, 1—14.

Boenigk, J., Beisser, D., Zimmermann, S., Bock, C., Jakobi, J., Grabner, D., Grobmann, L., Rahmann, S., Barcikowski, S., Sures, B., 2014. Effects of silver nitrate and silver nanoparticles on a planktonic community: general trends after short-term exposure. PLoS One 9, e107092.

Bonner, J.C., 2011. Carbon nanotubes as delivery systems for respiratory disease: do the dangers outweigh the potential benefits? Expert Rev. Respir. Med. 5, 779—787.

Bottini, M., Bruckner, S., Nika, K., Bottini, N., Bellocci, S., Magrini, A., Bergamaschi, A., Mustelin, T., 2006. Multi-walled carbon nanotubes induce T lymphocyte apoptosis. Toxicol. Lett. 160, 121—126.

Bourdon, J.A., Saber, A.T., Jacobsen, N.R., Jensen, K.A., Madsen, A.M., Lamson, J.S., Wallin, H., Pr, M., Loft, S., Yauk, C.L., Voge, U.B., 2012. Carbon black nanoparticle instillation induces sustained inflammation and genotoxicity in mouse lung and liver. Part. Fibre Toxicol. 9, 5.

Bouwmeester, H., Poortman, J., Peters, R.J., Wijma, E., Kramer, E., Makama, S., Puspitaninganindita, K., Marvin, H.J.P., Peijnenburg Ad, A.C.M., Hendriksen, P.J.M., 2011. Characterization of translocation of silver nanoparticles and effects on whole-genome gene expression using an in vitro intestinal epithelium coculture model. ACS Nano 5, 4091—4103.

Brannon-Peppas, L., Blanchette, J.O., 2004. Nanoparticle and targeted systems for cancer therapy. Adv. Drug Deliv. Rev. 56, 1649—1659.

Brook, R.D., Franklin, B., Cascio, W., Hong, Y., Howard, G., Lipsett, M., Luepker, R., Mittleman, M., Samet, J., Smith, S.C., Tager, I., 2004. Air pollution and cardiovascular disease: a statement for healthcare professionals from the expert panel on population and prevention science of the American Heart Association. Circulation 109, 2655—2671.

Brown, D.M., Beswick, P.H., Bell, K.S., Donaldson, K., 2000. Depletion of glutathione and ascorbate in lung lining fluid by respirable fibres. Ann. Occup. Hyg. 44, 101—108.

Busch, W., Kuhnel, D., Schirmer, K., Scholz, S., 2010. Tungsten carbide cobalt nanoparticles exert hypoxia-like effects on the gene expression level in human keratinocytes. BMC Genomics 11, 65.

de Boer, A.G., Gaillard, P.J., 2007a. Strategies to improve drug delivery across the blood—brain barrier. Clin. Pharmacokinet. 46, 553—576.

de Boer, A.G., Gaillard, P.J., 2007b. Drug targeting to the brain. Ann. Rev. Pharmacol. Toxicol. 47, 323—355.

Cai, D., Mataraza, J., Qin, Z., Huang, Z., Huang, J., Chiles, T., Carnahan, D., Kempa, K., Renet, Z., 2005. Highly efficient molecular delivery into mammalian cells using carbon nanotube spearing. Nat. Methods 2, 449—454.

Campagnolo, L., Massimiani, M., Magrini, A., Camaioni, A., Antonio Pietroiusti, A., 2012. Physico-chemical properties mediating reproductive and developmental toxicity of engineered nanomaterials. Curr. Med. Chem. 19, 4488—4494.

Campbell, W.J., Steel, E.B., Virta, R.L., Eisner, M.H., 1979. Relationship of Mineral Habit to Size Characteristics of Tremolite Cleavage Fragments and Fibers. Report of Investigations #8367. U.S. Department of the Interior, Bureau of Mines, Washington, DC.

Casado-Vela, J., Martínez-Esteso, M.J., Rodriguez, E., Borrás, E., Elortza, F., Bru-Martínez, R., 2010. iTRAQ-based quantitative analysis of protein mixtures with large fold change and dynamic range. Proteomics 10, 343—347.

Castranova, V., Kang, J.H., Moore, M.D., Pailes, W.H., Frazer, D.G., Schwegler-Berry, D., 1991. Inhibition of stimulant-induced activation of phagocytic cells with tetrandrine. J. Leukoc. Biol. 50, 412—422.

Castranova, V., 2004. Signaling pathways controlling the production of inflammatory mediators in response to crystalline silica exposure: role of reactive oxygen/nitrogen species. Free Radic. Biol. Med. 37, 916—925.

Cejas, P., Casado, E., Belda-Iniesta, C., De Castro, J., Espinosa, E., Redondo, A., Sereno, M., García-Cabezas, M.A., Vara, J.A.F., Domínguez-Cáceres, A., Perona, R., González-Barón, M., 2004. Implications of oxidative stress and cell membrane lipid peroxidation in human Cancer. Cancer Causes Control 15, 707—719.

Ceriello, A., Motz, E., 2004. Is oxidative stress the pathogenic mechanism underlying insulin resistance, diabetes, and cardiovascular disease? the common soil hypothesis revisited. Arterioscler. Thromb. Vasc. Biol. 24, 816–823.

Chairuangkitti, P., Lawanprasert, S., Roytrakul, S., Aueviriyavit, S., Phummiratch, D., Kulthong, K., et al., 2013. Silver nanoparticles induce toxicity in A549 cells via ROS-dependent and ROS-independent pathways. Toxicol. In Vitro 27, 330–338.

Chang, J., Jallouli, Y., Kroubi, M., Yuan, X.B., Feng, W., Kang, C.S., Pu, P.Y., Betbeder, D., 2009. Characterization of endocytosis of transferrin-coated PLGA nanoparticles by the blood–brain barrier. Int. J. Pharm. 379, 285–292.

Chen, B., Liu, Y., Song, W.M., Hayashi, Y., Ding, X.C., Li, W.H., 2011. In vitro evaluation of cytotoxicity and oxidative stress induced by multiwalled carbon nanotubes in murine RAW 264.7 macrophages and human A549 lung cells. Biomed. Environ. Sci. 24, 593–601.

Chen, P.S.M., Kanehira, K., Sonezaki, S., Taniguchi, A., 2011b. Development of sensor cells using NF-κB pathway activation for detection of nanoparticle-induced inflammation. Sensors 11, 7219–7230.

Chen, W.-F., Wu, J.-S., Kuo, P.-L., 2008. Poly(oxyalkylene) diamine-functionalized carbon nanotube/perfluorosulfonated polymer composites: synthesis, water state, and conductivity. Chem. Mater. 20, 5756–5767.

Chen, M., Zhang, M., Borlak, J., Tong, W., 2012. A decade of toxicogenomic research and its contribution to toxicological science. Toxicol. Sci. 130, 217–228.

Chen, H., Zhen, Z., Tang, W., Todd, T., Chuang, Y.-J., Wang, L., Pan, Z., Xie, J., 2013. Label-Free Luminescent Mesoporous Silica Nanoparticles for Imaging and Drug Delivery. Theranostics 3, 650–657.

Cherukuri, P., Bachilo, S., Litovsky, S., Weisman, R., 2004. Near-infrared fluorescence microscopy of single-walled carbon nanotubes in phagocytic cells. J. Am. Chem. Soc. 126, 15638–15639.

Cho, Y.J., Seo, M.S., Kim, J.K., Lim, Y., Chae, G., Ha, K.S., Lee, K.H., 1999. Silica-induced generation of reactive oxygen species in rat2 fibroblasts: role in activation of mitogen-activated protein kinase. Biochem. Biophys. Res. Commun. 262, 708–712.

Chou, C.-C., Hsiao, H.-Y., Hong, Q.-S., Chen, C.-H., Peng, Y.-W., Chen, H.-W., Yang, P.-C., 2008. Single-walled carbon nanotubes can induce pulmonary injury in mouse model. Nano Lett. 8, 437–445.

Clark, K.A., O'Driscoll, C., Cooke, C.A., Smith, B.A., Wepasnick, K., Fairbrother, D.H., Lees, P.S., Bressler, J.P., 2012. Evaluation of the interactions between multiwalled carbon nanotubes and Caco-2 cells. J. Toxicol. Environ. Health Part A 75, 25–35.

Clichici, S., Mocan, T., Filip, A., Biris, A., Simon, S., Daicoviciu, D., Decea, N., Parvu, A., Moldovan, R., Muresan, A., 2011. Blood oxidative stress generation after intraperitoneal administration of functionalized single-walled carbon nanotubes in rats. Acta Physiol. Hungarica 98, 231–241.

Collins, T., Read, M.A., Neish, A.S., Whitley, M.Z., Thanos, D., Maniatis, T., 1995. Transcriptional regulation of endothelial cell adhesion molecules: NF-kappa B and cytokine-inducible enhancers. FASEB J. 9, 899–909.

Coxon, A., Rieu, P., Barkalow, F.J., Askari, S., Sharpe, A.H., von Andrian, U.H., Arnaout, M.A., Mayadas, T.N., 1996. A novel role for the beta 2 integrin CD11b/CD18 in neutrophil apoptosis: a homeostatic mechanism in inflammation. Immunity 5, 653–666.

Coyuco, J.C., Liu, Y., Tan, B.J., Chiu, G.N., 2011. Functionalized carbon nanomaterials: exploring the interactions with Caco-2 cells for potential oral drug delivery. Int. J. Nanomed. 6, 2253–2263.

Crone, C., Christensen, O., 1981. Electrical resistance of a capillary endothelium. J. Gen. Physiol. 77, 349–371.

Crouzier, D., Follot, S., Gentilhomme, E., Emmanuel, F., Arnaud, R., Dabouis, V., Castellarin, C., Debouzy, J.C., 2010. Carbon nanotubes induce inflammation but decrease the production of reactive oxygen species in lung. Toxicology 272, 39–45.

Cui, D., Tian, F., Ozkan, C.S., Wang, M., Gao, H., 2005. Effect of single wall carbon nanotubes on human HEK293 cells. Toxicol. Lett. 155, 73–85.

Cui, H.F., Vashist, S.K., Al-Rubeaan, K., Luong, J.H.T., Sheu, F.S., 2010. Interfacing carbon nanotubes with living mammalian cells and cytotoxicity issues. Chem. Res. Toxicol. 23, 1131–1147.

Cui, W., Li, J., Zhang, Y., Rong, H., Lu, W., Jiang, L., 2012. Effects of aggregation and the surface properties of gold nanoparticles on cytotoxicity and cell growth. Nanomed. Nanotechnol. Biol. Med. 8, 46–53.

Davis, J.M., Addison, J., McIntosh, C., Miller, B.G., Niven, K., 1991. Variations in the carcinogenicity of tremolite dust samples of differing morphology. Ann. N.Y. Acad. Sci. 643, 473–490.

Deng, X.Y., Jia, G., Wang, H.F., Sun, H.F., Wang, X., Yang, S.T., Wang, T.C., Liu, Y.F., 2007. Translocation and fate of multi-walled carbon nanotubes in vivo. Carbon 45, 1419–1424.

Deng, X.Y., Yang, S.T., Nie, H.Y., Wang, H.F., Liu, Y.F., 2008. A generally adoptable radiotracing method for tracking carbon nanotubes in animals. Nanotechnology 19, 075101.

Deng, Zhou J., Monteiro, M., Toth, I., Minchin, Rodney F., 2011. Nanoparticle-induced unfolding of fibrinogen promotes Mac-1 receptor activation and inflammation. Nat. Nanotechnol. 6, 39–44.

Ding, L., Stilwell, J., Zhang, T., Elboudwarej, O., Jiang, H., Selegue, J.P., Cooke, P.A., Gray, J.W., Chen, F.F., 2005. Molecular characterization of the cytotoxic mechanism of multiwall carbon nanotubes and nano-onions on human skin fibroblast. Nano Lett. 5, 2448–2464.

Donaldson, K., Murphy, F.A., Duffin, R., Poland, C.A., 2010. Asbestos, carbon nanotubes and the pleural mesothelium: a review of the hypothesis regarding the role of long fibre retention in the parietal pleura, inflammation and mesothelioma. Part. Fiber Toxicol. 7, 5.

Donaldson, K., Aitken, R., Tran, L., Stone, V., Duffin, R., Forrest, G., Alexander, A., 2006. Carbon nanotubes: a review of their properties in relation to pulmonary toxicology and workplace safety. Toxicol. Sci. 92, 5–22.

Donaldson, K., Brown, G.M., Brown, D.M., Bolton, R.E., Davis, J.M., 1989. Inflammation generating potential of long and short fibre amosite asbestos samples. Br. J. Ind. Med. 46, 271–276.

Eckberg, D.L., Convertino, V.A., Fritsch, J.M., Doerr, D.F., 1991. Reproducibility of human vagal carotid baroreceptor-cardiac reflex responses. Am. J. Physiol. 263, R215–R220.

El-Ansary, A., Al-Daihan, S., 2009. On the toxicity of therapeutically used nanoparticles: an overview. J. Toxicol. 2009, 754810.

Elder, A., Gelein, R., Silva, V., Feikert, T., Opanashuk, L., Carter, J., Potter, R., Maynard, A., Ito, Y., Finkelstein, J., Oberdorster, G., 2006. Translocation of inhaled ultrafine manganese oxide particles to the central nervous system. Environ. Health Perspect. 114, 1172–1178.

Ellmeier, W., Aguzzi, A., Kleiner, E., Kurzbauer, R., Weith, A., 1992. Mutually exclusive expression of a helix-loop-helix gene and N-myc in human neuroblastomas and in normal development. EMBO J. 11, 2563–2571.

Eom, H., Choi, J., 2009. Oxidative stress of CeO_2 nanoparticles via p38-Nrf-2 signaling pathway in human bronchial epithelial cell, Beas-2B. Toxicol. Lett. 187, 77–83.

Eom, H.-J., 2009. Oxidative stress of silica nanoparticles in human bronchial epithelial cell, Beas-2B. Toxicol. In Vitro 23, 1326–1332.

Erdely, A., Hulderman, T., Salmen, R., Liston, A., Zeidler-Erdely, P.C., Schwegler-Berry, D., Castranova, V., Koyama, S., Kim, Y.-A., Endo, M., Simeonova, P.P., 2009. Cross-talk between lung and systemic circulation during carbon nanotube respiratory exposure. Potential biomarkers. Nano Lett. 9, 36–43.

Evans, P., Halliwll, B., 2001. Micronutrients: oxidant/antioxidant status. Br. J. Nutr. 85 (Suppl. 2), S67.

Fahmy, B., Cormier, S.A., 2009. Copper oxide nanoparticles induce oxidative stress and cytotoxicity in airway epithelial cells. Toxicol. In Vitro 23, 1365–1371.

Fang, Y.Z., Yang, S., Wu, G., 2002. Free radicals, antioxidants, and nutrition. Nutrition 18, 872–879.

Fang, Y.Z., Zheng, R.L. (Eds.), 2002. Theory and Application of Free Radical Biology. Scientific Press, Beijing, p. 647.

Farokhzad, O.C., Langer, R., 2006. Nanomedicine: developing smarter therapeutic and diagnostic modalities. Adv. Drug Deliv. Rev. 58, 1456–1459.

Fayol, D., Luciani, N., Lartigue, L., Gazeau, F., Wilhelm, C., 2012. Managing magnetic nanoparticle aggregation and cellular uptake: a precondition for efficient stem-cell differentiation and MRI tracking. Adv. Healthcare Mater. 2013 (2), 313–325.

Felgenhauer, K., 1980. Protein filtration and secretion at human body fluid barriers. Pflugers Arch. 384, 9–17.

Feng, J., Liu, H., Bhakoo, K.K., Lu, L., Chen, Z., 2011. A metabonomic analysis of organ specific response to USPIO administration. Biomaterials 32, 6558–6569.

Fenoglio, I., Tomatis, M., Lison, D., Muller, J., Fonseca, A., Nagy, J.B., Fubini, B., 2006. Reactivity of carbon nanotubes: free radical generation or scavenging activity? Free Radic. Biol. 40, 1227–1233.

Fenton, H.J.H., 1894. Oxidation of tartaric acid in presence of iron. J. Chem. Soc. 65, 899–911.

Fiorito, S., Serafino, A., Andreola, F., Bernier, P., 2006. Effects of fullerenes and single-wall carbon nanotubes on murine and human macrophages. Carbon 44, 1100–1105.

Fisichella, M., Berenguer, Frederic, Steinmetz, Gerard, Auffan, Melanie, Rose, Jerome, Prat, Odette, 2014. Toxicity evaluation of manufactured CeO_2 nanoparticles before and after alteration: combined physicochemical and whole-genome expression analysis in Caco-2 cells. BMC Genomics 15, 700.

Flahaut, E., Durrieu, M.C., Remy-Zolghadri, M., Bareille, R., Ch, B., 2006. Investigation of the cytotoxicity of CCVD carbon nanotubes towards human umbilical vein endothelial cells. Carbon 44, 1093–1099.

Foldvari, M., Bagonluri, M., 2008a. Carbon nanotubes as functional excipients for nanomed: I. Pharmaceutical properties. Nanomed. Nanotechnol. Biol. Med. 4, 173–182.

Foldvari, M., Bagonluri, M., 2008b. Carbon nanotubes as functional excipients for nanomedicines: II. Drug delivery and biocompatibility issues. Nanomedicine 4, 183–200.

Foncea, R., Carvajal, C., Almarza, C., Leighton, F., 2000. Endothelial cell oxidative stress and signal transduction. Biol. Res. 33, 2.

Fraczek, A., Menaszek, E., Paluszkiewicz, C., Blazewicz, M., 2008. Comparative in vivo biocompatibility study of single- and multi-wall carbon nanotubes. Acta Biomater. 4, 1593–1602.

Freidovich, I., 1999. Fundamental aspects of reactive oxygen species, or what's the matter with oxygen? Ann. N.Y. Acad. Sci. 893, 13.

Fu, K., Huang, W., Lin, Y., Riddle, L.A., Carroll, D.L., Sun, Y.-P., 2001. Defunctionalization of functionalized carbon nanotubes. Nano Lett. 1, 439–441.

Fubini, B., Hubbard, A., 2003. Reactive oxygen species (ROS) and reactive nitrogen species (RNS) generation by silica in inflammation and fibrosis. Free Rad. Biol. Med. 34, 1507–1516.

Fujita, K., Horie, M., Kato, H., Endoh, S., Suzuki, M., et al., 2009. Effects of ultrafine TiO_2 particles on gene expression profile in human keratinocytes without illumination: Involvement of extracellular matrix and cell adhesion. Toxicol. Lett. 191, 109–117.

Gabay, T., Ben-David, M., Kalifa, I., Sorkin, R., Abrams, Z.R., Ben-Jacob, E., Hanein, Y., 2007. Electro-chemical biological properties of carbon nanotube based multielectrode arrays. Nanotechnology 18, 035201.

Galano, A., 2008. Carbon nanotubes as free radical scavengers. J. Phys. Chem. C 112, 8922–8927.

Gan, C.W., Feng, S.S., 2010. Transferrin-conjugated nanoparticles of poly(lactide)-D-alpha-tocopheryl polyethylene glycol succinate diblock copolymer for targeted drug delivery across the blood–brain barrier. Biomaterials 31, 7748–7757.

Garcia, T., Shen, Y., Crawford, D., Oleksiak, M.F., Whitehead, A., Walter, R.B., 2012. RNA-Seq reveals complex genetic response to deepwater horizon oil release in Fundulus grandis. BMC Genomics 2012 (13), 474.

Gasser, M., Wick, P., Clift, M.J.D., Blank, F., Diener, L., Yan, B., Gehr, P., Krug, H.F., Rothen-Rutishauser, B., 2012. Pulmonary surfactant coating of multi-walled carbon nanotubes (MWCNTs) influences their oxidative and pro-inflammatory potential in vitro. Part. Fibre Toxicol. 9, 17.

Ge, L., Sethi, S., Ci, L., Ajayan, P.M., Dhinojwala, A., 2007. Carbon nanotube-based synthetic gecko tapes. PNAS 104, 10792–10795.

Ge, Y., Wallace, K., Winnik, W., Prasad, R.Y., 2011. Proteome profiling reveals potential toxicity and detoxification pathways following exposure of BEAS-2B cells to engineered nanoparticle titanium dioxide. Proteomics 11, 2406–2422.

Geiser, M., Rothen-Rutishauser, B., Kapp, N., Schurch, S., Kreyling, W., Schulz, H., Semmler, M., ImHof, V., Heyder, J., Gehr, P., 2005. Ultrafine particles cross cellular membranes by nonphagocytic mechanisms in lungs and in cultured cells. Environ. Health Perspect. 113, 1555–1560.

Gilbert, D.L., 2000. Fifty years of radical ideas. Ann. N.Y. Acad. Sci. 899, 1.

Gilmore, T.D., 2006. Introduction to NF-B, players, pathways, perspectives. Oncogene 25, 6680–6684.

Gomes, T., Chora, S., Pereira, C.G., Cardoso, C., Bebianno, M.J., 2014. Proteomic response of mussels *Mytilus galloprovincialis* exposed to CuO NPs and Cu^{2+}: an exploratory biomarker discovery. Aquat. Toxicol. 155, 327–336.

Gras, R., Almonacid, L., Ortega, P., Serramia, M.J., Gomez, R., et al., 2009. Changes in gene expression pattern of human primary macrophages induced by carbosilane dendrimer 2G-NN16. Pharm. Res. 26, 577–586.

Groneberg, D.A., Giersig, M., Welte, T., Pison, U., 2006. Nanoparticle-based diagnosis and therapy. Cur. Drug Targets 7, 643–648.

Gulati, N., Gupta, H., 2012. Two faces of carbon nanotube: toxicities and pharmaceutical applications. Crit. Rev. Ther. Drug Carrier Syst. 29, 65–88.

Guo, D., Bi, H., Liu, B., Wu, Q., Wang, D., Cui, Y., 2013. Reactive oxygen species-induced cytotoxic effects of zinc oxide nanoparticles in rat retinal ganglion cells. Toxicol. In Vitro 27, 731–738.

Gurr, J.R., Wang, A.S., Chen, C.H., Jan, K.Y., 2005. Ultrafine titanium dioxide particles in the absence of photoactivation can induce oxidative damage to human bronchial epithelial cells. Toxicology 213, 66–73.

Gwinn, M.R., Vallyathan, V., 2006. Nanoparticles: health effects—pros and cons. Environ. Health Perspect. 114, 1818–1825.

Hamilton Jr., R.F., Buford, M., Xiang, C., Wu, N., Holian, A., 2012. NLRP3 inflammasome activation in murine alveolar macrophages and related lung pathology is associated with MWCNT nickel contamination. Inhal. Toxicol. 24, 995–1008.

Hanagata, N., Zhuang, F., Connolly, S., Li, J., Ogawa, N., et al., 2011. Molecular responses of human lung epithelial cells to the toxicity of copper oxide nanoparticles inferred from whole genome expression analysis. ACS Nano 5, 9326–9338.

Haniu, H., Matsuda, Y., Usui, Y., Aoki, K., Shimizu, M., Ogihara, N., Hara, K., Okamoto, M., Takanashi, S., Ishigaki, N., Nakamura, K., Kato, H., Saito, N., 2011. Toxicoproteomic evaluation of carbon nanomaterials in vitro. J. Proteomics 74, 2703–2712.

Haseloff, R.F., Blasig, I.E., Bauer, H.C., Bauer, H., 2005. In search of the astrocytic factor(s) modulating blood–brain barrier functions in brain capillary endothelial cells in vitro. Cell Mol. Neurobiol. 25, 25–39.

He, X., Young, S., Schwegler-Berry, D., Chisholm, W.P., Fernback, J.E., Ma, Q., 2011. Multiwalled carbon nanotubes induce a fibrogenic response by stimulating reactive oxygen species production, activating NF-κB signaling, and promoting fibroblast-to-myofibroblast transformation. Chem. Res. Toxicol. 24, 2237–2241.

Herlaar, E., Brown, Z., 1999a. Molecular. Med. Today 5, 439–447.

Herlaar, E., Brown, Z., 1999b. p38 MAPK signalling cascades in inflammatory disease. Mol. Med. Today 5, 439–447.

Hofer, T.P., Frankenberger, M., Mages, J., Lang, R., Meyer, P., et al., 2008. Tissue-specific induction of ADAMTS2 in monocytes and macrophages by glucocorticoids. J. Mol. Med. (Berl.) 86, 323–332.

Holzinger, M., Vostrowsky, O., Hirsch, A., Hennrich, F., Kappes, M., Weiss, R., Jellen, F., 2001. Sidewall functionalization of carbon nanotubes. Angew. Chem. Int. Ed. 40, 4002–4005.

Hsieh, W.-Y., Chou, C.-C., Ho, C.-C., Yu, S.-L., Chen, H.-Y., Chou, H.-Y.E., Chen, J.J.W., Chen, H.-W., Yang, P.-C., 2012. Single-walled carbon nanotubes induce airway hyperreactivity and parenchymal injury in mice. Am. J. Respir. Cell Mol. Biol. 46, 257–267.

Huang, Y.C., Karoly, E.D., Dailey, L.A., Schmitt, M.T., Silbajoris, R., et al., 2011. Comparison of gene expression profiles induced by coarse, fine, and ultrafine particulate matter. J. Toxicol. Environ. Health A 74, 296–312.

Hubbard, A.K., Timblin, C.R., Rincon, M., Mossman, B.T., 2001. Use of transgenic luciferase reporter mice to determine activation of transcription factors and gene expression by fibrogenic particles. Chest 120, 24S–25S.

Hubbard, S.R., 2001. Theme and variations: juxtamembrane regulation of receptor protein kinases. Mol. Cell 8, 481–482.

Hubbard, A.K., Timblin, C.R., Shukla, A., Rincon, M., Mossman, B.T., 2002. Activation of NF-kB dependent gene expression by silica in lungs of luciferase reporter mice. Am. J. Physiol. Lung Cell. Mol. Physiol. 282, L968–L975.

Hubbard, S.R., 2002. Protein tyrosine kinases: autoregulation and small-molecule inhibition. Curr. Opin. Struct. Biol. 12, 735–741.

Huerta-García, E., Pérez-Arizti, J.A., Márquez-Ramírez, S.G., Delgado-Buenrostro, N.L., Chirino, Y.I., Iglesias, G.G., López-Marure, R., 2014. Titanium dioxide nanoparticles induce strong oxidative stress and mitochondrial damage in glial cells. Free Radic. Biol. Med. 73, 84–94.

Hunter, D.D., Undem, B.J., 1999. Identification and substance P content of vagal afferent neurons innervating the epithelium of the guinea pig trachea. Am. J. Respir. Crit. Care Med. 159, 1943–1948.

Hussain, S., Vanoirbeek, J.A., Luyts, K., De Vooght, V., Verbeken, E., Thomassen, L.C., Martens, J.A., Dinsdale, D., Boland, S., Marano, F., 2011. Lung exposure to nanoparticles modulates an asthmatic response in a mouse model. Eur. Respir. J. 37, 299–309.

Inoue, K., Takano, H., Ohnuki, M., Yanagisawa, R., Sakurai, M., Shimada, A., Mizushima, K., Yoshikawa, T., 2008. Size effects of nanomaterials on lung inflammation and coagulatory disturbance. Int. J. Immunopathol. Pharmacol. 21, 197–206.

Jaeger, A., Weiss, D.G., Jonas, L., Kriehuber, R., 2012. Oxidative stress-induced cytotoxic and genotoxic effects of nano-sized titanium dioxide particles in human HaCaT keratinocytes. Toxicology 296 (1–3), 27–36.

Jaeschke, H., Gores, G.J., Cederbaum, A.I., Hinson, J.A., Pessayre, D., Lemasters, J.L., 2002. Mechanisms of hepatotoxicity. Toxicol. Sci. 65, 166.

Jain, S., Thakare, V.S., Das, M., Godugu, C., Jain, A.K., Mathur, R., Chuttani, K., Mishra, A.K., 2011. Toxicity of multiwalled carbon nanotubes with end defects critically depends on their functionalization density. Chem. Res. Toxicol. 24, 2028–2039.

Ji, Z., Zhang, D., Li, L., Shen, X., Deng, X., Dong, L., Wu, M., Liu, Y., 2009. The hepatotoxicity of multi-walled carbon nanotubes in mice. Nanotechnology 20, 445101.

Jia, G., Wang, H.F., Yan, L., Wang, X., Pei, R.J., Yan, T., Zhao, Y.L., Guo, X.B., 2005. Cytotoxicity of carbon nanomaterials: single-wall nanotube, multi-wall nanotube, and fullerene. Environ. Sci. Technol. 39, 1378–1383.

Jin, Y.H., Dunlap, P.E., McBride, S.J., Al-Refai, H., Bushel, P.R., Freedman, J.H., 2008. Global transcriptome and deletome profiles of yeast exposed to transition metals. PLoS Genet. 4, e1000053.

Johnston, H.J., Hutchison, G.R., Christensen, F.M., Peters, S., Hankin, S., Aschberger, K., Stone, V., 2010. A critical review of the biological mechanisms underlying the in vivo and in vitro toxicity of carbon nanotubes: the contribution of physico-chemical characteristics. Nanotoxicology 4, 207–246.

Jonathan, P., 2009. Bioinformatics and Functional Genomics, second ed. Wiley-Blackwell, Hoboken, N.J. ISBN 9780470085851.

Jos, A., Pichardo, S., Puerto, M., Sánchez, E., Grilo, A., Cameán, A.M., 2009. Cytotoxicity of carboxylic acid functionalized single wall carbon nanotubes on the human intestinal cell line Caco-2. Toxicol. In Vitro 23, 1491–1496.

Juni, R.P., Duckers, H.J., Vanhoutte, P.M., Virmani, R., Moens, A.L., 2013. Oxidative stress and pathological changes after coronary artery interventions. J. Am. Coll. Cardiol. 61, 1471–1481.

Kagan, V.E., Tyurina, Y.Y., Tyurin, V.A., Konduru, N.V., Potapovich, A.I., Osipov, A.N., Kisin, E.R., Schwegler-Berry, D., Mercer, R., Castranova, V., Shvedova, A.A., 2006. Direct and indirect effects of single walled carbon nanotubes on RAW 264.7 macrophages: role of iron. Toxicol. Lett. 165, 88–100.

Kamp, D.W., Graceffa, P., Pryor, W.A., Weitzman, S.A., 1992. The role of free-radicals in asbestos induced diseases. Free Radic. Bio. Med. 12, 293–315.

Karlsson, H.L., Gustafsson, J., Cronholm, P., Moller, L., 2009. Size-dependent toxicity of metal oxide particles—a comparison between nano- and micrometer size. Tox. Lett. 188, 112–118.

Karoly, E.D., Li, Z., Dailey, L.A., Hyseni, X., Huang, Y.C., 2007. Up-regulation of tissue factor in human pulmonary artery endothelial cells after ultrafine particle exposure. Environ. Health Perspect. 115, 535–540.

Kasaliwal, G.R., Pegel, S., Göldel, A., Pötschke, P., Heinrich, G., 2010. Analysis of agglomerate dispersion mechanisms of multiwalled carbon nanotubes during melt mixing in polycarbonate. Polymer 51, 2708–2720.

Kawasaki, E.S., Player, A., 2005. Nanotechnology, nanomedicine, and the development of new, effective therapies for cancer. Nanomedicine 1, 101–109.

Kawata, K., Osawa, M., Okabe, S., 2009. In vitro toxicity of silver nanoparticles at noncytotoxic doses to HepG2 human hepatoma cells. Environ. Sci. Technol. 43, 6046–6051.

Khabashesku, V.N., Billups, W.E., Margrave, J.L., 2002. Fluorination of single-wall carbon nanotubes and subsequent derivatization reactions. Acc. Chem. Res. 35, 1087–1095.

Kim, E.Y., Schulz, R., Swantek, P., Kunstman, K., Malim, M.H., et al., 2012. Gold nanoparticle-mediated gene delivery induces widespread changes in the expression of innate immunity genes. Gene Ther. 19, 347–353.

Kim, S., Kim, S., Lee, S., Kwon, B., Choi, J., Hyun, J.W., Kim, S., 2011. Characterization of the effects of silver nanoparticles on liver cell using HR-MAS NMR spectroscopy. Bull. Korean Chem. Soc. 32, 2021–2026.

Knaapen, A.M., Borm, P.J.A., Albrecht, C., Schins, R.P.F., 2004. Inhaled particles and lung cancer, Part A: mechanisms. Int. J. Cancer 109, 799–809.

Kobayashi, N., Naya, M., Mizuno, K., Yamamoto, K., Ema, M., Nakanishi, J., 2012. Pulmonary and systemic responses of highly pure and well-dispersed single-wall carbon nanotubes after intratracheal instillation in rats. Inhal. Toxicol. 23, 814–828.

Kolosnjaj-Tabi, J., Hartman, K.B., Boudjemaa, S., Ananta, J.S., Morgant, G., Szwarc, H., Wilson, L.J., Moussa, F., 2010. In vivo behavior of large doses of ultrashort and full-length single-walled carbon nanotubes after oral and intraperitoneal administration to Swiss mice. ACS Nano. 4, 1481–1492.

Konduru, N.V., Tyurina, Y.Y., Feng, W., Basova, L.V., Belikova, N.A., Bayir, H., Clark, K., Rubin, M., Stolz, D., Vallhov, H., Scheynius, A., Witasp, E., Fadeel, B., Kichambare, P.D., Star, A., Kisin, E.R., Murray, A.R., Shvedova, A.A., Kagan, V.E., 2009. Phosphatidylserine targets single-walled carbon nanotubes to professional phagocytes in vitro and in vivo. PLoS One 4, e4398.

Kong, L., Tuomela, S., Hahne, L., Ahlfors, H., Yli-Harja, O., et al., 2013. NanoMiner — integrative human transcriptomics data resource for nanoparticle research. PLoS One 8, e68414.

Kostarelos, K., Lacerda, L., Pastorin, G., Wu, W., Wieckowski, S., Luangsivilay, J., Godefroy, S., Pantarotto, D., Briand, J.-P., Muller, S., Prato, M., Bianco, A., 2007. Cellular uptake of functionalized carbon nanotubes is independent of functional group and cell type. Nat. Nanotechnol. 2, 108–113.

Kovacic, P., Thurn, L.A., 2005. Cardiovascular toxicity from the perspective of oxidative stress, electron transfer, and prevention by antioxidants. Curr. Vasc. Pharmacol. 3, 107–117.

Kumar, A.A., Pandey, A.K., Singh, S.S., Shanker, R., Dhawan, A., 2011. Engineered ZnO and TiO_2 nanoparticles induce oxidative stress and DNA damage leading to reduced viability of *Escherichia coli*. Free Rad. Biol. Med. 51, 1872–1881.

Kuzmanya, H., Kukoveczb, A., Simona, F., Holzwebera, M., Krambergera, C., Pichler, T., 2004. Functionalization of carbon nanotubes. Synth. Metals 141, 113–122.

Kwon, Y.-K., Saito, S., Tománek, D., 1998. Effect of inter-tube coupling on the electronic. structure of carbon nanotube ropes. Phys. Rev. B 58, R13314.

Lacerda, L., Raffa, S., Prato, M., Bianco, A., Kostarelos, K., 2007. Cell-penetrating CNTs for delivery of therapeutics. Nano Today 2, 38–43.

Lam, C.W., James, J.T., McCluskey, R., Arepalli, S., Hunter, R.L., 2006. A review of carbon nanotube toxicity and assessment of potential occupational and environmental health risks. Crit. Rev. Toxicol. 36, 189–217.

Lam, C.W., James, J.T., McCluskey, R., Hunter, R.L., 2004. Pulmonary toxicity of single-wall carbon nanotubes in mice 7 and 90 days after intratracheal instillation. Toxicol. Sci. 77, 126.

Laskin, D.L., Laskin, J.D., 2001. Role of macrophages and inflammatory mediators in chemically induced toxicity. Toxicology 160, 111–118.

Laskin, D.L., Pendino, K.J., 1995. Macrophages and inflammatory mediators in tissue injury. Annu. Rev. Pharmacol. Toxicol. 35, 655–677.

Lee, H.J., Engelhardt, B., Lesley, J., Bickel, U., Pardridge, W.M., 2000. Targeting rat anti-mouse transferrin receptor monoclonal antibodies through blood–-brain barrier in mouse. J. Pharm. Exp. Ther. 292, 1048–1052.

Lee, H.-M., Song, H.-M., Yuk, J.-M., Lee, Z.-E., Lee, S.-H., Hwang, S.M., Lee, C.-S., Jo, E.-K., 2009. Nanoparticles up-regulate tumor necrosis factor-α and CXCL8 via reactive oxygen species and mitogen-activated protein kinase activation. Tox. Appl. Pharm. 238, 60–169.

Lee, T.L., Chan, W.Y., Rennert, O.M., Lee, T.L., Chan, W.Y., Rennert, O.M., 2009b. Assessing the safety of nanomaterials by genomic approach could be another alternative. ACS Nano 3, 3830.

Lee, Y., Geckeler, K.E., 2012. Cytotoxicity and cellular uptake of lysozyme-stabilized gold nanoparticles. J. Biomed. Mater. Res. Part A 100A, 848–855.

Lefer, D.J., Granger, D.N., 2000. Oxidative stress and cardiac disease. Am. J. Med. 109, 315–323.

Legramante, J.M., Valentini, F., Magrini, A., Palleschi, G., Sacco, S., Iavicoli, I., Pallante, M., Moscone, D., Galante, A., Bergamaschi, E., Bergamaschi, A., Pietroiusti, A., 2009. Cardiac autonomic regulation after lung exposure to carbon nanotubes. Hum. Expl. Toxicol. 28, 369–375.

Legramante, J.M., Sacco, S., Crobeddu, P., Magrini, A., Valentini, F., Palleschi, G., Pallante, M., Balocchi, R., Iavicoli, I., Bergamaschi, A., Galante, A., Campagnolo, L., Pietroiusti, A., 2012. Changes in cardiac autonomic regulation after acute lung exposure to carbon nanotubes: implications for occupational exposure. J. Nanomater. 2012. ID 397206.

Lei, R., Wu, C., Yang, B., et al., 2008. Integrated metabolomic analysis of the nano-sized copper particle-induced hepatotoxicity and nephrotoxicity in rats: a rapid in vivo screening method for nanotoxicity. Tox. Appl. Pharm. 232, 292–301.

Lenz, E.M., Wilson, I.D., 2007. Analytical strategies in metabonomics. J. Proteome Res. 26, 443–458.

Lerner, M.B., Daileya, J., Goldsmitha, B.R., Brissonb, D., Johnsona, A.T.C., 2013. Detecting Lyme disease using antibody-functionalized single-walled carbon nanotube transistors. Sens. Bioelectron. 45, 163–167.

Lewinski, N., Colvin, V., Drezek, R., 2008. Cytotoxicity of nanoparticles. Small 4, 26–49.

Li, J.J., Hartono, D., Ong, C., Bay, B., Yung, L.L., 2010. Autophagy and oxidative stress associated with gold nanoparticles. Biomaterials 31, 5996–6003.

Li, Z., Hulderman, T., Salmen, R., Chapman, R., Leonard, S.S., Young, S.H., Shvedova, A., Luster, M.I., Simeonova, P.P., 2007. Cardiovascular effects of pulmonary exposure to single-wall carbon nanotubes. Environ. Health Perspect. 115, 377–382.

Lin, Y.R., Kuo, C.J., Lin, H.Y., Wu, C.J., Liang, S.S., 2014. A proteomics analysis to evaluate cytotoxicity in NRK-52E cells caused by unmodified Nano-Fe_3O_4. Sci. World J. 2014, 754721.

Lindberg, H.K., Falck, G.C., Suhonen, S., Vippola, M., Vanhala, E., Catalán, J., Savolainen, K., Norppa, H., 2009. Genotoxicity of nanomaterials: DNA damage and micronuclei induced by carbon nanotubes and graphite nanofibres in human bronchial epithelial cells in vitro. Toxicol. Lett. 186, 166–173.

Liopo, A.V., Stewart, M.P., Hudson, J., Tour, J.M., Pappas, T.C., 2006. Biocompatibility of native functionalized singlewalled carbon nanotubes for neuronal interface. J. Nanosci. Nanotechnol. 6, 1365–1374.

Liu, S., Xu, L., Zhang, T., Ren, G., Yang, Z., 2010. Oxidative stress and apoptosis induced by nanosized titanium dioxide in PC12 cells. Toxicology 267, 172–177.

Liu, Y., Chen, Z., Gu, N., Wang, J., 2011. Effects of DMSA-coated Fe3O4 magnetic nanoparticles on global gene expression of mouse macrophage RAW264.7 cells. Tox. Lett. 205, 130–139.

Liu, Y., Guyton, K.Z., Gorospe, M., Xu, Q., Lee, J.C., Holbrook, N.J., 1996. Differential activation of ERK, JNK/SAPK and P38/CSBP/RK map kinase family members during the cellular response to arsenite. Free Radic. Biol. Med. 21, 771–781.

Love, S.A., Maurer-Jones, M.A., Thompson, J.W., Lin, Y.S., Haynes, C.L., 2012. Assessing nanoparticle toxicity. Ann. Rev. Anal. Chem. (Palo Alto Calif.) 5, 181–205.

Luyts, K., Napierska, D., Nemerya, B., Hoet, P.H.M., 2013. How physico-chemical characteristics of nanoparticles cause their toxicity: complex and unresolved interrelations? Environ. Sci. Process. Impacts 15, 23–38.

Ma, J.Y., Zhao, H., Mercer, R.R., Barger, M., Rao, M., Meighan, T., Schwegler-Berry, D., Castranova, V., Ma, J.K., 2011. Cerium oxide nanoparticle-induced pulmonary inflammation and alveolar macrophage functional change in rats. Nanotechnology 5, 312–325.

Madl, A.K., Pinkerton, K.E., 2009. Health effects of inhaled engineered and incidental nanoparticles. Crit. Rev. Toxicol. 39, 629–658.

Magrez, A., Kasas, S., Salicio, V., Pasquier, N., Seo, J.W., Celio, M., Catsicas, S., Schwaller, B., Forro, L., 2006. Cellular toxicity of carbon-based nanomaterials. Nano Lett. 6, 1121–1125.

Mahajan, S.D., Roy, I., Xu, G., Ken-Tye, Y., Hong, D., Aalinkeel, R., Reynolds, J., Sykes, D., Nair, B.B., Lin, E.Y., Elaine, Y., Prasad, P.N., Schwartz, S.A., 2010. Enhancing the delivery of anti retroviral drug "Saquinavir" across the blood–brain barrier using nanoparticles. Curr. HIV Res. 8, 396–404.

Mahajan, B., Shrestha, T.M., Gyawali, R., 2012. Antibacterial and cytotoxic activity of juniperus indica bertol from Napalese Hymalaya. IJPSR 3, 1104–1107.

Ma-Hock, L., Treumann, S., Strauss, V., Brill, S., Luizi, F., Mertler, M., Wiench, K., Gamer, A.O., van Ravenzwaay, B., Landsiedel, R., 2009. Inhalation toxicity of multiwall carbon nanotubes in rats exposed for 3 months. Toxicol. Sci. 112, 468–481.

Mangum, J.B., Turpin, E.A., Antao-Menezes, A., Cesta, M.F., Bermudez, E., Bonner, J.C., 2006. Single-walled carbon nanotube (SWCNT)-induced interstitial fibrosis in the lungs of rats is associated with increased levels of PDGF mRNA and the formation of unique intercellular carbon structures that bridge alveolar macrophages in situ. Part. Fibre Toxicol. 3, 15.

Manke, A., Wang, L., Rojanasakul, Y., 2013. Mechanisms of nanoparticle-induced oxidative stress and toxicity. BioMed. Res. Int. 2013 (942916), 15.

Manna, S.K., Sarkar, S., Barr, J., et al., 2005. Single-walled carbon nanotube induces oxidative stress and activates nuclear transcription factor-κB in human keratinocytes. Nano Lett. 5, 1676–1684.

Martinez-Morlanesa, M.J., Castellb, P., Alonsoc, P.J., Martinezb, M.T., Puértolasa, J.A., 2012. Multi-walled carbon nanotubes acting as free radical scavengers in gamma-irradiated ultrahigh molecular weight polyethylene composites. Carbon 50, 2442–2452.

Martins, J., Oliva Teles, L., Vasconcelos, V., 2007. Assays with Daphnia magna and Danio rerio as alert systems in aquatic toxicology. Environ. Int. 33, 414–425.

Masserini, M., 2013. Nanoparticles for brain drug delivery. ISRN Biochem. 2013, 238428.

Mercer, R.R., Hubbs, A.F., Scabilloni, J.F., Wang, L., Battelli, L.A., Schwegler-Berry, D., Castranova, V., Porter, D.W., 2010. Distribution and persistence of pleural penetrations by multi-walled carbon nanotubes. Part. Fibre Toxicol. 7, 28.

Meunier, E., Coste, A., Olagnier, D., Authier, H., Lefevre, L., Dardenne, C., Bernad, J., Beraud, M., Flahaut, E., Pipy, B., 2012. Double-walled carbon nanotubes trigger IL-1 release in human monocytes through Nlrp3 inflammasome activation. Nanomedicine 8, 987–995.

Mirzajani, F., Askari, H., Hamzelou, S., Schober, Y., Römpp, A., Ghassempour, A., Spengler, B., 2011. Proteomics study of silver nanoparticles toxicity on *Bacillus thuringiensis*. Proteomics 11, 4569–4577.

Moller, P., Jacobsen, N., Folkmann, J.K., Danielsen, P., Mikkelsen, L., Hemmingsen, J., Vesterdal, L.K., Forchhammer, L., Wallin, H., Loft, S., 2010. Role of oxidative damage in toxicity of particulates. Free Rad. Res. 44, 1–46.

Monteiro-Riviere, N.A., Nemanich, R.J., Inman, A.O., Wang, Y.Y., Riviere, J.E., 2005. Multi-walled carbon nanotube interactions with human epidermal keratinocytes. Toxicol. Lett. 155, 377–384.

Montiel-Davalos, A., Ventura-Gallegos, J.L., Alfaro-Moreno, E., Soria-Castro, E., García-Latorre, E., CabaasnMoreno, J.G., Ramos-Godínez, M.D., López-Marure, R., 2012. TiO_2 nanoparticles induce dysfunction and activation of human endothelial cells. Chem. Res. Toxicol. 25, 920–930.

Moos, P.J., Olszewski, K., Honeggar, M., Cassidy, P., Leachman, S., et al., 2011. Responses of human cells to ZnO nanoparticles: a gene transcription study. Metallomics 3, 1199–1211.

Mossman, B.T., Churg, A., 1998. Mechanisms in the pathogenesis of asbestosis and silicosis. Am. J. Respir. Crit. Care Med. 157, 1666–1680.

Mu, Q., Du, G., Chen, T., Zhang, B., Yan, B., 2009. Suppression of human bone morphogenetic protein signaling by carboxylated single-walled carbon nanotubes. ACS Nano 3, 1139–1144.

Mu, Q., Jiang, G., Chen, L., Zhou, H., Fourches, D., Tropsha, A., Yan, B., 2014. Chemical basis of interactions between engineered nanoparticles and biological systems. Chem. Rev. 114, 7740–7781.

Muller, J., Huaux, F., Heilier, J.F., Arras, M., Delos, M., Nagy, B.J., Lison, D., 2005. Respiratory toxicity of carbon nanotubes. Toxicol. Appl. Pharmacol. 207, 221–231.

Murakami, S., Okubo, K., Tsuji, Y., Sakata, H., Takahashi, T., Kikuchi, M., Hirayama, R., 2004. Changes in liver enzymes after surgery in anti-hepatitis C virus-positive patients. World J. Surg. 28, 671–674.

Murphy, F.A., Poland, C.A., Duffin, R., Al-Jamal, K.T., Ali-Boucetta, H., Nunes, A., Byrne, F., Prina-Mello, A., Volkov, Y., Li, S., Mather, S.J., Bianco, A., Prato, M., MacNee, W., Wallace, W.A., Kostarelos, K., Donaldson, H., 2011. Length-dependent retention of carbon nanotubes in the pleural space of mice initiates sustained inflammation and progressive fibrosis on the parietal pleura. Am. J. Pathol. 178, 2587–2600.

Murphy, F.A., Schinwald, A., Polant, C.A., Donaldson, K., 2012. The mechanism of pleural inflammation by long carbon nanotubes: interaction of long fibres with macrophages stimulates them to amplify pro-inflammatory responses in mesothelial cells.Part. Fibre Toxicol 9, 8.

Murray, A.R., Kisin, E., Leonard, S.S., Young, S.H., Kommineni, C., Kagan, V.E., Castranova, V., Shvedova, A.A., 2009. Oxidative stress and inflammatory response in dermal toxicity of single-walled carbon nanotubes. Toxicology 257, 161–171.

Mylvaganam, K., Zhang, L.C., 2004. Nanotube functionalization and polymer grafting: an ab initio study. J. Phys. Chem. B 108, 15009–15012.

Nafz, B., Just, A., Staub, H.M., Wagner, C.D., Ehmke, H., Kirchheim, H.R., Persson, P., 1986. Blood-pressure variability is buffered by nitric oxide. J. Auton. Nerv. Syst. 57, 181–183.

Nagai, H., Okazakia, Y., Chewa, S.H., Misawaa, N., Yamashitaa, Y., Akatsukaa, S., IshiharaaT, Y.K., Yoshikawac, Y., Yasuic, H., Jianga, L., Oharaa, H., Takahashid, T., Gaku, I.K., Kostarelosf, K., Miyatag, Y., Shinoharag, H., Toyokunia, S., 2011. Diameter and rigidity of multiwalled carbon nanotubes are critical factors in mesothelial injury and carcinogenesis. PNAS 108, E1330–E1338.

Nagai, H., Toyokuni, S., 2012. Differences and similarities between carbon nanotubes and asbestos fibers during mesothelial carcinogenesis: shedding light on fiber entry mechanism. Cancer Sci. 103, 1378–1390.

Naqvi, S., Samim, M., Abdin, M.Z., et al., 2010. Concentration-dependent toxicity of iron oxide nanoparticles mediated by increased oxidative stress. Int. J. Nanomed. 5, 983–989.

Nel, A., Xia, T., Madler, L., Li, N., 2006. Toxic potential of materials at the nanolevel. Science 311, 622–627.

Nemmar, A., Hoylaerts, M.F., Hoet, P.H., Nemery, B., 2004. Possible mechanisms of the cardiovascular effects of inhaled particles: systemic translocation and prothrombiotic effects of inhaled particels. Toxicol. Lett. 149, 243–253.

Nemmar, A., Hoet, P.H., Vanquickenborne, B., Dinsdale, D., Thomeer, M., Hoylaerts, M.F., Vanbilloen, H., Mortelmans, L., Nemery, B., 2002. Passage of inhaled particles into the blood circulation in humans. Circulation 105, 411–414.

Nemmar, A., Vanbilloen, H., Hoylaerts, M.F., Hoet, P.H., Verbruggen, A., Nemery, B., 2001. Passage of intratracheally instilled ultrafine particles from the lung into the systemic circulation in hamster. Am. J. Respir. Crit. Care Med. 164, 1665–1668.

Nerl, H.C., Cheng, C., Goode, A.E., Bergin, S.D., Lich, B., Gass, M., Porter, A.E., 2011. Imaging methods for determining uptake and toxicity of carbon nanotubes in vitro and in vivo. Nanomedicine 6, 849–865.

NIOSH, 2007. National Occupational Respiratory Mortality System (NORMS). http://webappa.cdc.gov/ords/norms.html.

NIOSH, 2011. Asbestos fibers and other elongate mineral particles: state of the science and roadmap for research. Bulletin 62, 2011–2159.

Ntim, S.A., Sae-Khow, O., Desai, C., Witzmann, F.A., Mitra, S., 2012. A size dependent aqueous dispersibility of carboxylated multiwall carbon Nanotubes. J. Environ. Monit. 14, 2772–2779.

Nunes, A., Al-Jamal, K., Nakajima, T., Hariz, M., Kostarelos, K., 2012. Application of carbon nanotubes in neurology: clinical perspectives and toxicological risks. Arch. Toxicol. 86, 1009–1020.

Oberdörster, G., Oberdörster, E., Oberdörster, J., June 2007. Concepts of nanoparticle dose metric and response metric. Environ. Health Perspect. 115 (6), A290.

Pacurari, M., Yin, X.J., Zhao, J., Ding, M., Leonard, S.S., Schwegler-Berry, D., Ducatman, B.S., Sbarra, D., Hoover, M.D., Castranova, V., Vallyathan, V., 2008. Raw single-wall carbon nanotubes induce oxidative stress and activate MAPKs, AP-1, NF-κB, and Akt in normal and malignant human mesothelial cells. Environ. Health Perspect. 116, 1211–1217.

Palmer, H.J., Paulson, K.E., 1997. Reactive oxygen species and antioxidants in signal transduction and gene expression. Nutr. Rev. 55, 353–361.

Palomaki, J., Valimaki, E., Sund, J., Vippola, M., Clausen, K.A., Jensen, K.A., Savolainen, K., Matikainen, S., Alenius, H., 2011. Long, needle-like carbon nanotubes and asbestos activite the NLRP3 inflammasome through a similar mechanism. Acs Nano 5, 6861–6870.

Pang, C.C.Y., 2001. Autonomic control of the venous system in health and disease. Effects of drugs. Pharmacol. Ther. 90, 179–230.

Pang, Z., Gao, H., Yu, Y., Chen, J., Guo, L., Ren, J., Wen, Z., Jiang, X., 2011. Brain delivery and cellular internalization mechanisms for transferrin conjugated biodegradable polymersomes. Int. J. Pharm. 415, 284–292.

Papageorgiou, I., Brown, C., Schins, R., et al., 2007. The effect of nano- and micron-sized particles of cobalt-chromium alloy on human fibroblasts in vitro. Biomaterials 28 (19), 2946–2958.

Park, K.H., Chhowalla, M., Iqbal, Z., Sesti, F., 2003. Single-walled carbon nanotubes are a new class of ion channel blockers. J. Biol. Chem. 278, 50212–50216.

Parkinson, H., Sarkans, U., Kolesnikov, N., Abeygunawardena, N., Burdett, T., et al., 2011. ArrayExpress update–an archive of microarray and high-throughput sequencing-based functional genomics experiments. Nucleic Acids Res. 39, D1002–D1004.

Patlolla, A., Knighten, B., Tchounwou, P., 2010. Multi-walled carbon nanotubes induce cytotoxicity, genotoxicity and apoptosis in normal human dermal fibroblast cells. Ethn. Dis. 20 (Suppl. 1), 65–72.

Patlolla, A.K., Mcginnis, B., Tchounwou, P.B., 2011. Biochemical and histo-pathological evaluation of functionalized single-walled carbon nanotube in Swiss-Webster mice. J. Appl. Toxicol. 31, 75–83.

Pauluhn, J., 2010. Subchronic 13-week inhalation exposure of rats to multiwalled carbon nanotubes, toxic effects are determined by density of agglomerate structures, not fibrillar structures. Toxicol. Sci. 113, 226–242.

Peng, H.P., Reverdy, P., Khabashesku, V.N., Margrave, J.L., 2003. Sidewall functionalization of single-walled carbon nanotubes with organic peroxides. Chem. Commun. 2003, 362–363.

Peretz, V., Motiei, M., Sukenik, C.N., Popovtzer, R., 2012. The effect of nanoparticle size on cellular binding probability. J. At. Mol. Opt. Phys. 2012, Article ID 404536.

Perez-Martınezc, F.C., Carrion, B., Cena, V., 2012. The use of nanoparticles for gene therapy in the nervous system. J. Alzheimer's Dis. 31, 697–710.

Petkovic, J., Zegura, B., Stevanović, M., Drnovšek, N., Uskoković, D., Novak, S., Filipič, M., 2011. DNA damage and alterations in expression of DNA damage responsive genes induced by TiO_2 nanoparticles in human hepatoma HepG2 cells. Nanotoxicology 5, 341–353.

Pichardo, S., Gutierrez-Praena, D., Puerto, M., Sanchez, E., Grilo, A., Camean, A.M., Jos, A., 2012. Oxidative stress responses to carboxylic acid functionalized single wall carbon nanotubes on the human intestinal cell line Caco-2. Toxicol. In Vitro 26, 672–677.

Poland, C.A., Duffin, R., Kinloch, I., Maynard, A., Wallace, W.A., Seaton, A., Stone, V., Brown, S., Macnee, W., Donaldson, K., 2008. Carbon nanotubes introduced into the abdominal cavity of mice show asbestos-like pathogenicity in a pilot study. Nat. Nanotechnol. 3, 423–428.

Pope 3rd, C.A., Burnett, R.T., Thurston, G.D., Thun, M.J., Calle, E.E., Krewski, D., Godleski, J.J., 2004. Cardiovascular mortality and long-term exposure to particulate air pollution: epidemiological evidence of general pathophysiological pathways of disease. Circulation 109, 71–77.

Porter, D.W., Hubbs, A.F., Mercer, R.R., Wu, N., Wolfarth, M.G., Sriram, K., Leonard, S., Battelli, L., Schwegler-Berry, D., Friend, S., Andrew, M., Chen, B.T., Tsuruoka, S., Endo M,Castranova, V., 2009. Mouse pulmonary dose- and time course-responses induced by exposure to multi-walled carbon nanotubes. Toxicology 10, 136–147.

Prousek, J., 2007. Fenton chemistry in biology and medicine. Pure Appl. Chem. 79, 2325–2338.

Pujalte, I., Passagne, I., Brouillaud, B., Tréguer, M., Durand, E., Ohayon-Courtès, C., L'Azou, B., 2011. Cytotoxicity and oxidative stress induced by different metallic nanoparticles on human kidney cells. Part. Fibre Toxicol. 8, 10.

Pulskamp, K., Diabate, S., Krug, H.F., 2007. Carbon nanotubes show no sign of acute toxicity but induce intracellular reactive oxygen species in dependence on contaminants. Toxicol. Lett. 168, 58–74.

Qi, W., Bi, J., Xiaoyong Zhang, X., Jing Wang, J., Jianjun Wang, J., Peng Liu, P., Zhan, Li, Wu, W., 2014. Damaging effects of multi-walled carbon nanotubes on pregnant mice with different pregnancy times. Sci. Rep. 4, 4352.

Qi, W., Bi, J., Zhang, X., Wang, J., Wang, J., LiuP, Li Z., Wu, W., 2005. Damaging effects of multi-walled carbon nanotubes on pregnant mice with different pregnancy times. Sci. Rep. 4, 4352.

Radomski, A., Jurasz, P., Alonso-Escolano, D., Drews, M., Morandi, M., Malinski, T., Radomski, M.W., 2005. Nanoparticle-induced platelet aggregation and vascular thrombosis. Br. J. Pharmacol. 146, 882–893.

Raghunathan, V.K., Devey, M., Hawkins, S., et al., 2013. Influence of particle size and reactive oxygen species on cobalt chrome nanoparticle-mediated genotoxicity. Biomaterials 34, 3559–3570.

Rallo, R.B.F., Liu, R., Nair, S., George, S., Damoiseaux, R., Giralt, F., Nel, A., Bradley, K., Cohen, Y., 2011. Self-organizing map analysis of toxicity-related cell signaling pathways for metal and metal oxide nanoparticles. Environ. Sci. Tech. 45 (4), 1695–1702.

Reddy, A.R., Krishna, D.R., Reddy, Y.N., Himabindu, V., 2010. Translocation and extra pulmonary toxicities of multi wall carbon nanotubes in rats. Toxicol. Mech. Methods 20, 267–272.

Reddy, A.R., Rao, M.V., Krishna, D.R., Himabindu, V., Reddy, Y.N., 2011. Evaluation of oxidative stress and anti oxidant status in rat serum following exposure of carbon nanotubes. Regul. Toxicol. Pharmacol. 59, 251–257.

Reddy, A.R., Reddy, Y.N., Krishna, D.R., Himabindu, V., 2012. Multi wall carbon nanotubes induce oxidative stress and cytotoxicity in human embryonic kidney (HEK293) cells. Toxicology 272, 11–16.

Ricci, R., 2008. Carbon Nanotubes for Neural Interfaces. 214th ECS Meeting, Abstract #2133. The Electrochemical Society, Honolulu, Hawaii.

Romoser, A.A., Berg, M., Seabury, C., Ivanov, I., Criscitiello, M.F., Sayes, V.M., 2011. Quantum dots trigger immunomodulation of the NFκB pathway in human skin cells. Mol. Immunol. 48, 1349–1359.

Rosen, G.M., Pou, S., Ramos, C.L., Cohen, M.S., Britigan, B.E., 1995. Free radicals and phagocytic cells. FASEB J. 9, 200–209.

Roux, P.P., Blenis, J., 2004. ERK and p38 MAPK-activated protein kinases: a family of protein kinases with diverse biological functions. Microbiol. Mol. Biol. Rev. 68, 320–344.

Ryman-Rasmussen, J.P., Tewksbury, E.W., Moss, O.R., Cesta, M.F., Wong, B.A., Bonner, J.C., 2009. Inhaled multi-walled carbon nanotubes potentiate airway fibrosis in murine allergic asthma. Am. J. Respir Cell Mol. Biol. 40, 349–358.

Ryman-Rasmussen, J.P., Cesta, M.F., Brody, A.R., Shipley-Phillips, J.K., Everitt, J.I., Tewksbury, E.W., Moss, O.R., Wong, B.A., Dodd, D.E., Andersen, M.E., Bonner, J.C., 2009. Inhaled carbon nanotubes reach the subpleural tissue in mice. Nat. Nanotechnol. 4, 747–751.

Samhan-Arias, A.K., Ji, J., Demidova, O.M., Sparvero, L.J., Feng, W., Tyurin, V., Tyurina, Y.Y., Epperly, M.W., Shvedova, A.A., Greenberger, J.S., Bayır, H., Kagan, V.E., Amoscato, A.A., 2012. Oxidized phospholipids as biomarkers of tissue and cell damage with a focus on cardiolipin. Biochim. Biophys. Acta 1818, 2413–2423.

Schroder, K., Tschopp, J., 2010. The inflammasomes. Cell 140, 821–832.

Sharma, C.S., Sarkar, S., Periyakaruppan, A., et al., 2007. Single-walled carbon nanotubes induces oxidative stress in rat lung epithelial cells. J. Nanosci. Nanotechnol. 7, 2466–2472.

Sharma, V., Anderson, D., Dhawan, A., 2012. Zinc oxide nanoparticles induce oxidative DNA damage and ROS-triggered mitochondria mediated apoptosis in human liver cells (HepG2). Apoptosis 17, 852–870.

Sharma, V.K., 2009. Aggregation and toxicity of titanium dioxide nanoparticles in aquatic environment—a review. J. Environ. Sci. Health A Tox. Hazard Subst. Environ. Eng. 44, 1485–1495.

Shimizu, M., Tainaka, H., Oba, T., Mizuo, K., Umezawa, M., Takeda, K., 2009. Maternal exposure to nanoparticulate titanium dioxide during the prenatal period alters gene expression related to brain development in the mouse. Part. Fibre Toxicol. 2009 (6), 20.

Shvedova, A.A., Kisin, E., Murray, A.R., Johnson, V.J., Gorelik, O., Arepalli, S., Hubbs, A.F., Mercer, R.R., Keohavong, P., Sussman, N., Jin, J., Yin, J., Stone, S., Chen, B.T., Deye, G., Maynard, A., Castranova, V., Baron, P.A., Kagan, V.E., 2008. Inhalation vs. aspiration of single-walled carbon nanotubes in C57BL/6 mice. Am. J. Physiol. Lung Cell Mol. Physiol. 295, L552–L565.

Shvedova, A.A., Fabisiak, J.P., Kisin, E.R., Murray, A.R., Roberts, J.R., Tyurina, Y.Y., Antonini, J.M., Feng, W.H., Kommineni, C., Reynolds, J., Barchowsky, A., Castranova, V., Kagan, V.E., 2008a. Sequential exposure to carbon nanotubes and bacteria enhances pulmonary Inflammation and infectivity. Am. J. Respir. Cell Mol. Biol. 38, 579–590.

Shvedova, A.A., Kagan, V.E., 2010. The role of nanotoxicology in realizing the 'helping without harm' paradigm of nanomedicine: lessons from studies of pulmonary effects of single-walled carbonnanotubes. J. Intern. Med. 267, 106–118.

Shvedova, A.A., Kisin, E., Murray, A.R., Johnson, V.J., Gorelik, O., Arepalli, S., Hubbs, A.F., Mercer, R.R., Keohavong, P., Sussman, N., Jin, J., Yin, J., Stone, S., Chen, B.T., Deye, G., Maynard, A., Castranova, V., Baron, P.A., Kagan, V.E., 2008b. Inhalation vs. aspiration of single-walled carbon nanotubes in C57BL/6 mice: inflammation, fibrosis, oxidative stress, and mutagenesis. Am. J. Physiol. Lung Cell Mol. Physiol. 295, L552–L565.

Shvedova, A.A., Kisin, E.R., Mercer, R., Murray, A.R., Johnson, V.J., Potapovich, A.I., Tyurina, Y.Y., Gorelik, O., Arepalli, S., Schwegler-Berry, D., Hubbs, A.F., Antonini, J., Evans, D.E., Ku, B.K., Ramsey, D., Maynard, M., Kagan, V.E., Castranova, V., Baron, P., 2005. Unusual inflammatory and fibrogenic pulmonary responses to single walled carbon nanotubes in mice. Am. J. Physiol. Lung Cell Mol. Physiol. 289, L698–L708.

Shvedova, A.A., Kisin, E.R., Porter, D., Schulte, P., Kagan, V.E., Fadeel, B., Castranova, V., 2009. Mechanisms of pulmonary toxicity and medical applications of carbon nanotubes: two faces of Janus? Pharmacol. Ther. 121, 192–204.

Shvedova, A.A., Kagan, V.E., Fadeel, B., 2010. Close encounters of the small kind: adverse effects of man-made materials interfacing with the nano-cosmos of biological systems. Annu Rev Pharmacol Toxicol. 50, 63–88.

Shvedova, A.A., Pietroiusti, A., Fadeel, B., Kagan, V.E., 2012. Mechanisms of carbon nanotube-induced toxicity: focus on oxidative stress. Toxicol. Appl. Pharm. 261, 121–133.

Shwe, T.T., Yamamoto, S., Kakeyama, M., Kobayashi, T., Fujimaki, H., 2005. Effect of intratracheal instillation of ultrafine carbon black on proinflammatory cytokine and chemokine release and mRNA expression in lung and lymph nodes of mice. Toxicol. Appl. Pharmacol. 209, 51–61.

Siddiqui, M.A., Ahamed, M., Ahmad, J., et al., 2012. Nickel oxide nanoparticles induce cytotoxicity, oxidative stress and apoptosis in cultured human cells that is abrogated by the dietary antioxidant curcumin. Food Chem. Tox. 50, 641–647.

Simon, D.F., Domingos, R.F., Hauser, C., Hutchins, C.M., Zerges, W., Wilkinson, K.J., 2013. Transcriptome sequencing (RNA-seq) analysis of the effects of metal nanoparticle exposure on the transcriptome of chlamydomonas reinhardtii appl. Environ. Microbiol. 79 (16), 4774–4785.

Singh, A.K., Jiang, Y., 2004. Differential activation of NF kappa B/RelA-p50 and NF kappa B/p50-p50 in control and alcohol-drinking rats subjected to carrageenin-induced pleurisy. Mediators Inflamm. 13, 255–262.

Sioutas, C., Delfino, R.J., Singh, M., 2005. Exposure assessment for atmospheric ultrafine particles (UFPs) and implications in epidemiologic research. Environ. Health Perspect. 113, 947–955.

Sleight, P., Eckberg, D.L., 1992. Human Baroreflexes in Health and Disease. Oxford University Press, Clarendon, ISBN 978-0198576938.

Smith, K.R., Klei, L.R., Barchowsky, A., 2001. Arsenite stimulates plasma membrane NADPH oxidase in vascular endothelial cells. Am. J. Physiol. 280, L442–L449.

Sohaebuddin, S.K., Thevenot, P.T., Baker, D., Eaton, J.W., Tang, L., 2010. Nanomaterial cytotoxicity is composition, size, and cell type dependent. Part. Fibre Toxicol. 7, 22.

Sohn, G.J., Choi, H.J., Jeon, I.Y., Chang, D.W., Dai, L., Baek, J.B., 2012. Water-dispersible, sulfonated hyperbranched poly(ether-ketone) grafted multiwalled carbon nanotubes as oxygen reduction catalysts. ACS Nano 6, 6345–6355.

Squadrito, G.L., Pryor, W.A., 1998. Oxidative chemistry of nitric acid: the roles of superoxide, peroxynitrate and carbon dioxide. Free Radic. Biol. Med. 25, 392–403.

Sriram, K., Porter, D., Tsuruoka, S., Endo, M., Jefferson, A., Wolfarth, W., Rogers, G.M., Castranova, V., Luster, M.I., 2007. Neuroinflammatory responses following exposure to engineered nanomaterials. Toxicologist 96, A1390.

Sriram, K., Porter, D.W., Jefferson, A.M., Lin, G.X., Wolfarth, M.G., Chen, B.T., McKinney, W., Frazer, D.G., Castranova, V., 2009. Neuroinflammation and blood–brain barrier changes following exposure to engineered nanomaterials. Toxicologist 108, A2197.

Stanton, M.F., Layard, M., Tegeris, A., Miller, E., May, M., Morgan, E., Smith, A., 1981. Relation of particle dimension to carcinogenicity in amphibole asbestoses and other fibrous minerals. J. Natl. Cancer Inst. 67, 965–975.

Stapleton, P.A., Minarchick, V., Cumpston, A., McKinney, W., Chen, B.T., Frazer, D., Castranova, V., Nurkiewicz, T.R., 2012. Impairment of coronary arteriolar endothelium-dependent dilation after multi-walled carbon nanotube inhalation: a time-course study. Int. J. Mol. Sci. 13, 13781–13803.

Stoeger, T.S.T., Frankenberger, B., Ritter, B., Karg, E., Maier, K., Schulz, H., Schmid, O., 2009. Deducing in vivo toxicity of combustion-derived nanoparticles from a cell-free oxidative potency assay and metabolic activation of organic compounds. Environ. Health Perspect. 117, 54–60.

Stone, V., Shaw, J., Brown, D.M., Macnee, M., Faux, S.P., Donaldson, K., 1998. The role of oxidative stress in the prolonged inhibitory effect of ultrafine carbon black on epithelial cell function. Toxicol. In Vitro 12, 649–659.

Sun, Y.P., Fu, K., Lin, Y., Huang, W., 2002. Functionalized carbon nanotubes: properties and applications. Acc. Chem. Res. 35, 1096–1104.

Sun, Q., Wang, A., Jin, X., Natanzon, A., Duquaine, D., Brook, R.D., Aguinaldo, J.G., Fayad, Z.A., Fuster, V., Lippmann, M., Chen, L.C., Rajagopalan, S., 2005. Long-term air pollution exposure and acceleration of atherosclerosis and vascular inflammation in an animal model. JAMA 294, 3003–3010.

Sund, J., Palomäki, J., Ahonen, N., Savolainen, K., Alenius, H., Puustinen, A., 2014. Phagocytosis of nano-sized titanium dioxide triggers changes in protein acetylation. J. Proteomics 108, 469–483.

Sundaresan, M., Yu, Z.X., Ferrans, V.J., Irani, K., Finkel, T., 1995. Requirements for generation of H_2O_2 for PDGF signal transduction. Science 270, 296–299.

Takagi, A., Hirose, A., Nishimura, T., Fukumori, N., Ogata, A., Ohashi, N., Kitajima, S., Kanno, J., February 2008. Induction of mesothelioma in p53+/− mouse by intraperitoneal application of multi-wall carbon nanotubes. J. Toxicol. Sci. ISSN: 1880-3989 33 (1), 105–116.

Takagi, A., Hirose, A., Futakuchi, M., Tsuda, H., Kanno, J., 2012. Dose-dependent mesothelioma induction by intraperitoneal administration of multi-wall carbon nanotubes in p53 heterozygous mice. Cancer Sci. 103, 1440–1444.

Takeda, K., Suzuki, K.-I., Ishihara, A., Kubo-Irie, M., Fujimoto, R., Tabata, M., Oshio, S., Nihei, Y., Ihara, T., Sugamata, M., 2009. Nanoparticles transferred from pregnant mice to their offspring can damage the genital and cranial nerve systems. J. Health Sci. 55, 95–102.

The *C. elegans* Sequencing Consortium, 1998. Platform for investigating biology, genome sequence of the nematode *C. elegans*: a platform for investigating biology. Science 282, 2012–2018.

Thess, A., Lee, R., Nikolaev, P., Dai, H., Petit, P., Robert, J., Xu, C., Lee, Y.H., Kim, S.G., Rinzler, A.G., Colbert, D.T., Scuseria, G.E., Tománek, D., Fischer, J.E., Smalley, R.E., 1996. Crystalline ropes of metallic carbon nanotubes. Science 273, 483–487.

Tian, F., Cui, D., Schwarz, H., Estrada, G.G., Kobayashi, H., 2006. Cytotoxicity of single-wall carbon nanotubes on human fibroblasts. Toxicol. In Vitro 20, 1202–1212.

Tin-Tin-Win-Shwe, Mitsushima, D., Yamamoto, S., Fukushima, A., Funabashi, T., Kobayashi, T., Fujimaki, H., 2007. Changes in neurotransmitter levels and proinflammatory cytokine mRNA expressions in the mice olfactory bulb following nanoparticle exposure. Toxicol. Appl. Pharmacol. 226, 192–198.

Tin-Tin-Win-Shwe, Yamamoto, S., Ahmed, S., Kakeyama, M., Kobayashi, T., Fujimaki, H., 2006. Brain cytokine and chemokine mRNA expression in mice induced by intranasal instillation with ultrafine carbon black. Toxicol. Lett. 163, 153–160.

Tran, P.A., Zhang, L., Webster, T.J., 2009. Carbon nanofibers and carbon nanotubes in regenerative medicine. Adv. Drug Deliv. Rev. 61, 1097–1114.

Tripathy, N., Hong, T.K., Ha, K.T., Jeong, H.S., Hahn, Y.B., 2014. Effect of ZnO nanoparticles aggregation on the toxicity in RAW 264.7 murine macrophage. J. Hazard Mater. 270, 110–117.

Tsai, Y.-Y., Huang, Y.-H., Chao, Y.-L., Hu, K.-Y., Chin, L.-T., Chou, S.-H., Hour, A.-L., Yao, Y.-D., Tu, C.-S., Liang, Y.-J., Tsai, C.-Y., Wu, H.Y., Tan, S.-W., Chen, H.-M., 2011. Identification of the nanogold particle-induced endoplasmic reticulum stress by omic techniques and systems biology analysis. ACS Nano 5, 9354–9369.

Tsuda, H., Xu, J., Sakai, Y., Futakuchi, M., Fukamachi, K., 2009. Toxicology of engineered nanomaterials - a review of carcinogenic potential. Asian Pac. J. Cancer Prev. 10, 975–980.

Tsukahara, T., Haniu, H., 2011. Cellular cytotoxic response induced by highly purified multi-wall carbon nanotube in human lung cells. Mol. Cell Biochem. 352, 57–63.

Tuomela, S., Autio, R., Buerki-Thurnherr, T., Arslan, O., Kunzmann, A., et al., 2013. Gene expression profiling of immune-competent human cells exposed to engineered zinc oxide or titanium dioxide nanoparticles. PloS One 8, e68415.

Tyurina, Y.Y., Kisin, E.R., Murray, A., Tyurin, V.A., Kapralova, V.I., Sparvero, L.J., Amoscato, A.A., Samhan-Arias, A.K., Swedin, L., Lahesmaa, R., Fadeel, B., Shvedova, A.A., Kagan, V.E., 2011. Global phospholipidomics analysis reveals selective pulmonary peroxidation profiles upon inhalation of single-walled carbon nanotubes. ACS Nano 5, 7342–7353.

Ulbrich, K., Hekmatara, T., Herbert, E., Kreuter, J., 2009. Transferrin- and transferrin-receptor-antibody-modified nanoparticles enable drug delivery across the blood–brain barrier (BBB). Eur. J. Pharm. Biopharma. 71, 251–256.

Unfried, K., Bierhals, K., Weissenberg, A., Abel, J., 2008. Carbon nanoparticle-induced lung epithelial cell proliferation is mediated by receptor-dependent Akt activation. Lung Cell Mol. Physiol. 294, L358–L367.

Uo, M., Akasaka, T., Watari, F., Sato, Y., Tohji, K., 2011. Toxicity evaluations of arious carbon nanomaterials. Dental Mater. J. 30, 245–263.

Uttara, B., Singh, A.V., Zamboni, P., Mahajan, R.T., 2009. Oxidative stress and neurodegenerative diseases: a review of upstream and down-stream antioxidant therapeutic options. Curr Neuropharmacol 7, 65–74.

Vakarelski, I.U., Brown, S.C., Higashitani, K., Moudgil, B.M., 2007. Penetration of living cell membranes with fortified carbon nanotube tips. Langmuir 23, 10893–10896.

Valko, M., Morris, H., Cronin, M.T.D., 2005. Metal, toxicity, and oxidative stress. Curr. Med. Chem. 12, 1161–1208.

Valko, M., Rhodes, C.J., Moncol, J., Izakovic, M., Mazur, M., 2006. Free radicals, metals and antioxidants in oxidative stress-induced cancer. Chem. Biol. Interact 160, 1–40.

Valko, M., Leibfritz, D., Moncol, J., Cronin, M.T., Mazur, M., Telser, J., 2007. Free radicals and antioxidants in normal physiological functions and human disease. Int. J. Biochem. Cell Biol. 39, 44–84.

Vallyathan, V., Shi, X., 1997. The role of oxygen free radicals in occupational and environmental lung diseases. Environ. Health Perspect. 105 (Suppl. 1), 165–177.

Vallyathan, V., Mega, J.F., Shi, X., Dalal, N.S., 1992. Enhanced generation of free radicals from phagocytes induced by mineral dusts. Am. J. Respir. Cell Mol. Biol. 6, 404–413.

de Vries, H.E., Kuiper, J., de Boer, A.G., Van Berkel, T.J.C., Breimer, D.D., 1997. The blood–brain barrier in neuroinflammatory diseases. Pharmacol. Rev. 49, 143–156.

Venter, J.C., Adams, M.D., Eugene, W., et al., 2001. The sequence of the human genome. Science 291, 1304–1351.

Vigolo, B., Pénicaud, A., Coulon, C., Sauder, C., Pailler, R., Journet, C., Bernier, P., Poulin, P., 2000. Macroscopic fibers and ribbons of oriented carbon nanotubes. Science 290, 1331–1335.

Vittorio, O., Raffa, V., Cuschieri, A., 2009. Influence of purity and surface oxidation on cytotoxicity of multiwalled carbon nanotubes with human. Neuroblastoma cells. Nanomedicine 5, 424–431.

Wagner, V., Dullaart, A., Bock, A.-K., Zweck, A., 2006. The emerging nanomedicine landscape. Nat. Biotech. 24, 1211–1217.

Wan, R., Mo, Y., Feng, L., Chien, S., Tollerud, D.J., Zhang, Q., 2012. DNA damage caused by metal nanoparticles: the involvement of oxidative stress and activation of ATM. Chem. Res. Toxicol. 25, 1402–1411.

Wang, R., Mikoryak, C., Li, S., Bushdiecker II, D., Musselman, I.H., Pantano, P., Drape, R.K., 2011. Cytotoxicity screening of single-walled carbon nanotubes: detection and removal of cytotoxic contaminants from carboxylated carbon nanotubes. Mol. Pharm. 8, 1351–1361.

Watts, P.C.P., Fearon, P.K., Hsu, W.K., Billingham, N.C., Kroto, H.W., Walton, D.R.M., 2003. Carbon nanotubes as polymer antioxidants. J. Mater. Chem. 13, 491–495.

Wick, P., Manser, P., Limbach, L.K., Dettlaff-Weglikowska, U., Krumeich, F., Roth, S., Stark, W.J., Bruinink, A., 2007. The degree and kind of agglomeration affect carbon nanotube toxicity. Toxicol. Lett. 168, 121–131.

Wolburg, H., Neuhaus, J., Kniesel, U., Krau, B., Schmid, E.M., Öcalan, M., Farrell, C., Risau, W., 1994. Modulation of tight junction structure in blood–brain barrier endothelial cells effects of tissue culture, second messengers and cocultured astrocytes. J. Cell Sci. 107, 1347–1357.

Wu, G., Morris Jr., S.M., 1998. Arginine metabolism: nitric oxide and beyond. Biochem. J. 336, 1.

Wu, W., Samet, J.M., Peden, D.B., Bromberg, P.A., 2010. Phosphorylation of p65 is required for zinc oxide nanoparticle–induced interleukin 8 expression in human bronchial epithelial cells. Environ. Health Perspect. 118, 982–987.

Xia, N.T., Mädler, L., Li, N., 2006a. Toxic potential of materials at the nanolevel. Sci. New Ser. 311, 622–627.

Xia, T., Kovochich, M., Brant, J., et al., 2006b. Comparison of the abilities of ambient and manufactured nanoparticles to induce cellular toxicity according to an oxidative stress paradigm. Nano Lett. 6, 1794–1807.

Xia, T., Kovochich, M., Liong, M., et al., 2008. Comparison of the mechanism of toxicity of zinc oxide and cerium oxide nanoparticles based on dissolution and oxidative stress properties. ACS Nano 2, 2121–2134.

Xiao, G., Gan, L.-S., 2013. Receptor-mediated endocytosis and brain delivery of therapeutic biologics. Int. J. Cell Biol. 2013, 703545.

Xu, H., Bai, J., Meng, J., Hao, W., Xu, H., Cao, J.M., 2009. Multi-walled carbon nanotubes suppress potassium channel activities in PC12 cells. Nanotechnology 2009 (20), 285102.

Xu, F.L., Li, Y.B., Deng, Y.P., Xiong, J., 2008. Porous nanohydroxyapatite/poly(vinylalcohol) composite hydrogel as artificial cornea fringe: characterization and evaluation in vitro. J. Biomater. Sci. Polymer E. 19, 431–439.

Xue, C., Wu, J., Lan, F., Liu, W., Yang, X., Zeng, F., Xu, H., 2010. Nano titanium dioxide induces the generation of ROS and potential damage in HaCaT cells under UVA irradiation. J. Nanosci. Nanotechnol. 10, 8500–8507.

Yamashita, K., Yoshioka, Y., Higashisaka, K., Mimura, K., Morishita, Y., Nozaki, M., Yoshida, T., Ogura, T., Nabeshi, H., Nagano, K., Abe, Y., Kamada, H., Monobe, Y., Imazawa, T., Aoshima, H., Shishido, K., Kawai, Y., Mayumi, T., Tsunoda, S., Itoh, N., Yoshikawa, T., Yanagihara, I., Saito, S., Tsutsumi, Y., 2010a. Silica and titanium dioxide nanoparticles cause pregnancy complications in mice. Nat. Nanotechnol. 6, 321–328.

Yamashita, K., Yoshioka, Y., Higashisaka, K., Morishita, Y., Yoshida, T., Fujimura, M., Kayamuro, H., Nabeshi, H., Yamashita, T., Nagano, K., Abe, Y., Kamada, H., Kawai, Y., Mayumi, T., Yoshikawa, T., Itoh, N., Tsunoda, S., Tsutsumi, Y., 2010b. Carbon nanotubes elicit DNA damage and inflammatory response relative to their size and shape. Inflammation 33, 276–280.

Yang, S.-T., Guo, W., Lin, Y., Deng, X.-Y., Wang, H.-F., Sun, H.-F., Liu, Y.-F., Wang, X., Wang, W., Chen, M., Huang, Y.-P., Sun, Y.-P., 2007. Biodistribution of pristine sin-gle-walled carbon nanotubes in vivo. J. Phys. Chem. C 111, 17761–17764.

Yang, S.-T., Jianbin Luo, J., Zhou, Q., Wang, Q., 2012. Pharmacokinetics, metabolism and toxicity of carbon nanotubes for biomedical purposes. Theranostics 2, 271–282.

Yang, S.-T., Wang, H., Meziani, M.J., Liu, Y., Wang, X., Sun, Y.-P., 2009. Bio-defunctionalization of functionalized single-walled carbon nanotubes in mice. Biomacromolecules 10, 2009–2012.

Yazdi, A.S., Drexler, S.K., Tschopp, J., 2010. The role of the inflammasome in nonmyeloid cells. J. Clin. Immunol. 30, 623–627.

Yazdi, A.S., Zhou, R., 2011. A role for mitochondria in NLRP3 inflammasome activation. Nature 469, 221–225.

Ye, Y.Y., Liu, J.W., Chen, M.C., Sun, L.J., Lan, M.B., 2010. In vitro toxicity of silica nanoparticles in myocardial cells. Environ. Toxicol. Pharmacol 29, 131–137.

Yoo, K.C., Yoon, C.H., Kwon, D., Hyun, K.H., Woo, S.J., Kim, R.K., et al., 2012. Titanium dioxide induces apoptotic cell death through reactive oxygen species-mediated fas upregulation and bax activation. Int. J. Nanomed. 7, 1203–1214.

Yuan, J., Gao, H., Sui, J., Chen, W.N., Ching, C.B., 2011. Cytotoxicity of single-walled carbon nanotubes on human hepatoma HepG2 cells: an iTRAQ-coupled 2D LC-MS/MS proteome analysis. Toxicol. In Vitro 25 (8), 1820–1827.

Zhang, H., Burnum, K.E., Luna, M.L., Petritis, B.O., Kim, J.S., Qian, W.J., Moore, R.J., Heredia-Langner, A., Webb-Robertson, B.J., Thrall, B.D., Camp 2nd, D.G., Smith, R.D., Pounds, J.G., Liu, T., 2014. Quantitative proteomics analysis of adsorbed plasma proteins classifies nanoparticles with different surface properties and size. Ecotoxicol. Environ. Saf. 100, 122–130.

Zhang, J., Song, W., Guo, J., Zhang, J., Sun, Z., Li, L., Ding, F., Gao, M., 2013. Cytotoxicity of different sized TiO_2 nanoparticles in mouse macrophages. Toxicol. Ind. Health 29, 523–533.

Zhang, T., Stilwell, J.L., Daniele, G., Lianghao, D., Omeed, E., Cooke, P.A., Gray, J.W., Alivisatos, A.P., Chen, F.F., 2006. Cellular effect of high doses of silica-coated quantum dot profiled with high throughput gene expression analysis and high content cellomics measurements. Nano Lett. 6, 800–808.

Zhang, Y., Xu, Y., Li, Z., Chen, T., Lantz, S.M., Howard, P.C., Paule, M.G., Slikker Jr., W., Watanabe, F., Mustafa, T., Biris, A.S., Ali, S.F., 2011. Mechanistic toxicity evaluation of uncoated and PEGylated single-walled carbon nanotubes in neuronal PC12 cells. ACS Nano 5, 7020–7033.

Zhao, X., Liu, R., 2012. Recent progress and perspectives on the toxicity of carbon nanotubes at organism, organ, cell, and biomacromolecule levels. Environ. Int. 40, 244–255.

van der Zande, M., Junker, R., Walboomers, X.F., Jansen, J.A., 2011. Carbon nanotubes in animal models: a systematic review on toxic potential. Tissue Eng. Part B 17, 57–69.

Zhu, X., Zhu, L., Duan, Z., Qi, R., Li, Y., Lang, Y., 2011. Comparative toxicity of several metal oxide nanoparticle aqueous suspensions to Zebrafish (Danio rerio) early developmental stage. J. Environ. Sci. Health Part A Toxcol. Hazard Subst. Environ. Eng. 43, 278–284.

Zhou, R., Yazdi, A.S., Menu, P., Tschopp, J., 2011. A role for mitochondria in NLRP3 inflammasome activation. Nature 469, 221–225.

Zuo, G., Kang, S.-G., Xiu, P., Zhao, Y., Zhou, R., 2013. Interactions between proteins and carbon-based nanoparticles: exploring the origin of nanotoxicity at the molecular level. Small 9, 1546–1556.

CHAPTER

8

Nanoparticle Ecotoxicology

OUTLINE

Engineered Nanoparticles
http://dx.doi.org/10.1016/B978-0-12-801406-6.00008-X

1. INTRODUCTION TO THE EARTH'S ENVIRONMENT

The environment consists of everything that surrounds us, ranging from the objects of nature such as soil, water, air, plants, and animals (Figure 1) to the objects of human activity such as physical structures, institutions, and industries (Figure 2). The key components of the environment are the atmosphere (air), hydrosphere (water), lithosphere (soil), and biosphere. All living forms interface the three abiotic components. Human activities such as construction of dams to control water flow, high-rise buildings, industries, corporate agriculture, excessive resource utilization, and pollution modulate the natural environment (Figure 2).

FIGURE 2 Effects of human activity. The natural environment is modified via urban and rural developments, such as building of dams to change a river's course.

1.1 Atmosphere

The atmosphere consists of five major layers. The *troposphere* contains about 80% of Earth's air and controls the weather. It extends to a height of about 11 miles at the equator. The *stratosphere* extends to a height of about 30 miles and includes the ozone layer, which blocks much of the sun's harmful ultraviolet (UV) rays. The *mesosphere* has a temperature of nearly −180°F at the top. Meteors generally burn up in the mesosphere. The *ionosphere* extends to about 430 miles and is home to the International Space Station. The *ionosphere* contains ions that form aurora. The outermost layer is the *exosphere,* extending outward in space. Currently, the atmosphere is being greatly affected by human activities that are increasing the amount of greenhouse gases, aerosols, and water vapors in the atmosphere. The largest contribution comes from the burning of fossil fuels, which releases carbon dioxide, methane, and nitrogen gases to the atmosphere. These greenhouse gases alter the incoming solar radiation and outgoing infrared (thermal) radiation, thus altering Earth's energy balance. A change in the atmospheric abundance or properties of these gases can lead to a warming or cooling of the climate system.

FIGURE 1 Natural ecosystem.

1.2 Lithosphere (Soil)

The *lithosphere* is the outermost solid layer of the earth. The surface is covered with nutrient-rich soil and a variety of living and dead life forms. Soil is composed of both organic and inorganic matter, including minerals (calcium, phosphorus, and potassium). The proportions of sand

(50 μm–2 mm), silt (2–50 μm), and clay (less than 2 μm) determine the soil's texture. Organic compounds include humus, which is decayed vegetation broken down by microorganisms. Humus binds the soil and improves its texture. The color of the soil is an indication of the amount of organic material it contains, with darker soils having greater organic content. Soil also contains micro- and macro-spaces and water channels (Figure 3(A)). Different soil samples may exhibit different pH values, which are measures of their acidity (or basicity). An acidic soil may lack calcium and potassium ions due to leaching. Brown soil with high organic content has a pH value of between 5 and 7—ideal values that are known to be very fertile, supporting a

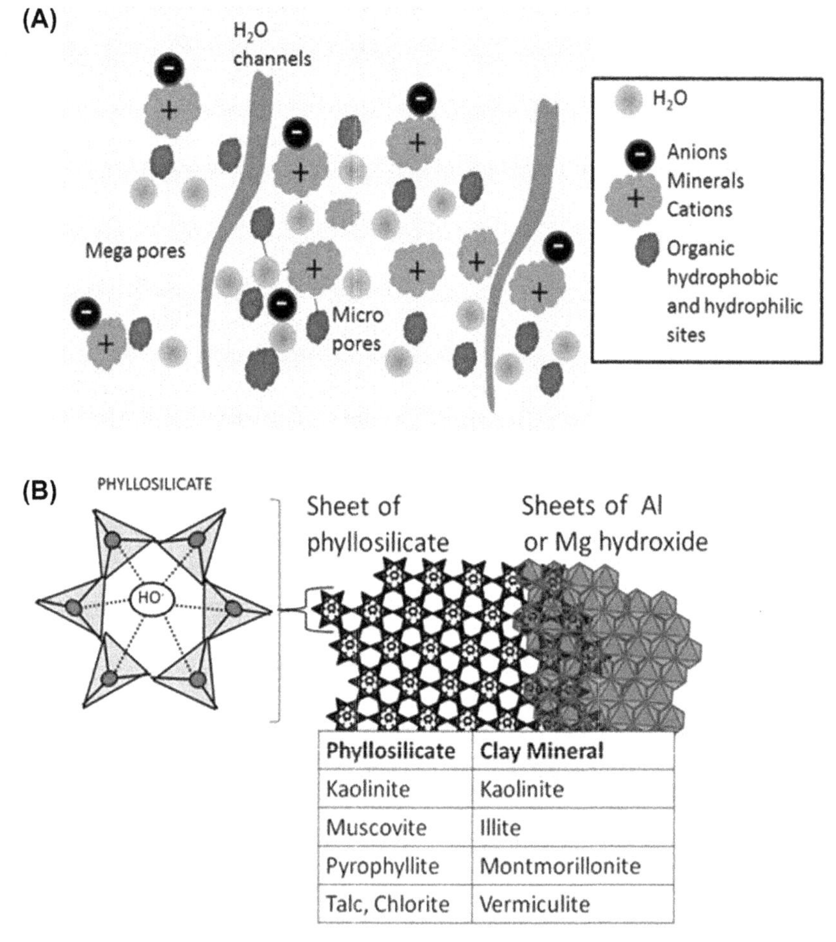

Phyllosilicate	Clay Mineral
Kaolinite	Kaolinite
Muscovite	Illite
Pyrophyllite	Montmorillonite
Talc, Chlorite	Vermiculite

FIGURE 3 (A) Soil structure. Soil consists of sand, clay, silt, minerals, ions, water, chemicals, and organic matter in different proportions. Soil also contains water channels and micropores for air. (B) Structure of clay minerals. In clay, the atoms are arranged in well-defined sheets and layers with the anions (O, OH, and occasionally F) in a phyllosilicate structuring. The anions are grouped tetrahedrally and oetahedrally around the cations. Si and Al are in tetrahedral positions and Al, Mg, and Fe in octahedral positions with interlayer cations such as K, Na, Ca, and Mg. Water layers exist between the silicate layers in such minerals as montmorillonite and vermiculite.

wide variety of plant life. Soil microorganisms such as bacteria, fungi, and protozoa are all vital to the reduction of organic and mineral wastes into plant nutrients.

Ideally, soil contains 45% minerals, 25% water, 25% air, and 5% organic matter, but the percentages of these components vary tremendously. In some cases, such as wetland soil, organic matter may be greater than 50% of the solid portion of the soil. In general, soil contains hydrophilic, hydrophobic, and organic sites that interact with different groups (Forouzangohar and Kookana, 2010). The mineral portion of soil is divided into three particle-size classes: sand (2–0.05 mm), silt (0.05–0.002 mm), and clay (<0.002 mm). Larger soil particles are called rock fragments. Sand (which does not retain water) and silt (retains water) are largely inert. Clay particles (kaolinite, illite, montmorillonite, and vermiculite) belong to a class of minerals called phyllosilicates and have sheet-like structures (Figure 3(B)). Clay particles are more reactive and undergo chemical, addition, and exchange reactions.

1.3 Hydrosphere

The hydrosphere include all the solid (ice), liquid, and gaseous (vapor, cloud) water of the planet, extending from the Earth's surface downward several kilometers into the lithosphere and upward about 12 km into the atmosphere. Only a small portion (about 3%) of the water in the hydrosphere is fresh (less than 0.25% salt), which flows as liquid or frozen. Ninety-seven percentage of Earth's water is salty (more than 35% salt). Water is unique because it exist in all the three states of matter—solid, liquid, and gas—within the temperature and pressure range of Earth. The uniqueness of water comes from its molecular structure: it is a polar covalent molecule having a slight positive and slight negative charge on opposite ends (Figure 4). Because water is a bent and partially polar molecule, it possesses the important characteristics, as described in the following sections.

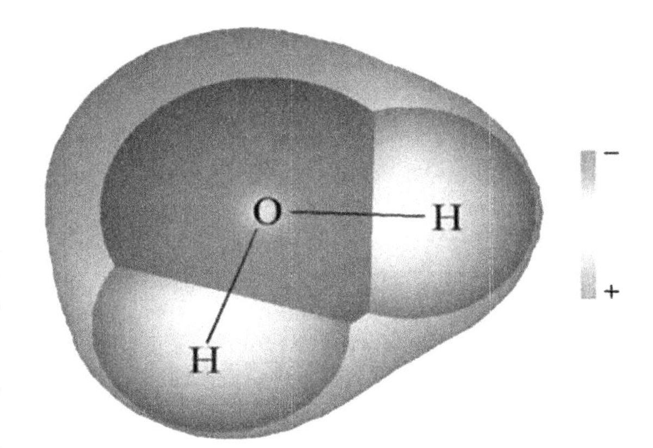

FIGURE 4 Structure of water.

1.3.1 *Polarity*

Water molecules have both a positively and negatively charged ends (Figure 4) that are responsible for effectively dissolving other polar molecules, such as sugars and ionic compounds. Water is considered to be a *universal solvent* because it can dissolve many common substances. Nonionized chemicals are not dissolved by water but dissolve in nonpolar solvents.

1.3.2 *Hydrogen Bonding*

The negatively charged oxygen forms a weak bond with the positively charged hydrogen ions of a neighboring water molecule via the *hydrogen bond*. Hydrogen bonds give water molecules two additional characteristics: cohesion and surface tension.

1.3.3 *Cohesion*

Because of the extensive hydrogen bonding in water, the molecules tend to stick together in a regular pattern. This phenomenon is called *cohesion*. The cohesive property of water allows tall trees to bring water to their highest leaves from sources below ground.

1.3.4 *Surface Tension*

The cohesive forces between molecules in a liquid are shared with all neighboring molecules.

The surface molecules have no neighboring molecules above; thus, they exhibit stronger attractive forces upon their nearest neighbors on and below the surface. Therefore, the surface of a liquid resists an external force.

1.3.5 *Conductivity*

Conductivity is a measure of the ability of water to pass an electrical current. Pure water is a nonconductor of electricity. However, water becomes conductive by the presence of inorganic dissolved salts resulting in the accumulation of positive and negative ions. Organic compounds such as oil, phenol, alcohol, and sugar do not conduct electrical currents very well in water.

1.3.6 *Water pH and Solubility*

pH ($-\log [H^+]$) is a measure of how acidic or basic water is. The pH range is 0 (highly acidic) to 14 (highly basic), with 7 being neutral. Water that has more free hydrogen ions is acidic, whereas water that has more free hydroxyl ions is basic. The pH of water determines the solubility and biological availability of chemicals. In the case of heavy metals, the degree to which they are soluble determines their toxicity. Metals tend to be more toxic at lower pH because they are more soluble. Figure 5 shows pH ranges from 0 to 14 and some of the possible effects. Eqns (1)–(3) define pH.

Ionization of acetic acid:

$$(CH_3COOH \rightleftharpoons CH_3COO^- + H^+).$$

$$K_a = \frac{[H^+][CH_3COO^-]}{[CH_3COOH]} \quad (1)$$

$$\frac{1}{[H]^+} = \frac{1}{K_a}\frac{[CH_3COO^-]}{[CH_3COOH]} \quad (2)$$

$$pH = pK_a + \log\frac{[CH_3COO^-]}{[CH_3COOH]} \quad (3)$$

The solubility of an acid or a base in water depends upon its K_a value, as shown in Eqns (1)–(3). An acid is soluble in water if water pH is higher than the acid's pKa (when the acid is ionized), while a base is soluble in water if water pH is lower than the base's pK (when the base is protonated). Hydrophobic chemicals poorly dissolve in water. In aquatic systems, colloid is the generic term applied to particles in the 1-nm to 1-μm size range.

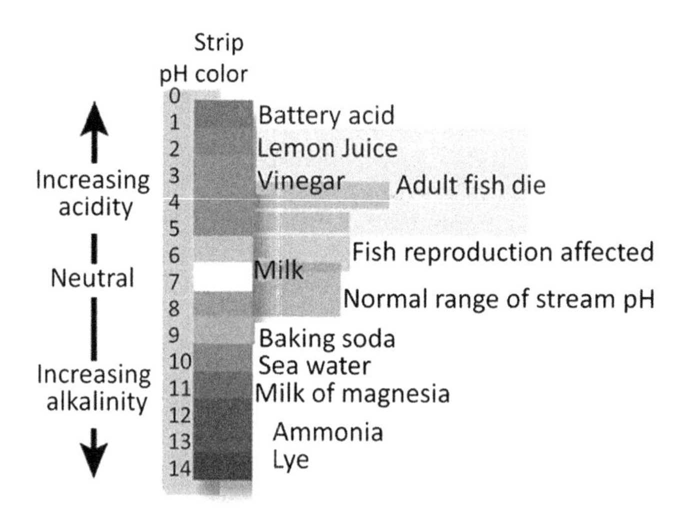

FIGURE 5 Environmental pH range.

1.4 Biosphere

The biosphere includes all living communities on Earth, including animals (the Animalia), plants, bacteria, fungi, and the photosynthetic microbes. The distribution of life on Earth reflects variations in the world's environments, principally in temperature and the availability of water. Organisms require proper temperature, moisture, nutrients, and habitat to live in an area. All the mineral nutrients necessary for life are contained within Earth's sphere, although regional differences are apparent. Nutrients contained in dead organisms or waste products of living cells are recycled so that other organisms can reuse them. This recycling of nutrients is necessary because there is no source of food outside the biosphere.

The regional variations are apparent in the satellite image in Figure 6, which shows the distribution of plant life on Earth. On land, the dark green colors show abundant vegetation and tan colors show relatively sparse plant cover. In the oceans, red, yellow, and green pixels show dense phytoplankton blooms, while blues and purples show where there is very little phytoplankton. It is difficult to image animal life on Earth from space, but it can be deduced from the distribution of photosynthetic plants.

1.5 Anthropogenic Spheres

Human activity has resulted in an artificial sphere including buildings, urban areas, dams, and waste. Gaseous waste directly enters the atmosphere. Solid waste is either dumped in a landfill or burned, while the liquid waste is processed through a water waste treatment facility.

1.6 Interaction Among the Four Spheres

Although the four spheres have distinct boundaries and characteristics, they exhibit high level of interaction (Figure 7) that regulates the climate and rain, recycle nutrients, and provide oxygen. The key interactions are described below.

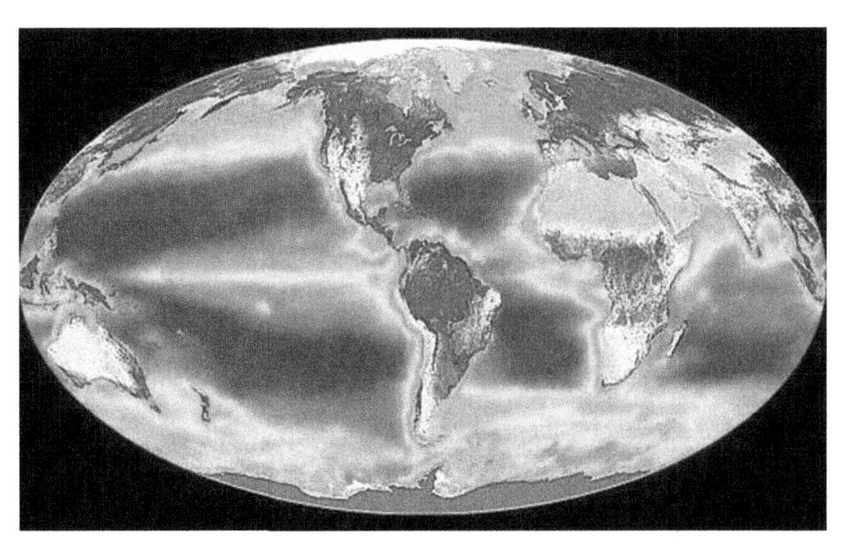

FIGURE 6 Earth from space. Green (gray in print versions) areas represent plants in the lithosphere and the dark blue (dark gray in print versions) areas represent algae in oceans. The yellow (light gray in print versions) areas are deserts. *From National Aeronautics and Space Administration files.*

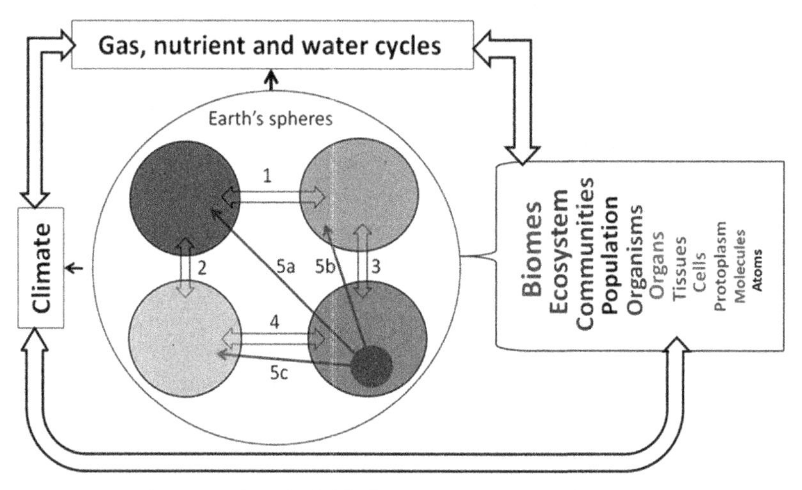

FIGURE 7 Interaction among the atmosphere (red (black in print versions) sphere), hydrosphere (blue (gray in print versions) sphere), lithosphere (yellow (light gray in print versions) sphere), biosphere (green (dark gray in print versions) sphere) including human activity (an inset in biosphere). Here, 1, 2, and 4 represent interactions among atmosphere with hydrosphere and biosphere, respectively; 3 and 4 represent interactions among biosphere, hydrosphere, and lithosphere; 5 a, b, and c represent the effects of human activity on other spheres.

1.6.1 *The Climate*

The climate is a composite of a region's temperature, air pressure, humidity, precipitation, sunshine, cloudiness, and winds throughout the year, averaged over a series of years. The climate of a region depends on many factors, including its height above sea level, the amount of sunlight it receives, the shape of the land, and how close it is to oceans. Because the equator receives more sunlight than the poles, the climate varies depending on distance from the equator. Overall, global climate also depends on the amount of energy received by the sun and the amount of energy that is trapped in the system. Human activity has manipulated and increased the energy trapped in the system, resulting in gradual warming of the environment. Overall, the following cycles regenerate a region's climate for its inhabitants.

1.6.2 *The Oxygen Cycle*

Most oxygen is stored in the oxide minerals of Earth's crust and mantle in the lithosphere. Because the minerals are bound to the rocks, oxygen is not available for animals and humans. Most oxygen comes from (1) photosynthesis by plants on land that release oxygen in the atmosphere and (2) phytoplankton on the ocean's surface that release oxygen in water. Animals and plants both use oxygen for energy production.

1.6.3 *Carbon, Mineral, and Nutrient Cycles*

CO_2 is released into the atmosphere by the terrestrial animals through energy utilization in which glucose is burned in the presence of O_2 and CO_2 is released. Aquatic animals release CO_2 in water, which reacts with water and forms carbonic acid. Rain dissolves the atmospheric CO_2 and the ensuing H_2CO_3 enters the hydrosphere.

$$C_6H_{12}O_6 + 6H_2O \rightarrow 6CO_2 \text{ (atmosphere)} + 6H_2O$$

$$CO_2 + H_2O \rightarrow H_2CO_3\ (H^+ + HCO_3^-) \text{ (water)}$$

$$CaSiO_3 + CO_2 \rightarrow CaCO_3 + SiO_2 \text{ (atmosphere and water)}$$

Plants and phytoplankton use CO_2 to make glucose. CO_2 is also removed when carbonic acid reacts with $CaSiO_3$ in rocks and forms $CaCO_3$ and SiO_3. The entire cycle removes CO_2 from the atmosphere.

The mineral cycles regulate the flow, distribution, and migration of mineral nutrients across Earth's surface. The nutrient cycle is the movement and exchange of organic and inorganic matter back into the production of living matter via the food web pathways that decompose matter into mineral nutrients. Nutrient cycles occur within ecosystems. Ecosystems are interconnected systems where matter and energy flows and is exchanged as organisms feed, digest, and migrate. Similar to carbon, biologically important elements such as oxygen, nitrogen, hydrogen, potassium, calcium, phosphorus, sulfur, and the trace elements cycle between living organisms, the soil, and, sometimes, the atmosphere.

1.6.4 Influence of Humans on the Environment

Although natural processes such as changes in the sun's energy, shifts in ocean currents, and other factors affect Earth's climate, they do not explain the warming that we have observed over the last half-century. Most of the warming of the past half century has been caused by human emissions of greenhouse gases coming from a variety of human activities, including but not limited to burning fossil fuels for heat and energy, clearing forests, fertilizing crops, storing waste in landfills, raising livestock, and producing some kinds of industrial products. Other activities such as agriculture or road construction can change the reflectivity of Earth's surface, leading to local warming or cooling. This effect is observed in urban centers, which are often warmer than surrounding, less populated areas. Emissions of small particles, known as aerosols, into the air can also lead to reflection or absorption of the sun's energy. This is relevant to the nanoparticles as described below.

2. SOURCES OF NANOPARTICLES IN THE ENVIRONMENT

2.1 Natural Sources of Nanoparticles

In nature, nanoparticles are produced in many processes, such as photochemical reactions, volcanic eruptions, forest fires (which can also be classified anthropogenic if the fire is intentional), and simple erosion. Natural events such as dust storms, volcanic eruptions, and forest fires can produce vast quantities of nanoparticles that may profoundly affect air quality worldwide. The aerosols generated by human activities are estimated to be only about 10% of the total; the remaining 90% have a natural origin (Taylor, 2002). Small particles suspended in the atmosphere (known as aerosols) absorb radiation from the sun and scatter it back to space, thus affecting the entire planet's energy balance (Houghton, 2005). Although the health effects of natural nanoparticles are not extensively studied, some studies have associated natural airborne particles with asthma, emphysema, and lung damage. Acute and chronic exposures to lunar dust irritated lungs and caused pneumoconiosis with fibrosis, respectively, in astronauts (Watson, 2005).

In atmosphere, cloud condensation (CC) is a natural mechanism for synthesis of nanoparticles using CC nuclei (CCN), a monomer particle that act as the initial site for condensation of water vapor into cloud droplets. Many of the atmospheric gases, such as nitrogen and sulfur gases or ozone, upon absorbing UV radiation, undergo oxidation and the oxidized products participate in the condensation reaction, resulting in the formation of nanosized aerosols on which vapor condensation takes place (Figure 8) (Westervelt et al., 2013; Kuang et al., 2009; Pierce and Adams, 2007; Kerminen et al., 2005).

In general, all cloud droplets or ice particles originate around some sort of CCN, which tend to attract water. Kavouras et al. (1998) have shown that forests, through their production of

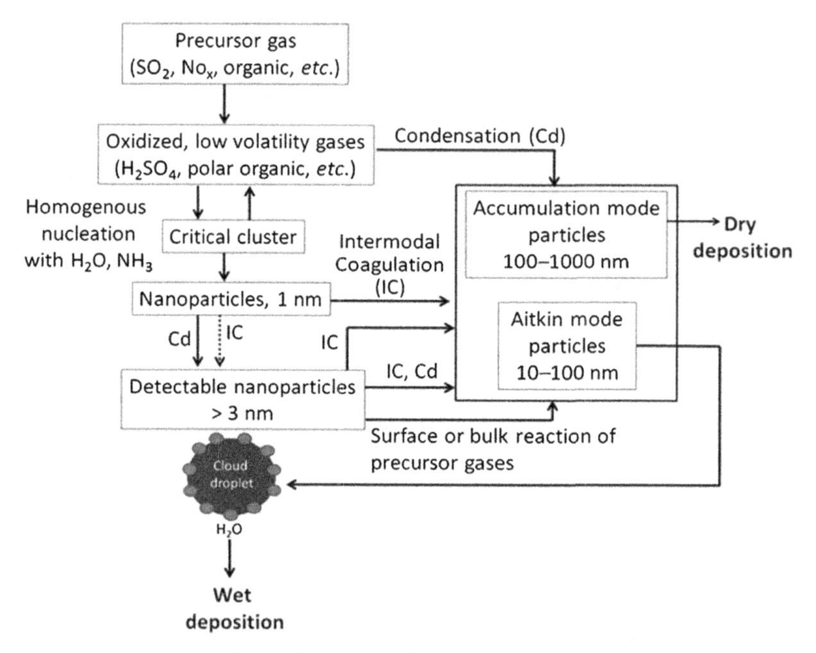

FIGURE 8 Diagram showing nanoparticle synthesis in clouds. Precursor particles facilitate the formation of cloud droplets that may include many toxins and even nanoparticles.

large quantities of organic aerosols, could be of considerable significance both for CCN formation and for heterogeneous atmospheric chemical processes. The size of the nuclei that help form clouds range from 1 to 3 nm in diameter. Condensation nuclei also come from the soil, the ocean (sea salt), and the burning of vegetation and fossil fuels, as well as nanoparticles.

Organic compounds constitute 20–90% of the submicron aerosol mass that contributes to the formation of new particles (Riipinen et al., 2012). Organic particles form in the atmosphere through the oxidation of volatile organic compounds (VOCs) by O_3, OH, and NO_3, followed by the conversion of the gaseous oxidation products to particles. The VOCs emitted by plants are probably the dominant source of secondary organic aerosol (SOA), although anthropogenic sources have also been reported (Hoyle et al., 2011; Jimenez et al., 2009, Guenther et al., 1995). An understanding of the role of SOA in nanoparticle formation is difficult to understand because SOA comprises thousands of compounds, including sulfuric acid, water, and basic compounds such as ammonia and amines (Sipila et al., 2010; Kuang et al., 2010, 2008; Kurten et al., 2008; Kirkby et al., 2011; Zhang et al., 2012b).

Soils contain inorganic and organic nanoparticles based on clay minerals, metal hydroxides, and humic substances. These nanoparticles are mostly associated with their inorganic counterparts, form coatings over mineral surfaces, and form electrostatic bonds with ions such as NH_4^+, Ca^{2+}, K^+, and Mg^{2+}, which contribute substantially to soil fertility by preventing the loss of these nutrients in groundwater. The naturally occurring nanoparticles in soil have important local, regional, and even global consequences. For example, airborne nanoparticles from the deserts of Africa are linked to the productivity of the oceans and lakes, which helps to sustain the nutrient demands of terrestrial ecosystems worldwide (Chadwick et al., 1999; Jickells et al., 2005; Prospero, 1999; Simonson, 1995). Nanoparticles are also synthesized by many organisms such as bacteria in sediments, which synthesize electrically conductive pili, called nanowires, for sensing neighbors or for transferring electrons and energy (Blango and Mulvey, 2009; Gorby et al., 2006). Dissimilarly, metal-reducing bacteria even respire on iron oxide nanoparticles in anaerobic environments (Bose et al., 2009).

The sources of natural nanoparticles in oceans and lakes are shown in Figure 9 and described in the following sections.

2.1.1 Release of Precursors from Hydrothermal Vents at the Ocean's Floor

Aqueous clusters of metal sulfides (FeS, ZnS, and CuS) constitute a major fraction of dissolved metal load in anoxic oceanic, sediment, freshwater, and deep ocean vent environments (Hartland et al., 2013; Yucel et al., 2011; Luther and Rickard, 2005). In water, Fe^{3+} exists as water-soluble $Fe(OH)_2^+$. When vent fluid (consisting of water-soluble H_2S, Fe^{2+}, FeS clusters; Zn^{2+}, Cu^{2+}, etc.) is mixed with cold, anoxic seawater, $Fe(OH)_2^+$ is reduced in the presence of H_2S into $Fe(OH)_3$,which precipitates with other polymetallic sulfides (the gray cloud in Figure 9). In

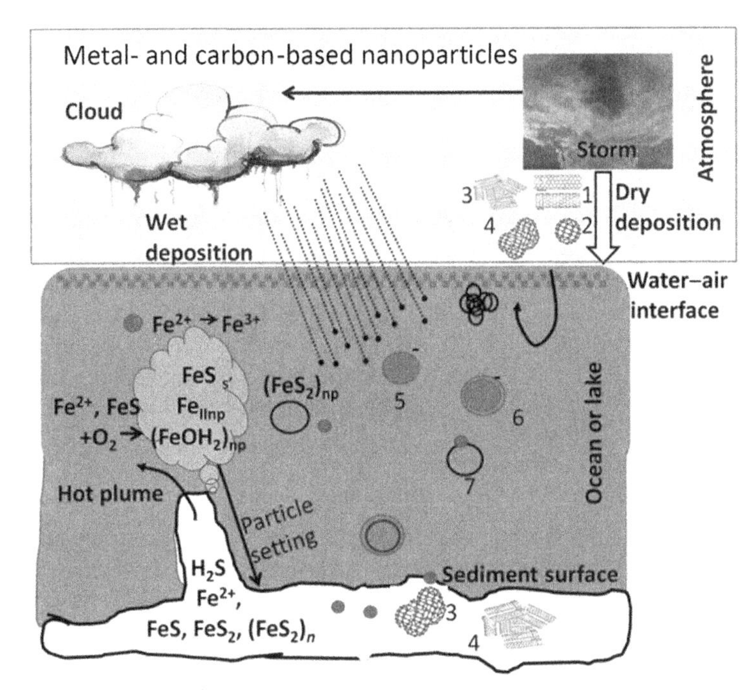

FIGURE 9 Natural sources of nanoparticles. Oceanic hot plumes (bottom) and storms, especially the dust/sand storms, are major natural sources of nanoparticles. The oceanic plumes contain high levels of sulfur and metal compounds that, once released in water, either precipitate ($FeOH_2$ and FeS) and settle down onto the ocean floor, or undergo synthesis of water-soluble nanoFeS_2, one of the most abundant nanoparticles on Earth. The dust/sand storms contain hydrophobic carbon-based (1, 2, and 4) or zero-valent metal, (3) nanoparticles that may get incorporated into water droplets. The hydrophobic nanoparticles from storm may get deposited onto the ocean surface and either aggregate and settle down at the floor or get incorporated into the surface water–air interface. Cloud nanoparticles undergo wet deposition via rain (5) and for adsorption by water particles (6). Nanoparticles may also trap different ions (7). (np: nanoparticles)

the presence of oxygen, the vent-derived Fe^{2+} is oxidized into organic Fe^{3+} complexes (in Figure 9, *L* stands for organic ligands) and the nanoparticle FeS_2 (pyrite or $(FeS_2)_{np}$ less than 200 nm in size. The pyrite nanoparticle, $(FeS_2)_{np}$, may account for about 10% of total discharge.

Because these nanoparticles exhibit relatively low settling (1.43 m/year compared to 143 m/year for 2-μm particles), they can travel long distances in water. Luther and Rickard (2005) and Luther et al. (2003, 1999) have shown neutral (Zn:S 1:1 such as ZnS or Zn_3S_3) as well as anionic (Zn:S 4:6 such as $Zn_4S_6{}^{4-}$) clusters present in ocean water as $Zn^3S^3(H_2O)_6$ and $[Zn_4S_6(H_2O)_4]^{4-}$, respectively. The densities of ZnS_{np} and ZnS_s are 0.057 ZnS_{np}/nm^3 and 1.024 ZnS_s/nm^3, respectively. ZnS_{np} also undergoes the following metal replacement reaction:

$$Cu^{2+} + ZnS_{aq} = Zn^{2+} + CuS_{aq}$$

$$Ag^{+} + ZnS_{aq} = Zn^{2+} + AgS^{-}_{aq}$$

Bianchi and Fortunati (1990) showed that Ag^+ was not toxic to *Daphnia magna* when water contained ZnS_{aq} but was toxic to the zooplankton when ZnS_{aq} was absent. This is because of a metal replacement reaction that releases nontoxic Zn^{2+}.

2.1.2 Dry Deposit from the Atmosphere During Stormy Periods, such as Hurricanes and Sand Storms in the Sahara

Nanoparticles present in sand storms of deserts have been characterized by Marconi et al. (2013), Wang et al. (2012a,b), and Goudie and Middleton (2001). In most cases, PM_{10} content accounted for more than 75% of the oxide content identified for several years. The PM_{10} content in Southern Italy's atmosphere ranges from 5.42 to 67.9 μg/m^3 when air currents are strong.

2.1.3 Nanoparticle Condensation in Clouds and Wet Deposition via Rain

Nanoparticles present in air, possibly due to storms, may serve as a nucleus for cloud or fog formation. This involves a change in gaseous water vapor to the condensed water phase (Figure 7). Wettability or hygroscopic agglomeration of suspended dry nanoparticles with already-formed water droplets facilitates condensation.

As nanotechnology advances, the likelihood of release of engineered nanoparticles into the environment also increases. Because the aqueous environment contains a huge excess of natural nanoparticles, the identification and quantification of engineered nanoparticles pose a big challenge.

2.2 Anthropogenic Sources of Nanoparticles

Anthropogenic means human made or arising, intentionally or unintentionally, from human activities, such as the commercial synthesis of nanoparticles, manufacturing, fossil fuel burning, etc.

The unintentional release of nanoparticles occurs from furnaces burning hydrocarbon gases, wood, or even wax candles (Kumar et al., 2013; Manoj et al., 2012; Dikio and Bixa, 2011; Shooto and Dikio, 2011). The pyrolysis of acetylene, methane, ethanol, benzene, and carbonization of synthetic polymers, such as polyvinyl alcohol, releases carbon nanotubes (CNTs) into the atmosphere (Shao et al., 2010, Dikio, 2011). The intentional release occurs primarily through emissions to wastewater treatment facilities, landfills, and soils. Some portion of these emissions may make their way into aquatic ecosystems, sewage, landfill, and drinking water (Figure 10).

Nanoparticles released from fossil-fuel burning are a mixture of many heterogeneous particles of different shapes and sizes. However, the nanoparticles used in medicine, drug delivery, and imaging are highly homogenous and engineered to perform specific functions. The most likely pathways of their release are accidental release into the environment, including the nearby natural

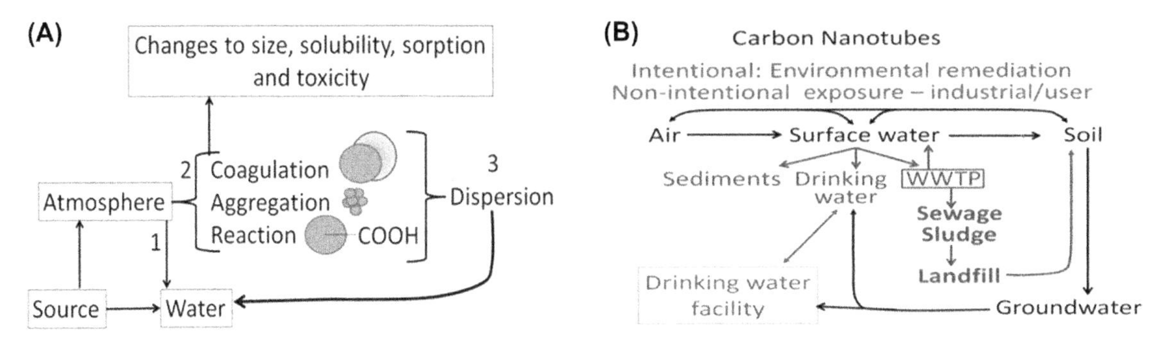

FIGURE 10 Fate of CNTs in the atmosphere (A) and surface water (B). CNTs are released into the atmosphere via combustion of fossil fuels, natural gases, coal, and wood (Murr et al., 2004, 2009). In the atmosphere, the nanotubes undergo chemical and physical changes, especially coagulation and aggregation, that increase their particle size. Dispersed or aggregated CNTs, depending upon the length and surface properties, distribute among air, surface/groundwater and soil. CNT may also reach the wastewater treatment plants (WWTP) and landfills. In fact, CNTs are currently being used in water treatment plants. If CNTs reach surface water, they will cause toxicity in aquatic animals and in animals drinking contaminated water. CNTs can also migrate to soil and taken up by plants.

water, during their production and application, through a factory's waste stream, and through other human activities (Muller and Nowack, 2008; Gottschalk et al., 2010). A brief description of various release pathways are described in the following sections.

2.2.1 *Accidental Release during Synthesis or Processing*

Although there is no reported case of nanoparticle escape from manufacturing factories, an exposure risk exists for the vast majority of nanoparticle production processes, including synthesis, recovery, and handling of synthesized materials. During synthesis, a leak or frequent starting and stopping of the process may lead to exposure (Biswas and Wu, 2005). Table 1 summarizes the main exposure risks during nanoparticle synthesis (Aitken et al., 2004). The means of prevention implemented must therefore take these variables into account.

In gas phase processes, if a reactor is operating under positive pressure, a gas leak may result in a nanoparticle leak (Aitken et al., 2004). If chemical processes involve colloid formation in solution, then inhalation exposure would occur mainly during agitation (Maynard and Kuempel, 2005). When nanoparticles are produced from bulk particles in dry conditions, significant quantities of inhalable nanoparticles will be generated and released during handling. Regardless of the synthetic approach, there is always a risk of exposure via the cutaneous route or ingestion when a worker touches contaminated surfaces following spills or atmospheric emissions. Airborne nanoparticles are difficult to contain and can rapidly spread to other ecosystems; in addition, contaminants cause many pulmonary conditions because their unique size makes it easy for them to pass deep into lung tissue.

2.2.2 *Human Activity*

Since the mid-twentieth century, there has been an exponential increase in the use of fossil fuel for energy, establishment of large-scale mining, and unpresented increase in urban growth. Many of the human-related activities have been shown to release nanoparticles into the environment.

Fossil fuel combustion (including gasoline, natural gas, or coal burning) typically produces nanoparticles such as CNTs and fullerenes (Mills et al., 2011; Hesterberg et al., 2010; Donaldson

TABLE 1 Risk of Exposure to Nanoparticles During Their Synthesis

Process	Exposure Risk
Gas phase condensation	Reactor leak, product recovery, post-production treatment, bagging, unbagging, shipping, maintenance/cleaning of production and ventilation sites, and equipment, accidental spill
Vapor deposition	Product recovery, post-production treatment, bagging, unbagging, shipping maintenance/cleaning of production and equipment, acciental spill.
Colloid formation and chemical precipitation	Product drying, spill/drying, agitation/transfilling, maintenance/cleaning of production sites and equipment
Mechanical attrition process	Product recovery, post-production treatment, product recovery, bagging, unbagging, shipping, maintenance/cleaning of production and ventilation sites and equipment, accidental spill

et al., 2005). Diesel and gasoline exhaust are important sources of atmospheric nanoparticles in heavily traveled areas (Kittelson, 1998, Burtscher, 2005). Carbon nanotubes and fibers are also found in engine exhaust as a byproduct of diesel combustion (Lagally et al., 2012; Kasper et al., 1999). The aspect ratio of these fibers is comparable to those of lung-retained asbestos, suggesting that strong carcinogens may exist in exhaust (Lagally et al. (2011).

Gomez et al. (2013) have shown intense generation of CNTs from a low-power soldering unit.

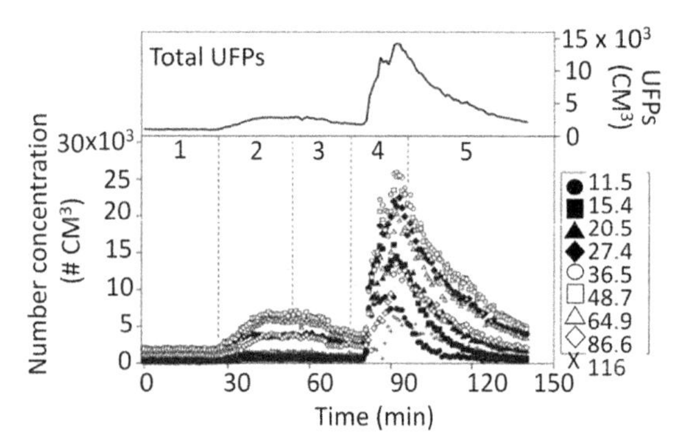

FIGURE 11 Three-dimensional (3D) printers are a significant source of nanoparticles, especially after prolonged use. The figure shows size-resolved and total (<100 nm) ultrafine particle (UP) concentrations measured in the office space. *Reprinted from Stephens et al. (2013) with permission.*

Many laser-printer inks have been shown to release CNTs in indoor air (He et al., 2007).

Stephens et al. (2013) have demonstrated ultrafine particle (UFP) emissions from commercially available desktop three-dimensional (3D) printers at extremely high rates, similar to those measured in previous studies of several other devices and indoor activities, including cooking on a gas or an electric stove, burning scented candles, operating laser printers, or even burning a cigarette. Although Stephens et al. (2013) for the first time showed nanoparticle release from 3D printers (Figure 11), previous studies have shown that moderately high temperature (e.g., 170–240 °C nozzle temperatures) thermal processing of thermoplastics emitted gaseous and particulates including carbon monoxides, CNTs, and other nanoparticles (Rutkowski and Levin, 1986).

Mining and metal refinery operations have been shown to generate metal and metal oxide nanoparticles (Miller et al., 2010a,b; Pfefferkorn et al., 2010; Harper et al., 2006). One particularly damaging aspect of nanoparticle release from combustion reactions or large-scale demolition is that the nanoparticles are often released directly into the air (although mine tailings may also introduce these nanoparticles into natural waters).

2.2.3 Nanoparticle-Enabled Consumer Products

Thousands of consumer products, including many items of clothing, personal care products, next-generation batteries, and sporting goods now contain nanomaterials. In future, nanoparticles will likely be incorporated into many more products, including pharmaceuticals and next-generation solar cells or batteries. The breakdown of these products at the end of their useful life also provides several key points of entry for manmade nanoparticles into the environment.

Nanoparticles disposed down the drain may eventually enter a wastewater treatment facility. Although the nanoparticle-contaminated water will undergo several purification processes, including mechanical filtering and settling treatments and digestion with microbes, none of these treatments remove nanoparticles from the waste stream. As a result, nanoparticles may remain in the purified water and released back into the environment. Use of contaminated sludge from wastewater purification may contaminate the environment. Similarly, products containing nanoparticles, at the end of their usefulness, are dumped in a landfill, where they can break down and release free nanoparticles into soil in and even groundwater sources. Gottschalk et al. (2010) identified the following nanomaterials in the environment: silver, titanium dioxide, zinc oxide, silica nanomaterials, single-walled carbon nanotubes (SWCNTs), multiwalled carbon nanotubes (MWCNTs), and fullerenes. However, it is difficult to estimate the relevant concentrations of nanoparticles released at any given time, possibly due to limited data and nanoparticle transformation (agglomeration, sedimentation, or change of surface moieties) (Klaine et al., 2008).

The atmosphere contains most anthropogenic nanoparticles, often referred to as ultrafine particles. The size range commonly found are particles $\geq 10\ \mu m$ (PM_{10}), fine particles $\leq 2.5\ \mu m$ ($PM_{2.5}$), and UFPs <100 nm (Vorbau et al., 2009). Smaller nanoparticles rapidly aggregate into larger ones; thus, they disappear rapidly (Wallace et al., 2008; He et al., 2007). Carbon nanotubes may be present in atmospheric dust (Kohler et al., 2008; Murr and Garza, 2009). In the atmosphere, nanoparticles are either released directly from a source (primary emissions) as nanoparticles or formed as a result of reactions in the atmosphere (secondary emissions). Natural sources of nanoparticles in the atmosphere include volcanic eruptions, physical and chemical weathering of rocks, precipitation reactions, and biological processes. Anthropogenic sources of nanoparticles can be from industrial, domestic, and fossil fuel sources, including the combustion of fuels by the vehicles.

As described, there are different methods by which a nanoparticle may grow in the atmosphere. Smaller particles are in a *nucleation mode* because they will rapidly grow due to coagulation and condensation (Sipila et al., 2010). Particles with diameters of $0.1-2\ \mu m$ are in the accumulation mode, and those greater than $2\ \mu m$ are in the coarse mode. These nanoparticles released into the atmosphere may remain in the atmosphere for years, or eventually be washed out by precipitation, which cleanses the air but contaminates water.

In an urban atmosphere, diesel- and gasoline-fueled vehicles and stationary combustion sources are key contributors of nanoparticles, amounting to more than 36% of the total particulate number concentrations (Burtscher, 2005; Shi et al., 2001). The total concentration of natural nanoparticles is considerably lower in comparison to the anthropogenic nanoparticles. Most research carried out in this area to date has focused on ultrafine particulate material, including nanoscale particles (Oberdorster et al., 2007). Comparatively little work has been done on ecological systems.

3. DISTRIBUTION OF NANOPARTICLES INTO THE ENVIRONMENT

As shown in Figure 12, nanoparticles, released from various applications, distribute among the different spheres of the environment (soil, water, air, sediments). Figure 13 shows the flow of nanoAg (a), nanoTiO_2 (b), and CNTs (c) in different environmental compartments (Mueller and Nowack, 2008). The predicted environmental concentrations (PEC) were nanoTiO_2 > nanoAg >> CNT. In the case of nanoAg and nanoTiO_2, the most prominent flows are between the products and the sewage treatment plant (STP) (3.3 t/a, 249 t/a), the STP and the waste isolation plant, WIP, (2.7 t/a, 202 t/a), and the WIP to the landfill (3.3 t/a, 231 t/a). Unlike nanoAg and nanoTiO_2, the most prominent flow for CNT is between the products and the WIP (1.75 t/a) and from the WIP to the landfill (1.30 t/a). However, small quantities of all three nanoparticles find their way to soil, water, and atmosphere.

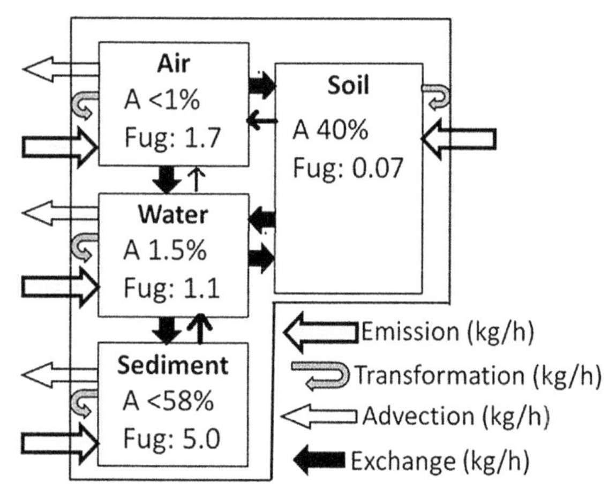

FIGURE 12 Possible distribution of nanoparticles in the environment. The arrow size indicates the direction of the nanoparticle movement. *Mueller and Nowack (2008) with permission.*

Compared to the metal nanoparticles, CNT exhibited poor PEC and distribution into the environment. This may be because a very small amount of CNTs are currently being used (in Switzerland) and find their way to contaminate the outdoor environment (Lam et al., 2006). In addition, the CNT-containing material ends almost exclusively in the WIP, if not recycled. Even within production facilities, the concentrations measured in the air and on gloves were small (Maynard et al., 2004). Because CNTs are partially burned in the WIP, the percentage of CNTs in the landfill is lower than that with the other two substances (one-fifth to about one-half of the total particle volume). However, with an increasing variety of products that contain CNTs, the CNT flows may change considerably.

The urban atmosphere, on average, contains 10^5 to 4×10^6 nanoparticles/cm^{-3}. Particle number concentrations (PNCs) in Beijing are shown to be generally higher than that in cities of developed countries, possibly due to concentrated industrial activity (Wu et al., 2008). The highest total particle number concentration occurred in spring due to the frequent nucleation events in the atmosphere. A study from Kuwait (Al-Dabbous and Kumar, 2014) reported that (1) the PNCs in Kuwait were up to 19 times higher (5.98×105 cm^{-3}) than those typically found in European roadside environments and (2) diurnal variations of PNCs coincided with the cyclic variations in CO, NOx, and traffic volume during morning and evening rush hours. Diesel exhaust and gas/wood combustion stoves increased atmospheric CNT density (Murr and Bang, 2003; Murr et al., 2004). However, in the atmosphere, CNTs may undergo aggregation into larger particles that, via Brownian diffusion, gravitational setting, or wet deposition, reach soil or surface water. These observations suggest that nanoparticles, depending upon their properties, distribute in different environmental spheres and may appear in freshwater supply.

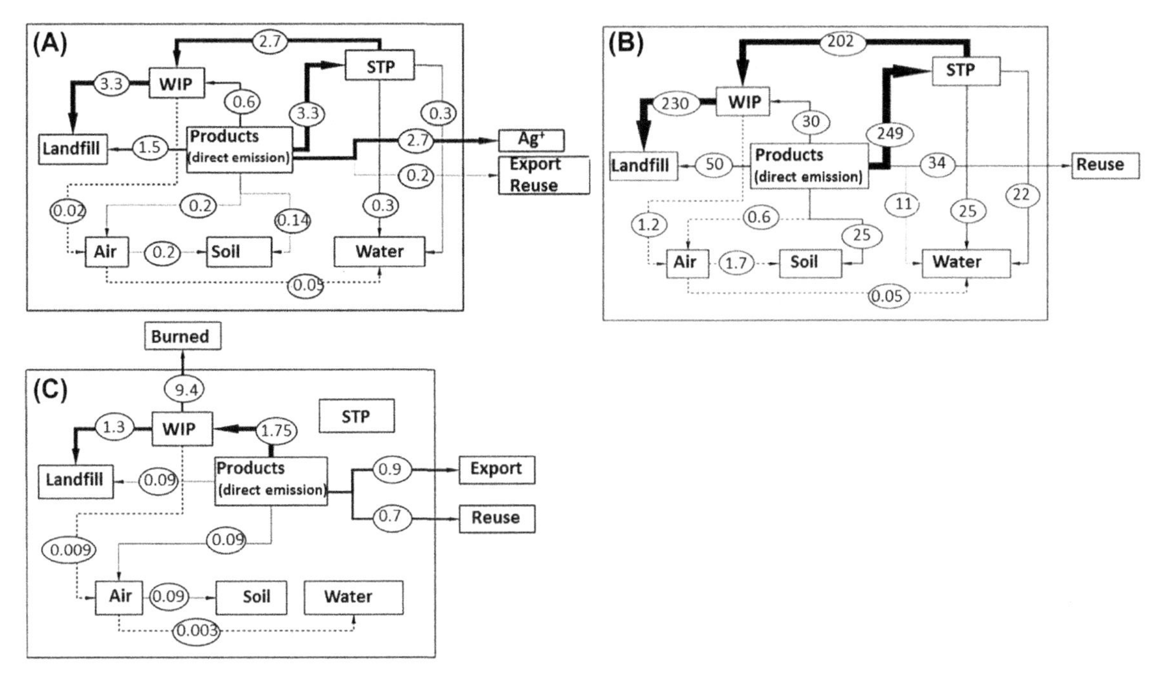

FIGURE 13 Flow of nanoAg (A), nanoTiO_2 (B), and CNT (C) from the products to the different environmental compartments, incinerators, and landfills. All flows are in tons/year. The thickness of the arrows is proportional to the amount of silver flowing between the compartments. Dashed arrows represent the lowest volume.

In atmosphere, soil, or water, nanoparticles may undergo physical and chemical changes that may further alter their fate in the environment.

4. FATE OF NANOPARTICLES IN THE ENVIRONMENT

The environmental fate is defined as the processes by which chemicals/nanoparticles move and are transformed in the environment. The fate processes include persistence in air, water, and soil; reactivity and degradation; migration in groundwater; removal from effluents by standard wastewater treatment methods; and bioaccumulation in aquatic or terrestrial organisms. Extensive research has been conducted on the characterization of the fate of bulk chemicals in the environment. However, the pivotal question is whether existing knowledge can be applied to predict the fate and behavior of nanoparticles, or if nanoparticles exhibit a distinct fate from bulk chemicals.

As discussed in earlier chapters, there is a distinct difference between the physicochemical, electronic, and magnetic properties of nanoparticles and bulk chemicals. One key difference is related to their size. The size of common engineered nanoparticles ranges from 2 to 100 nm in at least one dimension, while bulk particles (e.g., pesticides, herbicides, organic compounds, polymeric hydrocarbons, polyaromatic hydrocarbons (PAH), heavy metals) are >100 nm in all dimensions. As the particle size decreases below 100 nm, they acquire unique properties that are not present in their bulk counterparts. Although the basic principles of the environmental fate of nanoparticles and bulk chemicals are the same (water solubility, dissolution, soil sorption and desorption, migration, transformation, degradation, and bioaccumulation), the unique properties conferred to nanoparticles alter the mechanisms involved. In general, the fate of nanoparticles and bulk chemicals in the environment (atmosphere, water, and soil) is determined by their physicochemical and electronic properties, such as water solubility, hydrophilic-hydrophobic balance, vapor pressure, and steric and thermodynamic properties; as well as the soil properties including particle size, ionic composition, moisture and biota, especially microbial population. The fate of bulk chemicals in the environment is discussed in detail in Appendix 1. The following sections discuss the fate of nanoparticles in the environment.

Nanoparticles, upon their release into the environment, interact rapidly with soil particles; minerals and heavy metals, including lead, cadmium, and copper; dissolved organic matters; and toxic chemicals, such as pentafluooctane sulfonate, herbicides, and pesticides (Hyung and Kim, 2008; Hyung et al., 2007; Kwadijk et al., 2013). Dissolved organic matter may alter the environmental fate of nanoparticles as well as the fate of the chemicals adsorbed to it (Chen and Elimelech, 2009, Saleh et al., 2008, Gunasekara et al., 2003, Gunasekara and Xing, 2003). Pristine nanoparticles but not functioned nanoparticles aggregate into the environment (Jimenez and Madsen, 2003). In the environment, some nanoparticles, such as zero-valent metal nanoparticles and CNTs, may act similarly to charcoal by sequestering toxins and limiting their bioavailability and mobility. Conversely, these nanoparticles may concentrate and transport bioaccumulative chemicals from the environment into organisms, thus exacerbating their toxicity and transfer in the food chain, posing serious hazards to the environment and public health. The overall fate of nanoparticles in the environment is shown in Figure 14, and the details are described in the following sections.

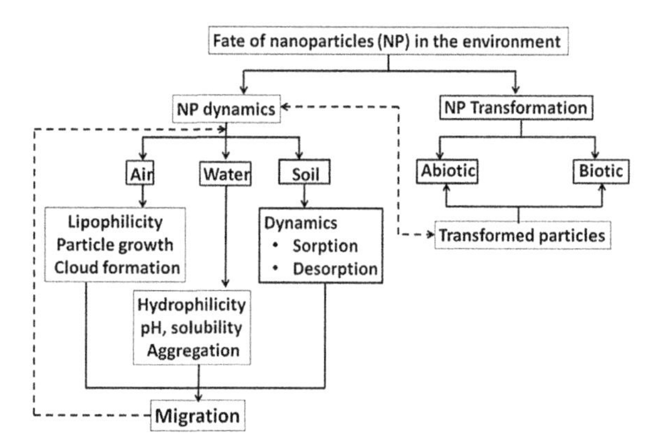

FIGURE 14 Overall fate of nanoparticles in the environment.

4.1 Fate of Nanoparticles in the Atmosphere

Nanoparticles are released into the atmosphere from natural and anthropogenic sources. The particles undergo aggregation, UV reaction, ozone reaction, oxidation, and condensation. Chianelli et al. (1998) and Zamaraev et al. (1994) have shown that atmospheric aerosols and smaller nanoparticles accelerate ozone-forming reactions photo- and thermo-catalytically. Chianelli et al. (1998) also showed that aerosols collected from Mexico City consisted of amorphous fullerene-like materials and inorganic materials consisting of nanoparticles of Fe_2O_3, MnO_2, and FeS_2 supported on clay minerals. In the presence of visible light and O_2, these inorganic components may partially oxidize hydrocarbons trapped in nanoparticles (Vinodgopal et al., 1996). The nanoparticles listed above exhibited band gaps that enhanced the photocatalytic adsorption of solar radiation (Rengifo-Herrera et al., 2011; Allen et al., 2009). As discussed above, critical cloud condensation results in the formation of hydrated particles also known as cloud particles (Biswas and Wu, 2005; Friedlander and Pui, 2004; Sioutas et al., 2005; Kulmala, 2003; Kulmala et al., 2004; Zhang and Wexler, 2004). The aggregated particles undergo dry deposition, while the hydrated particles contaminate soil and water via precipitation.

4.2 Fate of Nanoparticles in the Hydrosphere

Nanoparticles enter the hydrosphere via the natural and anthropogenic sources. The natural nanoparticles include aquatic colloids comprising of macromolecular organic materials, such as humic and fulvic acids, proteins, and peptides, as well as colloidal inorganic species (typically hydrous iron and manganese oxides). Their small size and large, highly active surface area per unit mass make them reactive to other ligands, forming both organic and inorganic contaminants. Their high surface energy, quantum confinement, and conformational behavior are likely to cause important environmental abnormalities. Colloidal nanoparticles less than 0.45 nm in diameter may exhibit different bioavailability from truly soluble organic or ionic metal bulk species (Lead et al., 1997). The sewage water is also a key source of manufactured nanoparticles because some nanoparticles may escape from treatment plants and are discharged into natural water bodies. The sewage nanoparticles can remain in the environment for long periods (Oberdorster, 2004; Velzeboer et al., 2008; Benn and Westerhoff, 2008; Kim et al., 2010). Upon release, nanomaterials may interact with aquatic surfaces and biological species and/or aggregate, depending on the interplay between electrostatic and van der Waals interactions (Thess et al., 1996; Saleh et al., 2008). Hydrophilic particles dissolved, while hydrophobic particles either settle down and accumulate at the sediment or bind to the suspended particles. The possible fate of hydrophilic and hydrophobic nanoparticles in aqueous environments is shown in Figure 15(A) and (B), respectively.

Water-soluble nanoparticles (Figure 15(A)) in aquatic environments release free ions (Figure 15(A3)), undergo aggregation (Figure 15(A4)), and/or adsorb foreign particles from the media (Figure 15(A5)). The large aggregates are water insoluble and settle onto the sediment surface. In addition, nanoparticles undergo biotic and abiotic transformations and the products are redistributed into the environment (Figure 15(A6–9)). Different mechanisms may be involved in determining the fate of water-insoluble nanoparticles in water. Hydrophobic

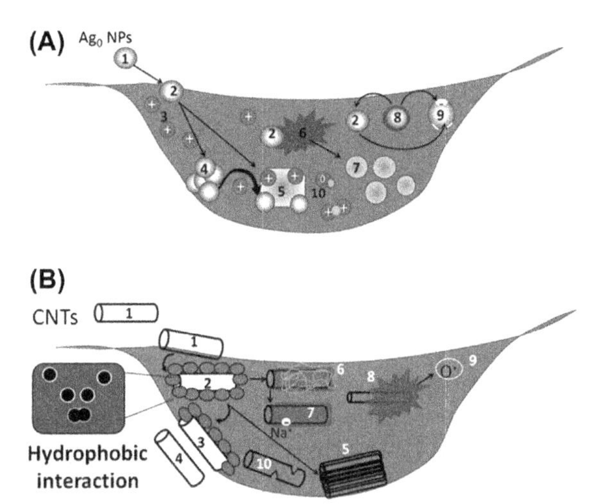

FIGURE 15 Diagrams showing the overall fate of nanoparticles in water. (A) Fate of zero-valent silver nanoparticles (Ag_0 NP) 1. In water Ag_0 NPs may undergo the following interactions: [i] Ag_0 NPs bind with cations (3) and form aggregates (4) that may further bind the suspended soil particles in water (5). Nanoparticles may be released into the water phase as monomers or dimers (10). [ii] Ag_0 NPs bind with organic matter (6) and form an organic corona (7). [iii] Different water-soluble chemicals including proteins, carbohydrates, fats and xenobiotics are adsorbed by the nanoparticles (8) that either desorbs under different conditions (2) or increases the particles hydrodynamic size (9). (B) Carbon nanotubes (CNTs) being highly hydrophobic have a different fate than those of the metals: [i] CNTs may adsorb hydrophobic chemicals (2) and settle at the bottom (3). In some cases, CNTs may dissociate (4) and bind soil particles. [2] CNTs may aggregate rapidly and settle down at the bottom (5) or bind organic carbon or proteins (5) that may further bind soluble ions (7). [3] CNTs may undergo oxidative stress and release reactive oxygen species (9) and/or attacked by the aqueous peroxidases that generate defects within the CNTs (10).

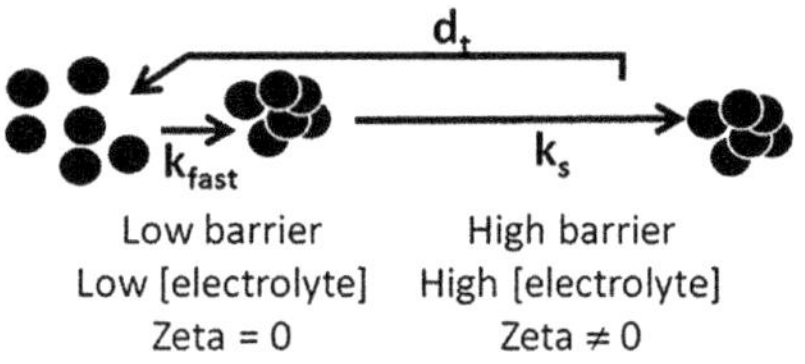

$$W = \frac{k_{fast}}{k_s} \qquad k_{fast} = \frac{8k_pT}{3\eta} \qquad \alpha = \frac{1}{W} = \frac{\left(\frac{dD_h(t)}{dt}\right)_{t\to 0}}{\left(\frac{dD_{h(t)}}{dt}\right)_{t\to fast}}$$

$$d_t = \left(d_0 - d_f\right) d_p^{\ -kt} + d_f$$

$$K_{cond.(nano)} = K_{cond} \cdot Exp\left(\frac{3M\gamma}{pr.RT}\right)$$

FIGURE 16 Aggregation-dissolution kinetics of nanoparticles in water. Details are described in the text.

particles do not disperse in water and undergo rapid aggregation or bind the hydrophobic moiety of the sediments (Figure 15(B) inset). Nanoparticle aggregation results in the formation of different sized aggregates that are insoluble in water and settle onto the sediments. Nanoparticles, depending upon their properties and the environmental conditions, undergo either fast aggregation characterized by k_{fast} or slow aggregation characterized by k_s, where the k_{fast}/k_s is called the stability ratio (W) (Figure 16). The kinetics of aggregation is characterized by the attachment efficiency, α (Figure 16). Under certain conditions, the aggregates disaggregate into smaller particles, characterized by the equation shown in Figure 16. The aggregation–disaggregation equilibrium is $K_{cond.\ (nano)}$ is shown in equation Figure 16. In addition to aggregate formation, nanoparticles may adsorb onto a solute particle and/or undergo chemical transformation.

4.2.1 *Factors Affecting the Fate of Nanoparticles in the Hydrosphere*

The fate of nanoparticles in the aqueous environment is determined by the physicochemical properties of both the nanoparticles (size, charge, aggregation, porosity, adsorption capacity, etc.) and the media (pH, ionic strength, soluble and particulate organic carbon, the water–sediment interface, etc.) (Filella, 2007), as described below.

4.2.1.1 SIZE

Nanoparticles, due to Brownian motion, collide with each other and, under certain circumstances, form bonds, resulting in the formation of larger particles, known as aggregates, of differing sizes. Larger aggregates do not dissolve in water and settle down at the sediment. Thus, larger particles are removed from the aquatic system. Because the collision frequency of nanoparticles is inversely related to the particle size, smaller particles may have a higher aggregation rate than larger particles. Aggregation may significantly affect the environmental fate and transport of nanoparticles in water and their adsorption onto the soil matrix (Areepitak and Ren, 2011; Godinez and Darnault, 2011). An increase in particle size leads to an increase in the sedimentation rate and ensuing decrease in their concentration in water (Zhang et al., 2009), as well as a decrease in their chemical reactivity and toxicity (Liu et al., 2009; Jiang et al., 2009).

4.2.1.2 SURFACE CHARGE AND IONIC STRENGTH

Nanoparticles' surface charge and/or ionic strength determine their zeta potential (a measure of the magnitude of the electrostatic or charge repulsion/attraction between particles), one of the fundamental parameters known to affect stability or aggregation (Figure 17(A)). The ion clouds around the nanoparticles (Figure 17(B)) with their net charge stabilize a colloid against agglomeration, while van der Waals forces always lead to an attractive interaction, as shown in Figure 17(B). The barrier against interparticle interaction can be achieved by introducing a charge onto the surface of the particle or by altering the pH of the solution. If the repulsive force surpasses the attractive force (e.g., zeta potential >20 nm), a stable system results. However, if the attractive force surpasses the repulsive force (e.g., zeta potential 0),

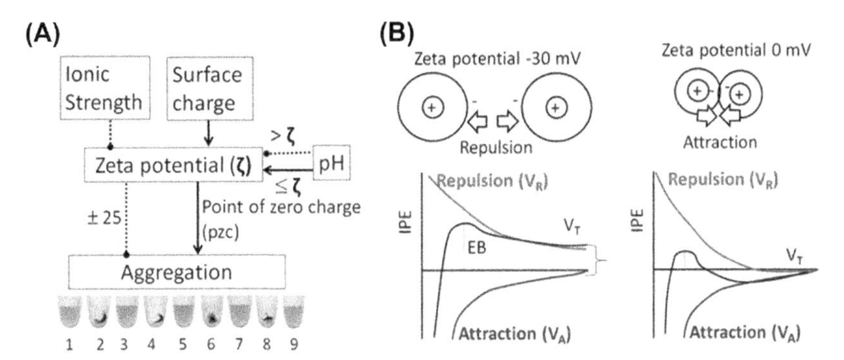

FIGURE 17 Possible mechanisms of aggregation. (A) The zeta potential (ξ) plays an important role in nanoparticles' aggregation. A ξ value of zero facilitates, while a ξ value of > ±25 mV opposes aggregation. 1: low ionic strength, 2: high ionic strength, 3: high surface charge, 4: low surface charge, 5: pH ≠ isoelectric point, 6: pH close isoelectric point, 7: high organic matter, 8: low organic matter and 9: pure water. (B) Energetics of interparticle interaction. At ξ > ±25 mV, the repulsion forces dominate, while at ξ close to 0, attractive forces dominate.

aggregation and instability may result. As shown in Figure 17(B), an energy barrier resulting from the repulsive forces inhibits two particles approaching and merging. The zeta potential indicates the size of this potential barrier. If the particles exhibit zero zeta potential (zeta potential = isoelectric point), then there will be inadequate repulsion to prevent the particles from coming together.

Figure 18 shows the effects of ionic strength (different concentrations of $NaNO_3$) on hydrodynamic diameter, particle attachment efficiency, and zeta potential (Bae et al., 2013). The time course of change in hydrodynamic

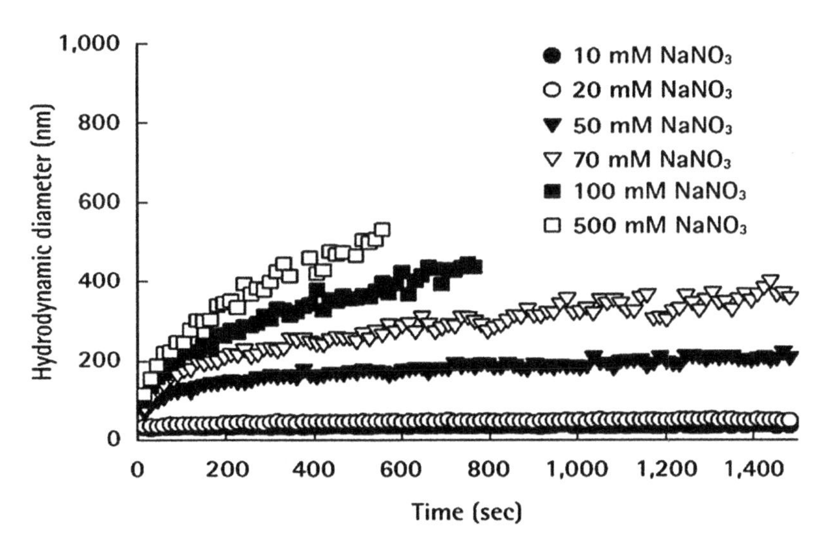

FIGURE 18 Effects of salt ($NaNO_3$) concentration on nanoparticles' hydrodynamic diameter. This figure demonstrates that an increase in salt concentrations increases aggregation, resulting in an increase of a nanoparticle's hydrodynamic diameter in water samples. *Reprinted from Bae et al. (2013) via open access.*

diameter of control nanoparticles and those exposed to 20 mM $NaNO_3$ did not differ significantly. A further increase in $NaNO_3$ concentration resulted in a concentration-dependent increase in the particles' hydrodynamic diameter. Thus, an increase in ionic strength results in the formation of nanoparticle aggregates that accumulate in sediments or soils rather than remain dispersed in the aqueous phase. Conversely, a decrease in ionic strength may disaggregate nanoparticles, resulting in an increase in their solubility.

4.2.1.3 ORGANIC MATTER AND pH OF WATER

The pH and organic matter are key determinants of nanoparticle sorption by soil (Omar et al., 2014). Figure 19 shows the effects of pH on the particles' zeta potential and hydrodynamic diameter.

The zeta potential values of nanoTiO_2 at pH 2, 6 (TiO_2's IEP), and 12 are +40 mV, 0 mV, and −40 mV, respectively (Loosli et al., 2013, Figure 19(A)). The hydrodynamic diameter of nanoparticles at pH 2 and 4 was <100 nm (dispersed) but, at pH 6, a sharp increase (due to aggregation) in size occurred. A further increase in pH resulted in a sharp decrease (disaggregation) in particle size, returning to the original size at pH 7.5 (Figure 19(B)). In another study, Baalousha (2009) and Baalousha et al. (2008) showed that nanoFe_2O_3, in a dose-dependent manner, modulates the effects of pH on the nanoparticle's zeta potential and z-average (average hydrodynamic diameter). The z-average values exhibited the following pattern: pH 10 (50 mg/L nanoFe_2O_3) > pH 6 (80 mg/L nanoFe_2O_3) > pH 11 (100 mg/L nanoFe_2O_3) > pH 12 (100 mg/L nanoFe_2O_3).

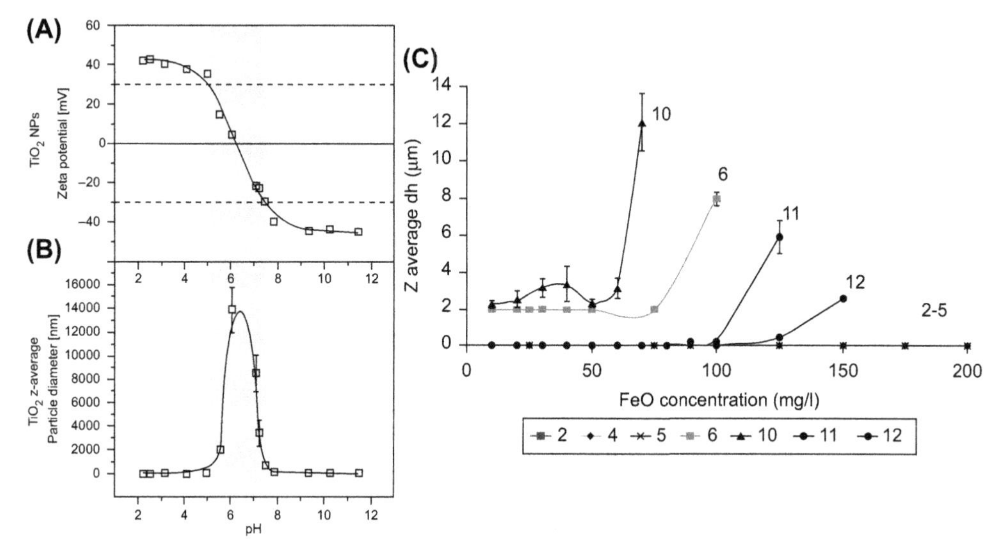

FIGURE 19 Effects of pH on ξ and z-average particle diameter. *(Reprinted from Baalousha (2009) with permission.)* (A) The ξ values of nanoTiO_2 is +40 mV at pH 2 that decreases with an increase in pH. ξ is 0 mV at pH 6.2. ξ becomes negative as the pH increases further, reaching −40 mV at pH 9. (B) The particle size remained unchanged for up to pH 5 then a sharp increase occurs at pH 6 followed by a sharp decrease, reaching the base level at pH 8. This sharp increase suggests aggregation. (C) Effects of FeO concentrations on the z-average diameter of nanoTiO_2 at different pH values. At pH 2–5, the diameter did not change at any FeO concentrations. The concentrations of FeO needed to aggregate TiO_2 at pH 10, 6, 11, and 12 were 60, 100, 125, and 150 mg/l, respectively. The particle size decreased as the FeO concentration increased.

The zeta potential is +40 mV at pH 2 and −40 mV at pH 12. At pH values in the range of 6−10, the nanoparticle (NP) surface charge decreases and particle aggregation increases gradually, resulting in a formation of larger particles (up to >2 μm). At pH 4 and 5, the particles possess charge, but lower than the particles at pH 2. The z-values of nanoFe_2O_3 at pH 4 and 5 were higher than the values at pH 2, but lower than the values at pH 10, 6, and 12. These pH-related changes may be attributed to the changes in the repulsive forces (according to DLVO theory, named after Boris Derjaguin, Lev Landau, Evert Verwey, and Theodoor Overbeek (Vervey and Overbeek, 1948)) between nanoparticles (Figure 20).

Similar to the change in pH, change in dissolved organic matter (DOM; such as humic acid) also modulated the nanoparticle stability in aqueous samples (Figure 21). The humic acid (HA)-zeta potential or diameter curve was similar to the pH-zeta potential or diameter curves (Figure 21(A)). The nanoparticles remained dispersed at HA concentrations 0−1.5 mg/L and 6−10 mg/L. Aggregation began at HA concentration >1.5 mg/L, peaked at HA concentration 3 mg/L, and returned to the original level at HA concentration >6 mg/L. Thus, aggregation, disaggregation, or stabilization of nanoTiO_2 (Loosli et al., 2013) or nanoZnO (Omar et al., 2014) depend on the solution pH, concentration ratio of HA to nanoparticles, and natural organic matter (NOM) physicochemical properties. Omar et al. (2014) suggests that typical environmental concentrations of HAs are sufficient to stabilize colloidal nanoparticles and that HAs can play major roles in the disaggregation and dispersion of already formed nanoparticle aggregates.

4.2.1.4 SEDIMENTATION UNDER GRAVITY

As nanoparticles aggregate, their density changes as water enters the pores. An increase in the cross-section increases the contact between particles as well as their aggregation. Sedimentation occurs when the gravitational force overcomes the surface charge and van der Waal interactions.

4.2.2 *Dynamics of Carbon Nanotubes in Water*

Pristine CNTs may remain aggregated, while COOH-functionalized CNTs, being relatively hydrophilic, remained dispersed in water (Stankovich et al., 2006). Fujigaya and Nakashima (2011) have suggested that, despite that pristine CNTs are hydrophobic and functionalized CNTs are hydrophilic, water can enter and

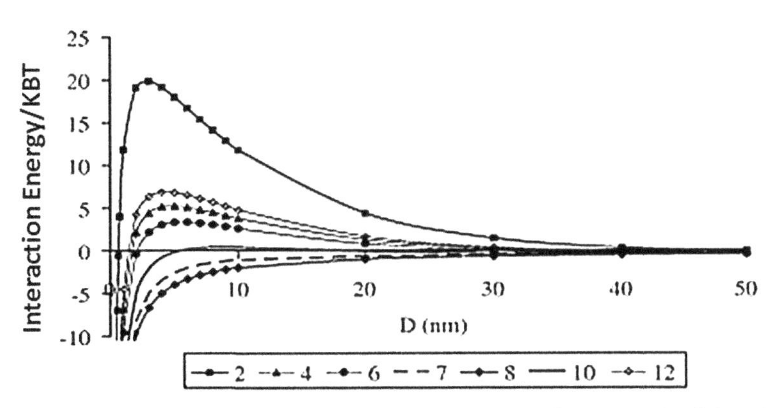

FIGURE 20 The attractive and repulsive forces according to the DLVO theory. *Reprinted from Baalousha (2009) with permission.*

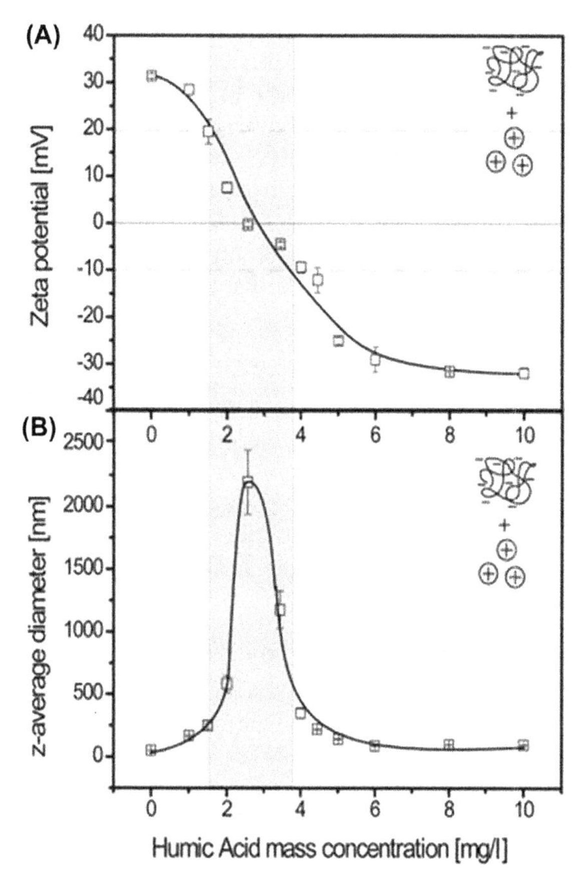

FIGURE 21 Effects of humic acid concentrations on ξ (A) and z-average diameter (B) of nanoparticles. The particle diameter increased sharply between the acid concentrations 2 and 4 mg/L. *Reprinted from Loosli et al. (2013) with permission.*

accumulate in hydrophobic CNTs <2 nm diameter. The transport properties of intra-CNT water are either isotropic or anisotropic, depending on the CNT's diameter (Rana and Chandra, 2007; Perez-Hernández and Schmidt, 2013). In anisotropic but not isotropic CNTs, water diffusivity, thermal conductivity, and viscosity in the axial direction are much larger than in the radial direction (Alexiadis and Stavros Kassino, 2008). The diffusivity of water in CNTs decreases as the diameter of the CNTs decreases. Contrary to the diffusivity, the axial thermal conductivity and shear viscosity in CNTs increase sharply as their diameter decreases (Liu and Wang, 2005). Water accumulated inside SWCNT(6, 6) formed an isotropic chain that spontaneously raptured into anisotropic-like fragments, SWCNT (7, 7) formed a stable trigonal helical chain, and SWCNT(8, 8) formed a pentagonal prismatic chain that may be due to water–water interaction, not water–CNT interaction (Perez-Hernández and Schmidt, 2013). Thus, if contaminated water enters the CNTs, it may remain inside CNTs or diffuse out depending on the nanotube's structure. However, the toxin-filled CNTs may transport the toxins across the food chain.

4.2.3 *Mechanisms and Kinetics of Nanoparticle Solubility in Water*

Solubility is defined as the maximum amount of a solute that dissolves in a solvent at equilibrium (where reactants and products reach a balance). In general, three major factors determine a particle's solubility: the properties related to the solute (surface charge, size, shape, etc.), properties related to the solvents (hydrophilicity, pH, surface charge, size, shape, etc.), and the properties of the interface between the solvent and the solute. One particulate descriptor that significantly influences the dissolution kinetics of nanoparticles is their size (Sotiriou and Pratsinis, 2010; Liu et al., 2010; Zhang et al., 2011; Elzey and Grassian, 2010; Ho et al., 2010). An increase in particle size correlates with a decrease in the particle's dissolution and the solubility product constant (k_{sp}, product of dissolved ions from a solid; Figure 22), defined for equilibrium between a solid and its respective ions in a solution.

The thermodynamic relationship between size and solubility is defined by Eqn (4).

$$\log K_{A\ nano} = \log K_{A\ bulk} + \frac{\gamma A_m}{3.45RT} \quad (4)$$

In this equation, $K_{A\ nano}$ is the solubility constant for nanoparticles, $K_{A\ bulk}$ is the solubility of the corresponding bulk particle, γ is surface tension, A_m is molar surface area, R is the gas

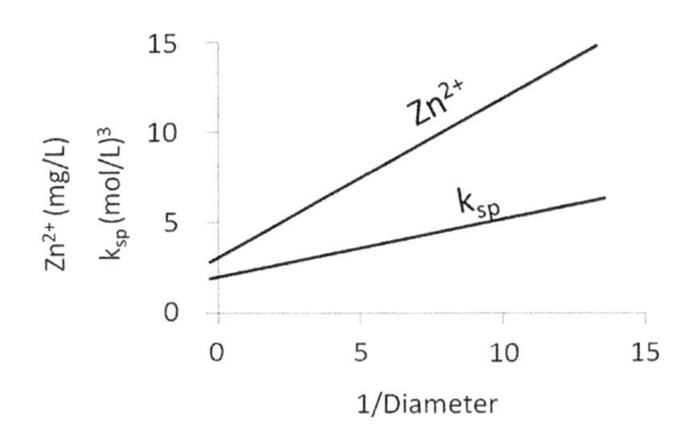

FIGURE 22 An inverse relationship between diameter and dissolution and k_{sp} of nanoparticles.

constant, and T is absolute temperature. Although the mechanism underlying the inverse relationship between size and solubility is not fully understood, the following possibilities have been proposed:

1. In general, metal/metal oxide nanoparticles are negatively charged, which facilitates their oxidation and dissolution (Liu et al., 2009; Plieth, 1982).
2. Aggregation decreases dissolution because an increase in size is associated with a lower surface-to-volume ratio (and thus less surface activity) and a lower rate of mass transfer (Hotze et al., 2010, 2008; Vikesland et al., 2007).

The dissolution of particles in water is determined by the classic Noyes and Whitney (1897) equation based on simple diffusion (Figure 23).

The rate at which a solid substance dissolves in its own solution is proportional to the difference between the concentration of that solution and the concentration of the saturated solution. Noyes and Whitney (1897), however, did not define the constant K in the equation. The Nernst-Brunner (Nernst, 1904) equation provides a physical meaning of K: (S.D)/δ, where S is the surface area available for dissolution, D is the diffusion coefficient of the drug within the liquid unstirred boundary layer, and δ is the boundary's thickness. The disadvantages of these equations are the following: (1) the equations do not take into account the particle's charge or shape and the solvent's viscosity, and (2) the equations assume that the particle size remains unchanged. Ostwald (1900) and Freundlich (1909) included surface tension (γ) in their calculation of particle solubility, while Lewis

$$\frac{dC}{dt} = \mathrm{K}\,(c_s - c_t)$$

Noyes and Whitney (1897)

$$\frac{dM}{dt} = \frac{S.D}{\delta}\,(c_s - c_t)$$

Nernst-Brunner (1904)

$$\frac{dM}{dt} = -K', S_t.\,(c_s - c_t)$$

Hixson and Crowell (1931abc)

$$\frac{RT}{V_m}\,\mathrm{Ln}\frac{S}{S_0} = \frac{2}{\gamma r}$$

Ostwald (1900)– Freundlich (1909)

$$r_{crit} = \sqrt[3]{\frac{q^2}{16\pi K\gamma'}}$$

Lewis (1909)

$$\frac{RT}{V_m}\,\mathrm{Ln}\frac{S}{S_0} = \frac{2}{\gamma r} - \frac{q^2}{8\pi K r^4}$$

Knapp (1922)

$$r = \sqrt[3]{\frac{q^2}{4\pi K\gamma'}}$$

Knapp (1922)

FIGURE 23 Different equations that characterize particle dissolution.

(1907) derived an equation that accounted for particle charge. Knapp (1922) showed that the inverse relationship between size and solubility is not infinite; below a critical size, the solubility decreases sharply (Figure 24).

These equations have one common flaw: all equations are designed for spherical particles. For nanoparticles, this is a major flaw because (1) nanoparticles consisting of the same number of units and volume may acquire many shapes, spheres, triangles, rods, etc.; (2) the size and shape of nanoparticle aggregates may change with time; and (3) polycrystalline nanoparticles are more soluble than monocrystalline nanoparticles in water (Pal et al., 2007; Yang et al., 2007, Elechiguerra et al., 2005; Wiley et al., 2004). However, the nanoparticle shape is rarely considered in the development of models for a nanoparticle's solubility in water. Mihranyan and Stromme (2007) developed a mathematical model that correlated the solubility with fractal surface dimensions of a nanoparticle. They showed that particles with fractal surface roughness may exhibit significantly higher solubility than predicted from the classical Ostwald–Freundlich equation.

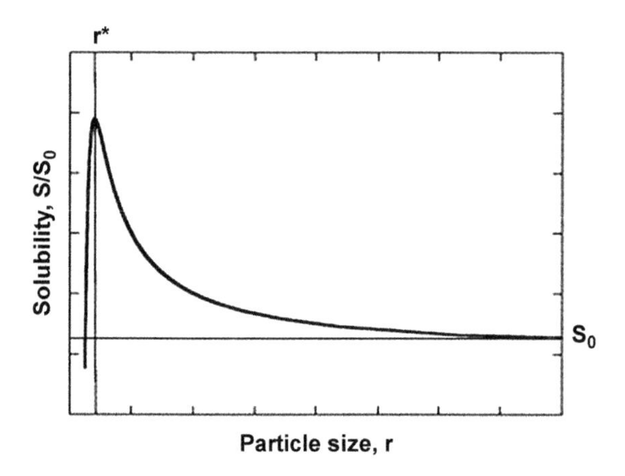

FIGURE 24 A theoretical nonlinear particle size (spherical)—solubility curve.

4.2.4 Nanoparticle Biotransformation in Water

The dissolved organic matter (DOM) in water can reduce ionic metals to nanometallic particles (nanoM) under natural sunlight—a reaction mediated by superoxide from photoirradiation of the phenol group in DOM (Yin et al., 2014). The dissolved O_2 significantly enhances the formation of nanoM. Ionic Hg is reduced by dissolved organic matter and sunlight, thus forming nanoHg. Natural reduction of silver ions (Ag^+) to nanoAg by DOM may play a crucial role in the transformation and transport of engineered NPs in aquatic environments. The light-induced reduction of metal ions depends highly on pH, concentration of metal ions, M_f-NOM (molecular weight fractionated NOM), light quality, and the molecular weight of M_f-NOM. The differential reduction of metal ions by NOM may not be due to their reductive ability but due to the differential light attenuation of Mf-NOM.

4.3 Fate of Nanoparticles in Soil

Natural nanoparticles in soil include nanoclays, organic matter, nano iron oxides, and nanoparticles of other minerals. These particles play an important role in biogeochemical processes. Soil nanoparticles influence soil's structural behavior and facilitate the movement of nanoparticles in soils and other porous media (Muller et al., 2007). Nanoparticle dynamics are determined by the following factors: sorption (adhering to the soil surface), desorption (dissociation), aggregation state (formation of larger and irregular aggregates), deposition (settling of the nanoparticles), interception (retention of the aggregates by soil particles), and detachment (release of the soil-bound particles) (Petosa et al., 2010; Phenrat et al., 2010, 2009a, b; 2007, and Raychoudhury et al., 2012, 2010) (Figure 25). These factors are regulated by the particle size, charge, and surface properties, as well as the

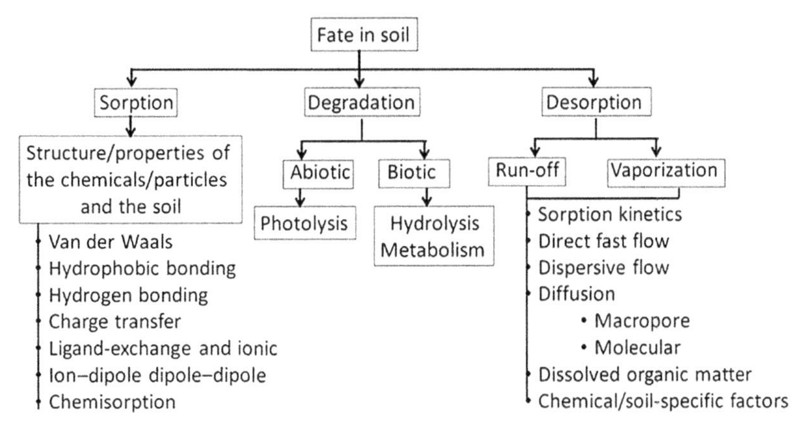

FIGURE 25 Fate of nanoparticles in soil.

soil's composition, organic content, particle size, electrolyte content, and pH.

Aggregation alters the nanoparticles' hydrodynamic size, thus influencing their deposition onto the soil surface and the magnitude of their collection by the soil. The collection efficiency (η_0) can be calculated using Eqn (5).

$$k_{dep,i} = \left[\frac{3(1-\varepsilon)V}{3d_c\varepsilon}\alpha p_c\right]\eta_{0,i} \quad (5)$$

In this equation, $k_{dep,i}$ is the particle deposition rate coefficient, V is the interstitial particle velocity, α is the attachment efficiency, ε is porosity, d_c is the average grain (soil) size, and η_0 is the collector efficiency. Chatterjee and Gupta (2009) have shown that larger particles were deposited at shallower bed depths compared to the smaller particles.

4.3.1 *Migration of Nanoparticles through Soil*

Migration occurs through the following mechanisms: advection, diffusion, and retardation mechanisms.

In *advection*, dissolved nanoparticles are transported via flowing groundwater at its average linear velocity. It is governed by Darcy's Law, with the Darcy flux, v_a, given by the equation $v_a = -k_i$ where k is the hydraulic conductivity (permeability coefficient) and i is the hydraulic gradient (Darcy, 1856). Variations in the magnitude and direction of local water flow determine the nanoparticles' spread in soil, while dispersion determines the nanoparticles' movement away from the mass.

Diffusion—the movement of nanoparticles from an area of high concentration to that of low concentration—can happen in the absence of any bulk air or water movement. Diffusion is generally governed by Fick's law, $f = -D\,(dc)/dz$, where D is the diffusion coefficient and dc/dz is the concentration gradient. In addition, diffusion is also influenced by other parameters, such as the tortuosity of the porous media, osmotic flow, electrical imbalance, and possible anion exclusion.

Retardation mechanisms include dilution, sorption, precipitation, volatilization, and decay, which may affect nanoparticle transport through soil. Sorption is the removal of nanoparticles from solution by soil particles or organic matter. Sorption can be further divided into adsorption and absorption. Adsorption refers to adhesion of nanoparticles to the surface of soil, while absorption is penetration of the nanoparticles in a soil matrix.

4.3.2 Adsorption, Desorption, and Migration

The mechanisms of adsorption, desorption, and transport of nanoparticles in soil are assessed using soil-breakthrough experiments (Appendix 1) consisting of (1) a column, coupled with temperature sensors and temperature control, loaded with desired soil; (2) a solvent delivery system including a pump, flow control, and a valve to control solvent delivery; and (3) a fraction collector. Depending upon the need, more complex units can be assembled with electronic controls and data processing. A tracer (chemical or nanoparticle) is applied from the top of the column and its concentration is measured in the effluent from the column bottom. The experiment is conducted in three stages (Figure 26), as described in the following sections.

4.3.2.1 ADSORPTION

The tracer's binding to the soil depends upon its physicochemical properties, as well as those of the soil, and the experiment's hydraulic conditions such as solvent's flow rate, composition, and pH. Thus, the solvent chemistry and flow rate must be controlled for accurate adsorption values. During this period, no tracer will appear in the efflux. The tracer may begin appearing in efflux due to desorption.

4.3.2.2 SATURATION

The column is considered saturated when all binding sites are occupied. This is when the tracer will remain consistently elevated (in continuous infusion) or begin decreasing in a single bolus mode (particles remain bound to the column).

4.3.2.3 DESORPTION

As discussed, particles dissociate from the binding site at relatively low pH, higher organic carbon content, and lower salt concentrations. In the case of hydrophobic binding, an organic solvent may be needed.

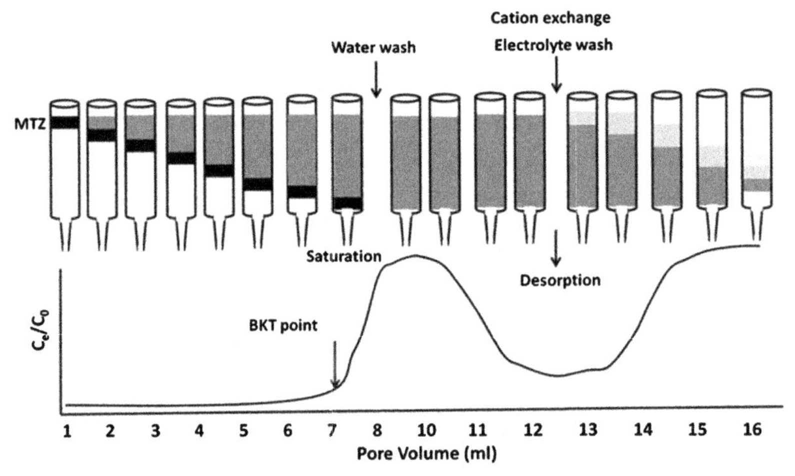

FIGURE 26 A tracer's binding to and desorption from a soil column. When water samples contaminated with a tracer are passed through a soil column, a wavefront known as the mass transfer zone (MTZ) is created. MTZ is where active adsorption happens in a packed column. As the soil reaches its equilibrium capacity, the MTZ travels further through the soil bed. When the MTZ moves across the soil bed, it leaves behind a section of soil bed completely saturated. The MTZ will continue to travel through the soil until it hits its breakthrough point when the tracer emerges in the effluent (bottom). If the tracer is stopped and bound tracer is not released, then the tracer concentration will decline gradually in efflux. Initiation of cation-exchange by passing KCl solution through the column will desorb the tracer, resulting in an increase in their appearance in the efflux.

4.3.3 *Interpretation of the Breakthrough Curve*

The shape of the breakthrough curve provides important information regarding the structure of soil pores and the tracer's interactions with soil (for details, see Appendix 1). The migration of nanoparticles through a soil column is determined by the column properties (e.g., flow rate, column length, solution pH, ionic strength), soil properties (e.g., soil composition, surface chemistry, soil particle size), and nanoparticle properties (e.g., size, shape, surface functionalization, zeta potential, and concentration).

4.3.3.1 EFFECTS OF COLUMN PROPERTIES ON THE BREAKTHROUGH POINT

Figure 27(A) shows the effects of flow rate on the nanoparticle's behavior in soil. The time taken for the breakthrough point and the flow rate are inversely related. The breakthrough point at 2, 5, and 10 ml/min flow rate occurred at 8, 15, and 36 h, respectively (Negrea et al., 2011). This suggests that the time required to reach the breakthrough decreased with an increased flow rate. This may be because an increase in flow rate may reduce contact time, resulting in a poor distribution of the liquid inside the column and a lower diffusivity of the solute. In addition to the flow rate, column length also modulated a nanoparticle's breakthrough point. An increase in column length increased the time required to reach the breakthrough point (Figure 27(B)).

4.3.3.2 EFFECTS OF pH, IONIC STRENGTH, AND CONCENTRATION ON BREAKTHROUGH POINT

The reactivity of nanoparticles in soil suspensions depends upon their zeta potential, a potential between the nanoparticle surface and the slippage plane (Figure 28(A)). A zeta potential of zero indicates uncharged nanoparticles that exhibit the highest aggregation (Figure 28(B)). An increase in zeta potential also increases nanoparticles' surface charge (+ve or −ve) and dispersion (Figure 28(C)).

A relationship between pH and zeta potential for nanoTiO_2 of different sizes (6−104 nm) is shown in Figure 28(D). An increase in particle size shifted the curves to the right. The zeta potential was zero at pH 4 for 6-nm particles and at pH 6 for 104-nm particles (at isoelectric points of the particles; Figure 28(D), inset). Figure 28(D) also shows the relationship of initial particle size with zeta potential and hydrodynamic diameter in aqueous samples. As the particle size increased, the zeta potential decreased and the hydrodynamic diameter increased (approximately 10-fold increase). This suggests that size

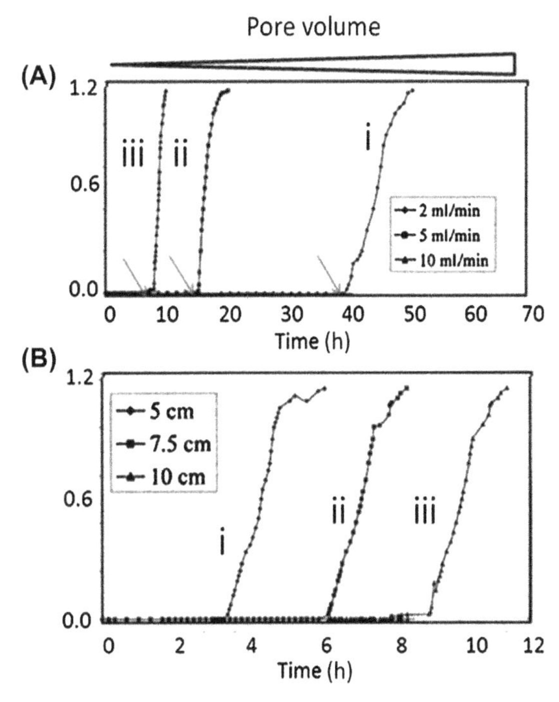

FIGURE 27 Effects of solvent flow rate (A) and column size (B) on soil breakthrough curves. An increase in flow rate accelerates breakthrough, while an increase in column height delays the breakthrough of particles from soil. *Reprinted from Negrea et al. (2011).*

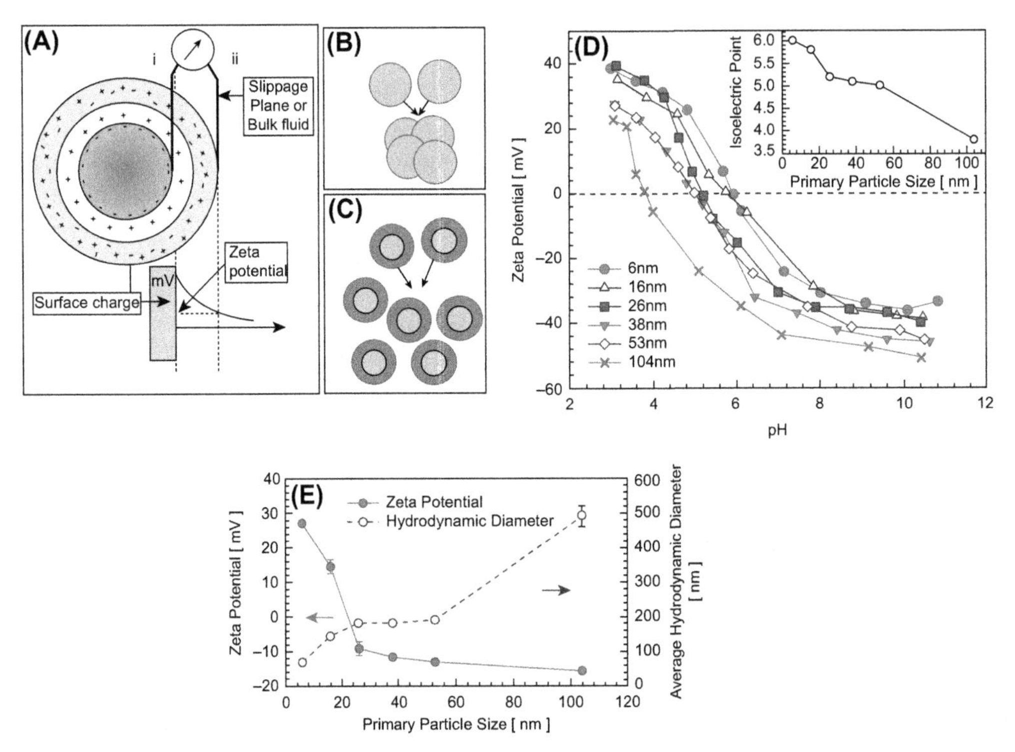

FIGURE 28 Principles of zeta potential (ξ). (A) ξ is defined as the potential difference between a particle's surface and the slippage plane of bulk fluid. A ξ value of 0 mV means lack of potential difference between the two planes. (B) Nanoparticles at ξ of 0 will aggregate rapidly. (C) Nanoparticles with ξ > ± 25 mV will remain dispersed. (D) This figure shows the effects of particle size and pH on ξ. An increase in pH decreased ξ of particles of all size studied (6–104 nm). However, the pH-ξ curve is displayed to the left as the particle size increases. For particle size of 6, 16, 26, 38, 53, and 104 nm, the zero point ξ occurred at pH 7, 6.8, 5.8, 5.5, 5.5, and 4, respectively. (E) Effects of primary particle size on ξ and hydrodynamic diameter (nm). As the particle size increases from 6 to 53 nm, the ξ value decreased from +28 to −15, but the hydrodynamic diameter changed only slightly (100–200 nm). When the particle size increased from 53 to 104 nm, the ξ values remained essentially unchanged but the hydrodynamic diameter increased from 200 to 600 nm. This suggests that ξ of a nanoparticle depends upon its size (primary and hydrodynamic diameter), ξ, and solution pH. *Reprinted from Suttiponparnit et al. (2011), open access.*

may play a key role in sorption and desorption of nanoparticles in soil.

Chowdhury et al. (2011) combined the effects of factors such as flow rate (2 ml/min (Figures 29 and 30(A) and (C)) and 10 ml/min (Figures 29 and 30(B) and (D))), the solution's pH (5 and 7), and ionic strength (1 and 10 nM KCl), and the nanoparticle's concentrations (100–800 mg/ml) on nanoTiO_2 breakthrough curve to study possible relationship between pH and zeta potential and hydrodynamic diameter. Figure 29 shows nanoparticle concentrations in column efflux, while Figure 30 shows nanoparticle distribution within the column.

Chowdhury et al. (2011) showed that nanoparticles at pH 7, irrespective of the KCl concentration or flow rate, adsorbed tightly to the top of the soil (Figure 30) and did not appear in the efflux (Figure 29). At pH 5, flow rate 2 ml/min (Figures 29 and 30(A)) or 10 ml/min (Figures 29 and 30(B)) and KCl 1 mM, the nanoparticles uniformly distributed in the column

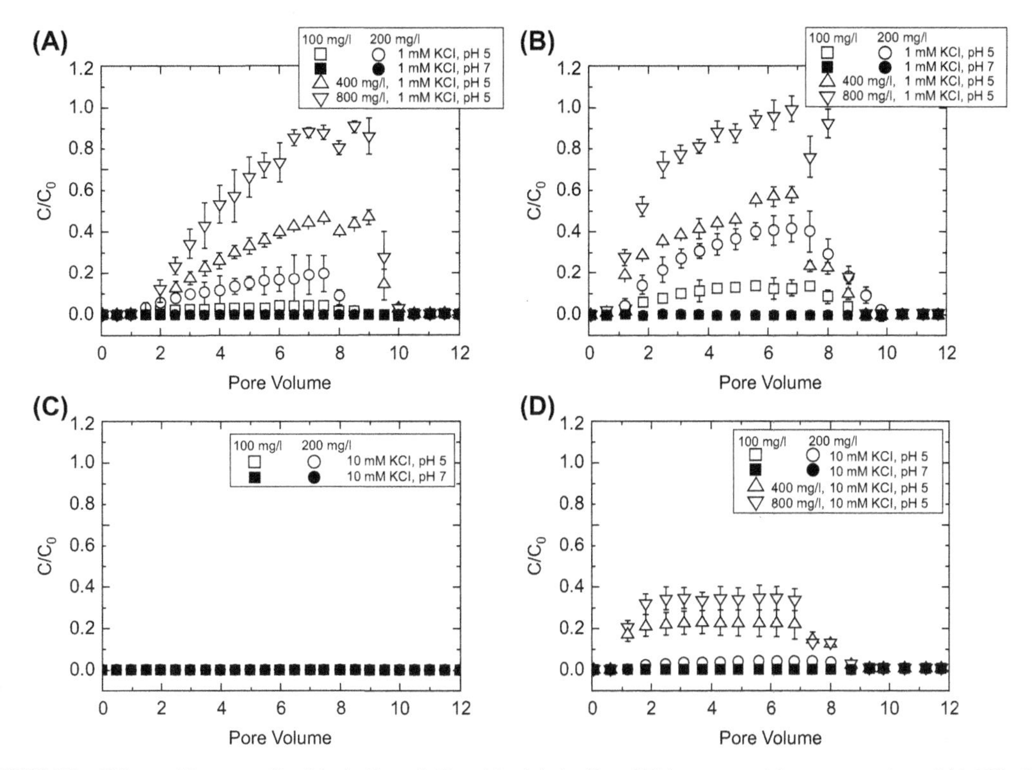

FIGURE 29 Effects of flow rate (2 ml/min (A and C) or 10 ml/min (B and D)), nanoparticle concentrations (100, 200, 400 and 800 mg/l), solution pH (5 and 7) and eluent KCl concentration on breakthrough properties of nanoTiO_2 in soil. Reprinted from Chowdhury et al. (2011) with permission. (A) At flow rate of 2 ml/min and KCl 1 mM, nanoTiO_2 A 100 mg/l and 200 mg/l solution of TiO_2 at pH 7.0 did not exhibit nanoparticles in the efflux. nanoTiO_2 solution (100 mg/l and 200 mg/l) at pH 5.0 exhibited trace levels of nanoparticles in the efflux. Considerable breakthrough occurred at 400 mg/l and 800 mg/l nanoparticle concentration (800 mg/l > 400 mg/l). (B) At flow rate 10 ml/min and KCl 1 mM, nanoTiO_2 breakthrough curve for 400 mg/l and 800 mg/l nanoparticle concentration were comparable to those shown in (A). However, nanoTiO_2 breakthrough curve for 100 mg/l nanoparticle concentration were higher than those shown in (A). This suggests that a higher flow rate increased sorption at lower nanoparticle concentrations. (C) At flow rate 2 ml/min and KCl 10 mM, nanoparticles, at all concentrations, did not appear in the efflux at all concentrations. (D) At flow rate 10 ml/min and KCl of 10 mM, nanoparticles at all concentrations, except 400 mg/l, did not appear in the efflux at all concentrations.

and appeared in the efflux, although the nanoparticle concentrations modulated the C/C_0 ratio (Figure 29(A)). However, at 10 mM KCl concentration, nanoparticles remained adsorbed to the column and poorly appeared in the efflux, irrespective of the pH (Figures 29 and 30(C) and (D)). This suggests that the solutions' flow rate, pH, and molarity and the nanoparticles' concentration interact with the soil and determine the nanoparticles' adsorption and movement.

4.3.3.3 DETERMINATION OF THE NANOPARTICLES' HYDRODYNAMIC SIZE

Fate and transport of nanoparticles in aqueous samples are significantly influenced by their aggregation in water and adsorption onto the soil matrix (Areepitak and Ren, 2011; Godinez and Darnault, 2011). Aggregation results in an uncontrolled increase in nanoparticles' size, resulting in their sedimentation and removal from aqueous samples and thereby a

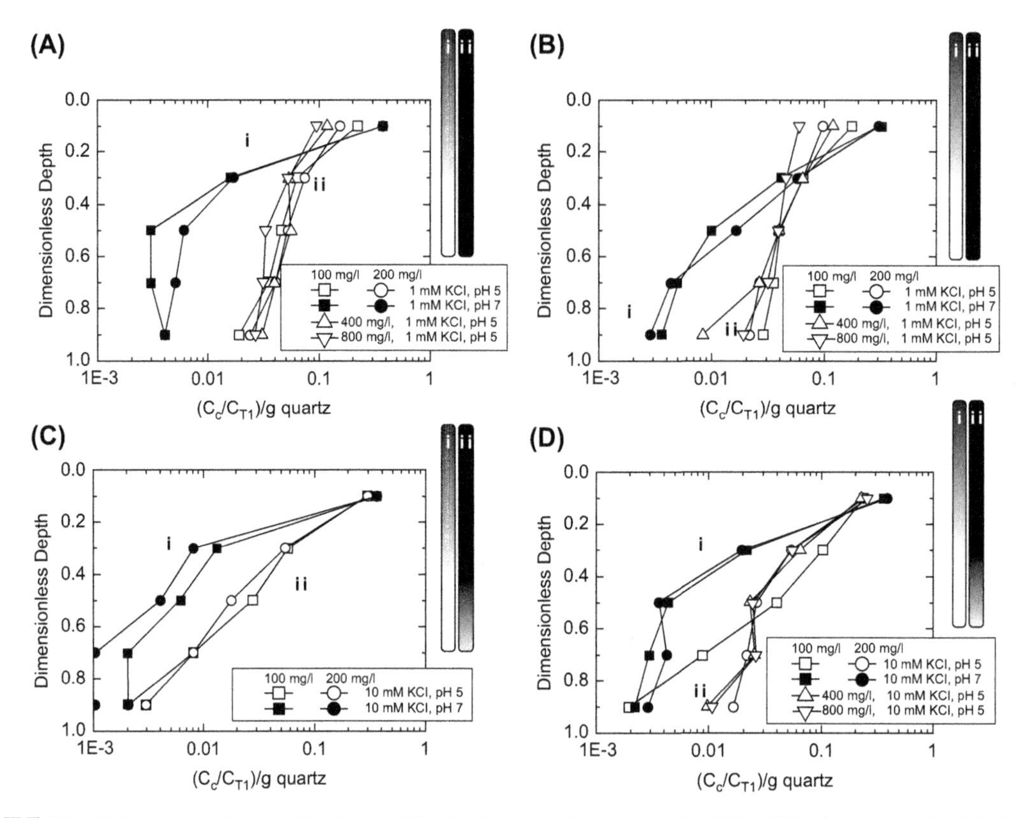

FIGURE 30 Column retention profile of nanoTiO_2. At the lower flow rate and pH 7, >90% of nanoparticles injected were retained at the column inlet (A[i]). At higher flow rate, the greatest retention was observed at the top of the column (B[i]). For the higher flow rate, greater KCl resulted in more retention at the inlet of the column at pH 7 as follows: 200 mg/l, 10 mM KCl >100 mg/l, 10 mM KCl >100 mg/l, 1 mM KCl >200 mg/l, 1 mM KCl (B). Retention levels corroborate the lack of elution observed in Figure 29. The retention profile trends for pH 5 differed from those at pH 7. At 10 mM KCl, >60% of nanoparticles were retained at the top of the column (compared to 90% at pH 7). At pH 5 and 2 mL/min, retention at the column inlet follows this order: 200 mg/l, 10 mM KCl >100 mg/l, 10 mM KCl >100 mg/l, 1 mM KCl >200 mg/l, 1 mM KCl. The higher flow rate experiments conducted at pH 5, retention was greatest in the first 1 cm of the column, and less substantial below the initial 1 cm past the inlet. In general, the nanoparticles remained mobile through the column and were not entirely retained under these conditions. This suggests that the mechanisms involved in particle retention likely include aggregation and straining, which are not considered in the traditional clean bed attachment theory and the contribution of each are discussed in the text. *Reprinted from Chowdhury et al. (2011) with permission.*

decrease of nanoparticle concentration in water (Zhang et al., 2009). Aggregation can also have a significant effect on their chemical reactivity (or toxicity) (Jiang et al., 2009; Liu et al., 2009). A number of studies have reported aggregation and soil adsorption behaviors of nanoparticles that may be involved in their transport (Fang et al., 2013; Bouchard et al., 2012; Huynh and Chen 2011, Bradford and Torkzaban, 2008; Chen and Elimelech, 2007; Bradford et al., 2006, 2003; Chen and Flury, 2005). Representative aggregation profiles as measured by time resolved dynamic light scattering (TR-DLS) at various $NaNO_3$ concentrations are presented in Figure 18 (Bae et al., 2013, http://www.ncbi.nlm.nih.gov/pmc/articles/PMC3657714/figure/F2/). At $NaNO_3$ concentrations below 20 mM, the size of nanoAg did not change and remained very stable through the measured duration. Above 20 mM $NaNO_3$, the aggregation rate increased with increasing $NaNO_3$ concentration. At approximately 100 mM $NaNO_3$, the aggregation rate of nanoAg was maximum and further increases in the $NaNO_3$ concentration had no appreciable effect.

4.3.4 *Effects of Nanoparticle Functionalization*

Raychoudhury et al. (2010) and He et al. (2009) described the breakthrough patterns of zero-valent iron (Fe^0) nanoparticles (NZVI) functionalized with either carboxymethyl cellulose (CMC-NZVI) or polyacrylic acid (PAA-NZVI) from packed sand columns. The nanoparticle diameters were 5.7 ± 0.9 nm and 4.6 ± 0.8 nm, respectively, for CMC-NZVI and PAA-NZVI. The pore water velocities were 0.02, 0.2, and 1 cm min^{-1} and the influent concentrations were 0.1, 0.5, and 3 g L^{-1}. They showed that the particle size changed significantly over time for both CMC-NZVI and PAA-NZVI. Based on the sedimentation test results, PAA-NZVI aggregated more rapidly and extensively than CMC-NZVI. Figure 31(A) and (B) show the breakthrough curves for adsorption of CMC-NZVI and PAA-NZVI, respectively, at 0.1 g/L, 0.5 g/L, and 3 g/L C_0 (initial nanoparticle concentration before the solution passed through the column) concentrations, named 0.1, 0.5, 3.0 C_0, respectively. CMC-and PAA-NZVI did not appear in efflux for up to 0.5 pore volume (PV). Thereafter, the breakthrough curve differed considerably for the two nanoparticles.

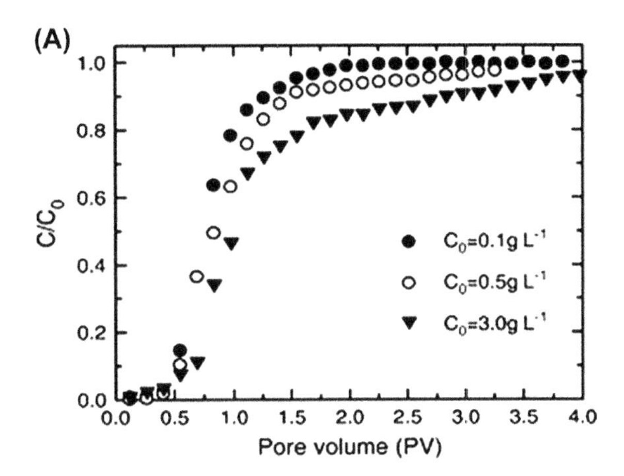

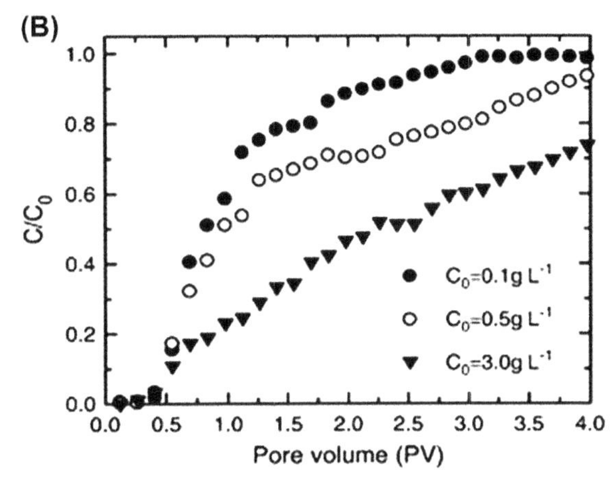

FIGURE 31 Breakthrough patterns of (A) carboxymethyl cellulose functionalized zero-valent iron nanoparticle (CMC-NZVI) and (B) polyacrylic acid functionalized zero-valent iron nanoparticle (PAA-NZVI) from sand column. Different C_0 (tracer concentration in starting solution) concentrations were 0.1, 0.5 and 3 μg/L. PAA-NZVI exhibits more prominent concentration-dependent curves than CMC-NZVI. *Reprinted from Raychoudhury et al. (2010) with permission.*

1. The breakthrough curve for three CMC-NZVI solutions yielded comparable curves with the breakthrough point occurring at 1–1.5 PV. The C/C_0 ratio exhibited the following pattern: $0.1C_0 > 0.5C_0 > 3.0C_0$ (Figure 31(A)). The curves were linear from 0.5 to 1.5 PV and then they flatten up. At 4 PV, the values for three solutions did not differ significantly. This suggests that all binding sites are saturated at 1.5 PV; thus, C approaches C_0 for all three samples.
2. For PAA-NZVI, the shape of the curve depended on the C_0 concentration. For $0.5C_0$ and $3.0C_0$ samples, the breakthrough curve was linear from 0.5 PV to 4 PV, although the C/C_0 ratio for $0.5C_0$ samples was several-fold greater than that for $3.0C_0$ samples. The

breakthrough point was not achieved for up to 4 PV. However, the C/C_0 ratios for $0.1C_0$ and $0.5C_0$ samples were comparable at 4 PV. This suggests that the PAA-NZVI's retention to the soil is lower than that of the CMC-NZVI's retention.

4.3.5 *Desorption of Nanoparticles*

Figure 32(A) shows adsorption isotherms of HA by nanoSiO_2 at pH 4.0. The adsorption of HA were fitted much better to Langmuir equation than Freundlich equation, particularly for the 20-nm SiO_2 particles. Figure 32(B) shows adsorption and sequential desorption of HA from 20 nm nanoSiO_2 at pH 4.0. The key observation is that the adsorption and desorption of HA followed different paths—a process known as hysteresis. The ratio of the hysteresis is the Figure 32: angle BAC that represents the extent of hysteresis or K. The values were 10°, 21° and 37° for first, second, and third desorption, respectively. This suggests that HA binds to the nanoSiO_3 with different mechanisms, and thus desorbed at different ionic strength.

Kanel and al-Abed (2011) have reported effects of pH on the zeta potential (Figure 33) and breakthrough curves (Figure 34) of bare (no functionalization) and carboxymethyl cellulose (CMC)-functionalized zinc oxide nanoparticles (nanoZnO) in soil. For bare nanoZnO, the zeta potentials at pH 7, 8, 9, and 11 were +20, +10, −10, and −20, respectively.

The zeta potential was zero at pH 8.9 for bare nanoZnO. For CMC-nanoZnO, the zeta potentials at pH 3, 5, 7, 9, and 11 were −10, −20, −30, −40, and −60, respectively (Figure 33). Thus, the CMC-nanoZnO were negatively charged at the entire pH range studied. Figure 34(A)–(D) showed the breakthrough curves for both nanoparticles at pH 3, 7, 9, and 11. At pH 3, both nanoparticles exhibited a comparable breakthrough curve (Figure 34(A)).

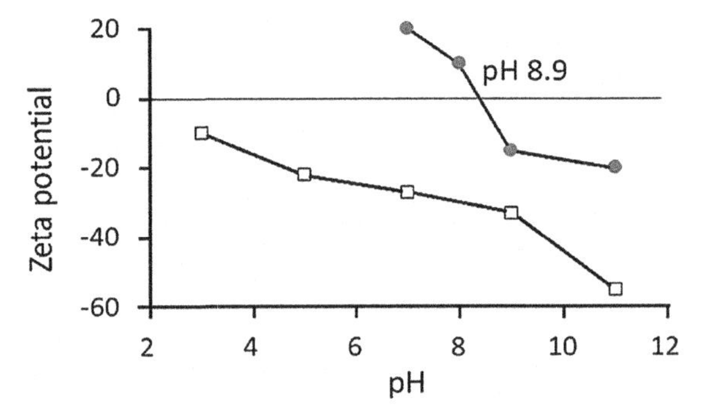

FIGURE 33 Effects of pH on the zeta potential (ξ) of bare (filled circle) or CMC functionalized (open circle) nanoZnO. *Reprinted from Kanel and Al-Abed (2011), open access.*

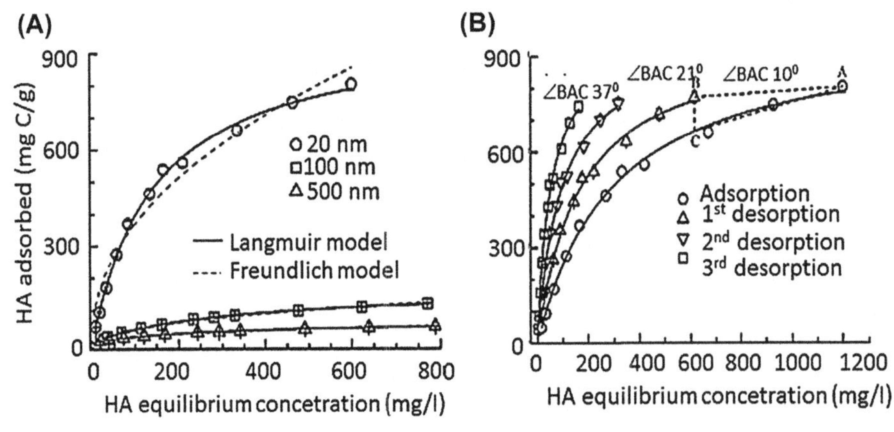

FIGURE 32 Desorption of humic acid from nanoSiO_2 exhibits hysteresis. (A) sorption and (B) sequential desorption. *Reprinted from Liang et al. (2011) with permission.*

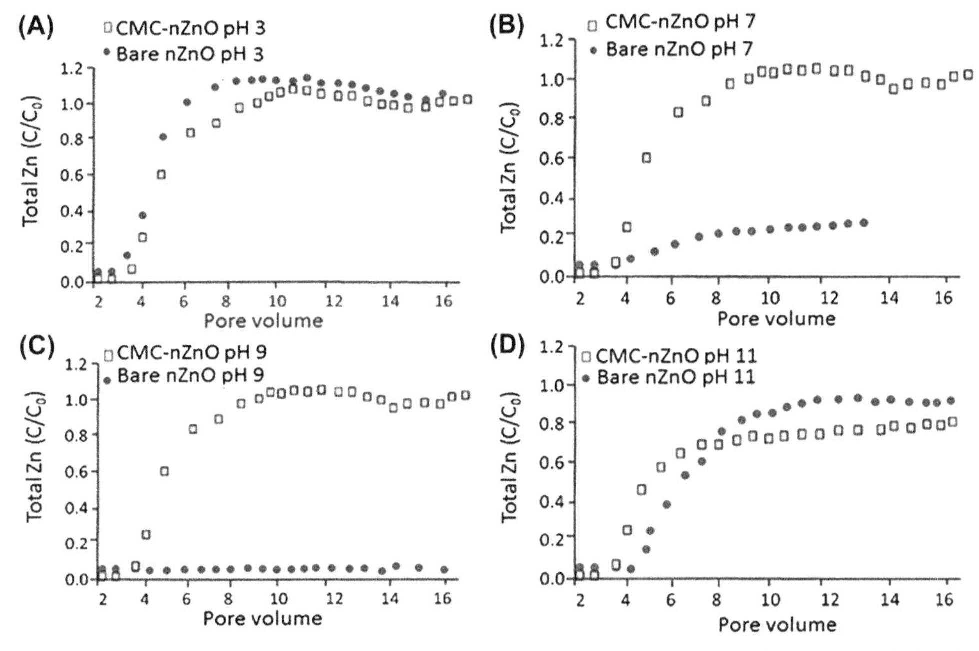

FIGURE 34 Effects of pH on the breakthrough curves for bare and CMC functionalized nanoZnO. Filled circle bare and open circle CMC-functionalized nanoparticles. (A–D) represents pH values 3, 7, 9, and 11, respectively. *Reprinted from Kanel and Al-Abed (2011), open access.*

Although possible causes of comparable breakthrough curves for bare- and CMC-nanoZnO are not known, earlier studies have suggested that, at pH 3, nanoZnO dissociates into Zn^{2+}, $Zn(OH)^{+}$, and $Zn(OH)_2$(aq) (Yamabi and Imai 2002, Stumm and Morgan 1996) and NaCl will remain as Na^{+} and Cl^{-}. He et al. (2009), using X-ray diffraction analysis, did not find ZnO for CMC–nanoZnO at lower pH (pH 5). Therefore, the Zn^{+} ions compete with the parent $nanoZnO^{+}$ for negatively charged sand particles (He et al., 2009). This phenomenon is called blocking effect (Ryan and Elimelech, 1996). At pH 7 (the zeta potentials for bare-nanoZnO and CMC-nanoZnOs were +20 and −20, respectively), about 20% of bare-nanoZnOs, while almost 90% of CMC-nanoZnOs appeared in the aqueous phase (Figure 34(B)). At pH 9 (the zeta potentials for bare-nanoZnO and CMC-nanoZnO were close to 0 mV and −40 mV, respectively), bare-nanoZnO did not appear in the efflux, while the entire CMC-nanoZnO appeared into the efflux. At pH 11 (both particles were negatively charged), CMC-nanoZnO concentration was higher than bare-nanoZnO in the efflux. These observations provide further support that charged particles are poorly adsorbed to soil than uncharged ones.

A conceptual model describing mechanisms of aggregation and deposition of surface functionalized nanoZVI in heterogeneous porous media is shown in Figure 34. Decreasing pH reduced the net charge on polymer-modified nanoZVI and subsequent flattening of the adsorbed layer, resulting in aggregation of nanoZVI (Figure 35(A)). At low pH, enhanced deposition of nanoZVI occurred on positively charged sites formed on kaolinite or metal oxide impurities on sand (Figure 35(B)). The effects of pH on nanoparticles' deposition is shown in Figure 35(C). At high pH, relatively low nanoparticle deposition occurred on sand. At low

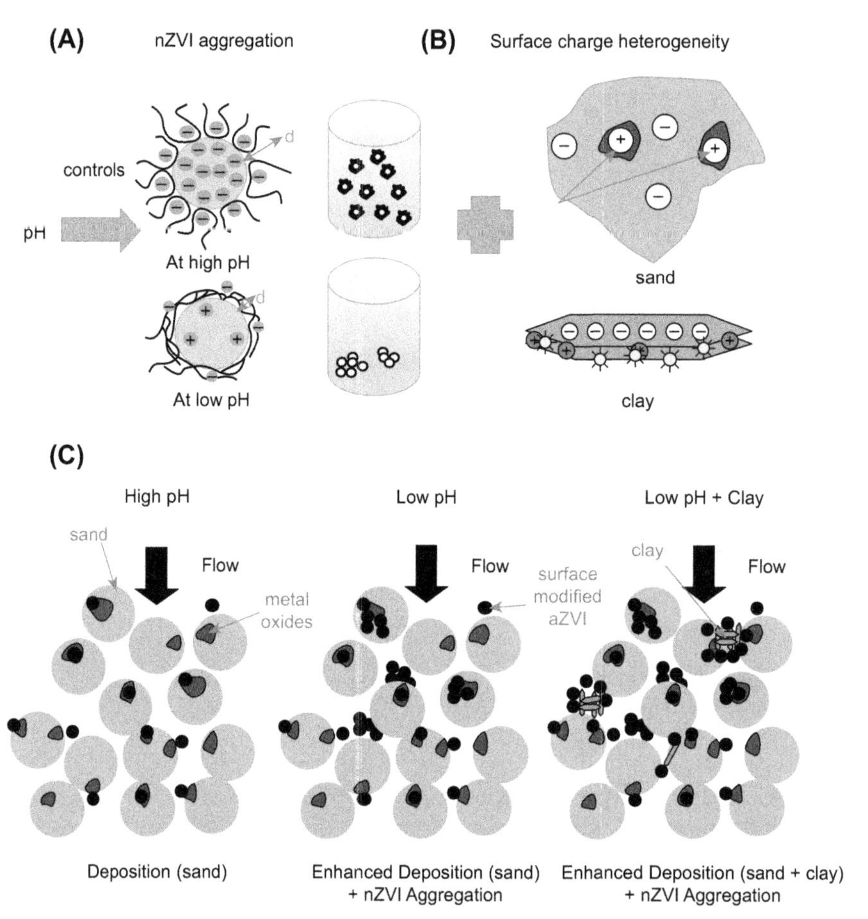

FIGURE 35 A model describing mechanisms of aggregation and deposition of surface functionalized zero-valent iron nanoparticles (nanoZVI) in heterogeneous porous media. (A) nZVI aggregation; (B) Surface charge heterogeneity; (C) Deposition. *Reprinted from Kim et al. (2012) with permission.*

pH, deposition and aggregation both enhanced. At low pH for sand and clay, deposition and mechanical filtration by clay aggregates formed the interstitial spaces.

Raychoudhury et al. (2010) and Kanel and al-Abed (2011) have described in detail the mechanisms underlying the fate of bare and functionalized nanoparticles in soil under different conditions. The key point here is that functionalization may have profound effects on their interaction with soil.

4.4 Nanoparticle Release from Soil via Cationic Exchange

Liang et al. (2013) have shown effects of salt concentrations (Figure 36: I, II, and III represent 1, 5, or 10 mM $Ca(NO_3)_2$, respectively) on the nanoAg breakthrough curves for soil. The experiment was conducted in three steps. In step A, a nanoparticle solution containing specific concentration (I, II, or III) of $Ca(NO_3)_2$ was allowed to pass through the soil column for nanoparticle

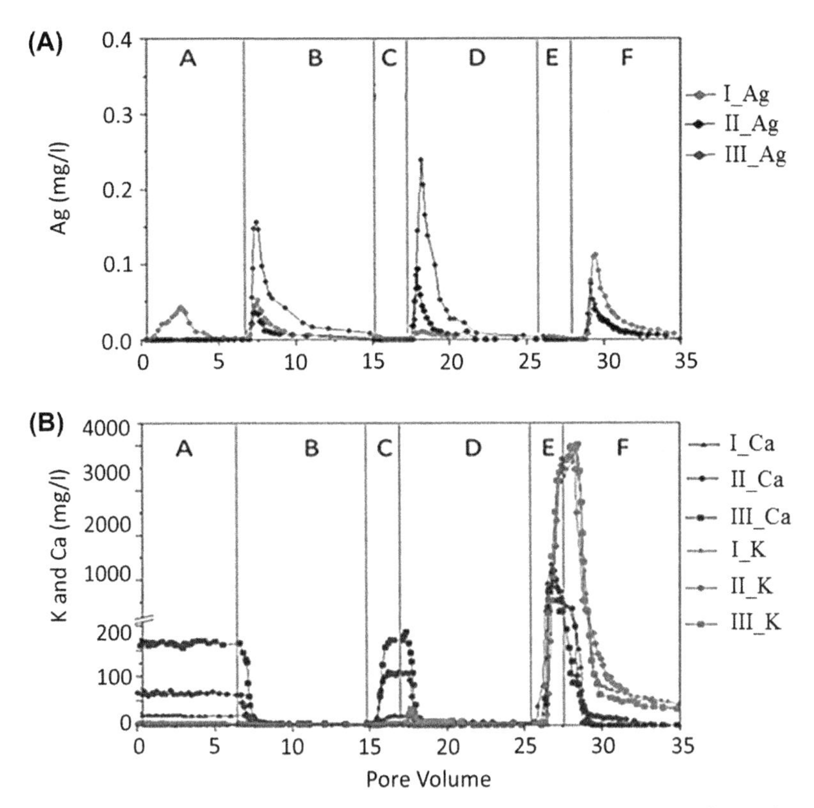

FIGURE 36 Gold nanoparticle breakthrough curves from different salt concentration conditions (I, II, and III represent different $Ca(NO_3)_2$ concentrations). (A) C/C_0 ratio and (B) K and Ca concentrations. (A–F) represent different elution profiles. *Reprinted from Liang et al. (2013) with permission.*

sorption. Then, nanoparticles were released by washing the columns with water (B), KCl low (1–10 mM) concentrations (C), water (D), KCl high (100 mM) concentrations (E), and water (F), as shown in Figure 36. The C/C_0 for nanoAg is shown in Figure 36(A), and the concentrations of K and Ca are shown in Figure 36(B). Although nanoAg remain adsorbed during step A in experiments I and II, a small amount of the injected nanoAg appeared in the efflux during step A in experiment I. The recoveries of nanoAg in step B (Milli-Q water) were 3.8%, 4.6%, and 16.0% for experiments I-III, respectively. KNO_3 was added in step C at the same ionic strength as in step A; further elution of nanoAg occurred in step D (III > II > I). Then, 100 mM of KNO_3 was added in step E (10 times greater than the initial ionic strength). Further elution of AgNPs occurred in step F (I > II and III). This release of the nanoparticles was initiated by the cation exchange during steps C and E, respectively.

Figure 36(B) shows plots of K^+ and Ca^{2+} concentrations in the column efflux from experiments I–III. Almost all of the injected K^+ mass and most of Ca^{2+} in step A remained in the column, even after flushing with Milli-Q water during step B. Addition of 1–10 mM KNO_3 resulted in the release of about 2–20% of bound Ca^{2+} in step C. Then, the addition of 100 mM KNO_3 further released about 30% of Ca^{2+} in step E, which was recovered by water in step F. This further indicates that excess amounts of K^+ in step E produced a substantial release of Ca^{2+} due to cation exchange. A comparison of Figure 36(A) and (B) reveals that very little nanoAg release occurred during cation exchange (C and E), while rather a significant release of nanoAg occurred during ionic strength reduction (D and F) immediately following cation exchange. These observations suggest that both cation exchange and ionic strength reduction were needed to release nanoAg from the soil grains in the presence of divalent cations.

4.5 Cotransport of NanoTiO_2 and Fullerene (C_{60})

Many previous studies have investigated the transport and deposition of nanoTiO_2 or nanoC_{60} under environmentally relevant conditions (Tripathi et al., 2012; Choudhury et al., 2012, 2011, Wang et al., 2012a,b; 2008a,b; Aiken et al., 2011; Chen et al., 2011, 2008a,b; Chen and Elimelech, 2009; Godinez and Darnault, 2011; Thio et al., 2011; Ben-Moshe and Dror, 2010; Tong et al., 2010; Li et al., 2008; Lecoanet and Wiesner, 2004). A key disadvantage of earlier studies is that they have mainly focused on understanding the transport of one type of nanoparticle, either nanoTiO_2 or nanoC_{60} in soil. Cai et al. (2013) have shown that nanoTiO_2 and nanoC_{60} exhibited cotransport and retention behaviors in saturated porous media under various ionic strength conditions. Figure 37 shows the breakthrough curves for (A) nanoTiO_2 in soil at pH 5 (Figure 37(A)) or pH 7 (Figure 37(B)) containing 0.1 to

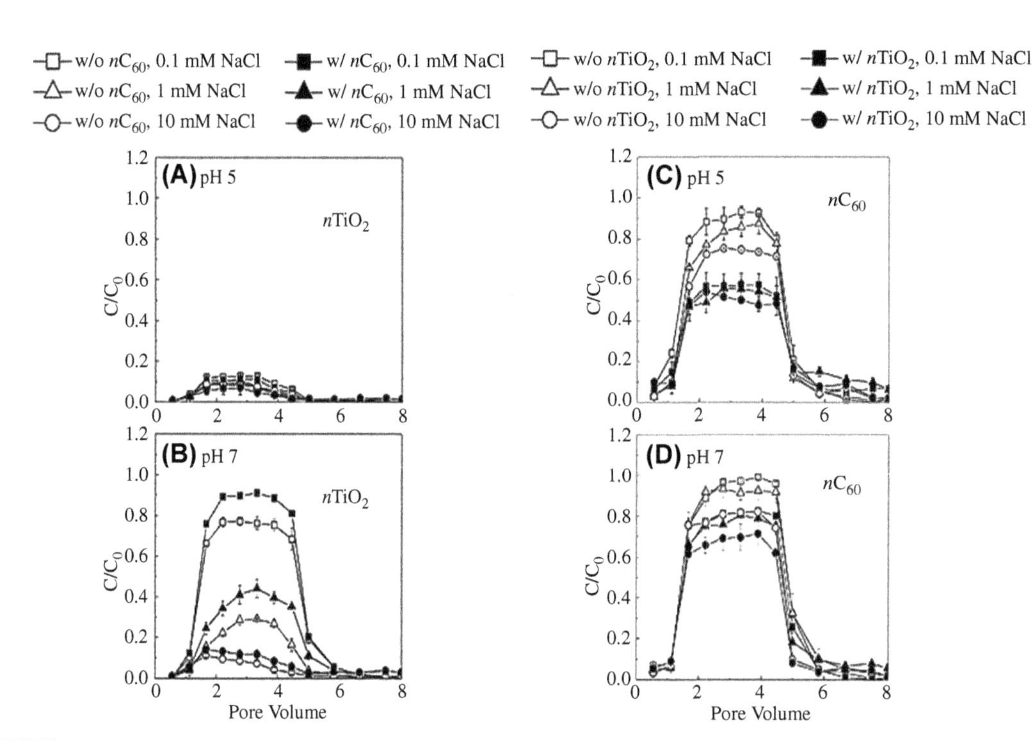

FIGURE 37 Effects of pH and nanoC_{60} on the breakthrough curves for (A) nanoTiO_2 in soil at pH 5 or 7 (B) containing 0.1–10 mM NaCl with or without nanoC_{60}, and nanoC_{60} in soil at pH 5 (C) or pH 7 (D) containing 0.1–10 mM NaCl with or without nanoTiO_2. *Reprinted from Cai et al. (2013) with permission. (n: nano)*

10 mM NaCl with or without nanoC_{60} and (B) nanoC_{60} in soil at pH 5 (Figure 37(C)) or pH 7 (Figure 37(D)) containing 0.1–10 mM NaCl with or without nanoTiO_2. Figure 38 shows the nanoparticle distribution in the column (experimental conditions identical to those described for Figure 37).

The results are summarized as follows:

1. The influence of solution ionic strength and/or nanoC_{60} on the transport of nanoTiO_2 in quartz sand was minimal at pH 5. The breakthrough curves of nanoTiO_2 in quartz sand ranged from 9 to 11% under all examined ionic strength conditions in the presence or absence of nanoC_{60} at pH 5.
2. At pH 7 without nanoC_{60}, relatively high breakthrough plateaus (11% and 78% at 0.1 mM for pH 5 and 7, respectively) of nanoTiO_2 were observed. The breakthrough plateau decreased with increasing ionic

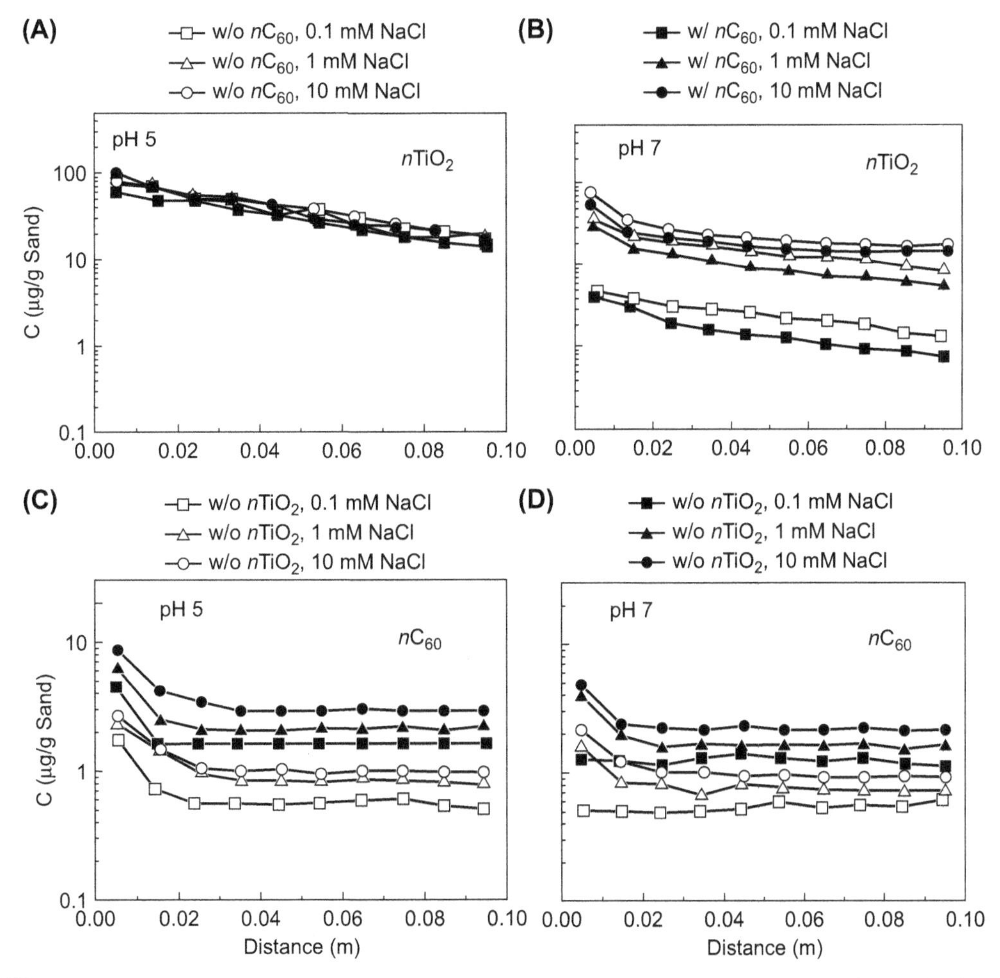

FIGURE 38 Nanoparticle distributions in the column described in Figure 37. *Reprinted from Cai et al. (2013) with permission. (n: nano).*

strength. The breakthrough plateau of nanoTiO_2 at pH 7 in the presence of nanoC_{60} was higher than those without nanoC_{60} under all examined ionic strength conditions (0.1–10 mM). The authors proposed that nanoC_{60} in suspension might compete with nanoTiO_2 for deposition sites on quartz sand, contributing to the enhanced transport of nanoTiO_2 in quartz sand at pH 7. In fact, the nanoC_{60}-nanoTiO_2 may be a major contributor to the enhanced nanoTiO_2transport in porous media at pH 7.

It is proposed that nanoC_{60} pre-exposure competitively suppresses sorption of nanoTiO_2, resulting in their mobilization in effluent (higher breakthrough plateau). However, nanoTiO_2 pre-exposure facilitates nanoC_{60} adsorption by binding with it and forming a soil-nanoTiO_2-nanoC_{60} complex. This results in a decrease in the breakthrough plateau.

4.6 Conclusion: Fate of Nanoparticles in the Environment

Recently, there has been a rapid increase in the application of different nanoparticles in medicine and environmental remediation. This has increased the likelihood of their release, either individually or in combination, into the environment. With rapidly increasing applications of various types of engineered nanoparticles, their simultaneous release into surface and subsurface environments is inevitable (Delay and Frimmel, 2013). Therefore, it is important to understand the sorption and mobilization of nanoparticles to assess the risk of water and food contaminations. The breakthrough experiments measure these parameters in soil. Although most studies have looked at individual nanoparticles, an understanding of cotransport of nanoparticles will help us to understand the fate and transport mechanisms of nanoparticles in complex and realistic environments. Cai et al. (2013) have suggested that the cotransport of TiO_2 and C_{60} nanoparticles in water-saturated porous media is far more complex than the transport of single-component nanoparticles. Further research is required to clarify the fate of nanoparticles in the environment.

5. NANOPARTICLE ECOTOXICOLOGY

One of the most alarming problems associated with the use, disposal, and manipulation of nanoparticles is their environmental contamination and ensuing toxic effects on the ecosystem (Monteiro-Riviere and Tran, 2007, 2005). A basic review of the ecosystem, principles of ecology, and ecotoxicology of classic bulk toxins (>500 nm) is discussed in Appendix 2 (readers are encouraged to review the Appendix section to fully appreciate nanoparticle ecotoxicology). Realizing the importance of ecological issues, research is being carried out in the area of ecotoxicity with regard to nanoparticles.

Different bioindicators include *Daniorerio* (zebrafish), *Pseudokirchneriella subcapitata* (microalgae), *Daphnia magna* (planktonic crustacean) and many other aquatic and soil dwellers (Bar-Ilan et al., 2009; Cheng et al., 2007; Gong et al., 2012; Griffitt et al., 2007; Hund-Rinke and Simon, 2006; Kasemets et al., 2009; Lin and Xing, 2007; Marsalek et al., 2012; Zhu and Cai, 2012). A disadvantage of these studies is that a limited number of species have been investigated, and then the results have been extrapolated to decipher the mechanisms of nanoparticle ecotoxicology. As shown in Appendix 2, ecosystems possess a complex but balanced environment in which a minor alteration may imbalance the ecosystem. Whether the ecosystem will recover or establish a new balance will depend upon its species–resource interaction. Many factors, such as speciation, diversity, trophic effect, competition for resources, etc. may play a role in the development of an ecosystem's homeostasis. In addition, the experimental data pertaining to ecotoxicological profiles are dispersed, inconsistent, and controversial. Therefore, many questions regarding the mechanism of ecotoxicity of nanoparticles remain unanswered.

Another problem associated with nanoparticles is that the research data collected for bulk particles cannot be extrapolated to estimate nanoparticle toxicity. This is because nanoparticles exhibit size- and surface-dependent properties not found in traditional bulk chemicals (discussed elsewhere in the book). Therefore, the aim of this section is to review the published work and propose a strategy to develop a unified hypothesis for ecological effects of nanoparticles.

5.1 Effects of Nanoparticles on Producers: Phytoplankton and Higher Plants

5.1.1 Phytoplankton

Phytoplanktons, a photosynthetic autotrophic microbial component of aquatic systems, are key primary producers in the marine food chain. Zooplanktons, including *Daphnia* (primary consumers in aquatic food chain), prey upon these photosynthetic microbes. The primary consumers in the marine ecosystem play an important role in transportation energy as well as aquatic pollutants across the marine food chain (Raisuddin et al., 2007; Ohman and Hirche, 2001). Because the copepod is both an omnivore and a filter-feeding organism (Lee et al., 2007; Ogawa, 1977; Ito, 1970), it may prove useful as a test organism for estimating the influence of nanoparticles in the marine environment (Lee et al., 2008). Studies have shown that phytoplanktons can be negatively affected by nanoparticles present in the aquatic ecosystem (Nel et al., 2006; Moore, 2006; Wigginton et al., 2007; Navarro et al., 2008; Cross et al., 2007; Mu and Sprando, 2010 Dıaz et al., 2008). However, these studies are limited to a few small-scale laboratory investigations heavily based on single-cultured species, with questionable applications to natural ecosystems. Consequently, there is a need to study nanoparticles' effects on natural algal communities that vary in species composition, physiological state, and experience in widely diverse ambient physicochemical conditions.

5.1.2 Higher Plants

Higher plants closely interact with soil, water, and the atmospheric compartments of the environment. Therefore, in addition to experiencing toxicity, plants also constitute one of the main routes of exposure to nanoparticles in the environment (Navarro et al., 2008). Plants, because of direct uptake and translocation of contaminants, are being used as one of the most important mechanisms for phytoremediation (Burken and Schnoor, 1999 and Collins, 2006). Unfortunately, only a few studies have considered interaction of nanoparticles with plants (Ma et al., 2010; Rico et al., 2011), and very little is known about the mechanisms of biological uptake and accumulation of nanoparticles in plants (Ju-Nam and Lead, 2008). In general, most research has studied edible plants to establish the plant–nanoparticle interaction (Nair et al., 2010; Rico et al., 2011), such as ryegrass (Lin and Xing, 2008), zucchini (Stampoulis et al., 2009), lettuce, crop plants, and wheat (Lee et al., 2012). Therefore, fundamental questions remain on the toxic effects of nanoparticles on higher plants and the impact of plant species and nanoparticles' physicochemical properties on the plant uptake potential.

A review of the literature revealed that the pioneer research on plant toxicity of nanoparticles was conducted by Yang and Watts (2005) and Lin and Xiang (2007). They showed that nanoAl_2O_3 and nanoZnO affected plant germination and that the largest effects in occurred in root elongation. Zn and nanoZnO changed the root shape and decreased plant biomass (Lin and Xing, 2008). Zhang et al. (2015) compared the rate of biomass increase in *Schoenoplectus tabernaemontani* exposed to Zn^{2+} ions (Figure 39(A)) or nanoZnO (Figure 39(B)). NanoZnO and Zn^{2+} both inhibited biomass growth depending upon the Zn^{2+} and nanoZnO concentrations as well as the exposure time. After 21 days, at concentrations of 10, 100, and 1000 mg L^{-1}, *S. tabernaemontani* was inhibited by 9%, 28%, and 54% by Zn^{2+} ions, while values were 6%, 13%, and 41%, respectively, for nanoZnO. Similarly, Lin and Xing (2008) reported a significant decrease in root elongation when *Lolium perenne* (ryegrass) was exposed to high concentrations of nanoZnO, with the biomass reduced by 50% after a 12-day exposure to nanoZnO at a level of 1000 mg/L.

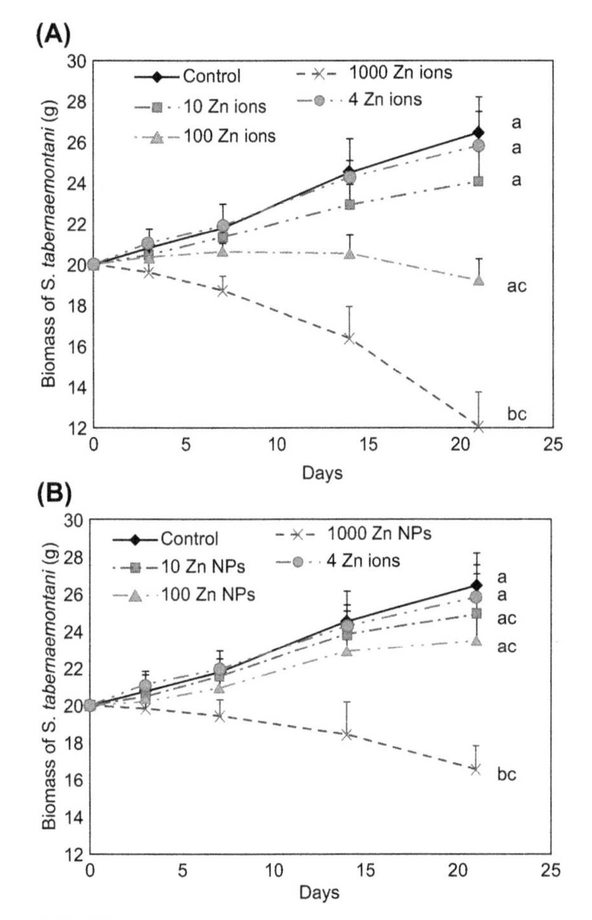

FIGURE 39 The rate of biomass increase in *S. tabernaemontani* exposed to Zn ions (A) or nanoZnO (B). *Reprinted from Zhang et al. (2015).*

An important question in determination of phytotoxicity of nanoZnO is this: whether the adverse effects of nanoZnO are due to its direct effect or to the potential dissolution of Zn^{2+} ions from the nanoZnO (Ma et al., 2010). Figure 39(B) shows no significant difference in plant growth between this treatment of 4.0 mg/L Zn^{2+} and the control, implying that the concentration of dissolved Zn^{2+} released from nanoZnO themselves was apparently not sufficient to cause significant toxicity. These observations concur with earlier studies by Lee et al. (2010) in *Arabidopsis thaliana* and Lin and Xing (2008) in ryegrass.

In addition to dissolution, chemical composition of nanoparticles also determine their phytotoxicity. Studies have shown the following:

1. Phytotoxicity of nanoZnO is greater than that of nanoAl_2O_3, nanoSiO_2, nanoNiO, and nanoFe_3O_4 (Ma et al., 2010; Parsons et al., 2010).
2. Nanoparticle accumulations exhibit species differences: nanoAl_2O_3 did not accumulate in *Phaseolus vulgaris*, but accumulated in *L. perenne* (Doshi et al., 2008). Canas et al. (2008, 2011) have shown that functionalized and non-functionalized SWCNTs penetrated the cell membranes and elongated the root systems in different crop species such as *Brassica oleracea, D. carota, C. sativus, Allium cepa, Lycopersicon esculentum, and L. sativa.* Non-functionalized SWCNTs were more effective on plants than functionalized SWCNTs.

Although possible mechanisms underlying phytotoxicity of nanoparticles are not yet known, studies (Grunathan et al., 2014, 2012; Von Moos and Slaveykova, 2014; Dimkpa et al., 2012; Lopez-Moreno et al., 2010) have proposed a key role of oxidative stress in the development of nanoparticle-induced adverse effects in plants. An increase in oxidative stress reduces the number of living cells in rice cells, causes cytotoxicity and genotoxicity in *Allium cepa* roots cell, and decreases the mitotic index with genotoxicity and DNA damage (Kumari and Mukherjee, 2009).

Table 2 shows the diversity in phytotoxicity of different nanoparticles (Rico et al., 2011). In fact, nanoparticle phytotoxicity studies are in their very early stages. Different studies have used different nanoparticles size and dose and diverse experimental designs. Future studies must involve multiple species and conduct studies in natural soil media. In addition to the germination and root/shoot growth assays, toxicity indicators based on biological markers, plant defense mechanism, and changes in plant integrity at the cellular or genetic level tested periodically during the plant's life cycle would be more appropriate. One of the urgent needs in plant–nanoparticle interaction studies is to determine the genetic response of plants, the genes that are upregulated, and those that are downregulated.

5.2 Nanoparticle Toxicity in Consumers: Soil Invertebrates

5.2.1 Soil Invertebrates

As shown in Figure 40, invertebrates play important roles in soil communities. Some directly consume detritus (detritivores that obtain nutrients by consuming detritus or decomposing plant and animal parts as well as feces), while others consume detritivores.

Soil invertebrates affect detritus decomposition rates, soil aeration, nutrient mineralization, and other services related to soil ecosystem (Six et al., 2002). Soil biota (Fortuna, 2012) also play critical roles in mobility and remediation of toxic chemicals, as well as mediate underground interactions among plants (Lal, 2002). Invertebrates are often prey for vertebrates such as birds and mammals; thus, they have vital roles in the food chains that include those animals. Despite diversified invertebrate species in soil, effects of nanoparticles have been studied only in a limited number of model species, as listed in Table 3. A comparison of the results obtained from different studies is difficult because of differences in the type of nanoparticles, soil samples, experimental protocols, and endpoints used. However, an incomplete picture of nanoparticles' phytotoxicity is emerging because of critical knowledge gaps and contradictory results.

Discrepancies in the literature that pose a serious challenge in the assessment of nanoparticles' health risk are primarily related to methodological and experimental shortcomings (inadequate nanoparticle characterization, aggregation or dissolution, lack of proper controls, and/or the use of environmentally irrelevant nanoparticle concentrations and/or exposure conditions). Table 3 also shows experimental conditions under which nanomaterials are bioavailable and cause toxicity in soil dwellers. The route of exposure to soil dwellers is species specific: collembolans will mainly be exposed to soil pore water, earthworms to both pore water and soil particles by dermal and oral contact, and woodlice to food (decaying leaf material) and soil particles by ingestion and to a limited extent to soil pore water (Tourinho et al., 2012). Despite the shortcomings, some definite conclusions can be made about nanoparticle toxicity in soils:

1. Almost all of the studies listed above showed the presence of nanoparticles in invertebrate tissue, suggesting that nanoparticles are bioavailable in soil dwellers.
2. The core elements of nanoparticles influence their toxicity (McShane et al., 2012; Canas et al., 2011).
3. For metal nanoparticles, toxicity is related to the particle itself, the dissolved ions, or both. Therefore, dissolution studies can help in deriving conclusions regarding the actual cause of toxic effects.

To achieve meaningful results, toxicity tests should be conducted in media that are as close

TABLE 2 Positive or No Consequential Effects of Nanoparticles on Food Crops

Nanoparticle	Particle Size (nm)	Plant	Concentration (mg L^{-1})	Observed Toxicity
nanoFe^0	-	G 1, G 2	100, 250, 500	No effect on germination
Al		G 3	2000	No effect on germination
	1–100	G 4	10, 100, 1000, 10000	No observed toxicity
	-	G 5	2000	Improved root growth
Ag	20	Flax	20, 40, 60, 80, 100	No effect on germination
Au	2	G 7	62, 100, 116	Low to zero toxicity
Au	10	G 7	62, 100, 116	Positive effect on germination index
Si	-	Zucchini	1000	No effect on germination
Cu	-	Lettuce	0.013% (w/w)	No effect on germination; improved shoot/root ratio
Dodecanethiol-functionalized Au	-	Lettuce	0.013% (w/w)	No effect on germination; improved shoot/root ratio
Pd entrapped in $Al(OH)_2$ matrix	-	Lettuce	0.013–0.066% (w/w)	No effect on germination; improved shoot/root ratio
3-amino-functionalized SiO_2	-	Lettuce	0.013–0.066% (w/w)	No effect on germination; improved shoot/root ratio
CeO_2	7	G 9	500, 1000, 2000, 4000	Significantly increased root and stem growth
CeO_2	<25	Wheat	100	
ZnO	8	Soybean	500	Increased root growth
Al_2O_3		Soybean	2000	No effect on germination
Fe_3O_4	20	Pumkin	500	No toxic effect
	7	Cucumber, lettuce	62, 100, 116	Low to zero toxicity
TiO_2	<100	Wheat	100	
Nanoanatase (TiO_2)	4–6	Spinach	0.25%	Enhanced protein levels (42%), Rubisco carboxylation, the rate of photosynthesis, single plant dry weight and chlorophyll content

TiO_2	5	Spinach	0.25%	Improved spinach growth related to N_2 fixation by TiO_2
TiO_2	5	Spinach	0.25%	Improved light absorbance, transformation from light energy to electron energy, and promoted carbon dioxide assimilation
Rutile (TiO_2)	-	Spinach (naturally aged)	0.25–4%	Increased germination and vigor indexes, plant dry weight, chlorophyll formation, ribulosebisphosphate carboxylase/ oxygenase activity, photosynthetic rate
Rutile (TiO_2)	-	Spinach	0.25–4%	Promoted photosynthesis, the rate of evolution of oxygen in the chloroplasts was accelerated
nanoNi$(OH)_2$	8.7	Mesquite	2	No effect
Mixture of nanoSiO_2/ nanoTiO_2		Soybean		Increased germination and shoot growth, nitrate reductase activity, absorption and utilization of water/fertilizer. Enhanced antioxidant system
Mixture of nanoAu/ nanoCu	-	Lettuce	0.013% (w/w)	No effect on germination; improved shoot/ root ratio
MWCNTs	-	Tomato	10–40	Significant increase in germination rate, fresh biomass, and length of stem; significantly enhanced moisture content inside tomato seeds
	-	G 10	2000	No effect on germination
	-	Ryegrass	2000	Increased root length
	-	Zucchini		No effect on germination
	Internal dimension: 110–170	Wheat	100	No significant effect on root or shoot growth
Single-walled carbon nanotube	8 8	G 11 G 12	104, 315, 1750	Significantly increased root length No effect
Functionalized single-walled carbon nanotube	8	G 13	9, 56, 315, 1750	No effect

Plant groups: G 1: Flax, Red clover, White clover, Meadow fescue; G 2: Barley, Ryegrass; G 3: Radish, Rape, Lettuce, Corn, Cucumber; G 4: Red kidney beans, Ryegrass; G 5: Radish, Rape; G 6: Flax; G 7: Cucumber, Lettuce; G 8: Corn, Alfalfa, Soybean; G 9: Radish, Rape, Ryegrass, Lettuce, Corn, Cucumber; G 10: Radish, Rape, Ryegrass, Lettuce, Corn, Cucumber; G 11: Onion, Cucumber; G 12: Cabbage, Carrot, Lettuce; G 13: Cabbage, Carrot, Tomato, Onion, Lettuce.
Data from Rico et al. (2011).

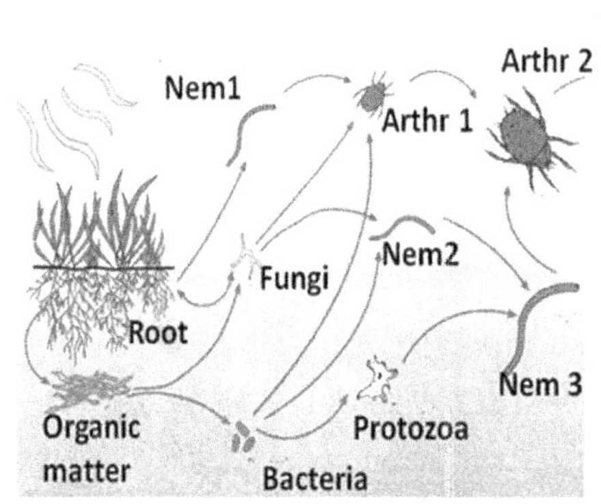

FIGURE 40 An example of the soil invertebrate network.

TABLE 3 Ecotoxicology of Nanoparticles (Tourinho et al., 2012)

NPs	Size (nm)	Species	Concentration Ranges	Exp. Media	Duration	EP	Outcomes
Ag	10 and 30–50 PC	*E fetida*	10–1000 mg Ag/kg dry soil	6	28 d	1	EC50 = 8.7 mg/kg
Ag	10 and 30–50 PC	*E fetida*	9.0–54 mg/kg (10) 0.3–27 mg/kg (30–50)	6	2 d	4	EC50 = 8.7 mg/kg (10) and 4.8 mg/kg (30–50)
Ag	50.3 (cit)	*C elegans*	10–1000 mg/L (survival) 1–10 mg/L (reproduction)	9	24 h (s) 48 h (r)	1	LC50 = 55 mg/L and EC50 = > 100 mg/L agar
Al_2O_3	60	*C elegans*	10.2–407.8 mg/L	10	24 h	1	LC50 = 82 mg/L
Cu	80	*E albidus*	130 and 230 mg Cu/kg	6	3 w 6 w	5	EC10 (lipid reduction) = 97 mg Cu NP/kg, EC10 (protein reduction) = 157 mg Cu NP/kg
TiO_2	50	*C elegans*	10	10	24 h	1	LC50 = 80 mg/L
ZnO	<100	*F candida*	100–6400 mg Zn/kg dry soil	8	28 d	1	LC50 = > 3086 mg/kg EC50 = 1964 mg/kg
ZnO	<100	*E veneta*	6–96 mg Zn/L DI water	10	24 h	1	LC50 = 1.75 mg/L (CI = 1.21–3.44 mg/L)
ZnO	<100	*E veneta*	250 and 750 mg/kg dry soil	7	21 d	2	750 mg/kg ~ reproduction EC30
ZnO	1–2.5	*C elegans*	10–1625 mg Zn/L	10	24 h 72 h	3	LC50 = 800 mg/L, EC50 (reproduction) = 46 mg/L, EC50 (mobility) = 600 mg/L
ZnO	20	*C elegans*	0.4–8.1 mg/L ZnO-NPs	10	24 h	1	LC50 = 2.2 mg/L

Experimental protocol: 1: Survival, growth, and reproduction, 2: Survival, immune activity, and life history trait, 3: Survival, reproduction, mobility, genetic, 4: voidance, 5: Energetic reserves.
Experimental media: 6: Artificial and natural soil, 7: Clay loam soil, 8: LUFA soil, 9: Agar and K-media, 10: water.

to realistic conditions. When nonsoil medium (e.g., liquids or agar) is used for testing, care should be taken in interpretating data because of the novel modes of action, such as photo-induced reactive oxygen species (ROS) generation for TiO_2, ZnO, and CeO_2 (Roh et al., 2010, 2009). Availability of RNAi, transgenic and mutant *C. elegans* strains, along with toxicogenomic approaches, may allow identification of the mechanisms of nanoparticle toxicity (Tsyusko et al., 2012). Important areas of future research include the following:

1. Investigation of additional potential mechanisms of toxicity
2. Further systematic probing of the effect of nanoparticle characteristics on toxicity to develop predictive models
3. Extension of molecular-level mechanistic toxicity to an understanding of cellular, tissue, and organism effects, and elucidation of multigenerational effects

5.2.2 Nanoparticles in Terrestrial Food Chains

Food chains (Figure 41) are linear links starting from a primary producer, such as plants that generate energy from the sun, and ends at species that are eaten by no other species in the web (mostly humans).

As discussed previously and described in earlier studies (Chithrani et al., 2006; Hillyer and Albrecht, 2002, Lin et al., 2009), plant uptake, gastrointestinal absorption, and transmembrane transport of nanomaterials occur at each trophic level in the environment. Terrestrial and aquatic plants absorbed functionalized MWCNTs and SWCNTs via their roots. The aggregated but not dispersed nanotubes caused phytotoxicity, including suppressed growth (Villagarcia et al., 2012; Rico et al., 2011; Baun et al., 2008). Because higher plants, algae, and planktons are primary producers in many food chains, the nanotubes may accumulate in consumers at different levels. This may pose

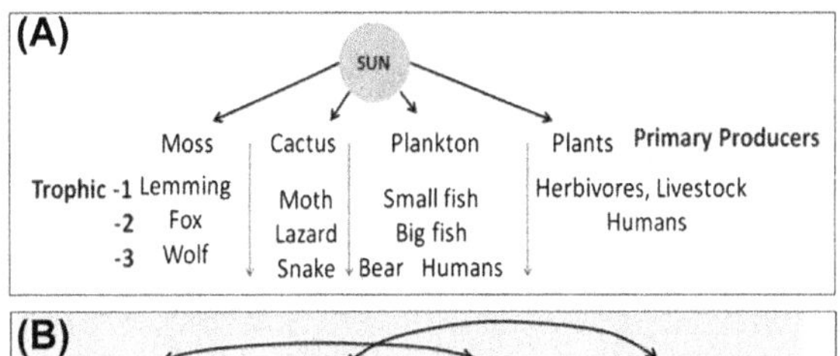

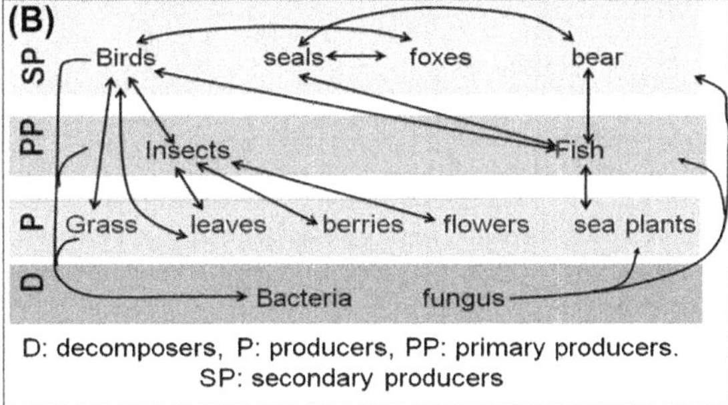

FIGURE 41 An example of food chain trophic level. (A) Relationship between the sun and different producers. (B) Different trophic levels.

significant risk to the consumers due to the nanoparticles' bioaccumulation in food chain (Judy et al., 2011; Unrine et al.,2008a,b, 2010, 2012; Wanga et al., 2008; Ferry et al., 2009).

Figure 42 shows three hypothetical food chains: soil (earthworms), soil (birds), and plants (insects). The primary producers in soil are higher plants and photosynthetic microbes, actinomycetes. Judy et al. (2011) provided evidence of trophic transfer and biomagnification of nanoparticles from a terrestrial primary producer to a primary consumer (Figure 43). Unrine et al. (2012) have shown trophic transfer and biodilution of nanoparticles from soil to earthworms (nematodes) (Figure 44). Holbrook et al. (2008) showed biomagnification of nanoparticles from soil into bacteria (Figure 45). These observations that nanoparticles can biomagnify highlights the importance of considering dietary uptake as a pathway for nanoparticle exposure and raises questions about potential health risks in humans. In fact, the risk may be magnified by the effects of human activity on the environment.

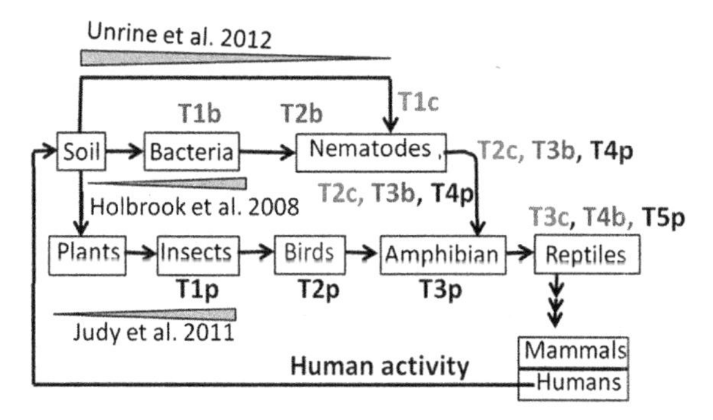

FIGURE 42 Examples of three hypothetical food chains (c: soil – earthworms, b: soil-birds and p: plants-insects, T: trophic levels (example T1b is trophic level 1 for food chain b)).

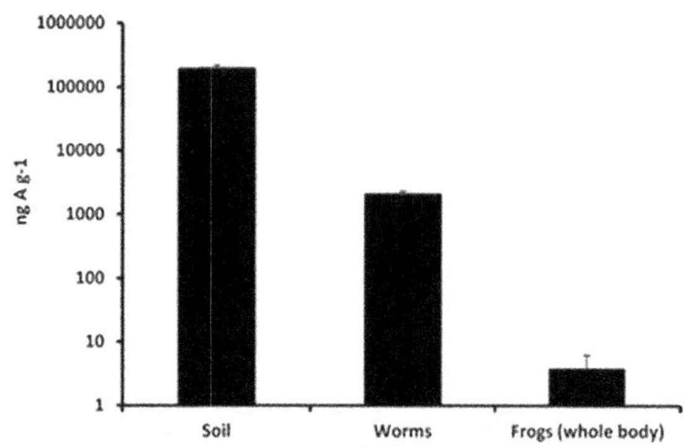

FIGURE 44 Transfer of $_{nano}$silver from soil (inhibiting primary producers) to worms (primary consumer) to frogs (secondary consumers or predator).

5.3 Aquatic Ecosystems

Overall characteristics of aquatic ecosystems are discussed in Appendix 3. In general, aquatic ecosystems include wetlands, rivers, lakes, oceans, and coastal estuaries. Aquatic ecosystems can be divided into three types: freshwater (rivers, lakes, and wetlands), salt water (oceans, sea, and estuaries), and polar water (Arctic and Antarctic oceans). Humans may have created a new type of water called *polluted* water.

5.3.1 Factors Determining the Distribution of Nanoparticles in Aquatic Ecosystems

Nanoparticles reach the aquatic environment via the following routes (Figure 46):

1. Direct exposure (determined by water to suspended sediment attachment rate, sedimentation, resuspension, and burial)
2. Leaching from soil (determined by runoff and transport, soil particle attachment rate, soil particle detachment rate, clay particle

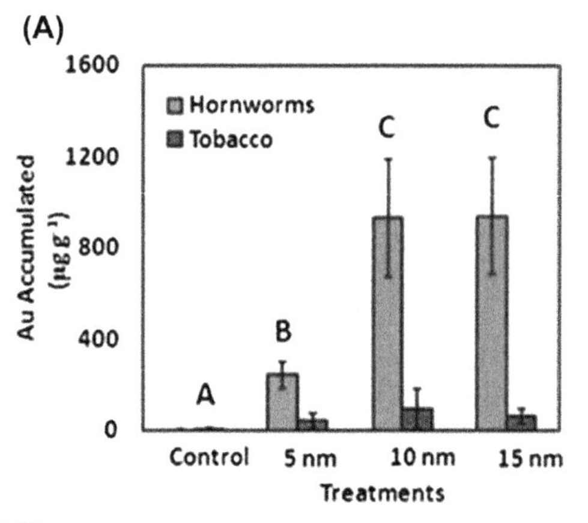

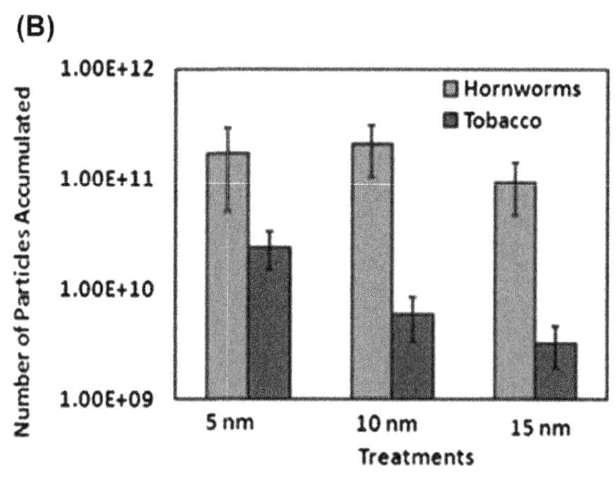

FIGURE 43 Transfer of gold nanoparticles between tobacco (primary producer, red (dark gray in print versions) filled) and hornworms (primary consumer, blue (gray in print versions) filled). (A) Au concentration and (B) $_{nano}$Au accumulated. *Reprinted from Judy et al. (2011) with permission.*

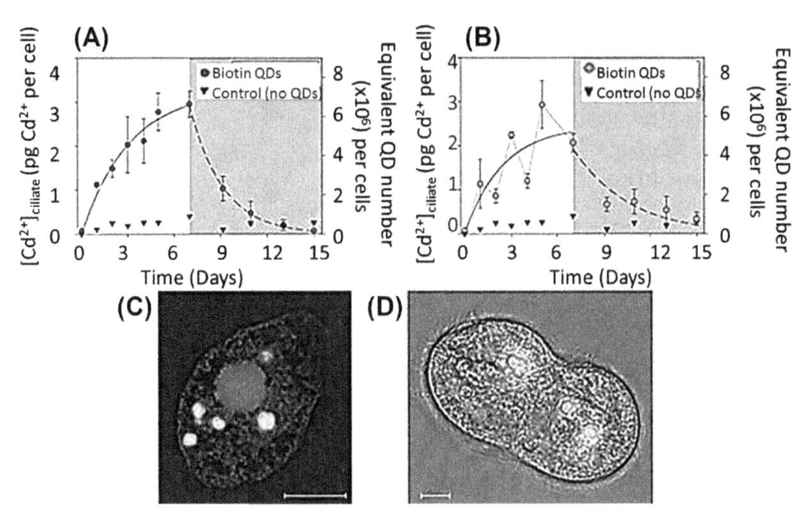

FIGURE 45 Biomagnification of nanoparticles from soil into bacteria. (A) COOH-functionalized quantum dots (QDs), (B) biotin-functionalized QDs (showing fluctuations), (C) QD accumulation in photosynthetic microbes and (D) QD accumulation in soil bacteria in which QDs were aggregated and toxic. *Reprinted from Holbrook et al. (2008) with permission.*

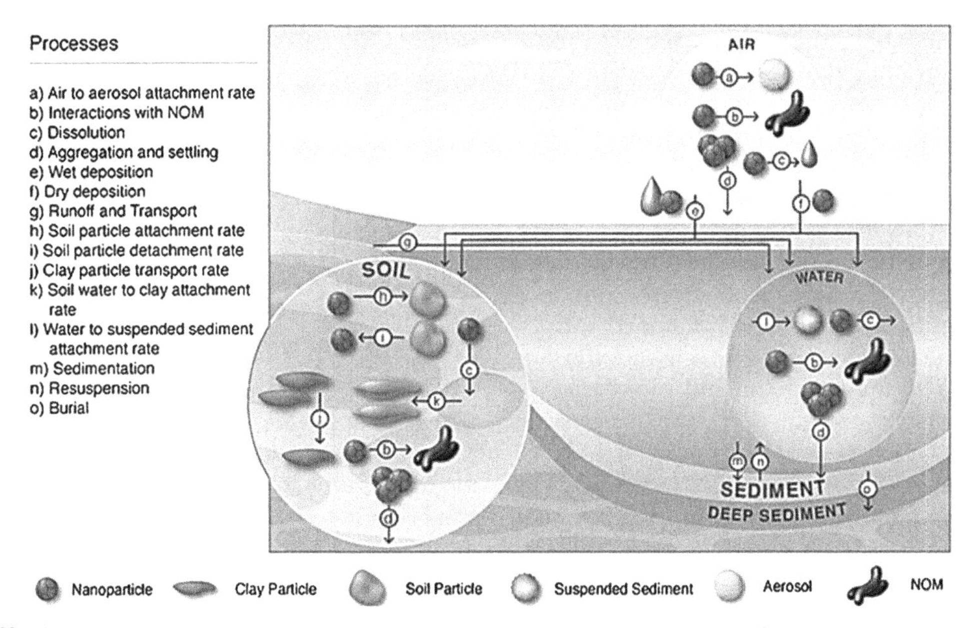

FIGURE 46 Distribution and fate of nanoparticles in the aquatic environment. *Reprinted from Garner and Keller (2014) with permission.*

transport rate, soil water to clay attachment rate)
3. Deposition from air (determined by aerosol attachment rate, interaction with organic materials, dissolution, aggregation and setting, wet deposition, and dry deposition)

In the aqueous environment, the fate of a nanoparticle is determined by its physicochemical properties as well as the properties of the environment (pH and ionic strength). Hydrophilic nanoparticles such as metal nanoparticles (charged), PEG-functionalized zero-charge metal nanoparticles, poly(amidoamine) dendrimers, etc., dissolve in water and translocate with water flow. Hydrophobic CNTs or polyaromatic dendrimers are water insoluble and rapidly aggregate, either binding to the suspended soil particles or settling down at the sediment surface, where they are ingested by the sediment dwellers. Another key factor determining the ecological effects of nanoparticles on oceans or lakes is the boundary layers between water and air, known as the surface–air interface, usually a water molecule monolayer. This represents a critical interface for biological, chemical, and physical processes. It is well established that water molecules form extensive intermolecular H bonds (-O…H-) that hold the mass of molecules together. Stiopkin et al. (2011) have characterized the water–air interface and have concluded that, unlike the subsurface water molecules, the surface water atoms contain a "dangling" bond (nonbonded H atom) protruding into the air (Figure 47). This makes the surface relatively hydrophobic and more reactive (Palacin and Barraud, 1991). This plays an important role in environmental toxicity of bulk toxicants and nanoparticle toxicity. The interface monolayer allows hydrophobic nanoparticles to remain at the surface and interact with water as well as air.

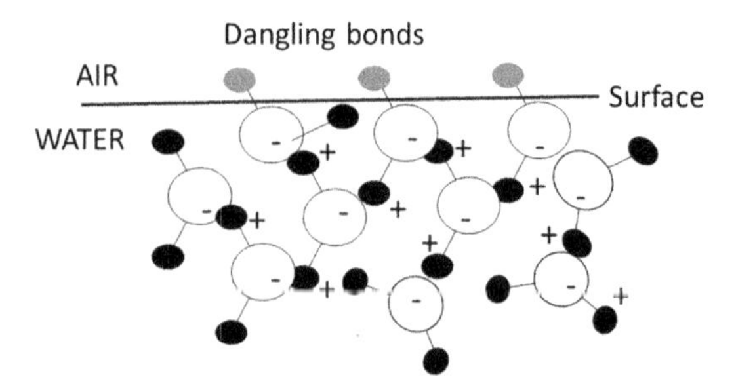

FIGURE 47 Water molecule at the surface–air interface. One hydrogen atom of surface water molecules may protrude in air (green (gray in print versions)). *Reprinted from Stooping et al. (2011) with permission.*

5.3.2 Sources of Nanoparticles in Lakes, Rivers, and Oceans

Two distinct sources are responsible for nanoparticles in aquatic environments: (1) natural nanoparticles synthesized in the aqueous environment and (2) engineered nanoparticles synthesized for medical or industrial applications that reach the aqueous environment via accidental/intentional release or end-of-life disposal. The ecological effects of natural nanoparticles have been discussed extensively elsewhere (Eriksson, 1952; Banfield and Navrotsky, 2001; Hochella et al., 2008; Gilbert and Banfield, 2005; Bargar et al., 2008; Manceau et al., 2008). However, the ecological effects of engineered nanoparticles have begun to receive attention recently.

5.3.3 Aquatic Ecotoxicology of Carbon-Based Nanoparticles

5.3.3.1 GRAPHENE AND GRAPHENE OXIDE

The projected rapid increases in the production and application of graphene may also increase the potential for their release into the environment. Graphene nano-preparations are used for environmental applications, such as adsorbents for wastewater and drinking water treatment (Upadhyay et al., 2014, Zhang et al.,

2011; Zhao et al., 2011); solid-phase extraction (Wang et al., 2012a,b); desalination (Mishra and Ramaprabhu, 2011); catalysts for aqueous organic pollutant oxidation and degradation (Sun et al., 2012); coating materials for filtration (Gao et al., 2011); and waste disposal. The graphene oxide nanocomposites, when subjected to photoactivation in the UV range, could release graphene oxide particles into the environment (Bernard et al., 2011). The released graphene may have significant adverse environmental impacts (Chowdhury et al., 2014, 2011; Zhao et al., 2014; Kotchey et al., 2011). One interesting effects of graphene is its antimicrobial activity comparable to or greater than kanamycin. The minimum inhibition values for graphene and kanamycin were 1–8 and 64–128 μg/mL, respectively (Krishnamoorthy et al., 2012). The gram-negative bacteria appears to be more sensitive to graphene than gram-positive bacteria, possibly due to the thinner peptidoglycan layer of gram-negative species (7–8 nm) as compared to gram-positive microbes (20–80 nm). However, Akhavan and Ghaderi (2010) showed the opposite result. In nematodes, crustaceans and fish, graphene exhibits poor toxicity (Pretti et al., 2014; Begum and Fugestu, 2013; Begum et al., 2011).

Algae may be one of most vulnerable organisms to graphene in the aquatic food chain (Gurunathan et al., 2012; Liu et al., 2011).The symptoms of phytotoxicity included necrotic lesions and membrane damage of leaf cells after graphene oxide exposure (dose: 500 mg/L to 2000 mg/L). Aquatic plants may be more sensitive to graphene as compared to terrestrial species. However, no comparative study has been performed to verify this hypothesis. The graphene-induced algae toxicity may have certain indirect consequences. A decrease in algal population may decrease pO_2, and an increase in pCO_2 with an ensuing decrease in pH. This will reduce the aquatic diversity and water quality.

Possible mechanisms for the relatively high phytotoxicity of graphene are shown in Figure 48. Graphene, after uptake, enters vesicles

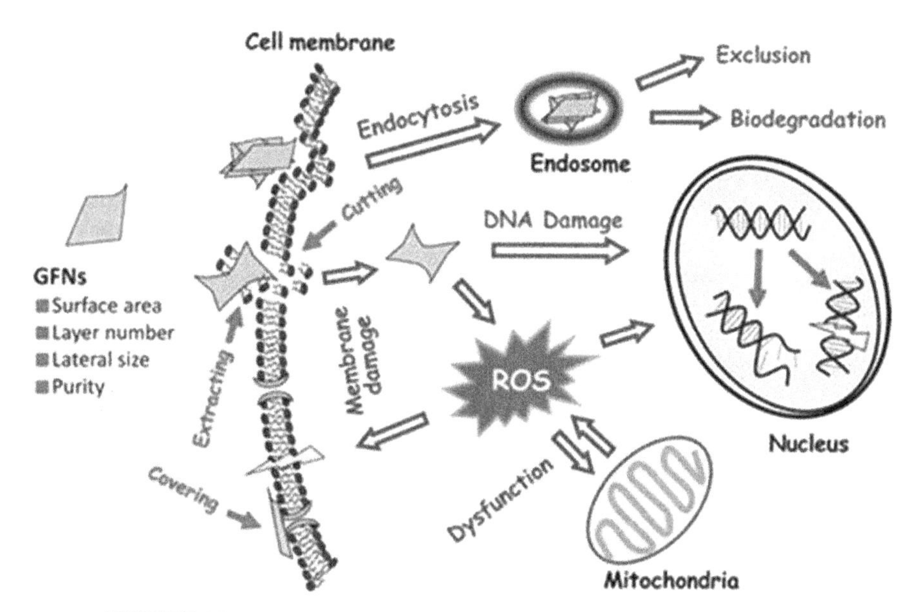

FIGURE 48 Mechanisms of phytotoxicity. Details are described in the text.

and then is deposited into the vacuoles (Begum and Fugetsu, 2013). The internalized cytoplasmic graphene induces oxidative stress (Gurunathan et al., 2014, 2012; Begum et al., 2011; Chang et al., 2011) by interacting with mitochondria, chloroplasts (for plant and alga cells), peroxisomes, and oxidases (Begum and Fugetsu, 2013). Oxidative stress may result in lipid peroxidation (Krishnamoorthi et al., 2012) and the loss of cell integrity. In addition, graphene may also cause DNA damage, including nucleic acid fragmentation and destruction independent of their effects on oxidative stress in human cells (Chang et al., 2011), but not in plant cells (Begum and Fugetsu, 2013). The significance of this needs further investigation. Although not yet recognized, graphene-induced DNA damage may have toxicological consequences.

5.3.3.2 FULLERENES (C_{60})

C_{60} is poorly soluble ($1.3 \times 10-14$ mg/L) in water, which makes C_{60} essentially irrelevant in the aquatic environment because it will rapidly bind soil particles or organic particles. However, studies have proposed that, in water, C_{60} aggregation results in formation of nanosized C_{60} (nanoC_{60}), which is negatively charged and relevant to the environment (Fortner et al., 2005). Two hypotheses have been proposed for nanoC_{60}'s negative charge: (1) molecular C_{60} is partially oxidized by water and (2) C_{60} accepts electrons from water molecules (Andrievsky et al., 1999; Deguchi et al., 2001). The latter is supported by ^{13}C-NMR results (Fortner et al., 2005). Thus, in experimental studies, formation of nanoC_{60} may depend upon the solubilization protocols used. The chemical properties and reactivity of nanoC_{60} are shown in Table 4.

Pristine C_{60} induces the production of ROS and ensuing toxicity (cell pathology and overloading the antioxidant defense system) (Oberdorster, 2004). An increase in ROS may lead to increased modification of pristine C_{60} molecules (Yamakoshi et al., 2003). A possible relationship between the band gap and the generation of ROS is shown in Figure 48. Lof et al. (1995) and Sergio et al. (2013) have shown the existence of a band gap of 2.3–2.75 eV in pristine C_{60}. The band gap decreased significantly (0.77–0.5 eV) upon the C60-Zn^{2+}/H^+ complex, making them more reactive and potent in forming ROS. Because

TABLE 4 Quantum Chemical Calculations of the Three Different Arrangements; C_{60} w/without Zn^{2+}, $C_{60}-H^+$ w/without Zn^{2+}, and $C_{60}-OH^-$ w/without Zn^{2+}. Their Difference and Band Gap Delineate the Reactivity of the Molecular Complex. The smaller the band gap, the higher the chemical reactivity (Sergio et al., 2013)

	C_{60}	$C_{60}-Zn^{2+}$	$C_{60}-H^+$	$C_{60}-H^+-Zn^{2+}$	$C_{60}-OH^-$	$C_{60}-OH^--Zn^{2+}$
E_{gap}(eV)	2.758	0.829	1.774	0.602	1.660	0.610
Reactivity	1	3.3×	1.5×	4.5×	1.7×	4.5×
Result	C_{60} binds Zn^{2+} spontaneously and increases its reactivity by a factor of 3.3. C_{60} may deplete ions such as zinc from the environment		Fullerene binds both the Zn and the H ions covalently and depletes zinc ions and protons in the cellular environment. The mitochondria, which rely on the high gradient of protons for energy generation, are particularly prone to toxicity of the C_{60}. Zn^{2+} increases the fullerenes' reactivity by 4.5 times		Fullerene binds the OH^- covalently and the zinc ion. Both the OH^- and the zinc ion increase the reactivity of the fullerene 4.5 times the pristine fullerene	

fullerenes are used in different modified and nonmodified forms, the molecules may have widely different absorption rates and periods of duration within the body (Table 4).

Oberdorster (2004) have shown that the LC_{50} values for nanoC_{60} (uncoated, water-soluble, colloidal fullerenes) in *Daphnia magna* were 800 ppb. Exposure of largemouth bass (*Micropterus salmoides*) to 0.5 ppm nanoC_{60} for 48 h did not cause mortality, but it resulted in lipid peroxidation in the brain and glutathione depletion in the gill. A cytochrome P450 (CYP2K4) involved in lipid metabolism was upregulated. Interestingly, the investigators later showed that daphnia 21-day exposures resulted in a significant delay of molting and reduction in offspring production at 2.5 and 5 ppm nanoC_{60}, which could possibly produce impacts at the population level (Zhu et al., 2006). Oberdorster et al (2006) have studies toxicity of water-soluble fullerenes (nC60) in the freshwater crustaceans Daphnia magna and Hyalella azteca, a marine harpacticoid copepod Heteropsyllus nr. nunni Coull, and two fish species, fathead minnow Pimephales promelas and Japanese medaka Oryzias latipes. In Daphnia, only 21-day exposures (at 2.5 and 5 ppm nC60) significantly delayed molting and reduced offspring production, which could possibly produce impacts at the population-level. The peroxisomal lipid transport protein PMP70 was significantly reduced in fathead minnow, but not medaka, indicating potential changes in acyl-CoA pathways. No significant toxicity occurred in other species. (Oberdorster et al., 2006). The toxicological significance of this is not fully understood.

5.3.3.3 CARBON NANOTUBES

Carbon nanotubes are hollow graphene cylinders of nanometers to millimeters in length (Eklund et al., 2007). CNTs can be SWCNTs with a diameter of 0.7–3 nm or MWCNTs with a diameter of 10–25 nm (Baughman et al., 2002). Metal catalyst residuals and amorphous carbon are the main impurities in the as-produced CNTs (Donaldson et al., 2006), which can be cleaned by oxidation, acid treatment, annealing, sonication, filtration, and functionalization processes. These cleaning processes often cause defects in CNTs, resulting in an increase in their reactivity.

CNTs are widely used in consumer products in the market, including sporting goods, textiles and shoes, vehicle fenders, electronics, X-ray tubes, and batteries (Dekkers et al., 2007). CNTs may enter aquatic environments through sources such as general weathering, disposal of CNT-containing consumer products, accidental spillages, and waste discharges. Although persistent in the aquatic environment, the CNTs' effects on aquatic organisms such as cladoceran (*Chydorus sphaericus*) and the salamander (*Ambystoma mexicanum*) have not been fully evaluated and the results of previous studies are not conclusive (Mwangi et al., 2012). No significant effects on mortality, development, and reproduction were observed in estuarine copepods (*Amphiascus tenuiremis*) exposed to purified SWCNTs, but mortality increased, fertilization rates decreased, and molting success decreased when the copepods were exposed to the as-produced SWCNTs that contained metal impurities (Templeton et al., 2006). Comparable results were obtained by other investigators (Cheng et al., 2007 and Kennedy et al., 2008) in other aquatic species. Taken together, metal impurities may play a key role in development of CNTs to ecotoxicity of aquatic organisms.

Mwang et al. (2012) have shown that exposure to MWCNTs or SWCNTs with and without sonication decreased the mean survival or biomass in the test organisms (Table 5). The survival of control amphipods was 88%, while the survival rates of amphipods exposure to nonsonicated and sonicated MWCNT-H were 5% and 2.5%, respectively. The mean survival of amphipods, midges, and mussels with nitric acid–cleaned MWCNT-S (metal depleted) was

TABLE 5 Mean Response of the Amphipod *Hyalella azteca*, the Midge *Chironomus dilutus*, the Mussel *Villosa iris*, and the Oligochaete *Lumbriculus variegatus* Exposed to Sonicated or Nonsonicated Single- or Multiwalled Carbon Nanotubes from Shenzhen (SWCNT-S or MWCNT-S) or from Helix (MWCNT-H) or Nitric Acid—Modified (NAM) MWCNT-S in 14-d Toxicity Tests[a] (Nwangi et al., 2012)

	Amphipod		Midge		Mussel		Oligochaete
Treatment	**Survival (%)**	**Biomass (mg)**	**Survival (%)**	**Biomass (mg)**	**Survival (%)**	**Length (mm)**	**Biomass (mg)**
Control	88 (5.0)x	1.2 (1.0)	80 (8.2)x	9.3 (2.6)x	98 (5.0)x	2.2 (0.1)	2.8 (0.2)x
Nonsonicated MWCNT-H	5.0 (10)y	NR	43 (9.6)y	3.7 (0.6)y	23 (17)y	NR	1.2 (1.0)z
Sonicated MWCNT-H	2.5 (5.0)y	NR	60 (8.2)z	1.4 (0.2)z	43 (19)y	NR	2.3 (0.4)y
Control	100 (0)x	NR	83 (5.0)x	NR	NT	NT	3.7 (0.8)x
Nonsonicated SWCNT-S	20 (12)y	NR	10 (8.2)y	NR	NT	NT	1.4 (0.4)z
Sonicated SWCNT-S	0 (0)z	NR	0 (0)z	NR	NT	NT	2.8 (0.2)y
Control	100 (0)x	NR	67 (7.5)[b]	NR	80 (28)x	NR	5.2 (4.0)x
Nonsonicated MWCNT-S	8.0 (10)y	NR	55 (3.0)	NR	35 (26)y	NR	0.8 (0.2)z
Sonicated MWCNT-S	5.0 (10)y	NR	8.0 (5.0)	NR	3.0 (10)z	NR	2.2 (0.6)y
Control	100 (0)x	0.8 (0.2)x	75 (19)x	5.5 (2.6)x	100 (0)x	1.3 (0.1)x	3.7 (1.2)x
Nonsonicated NAM MWCNT-S	95 (5.8)x	0.3 (0.1)y	60 (14)x	2.6 (0.9)y	98 (5.0)x	1.1 (0.04)y	1.3 (0.4)y

[a] *Standard deviations in parenthesis, n = 4. Different letters for survival (x), length (y), or biomass (z) in a column for a test indicate a significant difference among treatments ($p < 0.05$). The biomass is based on dry weight for the amphipod and midges and ash-free dry weight for the oligochaete.*
[b] *Control survival was less than test acceptability criteria 21. No statistical analysis was performed for this test NR = not reported because recovered organisms were used for photographs or transmission electron microscopy imaging or due to <50% survival in CNT treatments; NT = not tested.*

not significantly different from the controls, but the biomass of amphipods, midges, and oligochaetes was still significantly reduced (Table 5). The amphipods, midges, and oligochaetes also ingested the CNTs, which was confirmed with the transmission electron microscopy images. These observations suggest that aquatic organisms are affected by the metal ions present in unpurified CNTs. However, a direct intrinsic toxicity of CNTs cannot be ruled out because the acid-cleaned CNTs that did not affect the survival of amphipods, midges, and oligochaetes significantly reduced the biomass of the test organisms (Table 5). The possible mechanisms proposed for these observations are the following: (1) CNTs may form an outer surface coating, thus interfering with their respiratory systems; (2) CNTs may block the digestive track of the organisms, thus diminishing their nutrient uptake; and (3) CNTs may promote production of other chemicals, such as oxidative oxygen species, that may be toxic to organisms.

Zhu et al. (2007) conducted a comprehensive study on the 48-h acute toxicity of water suspensions of magnetic nanomaterials (MNMs) such as nano Zinc Oxide (nanoZnO), nano titanium oxide (nanoTiO_2), nano aluminum oxide (nanoAl_2O_3), C_{60}, SWCNTs, and MWCNTs to *Daphnia magna* (Table 6). The effects of such nanoparticles on the immobilization of *D. magna* were in the order: nanoZnO > SWCNTs > C_{60} > MWCNTs > nanoTiO_2 > nanoAl_2O_3, while their effects on mortality showed the following pattern: nanoZnO, SWCNTs > C_{60} > MWCNTs > nanoTiO_2, nanoAl_2O_3. The acute toxicities of all nanoparticles tested are dose dependent. *Daphnia magna* ingested nanomaterials from the test solutions, which indicates that the potential ecotoxicities and environmental health effects of carbon-based nanoparticles cannot be neglected.

In addition to aquatic ecosystem, CNTs have also been detected in municipal wastewater. Figure 49 shows the anticipated lifecycle of CNTs used in the development of commercial products. At the end of their life cycle, CNTs may eventually end up in waste treatment plants, landfills, and, to some extent, the incineration facilities. In certain cases, CNTs may enter the waste at earlier stages of their life cycle (NIOSH, 2011). For example, CNTs may leach from the consumer products and enter the environment during their use. As the production of engineered nanomaterials and their subsequent incorporation into the consumer products increase, the likelihood of the amount of nanomaterials in the waste will also increase. Once in the municipal waste, the organic contents of the waste may alter the physicochemical and biological properties of CNTs (Tong et al., 2007; Kang et al., 2009), while the antibacterial properties of CNTs (Kang et al., 2007; Ghafari et al., 2008) may alter the bacterial composition of the sludge. In landfills, high-molecular-weight organics, such as humic acid, may stabilize CNTs and increase their mobility (Lozano and Berge, 2012). However, high electrolyte concentrations may greatly reduce the mobility of

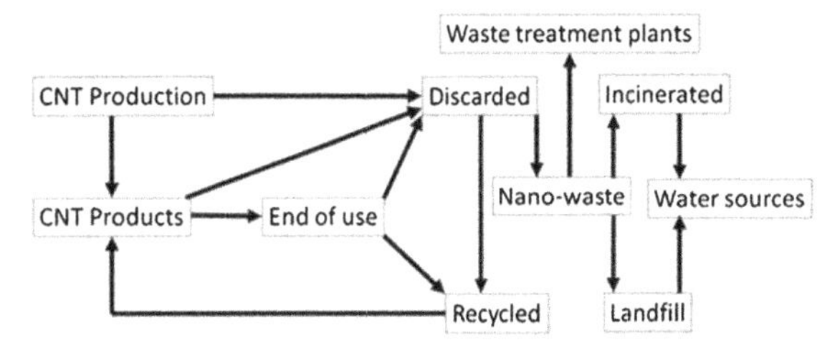

FIGURE 49 A life cycle of carbon nanotubes from production to disposal.

TABLE 6 The 48-h EC_{50} and LC_{50} to *D. Magna* of Water Suspensions of the Tested Materials (Zhu et al., 2009a)

Material suspensions	EC_{50}(mg/L)	95% CI	LC_{50}(mg/L)	95% CI
nanoAl_2O_3	114.357	111.232–191.100	162.392	124.325–214.803
Al_2O_3/bulk	>500	n.d.	>500	n.d.
nanoTiO_2	35.306	25.627–48.928	143.387	106.466–202.818
TiO_2/bulk	275.277	170.661–570.045	>500	n.d.
nanoZnO	0.622	0.411–0.805	1.511	1.120–2.108
ZnO/bulk	0.481	0.301–0.667	1.250	0.985–1.848
SWCNTs	1.306	0.821–1.994	2.425	1.639–3.550
MWCNTs	8.723	6.284–12.128	22.751	15.678–34.388
C_{60}	9.344	7.757–11.262	10.515	8.658–12.757
Carbon black	37.563	33.076–41.968	61.547	54.546–68.232

CNTs (Chen and Elimelech, 2006; Wang et al., 2008a,b). Potential health risks may result if CNTs are mobile and subsequently released to the surrounding environment, including surface and subsurface waters. CNTs are highly thermostable, but conventional incineration at 800–900 °C may degrade CNTs into biodegradable carbon products (Manzetti and Andersen, 2013). Earlier studies have shown that CNTs may partially degrade above 500 °C, but part of their molecular structure remains intact even at 1300 °C (Bom et al., 2002; Sobek and Bucheli, 2009). Thus, incineration may not be a viable option to handle CNT-contaminated waste. Development of new techniques for CNT-contaminated waste treatment is essential to handle the rapid increase in the use of CNTs in consumer products (Lam et al., 2004; Muller et al., 2008). The practice of converting polyethylene-based plastics into CNTs aggravates the dangers from the increased presence of CNTs in society (Pol and Thivagarajan, 2009).

5.3.4 *Aquatic Ecotoxicology of Metal Nanoparticles*

5.3.4.1 BEHAVIOR OF NANOPARTICLES IN AQUATIC ENVIRONMENTS

Metal oxide nanoparticles such as TiO_2, ZnO, and CeO_2 are extensively used in medicine, environmental remediation, and industry. Some commonly used applications include sunscreen, cosmetics, paint, medicine, and pharmacology. Because of their extensive use, metal/metal oxide nanoparticles are also finding their way into the aquatic environment (Garner and Keller, 2014). The overall behavior of nanoparticles in the aquatic environment is shown in Figure 50.

In water, nanoparticles' charge status and water's pH, salinity, and organic matter determine the fate of nanoparticles in water. Hydrophilic (charged particles such as ions and dipoles) but not hydrophobic metal particles (zero-charge metal nanoparticles or conjugated acids or bases; e.g., acids such as CH_3COOH at

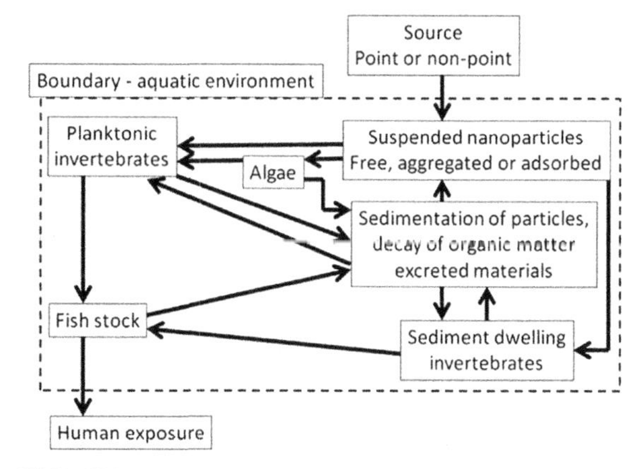

FIGURE 50 Behavior of nanoparticles in aquatic environment.

pH < pKa or bases such as $C_6H_5\text{-}NH_3$ at pH > pKa) are not miscible with water. Water-insoluble particles (1) rapidly aggregate and settle down onto the sediment surface, (2) bind the suspended soil particles and/or organic matter, and/or (3) distribute within the surface monolayer that has relatively hydrophobic characteristics. The sediment dwellers ingest nanoparticles onto the sediment surface. Many aquatic organisms, especially filter feeders, ingest nanoparticles bound to the suspended soil or organic matter. The surface monolayer, however, receives a rich supply of light that results in phototransformation of metal nanoparticles.

Water-soluble particles may remain soluble or precipitate when water properties change (increase in salinity or pH). The water-insoluble nanoparticles, such as nanoFe^0, may undergo an oxidation reaction resulting in the formation of nanoFe^{2+}.

$$Fe^{2+} + RH + X^- \Leftrightarrow Fe^0 + RX \text{ (alkyl halide)} + H^+$$

The above equation is an example of net reductive dehalogenation by iron. It is equivalent to iron corrosion, with the alkyl halide acting as a strong oxidizing agent. When there are no strong oxidizing agents present, the following modifications occur:

$$2Fe^{2+} + 4OH^- \Leftrightarrow 2Fe^0 + O_2 + 2H_2O$$

The metal transformation may also occur under anaerobic conditions, as shown below:

$$Fe^{2+} + H_2 + 2OH^- \Leftrightarrow Fe^0 + 2H_2O$$

An increase in pH of the aqueous media may precipitate zero-valent (M^0) nanoparticles. As shown in Figure 51, ecotoxicity of metal nanoparticles depends upon their stability in water. Dispersed nanoparticles are bioavailable but generally nontoxic (there are important exceptions to this, as discussed later).

If nanoparticles aggregate in water, their bioavailability may decrease; thus, they may not exhibit ecotoxicity. Metal oxide nanoparticles that release dissolved metal ions can potentially cause harm to aquatic organisms (Miller et al., 2010b; Zhu et al., 2009a,b; Xia et al., 2008). Figure 52 shows the behavior (aggregation and sedimentation) of metal oxide nanoparticles (nanoTiO_2, nanoCeO_2, and nanoZnO) in seawater, freshwater and different natural waters.

In seawater, but not freshwater, the nanoparticles' size increased gradually and the increase showed the following pattern: nanoTiO_2 (greater than 1000 nm) > nanoCeO_2 (greater than 800 nm) >> nanoZnO (less than 400 nm) (Figure 52(A-i)) (Keller et al., 2010). In freshwater, the size stabilized at about 300–350 nm for all three nanoparticles (Figure 52(A-ii)). Figure 52(B) shows the time-course of change in C/C_0 ratio, where C_0 is initial dispersed nanoparticle concentration in pure water and C is nanoparticle concentration at different time intervals after incubation in the

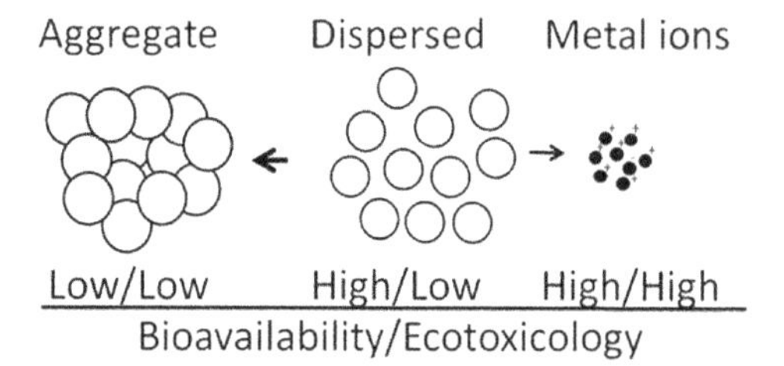

FIGURE 51 Possible toxicity of aggregated and dispersed metal nanoparticles and metal ions.

test-water samples. Sedimentation of the aggregated nanoparticles will decrease the C/C_0 ratio. As shown in Figure 52(B), the C/C_0 ratio decreased rapidly in seawater (decrease in nanoTiO_2 > nanoCeO_2 > nanoZnO) but did not change in freshwater. This suggests that under low organic carbon and high ionic strength conditions of seawater, the rate of sedimentation was very high for the three metal oxide nanoparticles. Metal nanoparticles entering seawater may be removed from the water column in a few hours. Aquatic organisms in the seawater column may have lower toxicity risk unless the exposure is frequent or continuous. Benthic organisms may be exposed to relatively large aggregates (>1 μm) of nanoparticles.

As discussed, nanoparticles are most soluble at zeta potential greater than ±25 mV and aggregate rapidly at zeta potential close to zero. pH is a key determinant of nanoparticles' zeta potential in aqueous samples. In general, an increase in pH is associated with a decrease in zeta potential that attains a value of zero (pHzpc where zpz is zero particle charge) at nanoparticles' isoelectric point. pHzpc for nanoGold, nanoSiO_2, nanoTiO_2, nanoFe_2O_3, and nanoZnO is 2, 2.2, 5.19, 4.24, and 8.4–9, respectively. Therefore, these nanoparticles, at their respective pHzpc will exhibit highest aggregation. At pH higher or lower than the nanoparticles' isoelectric point, when the zeta potential is over ±25 mV, they remain dispersed. However, Tso et al. (2010), while studying the effects of pH changes (using NaOH for increasing pH and HCl for decreasing pH), found that nanoTiO_2 remained dispersed at pH 2–8, but aggregated when the pH was adjusted to 10.4 using NaOH, size increased from 300 nm to almost 3000 nm. They attributed this to the effects of Na^+ from pH adjustment. The effect of cations in water is more important than that of pH. NanoZnO was unique because it possessed a high degree of pH buffer capacity. Initially, nanoZnO neutralized the incubation pH, then lowered it to 5.1 close to its pHzpc, resulting information of nanoZnO aggregates

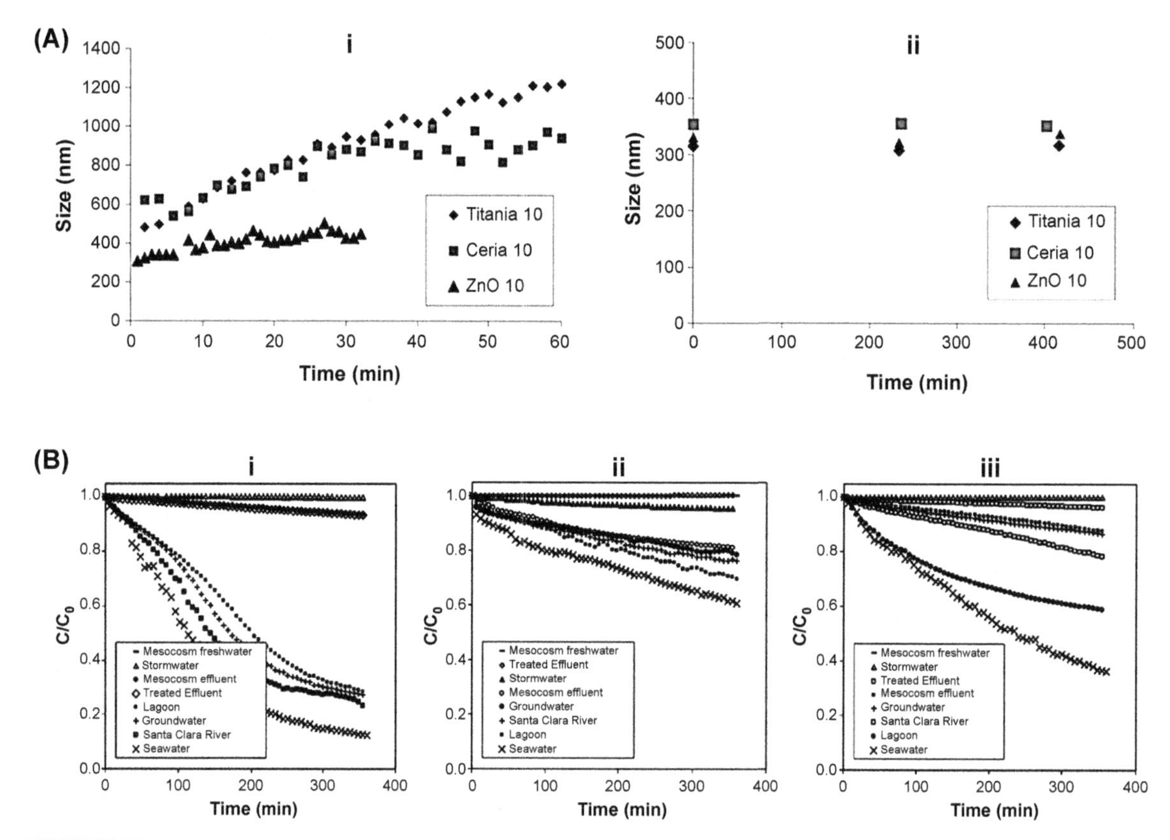

FIGURE 52 Behavior of metal nanoparticles in different water types ranging from sea water to freshwater. (A) Aggregation of nanoTiO_2, nanoZnO, and nanoCeO_2 at 10 mg L^{-1} (i) in seawater; and (ii) in freshwater. (B) Sedimentation in different natural aqueous media of (i) nanoTiO_2; (ii) nanoZnO; and (iii) nanoCeO_2. The order of the media in the legend reflects sedimentation rate from lowest to highest. *Reprinted from Keller et al. (2010) with permission.*

(>1000 nm), possibly due to dissolved zinc species. Taken together, these observations suggest considerable diversity in stability of metal nanoparticles in aquatic environment. Depending upon the pH and ionic strength, nanoparticles in water may remain dispersed (±25 mV zeta potential) or aggregate and settled onto the sediment surface. However, in the presence of cations, nanoparticles may remain aggregated even at −25 mV zeta potential (Figure 53). This may explain the observations that metal oxide nanoparticles induce a range of biological responses varying from cytotoxicity to cytoprotection (Xia et al., 2008; Schubert et al., 2006; Tarnuzzer et al., 2005). Further research is needed to fully understand the seemingly opposing effects of nanoparticle toxicity. In the following sections, the effects of metal nanoparticles on different ecosystems are discussed.

5.3.4.2 BIOACCUMULATION OF METAL NANOPARTICLES IN AQUATIC ORGANISMS

Metal nanoparticles interact with different components of the natural water ecosystem that may change the nanoparticles' physical or chemical properties (e.g., size, surface charge), resulting in modulation of their environmental behavior and toxicity Kennedy et al., 2008). Zhang et al. (2012a,b,c) have shown that, in the aquatic system, the zeta potential of ceria

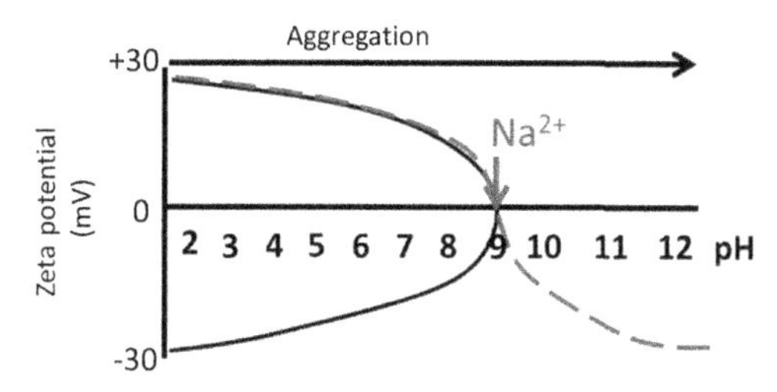

FIGURE 53 A diagram showing the possible mechanisms underlying the effects of pH without or with cations on metal nanoparticle stability. At relatively low cation concentration, metal nanoparticles may remain aggregated at pH near the particles' isoelectric point (pH 9, in this case), but gradually dissolve at the pH higher or lower than their isoelectric point. In the presence of high-cation concentrations (red (gray in print versions) line), nanoparticles remain aggregated at higher pH. Thus, the nanoparticle stability is determined by the pH and cation concentration of water.

nanoparticles changed from 30.9 mV to near the zero charge, which may weaken the interparticle electrostatic repulsion and ultimate aggregation. The dissolved nanoparticles are available to the ecosystem organisms such as hornwort, mud snails, and fish.

Figure 54 shows the accumulation of nanoparticles in fish. The nanoparticle concentration increased and peaked to about 700 ng/g early, then decreased and remained stable at around 300 ng/g throughout the experiment. This may account for various toxic effects of nanoparticles to fish (Handy et al., 2008). Gill, gut, liver, and brain tissues are identified as the target organs. Ostroumov and Kolesov (2010) has shown that gold nanoparticles with or without protein modification can accumulate in hornworts—a

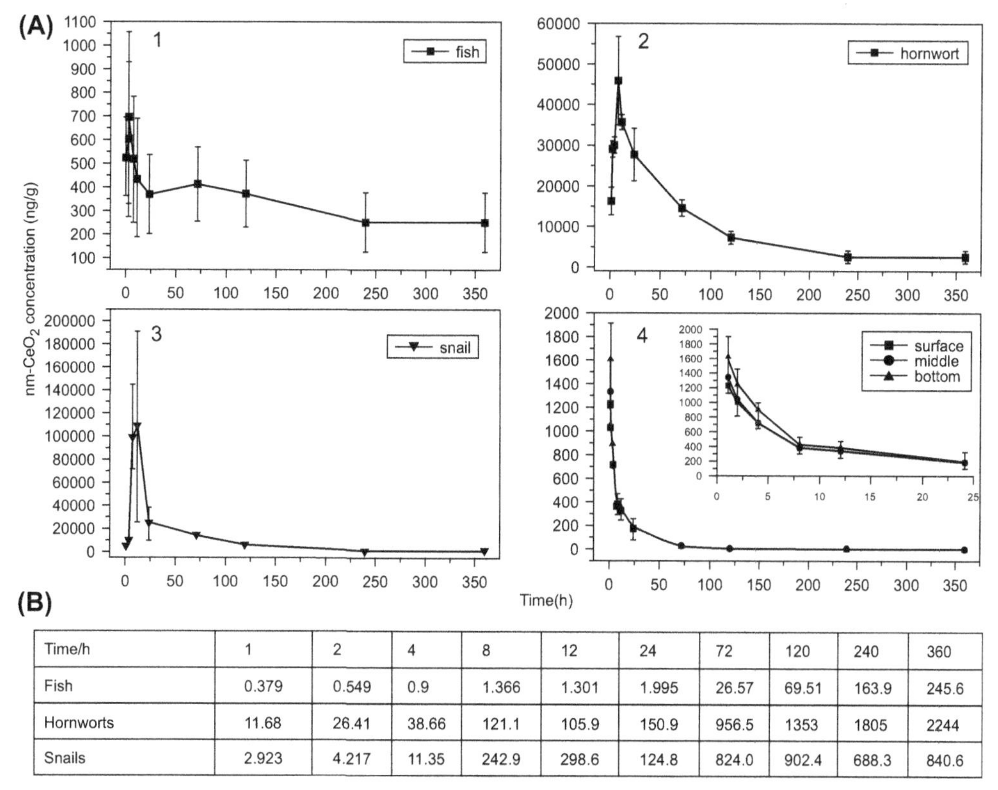

Time/h	1	2	4	8	12	24	72	120	240	360
Fish	0.379	0.549	0.9	1.366	1.301	1.995	26.57	69.51	163.9	245.6
Hornworts	11.68	26.41	38.66	121.1	105.9	150.9	956.5	1353	1805	2244
Snails	2.923	4.217	11.35	242.9	298.6	124.8	824.0	902.4	688.3	840.6

FIGURE 54 (A) Accumulation of $_{nano}CeO_2$ in fish (1), hornwort (2), snail (3), and water (4). (B) The time course of bioaccumulation (as bioaccumulation factor) is shown in different species. *Reprinted from Zhang et al. (2012a) with permission.*

submerged, free-floating rootless aquatic macrophyte in fresh water.

Zhang et al. (2012a) have shown fast absorption and desorption of ceria nanoparticles in hornworts. The nanoparticle concentration quickly increased, peaked at about 45,000 ng/g, and then decreased and remained stable at around 1000 ng/g throughout the experiment (Figure 54(A2)). Direct and quick absorption of nanoparticles from the sediment surface might result in the high bioaccumulation (Figure 54(B)) in snails. In water, the ceria-nanoparticle concentration decreased rapidly (Figure 54(A4)). After 15 days, only 0.095% of the total ceria was found in the water, indicating that almost all of the nanoparticles have been thoroughly removed from water. Because most of the applied ceria nanoparticles were recovered from the sediments, the decrease in the nanoparticle concentrations is due to their aggregation.

Zhang et al. (2012a) used a static freshwater ecosystem, representing lakes and reservoirs that are usually stratified and experience little mixing. The hydrodynamic and morphological characteristics of static ecosystem may be different from those of marine or estuarine systems. Estuarine and marine systems are continuous flowing systems with changing salinity. Colloids will precipitate when entering into an estuarine system due to the change of salinity. Therefore, in a realistic aquatic ecosystem, the behavior and fate of nanoparticles will be more complicated than the present simulated system.

5.3.4.3 TOXICITY MECHANISMS

As water travels from low to high salinity, low to high temperature, and/or different pH values, metal nanoparticles may aggregate to form colloidal suspensions (Petosa et al., 2010). Nanoparticle aggregation may lead to sedimentation and ensuing accumulation of the colloids on sediments' surface. This may also transfer nanoparticles to the sediments where they can be ingested by organisms. Toxicity of metal nanoparticles may depend on (1) the physicochemical properties of the nanoparticles (composition, size, charge, specific surface area, shape, and the saturation magnetization for magnetic particles) and (2) the characteristics of the water column such as pH, ionic strength, hardness, salinity, temperature, flow velocity, organic matter, and other particulate constituents (Figure 55). The detailed mechanisms of toxicity are shown in Figures 55 and 56.

In general, the effects of metal nanoparticles on aquatic ecosystems are produced via accumulation of ROS that cause oxidative damage (internalization by cells) and interaction between nanoparticles and the cell membrane (without internalization), possibly due to electrostatic attraction and/or the presence of thiol groups localized in the cell-wall proteins. Both interactions destabilize the outer membrane, resulting in the disruption of their

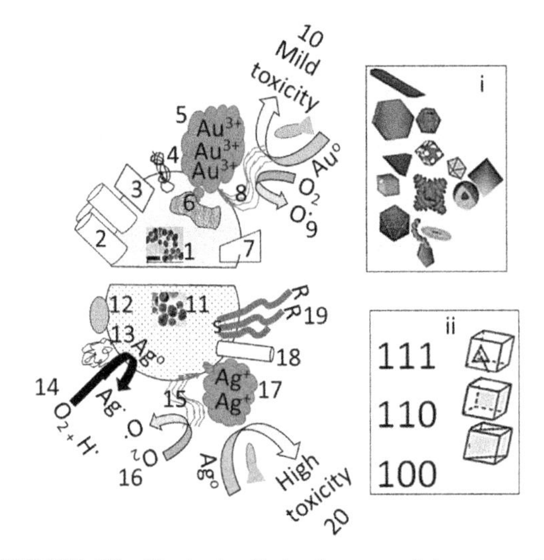

FIGURE 55 The toxic effects of nanoparticles are modulated by their physicochemical properties including surface area, composition, charge, crystal structure, solubility, and morphology (size, shape, and topology); interaction with natural organic matter (NOM), dispersed chemicals or other surrounding nanoparticles; and surface modification. nanoAg exhibits more severe adverse effects on fresh and estuarine organisms than $_{nano}$Au. Surface modification such as sulfidation and PEG coating may reduce the oxidation and subsequently the ion release, thus modulating toxicity. The facet {1,1,1} of the particle exhibits the most reactive (biocidal) properties. *Reprinted from Lapresta-Fernánde et al. (2012) with permission.*

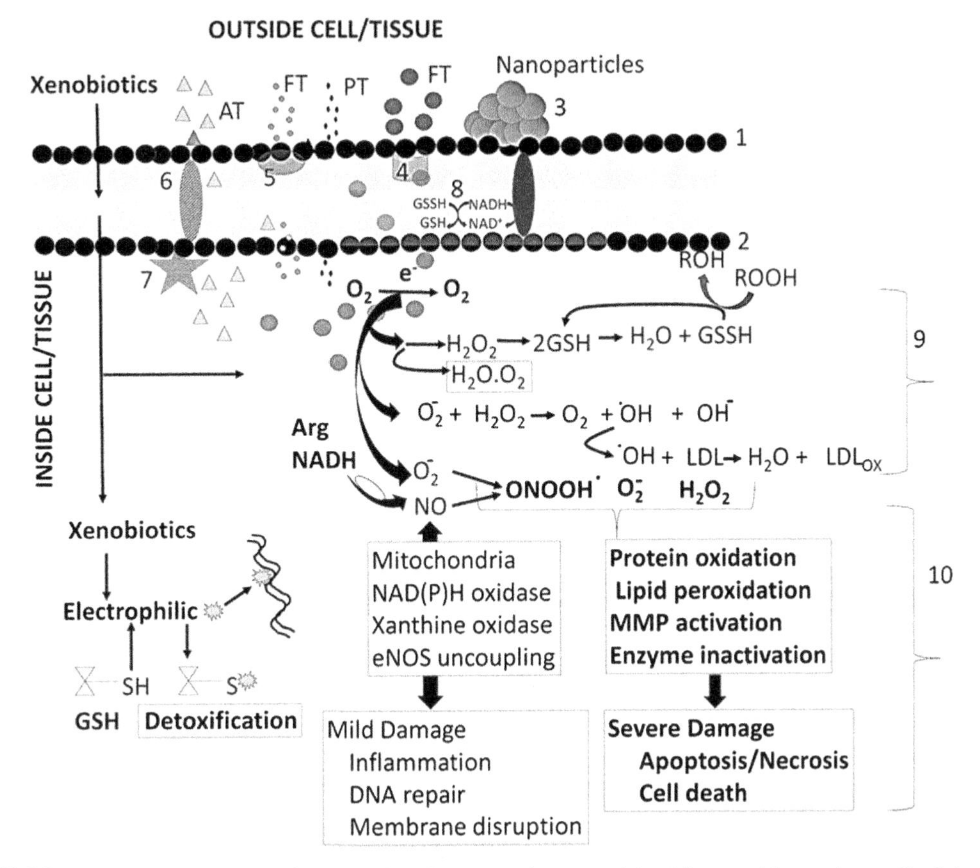

FIGURE 56 Mechanism of cell-membrane-mediated toxicity of nanoparticles. Nanoparticles are internalized through the cell membrane by diffusion and/or by active transport. The interaction between cell membrane and nanoparticles may destabilize the membrane (1) through the thiol groups containing protein and/or (2) by electrostatic interaction. This may interfere with the electrochemical gradient of the transmembrane, involving a disturbance in the electron flow generated in the respiratory chain, which may cause the generation of reactive oxygen species (ROS). The sources of nitric oxide (NO) are restricted to various NO synthases (NOS). O_2 may be produced by NADPH oxidase or xanthene oxidase, processes that disrupt the mitochondrial electron-transport system. The cytotoxicity may be due to protein nitration and oxidation, lipid peroxidation, activation of matrix metalloproteins, and inactivation of a series of enzymes. *Reprinted from Lapresta-Fernánde et al. (2012) with permission.*

electrochemical gradients. In freshwater, nanoAg (1) inhibits Na^+ K^+ -ATPase activity through a noncompetitive inhibition of Na^+ and Cl^- transport in fish gills and (2) inhibits carbonic anhydrase that increases the intracellular acidity. High levels of Na^+ may counteract the inhibition of the Na^+ K^+ -ATPase activity. These pathways, taken together, participate in the development of nanoparticles as follows:

1. Nanoparticle size, shape, chemistry, and capping agents determine their stability, and thus bioavailability within aqueous media.
2. Nanoparticles form heteroaggregates determined by the molecular weight of NOM in the water. The release kinetics may determine the nanoparticle toxicity.
3. An increase in ionic strength decreases the nanoparticles' zeta potential, causing

aggregation and promoting rapid sedimentation.

4. In dynamic environments such as estuaries, aggregated nanoparticles could solubilize and become available to the organisms in the water column. Filter, suspension, and deposit feeders take up both nano-sized and micron-size particles during feeding; however, other invertebrates may suffer from clogged gills during ventilation. Fish will take up nanoparticles in their gut as an artefact of drinking.
5. The degree of nanoparticle-induced oxidative stress may depend upon the antioxidative capacity of an organism: the greater the antioxidation capacity, the lesser the toxicity.
6. Some metal nanoparticles, such as TiO_2, exhibit phototoxicity—that is, the nanoparticle absorbs light of a certain wavelength and causes adverse effects (Li et al., 2014b). Figure 57 shows the interaction between aggregated nanoparticles and surface of three benthic organisms—*H. azteca*, *L. variegatus*, and *C. dilutes*—which were exposed to four preparations of nanoparticles: spike + thick sand, disperse + thick sand, spike + thin sand, and spike + thin sand. Minimal surface attachment of was observed for *H. Azteca* (Figure 57 A-D) and L.

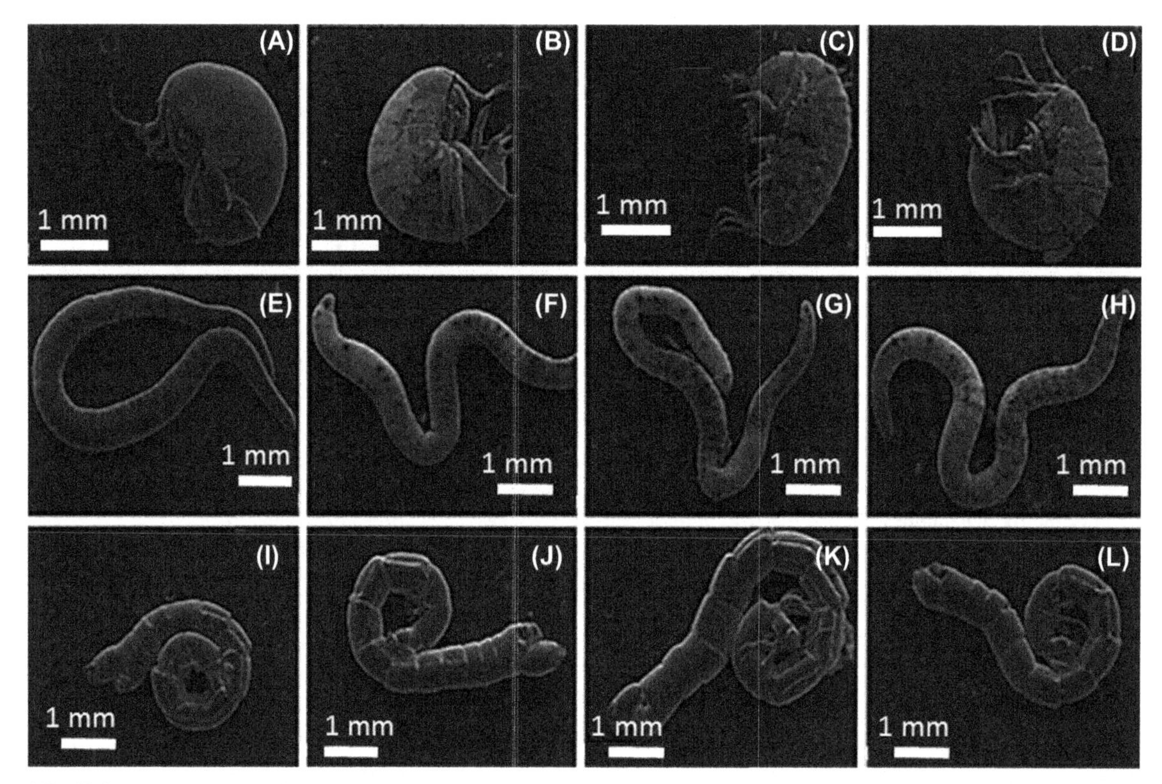

FIGURE 57 Interaction between aggregated nanoparticles and surface of three benthic organisms *H. Azteca* (A–D), *L. variegatus* (E–H), and *C. dilutes* (I–L), which were exposed to four preparations of nanoparticles: spike + thick sand (A, E, I), disperse + thick sand (B, F, J), spike + thin sand (C, G, K) and disperse + thin sand (D, H, L). *Reprinted from Li (2014b) with permission.*

variegatus (Figure 57 E-L) when nano-TiO_2 was spiked into thick sand (Figure 57 A and E), whereas extensive surface attachment occurred in the other 3 exposure scenarios because more nano-TiO_2 was bio-accessible to these epibenthic organisms (Figure 57 B–D, F–H). No apparent attachment was observed for *C. dilutus* in all exposure scenarios. This suggests that adsorption of metal nanoparticles onto the aquatic organisms' surface depended upon the nanoparticles' properties and the soils' characteristics.

Figure 58(A) shows the effects of UV intensity on the survival of three species: *Hyalella azteca*, *L. variegatus*, and *C. dilutes*. No mortality was observed in *L. variegatus* exposed to UV intensity ranging from 0 to 41 W/m^2. Survival of *C. dilutus* was between 75% and 85% after exposure to the same UV intensity range.

A sharp drop in *Hyalella azteca* survival was observed when UV intensity was higher than 9.4 W/m^2 (Figure 58(A)). A combination of UV and nanoTiO_2 caused most toxicity in *H. azteca* (the most sensitive species). Infaunal invertebrates, such as *C. dilutes*, were shielded from UV exposure when they hid inside a matrix. Without UV activation, photoactive chemicals showed no toxicity to invertebrates (Ma et al., 2012; Dabrunz et al., 2011). In addition, different degrees of nanoTiO_2 attachment to the organism's surface could lead to different susceptibilities. Compared with *H. azteca* and *L. variegatus*,

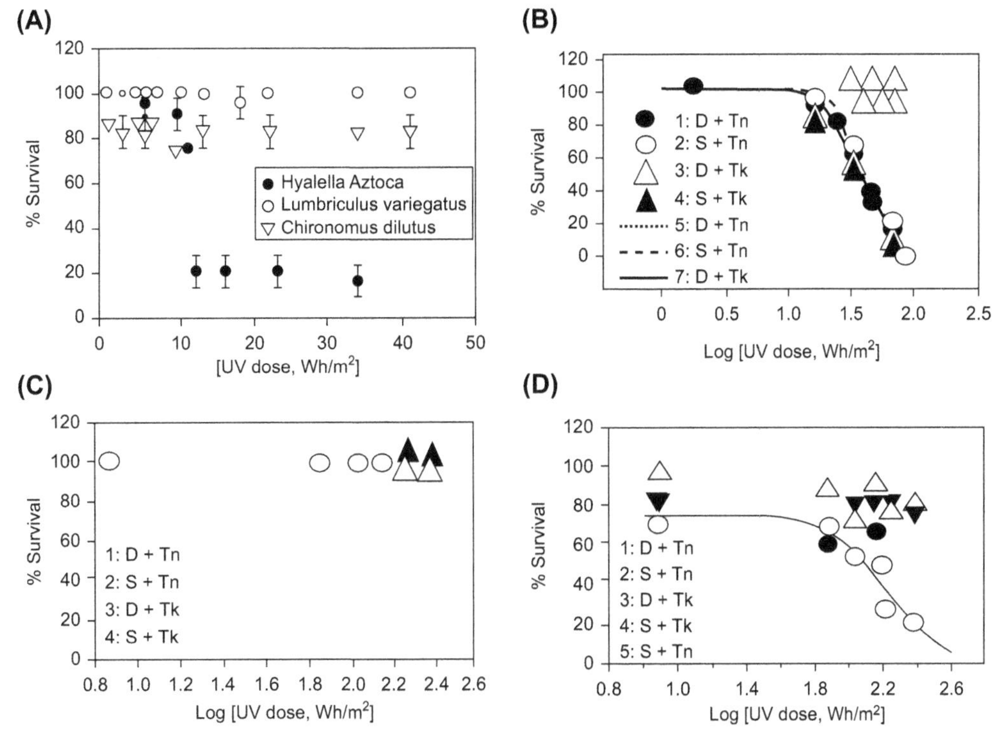

FIGURE 58 Effects of UV intensity alone (A) or in combination with nanoTiO_2 (58B–D) on survival of the species listed in Figure 57. (B) Phototoxic responses for *Hyallela azteca*; (C) phototoxic responses for *Lumbriculus variegatus*; and (D) phototoxic responses for *Chironomus dilutus*. (D: dispersed, S: spiked, Tn: thin sand, Tk: thick sand and lines are fitted data.) *Reprinted from Li (2014b) with permission.*

C. dilutus had minimal nanoTiO_2 attachment and hence were less affected (Figure 57). Compared with *H. azteca*, *L. variegatus* is more tolerant to nanoTiO_2 phototoxicity, although both species had large amounts of nanoTiO_2 on their surfaces (Figures 57 and 58). Studies have suggested that the less susceptible benthic organisms might also have a larger amount of pigmentations or UV-absorbing compounds, such as mycosporine-like amino acids (McClintock, 1997; Dunlap, 1995).

Taken together, the ecotoxicity of metal nanoparticles is determined by a complex interaction between the nanoparticle and the environment having numerous compounding factors (Figures 54 and 55). Major changes in toxicity may occur with a minor change in either the particles' or the environment's properties.

5.4 Nanoplastic Ecotoxicology

Worldwide annual production of plastics reached about 280 million tons in 2012 (Rochman et al., 2013) and is expected to increase in the future. Plastic products have a short useful life, then they are discarded in landfills or oceans. Plastics constitute 60–80% of marine litter (Moore, 2008), estimated between 7,000 and 35,000 metric tons. Plastic in the ocean is degraded into smaller pieces through UV radiation, mechanical abrasion, biological degradation, and disintegration (Wegner et al., 2012; Imhof et al., 2012), resulting in formation of plastic nanoparticles called nanoplastics (NPl), which are expected to persist in the marine environment for centuries (Barnes et al., 2009; Thompson, 1996; Thompson et al., 2009). Polystyrene is one of the most common nanopolymers found in marine plastic debris (Hidalgo-Ruz et al., 2012). NPl are easily ingested by marine wildlife, such as mussels (Browne et al., 2008), fish (Boerger et al., 2010; Carpenter et al., 1999, 1993, 1972), seabirds (Andrady, 2011), and whales (Fossi et al., 2012). Ingestion of plastic microparticles can harm animals via the release of plastic monomers and toxic chemical additives, such as phthalates (Cole et al., 2011). Despite the abundance of NPl, the knowledge of their mechanisms of toxicity is limited.

Earlier studies have shown that mussels can ingest NPls by translocation from the gut to the circulatory system, where the particles may be retained for more than a month (Browne et al., 2008). As particle size decreases to the nanometer range, they could partition into cell membranes and diffuse through them (Radlinska et al., 1995). Recently, Mattsson et al. (2015) exposed fish to NPls through the aquatic food chain (algae, phytoplankton > *Daphnia* > fish) and showed that the uptake of NPls through a food chain strongly affects behavior of top consumers by reducing their activity, feeding rate, and changing their social behavior. In addition, they also showed genetic changes in liver and muscle metabolism as well as morphological alterations in the brain and muscles. Besseling et al. (2014) have demonstrated that NPls (1) reduced population growth and reduced chlorophyll concentrations in algae and (2) suppressed body size and cause severe alterations in reproduction of *Daphnia*: numbers and body size of neonates were lower, while the number of neonate malformations rose to 68%. Rossi et al. (2014) have shown that NPl permeate easily into lipid membranes and alter membrane structure (Giulia et al., 2014), thus, significantly reduing molecular diffusion. NPls also severely affects membrane's lateral organization by stabilizing raft-like domains. Thus, changes in membrane properties and lateral organization can severely affect the activity of membrane proteins and thereby cellular function.

6. CONCLUSIONS

A rapid increase in production of engineered nanoparticles is also increasing the potential for release of nanoparticles into the environments and causing damage to the environment. Therefore, there is a rising need for monitoring nanoparticles' potential environmental toxicity. This chapter reviewed the previous and latest research results, including the fate and ecotoxicology of nanoparticles. Considering the diversity in different components of the environment and structural and physicochemical properties of nanoparticles, a single mechanism or pathway may not characterize their environmental toxicity. Particle size, shape, surface charge, and the presence of other materials in the environment influence nanoparticle behavior in the environment. Nanoparticles, as they flow from freshwater to saltwater, tend to aggregate with uncontrolled increase in particle size. Aggregation modulates their toxicity. In this chapter, these variables have been brought together to explain overall ecotoxicity of nanoparticles. The biology of nanoparticle-induced oxidative stress is an important mechanistic paradigm, and so are solvent effects and the effects of nanoparticles on other substances.

Despite extensive research, a remarkable lack of information on some key aspects prevents a better understanding and assessment of ecotoxicity of nanoparticles to ecosystem organisms. The following are important future research directions:

1. Basic mechanisms of toxicity of nanoparticles of different composition, size, shape, and functionalization in air, water, and soil
2. Effects of human activity on nanoparticles' ecotoxicity
3. Systematic investigation of nanoparticles' ecotoxicology in diverse ecosystems
4. Modulation of nanoparticles toxicity by other environmental pollutants in organisms
5. Development of new biological indices and analytical tests to assess nanoparticle ecotoxicology

A greater understanding of the knowledge gaps identified above may help develop strategies for safer application of nanoparticles.

7. APPENDICES

7.1 Appendix 1: Physicochemical Basis of Adsorption, Distribution, and Migration

The overall physicochemical processes (adsorption, distribution, translocation, and biotic or abiotic transformation) that a chemical or particle endures determine their fate in the environment. Figure 59 shows a fugacity model describing the fate of a hydrophobic chemical in the environment (atmosphere, hydrosphere, sediments, and soil). The model provides an overall view of the chemical's fate including emission, efflux, chemical modification, and intermedia exchange. The sections below describe the mechanisms regulating the environmental fate of chemicals or particles.

7.1.1 Sorption

The factors determining soil sorption of bulk chemicals or nanoparticles are Van der Waal interactions, hydrophobic interactions, hydrogen bonding, charge transfer (electron acceptor (aromatic), humic acid, ligand-exchange

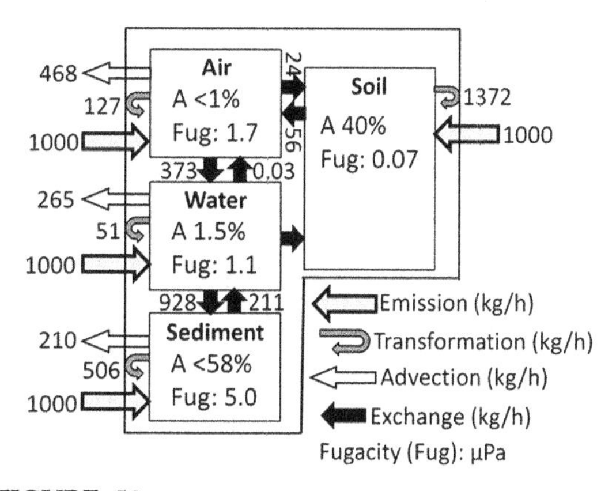

FIGURE 59 Fugacity plot showing distribution of the polybrominated diphenyl ethers in different components of the environment. The unit of input and output data is Kg/H. Possible exchanges are shown as filled arrows. *Reprinted from Palm et al. (2002) with permission.*

(electrostatic or ion specific), ion-dipole or dipole–dipole interaction (induced or permanent), and chemisorption (chemical bonding such as covalent bonding). Soil sorption kinetics influence (1) the extent of a compounds' removal from the soil surface and (2) their degradation and migration, that in itself is determined by a number of processes such as fast flow, dispersive flow, diffusion through macropores, and molecular diffusion. Therefore, to understand the fate of a chemical or particle in soil, it is necessary to understand the above processes as well as the variables such as molecular and structural properties of the compounds concerned, the heterogeneous nature and composition of the soil including its minerals, organic matter, and microbes. Normalization of the organic carbon content results in the sorption coefficient K_{oc}, an important parameter to understand the chemical's fate in soil.

7.1.1.1 MECHANISMS OF SORPTION

7.1.1.1.1 HYDROPHOBIC INTERACTION Hydrophobic (nonpolar) chemicals partition from aqueous phase into the soil phase via hydrophobic interactions. If a hydrophobic particle and the hydrophobic moiety of soil approach each other, the space between these molecules will diminish due to water displacement. This results in an increase in entropy, a thermodynamically favored situation. Hydrophobic binding is described for PAHs, aromatic hydrocarbons, and pesticides with large hydrophobic regions. Hydrophobic binding is responsible for the strong sorption of DDT or other organochlorine insecticides to soil (Hamakcr and Thompson, 1972). Karickhof et al. (1979) have shown that sorption isotherms for pyrene, methoxychlor, naphthalene, 2-methylnaphthalene, anthracene, 9-methylanthracene, phenanthrene, tetracene, hexachlorobiphenyl benzene (water solubility from 500 ppt to 1800 ppm) were similar and linear over a broad range of aqueous phase compound concentrations. The linear partition coefficients (K_p) were relatively independent of

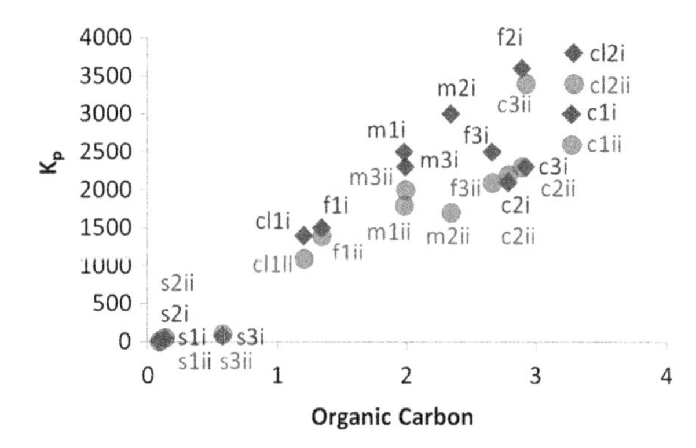

FIGURE 60 Plot of soil organic carbon versus soil/water partition coefficient (Kp). s: sand, c: coarse silt, m: medium silt, f: fine silt, cl: clay, i (green (black in print versions) font): pyrene, ii (red (gray in print versions) font): methoxychlor. *Data from EPA (epa.gov/superfund/health/conmedia/ soil/ pdfs/appd_k.pdf).*

sediment concentration and salt content in the suspensions (Figure 60). The K_p's for a given compound were directly related to organic carbon content of sediments. Based on the K_{oc} values, the sand fraction was a considerably less effective sorbent (50–90% reduction in K_{oc}) than the fine fraction (sediment particles <50 pm). Differences in sorption within the silt and clay fractions were largely the result of differences in organic carbon content.

7.1.1.1.2 VAN DER WAALS INTERACTIONS Van der Waals interactions are common to both hydrophobic and hydrophilic molecules. A Van der Waals interaction results from the formation of transient dipole moments in atoms or molecules, thus allowing weak interaction between molecules having transient opposite charges. This attraction has an energy of about 1–2 kcal/mol for atoms or small molecules. The interaction decreases with increasing distance. These effects, being additive, can contribute to the sorption of large molecules to a greater extent.

7.1.1.1.3 HYDROGEN BONDING A hydrogen bond is a weak type of force (stronger then the Van der Waal but weaker than the covalent bonds) that forms when a hydrogen atom bonded to a strongly electronegative atom exists in the vicinity of another electronegative atom with a lone pair of electrons. Hydrogen bond formation required a hydrogen donor (hydrogen bonded to atoms such as O, N or F) and an acceptor (must possess a lone electron pair) present. Earlier studies using infrared spectroscopy showed hydrogen bonds between humic substances and s-triazines or 2, 4-D (Pignatello, 1989; Sullivan and Fellbcck, 1968; Senesi and Testini, 1982, 1983; Khan, 1973, 1974). Hydrogen bonding also plays a key role in adsorption of water to soil, especially clays. Soil minerals are essentially made up of oxygen or hydroxyl groups, facilitating easy formation of hydrogen bonds. Surface oxygen can attract positive corner of water molecules (H^+) as depicted in Figure 61.

7.1.1.1.4 CHARGE TRANSFER Charge-transfer interactions occur when a fraction of electronic charge is transferred between two molecular entities, resulting in formation of a donor–acceptor complex. These complexes are divided into two classes: the electron-rich H-donors, such as aromatics or alkynes, and the lone pair donors, such as e.g. alcohols or amines and weak acids. In general, charge-transfer seems to be a likely mechanism for sorption to humic acid (Hamaker, 1972; Pignatello, 1989).

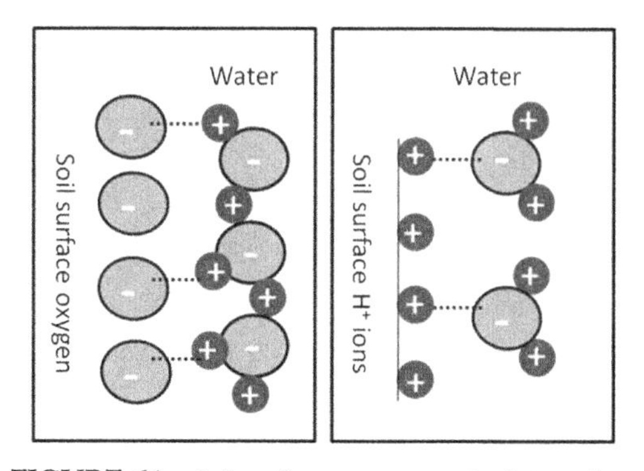

FIGURE 61 Soil surface oxygen–water interaction.

7.1.1.1.5 LIGAND EXCHANGE AND ION BONDING A *ligand exchange reaction* involves the substitution of one or more ligands in a complex ion with one or more different ligands. Ion sorption to soil is either specific or nonspecific. The nonspecific sorption involves electrostatic interactions, whereas specific ionic sorption refers to the positive or negative charged sites of the soil. The specific sorption of cations results in a substitution of central cations in clay minerals. Cation exchange interactions with clay minerals have been observed for parquet and piquet (Hayes and Himes, 1986; Hayes et al., 1975; Burns and Hayes, 1974). Nonspecific anion sorption refers to a binding of anions to the positively charged sites of clay minerals (Huang, 1980).

Isomorphic substitution and terminal bonds are two primary processes by which minerals exert a charge. Isomorphic substitution is the replacement of one ion with another having a different charge (e.g., substitution of Al^{3+} for Si^{4+} or Al^{3+} for Mg^{2+}). This charge is permanent because it will not vary as a function of the solution composition. Terminal bonds result from coordinately unsaturated central captions or ligands that occur at terminal ends of many minerals (Figure 62).

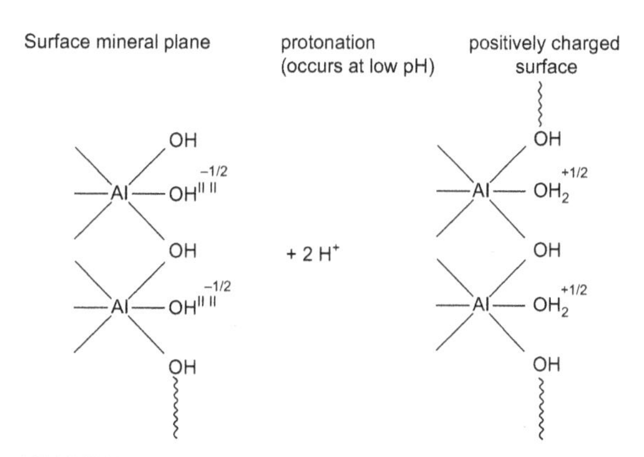

FIGURE 62 Terminal (broken) bonds that result from coordinately unsaturated central cations or ligands that occur at terminal ends of many minerals.

7.1.1.1.6 DIRECT AND INDUCED ION–DIPOLE AND DIPOLE–DIPOLE INTERACTION These interactions depend on the charge of the soil surface, the strength of the dipole of the sorbate, and the orientation of the dipoles in space (Voice and Weber, 1983). The direct or induced dipole–dipole interactions result from the reciprocal attraction of two dipoles. This type of bonding may contribute to the sorption of polar or ionic compounds.

7.1.1.1.6 CHEMISORPTION Chemisorption involves formation of a bond between the adsorbate molecule and the surface atoms. The bond is usually a covalent bond but is sometimes ionic (Hamaker, 1972).

7.1.1.1.7 CATION EXCHANGE CAPACITY A common property of clay and DOM both is their *net negative charge* due to the proton-electron imbalance and ionization of hydroxyl groups at the edge of a crystal (called *broken edge* charge). The net negative charge is responsible for their cation-exchange capacity (CEC). When placed in a solution, the negative ions of clay or DOM attract positively charged particles to their surfaces. The bonds between the clay mineral surface and the cations are relatively weak, and these cations can be exchanged for other cations that are dissolved in the solution. Studies have shown that cation exchange is an important property that allows soil particles to retain or exchange positively charged particles onto the colloids. Described below is a generic reaction of exchange between the cations A^{u+} and B^{v+} on a soil:

$$vAX_u(s) + uB^{v+}(aq) = uBX_v(S) + vA^{u+}(aq)$$

In this equation, X^{-1} represents one equivalent of the soil's exchange complex; this allows the same exchange isotherm for forward (starting from fully saturated with A^{u+}) or backward (starting from fully saturated B^{v+}). This thermodynamically reversible equation suggests that performing exchange experiments in only one direction, either the forward or the backward reaction, is necessary to study the exchange process.

7.1.1.2 SORPTION ISOTHERMS

Because of structural diversity, soil may bind different nanoparticles differently (Figure 63) depending upon the binding sites (I: binding to one type of site, II and IV: binding to more than one sites), attractive or repulsive interaction (III), and phase transition (V).

One of the methods used for sorption characterization is construction of adsorption isotherm, a mathematical equation describing the relationship between the adsorbate concentrations and the extent of adsorption at any fixed temperature. Various adsorption isotherms commonly employed in describing the adsorption data are described in the following sections.

7.1.1.2.1 FREUNDLICH ADSORPTION ISOTHERM In a Freundlich adsorption isotherm, the adsorbate forms a monomolecular layer on the surface of the adsorbent (Eqn (1)) for nonlinear and Eqn (2) for linear curves (Table 7). In these equations, K_f is the Freundlich constant or maximum absorption capacity, C_e is the solution concentration equilibrium (μ mol L^{-1}), and x is the weight of the particles adsorbed by 'm'

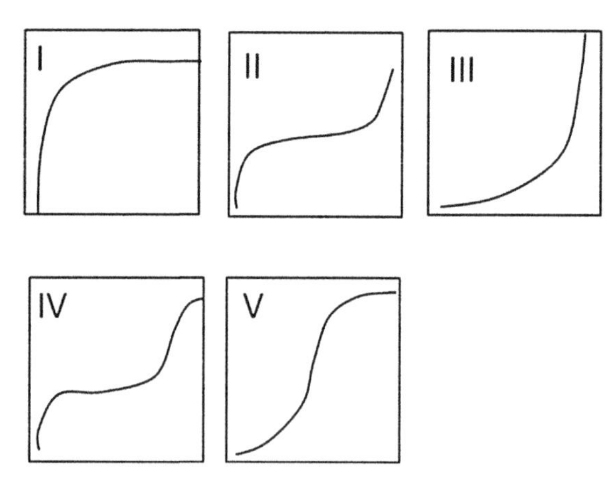

FIGURE 63 The nanoparticle-soil sorption curves reveal the properties of the binding sites.

grams of the adsorbent ($x/m = q_e$ absorbed solute). The ratio, *x/m*, represents the amount of particles adsorbed by the adsorbents per gram (unit mass). The index $1/n$ is the measure of intensity. The shape of C_e vs q_e plots is determined by the $1/n$ values (Figure 64: $1/n = 1$ yields a linear plot, while $n \neq 1$ yields a nonlinear plot).

The Freundlich equation in a linear mode yields a linear plot in which the slope is $1/n$. The K_f values can be determined by fitting the $1/n$ values in Eqn (7). In a nonlinear mode, the plot is linear for up to 50% max saturation and then becomes nonlinear (Figure 63, solid line). Although the Freundlich equation provides important information regarding sorption of particles, it has certain limitations:

1. The Freundlich equation is purely empirical.
2. The equation is valid only up to a certain concentrations, above which it becomes nonlinear.

TABLE 7 Freundlich, Langmuir, and Redlich–Peterson Equations (Brdar et al., 2012)

Equation	Isotherm	Equation#
$q_e = K_f C^{\frac{1}{n}}$	Freundlich nonlinear	6
$\log(q_e) = \log K_f + \frac{1}{n}\log C$	Freundlich linear	7
$\frac{C_e}{q_e} = \frac{1}{q_m}C_e + \frac{1}{K_a q_m}$	Langmuir linear 1	8
$\frac{1}{q_e} = \left(\frac{1}{K_a q_m}\right)\frac{q_e}{C_e}$	Langmuir linear 2	9
$q_e = q_m - \left(\frac{1}{K_a}\right)\frac{q_e}{C_e}$	Langmuir linear 3	10
$\frac{q_e}{C_e} = K_a\, q_m - K_a q_e$	Langmuir linear 4	11
$q_e = \frac{Q_0 K_L C_e}{1 + K_L C_e}$	Langmuir nonlinear	12
$R_L = \frac{1}{1 + K_L C_0}$		13
$q_e = \frac{K_r C_e}{1 + \alpha_r C_e^{\beta}}$	Redlich–Peterson nonlinear	14
$\log\left(K_r \frac{C_e}{q_e} - 1\right) = \beta \log(C_s) + \ln(\alpha_r)$	Redlich–Peterson linear	15

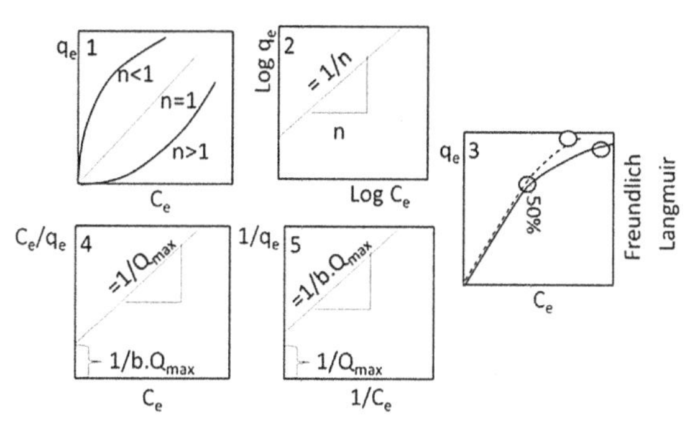

FIGURE 64 Different types of Freundlich and/or Langmuir equations (see the text for details).

3. The constant *K* changes with change in temperature.

7.1.1.2.2 LANGMUIR EQUATION The Langmuir theory correlates the number of active sites of the surface undergoing adsorption with C_e (concentration at equilibrium). Equation (12) defines the nonlinear isotherm, while Eqns (8)–(11) represent linear plots for one to four binding sites, respectively. In these equations, C_e is the concentration of metal in solution at equilibrium (μmol L^{-1}), *x* is the amount of adsorbed metal (μmol), m is the mass of sorbent (g), and B_L and K_L are the Langmuir parameters that characterize the sorption process. It is assumed that each particle independently binds to each binding site, which is unaffected by the sites already occupied. The parameter B_L is the maximum sorption capacity (μmol g^{-1}), which measures the number of sorption sites per gram of sorbent.

The Langmuir constant K_L (L μmol^{-1}) is an equilibrium constant that increases exponentially with the sorption energy. The values of *x/m* and K_L can be computed from the slope and intercept of the Langmuir plot of C_e versus q_e (Figure 64(3)). When C_e is plotted against C_e/q_e, the slope is $1/Q_{max}$ (in some equations, Q_{max} is represented as Q^0) and the intercept is $1/bQ_{max}$ *b* is the absorption maximum. The essential features of the Langmuir isotherm may be expressed in terms of equilibrium parameter R_L, which is a dimensionless constant referred to as separation factor or equilibrium parameter. For a Langmuir type adsorption process, whether adsorption is favorable or unfavorable can be classified by a term R_L (Eqn(13), El Qada et al., 2006; McKay et al., 1987).

7.1.1.2.3 REDLICH–PETERSON ISOTHERM

The Redlich–Peterson isotherm (Redlich and Peterson, 1959) is a hybrid of the Langmuir and Freundlich isotherms. The equation amends inaccuracies of Langmuir and Freundlich isotherm equations, thus providing more accurate information regarding adsorption systems. The Redlich–Peterson isotherm (Eqns (13) and (14)) contains three parameters: K_r, α_r, and β. K_r is the modified Langmuir constant (dm3/g); thus, for simplicity, $Kr = K_L$. The α_r (dm^3/mg) and β are the constants. The model has a linear dependence on concentration in the numerator and an exponential function in the denominator. When $\beta = 1$, the Redlich–Peterson equation becomes the Langmuir equation; when $\beta = 0$, the equation represents Henry's law.

7.1.1.2.4 INTERPRETATION OF THE ISOTHERMS Linearization of a nonlinear equation adds the inherent biases. The optimization procedure requires the selection of an error function in order to evaluate the fit of the isotherm to the experimental data. The commonly used error functions are ERRSQ (sum of the squares of errors), HYBRD (a composite fractional error function), MPSD (a derivative of Marquardt's prevent standard deviation), ARE (the average relative error), and EABS (the sum of the absolute errors) (Brdar et al., 2012). Figure 65 shows the isotherms for binding of Cu ions to lignin (complex plant polymer). The Redlich–Peterson model exhibited the best fit for absorption data (red circles), while ERRSQ, HYBRD, and MPSD exhibited the highest r^2 and lowest χ^2.

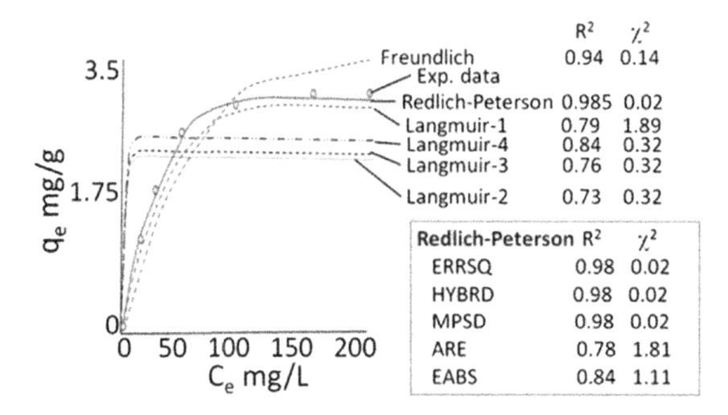

FIGURE 65 Examples of Freundlich, Langmuir, and Redlich–Peterson plots (see the text for details). For the dataset shown, the Redlich–Peterson and Freundlich plots best characterize the sorption process. *Reprinted from Brdar et al. (2012).*

7.1.2 Desorption

When a chemical bind to a true ion-exchanger such as cationic- or anionic-resins, its sorption and desorption pathways are traceable However, in soil, sorption and desorption are not traceable and follow different paths (Figures 66 (3) and 66 (4) - plots (a) representing adsorption and (d) representing desorption). The situation when desorption does not follow that same path as adsorption is called hysteresis (Verburg

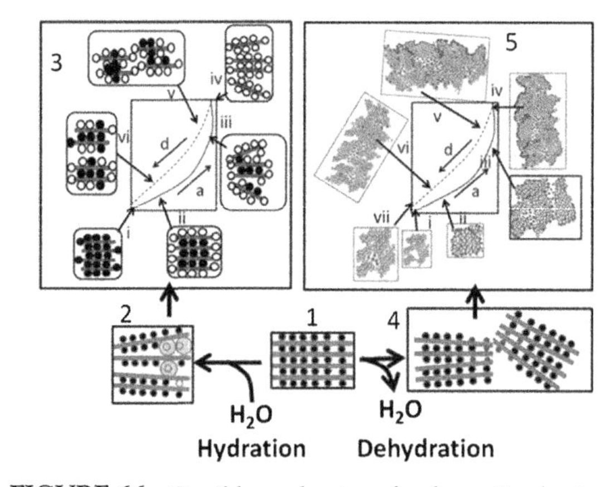

FIGURE 66 Possible mechanisms for desorption hysteresis (when sorption and desorption follow different paths) in soil samples (see the text for detailed information).

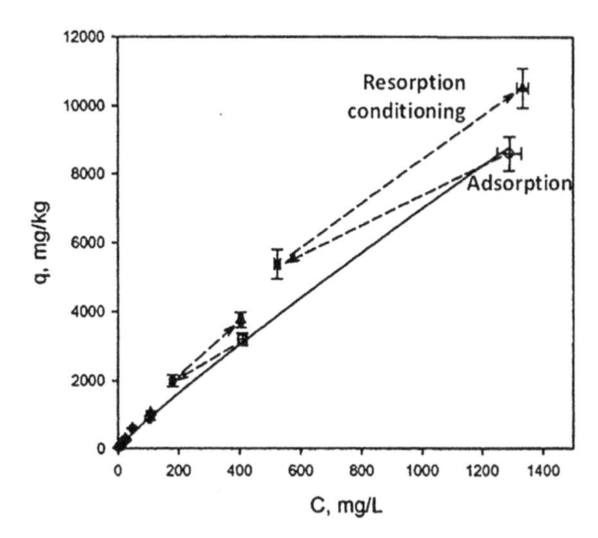

FIGURE 67 An example of conditioning due to hysteresis (Lu and Pignatello, 2002; see the text for details).

and Baveye, 1994). In Figure 67, the angle BAC1 or BAC2 is the angle of hysteresis, a measure of irreversible binding (part of the particles that remained bound to the soil and did not desorb) between chemicals/nanoparticles and soil. Fripiat et al. (1965) who showed that a homogenous ion exchange resin, Amberlite, exchanged NH_4^+ with Li, Na, K, Rb, Cs, Mg, Ca, Sr, and Ba in a reversible manner—that is, the kinetics of adsorption and desorption were identical. However, on different types of soil including montmorillonite, the exchanges between NH_4^+ and Li, Na, K, Rb, Cs, Mg, Ca, Sr, or Ba exhibited significant hysteresis—that is, the kinetics of adsorption and desorption were different and desorption did not follow the adsorption path (Figure 67).

Although the mechanism of hysteresis in soil are not fully understood, the following possibilities have been proposed: charge or site heterogeneity at the surface of the soil exchanger, soil aggregation mediated entrapment, differential hydration of the exchanging cations, dehydration of the soil or clay, and/or inaccessibility of sites caused by domain or quasi-crystal formation. Figure 66 shows the following pathways for development of hysteresis:

1. When a monovalent ion such as Li^+ replaces a divalent ion Ca^{2+}, because of their lower charge, water would enter the lattice and disrupt the interplatelet regions (Figure 66(2)). This allows the exchange reaction to proceeds with very fast kinetics until the exchanger is fully saturated with Li^+ (Figure 66(3-iv)). Figure 66(3) shows sorption and desorption of chemicals when Li^+ replaces a divalent ion Ca^{2+}. As the density of Li^+ increases, soil exhibits greater deformation (Figure 66(3-(ii−iv))). In the backward reaction, divalent ions such as Ca^{2+} replace K^+ (Figure 66(3-(v−i))) and the suspension reverts to its initial state. However, the adsorption and desorption take different paths. This is because of the heterogeneity in soil at stage iv.
2. When a monovalent ion such as NH_4^+, Rb^+, Cs^+ or K^+ replaces a divalent ion Ca^{2+}, the monovalent ions form a partially aggregated suspension (Figure 66(4)), instead of a fully dispersed one formed by Li^+. This may result in quasi-crystals with interplate spacing smaller than in the Ca^{2+} (Figure 66(4)). Because of aggregation, the soil's heterogeneity may increase and the monovalent ions that may be trapped in soil cavities. During desorption (Figure 66(5-(i−vii)), the exchanger may replace the surface monovalent ions with bivalent ions much easier than the monovalent ions in cavities, resulting in the development of hysteresis. In fact, the exchanger may not return to its original state (i).

Earlier studies (Selim et al., 1976; Gamble and Khan, 1990; Ma and Selim, 1994) have also shown that sorption and desorption of chemicals and particles by soil exhibits considerable hysteresis. Hysteresis results in a phenomenon known as *conditioning*, which is defined as enhancement in sorption of a compound following brief exposure of the sorbent to a high concentrations of the same or a similar compound. Typical curves for adsorption, desorption, and conditioning effects are shown in Figure 67 (Li and Pignatello, 2002).

The adsorption curve was linear from 1 to 1400 mg/L. For samples exposed to 1200 mg/L ligand, the adsorption and desorption paths were different, revealing hysteresis. The Kd values of the desorption points are 30–40% above the initial sorption values. Based on these observations, Lu and Pignatello (2002) proposed the following:

1. Conditioning is accompanied by an enhancement in sorption and a decrease in linearity.
2. The conditioning effect is inversely related to the initial concentration and the effect diminishes as the solute concentration increases.
3. The memory of the conditioning effect may be eliminated after the soil is heated.
4. Sorption-desorption hysteresis is accompanied by an enhanced resorption (following desorption).

In the environment, soil samples may adsorb contaminants rapidly but release them into the water phase relatively slowly, depending upon the chemical nature of water flowing through them. The sorbate accumulates in soil cavities or within layers; it may become immobilized and remain trapped for longer time.

7.1.3 Transport

In general, there are three types of pollutants: water insoluble or hydrophobic, water soluble or hydrophilic, and those slightly soluble and heavier than water (Figure 68). Hydrophobic particles either bind to the organic carbon fraction of soil particles or migrate to the water level and settle as sediment (Figure 68(1)). Hydrophobic particles remain bound to the soil particles and migrate with soil migration. Hydrophilic

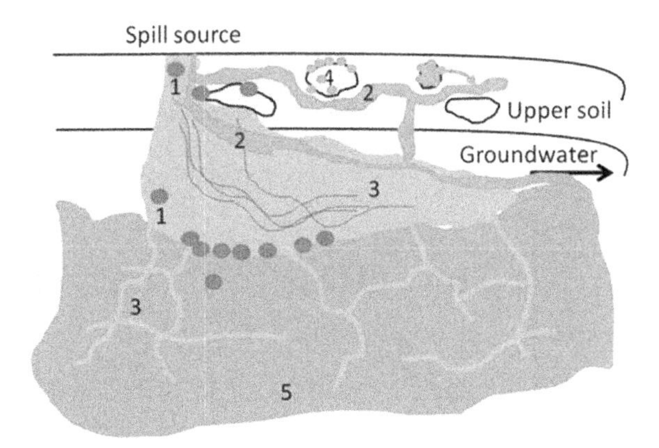

FIGURE 68 Distribution of chemicals with different water solubility in soil (see the text for details).

particles either bind to the ions adsorbed onto the soil surface and undergo ion exchange or dissolve in interstitial water and migrate (Figure 68(2–4)). The partially water-soluble, denser, particles may sink through groundwater and permeate into fine-grained soil units, such as silt and clay (Figure 68(5)). They can migrate in multiple directions through fractures in bedrock. The fate of heavier pollutants is difficult to predict. In bedrock and fine-grained soils, particles can act as continuing sources of contamination, which may cause long-term impacts on ground water and pose a significant challenge. In soil and water, chemicals and particles may undergo chemical and biological transformation and the products will undergo the same fate as the parents.

Different physical, chemical, and biological processes, including sorption, desorption, volatilization, chemical and biological degradation, uptake by plants, run-offs, and leaching govern dissipation of chemicals and particles in soil (Figure 69). Distribution of chemicals occurs either horizontally through the soil or vertically from plants into the atmosphere, depending upon the weather conditions.

Plants play an important role in mobilization of contaminants (Figure 69). Water-soluble contaminants are absorbed by the roots and

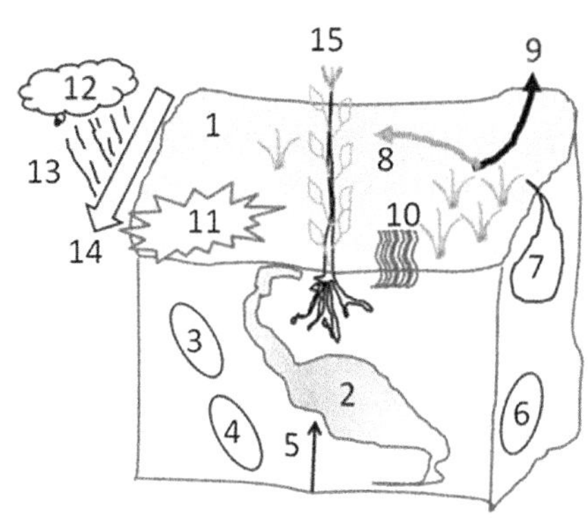

1. Surface soil
2. Adsorption
3. Chemical decomposition
4. Leaching
5. Capillary flow
6. Biodegradation
7. Crop adsorption/fixation
8. Horizontal transport
9. Crop removal
10. Volatilization
11. Photodecomposition
12. Clouds
13. Rain
14. Runoff

FIGURE 69 Dissipation of chemicals and particles in soil.

transported to the leaves. If contaminants fall onto the plant's surface, the following scenarios may materialize (Figure 70):

1. Depending upon its vapor pressure, the contaminant may volatilize translocate into the atmosphere or washed off by rain or irrigation.
2. In dry weather, pollutants may remain on the outer surface of the trunk or leaves in a viscous liquid or crystalline form, and undergo photodegradation.
3. Hydrophobic contaminants may penetrate the cuticle but remain absorbed in the lipoid components of the cuticle. If dislodged, they may migrate to another site and bind to the lipoid component.
4. Hydrophilic contaminants may also penetrate the cuticle but enter the cell walls and then translocate (apoplastic translocation) prior to entering the symplasm.
5. Some contaminants may penetrate the cuticle, enter the cell walls, and then move into the cellular system through the plasma lemma for symplastic translocation via the phloem movement.

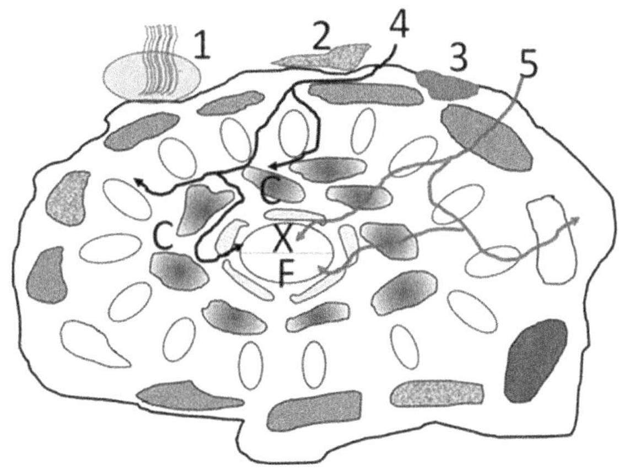

FIGURE 70 Distribution of chemicals in plants' inner structures.

7.1.4 *Transformation and Degradation*

Xenobiotic chemicals in the environment, dissolved in water, adsorbed by soil, or absorbed by organisms undergo transformation and degradation via abiotic reactions and/or biotic processes as described below.

7.1.4.1 ABIOTIC TRANSFORMATIONS

Abiotic transformations are relevant to aquatic systems, water-soil interphase receiving direct sunlight, and/or moist soil surface. Photo-oxidation of pollutants, such as PAHs and pesticides, occur in water or soil in the presence or absence of catalysts (Wen et al., 2002). When aromatic compounds are exposed to UV light, partially oxidized intermediates are produced, which are more susceptible to degradation than their parent compounds. The polarity of the solvent is directly proportional to the rate of the degradation process; hence, the higher the polarity of the solvent, the faster the degradation process. The key abiotic transformation processes are photo-oxidation and photolysis (Figure 71).

7.1.4.1.1 ABIOTIC OXIDATION All pollutant molecules in the environment are subjected to oxidation and hydrolysis. An oxidation reaction occurs between an electron donor and an electron acceptor according to $AH_2 + X = A + XH_2$, where AH_2 is the reducing agent or electron donor (phenols, aldehydes, aromatic amines, or thiols) and X the oxidant or electron acceptor (molecular oxygen or super oxide radicals, hydroxyl radical or peroxy radicals). Oxidation reactions occur in the atmosphere-lithosphere interface (the surface soil) where oxygen and other oxidizing agents are present in large concentrations. In aquatic systems, however, oxygen and oxidizing agents are present in much lower concentrations; thus, there is a relatively poor rate of abiotic oxidation depending upon the pH and redox states. At water–air interfaces that are rich in light and oxygen, chemicals may undergo photo-oxidation occurring via two mechanisms: type I using radical chemistry and type II using the transfer of excitation energy (Foote, 1968; Kramer and Maute, 1972; Nilsson and Kearns, 1973; Silva, 1979; Sconfienza et al., 1980; Tsai et al., 1985) (Figure 72).

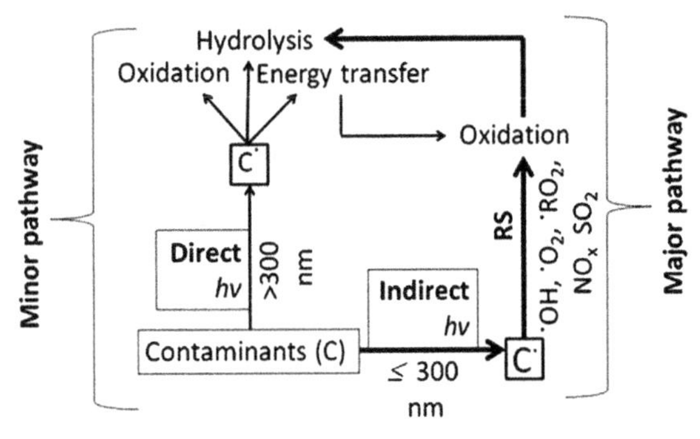

FIGURE 71 The key abiotic transformation processes, photo-oxidation, and photolysis (RS: reactive species, free radicals).

Both mechanisms share a common initial step in which a sensitizer absorbs photons and then transfers the energy to a ground-state substrate (S), converting it to a triplet-state substrate (^3S). In a type I mechanism, ^3S reacts with molecular oxygen forming an oxidized product (AO_2) via radical intermediates (Figure 72). In a type II mechanism, the excitation energy from ^3S is transferred from the sensitizer in the triplet state to molecular oxygen, giving rise to $^1O_2$ ($O_2(^1\Delta_g)$) which, in turn, reacts with the acceptor forming oxidized product. It should be noted that there are two types of singlet oxygen, $[O_2(^3\Sigma_g^-)]$ and $[O_2(^1\Delta_g)]$. Of these, only $[O_2(^1\Delta_g)]$ can effectively initiate photo-oxidation. The importance of $[O_2(^1\Delta_g)]$ as an oxidant in a given reaction medium can be expressed as the half-life of the substrate, a value which can be deduced from the rate of substrate photo-oxidation (Eqn (16))

$$-\frac{d[P]}{dt} = k_r[P]\left[O_2\left(^1\Delta_g\right)\right] \quad (16)$$

The actual significance of $O_2(^1\Delta_g)$-mediated oxidation in terms of photodegradation is given by the effective quantum yield of photo-oxidation (Φ_r) defined as $\Phi_r = \Phi_{\Delta r}[[P]/k_D + k_t[P]$. In this equation, $\Phi_{\Delta r}$ is the quantum yield of $O_2(^1\Delta_g)$ generation by the sensitizer and k_r is the reactive rate constant. The factors that

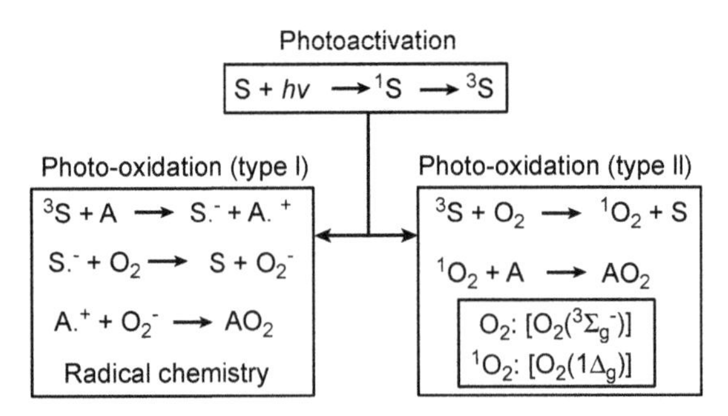

FIGURE 72 Mechanisms of photoactivation and photo-oxidation.

influence the oxidation reaction are (1) the presence of OH groups and their position on the aromatic ring and (2) the nature of substituents and their position on the aromatic ring (electron-donating substituents, $-CH_3$ or $-CH_2CH_3$, donate electrons much easier than the electron-withdrawing substituents $-NO_2$ or $-Cl$). In the environment, phenolic compounds in the presence of sensitizers (Fe_2O_3, MnO_2, soil and clays) or the enzyme phenol-oxidases and peroxidases are transformed into polymeric or simpler products.

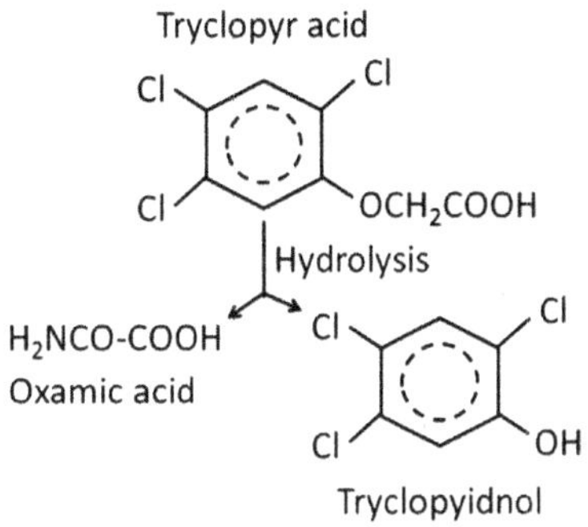

Tryclopyr acid
Cl
Cl
Cl
OCH_2COOH
Hydrolysis
$H_2NCO\text{-}COOH$
Oxamic acid
Cl
Cl
Cl
OH
Tryclopyidnol

FIGURE 73 Photo-hydrolysis of tryclopyr acid.

7.1.4.1.2 ABIOTIC HYDROLYSIS As shown in Figure 73, bulk chemicals react with H^+ or OH^- groups that split them by addition of a water molecule (Burrows et al., 2002). The rate of hydrolysis reaction is very sensitive to pH. Degradation of compounds in the environment via direct hydrolysis is limited. Rates of degradation throughout other processes (e.g. biological degradation) are typically much faster than direct hydrolysis.

7.1.4.2 BIOTIC TRANSFORMATION

The habitat for soil organisms is determined by soil texture, the pore-size distribution, soil water holding capacity, the amount of water, and air-filled pore space in soil aggregates. Various soil components may form varying size aggregates (macroaggregates: >250 μm and microaggregates: 20–250 μm) consisting of sand, silt, clay, organic matter, root hairs, microorganisms, mucilages, extracellular polysaccharides, and hyphae (filaments) of fungi as well as the resulting pores. The macroaggregates constitute the ecosystems for soil organisms. The biotic population can be divided into four groups: (1) surface and upper microfona— bacteria, algae, fungi, and mycorrhizae (plant root fungi); (2) lower microfona—bacteria or fungi feeding protozoa and nematodes; (3) mesofona (mm depth)—microarthropods including mites and collemobola; and (4) macrofona (cm depth)—including, but not limited to, earthworms, macroarthropods, and enchytraeids. Realizing the importance of microorganisms' role in chemical composition of soil as well as transformation of xenobiotic, there has been major research effort in this area that has generated considerable amounts of new data. However, as shown in Figure 74 (Madsen, 2005), the information still awaiting discovery greatly exceeds our current knowledge, given that nearly all current information about prokaryotes is based on 0.1% of the total estimated diversity of prokaryotes in the biosphere.

In this section, we review the published information to assess the possible roles of soil organisms in xenobiotic transformations. The biotic fate of xenobiotic in soil includes (1) mineralization (complete degradation of organic molecules into component minerals that can be used by other organisms) and complete degradation of the parent or transformed product of the compound in soil; (2) metabolic degradation; and (3) biotransformation. A key problem associated with biotic transformation is that the catabolic enzymes (enzymes that cause degradation) are substrate-specific; thus, they can act only on their natural substrate. The enzymes will act on a xenobiotic if there is reasonable structural similarity between the substrate and the xenobiotic. In certain cases, organisms are able to mineralize and completely degrade (productive degradation) the xenobiotic (Figure 75, organism 1). However, more often, an initial microbe or a group of microbes convert a chemical to a product that is

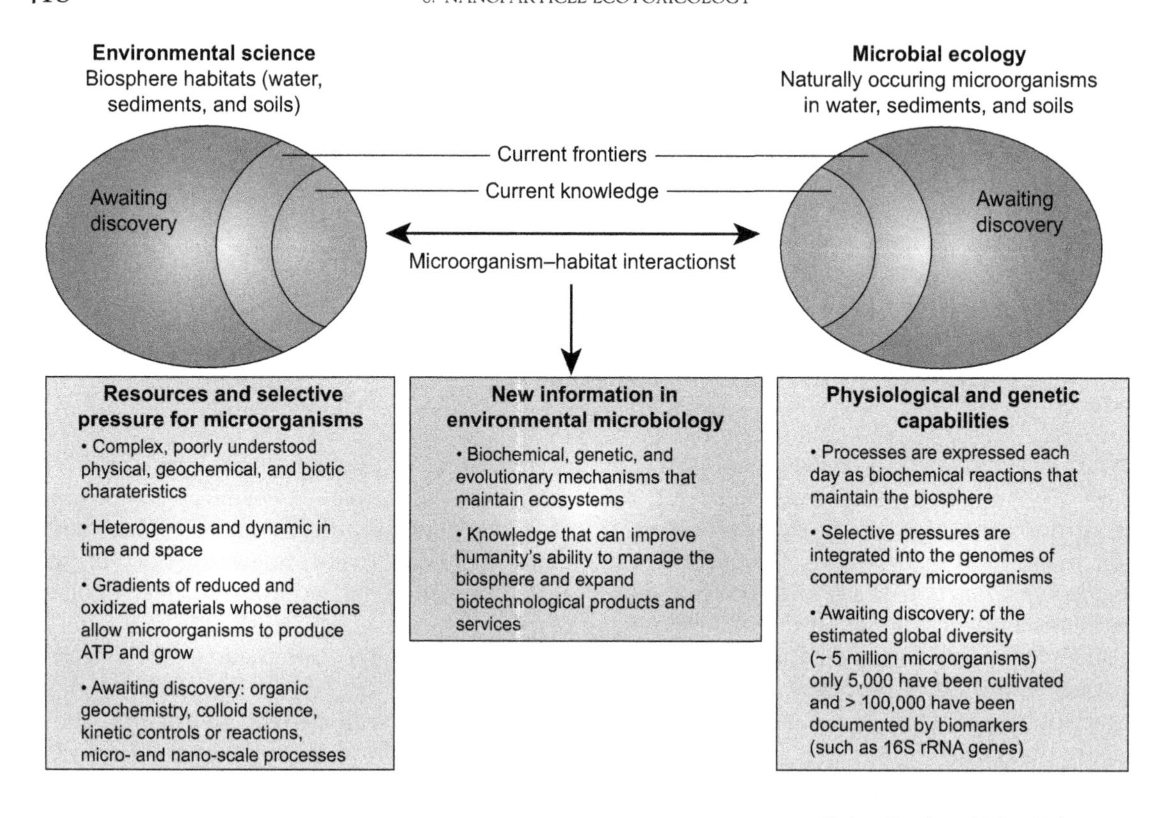

FIGURE 74 The information still awaiting discovery. Nearly all current information about prokaryotes is based on 0.1% of the total estimated diversity of prokaryotes in the biosphere. *Reprinted from Madsen (2005) with permission.*

transformed, but not mineralized—a process known as unproductive metabolism (Figure 75, organism 2).

Another group of bacteria may internalize and mineralize the transformed products (Figure 75, organism 3). In some cases, a co-metabolically active organism may further transform the chemical, forming a stable product (dead-end products, the process known as unproductive) (Figure 75, organism 4). This process is known as an unproductive process because it is not complete degradation to the elements. Therefore, a wide variety of dead-end products may accumulate in soil exposed to xenobiotic chemicals. The soil microbes may mineralize the partially degraded (transformed) products of abiotic process (Figure 75, organism 4).

7.1.4.2.1 SOIL REDUCTION–OXIDATION (REDOX) POTENTIAL AND BIOTIC PROCESSING OF XENOBIOTIC CHEMICALS Xenobiotic compounds, once released into the environment, are transported and distributed within different components of the environment (soil, air, surface water, sediment, and biota). The factors that determine distribution are dosage, the physicochemical properties of the chemical or particle, partitioning to soil and sediment, abiotic and biotic degradation, environmental characteristics including soil type and climatic conditions.

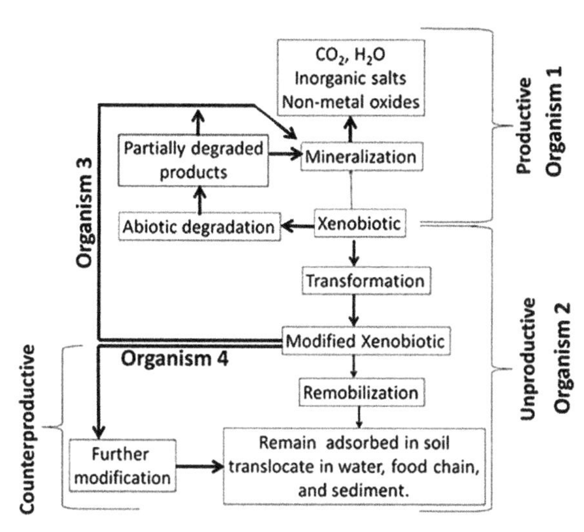

FIGURE 75 Biotic transformation of xenobiotic in soil. Productive organism 1 causes complete mineralization of toxins. Unproductive organism 2 partially degrades a xenobiotic. Organisms 3 and 4 mineralize the partially degraded chemicals.

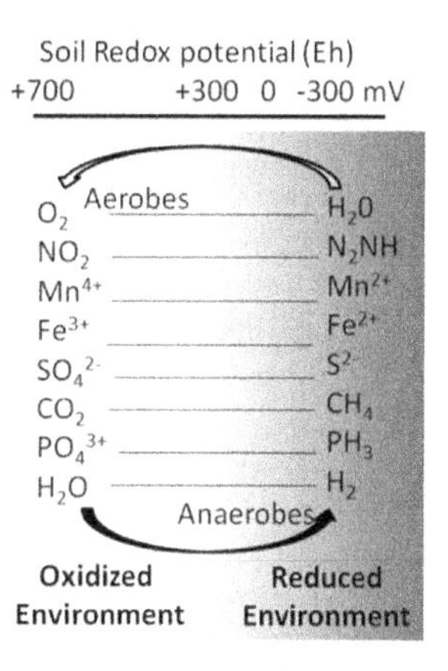

FIGURE 76 Redox potential of soil and its aerobic and anaerobic processes. Oxidative soil (Eh value around 600–700 mV) contain aerobic microbes, while the reductive soil (Eh 300 to −300 mV) contains anaerobic microbes.

Aerobic and anaerobic processes, maintained by the respective bacterial populations, play a critical role in biotic transformation of xenobiotic compounds. Aerobic organisms such as aerobic bacteria, fungi, and actinobacteria require aeration, critical moisture, near neutral pH, and organic matter; thus, they inhibit the surface and upper soil. Aerobic organisms use oxygen as the electron acceptor (reduction of O_2 to H_2O) during energy production. Anaerobic bacteria resides in deep soil or flooded surface soil lacking oxygen and use other sources for energy production (Figure 76). The distribution of aerobic and anaerobic organisms in soil depends upon the soil ecology and redox state measured as the soil's REDOX potential (Eh), characterized by Eqn (17) (Figure 77).

$$Eh = Eh^0 - \left(\frac{0.059}{n}\right)\log\frac{\{\text{Reduced species}\}}{\{\text{oxidized specied}\}[H^+]} \tag{17}$$

In this equation, Eh^0 is standard Eh, 0.059 is a constant (RT/F), and n is the number of electrons involved. In general, Eh values range from −300 mV to +700 mV. As shown in Figure 76, Eh for reduced soil ranges from 300 to −300, and Eh for oxidized soil is > 300 (up to 700). Aerobic bacteria that oxidize the reduced substrates reside in the oxidized soil ($Eh > 300$), while anaerobic bacteria that reduce the oxidized substrates reside in the reduced soil.

7.1.4.2.2 DISTRIBUTION OF BACTERIA IN SOIL Effects of soil depth on aerobic and anaerobic bacterial populations have been shown by Xu et al. (2014), who measured vertical distribution of methane-oxidizing bacteria (MOBs) expressing methane monooxygenase and sulfate-reducing bacteria (SRBs) expressing α-subunit of adenosine-5′-phosphosulfate (APS) reductase in soil from surface to 738 mm depth. As shown in Figure 76, from surface to 160 cm depth, MOB's population was several thousand-fold higher than APS's population, indicating mostly aerobic oxidation of methane to methanol.

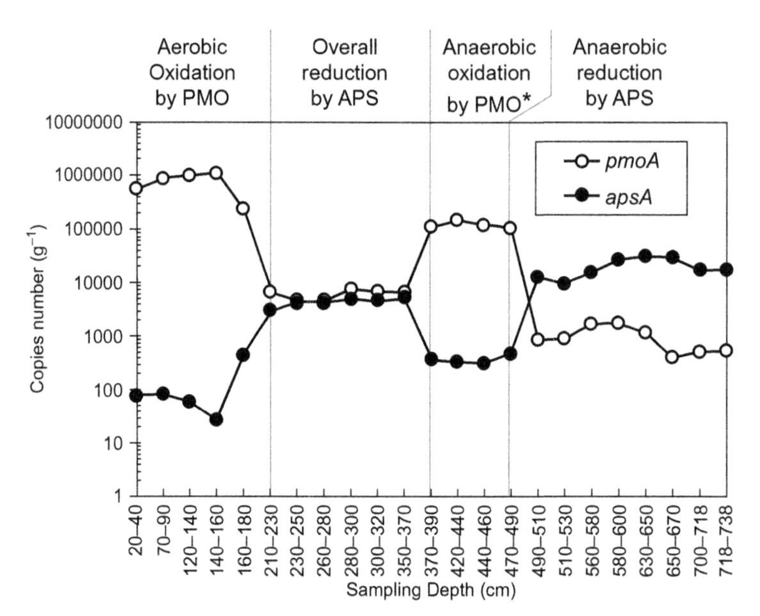

FIGURE 77 Vertical distributions of methane-oxidizing bacteria (MOBs) and sulfate-reducing bacteria (SRBs) in the marine sediment core of DH-CL14 by quantification of pmoA and apsA, respectively. *Reprinted from Xu et al. (2014) open access.*

Then, the MOB's population decreased, while the APS's population increased, reaching comparable populations from 230 to 370 cm depths, indicating a balance between methane oxidation and reduction of sulfate. As the depth increased, a proportional increase in MOB and a decrease in APS populations occurred. This may reflect anaerobic oxidation of methane using another acceptor. Then, a proportional increase in APS and a decrease in MOB occurred and the APS population surpassed the MOB population, indication of sulfate reduction. This suggests that depth plays a critical role in the distribution of aerobic and anaerobic bacteria in soil.

7.1.4.2.3 AEROBIC PROCESSING OF XENOBIOTIC CHEMICALS Aerobic bacteria require oxygen for survival. They are present in aerated moist soil containing organic carbon sources. There are two main types of aerobic bacteria:

1. The *obligate aerobes* that compulsorily require oxygen for deriving energy, growth, reproduction, and cellular respiration. These organisms do not survive in the absence of oxygen or flooding.
2. *Facultative and Microaerophile aerobes*: Facultative bacteria behave both aerobically and anaerobically, according to the prevailing conditions. In reduced environments, they acquire energy via anaerobic pathways, whereas in oxidative environments, they develop aerobic pathways. The microaerophilic bacteria require oxygen but in very low concentrations.

Studies have shown that (1) microbial community composition may play an important role in xenobiotic processing in soil and other soil processes (Cavigelli and Robertson, 2000; Balser et al., 2002) and (2) the microbial communities residing at depth are not simply diluted analogs of the surface microbial communities, but exhibit considerable differentiation (Ghiorse and Wilson, 1988; Fritze et al., 2000; Blume et al., 2002). In fact, the microbial communities in the soil subsurface may function differently from those at different depths. Aerobic microbes predominated the surface soil or soil rich in moisture and oxygen, whereas anaerobic microbes predominated the deep soil (Li et al., 2014a; Fierera et al., 2003; Blume et al., 2002).

Figure 78 shows the distribution of total aerobic (T. aerobic) organisms, aerobic bacteria, fungi, facultative anaerobic (Fa. aerobic) bacteria, and total anaerobic (T. anaerobic) bacteria in unplowed and plowed soil (Linn and Doran, 1984). This study is based on the hypothesis that plowing may increase oxygen and moisture, thus increasing that aerobic capacity and decreasing the ratio of population in nonplowed (PnP) soil/population in plowed (PP) soil for soil samples. In samples collected from 0 to 75 mm depth, the PnP/PP ratio ranged from 1.27 to 1.35, indicating that plowing reduced bacterial (aerobic or anaerobic) populations in top soil samples, possibly due to the lost moisture. In samples collected from deep soil, the PnP/PP ratio ranged from 0.6 to 0.8 for aerobic microbes, while the ratio ranged from 0.9 to 1.1 for anaerobic microbes. This suggests that although plowing significantly increased aerobic microbial population and/or decreased anaerobic populations, the surface microbes were more susceptible to plowing than were deep-soil microbes (Li et al., 2014a).

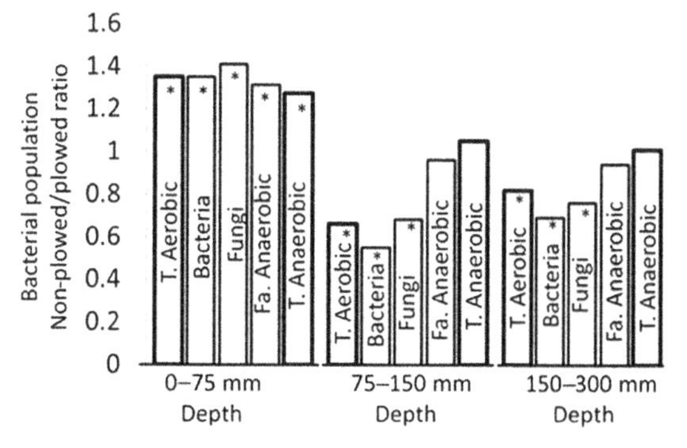

FIGURE 78 Distribution of microbes at different depths. *Reprinted from Xu et al. (2014), open access.*

Structurally diverse aromatic chemicals are degraded via the common products such as catechol, protocatechuate, gentisate, hydroquinone (benzene-1,4-diol), homoprotocatechuate, dihydroxyphenyl propionates, and homogentisate (Vaillancourt et al., 2006). The ring cleavage of catechol is performed by two distinct enzymes: intradiol oxygenases (utilize nonheme Fe(III) to cleave the aromatic nucleus ortho to (between) the hydroxyl substituents) and extradiol oxygenases (extradiol dioxygenases utilize nonheme Fe(II) to cleave the aromatic nucleus meta (adjacent) to the hydroxyl substituents) (Harayama and Rekik, 1989).

Figure 79(A)–(C) summarize the significance of the three aerobic processes. In the growth pathway (Figure 79A), the substrate is processed by an oxygenase that uses NAD(P)H and forms $NAD(P)^+$, while the intermediate is processed by a reductase that reduces $NAD(P)^+$ to NAD(P)H. This regeneration of NAD(P)H generates sufficient energy and precursors to sustain growth. In the nongrowth process (Figure 79B), the initial processing of the nongrowth substrate may generate dead-end products (not degraded by the bacterial populations) that may have adverse (toxic) effects. When a soil sample contains both growth and nongrowth substrates (Figure 79C), they compete for the enzyme, resulting in degradation of the nongrowth substrate. This strategy is commonly used in bioremediation processes.

7.1.4.2.4 ANAEROBIC PROCESSING OF XENOBIOTIC CHEMICALS Anaerobic conditions such as in groundwater, sediments, landfill, and sludge have been shown to transform and degrade diverse groups of pollutants (Janke and Fritsche, 1985; Mogensen et al., 2003a, b; Schink, 2002; Heider and Fuchs 1997, Prince, 1993; Spormann and Widdel, 2000; Widdel and Rabus, 2001; Ahmad and Hughes, 2000; Gorontzy et al., 1994; Marvin-Sikkema and de Bont, 1994; Peres and Agathos, 2000; Abramowicz, 1990; Bedard, 2003; El Fantroussi et al., 1998;

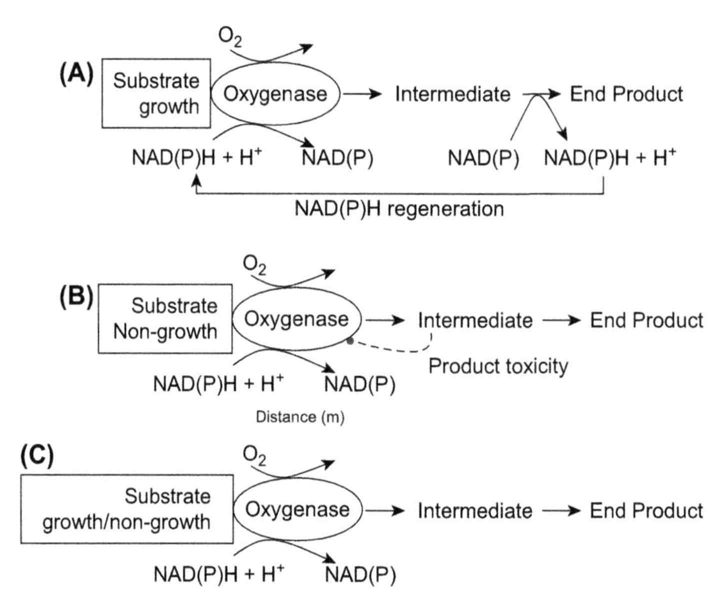

FIGURE 79 Growth and nongrowth metabolism of xenobiotics: (A) growth pathway, (B) non-growth pathway and (C) both growth and non-growth pathways.

Haggblom and Young, 1995, 1990; Haggblom et al., 2000; Ohtsubo et al., 2004; Williams, 1977). Figure 80(A) shows a typical anaerobic dechlorination pathway for the biodegradation of chlorinated aliphatic tetrachloroethylene (PCE) into ethene (Cookson, 1995).

The reaction is regulated by two enzymes: PCE reductive dehydrogenase (PCE-RDase or

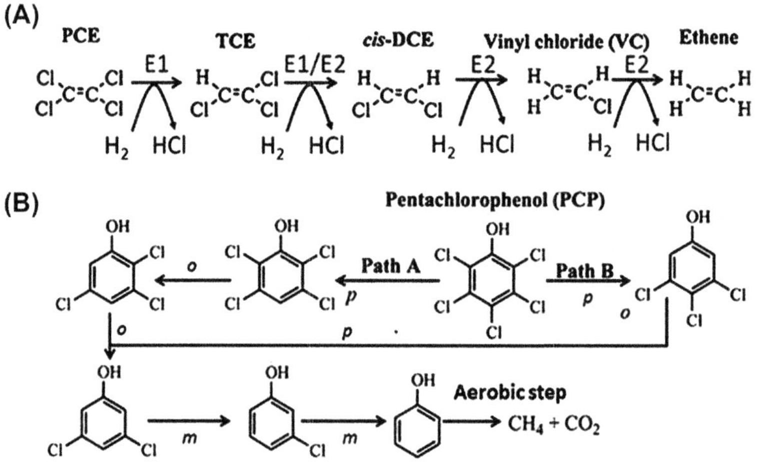

FIGURE 80 Biotransformation of (A) polychloroethane (PCE) and (B) pentachlorophenol (PCP).

E1) and trichloroethene reductive dehydrogenase (TCE-RDase or E2). Figure 80(B) shows the anaerobic pathways for degradation of pentachlorophenol (PCP). Although dechlorination is the key step, certain bacteria may preferentially remove chlorines in the order of para > ortho > meta (Bryant et al., 1991), whereas in others an ortho > para > meta order of chlorine removal has been reported (Mikesell and Boyd, 1986). The common reactions responsible for anaerobic degradation of pesticides are dechlorination (chlorinated pesticides), hydrolysis, nitro-reduction, and dealkylation (Cookson, 1995). For persistent compounds including polychlorinated biphenyls, dioxins, and DDT, anaerobic processes are slow for remedial applications but can be a significant long-term avenue for natural attenuation.

7.1.4.2.5 SEQUENTIAL ANAEROBIC-AEROBIC PROCESSING In some cases, a sequential anaerobic-aerobic strategy is needed for total mineralization (destruction) of xenobiotic compounds (Coleman et al., 2002a,b). The aerobic bacteria ensure completely optimal conditions such as redox, electron donors (normally H_2), and competing electron acceptors (e.g., nitrate, sulfate) for anaerobic bacteria to function. Conversely, anaerobic bacteria generate reduced substrates for aerobic bacteria. Mineralization of PCP or related compounds may be achieved via anaerobic-aerobic coupling. The degradation of atrazine, a nitrogen-containing pesticide, involves hydrolytic dechlorination, dealkylation, and the cleavage of C—N in the cyclic ring, yielding ultimate mineralization to CO_2 and NH_3. As shown in Figure 81, anaerobic degradation of trichloroethylene results in dead-end products ethane and ethane. The aerobic degradation of trichloroethylene (a cosubatrate for CH_4 degradation) results in accumulation of mineralized carbon and dead-end products (trichloroethanol, dichloroecataldehyde, and glycoaldehyde). The anaerobic metabolite of trichloroethylene, vinyl

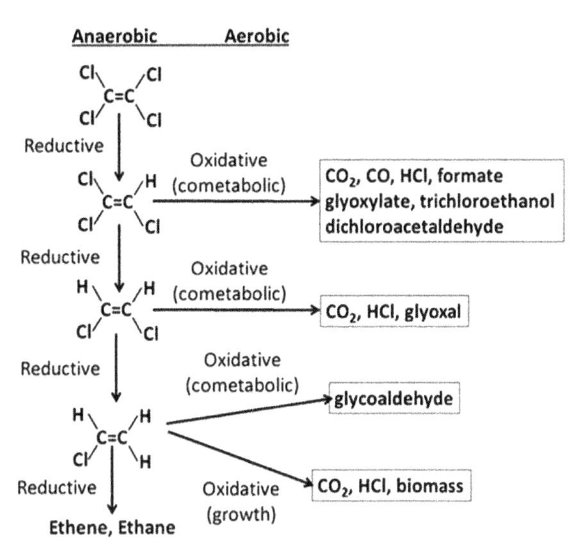

FIGURE 81 Sequential aerobic and anaerobic processing of xenobiotics.

chloride, is a growth substrate. Its aerobic metabolites are CO_2 and HCl (Figure 81).

7.2 Appendix 2: Principles of Ecology

The following principles are at play in every ecosystem formed by the interaction of a community of organisms and their environment (Figure 82). Ecosystems are nested at many

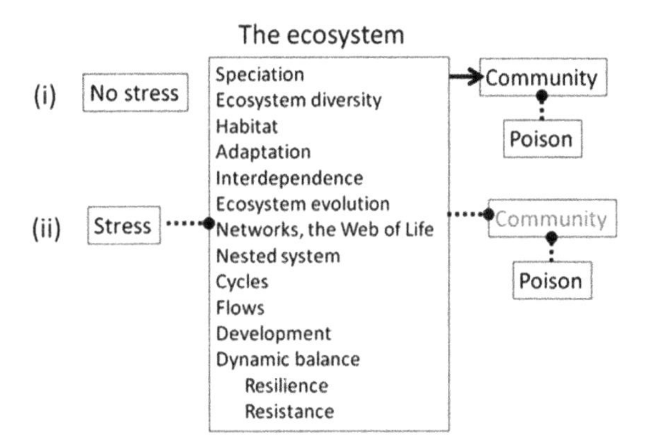

FIGURE 82 Components of an ecosystem.

scales from a small pond to the entire planet. The human species, just like any other species, is entirely interdependent on the ecosystems in which we live.

Some of the principles of an ecosystem is described below:

7.2.1 *Speciation or Formation of New Species*

Organisms may change to the point where they are different enough to be considered separate species. This occurs when populations of one species are separated and adapt to their new environment or conditions (physiological, geographic, or behavioral). The key to speciation is the evolution of genetic differences between the incipient species (a group of organisms that is about to become a separate species from other, related individuals). For a lineage to split, mating between certain groups either do not happen or are not successful because of a change in the timing, location, or rituals of mating. This change might evolve by natural selection or genetic drift. Table 8 compares some of these speciation modes.

7.2.2 *Ecosystem Diversity*

There are many different kinds of ecosystems on Earth, ranging from deserts to mountain slopes and from the ocean floor to the Antarctic. Coral reefs and rainforests are among the richest of these ecosystems. Each ecosystem provides many different kinds of habitats or living places. The living things and the nonliving environment (earth forms, soil, rocks, and water) interact constantly and in complex ways that change over time, with no two ecosystems being the same. Although ecosystems are ever-changing and complex, some universal principles apply:

1. Matter constantly cycles and recycles.
2. Energy moves through the cycle, being used, absorbed, and stored. The Sun is a constant source of Earth's energy.

Humans have to appreciate how the elements in each ecosystem are connected to each other and the diversity that exists among Earth's ecosystems. Maintaining this ecological diversity is important for the health of the planet.

TABLE 8 Mode of speciation

Mode	
Allopatric: Geographically isolated populations	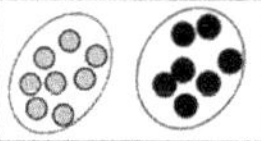
Peripatric: A small species at edge of a larger population.	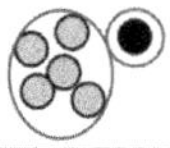
Parapatric: A continuously distributed population	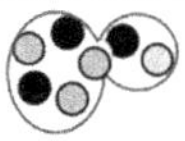
Sympatric: Within a range of ancestral population.	

7.2.3 *Habitat*

A habitat is the place in the environment where a population (group of living organisms of the same kind living in the same place at the same time) resides. Individual species have adapted to the conditions of unique niches, territories, watersheds, and climates. All of the populations interact and form a community. The community of living things interacts with the nonliving world around it to form the ecosystem. The habitat must supply the needs of organisms, such as food, water, temperature, oxygen, and minerals. If the population's needs are not met, it will move to a better habitat.

Habitat loss and its effects on biodiversity are growing global concerns. Loss of habitat is a major cause of the decline of a species. Habitat is lost when land changes due to human activity. Humans are progressively taking over natural areas for their own use, thus encroaching on the habitat of other creatures. In addition to the physical changes, pollution is also an increasingly important factor determining the health and distribution of wildlife and biodiversity. Pollution arises from industrial, agricultural, and transportation industries. Often, these polluted habitats support key communities within estuarine, coastal, and/or marine subsystems and have a high biodiversity. The potential impacts of negative changes on the habitat include erosion, sediment deposition, poor water quality, loss of habitat-dependent species, and a loss in visual amenity.

7.2.4 *Adaptation*

Adaptation means the adjustment of organisms to environmental conditions and to other living things, either in an organism's lifetime (physiological adaptation) or in a population over many generations (evolutionary adaptation). Individual species change or adapt to changing conditions in their habitat. If the features and behaviors are successful in adapting to the changes, the particular organisms survive and reproduce. If the features and behaviors are not successful, the organisms will not survive or reproduce.

7.2.5 *Interdependence*

All of life is a web of interactions and evolving adaptations between species and their habitats. No species, including humans, can survive separately from this web. Figure 83 shows interdependence among plants, caterpillar, blue tit, and kestrel. Although kestrel does not feed on leaves, it will not survive without one.

7.2.6 *Evolution*

The interdependent adaptations between species and habitat create biological change over time. New forms of life are always emerging, converging, and diverging, pursuing the greatest health and flexibility for the entire system.

FIGURE 83 An example of interdependence.

7.2.7 *Networks or the Web of Life*

An ecological community is a group of interacting species living in the same place. A community is bound together by the network of influences that species have on one another. Inherent in this view is the notion of the *balance of nature*—that is, whatever affects one species also affects many others. There are usually four types of interactions, as shown in Table 9.

Food webs are graphical depictions of the interconnections among species based on energy flow (Figure 41) (Odum, 1968). Energy from the sun enters a biological web of life at the bottom of the producer level through the green plants. Many food webs also gain energy inputs through the decomposition of organic matter aided by microbes. Energy moves from

TABLE 9 Description of Species Interaction in an Ecosystem

Type of Interaction	Effects	Examples
Mutualism	Both species benefit from interaction	Pollination
Commensalism	One species benefits, one unaffected	Egrets and cattle
Competition	Each species affected negatively	Resource overlap
Predation, parasitism, herbivory	One species benefits, one is disadvantaged	Antagonistic interactions

lower to higher trophic (feeding) levels by consumption: herbivores consume plants (producers), predators consume herbivores, which may in turn be eaten by top predators. Some species feed at more than one tropic level and hence are termed omnivores.

7.2.8 *Nested Systems*

Often, we find multileveled structures of systems nesting within systems. Each forms an integrated whole within a boundary, while at the same time being a part of a larger whole (Figure 84). Each nested ecosystem may be diverse in terms of flora and fauna, but it interacts with others. Nested dynamic systems are a real feature of life on Earth.

7.2.9 *Cycles*

The interactions among the members of an ecological community involve the exchange of energy and resources in continual cycles such as water, carbon, and nitrogen cycles (Figure 85). The cycles in an ecosystem intersect with larger cycles in the bioregion and in the planetary biosphere.

Water cycle: In the water cycle, energy from the sun evaporates water from ocean surfaces or from treetops and then the weather systems move the water vapor (clouds) from one place to another. Precipitation occurs when water condenses from a gaseous state in the atmosphere and falls to Earth. Gravity either pulls water from the surface into the underground reserves or runoff. Frozen water may be trapped in cooler regions of Earth (the poles, glaciers

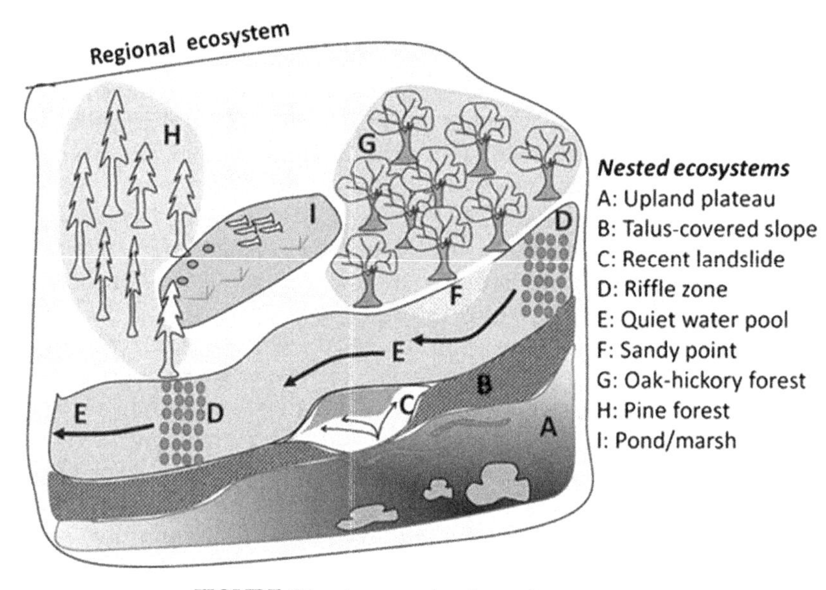

FIGURE 84 An example of nested ecosystem.

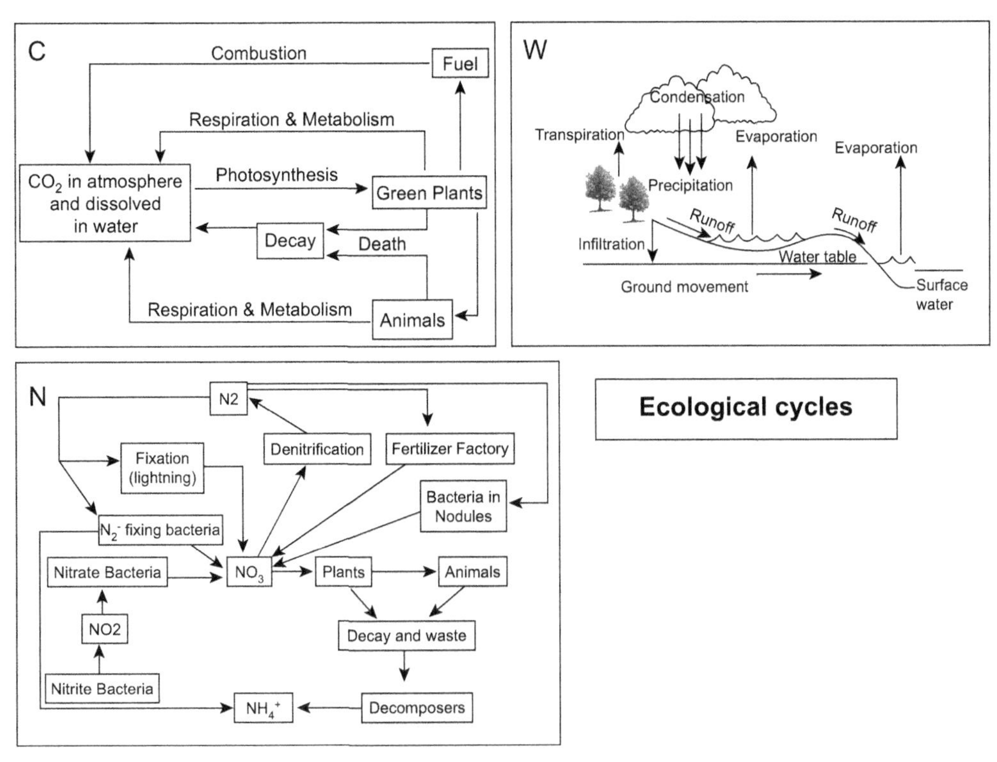

FIGURE 85 Ecological cycles. C: carbon, N: nitrogen, and W: water cycles.

on mountaintops, etc.) as snow or ice. The oceans are salty and not drinkable. Water leaves the oceans by evaporation that leaves the minerals behind. Rainfall and snowfall are comprised of relatively clean water, with the exception of pollutants (such as acids) picked up as the water falls through the atmosphere. Plants absorb water via the roots and move it to the gas exchange organs, the leaves, where it evaporates quickly (transpiration). In both plants and animals, the breakdown of carbohydrates (sugars) to produce energy (respiration) also produces both carbon dioxide and water as waste products.

Nitrogen cycle: Nitrogen exists in the atmosphere as nitrogen gas (N_2) that is biochemically inert and plants and animals are unable to use it as nutrient. Plants can only take up nitrogen in the form of ammonium ion (NH_4^+) and nitrate ion (NO_3^-), while animals receive the nitrogen in the form of proteins and amino acids. Atmospheric N_2 is converted into NH_4^+ ions that is further modified into nitrite (NO_2^-) and nitrate (NO_3^-) via nitrifying bacteria. Nitrates and nitrites are highly mobile because they do not bind soil particles. In the ocean, nitrates are converted into nitrous oxide gas, which is released into the atmosphere.

Carbon cycle: The carbon cycle refers to the biogeochemical cycle by which carbon is exchanged between the biosphere (which usually includes animals, plants and bacteria), geosphere (soil and soil bacteria), hydrosphere (which includes dissolved inorganic carbon and living and non-living marine biota) and atmosphere (mostly CO_2). The ocean contains the largest active pool of carbon near the surface of Earth, but the deep ocean part of this pool does not rapidly exchange with the atmosphere.

7.2.10 *Flow*

All organisms are open systems, which means that they need to feed on a continual flow of energy and resources to stay alive. The constant flow of solar energy sustains life and drives all ecological cycles.

7.2.11 *Development*

On Earth, every life form undergoes development and learning at the individual level and evolution at the species level. This involves interplay of creativity and mutual adaptation in which organisms and environment coevolve.

7.2.12 *Dynamic Balance*

In a stable ecosystem, the biotic (living) and abiotic (nonliving) parts are in equilibrium, characterized by continual fluctuations. All ecological cycles act as feedback loops, so that the ecological community regulates and organizes itself, maintaining a state of dynamic balance. This means that the nutrients are able to cycle efficiently, and no community of organisms or natural phenomena is interrupting the flow of energy and nutrients to other parts of the ecosystem. Resilience and resistance characterize the ecosystem stability. Resilience refers to the rate at which the density of a population in an ecosystem comes back to the equilibrium after a particular disturbance. Resistance stands for the potential of an ecosystem to prevent the animal population and trees from succumbing to stresses such as high pollution or drought.

7.3 Appendix 3: Aquatic Ecosystems

Aquatic ecosystems include wetlands, rivers, lakes, oceans, and coastal estuaries. These ecosystems are critical elements of Earth's dynamic processes and are essential to human economies and health. Wetlands connect land and water, serving as natural filters, reducing pollution, controlling floods, and acting as nurseries for many aquatic species. Rivers, lakes, and estuaries serve as important transportation, recreation, and wildlife hubs. As shown in Figure 86, aquatic ecosystems can be divided into three types: freshwater (rivers, lakes and wetlands), salt water (oceans, seas and estuaries), and polar water (Arctic and Antarctic oceans). Although Arctic and Antarctic oceans also have salt water, it is covered with a thick layer of ice and thus lacks a water–air interface.

7.3.1 *Freshwater Ecosystems*

7.3.1.1 LAKES

These regions range in size from less than one square mile to thousands of square miles. Ponds are smaller and usually seasonal, lasting just a couple of months. Lakes are divided into three different zones, usually determined by depth and distance from the shoreline:

1. *Littoral zone:* This warmest, lighted, and oxygen-rich zone is the topmost zone near the shore of a lake. It sustains fairly diverse

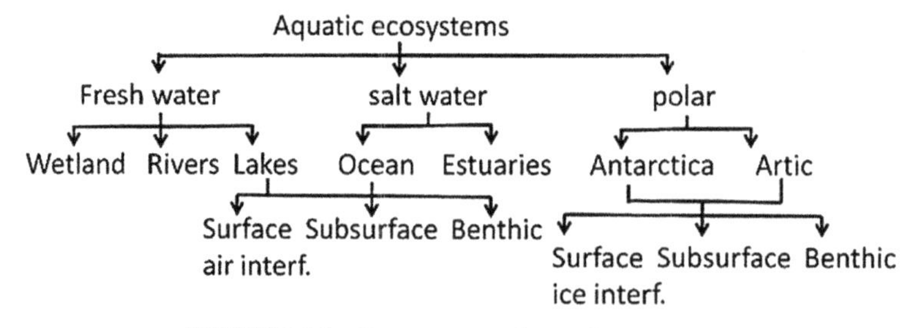

FIGURE 86 Components of aquatic ecosystem.

communities, including algae (like diatoms), rooted and floating aquatic plants, grazing snails, clams, insects, crustaceans, fishes, and amphibians. The vegetation and animals living in the littoral zone are food for other creatures such as turtles, snakes, and ducks (Carpenter and Kitchell, 1993).

2. *Limnetic zone:* This is near-surface open water zone surrounded by the littoral zone. The limnetic zone is well-lighted and is dominated by plankton, both phytoplankton and zooplankton. A variety of freshwater fish also occupy this zone.
3. *Profumdal zone:* This is the deep-water part of the lake/pond, much colder and denser than the other two. Little light penetrates all the way through the limnetic zone into the profundal zone. The fauna are heterotrophs, meaning that they eat diverse organisms and use oxygen for cellular respiration.

7.3.1.2 RIVERS

These are bodies of flowing water moving in one direction from a source to the ocean or another river. The characteristics of a river or stream and its flora change during the journey from the source to the mouth.

7.3.1.3 WETLANDS

These are areas of standing water supporting aquatic plants. Marshes, swamps, and bogs are all considered wetlands. Plant (pond lilies, cattails, sedges, tamarack, and black spruce) species adapted to the very moist and humid conditions. Wetlands have the highest species diversity of all ecosystems.

7.3.1.4 SALTWATER ECOSYSTEMS

Saltwater ecosystems can be divided into oceans, estuaries (where the saltwater and freshwater mix), coral reefs, and shorelines, each having diverse characteristics. The plant life, animal life, and land structures that make up saltwater ecosystems also are varied. All saltwater ecosystems have landforms, plant life, and animal life, which may be unique to their location. Although these saltwater ecosystems are very different, they share a high degree of interdependency by the nature of their association. In ocean, the animal and plant life is diverse and, for the most part, is limited to the first few hundred feet in depth. Plant and animal lives in the oceans range in size from microscopic organisms to the largest mammals on Earth, the blue whale. In an estuary, plant life includes kelp forests, sea grass, and woody swamps, and the animal life includes birds, microscopic organisms, crustaceans, sea reptiles, and amphibians. The coral reef includes sponges, corals, snails, fishes, invertebrates, and other marine life. Seagrass, mangroves, and algae compose the plant life.

7.3.1.5 POLAR ECOSYSTEMS

Although the Arctic and Antarctic oceans are capped with ice, plunged into total darkness during the winter, buffeted by blizzard winds and bitterly cold, the oceans are yet beautiful environments on Earth. The Arctic and Antarctic oceans are teeming with life. Great polar bears roam the Arctic ice and swim the Arctic seas. Supporting these top predators is a complex ecosystem that includes plankton, fish, birds, seals, walruses, and even whales. At the center of this food web, supporting all of this life, are phytoplankton and algae that produce organic material using energy from the sun.

7.3.2 *Ecotoxicology*

7.3.2.1 INTRODUCTION

Toxicology is the study of poison and poisoning, including the nature, effects, and detection of poisons and the treatment of poisoning. As discussed earlier, the degree of poisoning is related to the exposure dose in terms of mass/unit weight. The adverse effects of toxins or poisons are studies in whole animals and in vitro methods to decipher how various chemicals and dosages interact with living systems. In classic toxicology, the focus is mostly on the organisms including humans, experimental animals and companion and food animals under controlled ecosystems such as human settlements or animal farms. Ecotoxicology, on the other hand, is study of the effects of contaminants on the ecosystems such as species abundance, diversity, community composition, and species interactions. Ecotoxicol ogy is based on the principles of ecology (Appendix 2) where the adverse effects of chemical pollutants may depend not only on their concentrations and mechanism of toxicity, but also on the prevailing environmental conditions Noyes et al., 2009 and Schiedek et al., 2007). Environmental factors influence the toxic effects of pollutants, resulting in multiple stressor effects in a community Clements and Newman, (2002) and Clements and Rohr, (2009). Schnitt-Jansen et al. (2008) have proposed two basic assumptions in applying concepts of community ecology to ecotoxicology:

1. Communities are more than the sum of individual populations.
2. The protection of community structure maintains ecosystem structure and function (Calow and Forbes, 2003).

7.3.2.2 COMMUNITY AND TROPHIC LEVELS

In general, the health of an ecosystem is assessed by using a single species tests and then extrapolating the effects of chemicals on entire ecosystems. Although the single-species tests are well reproducible, they may not accurately reflect the status of the ecosystem community because the tests do not account for indirect effects like species interactions and trophic-based accumulation of toxin. Sommer (1986) and Carpenter et al. (1987) have shown that changes occurring in one trophic level may propagate to other levels by top-down or bottom-up regulation. Ecosystems may shift to different stable states after disturbance. Trophic interactions may define the sensitivity of a species to a toxicant. Figure 87 described a hypothetical situation of classic toxicology (Figure 87(A) and (B)) and ecotoxicology (C-E) that provides some information regarding the direct and indirect effects of toxicants and effect propagation.

As shown in Figures 87(A–E), animals exposed to insecticides, but not herbicides, experience toxicity, while phytoplankton exposed to herbicide, but not insecticide, experiences toxicity (usually lethal). If this information is extrapolated, one may conclude that in the presence of herbicides, animals will not experience toxicity. However, when the laws of ecology are applied, either an herbicide or a pesticide will disrupt the ecosystem balance (shown in Figure 87(C)) that will adversely affect the ecosystem animal community. An herbicide will destroy the phytoplankton that will reduce the herbivores population, resulting in a reduction of the predators' population (Figure 87(D)). In long term, other species may takeover that ecosystem. The herbivores and the predators of the herbivores will take up insecticide from water, but, according to the laws of ecology, the predator will accumulate more toxin than the prey. Thus, eventually, there will be more prey than predators. A reduced pressure will initially increase the phytoplankton population. Thus, the herbivore population will increase at a higher rate than the predator population. The ecosystem is imbalanced but different than the system shown in Figure 87(D).

According to the information provided above and the results of other published studies (Ashauer, 2010 and Rozman and Doull, 2000), two possible pathways that lead to the development of ecotoxicology have been proposed:

1. Direct effects of the toxin on an ecosystem community: In this case, the adverse effects are determined by the toxin's toxicokinetics (TC), which define the time course of the toxins reaching the target site, and the toxin's toxicodynamics (TD), which define the target-site characteristics and the outcome of the

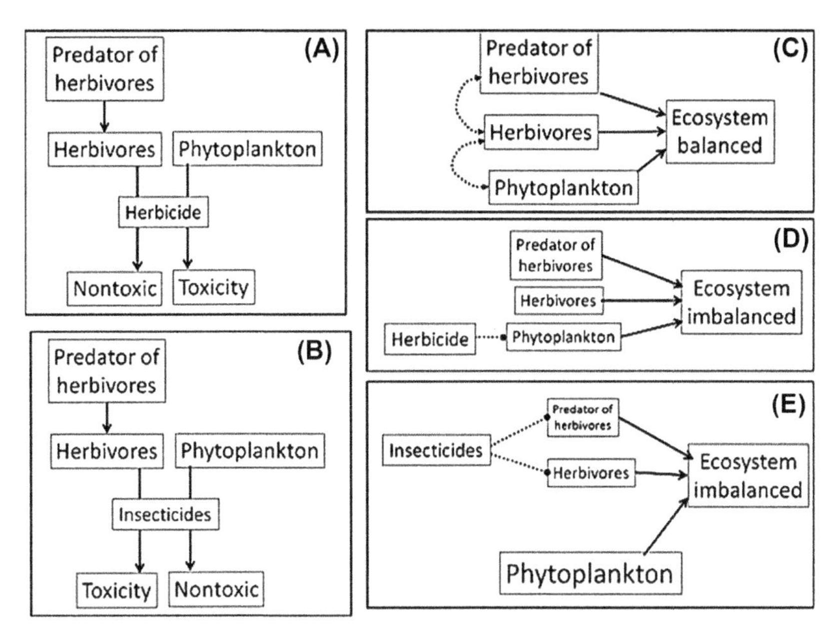

FIGURE 87 A hypothetical situation of classic toxicology (A) and (B) and ecotoxicology (C–E) that provides important information regarding the direct and indirect effects of toxicants and effect propagation. (A) Herbicides directly affect phytoplankton and not animals that remain healthy. (B) Insecticides directly affect all animals and phytoplankton remains unaffected. (C) A balance exists among the producer, herbivore, and predator. (D) Herbicides that kill the phytoplankton stress the herbivores, resulting in a decrease in their population. This may also decrease the predator population. The ecosystem is imbalanced. (E) An insecticide may kill both the herbivores and the predator, resulting in uncontrolled growth of the producers. The ecosystem is imbalanced.

chemical reaching the target site. The toxicity potential is related to the consumer's trophic levels—the higher the trophic level, the greater the accumulation and toxicity.

2. Ecosystem stress (due to the toxins independent factors) modulates the community toxicity by interfering with the toxin's TC and TD. That is, the stressors either modulate the toxins fate in body and/or modulate the characteristics of the target site.

7.3.2.3 HOW DO TOXINS AFFECT INTERSPECIES INTERACTION?

Toxins, usually arising from some form of human activity, become available to an ecosystem community via food or direct contact through skin or cell wall and cause toxicity. Even if only a few components are affected by the contaminants, the entire ecosystem may suffer due to direct and indirect effects. Some of the key mechanisms are described as follows:

- *Keystone effects*: A keystone species is defined as a predator (P) that enhances the abundance of one or more inferior competitors (C_I) by reducing the abundance of a dominant (superior) competitor (C_D) (Figure 88(A)) (Paine, 1966). A toxicant can increase diversity by targeting C_D.
- *Consumptive competition*: Competition between consumers occurs when the predator or consumer species inhibit one another by consuming a shared resource. Some contaminants such as herbicides can have

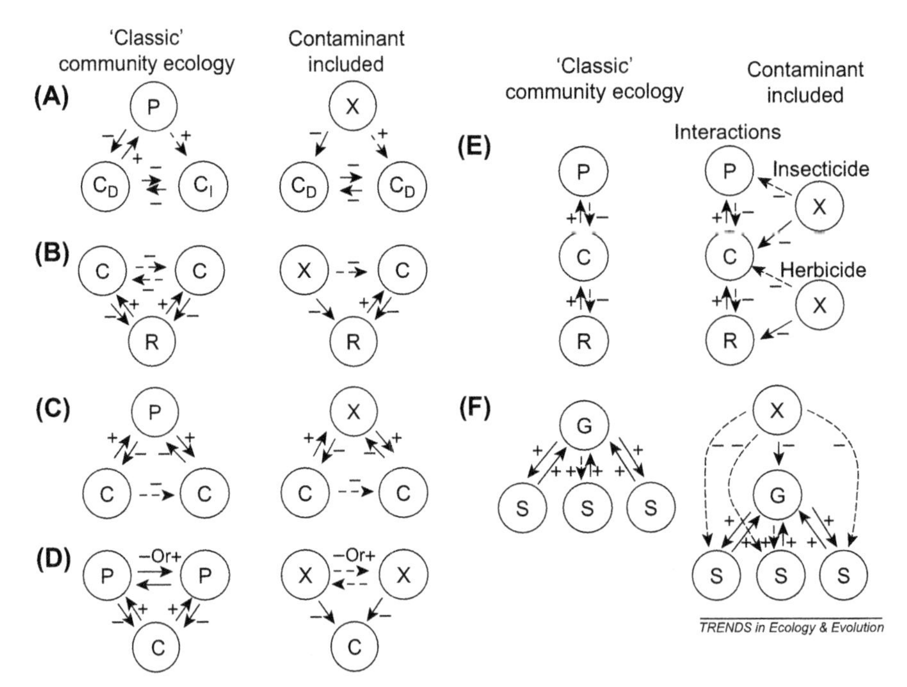

FIGURE 88 Possible effects of poisons on species interaction. (A) Keystone predation: contaminant impacts on competitive dominant species (C_D) enable competitive inferior species (C_I) to persist. (B) Consumptive competition: contaminant impacts on prey negatively impact predators of that prey. (C) Apparent competition or shared sensitivity to a pesticide: pesticide used to control a pest species negatively affects other species that do not compete with the pest. (D) Multiple predators: contaminants can interact additively, synergistically, or antagonistically with one another. (E) Tri-trophic interactions: contaminants can have direct and indirect effects at milti-trophic levels. (F) Direct mutualism: contaminants can affect the central hub of generalists by having indirect effects on the specialists that rely heavily on these generalists. (C, consumer; G, generalist; P, predator; R, resource; S, specialist; X, contaminant. Dashed lines represent indirect effects and solid lines are direct effects. Positive and negative effects are signified by + and −, respectively.)

similar negative effects by reducing plant availability (Rohr et al., 2008, 2006; Rohr and Crumrine, 2005; Fleeger et al., 2003). Unlike the competing consumer, the toxin−resource relationship is unidirectional and thus does not strictly fit the definition of competition.

- *Apparent competition*: Apparent competition involves negative interactions between prey species (C) through a shared predator (*P*) rather than shared resources (*R*). An increase in either prey species increases the abundance of a shared predator, resulting in increased predation on the other species (Figure 88) (Abrams and Matsuda, 1996). Here, we view pesticides as being analogous to the shared predator.
- *Multiple predator effects:* In an ecosystem, organisms are often exposed to multiple predators and, similarly, they may also be exposed to multiple contaminants (Thompson, 1996). As shown in Table 10 (Sih et al., 1998), the combined presence of two predator taxa results in greater than expected prey mortality, whereas in other cases, two predator species result in less than expected prey mortality.
- *Tri-trophic interactions*: Consumer and resource interactions can have indirect effects that

TABLE 10 Studies that Examined Multiple Predator Effects (MPEs) on Prey Survival and/or Mortality Grouped by Type of Effect Observed. (Sih et al., 1998)

Effect	Habitat	Prey	Predators
Additive $p = p_a + p_b$	Terrestrial	Arthropods	Spiders, lizards
			Mantids, spiders
	Marine	Barnacles, mussels	Snails, birds
		Invertebrates	Crabs, fish
Risk reduction $p < p_a + p_b$	Terrestrial	Arthropods	Mantids, spiders
		Aphids	Parasitoids, coccinellids
			Lacewings, hemipterans
	Marine	Barnacles, mussels	Snails, birds
		Fish	Fish, birds
Risk enhancement $p > p_a + p_b$	Lotic	Mayflies	Stoneflies, fish
			Dragonflies, fish
	Lentic	Tadpoles	Two species of dragonflies
Substitutable p mean of p_a and p_b	Terrestrial	Aphids	Coccinellids
	Lotic	Mayflies	Two species of stoneflies

p: predation, a and b: species A and species B.

proliferate across three or more trophic levels. As shown in Figure 88(E), pesticides, herbicides and hydrocarbons demonstrate tri-trophic indirect effects generated by top-down or bottom-up effects (Rohr and Crumrine, 2005; Fleeger et al., 2003; Brock, 2000).

- *Direct mutualism:* Mutualism refers to mutually beneficial interaction such as an interaction between a tree and its pollinator. In mutualism, there are two groups of species: generalists (*G*) that interact with each other and specialists (*S*) that do not interact with each other but interact with generalists (Figure 88(F)). Contaminants that have substantial direct or indirect deleterious effects on the species of the central hub of mutualistic networks could be catastrophic to the many specialists that rely on this hub and could lead to system collapse (Figure 88).

7.3.2.4 EFFECTS OF TOXINS ON ECOSYSTEM INTEGRITY

Ecosystems consist of highly interactive abiotic and biotic systems in which diverse communities live in balance. A healthy ecosystem is highly diverse, in which each member (from producer to consumer) performs its specific function. Even a minor imbalance may affect entire ecosystem via bottom-up or top-down manner. The health of an ecosystem is usually assessed using computer-based modeling. Some of the approaches are described in the following sections. A simplified model showing interaction of toxins with ecosystem is shown in Figure 89.

Prediction of the effects of contaminants on communities and ecosystem properties poses many more challenges than predicting the responses of individual taxa in isolation. This requires an integration of (1) both direct and indirect density and trait-mediated effects of contaminants, (2) ecosystem community composition, and (3) nonadditive effects (synergisms and antagonisms), which are more likely with species interactions. Many approaches have been used to assess effects of contaminants on ecosystems. Earlier studies (Halstead et al., 2014; Rohr et al., 2006; Clements and Rohr, 2009) have proposed that the food web paradigm provides a key framework for predicting ecosystem health by using null hypotheses regarding antagonistic, additive, or synergistic interactions between chemicals (Paine, 1966). This approach is based on the following principles:

1. Only certain species are sensitive to direct effects by particular chemical classes.
2. Their reproductive system allows recovery after a chemical exposure.

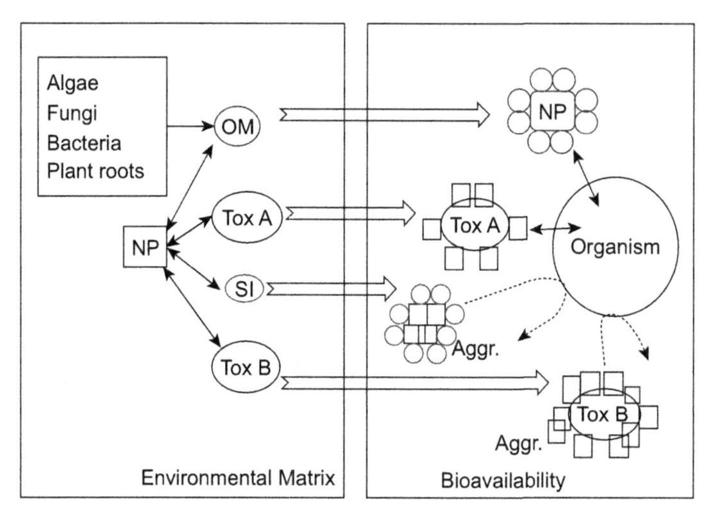

FIGURE 89 A diagram showing possible interaction of nanoparticles with toxicants (Tox A and B), salt ions (SI), organic matter (OM), or compounds released by plants, fungi, bacteria, and algae. OM and some of the phytochemicals increase the nanoparticles' bioavailability (represented as solid arrows entering organisms), whereas salt ions facilitate their aggregation, thus reducing their bioavailability (represented as dotted arrows not entering organisms). In other cases, nanoparticles' bioavailability might be either increased or decreased.

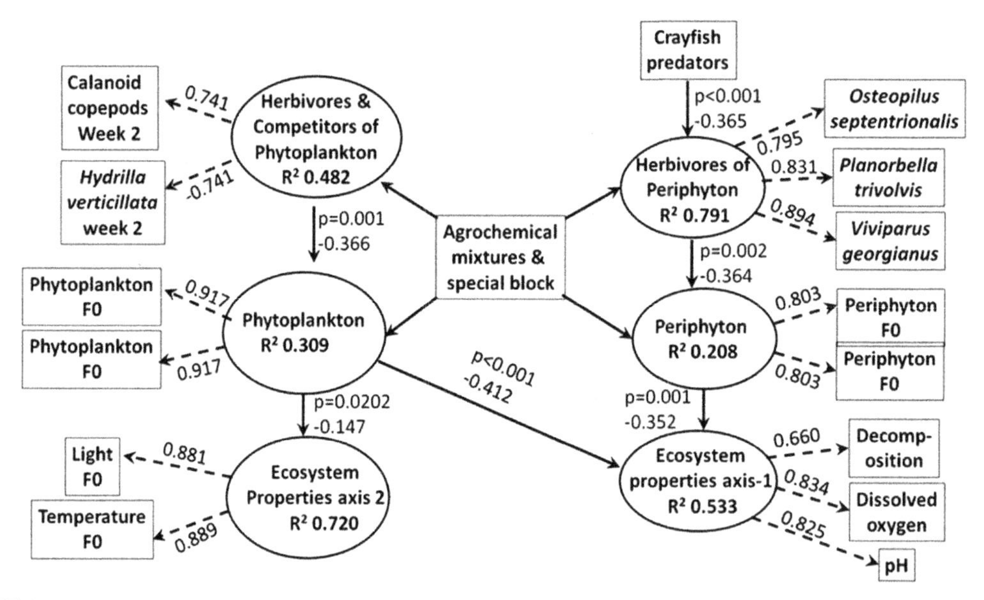

FIGURE 90 Effects of agriculture chemicals on ecosystem properties and biodiversity. *Reprinted from Halstead et al. (2014) with permission.*

3. Strong interspecies interactions in food webs that account for indirect effects.
4. Integration of the information collected in steps 1–3 with the effects that functional groups have on ecosystem properties and community.

Halstead et al. (2014) tested this general hypothesis by conducting an outdoor mesocosm experiment in which the effects of a fertilizer, herbicide (atrazine), insecticide (malathion), and fungicide (chlorothalonil) were quantified in isolation and in all pairwise combinations, on 24 species and seven ecosystem-level responses. They evaluated two ecosystems shown in Figure 90.

- *Axis 1,* consisting of periphyton (F_0 measure of chlorophyll a and QY is the measure of photosynthetic efficiency) as the producer, herbivores of periphyton (*Osteopilus septentionalis, Planorbella trivolvis and Viciparys grorgianus*) and predators. This ecosystem correlated with dissolved oxygen, pH, and decomposed leaf litter
- *Axis 2,* consisting of phytoplankton as producers and herbivores (Calanoid copepod) as the consumer and *Hydrilla verticillata* as the competitors of phytoplankton. The ecosystem correlated with light and temperature.

As shown in Figure 89, none of the agrochemicals have a direct effect on the ecosystem properties. Instead, the ecosystem properties were modulated indirectly via the effects of contaminants on herbivores. A decrease in herbivore population (1) reduced pressure on producers, resulting in an increase in producers and (2) decreased the predator population. An increase in the producers increased dissolved oxygen concentration through photosynthesis. However, an increase in phytoplankton may decrease light penetration through the water column, thus decreasing temperature. These conditions may allow recovery of the herbivore population followed by the recovery of the predator population. The recovery will be hindered if pesticides directly reduce the predator populations.

References

Abramowicz, D.A., 1990. Aerobic and anaerobic biodegradation of PCBs: a review. Crit. Rev. Biotechnol. 10, 241–251.

Abrams, P.A., Matsuda, H., 1996. Positive indirect effects between prey species that share predators. Ecology 77, 610–616.

Ahmad, F., Hughes, J.B., 2000. Anaerobic transformation of TNT by *Clostridium*. In: Spain, J.C., Hughes, J.B., Knackmuss, H.-J. (Eds.), Biodegradation of Nitroaromatic Compounds and Explosives. Lewis, Boca Raton, Fla, pp. 185–212.

Aiken, G.R., Hsu-Kim, H., Ryan, J.N., 2011. Influence of dissolved organic matter on the environmental fate of metals, nanoparticles, and colloids. Environ. Sci. Technol. 45, 3196–3201.

Aitken, R.J., Creely, K.S., Tran, C.L., 2004. Nanoparticles: An Occupational Hygiene Review. HSE Books RR274, Sudbury, UK.

Akhavan, O., Ghaderi, E., 2010. Toxicity of graphene and graphene oxide nanowalls against bacteria. ACS Nano 4, 5731–5736.

Al-Dabbous, A.N., Kumar, P., 2014. Number and size distribution of airborne nanoparticles during summertime in Kuwait: first observations from the Middle East. Environ. Sci. Technol. 2014, 13634–13643.

Alexiadis, A., Kassino, S., 2008. Molecular simulation of water in carbon nanotubes. Chem. Rev. 108, 5014–5034.

Allen, N.S., Edge, M., Verran, J., Caballero, L., Abrusci, C., Stratton, J., Maltby, J., Bygott, C., 2009. Photocatalytic surfaces: environmental benefits of Nanotitania. Open Mate. Sci. J. 3, 6–27.

Andrady, A.L., 2011. Microplastics in the Marine environment. Mar. Pollut. Bull. 62, 1596–1605.

Andrievsky, G.V., Klochkov, V.K., Karyakina, E.L., McHedlov-Petrossyan, N.O., 1999. Studies of aqueous colloidal solutions of fullerene C-60 by electron microscopy. Chem. Phys. Lett. 300, 392–396.

Areepitak, T., Ren, J., 2011. Model simulations of particle aggregation effect on colloid exchange between streams and streambeds. Environ. Sci. Technol. 45, 5614–5621.

Ashauer, R., 2010. Toxicokinetic–toxicodynamic modelling in an individual based context—consequences of parameter variability. Ecol. Model 221, 1325–1328.

Baalousha, M., Manciulea, A., Cumberland, S., Kendall, K., Lead, J.R., 2008. Aggregation and surface properties of iron oxide nanoparticles: influence of ph and natural organic matter. Environ. Tox. Chem. 27, 1875–1882.

Baalousha, M., 2009. Aggregation and disaggregation of iron oxide nanoparticles: influence of particle concentration, pH and natural organic matter. Sci. Total Environ. 407, 2093–2101.

Bae, S., Hwang, U.S., Lee, Y.-J., Lee, S.-K., 2013. Effects of water chemistry on aggregation and soil adsorption of silver nanoparticles. Environ. Health Toxicol. 28, e2013006.

Balser, T., Kinzig, A., Firestone, M., 2002. The functional consequences of biodiversity. In: Kinzig, A., Pacala, S., Tilman, D. (Eds.), The Functional Consequences of Biodiversity. Princeton University Press, Princeton, pp. 265–293.

Nanoparticles and the environment. In: Banfield, J., Navrotsky, A. (Eds.), Rev. Mineral. Geochem. 44. Mineral. Soc. Am., (Washington, DC).

Bargar, J.R., Bernier-Latmani, R., Giammar, D.E., Tebo, B.M., 2008. Biogenic uraninite nanoparticles and their importance for uranium remediation. Elements (Chantilly, VA, USA) 4, 407–412.

Bar-Ilan, O., Albrecht, R.M., Fako, V.E., Furgeson, D.Y., 2009. Toxicity assessments of multisized gold and silver nanoparticles in zebrafish embryos. Small 5, 1897–1920.

Barnes, D.K.A., Galgani, F., Thompson, R.C., Barlaz, M., 2009. Accumulation and fragmentation of plastic debris in global environments. Phil Trans. Royal Soc. B: Biol. Sci. 364, 1985–1998.

Baughman, R.H., Zakhidov, A.A., de Heer, W.A., 2002. Carbon nanotubes-the route toward applications. Science 297, 593–596.

Baun, A., Hartmann, N.B., Grieger, K., 2008. Ecotoxicity of engineered nanoparticles to aquatic invertebrates: a brief review and recommendations for future toxicity testing. Ecotoxicology 17, 387–395.

Bedard, D.L., 2003. Polychlorinated biphenyls in aquatic sediments: environmental fate and outlook for biological treatment. In: Bossert, I.D., Haggblom, M.M. (Eds.), Dehalogenation. Kluwer, Norwell, Mass, pp. 443–465.

Begum, P., Fugetsu, B., 2013. Induction of cell death by graphene in *Arabidopsis thaliana* (Columbia ecotype) T87 cell suspensions. J. Hazard. Mater. 260, 1032–1041.

Begum, P., Ikhtiari, R., Fugetsu, B., 2011. Graphene phytotoxicity in the seedling stage of cabbage, tomato, red spinach, and lettuce. Carbon 49, 3907–3919.

Ben-Moshe, T., Dror, I., Berkowitz, B., 2010. Transport of metal oxide nanoparticles in saturated porous media. Chemosphere 81, 387–393.

Benn, T., Westerhoff, P., 2008. Nanoparticle silver released into water from commercially available sock fabrics. Environ. Sci. Technol. 42, 4133–4139.

Bernard, C., Nguyen, T., Pelligrin, B., Holbrook, R.D., Zhao, M., Chin, J., 2011. Fate of graphene in polymer nanocomposite exposed to UV radiation. J. Phys. Conf. Ser. 304, 012063.

Besseling, E., Wang, B., Lürling, M., Koelmans, A.K., 2014. Nanoplastic affects growth of *S. obliquus* and reproduction of *D. magna*. Environ. Sci. Technol. 48, 12336–12343.

Bianchi, V., Fortunati, E., 1990. Cellular effects of an anionic surfactant detected in V79 fibroblasts by different cytotoxicity tests. Toxicol. In Vitro 4, 9–16.

Biswas, P., Wu, C.Y., 2005. Critical review: nanoparticles and the environment. J. Air Waste Manage Assoc. 55, 708–746.

Blango, M.G., Mulvey, M.A., 2009. Bacterial landlines: contact-dependent signaling in bacterial populations. Curr. Opin. Microbiol. 12, 177–181.

Blume, E., Bischoff, M., Reichert, J., Moorman, T., Konopka, A., Turco, R., 2002. Surface and subsurface microbial biomass, community structure and metabolic activity as a function of soil depth and season. App. Soil Ecol. 592, 1–11.

Boerger, C.M., Lattin, G.L., Moore, S.L., Moore, C.J., 2010. Plastic ingestion by Planktivorous fishes in the North Pacific Central Gyre. Mar. Pollut. Bull. 60, 2275–2278.

Bom, D., Andrews, R., Jacques, D., Anthony, J., Bailin Chen, B., Meier, M., Selegue, J., 2002. Thermogravimetric analysis of the oxidation of multiwalled carbon nanotubes: evidence for therole of defect sites in carbon nanotube chemistry. Nano Lett. 2, 615–619.

Bose, S., Hochella, M.F., Gorby, Y.A., Kennedy, D.W., McCready, D.E., Madden, A.S., Lower, B.H., 2009. Bioreduction of hematite nanoparticles by the dissimilatory iron reducing bacterium *Shewanella oneidensis* MR-1. Geochim. Cosmochim. Acta 73, 962–976.

Bouchard, D., Zhang, W., Powell, T., Rattanaudompol, U., 2012. Aggregation kinetics and transport of single-walled carbon nanotubes at low surfactant concentrations. Environ. Sci. Technol. 46, 4458–4465.

Bradford, S.A., Torkzaban, S., 2008. Colloid transport and retention in unsaturated porous media: a review of interface-, collectors-, and pore-scale processes and models. Vadose Zone J. 7, 667–681.

Bradford, S.A., Simunek, J., Bettahar, M., van Genuchten, M.T., Yates, S.R., 2003. Modeling colloid attachment, straining, and exclusion in saturated porous media. Environ. Sci. Technol. 37, 2242–2250.

Bradford, S.A., Simunek, J., Bettahar, M., van Genuchten, M.T., Yates, S.R., 2006. Significance of straining in colloid deposition: evidence and implications. Water Resour. Res. 42, W12S15.

Brdar, M., Sciban, M., Takaci, A., Dosenovi, T., 2012. Comparison of two and three parameters adsorption isotherm for Cr(VI) onto Kraft lignin. Chem. Eng. J. 183, 108–111.

Brdar, M.M., Takači, A.A., Sciban, M.B., Rakic, D.J., 2012. Isotherms for the adsorption of Cu(II) onto lignin – comparison of linear and non-linear methods. Hem. Ind. 66, 497–503.

Brock, T.C.M., 2000. Ecological Risks of Pesticides in Freshwater Ecosystems Part 1. Herbicides,Green World Research, Alterra.

Browne, M.A., Dissanayake, A., Galloway, T.S., Lowe, D.M., Thompson, R.C., 2008. Ingested microscopic plastic translocates to the circulatory system of the mussel, *Mytilus edulis*(L.). Environ. Sci. Technol. 42, 5026–5031.

Bryant, F.O., Hale, D.D., Rogers, J.E., 1991. Regiospecific dechlorination of pentachlorophenol by dichlorophenol-adapted microorganisms in freshwater, anaerobic sediment slurries. Appl. Environ. Microbiol. 57, 2293–2301.

Burken, J.G., Schnoor, J.L., 1999. Distribution and volatilization of organic contaminants following uptake by hybrid poplar trees. Int. J. Phytoremed. 1, 139–152.

Burns, I.G., Hayes, H.B., 1974. Some physico-chemical principles involved in the sorption of the organic cation paraquat by soil humic materials. Residue Rev. 52, 117–146.

Burrows, H.D., Conle, M., Santaballa, J.A., Steenken, S., 2002. Reaction pathways and mechanisms of photodegradation of pesticides. J. Photochem. Photobiol. B 67, 71–108.

Burtscher, J., 2005. Physical characterization of particulate emissions from diesel engines: a review. Aerosol Sci. 36, 896–932.

Cai, L., Tong, M., Ma, H., Kim, H., 2013. Cotransport of titanium dioxide and fullerene nanoparticles in saturated porous Media. Environ. Sci. Technol. 47, 5703–5710.

Calow, P., Forbes, V.E., 2003. Does ecotoxicology inform ecological risk assessment? Environ. Sci. Technol. 37, 147–151.

Canas, J.E., Long, M., Nations, S., Vadan, R., Dai, L., Luo, M., Ambikapathi, R., Lee, E.H., Olszyk, D., 2008. Effects of functionalized and nonfunctionalized single-walled carbon-nanotubes on root elongation of select crop species. Nanomat. Environ. 27, 1922–1931.

Canas, J.E., Qi, B., Li, S., Maul, J.D., Cox, S.B., Das, S., Green, M.J., 2011. Acute and reproductive toxicity of nano-sized metal oxides (ZnO and TiO_2) to earthworms (*Eisenia fetida*). J. Environ. Monit. 13, 3351–3357.

Carpenter, S.R., Kitchell, J.F., 1993. The Trophic Cascade in Lakes. Cambridge University Press, New York, NY, USA.

Carpenter, E.J., Anderson, S.J., Harvey, G.R., Miklas, H.P., Peck, B.B., 1972. Polystyrene spherules in coastal waters. Science 178, 749–750.

Carpenter, S.R., Kitchell, J.F., Hodgson, J.R., Cochran, P.A., Elser, J.J., Elser, M.M., Lodge, D.M., Kretchmer, D., He, X., Ende, C.N., 1987. Regulation of lake primary productivity by food web structure. Ecology 68, 1863–1876.

Carpenter, S.R., Ludwig, D., Brock, W.A., 1999. Management of eutrophication for lakes subject to potentially irreversible change. Ecol. Appl. 9, 751–771.

Cavigelli, M.A., Robertson, G.P., 2000. The functional significance of denitrifier community composition in a terrestrial ecosystem. Ecology 81, 1402–1414.

Chadwick, O.A., Derry, L.A., Vitousek, P.M., Huebert, B.J., Hedin, L.O., 1999. Changing sources of nutrients during four million years of ecosystem development. Nature 397, 491–497.

Chang, Y., Yang, S.T., Liu, J.H., Dong, E., Wang, Y., Cao, A., Liu, Y., Wang, H., 2011. In vitro toxicity evaluation of graphene oxide on A549 cells. Toxicol. Lett. 200, 201–210.

Chatterjee, J., Gupta, S.K., 2009. An agglomeration-based model for colloid filtration. Environ. Sci. Technol. 43, 3694–3699.

Chen, K., Elimelech, M., 2006. Aggregation and deposition kinetics of fullerene (C-60) nanoparticles. Langmuir 22, 10994–11001.

Chen, K.L., Elimelech, M., 2007. Influence of humic acid on the aggregation kinetics of fullerene (C_{60}) nanoparticles in monovalent and divalent electrolyte solutions. J. Colloid Interface Sci. 309, 126–134.

Chen, K.L., Elimelech, M., 2009. Relating colloidal stability of fullerene (C_{-60}) nanoparticles to nanoparticle charge and electrokinetic properties. Environ. Sci. Technol. 43, 7270–7276.

Chen, G., Flury, M., 2005. Retention of mineral colloids in unsaturated porous media as related to their surface properties. Colloids Surf. A: Physicochem. Eng. Aspects 256, 207–216.

Chen, Y., Westerhoff, P., Hristovski, K., Crittenden, J.C., 2008a. Stability of commercial metal oxide nanoparticles in water. Water Res. 42, 2204–2212.

Chen, Z., Westerhoff, P., Herckes, P., 2008b. Quantification of $C_{(60)}$ fullerene concentrations in water. Environ. Toxicol. Chem. 27, 1852–1859.

Chen, G.X., Liu, X.Y., Su, C.M., 2011. Transport and retention of TiO_2 rutile nanoparticles in saturated porous media under low-ionic-strength conditions: measurements and mechanisms. Langmuir 27, 5393–5402.

Cheng, J., Flahaut, E., Shuk, H.C., 2007. Effect of carbon nanotubes on developing zebrafish (*Danio rerio*) embryos. Environ. Toxicol. Chem. 26, 708–716.

Chianelli, R.R., Yácaman, M.J., Arenas, J., Aldape, F., 1998. Atmospheric nanoparticles in photocatalytic and thermal production of atmospheric pollutants. J. Hazard. Subst. Res. 1, 1–17.

Chithrani, B., Ghazani, A.A., Chan, W.C.W., 2006. Determining the size and shape dependence of gold nanoparticle uptake into mammalian cells. Nano Lett. 6, 662–668.

Chowdhury, I., Hong, Y., Honda, R.J., Walker, S.L., 2011a. Mechanisms of TiO(2) nanoparticle transport in porous media: role of solution chemistry, nanoparticle concentration, and flow rate. J. Colloid Interface Sci. 360, 548–555.

Chowdhury, I., Cwiertny, D.M., Walker, S.L., 2012. Combined factors influencing the aggregation and deposition of nanoTiO_2 in the presence of humic acid and bacteria. Environ. Sci. Technol. 46, 6968–6976.

Chowdhury, I., Duch, M.C., Mansukhani, N.D., Hersam, M.C., Bouchard, D., 2014. Deposition and release of graphene oxide nanomaterials using a quartz crystal microbalance. Environ. Sci. Technol. 48, 961–969.

Clements, W.H., Newman, M.C., 2002. Community Ecotoxicology. John Wiley, Chichester, West Sussex, UK.

Clements, W.H., Rohr Jr., J.R., 2009. Community response to contaminants: using basic ecological principles to predict ecotoxicological effects. Environ. Toxicol. Chem. 28, 1789–1800.

Cole, M., Lindeque, P., Halsband, C., Galloway, T.S., 2011. Microplastics as contaminants in the marine environment: a review. Mar. Pollut. Bull. 62, 2588–2597.

Coleman, N.V., Mattes, T.E., Gossett, J.M., Spain, J.C., 2002a. Biodegradation of *cis*-dichloroethene as the sole carbon source by a beta-proteobacterium. Appl. Environ. Microbiol. 68, 2726–2730.

Coleman, N.V., Mattes, T.E., Gossett, J.M., Spain, J.C., 2002b. Phylogenetic and kinetic diversity of aerobic vinyl chloride-assimilating bacteria from contaminated sites. Appl. Environ. Microbiol. 68, 6162–6171.

Collins, C.D., 2006. Implementing phytoremediation of petroleum hydrocarbons. In: Willey, N. (Ed.), Methods in Biotechnology, Phytoremediation: Methods and Reviews, vol. 23. © Humana Press, Inc., Totowa, NJ.

Cookson Jr., J.T., 1995. Bioremediation Engineering: Design and Application. McGraw-Hill, New York, NY.

Cross, S.E., Innes, B., Roberts, M.S., Tsuzuki, T., Robertson, T.A., McCormick, P., 2007. Human skin penetration of sunscreen nanoparticles: in-vitro assessment of a novel micronized zinc oxide formulation. Skin Pharmacol. Physiol. 20, 148–154.

Dabrunz, A., Duester, L., Prasse, C., Seitz, F., Rosenfeldt, R., Schilde, C., Schaumann, G.E., Schulz, R., 2011. Biological surface coating and molting inhibition as mechanisms of TiO_2 nanoparticle toxicity in Daphnia magna. PlOS One 6, e20112.

Darcy, H., 1856. Les Fontaines Publiques de la Ville de Dijon. Dalmont, Paris.

Deguchi, S., Alargova, R.G., Tsujii, K., 2001. Stable dispersions of fullerenes, C-60 and C-70, in water. Preparation and characterization. Langmuir 17, 6013–6017.

Dekkers, S., de Heer, C., de Jong, W.H., Sips, A.J.A.M., van Engelen, J.G.M., Kampers, F.W.H., 2007. Nanomaterials in Consumer Products. Availability on the European market and adequacy of the regulatory framework. RIVM/SIR Advisory Report 11014 (IP/A/ENVI/IC/2006-193):49. Economic and Scientific Policy, Policy Department, European Parliament, Brussels, Belgium.

Delay, M., Frimmer, F.H., 2013. Nanoparticles in aquatic systems. Anal. Bioanal. Chem. 402, 583–592.

Dıaz, B., Sánchez-Espinel, C., Arruebo, M., Faro, J., de Miguel, E., Magadán, S., et al., 2008. Assessing methods for blood cell cytotoxic responses to inorganic nanoparticles and nanoparticle aggregates. Small 4, 2025–2034.

Dikio, E.D., 2011. Morphological characterization of soot from the atmospheric combustion of kerosene. E-J. Chem. 8, 1068–1073.

Dikio, D., Bixa, N., 2011. Carbon nanotubes synthesis by catalytic decomposition of ethyne using Fe/Ni catalyst on aluminium oxide support. Int. J. App. Chem. 7, 35–42.

Dimkpa, C.O., McLean, J.E., Latta, D.E., Manangón, E., Britt, D.W., Johnson, W.P., Boyanov, M.I., Anderson, A.J., 2012. CuO and ZnO nanoparticles: phytotoxicity, metal speciation, and induction of oxidative stress in sand-grown wheat. J. Nanopart. Res. 14, 1125.

Donaldson, K., Tran, L., Jimenez, L.A., et al., 2005. Combustion-derived nanoparticles: a review of their toxicology following inhalation exposure. Part. Fibre Tox. 2, 10.

Donaldson, K., Aitken, R., Stone, V., Duffin, R., Forrest, G., Alexander, A.Y., 2006. Carbon nanotubes: a review of their properties in relation to pulmonary toxicology and workplace safety. Toxicol. Sci. 92, 5–22.

Doshi, R., Braida, W., Christodoulatos, C., Wazne, M., O'Connor, G., 2008. Nano-aluminum: transport through sand columns and environmental effects on plants and soil communities. Environ. Res. 106, 296–303.

Dunlap, W.C.Y.Y., 1995. Small-molecule antioxidants in marine organisms: antioxidant activity of mycosporine-glycine. Comp. Biochem. Physiol. B 112, 105.

Eklund, P., Ajayan, P., Blackmon, R., Hart, A.J., Kong, J., Pradham, B., Rao, A., Rinzler, A., 2007. International Assessment of Research and Development of Carbon Nanotube Manufacturing and Applications. World Technology Evaluation Center, Baltimore, MD, USA pp. 1–120.

El Fantroussi, S., Naveau, H., Agathos, S.N., 1998. Anaerobic dechlorinating bacteria. Biotechnol. Prog. 14, 167–188.

El Qada, E.N., Allen, S.J., Walker, G.M., 2006. Adsorption of methylene blue onto activated carbon produced from steam activated bituminous coal: a study of equilibrium adsorption isotherm. Chem. Engineer J. 124, 103–110.

Elechiguerra, J.L., Larios-Lopez, L., Liu, C., Garcia-Gutierrez, D., Camacho-Bragado, A., Yacaman, M.J., 2005. Corrosion at the nanoscale: the case of silver nanowires and nanoparticles. Chem. Mater. 17, 6042–6052.

Elzey, S., Grassian, V.H., 2010. Agglomeration, isolation and dissolution of commercially manufactured silver nanoparticles in aqueous environments. J. Nanopart. Res. 12, 1945–1958.

Eriksson, E., 1952. Cation-exchange equilibria on clay minerals. Soil Sci. 74, 103–113.

Fang, J., X.u, M-j, Wangb, D-j, Wenc, B., Han, J-y, 2013. Modeling the transport of TiO_2 nanoparticle aggregates in saturated and unsaturated granular media: effects of ionic strength and pH. Eater Res. 47, 1399–1408.

Ferry, J., Craig, P., Hexel, C., Sisco, P., Frey, R., Pennington, P.L., Fulton, M.H., Scott, I.G., Decho, A.W., Kashiwada, S., Murphy, C.J., Shaw, T.J., 2009. Transfer of gold nanoparticles from the water column to the estuarine food web. Nat. Nanotechnol. 4, 441–444.

Fierera, N., Schimela, J.P., Holdenb, P.A., 2003. Variations in microbial community composition through two soil depth profiles. Soil Biol. Biochem. 35, 167–176.

Filella, M., 2007. Colloidal properties of submicron particles in natural waters. In: Wilkinson, K.J., Lead, J.R. (Eds.), Environmental Colloids and Particles: Behaviour, Structure and Characterization. IUPAC Series on Analytical and Physical Chemistry of Environmental Systems. John Wiley and Sons, Chichester, pp. 17–93.

Fleeger, J.W., Carman, K.R., Nisbet, R.M., 2003. Indirect effects of contaminants in aquatic ecosystems. Sci. Total Environ. 317, 207–233.

Foote, C.S., 1968. Photosensitized oxygenation and the role of singlet oxygen. Acc. Chem. Res. 1, 104–110.

Forouzangohar, M., Kookana, R.S., 2010. Sorption of nano-C60 clusters in soil: hydrophilic or hydrophobic interactions? J. Environ. Monit. 13, 1190–1994.

Fortner, D., Lyon, D.Y., Sayes, C.M., Boyd, A.M., Falkner, J.C., Hotze, E.M., Alemany, L.B., Tao, Y.J., Guo, W., Ausman, K.D., Colvin, V.L., Hughes, J.B., 2005. C60 in water: nanocrystal formation and microbial response. Environ. Sci. Technol. 39, 4307–4316.

Fortuna, A., 2012. The soil biota. Nat. Educ. Knowledge 3, 1.

Fossi, M.C., Panti, C., Guerranti, C., Coppola, D., Giannetti, M., Marsili, L., Minutoli, R., 2012. Are baleen whales exposed to the threat of microplastics? A case study of the Mediterranean fin whale (Balaenoptera physalus). Mar. Pollut. Bull. 64, 2374–2379.

Freundlich, H., 1909. Kapillarchemie: Eine Darstellung der Chemie der Kolloide und verwandter Gebiete [Translation: Capillary Chemistry: A Presentation of Colloid Chemistry and Related Fields]. Leipzig, Germany. Akademische Verlagsgesellschaft, 144.

Friedlander, S.K., Pui, D.Y.H., 2004. Emerging issues in nanoparticle aerosol science and technology. J. Nanopart Res. 6, 313–314.

Fripiat, J.J., Lronard, A., Uytterhoeven, J.B., 1965. Structure and properties of amorphous silicoaluminas. II. Lewis and bronsted acid sites. J. Phys. Chem. 69, 3274–3279.

Fritze, H., Pietikainen, J., Pennanen, T., 2000. Distribution of microbial biomass and phospholipid fatty acids in podzol profiles under coniferous forest. Eur. J. Soil Sci. 51, 565–573.

Fujigaya, T., Nakashima, N., 2011. Single-walled carbon nanotubes as a molecular heater for thermoresponsive polymer Gel composite. In: Yellampalli, S. (Ed.), Carbon Nanotubes – Polymer Nanocomposites. ISBN:978-953-307-498-6.

Gamble, D.S., Khan, S.U., 1990. Atrazine in organic soil: chemical speciation during heterogeneous catalysis. J. Agric. Food Chem. 38, 297–308.

Gao, W., Majumder, M., Alemany, L.B., Narayanan, T.N., Ibarra, M.A., Pradhan, B.K., Ajayan, P.M., 2011. Engineered graphite oxide materials for application in water purification. ACS Appl. Mater. Interfaces 3, 1821–1826.

Garner, K.L., Keller, A.A., 2014. Emerging patterns for engineered nanomaterials in the environment: a review of fate and toxicity studies. J. Nanopart Res. 16, 2503.

Ghafari, P., St-Denis, C.H., Power, M.E., Jin, X., Tsou, V., Mandal, H.S., Bols, N.C., Tang, X.S., 2008. Impact of carbon nanotubes on the ingestion and digestion of bacteria by ciliated protozoa. Nat. Nanotechnol. 3, 347–351.

Ghiorse, W., Wilson, J., 1988. Microbial ecology of the terrestrial subsurface. In: Laskin, A. (Ed.), Advances in Applied Microbiology. Academic Press, New York, pp. 107–172.

Gilbert, B., Banfield, J.F., 2005. Molecular-scale processes involving nanoparticulate minerals in biogeochemical systems. Rev. Mineral. Geochem. 59, 109–155.

Giulia, R., Jonathan, B., Luca, M., 2014. Polystyrene nanoparticles perturb lipid membranes. J. Phys. Chem. Lett. 5, 241–246.

Godinez, I.G., Darnault, C.J., 2011. Aggregation and transport of nanoTiO_2 in saturated porous media: effects of pH, surfactants and flow velocity. Water Res. 45, 839–851.

Gomez, V., Irusta, S., Balas, F., Santamaria, J., 2013. Intense generation of respirable metal nanoparticles from a low-power soldering unit. J. Hazard. Mat. 256–257, 84–89.

Gong, P., Loh, P.R., Barker, N.D., Tucker, G., Wang, N., Zhang, C., Escalon, B.L., 2012. Building quantitative prediction models for tissue residue of two explosives compounds in earthworms from microarray gene expression data. Environ. Sci. Technol. 46, 19–26.

Gorby YA, S., Yanina McLean, J.S., Rosso, K.M., Moyles, D., Dohnalkova, A., Beveridge, T.J., Chang, I.S., Kim, B.H., Kim, K.S., Culley, D.E., Reed, S.B., Romine, M.F., Saffarini, D.A., Hill, E.A., Shi, L., Elias, D.A., Kennedy, D.W., Pinchuk, G., Watanabe, K., Ishii, S., Logan, B., Nealson, K.H., Fredrickson, J.K., 2006. Electrically conductive bacterial nanowires produced by Shewanella oneidensistrain MR-1 and other microorganisms. PNAS USA 103, 11358–11363.

Gorontzy, T., Drzyzga, O., Kahl, M.W., Bruns-Nagel, D., Breitung, J., von Loew, E., Blotevogel, K.H., 1994. Microbial degradation of explosives and related compounds. Crit. Rev. Microbiol. 20, 265–284.

Gottschalk, F., Sonderer, T., Scholz, R.W., Nowack, B., 2010. Possibilities and limitations of modeling environmental exposure to engineered nanomaterials by probabilistic material flow analysis. Environ. Toxicol. Chem. 29, 1036–1048.

Goudie, A.S., Middleton, N.J., 2001. Saharan dust storms: nature and consequences. Earth Sci. Rev. 56, 179–204.

Griffitt, R.J., Weil, R., Hyndman, K.A., Denslow, N.D., Powers, K., Taylor, D., Barber, D.S., 2007. Exposure to copper nanoparticles causes gill injury and acute lethality in zebrafish (*Danio rerio*). Environ. Sci. Technol. 41, 8178–8186.

Guenther, A., Hewitt, C.N., Erickson, D., Fall, R., Geron, C., Graedel, T., Harley, P., Klinger, L., Lerdau, M., Mckay, W.A., Pierce, T., Scholes, B., Steinbrecher, R., Tallamraju, R., Taylor, J., Zimmerman, P., 1995. A global-model of natural volatile organic-compound emissions. J. Geophys. Res. Atmos. 100, 8873–8892.

Gunasekara, A.S., Xing, B., 2003. Sorption and desorption of naphthalene by soil organic matter: importance of aromatic and aliphatic components. J. Environ. Qual. 32, 240–246.
Gunasekara, A.S., Simpson, M.J., Xing, B., 2003. Identification and characterization of sorption domains in soil organic matter using structurally modified humic acids. Environ. Sci. Technol. 37, 852–858.
Gurunathan, S., Han, J.W., Dayem, A.A., Eppakayala, V., Kim, J.H., 2012. Oxidative stress-mediated antibacterial activity of graphene oxide and reduced graphene oxide in *Pseudomonas aeruginosa*. Int. J. Nanomedicine 7, 5901–5914.
Gurunathan, S., Han, J.W., Dayem, A.A., Eppakayala, V., Kim, J.H., 2014. Oxidative stress-mediated antibacterial activity of graphene oxide: membrane and oxidative stress. Environ. Sci. Technol. 48, 9995–10009.
Haggblom, M.M., Young, L.Y., 1990. Chlorophenol degradation coupled to sulfate reduction. Appl. Environ. Microbiol. 56, 3255–3260.
Haggblom, M.M., Young, L.Y., 1995. Anaerobic degradation of halogenated phenols by sulfate-reducing consortia. Appl. Environ. Microbiol. 61, 1546–1550.
Haggblom, M.M., Knight, V.K., Kerkhof, L.J., 2000. Anaerobic decomposition of halogenated aromatic compounds. Environ. Pollut. 107, 199–207.
Halstead, B.J., Wylie, G.D., Casazza, M.L., 2014. Ghost of habitat past: historic habitat affects the contemporary distribution of giant garter snakes in a modified landscape. Anim. Conserv. 17, 144–153.
Hamaker, J.W., Thompson, J.M., 1972. In: Goring, C.A.I., Hamaker, J.W. (Eds.), Adsorption in Organic Chemicals in the Soil Environment, Vol 1. Marcel Dekker Inc., New York.
Handy, R.D., Owen, R., Valsami-Jones, E., 2008. The ecotoxicology of nanoparticles and nanomaterials: current status, knowledge gaps, challenges, and future needs. Ecotoxicology 17, 315–325.
Harayama, S., Rekik, M., 1989. Bacterial aromatic ring-cleavage enzymes are classified into two different gene families. J. Biol. Chem. 264, 15328–15333.
Harper, M., Pacolay, B., Hintz, P., Andrew, M.E., 2006. A comparison of portable XRF and ICP-OES analysis for lead on air filter samples from a lead ore concentrator mill and a lead-acid battery recycler. J. Environ. Monit. 8, 384–392.
Hartland, A., Lead, J.R., Slaveykova, V.I., O'Carroll, D., Valsami-Jones, E., 2013. The environmental significance of natural nanoparticles. Nat. Educ. Knowledge 4 (8), 7.
Hayes, M.H.B., Himes, F.L., 1986. Nature and properties of humus-mineral complexes. In: Huang, P.M., Schnitzer, M. (Eds.), Interaction of Soil Minerals with Natural Organics and Microbes. SSSA, Madison WI, pp. 103–158.
Hayes, M.H.B., Pick, M.E.M., Toms, B.A., 1975. Interactions between clay minerals and bipyridylium herbicides. Residue Rev. 57, 1–25.
He, C., Morawska, L., Taplin, L., 2007. Particle emission characteristics of office printers. Environ. Sci. Technol. 41, 6039–6045.
He, F., Zhang, M., Qian, T.W., Zhao, D.Y., 2009. Transport of carboxymethyl cellulose stabilized iron nanoparticles inporous media: column experiments and modeling. J. Colloid Interface Sci. 334, 96–102.
Heider, J., Fuchs, G., 1997. Anaerobic metabolism of aromatic compounds. Eur. J. Biochem. 243, 577–596.
Hesterberg, T.W., Long, C.M., Lapin, C.A., Hamade, A.K., Valberg, P.A.D., 2010. Diesel exhaust particulate (DEP) and nanoparticle exposures: what do DEP human clinical studies tell us about potential human health hazards of nanoparticles? Inhal Toxicol. 22, 679–694.
Hidalgo-Ruz, V., Gutow, L., Thompson, R.C., 2012. Thiel, M. Microplastics in the Marine environment: a review of the methods used for identification and quantification. Environ. Sci. Technol. 46, 3060–3075.
Hillyer, J.F., Albrecht, R.M., 2002. Gastrointestinal persorption and tissue distribution of differently sized colloidal gold nanoparticles. J. Pharm. Sci. 90, 1927–1936.
Ho, C., Yau, S., Lok, C., So, M., Che, C., 2010. Oxidative dissolution of silver nanoparticles by biologically relevant oxidants: a kinetic and mechanistic study. Chem. Asian J. 5, 285–293.
Hochella, M.F., Lower, S.K., Maurice, P.A., Penn, R.L., Sahai, N., Sparks, D.L., Twining, B.S., 2008. Nanominerals, mineral nanoparticles, and earth systems. Science 319, 1631–1635.
Holbrook, R.D., Murphy, K.E., Morrow, J.B., Cole, K.D., 2008. Trophic transfer of nanoparticles in a simplified invertebrate food web. Nat. Nanotechnol. 3, 352–355.
Hotze, E.M., Labille, J., Alvarez, P., Wiesner, M.R., 2008. Mechanisms of photochemistry and reactive oxygen production by fullerene suspensions in water. Environ. Sci. Technol. 42, 4175–4180.
Hotze, E.M., Bottero, J.Y., Wiesner, M.R., 2010. Theoretical framework for nanoparticle reactivity as a function of aggregation state. Langmuir 26, 11170–11175.
Houghton, J., 2005. Global warming. Rep. Prog. Phys. 68, 1343–1403.
Hoyle, C.R., Boy, M., Donahue, M.M., Fry, J.L., Glasius, M., Guenther, A., Hallar, A.G., Hartz, K.H., Petters, M.D., Petäjä, T., Rosenoern, T., Sullivan, T.P., 2011. A review of the anthropogenic influence on biogenic secondary organic aerosol. Atmos. Chem. Phys. 11, 321–343.
Huang, P.M., 1980. Adsorption Processes in Soil. In: Hutzinger, O. (Ed.), The Handbook of Environmental Chemistry 2 A, Reactions and Processes, pp. 47–59.
Hund-Rinke, K., Simon, M., 2006. Ecotoxic effect of photocatalytic active nanoparticles (TiO_2) on algae and daphnids. Environ. Sci. Pollut. Res. Int. 13, 225–232.

Huynh, K.A., Chen, K.L., 2011. Aggregation kinetics of citrate and polyvinylpyrrolidone coated silver nanoparticles in monovalent and divalent electrolyte solutions. Environ. Sci. Technol. 45, 5564–5571.

Hyung, H., Kim, J.-H., 2008. Natural organic matter (NOM) adsorption to multi-walled carbon nanotubes: effect of NOM characteristics and water quality parameters. Environ. Sci. Technol. 42, 4416–4421.

Hyung, H., Fortner, J.D., Hughes, J.B., Kim, J.H., 2007. Natural organic matter stabilizescarbon nanotubes in the aqueous phase. Environ. Sci. Technol. 41, 179–184.

Imhof, H.K., Schmid, Niessner, R., Ivleva, N.P., Laforsch, C., 2012. A novel, highly efficient method for the separation and quantification of plastic particles in sediments of aquatic environments. Limnol. Oceanogr. Meth. 10, 524–537.

Ito, T., 1970. The biology of a harpacticoid copepod, Tigriopus japonicus Mori. J. Fac. Sci. Hokkaido Univ. 17, 474–500.

Janke, D., Fritsche, W., 1985. Nature and significance of microbial cometabolism of xenobiotics. J. Basic Microbiol. 25, 603–619.

Jiang, J., Oberdorster, G., Biswas, P., 2009. Characterization of size, surface charge, and agglomeration state of nanoparticle dispersions for toxicological studies. J. Nanopart Res. 11, 77–89.

Jickells, T.D., An, J.S., Andersen, K.K., Baker, A.R., Bergametti, G., Brooks, N., Cao, J.J., Boyd, P.W., Duce, R.A., Hunter, K.A., Kawahata, H., Kubilay, N., laRoche, J., Liss, P.S., Mahowald, N., Prospero, J.M., Ridgwell, A.J., Tegen, I., Torres, R., 2005. Global iron connections between desert dust, ocean biogeochemistry, and climate. Science 308, 67–71.

Jimenez, L.A., Madsen, O.S., 2003. A simple Formula to estimate settling velocity of natural sediments. J. Waterway Port, Coastal, Ocean Eng. 129, 70–78.

Jimenez, J.L., Canagaratna, M.R., Donahue, N.M., Prevot, A.S.H., Zhang, Q., Kroll, J.H., DeCarlo, P.F.P., Robinson, A.L., Duplissy, J., Smith, J.D., Wilson, K.R., Lanz, V.A., Hueglin, C., Sun, Y.L., Tian, J., Laaksonen, A., Raatikainen, T., Rautiainen, J., Vaattovaara, P., Ehn, M., Kulmala, M., Tomlinson, J.M., Collins, D.R., Cubison, M.J., Dunlea, E.J., Huffman, J.A., Onasch, T.B., Alfarra, M.R., Williams, P.I., Bower, K., Kondo, Y., Schneider, J., Drewnick, F., Borrmann, S., Weimer, S., Demerjian, K., Salcedo, D., Cottrell, M., Dzepina, K., Kimmel, J.R., Sueper, D., Jayne, J.T., Herndon, S.C., Trimborn, A.M., Williams, L.R., Wood, E.C., Middlebrook, A.M., Kolb, C.E., Baltensperger, U., Worsnop, D.R., 2009. Evolution of organic aerosols in the atmosphere. Science 326, 1525–1529.

Judy, J.D., Unrine, J.M., Bertsch, P.M., 2011. Evidence for biomagnification of gold nanoparticles within a terrestrial food chain. Environ. Sci. Technol. 45, 776–781.

Ju-Nam, Y., Lead, J.R., 2008. Manufactured nanoparticles: an overview of their chemistry, interactions and potential environmental implications. Sci. Total Environ. 400, 396–414.

Kanel, S.K., Al-Abed, S.R., 2011. Influence of pH on the transport of nanoscale zinc oxide in saturated porous media. J. Nanopart. Res. 13, 4035–4047.

Kang, S., Pinault, M., Pfefferle, L.D., Elimelech, M., 2007. Single-walled carbon nanotubes exhibit strong antimicrobial activity. Langmuir 23, 8670–8673.

Kang, S., Mauter, M.S., Elimelech, M., 2009. Microbial cytotoxicity of carbon-based nanomaterials: implications for river water and wastewater effluent. Environ. Sci. Technol. 43, 2648–2653.

Karickhoff, S.W., Brown, D.S., Scott, T.A., 1979. Sorption of hydrophobic pollutants on natural sediments. Water Res. 13, 241–248.

Kasemets, K., Ivask, A., Dubourguier, H.C., Kahru, A., 2009. Toxicity of nanoparticles of ZnO, CuO and TiO_2 to yeast *Saccharomyces cerevisiae*. Toxicol. In Vitro 23, 1116–1122.

Kasper, M., Sattler, K., Siegmann, K., Matter, U., Siegmann, H.C., 1999. The influence of fuel additives on the formation of carbon during combustion. J. Aerosol Sci. 30, 217–225.

Kavouras, I.G., Mihalopoulos, N., Stephanou, E.G., 1998. Formation of atmospheric particles from organic acids produced by forests. Nature 395, 683–686.

Keller, A.A., Wang, H., Zhou, D., Lenihan, H.S., Cherr, G., Cardinale, B.J., Miller, R., Ji, Z., 2010. Stability and aggregation of metal oxide nanoparticles in natural aqueous matrices. Environ. Sci. Technol. 44, 1962–1967.

Kennedy, A.J., Hull, M.S., Steevens, J.A., Dontsova, K.M., Chappell, M.A., Gunter, J.C., Weiss, C.A., 2008. Factors influencing the partitioning and toxicity of nanotubes in the aquatic environment. Environ. Toxicol. Chem. 27, 1932–1941.

Kerminen, V.-M., Lihavainen, H., Komppula, M., Viisanen, Y., Kulmala, M., 2005. Direct observational evidence linking atmospheric aerosol formation and cloud droplet activation. Geophys. Res. Lett. 32, L14803.

Khan, S.U., 1973. Equilibrium and kinetic studies of the adsorption of 2,4-D and picloram on humic acid. Can. J. Soil Sci. 53, 429–434.

Khan, S.U., 1974. Adsorption of 2,4-D from aqueous solution by fulvic acid-clay complex. Environ. Sci. Technol. 8, 236–238.

Kim, B., Park, C.S., Murayama, M., Hochella, M.F., 2010. Discovery and characterization of silver sulfide nanoparticles in final sewage sludge products. Environ. Sci. Technol. 4, 7509–7514.

Kim, S.W., Nam, S.-H., An, Y.-J., 2012. Interaction of silver nanoparticles with biological surfacesof *Caenorhabditis elegans*. Ecotoxicol Environ. Saf. 77, 64–70.

Kirkby, J., et al., 2011. Role of sulphuric acid, ammonia and galactic cosmic rays in atmospheric aerosol nucleation. Nature 476, 429–433.

Kittelson, D.B., 1998. Engines and nanoparticles: a review. J. Aerosol Sci. 29, 575–588.

Klaine, S.J., Alvarez, P.J.J., Batley, G.E., Fernandes, T.F., Handy, R.D., Lyon, D.Y., Mahendra, S., McLaughlin, M.J., Lead, J.R., 2008. Nanomaterials in the environment: behavior, fate, bioavailability, and effects. Environ. Toxicol. Chem. 2008 (27), 1825–1851. [Erratum appears in Environ Toxicol Chem. (2012). 31:2893].

Knapp, L.F., 1922. The solubility of small particles and the stability of colloids. Trans. Faraday Soc. 17, 457–465.

Kohler, A., Som, C., Helland, A., Gottschalk, F., 2008. Studying the potential release of carbon nanotubes throughout the application lifecycle. J. Cleaner Production 16, 927–937.

Kotchey, G.P., Allen, B.L., Vedala, H., Yanamala, N., Kapralov, A.A., Tyurina, Y.Y., Klein-Seetharaman, L., Kagan, V.E., Star, A., 2011. The enzymatic oxidation of graphene oxide. ACS Nano 5, 2098–2108.

Kramer, H.E., Maute, A., 1972. Sensitized photooxygenation according to type I mechanism (radical mechanism). I. Flash photolysis experiments. Photochem. Photobiol. 15, 7–23.

Krishnamoorthy, K., Veerapandian, M., Zhang, L.H., Yun, K., Kim, S.J., 2012. Antibacterial efficiency of graphene nanosheets against pathogenic bacteria via lipid peroxidation. J. Phys. Chem. C 116, 17280–17287.

Kuang, C., McMurry, P.H., McCormick, A.V., Eisele, F.L., 2008. Dependence of nucleation rates on sulfuric acid vapor concetration in diverse atmospheric locations. J. Geophys. Res. 113 (D10), 209.

Kuang, C., McMurry, P.H., McCormick, A.V., 2009. Determina tion of cloud condensation nuclei production from measured new particle formation events. Geophys. Res. Lett. 36, L09822.

Kuang, C., Riipinen, I., Sihto, S.L., Kulmala, M., McCormick, A.V., McMurry, P.H., 2010. An improved criterion for new particle formation in diverse atmospheric environments. Atmos. Chem. Phys. 10, 8469–8480.

Kulmala, M., Vehkamaki, H., Petaja, T., Dal Maso, M., Lauri, A., Kerminen, V.M., Birmili, W., McMurry, P.H., 2004. Formation and growth rates of ultrafine atmospheric particles: a review of observations. J. Aerosol Sci. 35, 143–176.

Kulmala, M., 2003. How particles nucleate and grow. Science 302, 1000–1001.

Kumar, P., Pirjolac, L., Ketzele, M., Harrisonf, R.M., 2013. Nanoparticle emissions from 11 non-vehical exhaust cources – a review. Atmos. Env. 67, 252–277.

Kumari, M., Mukherjee, A., 2009. Chadrasekaran N. Genotoxicity of silver nanoparticle in Allium cepa. Sci. Total Environ. 407, 5243–5246.

Kurten, T., Loukonen, V., Vehkamäki, H., Kulmala, M., 2008. Amines are likely to enhance neutral and ion-induced sulfuric acid-water nucleation in the atmosphere more effectively than ammonia. Atmos. Chem. Phys. 8, 4095–4103.

Kwadijk, C.J.A.F., Velzeboer, I., Koelmans, A.A., 2013. Sorption of perfluorooctane sulfonate to carbon nanotubes in aquatic sediments. Chemosphere 90, 1631–1636.

Lagally, C.D., Reynolds, C.C.O., Grieshop, A.P., Kandlikar, M., Rogak, S.N., 2011. Carbon nanotube and fullerene emissions from spark-ignited engines. Aerosol Sci. Technol. 42, 15–164.

Lagally, C.D., Reynolds, C.C.O., Grieshop, A.P., Kandlikar, M., Rogak, S.N., 2012. Carbon nanotube and fullerene emissions from spark-ignited engines. Aerosol Sci. Technol. 46, 156–164.

Lal, R., 2002. Soil carbon sequestration in China through agricultural intensification, and restoration of degraded and desertified ecosystems. Land Degrad. Dev. 13, 469–478.

Lam, C.W., James, J.L., McCluskey, R., Hunter, R.L., 2004. Pulmonary toxicity of single-wall carbon nanotubes in mice 7 and 90 days after intratracheal instillation. Tox. Sci. 77, 126–134.

Lam, C.W., James, J.T., McCluskey, R., Arepalli, S., Hunter, R.L., 2006. A review of carbon nanotube toxicity and assessment of potential occupational and environmental health risks. Crit. Rev. Toxicol. 36, 189–217.

Lapresta-Fernandez, A., Fernandez, A., Blasco, J., 2012. Nanoecotoxicity effects of engineered silver and gold nanoparticles in aquatic organisms. TrAC 32, 40–59.

Lead, J.R., Davison, W., Hamilton-Taylor, J., Buffle, J., 1997. Characterizing colloidal material in natural waters. Aquatic Geochem. 3, 213–232.

Lecoanet, H.F., Wiesner, M.R., 2004. Velocity effects on fullerene and oxide nanoparticle deposition in porous media. Environ. Sci. Technol. 38, 4377–4382.

Lee, K.W., Raisuddin, S., Hwang, D.S., Park, H.G., Lee, J.S., 2007. Acute toxicities of trace metals and common xenobiotics to the marine copepod Tigriopus japonicus: evaluation of its use as a benchmark species for routine ecotoxicity tests in Western Pacific coastal regions. Environ. Toxicol. 22, 532–538.

Lee, K.W., Raisuddin, S., Hwang, D.S., Park, H.G., Dahms, H.U., Ahn, I.Y., Lee, J.S., 2008. Two-generation toxicity study on the copepod model species *Tigriopus japonicas*. Chemosphere 72, 1359–1365.

Lee, W.M., Kim, S.W., Kwak, J.I., Nam, S.-H., Shin, Y.-J., An, Y.-J., 2010. Research trends of ecotoxicity of nanoparticles in soil environment. Toxicol. Res. 26 (4), 253–259.

Lee, W.M., Kwak, J.I., An, Y.J., 2012. Effect of silver nanoparticles in crop plants *Phaseolus radiatus* and *Sorghum bicolor*: media effect on phytotoxicity. Chemosphere 86, 491–499.

Lewis, G.N., 1907. Outlines of a new system of thermodynamic chemistry. Proc. Am. Acad. Arts Sci. 43, 259–293.

Li, Y., Pignatello, J.J., 2002. Demonstration of the "Conditioning effect" in soil organic matter in support of a pore deformation mechanism for sorption hysteresis. Environ. Sci. Technol. 36, 4553–4561.

Li, Y., Wang, Y., Pennell, K.D., Abriola, L.M., 2008. Investigation of the transport and deposition of fullerene (C_{60}) nanoparticles in quartz sands under varying flow conditions. Environ. Sci. Technol. 42, 7174–7180.

Li, N., Yao, S.-H., You, M.-Y., Zhang, Y.-L., Qiao, Y.-F., Zou, W.-X., Han, X.-Z., Zhang, B., 2014a. Contrasting development of soil microbial community structure under no-tilled perennial and tilled cropping during early pedogenesis of a Mollisol. Soil Biol. Biochem. 77, 221–232.

Li, S., Wallis, L.K., Diamond, S.A., Ma, H., Hoff, D.J., 2014b. Species sensitivity and dependence on exposure conditions impacting the phototoxicity of TiO_2 nanoparticles to benthic organisms. Env. Tox. Chem. 33, 1563–1569.

Liang, Y., Bradford, S.A., Simunek, J., Heggen, M., Vereecken, H., Klumpp, E., 2013. Retention and remobilization of stabilized silver nanoparticles in an undisturbed loamy sand soil. Environ. Sci. Technol. 47, 12229–12237.

Lin, D., Xing, B., 2007. Phytotoxicity of nanoparticles: inhibition of seed germination and root growth. Environ. Pollut. 150, 243–250.

Lin, D., Xing, B., 2008. Root uptake and phytotoxicity of ZnO nanoparticles. Environ. Sci. Technol. 42, 5580–5585.

Lin, S., Reppert, J., Hu, Q., Hudson, J.S., Reid, M.L., Ratnikova, T.A., Rao, A.M., Luo, H., Ke, P.C., 2009. Uptake, translocation, and transmission of carbon nanomaterials in rice plants. Small 5, 1128–1132.

Linn, D.M., Doran, J.W., 1984. Aerobic and anaerobic microbial populations in no-till and plowed soils. Soil Sci. Soc. Am. J. 48, 794–799.

Liu, J., Aruguete, D.M., Murayama, M., Hochella Jr., M.F., 2009. Influence of size and aggregation on the reactivity of an environmentally and industrially relevant nanomaterial (PbS). Environ. Sci. Technol. 43, 8178–8183.

Liu, J., Sonshine, D.A., Shervani, S., Hurt, R.H., 2010. Controlled release of biologically active silver from nanosilver surfaces. ACS Nano 4, 6903–6913.

Liu, S., Zeng, T.H., Hofmann, M., Burcombe, E., Wei, J., Jiang, R., Kong, J., Chen, Y., 2011. Antibacterial activity of graphite, graphite oxide, graphene oxide, and reduced graphene oxide: membrane and oxidative stress. ACS Nano 5, 6971–6980.

Liu, Y., Wang, Q., 2005. Transport behavior of water confined in carbon nanotubes. Phys. Rev. B 72, 085420.

Lof, R.W., vanVeenendaal, M.A., Jonkman, H.T., Sawatzky, G.A., 1995. Bang gap, excitations and coulomb interaction in solid C_{60}. J. Electron Spectr. Relat. Phenomena 72, 83–87.

Loosli, F., Mueller, N.C., Nowack, B., Coustumer, P.L., Stoll, S., 2013. TiO_2 nanoparticles aggregation and disaggregation in presence of alginate and Suwannee river humic acids. pH and concentration effects on nanoparticle stability. Water Res. 47, 6052–6063.

Lopez-Moreno, M.L., De La Rosa, G., Hernandez-Viezcas, J.A., Peralta-Videa, J.R., Gardea-Torresdey, J.L., 2010. X-ray absorption spectroscopy (XAS) corroboration of the uptake and storage of CeO_2 nanoparticles and assessment of their differential toxicity in four edible plant species. J. Agric. Food Chem. 58, 3689–3693.

Lozano, P., Berge, N.D., 2012. Single-walled carbon nanotube behavior in representative mature leachate. Waste Manage. 32, 1699–1711.

Lu, Y., Pignatello, J.J., 2002. Demonstration of the "Conditioning effect" in soil organic matter in support of a pore deformation mechanism for sorption hysteresis. Environ. Sci. Technol. 36, 4553–4561.

Luther III, G.W., Rickard, D.T., 2005. Metal sulfide cluster complexes and their biogeochemical importance in the environment. J. Nanopart Res. 7, 389–407.

Luther, G.W., Theberge, S.M., Rickard, D.T., 1999. Evidence for aqueous clusters as intermediates during zinc sulfide formation. Geochim. Cosmochim. Acta 63, 3159–3169.

Luther, G.W., Glazer, B., Shufeb, M., Trouwborst, R., Schultz, B.R., Druschel, G., Kraiya, C., 2003. Iron and sulfur chemistry in a stratified lake: evidence for iron-rich sulfide complexes. Aquat. Geochem. 9, 87–110.

Ma, L.W., Selim, H.M., 1994. Predicting atrazine adsorption-desorption in soils: a modified second-order kinetic model. Water Resour. Res. 30, 447–456.

Ma, X., Geiser-Lee, J., Deng, Y., Kolmakov, A., 2010. Interactions between engineered nanoparticles (ENPs) and plants: phytotoxicity, uptake and accumulation. Sci. Total Environ. 408, 3053–3061.

Ma, H., Brennan, A., Diamond, S.A., 2012. Phototoxicity of TiO_2 nanoparticles under solar radiation to two aquatic species: *Daphnia magna* and *Japanese medaka*. Environ. Toxicol. Chem. 31, 1621–1629.

Madsen, E.L., 2005. Identifying microorganisms responsible for ecologically significant biogeochemical processes. Nat. Rev Microbiol. 3, 439–446.

Manceau, A., Nagy, K.L., Marcus, M.A., Lanson, M., Geoffroy, N., Jacquet, T., Kirpichtchikova, T., 2008. Formation of metallic copper nanoparticles at the soil–root interface. Environ. Sci. Technol. 42, 1766–1772.

Manoj, B., Sreelaksmi, S., Mohan, A.N., Kunjomana, A.G., 2012. Characterization of diesel soot from the combustion in engine by x-ray and spectroscopic techniques. Int. J. Electrochem. Sci. 7, 3215–3221.

Manzetti, S., Andersen, O., 2013. Carbon nanotubes in electronics: background and discussion for waste-handling strategies. Challenges 4, 75–85.

Marconi, M., Sferlazzo, D.M., Becagli, S., Bommarito, C., Calzolai, G., Chiari, M., di Sarra, A., Ghedini, C., Gómez-Amo, L., Lucarelli, F., Meloni, D., Monteleone, F., Nava, S., Pace, G., Piacentino, S., Rugi, F., Severi, M., Traversi, R., Udisti, R., 2013. Saharan dust aerosol over the central mediterranean sea: optical columnar measurements vs. aerosol load, chemical composition and marker solubility at ground level. Atmos. Chem. Phys. Discuss 13, 21259–21299.

Marsalek, B., Jancula, D., Marsalkova, E., Mashlan, M., Safarova, K., Tucek, J., Zboril, R., 2012. Multimodal action and selective toxicity of zerovalent iron nanoparticles against cyanobacteria. Environ. Sci. Technol. 46, 2316–2323.

Marvin-Sikkema, F.D., de Bont, J.A., 1994. Degradation of nitroaromatic compounds by microorganisms. Appl. Microbiol. Biotechnol. 42, 499–507.

Mattsson, K., Ekvall, M.T., Hansson, L.A., Linse, S., Malmendal, A., Cedervall, T., 2015. Altered behavior, physiology, and metabolism in fish exposed to polystyrene nanoparticles. Environ. Sci. Technol. 49, 553–561.

Maynard, A.D., Kuempel, E.D., 2005. Airborne nanostructured particles and occupational health. J. Nanoparticle Res. 7, 587–614.

Maynard, A.D., Baron, P.A., Foley, M., Shvedova, A.A., Kisin, E.R., Castranova, V., 2004. Exposure to carbon nanotube material: aerosol release during the handling of unrefined single-walled carbon nanotube material. J. Toxicol. Environ. Health Part A 67, 87107.

McClintock, J.B.K.D., 1997. Mycosporine-like amino acids in 38 species of subtidal marine organisms from McMurdo Sound, Antarctica. Antarct Sci. 9, 392–398.

McKay, G., El Guendi, M., Nassar, M., 1987. Equilibrium studies during the removal of dyestuffs from aqueous solutions using bagasse pith. Water Res. 21, 1513–1520.

McShane, H., Sarrazin, M., Whalen, J.K., Hendershot, W.H., Sunahara, G.I., 2012. Reproductive and behavioral responses of earthworms exposed to nano-sized titanium dioxide in soil. Environ. Toxicol. Chem. 31, 184–193.

Mihranyan, A., Strømme, M., 2007. Solubility of fractal nanoparticles. Surf. Sci. 601 (2), 315–319.

Mikesell, M.D., Boyd, S.A., 1986. Complete reductive dechlorination and mineralization of pentachlorophenol by anaerobic microorganisms. Appl. Environ. Microbiol. 52, 861–865.

Miller, A., Drake, P.L., Hintz, P., Habjan, M., 2010a. Characterizing exposures to airborne metals and nanoparticle emissions in a refinery. Ann. Occup. Hyg. 54, 504–513.

Miller, R.J., Lenihan, H.S., Muller, E.B., Tseng, N., Hanna, S.K., Keller, A.A., 2010b. Impacts of metal oxide nanoparticles on marine phytoplankton. Environ. Sci. Technol. 44, 7329.

Mills, N.L., Miller, M.R., Lucking, A.L., Beveridge, J., Flint, L., Boere, A.J.F., Fokkens, P.H., Boon, N.A., Sandstrom, T., Blomberg, A., Duffin, R., Donaldson, K., Hadoke, P.W.F., Cassee, F.R., Newby, D.E., 2011. Combustion-derived nanoparticulate induces the adverse vascular effects of diesel exhaust inhalation. Eur. Heart J. 32, 2660–2671.

Mishra, A.K., Ramaprabhu, S., 2011. Functionalized graphene sheets for arsenic removal and desalination of sea water. Desalination 282, 39–45.

Mogensen, A.S., Dolfing, J., Haagensen, F., Ahring, B.K., 2003a. Potential for anaerobic conversion of xenobiotics. Adv. Biochem. Eng. Biotechnol. 82, 69–134.

Mogensen, A.S., Haagensen, F., Ahring, B.K., 2003b. Anaerobic degradation of linear alkylbenzene sulfonate. Environ. Toxicol. Chem. 22, 706–711.

Monteiro-Riviere, N.A., Tran, C.L., 2007. Nanotoxicology: Characterization, Dosing and Health Effects. Informa Healthcare, 1420045148.

Monteiro-Riviere, N.A., Nemanich, R.J., Inman, A.O., Wang, Y.Y.Y., Riviere, J.E., 2005. Multi-walled carbon nanotube interactions with human epidermal keratinocytes. Toxicol. Lett. 155, 377–384.

Moore, M.N., 2006. Do nanoparticles present ecotoxicological risks for the health of the aquatic environment? Environ. Int. 32, 967–976.

Moore, C.J., 2008. Synthetic polymers in the marine environment: a rapidly increasing long-term threat. Env. Res. 108, 131–139.

Mu, L., Sprando, R.L., 2010. Application of nanotechnology in cosmetics. Pharmacol. Res. 27, 1746–1749.

Mueller, N.C., Nowack, B., 2008. Exposure modeling of engineered nanoparticles in the environment. Environ. Sci. Technol. 42, 4447–4453.

Muller, K., Skepper, J.N., Posfai, M., Trivedi, R., Howarth, S., Corot, C., Lancelot, E., Thompson, P.W., Brown, A.P., Gillard, J.H., 2007. Effect of ultrasmall supraparamagnetic iron oxide nanoparticles (Ferumoxtran-10) on human monocyte-macrophages in vitro. Biomaterials 28, 1629–1642.

Muller, J., Huaux, F., Fonseca, A., Nagy, J.B., Moreau, N., Delos, M., Raymundo-Piñero, E., Béguin, F., Kirsch-Volders, M., Fenoglio, I., 2008. Structural defects play a major role in the acute lung toxicity of multiwall carbon nanotubes: toxicological aspects. Chem. Res. Toxicol. 21, 1698–1705.

Murr, L.E., Bang, J.J., 2003. Electron microscopy comparisons of fine and ultrafine carbonaceous and nano-carbonaceous, airborne particulates. Atmos. Environ. 37, 4795–4806.

Murr, L.E., Garza, K.M., 2009. Natural and anthropogenic environmental nanoparticulates: their microstructural characterization and respiratory health implications. Atmos. Environ. 43, 2683–2692.

Murr, L.E., Bang, J.J., Esquivel, E.A., 2004. Carbon nanotubes, nanocrystal forms, and complex nanoparticle aggregates in common fuel gas combustion streams. J. Nanoparticle Res. 6, 241–251.

Mwangi, J.N., Wang, N., Ingersoll, C.G., Hardesty, D.K., Brunson, E.L., Li, H., Deng, B., 2012. Toxicity of carbon nanotubes to freshwater aquatic invertebrates. Env. Tox. Chem. 31, 1823–1830.

Nair, R., Varghese, S.H., Nair, B.G., Maekawa, T., Yoshida, Y., Kumar, S., 2010. Nanoparticle material delivery to plants. Plant Sci. 179, 154–163.

Navarro, E., Baun, A., Behra, R., Hartmann, N.B., Filser, J., Miao, A.J., Quigg, A., Santschi, P.H., Sigg, L., 2008. Environmental behavior and ecotoxicity of engineered nanoparticles to algae, plants, and fungi. Ecotoxicology 17, 372–386.

Negrea, A., Lupa, L., Ciopec, M., Negrea, p., 2011. Experimental and modelling studies on As(III) removal from aqueous medium on fixed bed. Column. Chem. Bull. "Politehnica" Univ. (Timisoara) 56, 89–93.

Nel, A., Xia, T., Madler, L., Li, N., 2006. Toxic potential of materials at the nanolevel. Science 311, 622–627.

Nernst, W., 1909. Theorie der Reaktionsgeschwindigkeit in heterogenen Systemen. Z. Phys. Chem. 47, 52–55.

Nilsson, R., Kearns, D.R., 1973. Remarkable deuterium effect on the rate of photosensitized oxidation of alcohol dehydrogenase and trypsin. Photochem. Photobiol. 17, 65–68.

NIOSH, 2011. Asbestos fibers and other elongate mineral particles: State of the Science and Roadmap for Research. Bulletin 62, 2011–2159.

Noyes, A.A., Whitney, W.R., 1897. The rate of solution of solid substances in their own solutions. J. Am. Chem. Soc. 19, 930–934.

Noyes, P.D., McElwee, M.K., Miller, H.D., Clark, B.W., Van Tiem, L.A., Walcott, K.C., Erwing, K.N., Levin, E.D., 2009. The toxicology of climate change: Environmental contaminants in a warming world. Environ. Int. 35, 971–986.

Nwangi, J.N., Wang, N., Ingersoll, C.G., Hardesty, D.K., Brunson, E.L., Li, H., Deng, B., 2012. Toxicity of carbon nanotubes to freshwater aquatic invertebrates. Environ. Toxicol. Chem. 1823–1830.

Oberdorster, E., Zhu, S., Blickley, T.M., McClellan-Green, M., Haasch, M.L., 2006. Ecotoxicology of carbon-based engineered nanoparticles: effects of fullerene (C60) on aquatic organisms. Carbon 44, 1112–1120.

Oberdorster, G., Oberdorster, E., Oberdorster, J., 2007. Concepts of nanoparticle dose metric and response metric. Environ. Health Perspect. 115, 290.

Oberdorster, E., 2004. Manufactured nanomaterials (fullerenes, C60) induce oxidative stress in the brain of juvenile largemouth bass. Environ. Health Perspect. 112, 1058–1062.

Odum, E.P., 1968. Energy flow in ecosystems: a historical review. Am. Zool. 8, 11–18.

Ogawa, K., 1977. The role of bacterial floc as food for zooplankton in the sea. Nippon Suis. Gak 43, 395–407.

Ohman, M.D., Hirche, H.J., 2001. Density-dependent mortality in an oceanic copepod population. Nature 412, 638–64119.

Ohtsubo, Y., Kudo, T., Tsuda, M., Nagata, Y., 2004. Strategies for bioremediation of polychlorinated biphenyls. Appl. Microbiol. Biotechnol. 65, 250–258.

Omar, F.M., Aziz, H.A., Stoll, S., 2014. Aggregation and disaggregation of ZnO nanoparticles: influence of pH and adsorption of Suwannee River humic acid. Sci. Total Environ. 468–469, 195–201.

Ostroumov, S.A., Kolesov, G.M., 2010. The aquatic macrophyte *Ceratophyllum demersum* immobilizes Au nanoparticles after their addition to water. Doklady Biol. Sci. 431, 124–127.

Ostwald, W., 1900. Über die vermeintliche Isomerie des roten und gelben Quecksilbersoxyds und die Oberflächenspannung fester Körper (Translation: On the supposed isomerism of red and yellow mercury oxide and the surface tension of solid bodies). Z. Physikalische Chem. 34, 495–503.

Paine, R.T., 1966. Food web complexity and species diversity. Am. Nat. 100, 65–75.

Pal, S., Tak, Y.K., Song, J.M., 2007. Does the antibacterial activity of silver nanoparticles depend on the shape of the nanoparticle? A study of the gram-negative bacterium Escherichia Coli. Appl. Environ. Microbiol. 73, 1712–1720.

Palacin, S., Barraud, A., 1991. Chemical reactivity at the air-water interface: redox properties of the tetrapyridino porphyraziniumring. Colloids Surf. 52, 123–147.

Palm, A., Cousins, I.T., Mackay, D., Tysklind, M., Metcalfe, C., Alaee, M., 2002. Assessing the environmental fate of chemicals of emerging concern: a case study of the polybrominated diphenyl ethers. Env. Pol. 117, 195–213.

Parsons, J.G., Lopez, M.L., Gonzalez, C.M., Peralta-Videa, J.R., Gardea-Torresdey, J.L., 2010. Toxicity and biotransformation of uncoated and coated nickel hydroxide nanoparticles on mesquite plants. Environ. Toxicol. Chem. 29, 1146–1154.

Peres, C.M., Agathos, S.N., 2000. Biodegradation of nitroaromatic pollutants: from pathways to remediation. Biotechnol. Annu. Rev. 6, 197–220.

Perez-Hernández, G., Schmidt, B., 2013. Anisotropy of the water–carbon interaction: molecular simulations of water in low-diameter carbon nanotubes. Phys. Chem. Chem. Phys. 15, 4995–5006.

Petosa, A.R., Jaisi, D.P., Quevedo, I.R., Elimelech, M., Tufenkji, N., 2010. Aggregation and deposition of engineered nanomaterials in aquatic environments: role of physicochemical interactions. Environ. Sci. Technol. 44, 6532–6549.

Pfefferkorn, F.E., Bello, D., Haddad, G., Park, J.Y., Powell, M., McCarthy, J., Bunker, K.L., Fehrenbacher, A., Jeon, Y., Virji, M.A., Gruetzmacher, G., Hoover, M.D., 2010. Characterization of exposures to airborne nanoscale particles during friction stir welding of aluminum. Ann. Occup. Hyg. 54, 486–503.

Phenrat, T., Saleh, N., Sirk, K., Tilton, R.D., Lowry, G.V., 2007. Aggregation and sedimentation of aqueous nanoscale zerovalent iron dispersions. Environ. Sci. Technol. 41, 284–290.

Phenrat, T., Kim, H.-J., Fagerlund, F., Illangasekare, T., Tilton, R.D., Lowry, G.V., 2009a. Particle size distribution, concentration, and magnetic attraction affect transport of polymer-modified Fe^0 nanoparticles in sand columns. Env Sci.Tech. 43, 5079–5085.

Phenrat, T., Long, T.C., Lowry, G.V., Veronesi, B., 2009b. Partial oxidation ("Aging") and surface modification decrease the toxicity of nanosized zerovalent iron. Environ. Sci. Technol. 43, 195–200.

Phenrat, T., Kim, H.-J., Fagerlund, F., Illangasekare, T., Lowry, G.V., 2010. Empirical correlations to estimate agglomerate size and deposition during injection of a polyelectrolyte-modified Fe^0 nanoparticle at high particle concentration in saturated sand. J. Contam. Hydrol. 118, 152–164.

Pierce, J.R., Adams, P.J., 2007. Efficiency of cloud condensation nuclei formation from ultrafine particles. Atmos. Chem. Phys. 7, 1367–1379.

Pignatello, J.J., 1989. Sorption dynamics of organic compounds in soils and sediments. In: Reactions and Movements of Organic Chemicals in Soils. SSSA Special Publicat No. 22, pp. 45–80.

Plieth, W.J., 1982. Electrochemical properties of small clusters of metal atoms and their role in surface enhanced Raman scattering. J. Phys. Chem. 86, 3166–3170.

Pol, V., Thivagarajan, P., 2009. Remediating plastic waste into carbon nanotubes. J. Environ. Monit. 12, 455–459.

Pretti, C., Oliva, M., Di Pietro, R., Monni, G., Cevasco, G., Chiellini, F., Pomelli, C., Chiappe, C., 2014. Ecotoxicity of pristine graphene to marine organisms. Ecotox. Environ. Safe 101, 138–145.

Prince, R.C., 1993. Petroleum spill bioremediation in marine environments. Crit. Rev. Microbiol. 19, 217–242.

Prospero, J.M., 1999. Long-range transport of mineral dust in the global atmosphere: impact of African dust on the environment of the southeastern United States. PNAS. USA 96, 3396–3403.

Radlinska, E.Z., Gulik-Krzywicki, T., Lafuma, F., Langevin, D., Urbach, W., Williams, C.E., Ober, R., 1995. Polymer confinement in surfactant bilayers of a lyotropic lamellar phase. Phys. Rev. Lett. 74, 4237–4240.

Raisuddin, S., Kwok, K.W.H., Leung, K.M.Y., Schlenk, D., Lee, J.S., 2007. The copepod Tigriopus: a promising marine model organism for ecotoxicology and environmental genomics. Aquat. Toxicol. 83, 161–173.

Rana, M., Chandra, A., 2007. Filled and empty states of carbon nanotubes in water: dependence on nanotube diameter, wall thickness and dispersion interactions. J. Chem. Sci. 119, 367–376.

Raychoudhury, T., Naja, G., Ghoshal, S., 2010. Assessment of transport of two polyelectrolyte-stabilized zero-valent iron nanoparticles in porous media. J. Contaminant Hydrol. 118, 143–151.

Raychoudhury, T., Tufenkji, N., Ghoshal, S., 2012. Aggregation and deposition kinetics of carboxymethyl cellulose-modified zero-valent iron nanoparticles in porous media. Water Res. 46, 1735–1744.

Redlich, O., Peterson, D.L., 1959. A useful adsorption isotherm. J. Phys. Chem. 63, 1024.

Rengifo-Herrera, J.A., Pizzio, L.R., Blanco, M.N., Roussel, C., Pulgarin, C., 2011. Photocatalytic discoloration of aqueous malachite green solutions by UV-illuminated TiO_2 nanoparticles under air and nitrogen atmospheres: effects of counter-ions and pH. Photochem. Photobiol. Sci. 10, 29–34.

Rico, C.M., Majumdar, S., Duarte-Gardea, M., Peralta-Videa, J.R., Gardea-Torresdey, J.L., 2011. Interaction of nanoparticles with edible plants and their possible implications in the food chain. J. Agric. Food Chem. 59, 3485–3498.

Riipinen, I., Yli-Juuti, T., Pierce, J.R., Petaja, T., Worsnop, D.R., Kulmala, M., Donahue, N.M., 2012. Contribution of organics to atmospheric nanoparticle growth. Nat. Geosci. 5, 453–485.

Rochman, C.M., Browne, M.A., Halpern, B.S., Hentschel, B.T., Hoh, E., Karapanagioti, H.K., Rios-Mendoza, L.M., Takada, H., Teh, S., Thompson, R.C., 2013. Policy classify plastic waste a hazardous. Nature (London, United Kingdom) 494, 169–171.

Roh, J.-Y., Sim, S.-J., Park, Y.K., Chung, K.H., Ryu, D.-Y., Choi, J., 2009. Ecotoxicity of silver nanoparticles on the soil nematode *Caenorhabditis elegans* using functional ecotoxicogenomics. Environ. Sci. Technol. 43, 3933–3940.

Roh, J.-Y., Park, Y.-K., Park, K., Choi, J., 2010. Ecotoxicological investigation of CeO_2 and TiO_2 nanoparticles on the soil nematode *Caenorhabditis elegans* using gene expression, growth, fertility, and survival as endpoints. Environ. Toxicol. Pharmacol. 29, 167–172.

Rohr, J.R., Crumrine, P.W., 2005. Effects of an herbicide and an insecticide on pond community structure and processes. Ecol. Appl. 15, 1135–1147.

Rohr, J.R., Kerby, J.L., Sih, A., 2006. Community ecology as a framework for predicting contaminant effects. Trends Ecol. Evol. 21, 606–613.

Rohr, J.R., Schotthoefer, A.M., Raffel, T.R., Carrick, H.J., Halstead, N., Hoverman, J.T., Johnson, C.M., Johnson, L.B., Lieske, C., Piwoni, M.D., Schoff, P.K., Beasley, V.R., 2008. Agrochemicals increase trematode infections in a declining amphibian species. Nature 455, 1235–1239.

role in phytotoxicity of alumina nanoparticles. Toxicol. Lett. 14:158:122–132.

Rossi, J., Barnoud, J., Monticelli, L., 2014. Polystyrene nanoparticles perturb lipid membranes. J. Phys. Chem. Lett. 5 (1), 241–246.

Rozman, K.K., Doull, J., 2000. Dose and time as variables of toxicity. Toxicology 144, 169–178.

Rutkowski, J.V., Levin, B.C., 1986. Acrylonitrile-butadiene-styrene copolymers (ABS): pyrolysis and combustion products and their toxicity? a review of the literature. Fire Mater. 10, 93.

Ryan, J.N., Elimelech, M., 1996. Colloid mobilization and transport in groundwater. Colloids Surf. A: Physicochem. Eng. Aspects 107, 1–56.

Saleh, N.B., Pefferle, L.D., Elimelech, M., 2008. Aggregation kinetics of multiwalled carbon nanotubes in aquatic systems: measurements and environmental implications. Environ. Sci. Technol. 42, 7963–7969.

Schiedek, D., Sundelin, B., Readman, J.W., Macdonald, R.W., 2007. Interactions between climate change and contaminants. Mar. Pollut. Bull. 54, 1845–1856.

Schink, B., 2002. Anaerobic digestion: concepts, limits and perspectives. Water Sci. Technol. 45, 1–8.

Schmitt-Jansen, M., Veit, U., Dudel, G., Altenburger, R., 2008. An ecological perspective in aquatic ecotoxicology: approaches and challenges. Basic Appl. Ecol. 9, 337–345.

Schubert, D., Dargusch, R., Raitano, J., Chan, S.W., 2006. Cerium and Yttrium oxide nanoparticles are neuroprotective. Biochem. Biophys. Res. Commun. 342, 86–91.

Sconfienza, C., van de Vorst, Jori, G., 1980. Type I and type II mechanisms in the photooxidation of L-tryptophan and tryptamine sensitized by hematoporphyrin in the presence and absence of sodium dodecyl sulfate micelles. Photochem. Photobiol. 31, 351–357.

Selim, H.M., Davidson, J.M., Mansell, R.S., 1976. Evaluation of a two-site adsorption–desorption model for describing solute transport in soil. In: Proc. of the Computer Simulation Conf., J. Washington, DC. Am. Inst. of Chem. Eng., Washington, DC, pp. 444–448.

Senesi, N., Testini, C., 1982. Physico-chemical investigations of interaction mechanism between striazine herbicides and soil-humic acids. Geoderma 28, 129–146.

Senesi, Testini, C., 1983. Spectroscopic investigations of electron donor-acceptor processes involving organic free radicals in the adsorption of substituted urea herbicides by humic acids. Pest Sci. 14, 79–89.

Sergio, M., Behzadi, H., Otto, A., van der Spoel, D., 2013. Fullerenes toxicity and electronic properties. Environ. Chem. Lett. 11, 105–118.

Shao, D.D., Ren, X.M., Hu, J., Chen, Y.X., Wang, X.K., 2010. Preconcentration of $Pb2^{+}$ from aqueous solution using poly(acrylamide) and poly(N,N-dimethylacrylamide) grafted multi walled carbon nanotubes. Colloids Surf. A 360, 74–84.

Shi, J.P., Evans, D.E., Khan, A.A., Harrison, R.M., 2001. Sources and concentration of nanoparticles (<10 nm diameter) in the urban atmosphere. Atmos. Environ. 35, 1193–1202.

Shooto, D.N., Dikio, E.D., 2011. Morphological characterization of soot from the combustion of candle wax. Int. J. Electrochem. Sci. 6, 1269–1276.

Sih, A., Englund, G., Wooster, D., 1998. Emergent impacts of multiple predators on prey. TREE 13, 9.

Silva, E., 1979. Rate constants studies of the dye-sensitized photoinactivation of lysozyme. Radiat. Environm Biophys. 16, 71–79.

Simonson, R.W., 1995. Airborne dust and its significance to soils. Geoderma 65, 1–43.

Sioutas, C., Delfino, R.J., Singh, M., 2005. Exposure assessment for atmospheric ultrafine particles (UFPs) and implications in epidemiologic research. Environ. Health Perspect. 113, 947–955.

Sipila, M., et al., 2010. Role of sulphuric acid in atmospheric nucleation. Science 327, 1243–1246.

Six, J., Feller, C., Denef, K., Ogle, S.M., de Moraes Sa, J.C., Albrecht, A., 2002. Soil organic matter, biota and aggregation in temperate and tropical soils - effects of no-tillage. Agronomie 22, 755–775.

Sobek, A., Bucheli, T., 2009. Testing the resistance of single- and multi-walled carbon nanotubes to chemothermal oxidation used to isolate soots from environmental samples. Environ. Pollut. 157, 1065–1071.

Sommer, T.J., 1986. The periodicity of phytoplankton in Lake Con stance (Bodensee) in comparison to other deep lakes of central Europe. Hydrobiologia 138, 1–7.

Sotiriou, G.A., Pratsinis, S.E., 2010. Antibacterial activity of nanosilver ions and particles. Environ. Sci. Technol. 44, 5649–5654.

Spormann, A.M., Widdel, F., 2000. Metabolism of alkylbenzenes, alkanes, and other hydrocarbons in anaerobic bacteria. Biodegradation 11, 85–105.

Stampoulis, D., Sinha, S.K., White, J.C., 2009. Assay-dependent phytotoxicity of nanoparticles to plants. Environ. Sci. Technol. 43, 9473–9479.

Stankovich, S., Piner, R.D., Chen, X., Wu, N., Nguyen, S.T., Ruoff, R.S., 2006. Enhanced mechanical properties of Nanocomposites at low graphene content. J. Mater. Chem. 16, 155–158.

Stephens, B., Azimi, P., El Orch, Z., Ramos, T., 2013. Ultrafine particle emissions from desktop 3D printers. Atm. Environ. 79, 334–339.

Stiopkin, I.V., Weeraman, C., Pieniazek, P.A., Shalhout, F.Y., Skinner, J.L., Benderskii, A.V., 2011. Hydrogen bonding at the water surface revealed by isotopic dilution spectroscopy. Nature 474, 192–195.

Stumm, W., Morgan, J.J., 1996. Aquatic Chemistry, 3rd edn. Wiley, New York.

Sullivan Jr., J.D., Felbeck, G.T., 1968. The interaction of s-triazine herbicides with humic acids from three different soils. Soil Sci. 106, 45–52.

Sun, H., Liu, S., Zhou, G., Ang, H.M., Tade, M.O., Wang, S., 2012. Reduced graphene oxide for catalytic oxidation of aqueous organic pollutants. ACS Appl. Mater. Interfaces 4, 5466–5471.

Suttiponparnit, K., Jiang, J., Sahu, M., Suvachittanont, S., Charinpanitkul, T., Biswas, P., 2011. Role of surface area, primary particle size, and crystal phase on titanium dioxide nanoparticle dispersion properties. Nanoscale Res. Lett. 6, 27.

Tarnuzzer, R.W., Colon, J., Patil, S., Seal, S., 2005. Vacancy engineered ceria nanostructures for Protection from radiation-induced cellular damage. Nano Lett. 5, 2573–2577.

Taylor, D.A., 2002. Dust in the wind Environ. Health Persp. 110, A80–A87.

Templeton, R.C., Ferguson, P.L., Washburn, K.M., Scrivens, W.A., Chandler, G.T., 2006. Life-cycle effects of single-walled carbon nanotubes (SWCNTs) on an estuarine meiobenthic copepod. Environ. Sci. Technol. 40, 7387–7393.

Thess, A., Lee, R., Nikolaev, P., Dai, H.J., Petit, P., Robert, J., Xu, C.H., Lee, Y.H., Kim, S.G., Rinzler, A.G., Colbert, D.T., Scuseria, G.E., Tomanek, D., Fischer, J.E., Smalley, R.E., 1996. Crystalline ropes of metallic carbon nanotubes. Science 273, 483–487.

Thio, B.J.R., Zhou, D.X., Keller, A.A., 2011. Influence of natural organic matter on the aggregation and deposition of titanium dioxide nanoparticles. J. Hazard. Mater. 189, 556–563.

Thompson, R.C., Moore, C.J., vom Saal, F.S., Swan, S.H., 2009. Plastics, the environment and human health: current consensus and future trends. Phil Trans. Royal Soc. B: Biol. Sci. 364, 2153–2166.

Thompson, H.M., 1996. Interactions between pesticides: a review of reported effects and their implications for wildlife risk assessment. Ecotoxicology 5, 59–81.

Tong, Z., Bischoff, M., Nies, L., Applegate, B., Turco, R.F., 2007. Impact of fullerene (C60) on a soil microbial community. Environ. Sci. Technol. 41, 2985–2991.

Tong, M.P., Ding, J.L., Shen, Y., Zhu, P.T., 2010. Influence of biofilm on the transport of fullerene (C_{-60}) nanoparticles in porous media. Water Res. 44, 1094–1103.

Tourinho, P.S., van Gestel, C.A.M., Lofts, S., Svendsen, C., Soares, A.M.V.M., Loureiro, S., 2012. Metal-based nanoparticles in soil: fate, behavior and effects on soil invertebrates. Environ. Toxicol. Chem. 31, 1–14.

Tripathi, S., Champagne, D., Tufenkji, N., 2012. Transport behavior of selected nanoparticles with different surface coatings in granular porous media coated with *Pseudomonas aeruginosa* biofilm. Environ. Sci. Technol. 46, 6942–6949.

Tsai, C.S., Gidin, J.R.P., Wandn, A.J., 1985. Dye-sensitized photooxidation of enzymes. Biochem. J. 225, 203–208.

Tso, C.P., Zhung, C.M., Shih, Y.H., Tseng, Y.M., Wu, S.C., Doong, R.A., 2010. Stability of metal oxide nanoparticles in aqueous solutions. Water Sci. Technol. 61, 127–213.

Tsyusko, O., Unrine, J.M., Spurgeon, D.J., Blalock, E.M., Starnes, D., Tseng, T.M., Joice, G., Bertsch, P.M., 2012. Toxicogenomic responses of the model organism *Caenorhabditis elegans* to gold nanoparticles. Environ. Sci. Technol. 46, 4115–4124.

Unrine, J.,P., Hunyadi, B.S., 2008a. Bioavailability, trophic transfer, and toxicity of manufactured metal and metal oxide nanoparticles in terrestrial environments. In: Grassian, V. (Ed.), Nanoscience and Nanotechnology: Environmental and Health Impacts. Wiley, Hoboken, NJ, USA, pp. 345–366.

Unrine, J.M., Bertsch, P.M., Hunyadi, S.E., 2008b. In: Grassian, V.H. (Ed.), Nanoscience and Nanotechnology: Environmental and Health Impacts. John Wiley and Sons, Inc., Hoboken.

Unrine, J.M., Tsyusko, O.V., Hunyadi, S.E., Judy, J.D., Bertsch, P.M., 2010. Effects of particle size on chemical speciation and bioavailability of copper to earthworms (*Eisenia fetida*) exposed to copper nanoparticles. J. Environ. Qual. 39, 1942–1953.

Unrine, J.M., Shoults-Wilson, W.A., Zhurbich, O., Bertsch, P.M., Tsyusko, O.V., 2012. Trophic transfer of Au nanoparticles from soil along a simulated terrestrial food chain. Environ. Sci. Technol. 46, 9753–9760.

Upadhyay, R.K., Soin, N., Roy, S.S., 2014. Role of graphene/metal oxide composites as photocatalysts, adsorbents and disinfectants in water treatment: a review. RSC Adv. 4, 3823–3851.

Vaillancourt, F.H., Bolin, J.T., Eltis, L.D., 2006. The ins and outs of ring-cleaving dioxygenases. Crit. Rev. Biochem. Mol. Biol. 41, 241–267.

Velzeboer, I., Hendriks, J., Ragas, M.J., van de Meent, D., 2008. Aquatic ecotoxicity tests of some nanomaterials. Environ. Toxicol. Chem. 27, 1942–1947.

Verburg, K., Baveye, P., 1994. Hysteresis in the binary exchange of cations on 2:1 clay minerals: a critical review. Clays and Clay Minerals 42, 207–220.

Vervey, E.J.W., Overbeek, J.Th.G., 1948. Theory of the Stability of Lyophobic Colloids. Elsevier, Amsterdam, New York, 1948.

Vikesland, P.J., Heathcock, A.M., Rebodos, R.L., Makus, K.E., 2007. Particle size and aggregation effects on magnetite reactivity toward carbon tetrachloride. Environ. Sci. Technol. 41, 5277–5283.

Villagarcia, H., Dervishi, E., de Silva, K., Biris, A.S., Khodakovskaya, M.V., 2012. Surface chemistry of carbon nanotubes impacts the growth and expression of water channel protein in tomato plants. Small 8, 2328–2334.

Vinodgopal, K., Wynkoop, D.E., Kamat, P.V., 1996. Environmental photochemistry on semiconductor surfaces. Environ. Sci. Technol. 30, 1660.

Voice, T.C., Weber Jr., W.J., 1983. Sorption of hydrophobic compounds by sediments, soils and suspended solids – I theory and background. Water Res. 17, 1433–1441.

Von Moos, N., Slaveykova, V.I., 2014. Oxidative stress induced by inorganic nanoparticles in bacteria and aquatic microalgae–state of the art and knowledge gaps [Review]. Nanotoxicology 8, 605–630.

Vorbau, M., Hillemann, L., Stintz, M., 2009. Method for the characterization of the abrasion induced nanoparticle release into air from surface coatings. Aerosol Sci. 40, 209–217.

Wallace, L., Wang, F., Howard-Reed, C., Persily, A., 2008. Contribution of Gas and electric stoves to residential ultrafine particle concentrations between 2 and 64 nm: size distributions and emissions and coagulation rates. Environ. Sci. Technol. 42, 8641–8647.

Wang, Y.G., Li, Y.S., Fortner, J.D., Hughes, J.B., Abriola, L.M., Pennell, K.D., 2008a. Transport and retention of nanoscale C-60 aggregates in water-saturated porous media. Env. Sci. Tech. 42, 3588–3594.

Wang, Y.G., Li, Y.S., Pennell, K.D., 2008b. Influence of electrolyte species and concentration on the aggregation and transport of fullerene nanoparticles in quartz sands. Env. Tox. Chem. 27, 1860–1867.

Wang, G.H., Zhou, B.H., Cheng, C.L., Cao, J.J., Meng, J.J., Li, J.J., Tao, J., Zhang, R.J., Fu, P.Q., 2012a. Impact of Gobi desert dust on aerosol chemistry of Xi'an, inland China during spring 2009: differences in composition and size distribution between the urban ground surface and the mountain atmosphere. Atmos. Chem. Phys. Discuss. 12, 21355–21397.

Wang, Y., Li, Y., Costanza, J., Abriola, L.M., Pennell, K.D., 2012b. Enhanced mobility of fullerene (C_{60}) nanoparticles in the presence of stabilizing agents. Environ. Sci. Technol. 46, 11761–11769.

Wanga, H., Wicka, R., Xing, B., 2008. Toxicity of nanoparticulate and bulk ZnO, Al_2O_3 and TiO_2 to the nematode *Caenorhabditis elegans*. Environ. Pollut. 154, 1171–1177.

Watson, T., 2005. NASA to Detail Plans for Trip to Moon. USA Today 18 September.

Wegner, A., Besseling, E., Foekema, E.M., Kamermans, P., Koelmans, A.A., 2012. Effects of nanopolystyrene on the feeding behavior of the blue mussel (*Mytilus edulis* L.). Environ. Toxicol. Chem. 31, 2490–2497.

Wen, S., Zhao, J., Sheng, G., Fu, J., Peng, P., 2002. Photocatlytic reactions of phenanthrene at TiO_2/water interfaces. Chemosphere 46, 871–877.

Westervelt, D.M., Pierce, J.R., Riipinen, I., Trivitayanurak, W., Hamed, A., Ulmala, M., Laaksonen, A., Decesari, S., Adams, P.J., 2013. Formation and growth of nucleated particles into cloud condensation nuclei: model–measurement comparison. Atmos. Chem. Phys. 13, 7645–7663.

Widdel, F., Rabus, R., 2001. Anaerobic biodegradation of saturated and aromatic hydrocarbons. Curr. Opin. Biotechnol. 12, 259–276.

Wigginton, N.S., Haus, K.L., Hochella Jr., M.F., 2007. Aquatic environmental nanoparticles. J. Environ. Monit. 9, 1306–1316.

Wiley, B., Herricks, T., Sun, Y., Xia, Y., 2004. Polyol synthesis of silver nanoparticles: use of chloride and oxygen to promote the formation of single-crystal, truncated cubes and tetrahedrons. Nano Lett. 4, 1733–1739.

Williams, P.P., 1977. Metabolism of synthetic organic pesticides by anaerobic microorganisms. Residue Rev. 66, 63–135.

Wu, Z., Hu, M., Lin, P., Liu, S., Wehner, B., Wiedensohler, A., 2008. Particle number size distribution in the urban atmosphere of Beijing, China. Atmos. Environ. 42, 7967–7980.

Xia, T., Kovochich, M., Liong, M., Mdler, L., Gilbert, B., Shi, H., Yeh, J.I., Zink, J.I., Nel, A.E., 2008. Comparison of the mechanism of toxicity of zinc oxide and cerium oxide nanoparticles based on dissolution and oxidative stress properties. ACS Nano 2, 2121–2134 [Erratum appears in ACS Nano. 2008 Dec 23;2(12):2592].

Xu, X.-M., Fu, S.-Y., Zhu, Q., Xiao, X., Yuan, J.-P., Peng, J., Wu, C.-F., Wang, J.H., 2014. Depth-related coupling relation between methane-oxidizing bacteria (MOBs) and sulfate-reducing bacteria (SRBs) in a marine sediment core from the Dongsha region, the South China Sea. Appl. Microbiol. Biotechnol. 98, 10223–10230.

Yamabi, S., Imai, H., 2002. Growth conditions for Wurtzite zinc oxide films in aqueous solutions. J. Mater. Chem. 12, 3773–3778.

Yamakoshi, Y., Umezawa, N., Ryu, A., Arakane, K., Miyata, N., Goda, Y., Masumizu, T., Nagano, T., 2003. Active oxygen species generated from Photoexcited fullerene (C_{60}) as potential medicines: O^{2-} versus $1O^2$. J. Am. Chem. Soc. 125, 12803–12809.

Yang, L., Watts, D.J., 2005. Particle Surface Characteristics May Play an Important.

Yang, F., Liu, C., Gao, F., Su, M., Wu, X., Zheng, L., Hong, F., Yang, P., 2007. The improvement of spinach growth by nano-anatase TiO_2 treatment is related to nitrogen photoreduction. Biol. Trace Elem. Res. 119, 77–88.

Yin, Y., Shen, M., Zhou, X., Yu, S., Chao, J., Liu, J., Jiang, G., 2014. Photoreduction and stabilization capability of molecular weight fractionated natural organic matter in transformation of silver ion to metallic nanoparticle. Environ. Sci. Tech. 48, 9366–9373.

Yucel, M., Gartman, A., Chan, C.S., Luther III, G.W., 2011. Hydrothermal vents as a kinetically stable source of iron-sulphide-bearing nanoparticles to the ocean. Nat. Geosci. 4, 367–371.

Zamaraev, K.I., Khramov, M.I., Parmon, V.N., 1994. Possible impact of Heterogeneous Photocatalysis on the global chemistry of the Earth's atmosphere. Catal. Rev. Sci. Eng. 36, 617.

Zhang, K.M., Wexler, A.S., 2004. Evolution of particle number distribution near roadways — Part I: analysis of aerosol dynamics and its implications for engine emission measurement. Atmos. Environ. 38, 6643–6653.

Zhang, Y., Chen, Y., Westerhoff, P., Crittenden, J., 2009. Impact of natural organic matter and divalent cations on the stability of aqueous nanoparticles. Water Res. 43, 4249–4257.

Zhang, W., Yao, Y., Sullivan, N., Chen, Y.S., 2011. Modeling the primary size effects of citrate coated silver nanoparticles on their ion release kinetics. Environ. Sci. Technol. 45, 4422–4428.

Zhang, P., He, X., Ma, Y., Lu, K., Zhao, Y., Zhang, Z., 2012a. Distribution and bioavailability of ceria nanoparticles in an aquatic ecosystem model. Chemosphere 89, 530–535.

Zhang, R.A., Khalizov, A., Wang, L., Hu, M., Xu, W., 2012b. Nucleation and growth of nanoparticles in the atmosphere. Chem. Rev. 112, 1957–2011.

Zhang, L., Hou, L., Wang, L., Kan, A.T., Chen, W., Tomson, M.B., 2012c. Transport of fullerene nanoparticles (nC_{60}) in saturated sand and sandy soil: controlling factors and modeling. Environ. Sci. Technol. 46, 7230–7238.

Zhang, D., Hua, T., Xiao, F., Chen, C., Gersberg, R.M., Liu, Y., Stuckey, D., Ng, W.J., Tan, S.K., 2015. Phytotoxicity and bioaccumulation of ZnO nanoparticles in *Schoenoplectus tabernaemontani*. Chemosphere 120, 211–219.

Zhao, G., Li, J., Ren, X., Chen, C., Wang, X., 2011. Few-layered graphene oxide nanosheets as superior sorbents for heavy metal ion pollution management. Environ. Sci. Technol. 45, 10454–10462.

Zhao, J., Wang, Z., White, J.C., Xing, B., 2014. Graphene in the aquatic environment: adsorption, dispersion, toxicity and transformation. Environ. Sci. Technol. 48, 9995–10009.

Zhu, X., Cai, Z., 2012. Behavior and effect of manufactured nanomaterials in the marine environment. Integr. Environ. Assess. Manage. 8, 566–567.

Zhu, S., Oberdorster, E., Haasch, M.L., 2006. Toxicity of an engineered nanoparticle (fullerene, C_{60}) in two aquatic species, Daphnia and fathead minnow. Mar Environ. Res. 62 (Suppl), S5–S9.

Zhu, X., Zhu, L., Li, Y., Duan, Z., Chen, W., Alvarez, P.J., 2007. Developmental toxicity in zebrafish embryos after exposure to manufactured nanomaterials: buckminsterfullerene aggregates (nC_{60}) and fullerol. Environ. Toxicol. Chem. 26, 976–979.

Zhu, X., Zhu, L., Chen, Y., Tian, S., 2009a. Acute toxicities of six manufactured nanomaterial suspensions to *Daphnia magna*. J. Nanopart Res. 11, 67–75.

Zhu, X., Wang, J.X., Zhang, X.Z., Chang, Y., Chen, Y., 2009b. The impact of ZnO nanoparticle aggregates on the embryonic development of zebrafish (*Danio rerio*). Nanotechnology 20, 195103.

CHAPTER

9

Human and Environmental Risk Characterization of Nanoparticles

OUTLINE

Engineered Nanoparticles
http://dx.doi.org/10.1016/B978-0-12-801406-6.00009-1

1. INTRODUCTION

As discussed in earlier chapters and shown in Figure 1, the physicochemical and biological properties of nanoparticles (1–100 nm diameter) and bulk particles (>100 nm) are different. For nanoparticles, an increase in particle size is associated with the following:

1. A transition from quantum properties (<100 nm) to linear properties (as found in bulk particles)
2. A decrease in the surface area-to-volume ratio, surface reactivity, specific heat capacity, and band gap (highest occupied molecular orbital HOMO-lowest unoccupied molecular orbital LUMO) energy
3. An increase in electron conductivity, cohesive energy, enthalpy, and Curie temperature
4. A change in emission wavelength associated with color change

Carbon nanotubes (CNTs), depending upon the diameter or graphene-folding configuration, exhibit metallic or semiconductor properties.

The size dependence of unique properties of nanoparticles is currently being exploited in nanoparticle-based products. However, the very properties responsible for nanoparticles' beneficial properties also pose serious health hazards to humans and the environment. Of special concern is occupational exposure to free nanoparticles of individuals in the workplaces

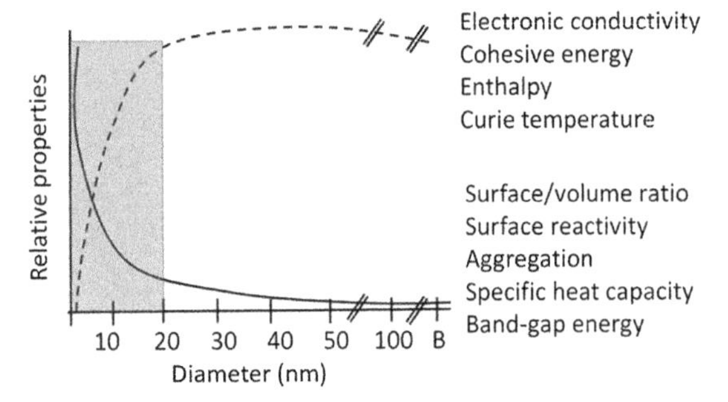

FIGURE 1 Side-dependent physiochemical, thermodynamic, and electronic properties of nanoparticles.

(Kuhlbusch et al., 2011). Prolonged exposure to nanoparticles having characteristics not previously encountered may compromise human health by either directly challenging the normal defense mechanisms in humans and animals—especially their immune and inflammatory systems—or indirectly by impairing the environment and ecosystems. The environmental impact of nanoparticles may be related to their dispersion and persistence in the environment.

Nanoparticles, for a desired medicinal or commercial outcome, are functionalized with specific groups performing unique functions; for example, the attachment of a hydrophilic group to the surface of a hydrophobic particle may increase their water solubility, functionalization with a fluorescent group may facilitate imaging, and/or drug loading may improve their pharmaceutical capacity. However, if functionalization alters a nanoparticle's size, their physiochemical properties may also change accordingly; for example, bare 10-nm gold and functionalized 10-nm gold nanoparticles may exhibit relatively different physicochemical properties. The surface of nanoparticles in tissue culture media, biological systems, or the environment may form a reversible protein/lipid corona that further modulates their properties. These observations, taken together, raise the following important questions:

1. Do the unique features of nanoparticles pose any inherent threats (different from their bulk counterparts) to humans and the environment?
2. Are existing risk-assessment methodologies developed for bulk particles appropriate to assess potential and plausible risks associated with nanoparticles?

Unfortunately, at present, there are no paradigms to anticipate the significance of these characteristics. On face value, the safety evaluation of nanoparticles cannot rely on the toxicological and ecotoxicological profiles of the bulk material. Therefore, an understanding of the strategies commonly used to assess nanoparticles' risk is needed to develop protocols for their safe use in daily-use products. The aims of this chapter are to discuss the basic principles of risk assessment, management, and communication, as well as the commonly used risk-assessment strategies for bulk particles (discussed in the Appendix section) and nanoparticles.

2. INTRODUCTION TO RISK ASSESSMENT, RISK MANAGEMENT, AND RISK COMMUNICATION

According to the US Federal Regulations (42 Code of, Part 90, published in 55 Federal Register 5136, February 13, 1990A), public health assessment is defined as the evaluation of data and information on the release of hazardous substances into the environment in order to (1) assess any past, current, or future impact on public health; (2) develop health advisories or other recommendations; and (3) identify studies or actions needed to evaluate and mitigate or prevent human health effects. There are two approaches commonly used to secure public health: the approach used by the Agency for Toxic Substances and Disease Registry (ATSDR) and the qualitative/quantitative risk assessments conducted by regulatory agencies such as US Environmental Protection Agency (EPA) and the European Union (EU).

2.1 ATSDR Approach

The ATSDR approach focuses more closely on site-specific exposure conditions and specific community health concerns using any available health outcome data to provide a more qualitative, less theoretical evaluation of possible public health hazards. It considers past exposures in addition to current and potential future exposures. The ATSDR approach may not go into the environmental aspects of pollution. Figure 2(A) shows the overall ATSDR approach, and Figure 2(B) shows the information needed to conduct the public health assessment. Communities (people who live and work at or around the site) often play an important role in the public health assessment process. The evaluation process is iterative and dynamic, considering all available data from varying perspectives (Table 1). It is important to identify data gaps and limitations, such as the need for further environmental sampling. Detailed information regarding the ATSDR approach can be obtained from the following site: http://www.atsdr.cdc.gov/HAC/PHAManual/ch2.html.

2.2 Qualitative/Quantitative Risk Assessments Based on the US EPA and EU Protocols

The protocols developed by the US EPA and the EU's public health departments are based on the qualitative and quantitative risk-assessment strategies that provides a numeric estimate of theoretical risk or hazard, assuming no cleanup has taken place. It focuses on current and potential future exposures and considers all contaminated media, regardless of the mode of exposure. By design, it generally uses standard protective exposure assumptions when evaluating the site risk.

2.2.1 Basic Concepts of Risk Assessment

This section provides a brief definition of risk and hazard, and then discusses basic principles of risk determination strategies (http://epa.gov/riskassessment/guidance.htm).

2.2.1.1 RISK AND HAZARD

Risk is the chance (probability) that harmful effects may occur to human health or to ecosystems resulting from exposure to any stressor (physical, chemical, or biological entity) that can induce an adverse response. Hazard is the potential for a stressor to cause harm (toxicity). Hazard is an intrinsic property of a stressor.

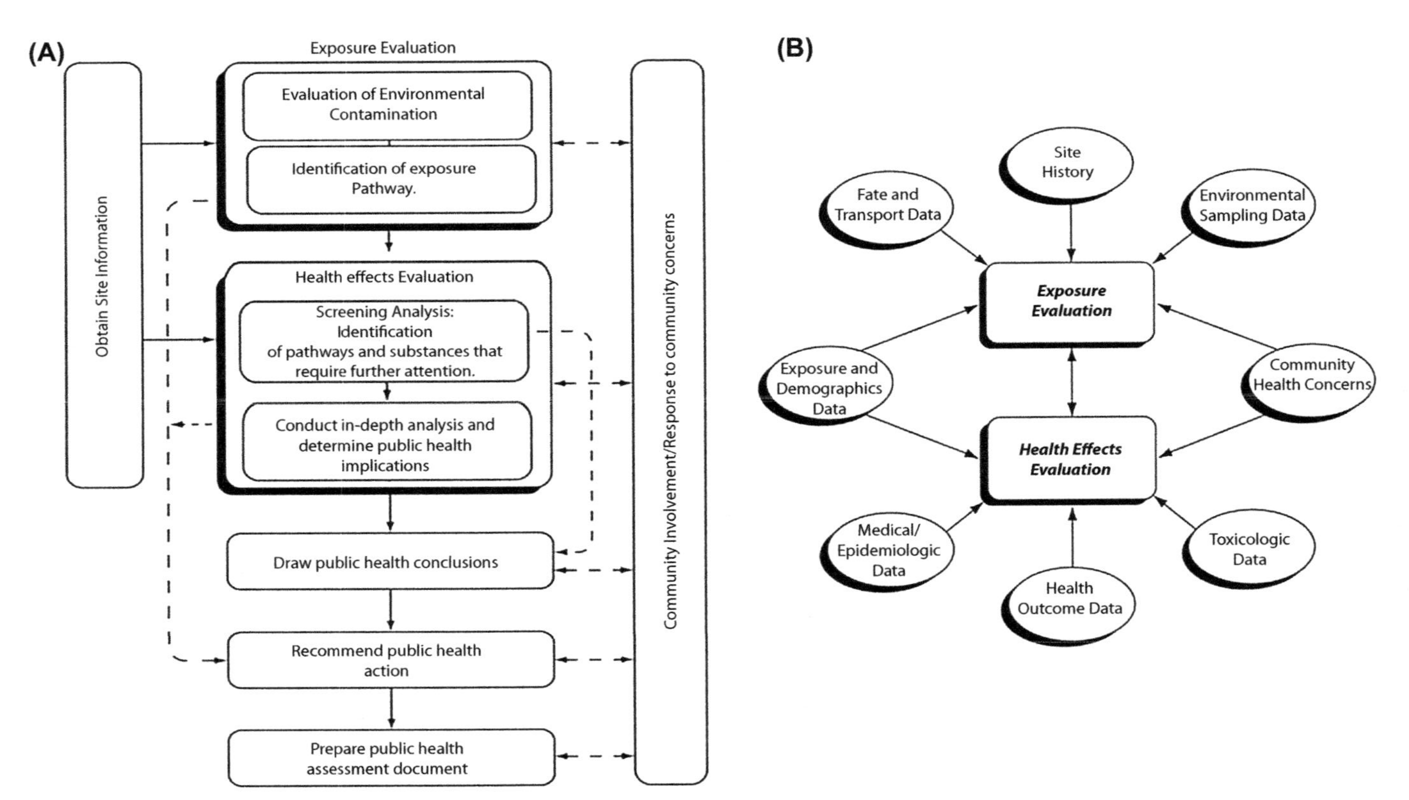

FIGURE 2 (A) Typical approach used by the Agency for Toxic Substances and Disease Registry (ATSDR) for determining human risk focusing more closely on site-specific exposure conditions, specific community health concerns, and any available health outcome data. ATSDR provides a more qualitative, less theoretical evaluation of possible public health hazards. (B) Information needed to evaluate ATSDR risk assessment (EPA Website, public access).

TABLE 1 Basic Data Needs for Conducting a Public Health Assessment

Background information	Community health concerns and information
Site Description • Site name(s) and address. • Site boundaries. • Site maps—current and historical. • Physical hazards. • Contact person(s) (local/county, state, tribal, and federal). *Site Operations and History* • Current and past site-related activities. • Current and past hazardous waste treatment, storage, and disposal practices. • Current and past site use. *Regulatory History and Activities* • Current CERCLA or RCRA status of the site. • Site investigation results. • Permit and compliance information. • Site remedial activities and actions taken to address contaminant releases. • Types of institutional controls planned or in place. *Land Use and Natural Resources Information* • Control of public access. • Land use on or near the site, including schools. • Estimated frequency and types of recreational activities on or near the site. • Children's play areas on or near site. • Planned or proposed land use or development. • Location and purveyors of public water supplies. • Surface water uses downstream of the site. • Drainage systems on and in the site and vicinity. • Agriculture, aquaculture, animal husbandry, hunting, fishing, and tribal activities near the site. *Demographic Information* • Population activities on or near the site. • Sensitive populations in the vicinity of the site.	• Records of complaints by the public about the site. • Logs of actions taken by federal, state, or local agencies site in response to health complaints. • Information from the community, gathered during meetings or health studies. • Environmental justice, tribal member concerns, or cultural issues. *Environmental Contamination Information* • Summary of current and historical sampling data. • List of substances analyzed for, tested for, and not found and detected. • Range of detected concentrations. • Sampling and analytical methods used. • QA/QC documentation. *Exposure Pathway Information* • Contaminant sources. • Description of physical barriers to prevent the pollutant transport. • Topography, geology, and hydrogeology information. • Description of upstream (surface water) or other nearby activities contributing to contamination. • Affected media, including groundwater, surface water, soil/sediment, air, and food chain (biota). • Exposure point and exposure route. *Health Outcome Data* • Relevant health outcome databases (e.g., mortality data, cancer incidence, birth defect data). • Any site-specific community health records. *Substance-Specific Information* • Information on chemical and physical properties of environmental contaminants of concern. • Toxicologic and epidemiologic data. • Biologica.

www.atsdr.cdc.gov/HAC/PHAManual/ch3.html

The following examples may explain the difference between the risk and the hazard.

- If oil is spilled on steps, then that oil presents a slipping risk to people trying to use the steps. However, if a physical barrier is placed to prevent people using the steps, then, although the hazard of slippage is still there, people are not at risk of getting hurt by slipping. Therefore, people are at risk if they are *exposed* to the hazard.
- Hydrochloric acid is highly corrosive or irritating to humans. It can burn the skin upon contact. However, the acid is only a risk to human health if humans are exposed to it. The degree of harm caused by the exposure will depend on the specific exposure scenario. A diluted acid may not cause any harm to humans, although the hazardous property of the acid will remain unchanged.

- Anticoagulant rodenticides are highly hazardous (dangerous), but in patients with heart conditions, very low doses of these chemicals are used to prevent blood clots.

Taken together, these scenarios suggest that a hazard may pose a risk when there is exposure and the dose is sufficiently high to cause an adverse effect. This brings up the concept of a *safe dose*, defined as a dose safely tolerated by the body without any harmful effects. Risk and hazard are related by the following equation: Risk = Hazard × Exposure (dose, mass/unit - weight). This equation defines the probability that exposure (in terms of dose) to a hazard will lead to a negative consequence. Thus, a hazard poses no risk if there is no exposure to that hazard. In general, risk is a combination of the following:

1. Probability that an adverse effect will occur
2. The consequences of the effect

For example, a 10 mg/kg exposure of an organophosphate toxin results in blood insecticide concentration ranging from 100 to 300 μg/100 ml, which may cause a 30–60% decrease in blood cholinesterase activity. The probability is that a 30% decrease blood cholinesterase activity causes seizures in 10% of the cases, while a 60% decrease in blood cholinesterase activity may be fatal in 20% of the exposed population but cause seizures in 60% of the exposed population. If the dose used (say 5 mg/kg) does not inhibit blood cholinesterase activity, then it is assumed that no adverse effects will occur.

2.2.1.2 RISK CHARACTERIZATION

Risk characterizes the nature and magnitude of health risks to humans, animals, and/or ecosystems (e.g., plants, phytoplankton, birds, fish, wildlife) from a stressor present in food or the environment (Sand 1987). This information is used to protect humans and the environment from stressors or contaminants. Risk characterization is a scientific process that depends on the following factors (Figure 3):

1. Planning and scoping stage, where the purpose and scope of a risk assessment are decided
2. The risk-assessment process, consisting of hazard identification and characterization, exposure assessment, dose–response assessment, and risk analysis
3. The risk management and communication processes

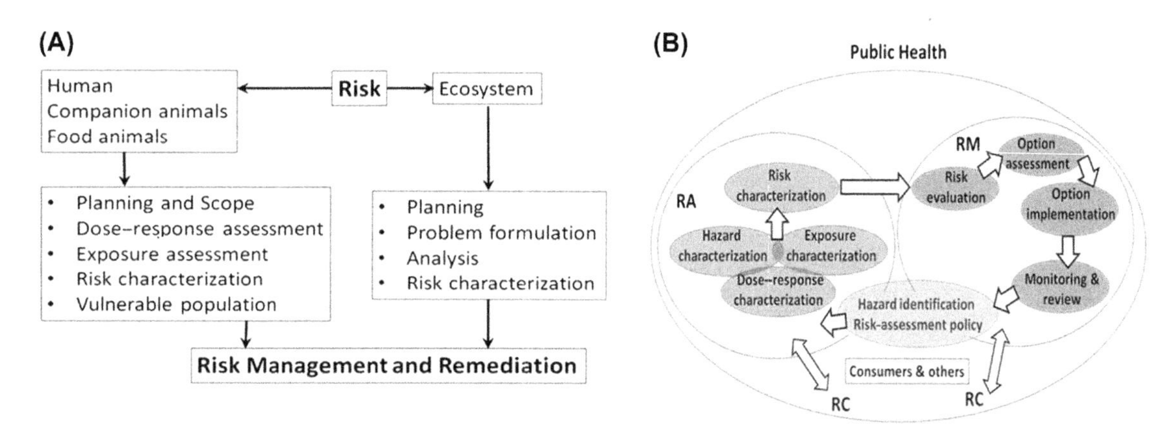

FIGURE 3 (A) Overall approach used by the USEPA for human and environment risk assessment (RA) and management (RM). (B) Components of risk assessment, management, and communication (RC).

The EPA, since its inception, was concerned with risk characterization practices. The agency has developed and conducted risk assessments of numerous environmental contaminants, including nanoparticles. Figure 3(A) and (B) shows the key aspects of the EPA's risk assessment strategies (http://epa.gov/riskassessment/ecological-risk.htm, ECB, 2003)

2.2.1.2.1 STAGES OF RISK CHARACTERIZATION

- *The planning and scoping stage*: The investigator identifies the hazard and relevant health impacts that need to be addressed. Scoping includes characterization of the potential hazard; setting boundaries such as the time scale, geographical area, and the population that could be affected; noting any sectors of special concern or vulnerability, identifying the authorities that need to be informed; classifying the situation, ranging from catastrophic to negligible; and preparing a proposal (Figure 4).
- *Hazard identification and characterization*: Ideally, hazard identification starts before there is significant use of the agent. As shown in Figure 5, physicochemical, biological, and toxicological properties are part of hazard characterization.
  Toxicity of a hazard is characterized using cell-based in vitro studies, animal bioassays, human studies, and multiple health-related endpoints. Common toxicity endpoints for hazard identification are carcinogenicity, mutations, altered immune function, teratogenicity, altered reproductive function, neurobehavioral toxicity, and organ-specific effects. Human studies have the obvious advantage of being done on the subject of most interest, but they are time-consuming, expensive, and often have many variables that are difficult to control.

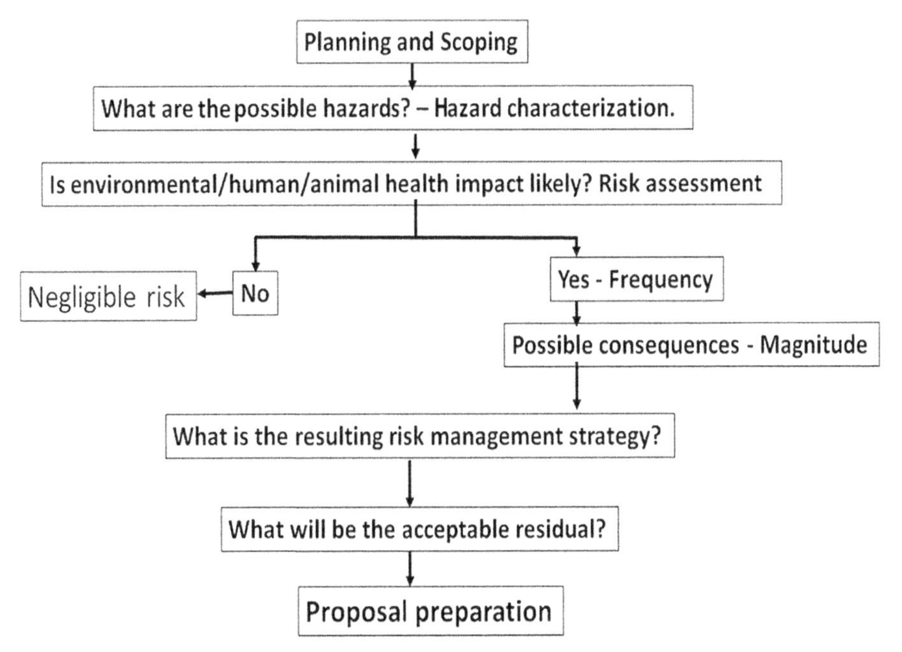

FIGURE 4 A flow diagram for conducting risk assessment from planning to proposal preparation.

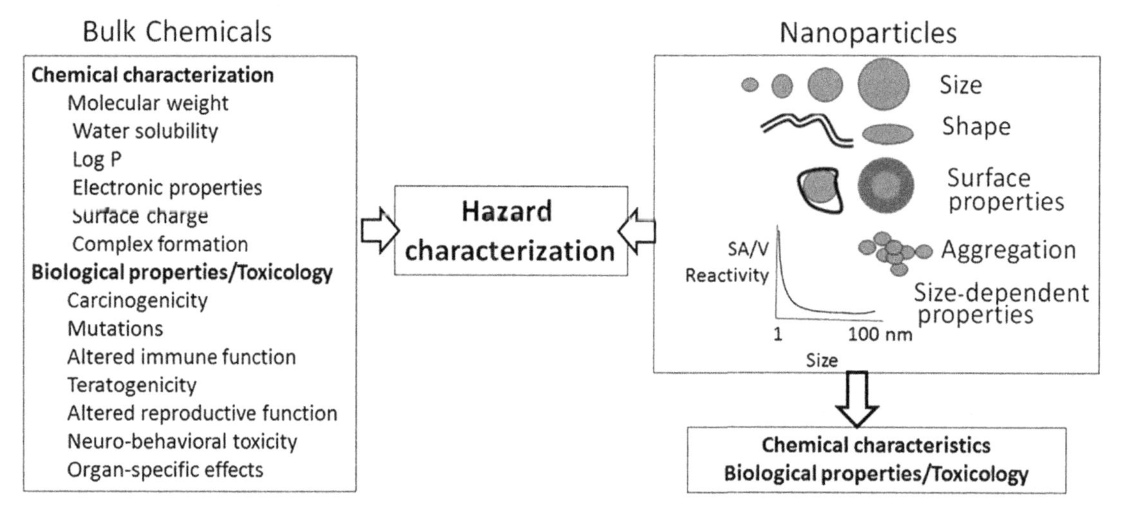

FIGURE 5 Comparison of hazards posed by bulk particles and nanoparticles including selected indices used to measured hazard. Size, shape and surface determine the Nanoparticles', but not the bulk particles', hazard and toxicity.

- *Exposure*: If the hazard assessment indicates that the compound is potentially hazardous, the next step is to evaluate various possibilities for exposure. This includes the routes of exposure (oral, inhalation, or skin), dose, site (home, workplace, school, or other areas), duration, and frequency of exposure. This information helps to define the population of concern. Exposure information may also be important for designing appropriate studies on hazard assessment and certainly for the next step of establishing dose–response relationships (MacPhail et al., 2013).
- *Dose–response relationship* (also known as the concentration–response or exposure–response relationship): The dose–response relationship describes how the severity of adverse health effects (the responses) is related to the dose for a given mode of exposure. Typically, as the dose increases, the measured response also increases. At low doses, there may be no response. At some level of dose, the responses begin to occur; at a certain dose, maximum response occurs (Figure 6). In general, the response that occurs at the lowest dose is selected as the critical effect for risk assessment. The underlying assumption is that if the critical effect is prevented from occurring, then no other toxic effects will occur.

  Based on this mode of action, the nature of the extrapolation is determined either through nonlinear or linear dose–response assessment.
  - *Nonlinear dose–response* assessment holds that a range of exposures from zero to some finite value can be tolerated by the organism with essentially no toxic effect (no observed adverse effect level, or NOAEL). The threshold of toxicity is where the effects (or their precursors) begin to occur (the lowest observed adverse effect level, or LOAEL). Using mathematical modeling, an alternative of NOAEL, the benchmark dose lower-confidence limit (BMDL), is determined. BMDL represents a statistical lower confidence limit for the dose that produces a selected response.

    Another index, the reference dose (RfD) or acceptable daily intake (ADI), is an

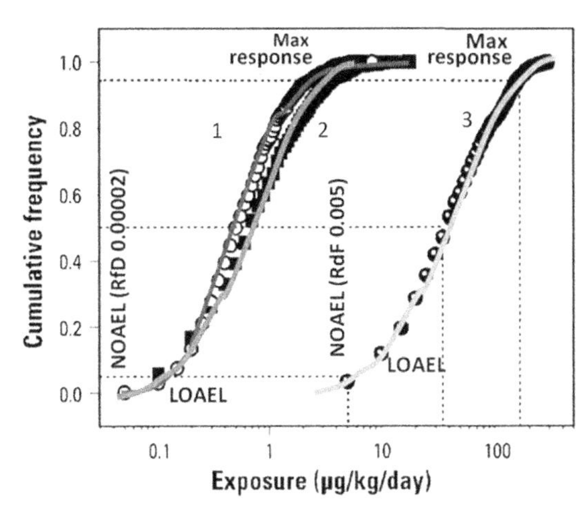

FIGURE 6 Dose (mass/unit weight)-response curves following inhalation (plot 1), dermal (plot 2) and oral (plot 3) exposure. The deduced NOAEL, LOAEL, and RfD depend upon the dose as well as the exposure mode.

estimate of the daily oral exposure to the human population (including sensitive groups, such as asthmatics, or life stages, such as children or the elderly) that is likely to be without an appreciable risk of deleterious effects during a lifetime. RfD is calculated by the equation RfD = NOAEL/UF, where UF is the uncertainty factor (discussed in detail later in this chapter). For example, an uncertainty factor of 1000 (10 each to account for inadequate animal data, animal to human extrapolation, and human variability or individual sensitivity) is used for converting the animal NOAEL data to human RfD or ADI.

In a nonlinear dose–response curve, there is enormous inter- and intraspecies variability in the extent and nature of different databases used for risk assessment. Figure 8 shows dose–response curves prepared for inhalation, dermal, and oral exposure, and the respective NOAEL values (0.02 μg/kg/day for inhalation and dermal exposure but 5 μg/kg/day for oral exposure) exhibiting about 250-fold variation.

- The *linear dose–response* curve assumes that there is no tolerated dose. The slope of the straight line estimates the risk at the exposure level that falls along the line. The linear dose–response curve is relevant to the development of cancer.

2.2.1.3 RISK ASSESSMENT

As a probability function, risk assessment describes the likelihood of dose-dependent adverse effects in humans or animals exposed to a stressor (bulk chemicals and nanoparticles). The first step in risk assessment is the identification of an adverse event. The next step is identification and characterization of hazard including dose, frequency, and magnitude of exposure as a consequence of contact with the contaminated medium. Then, information on the inherent toxicity of the chemical (i.e., the expected response to a given level of exposure) is collected. The collected data are combined to predict the probability, nature, and magnitude of the adverse health effects that may occur. For quantitative risk assessment, one would require reliable and complete data on the nature and extent of contamination, fate and transport processes, the magnitude and frequency of human exposure, toxicokinetics, and toxicodynamics. Unfortunately, such complete data are not yet available for nanoparticles. The lack of data results in knowledge gaps or uncertainties that need to be addressed in risk characterization (discussed below). It is important that the uncertainties in the calculations are clearly defined to establish the reliability of the risk estimates.

2.2.1.4 UNCERTAINTY FACTOR

The uncertainty factor deals with gaps in knowledge, such as variations in the database. Some databases have limited or barely adequate animal data, while others may contain detailed information on the mechanism of toxicity and/or toxicokinetics. Some databases may include animal data, whereas others may include both animal and human data. Thus, the risk evaluation based on different databases may encounter different degrees of uncertainties.

To account for these uncertainties, a flexible but structured approach to the selection of uncertainty factors was developed by Renwick (1993), which retains the two 10-fold uncertainty for interspecies and intraspecies differences, respectively. In the interspecies 10-fold factors, 2.5-fold is assigned for toxicokinetic and 4-fold is assigned for toxicodynamic aspects (WHO, 1987). In the intraspecies 10-fold factors, Renwick (Renwick and Lazarus, 1998: Renwick et al., 2003) assigned 2.5-fold for toxicokinetic and 4-fold is for toxicodynamic aspects, whereas WHO (1994) assigned equal weight (3.2 each) to both. This allows the incorporation of appropriate data on the compound of interest in one or both of these aspects where they exist, thereby reducing the extent of uncertainty (Figure 7).

In some cases, such as the linear dose–response relationship, the application of an additional uncertainty factor of 10–100 applied to the NOAEL (in addition to the 100 normally used to allow for interspecies and intraindividual differences), as proposed by Weil (1972). The magnitude of an additional factor could be made more logical and consistent with the weight of evidence of carcinogenicity. Such an approach has not been adopted in previous risk assessments but may be a reasonable method, especially in the numerous cases where tumors are detected only at the top dose tested.

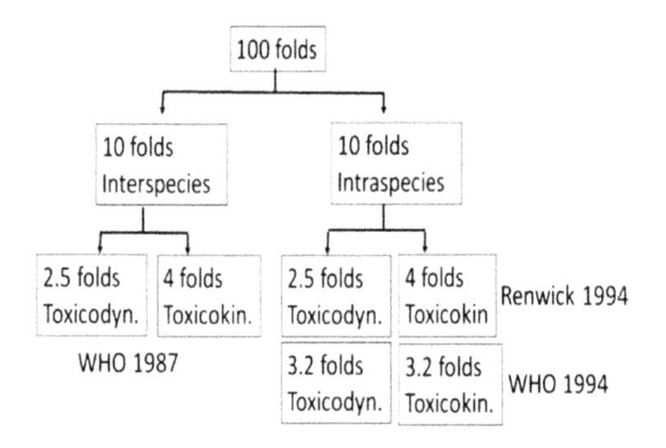

FIGURE 7 Uncertainty determination in risk analysis.

2.2.1.5 RISK MANAGEMENT

Following risk assessment, which provides information on potential health risks, risk management evaluates the recommendations and determines how to protect public or environmental health. Risk management strategies/or options can be broadly classified as regulatory, nonregulatory, economic, advisory, or technological, which are not mutually exclusive. In addition to the scientific factors that provide the basis for the risk assessment, risk management may deal with the following:

- The economic factors, such as the cost of risks, the benefits of reducing them, the costs of risk mitigation or remediation options, and the distributional effects.
- Legal issues such as federal, state, and local laws that must be followed.
- Social factors, such as the communities' income level, ethnic background and values, land use and zoning, and availability of health care, lifestyle, and psychological conditions of the affected populations.
- Technological factors include the feasibility, impacts, and range of risk management options.
- Political factors are based on the interactions among branches of the federal government, with other federal, state, and local government entities, and even with foreign governments; these may range from practices defined by agency policy and political administrations through inquiries from the members of Congress, special interest groups, or concerned citizens.

2.2.1.6 RISK PERCEPTION AND COMMUNICATION

In contrast to risk assessment and risk management, which are based on science and experimental data, the majority of individuals rely on intuitive judgment typically called "risk perception." This may be based on information coming from the news media, which principally document mishaps and threats occurring globally (Slovic, 1987, 1993; Kraus and Slovic, 1988; Cohrssen and Covello, 1989; Sandman et al., 1993; Van Eijndhoven et al., 1994). A major factor that influences the complexity of the social debate over appropriate laws and regulations is the nature and extent of the perceived threat. People perceive risks differently, depending on the likelihood of a hazard having adverse effects; whom it affects; how familiar, widespread, and dreaded the effects are; and the possible effect of a hazard on an individual. Perceptions of risk are also influenced to a large degree by the supposed benefits derived from accepting the risk. Figure 8 (Morgan, 1993; Morgan and Henrion, 1992, 1990) illustrates a mosaic of public perception of risks in terms of risk space quadrants; the upper right quadrant of this space captures uncontrollable risks that are most likely to provoke calls for government regulation.

Risks perceived as potentially uncontrollable, capable of causing catastrophe on a global scale, and affects future generations leading to public anxiety. While different people weigh these factors differently in reaching their overall perceptions of the riskiness of a hazard, the set of factors that are important in determining the relative perceptions of risk go well beyond the statistical frequency, magnitude, and uncertainty of effects. Public opinion on acceptable risk constantly changes, usually in the direction of further risk reduction, which provides further impetus for additional legislation and regulation in many quarters. Risk communication is an interactive process of exchange of information and opinion among individuals, groups, and institutions. The topics usually discussed are the nature of risk and other messages not strictly about risk but expressing concerns, opinions, or reactions to risk messages or to legal and institutional arrangements for risk management. Risk communication deals with both the exchanges of information and views that actually occur in relation to any risk or hazard and how these exchanges ought to occur (Kandlikar et al., 2005).

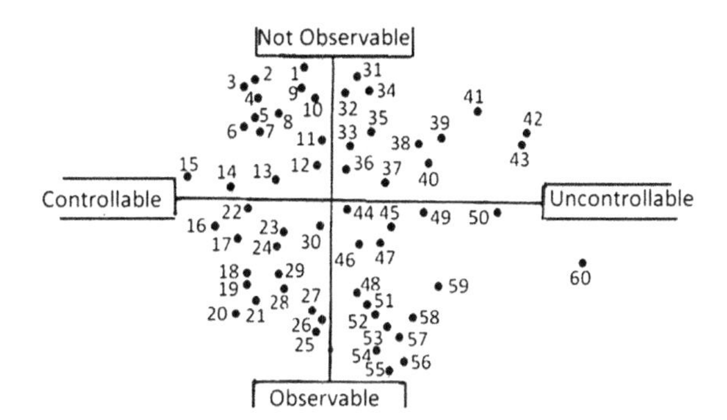

FIGURE 8 Comparative public risk perception. Details are described in the text. **Not observable–controllable axis**: (1) Microwave oven, (2) water fluoridation, (3) saccharin, (4) unidentified, (5) oral contraceptive, (6) valium, (7) IUDs, (8) X-ray, (9) nitrites, (10) polyvinyl chloride, (11) antibiotics, (12) lead (autos), (13) lead paint, (14) vaccines, and (15) aspirin. **Controllable–observable axis**: (16) snowmobile, (17) trampolines, (18) swimming pool, (19) downhill skiing, (20) bicycles, (21) boating, (22) skateboard, (23) snowmobile, (24) tractors, (25) fireworks, (26) alcohol-related accidents, (27) motorcycles, (28) elevators, (29) chain saws, and (30) smoking (disease). **Not observable–uncontrollable axis**: (31) electric fields, (32) fertilizers, (33) asbestos, (34) DES, (35) pesticides, (36) mercury, (37) coal-burning pollution, (38) three nuclear weapons, (39) uranium mining, (40) satellite crashes, (41) radioactive waste, (42) nuclear accidents, and (43) nuclear fallout. **Uncontrollable–observable axis**: (44) carbon monoxide (auto), (45) black lung, (46) skyscraper fire, (47) large dams, (48) underwater construction, (49) liquid natural gas storage and transportation, (50) gas accidents, (51) sports parachutes, (52) general aviation, (53) railroad crossing, (54) auto racing, (55) dynamite, (56) handguns, (57) commercial aviation, (58) high construction, (59) coal mining accidents, and (60) nuclear weapons (war).

2.2.2 *Ecological versus Human Risk Assessment*

Ecosystems consist of *producers* that convert light into chemical energy via photosynthetic conversion of environmental carbon dioxide into glucose, *primary consumers* that feed on producers to generate energy for their survival, *consumers* of higher trophic levels of the food chain, and sufficient *abiotic nutrients* for its inhabitants. A stable ecosystem maintains relatively stable abiotic and biotic components with resistance (ecosystem resists change) and resilience (system destabilizes but then recovers to the original level). The diversity redundancy of an ecosystem determines its resistance.

Although the basic premises of human and ecological risk assessments are the same (Figure 9), there are some major operational differences, as described in the EPA frameworks and other authors (Ferenc and Foran, 2000). For humans (or companion/food animals), the environment is usually controlled; thus, it is not a variable. In the case of ecological risk, environmental conditions are a key variable. According to the EPA, "a human health risk assessment is the process to estimate the nature and probability of adverse health effects in humans who may be exposed to chemicals in contaminated environmental media, now or in the future." An ecological risk assessment is a considerably more complex process because it involves evaluation of the likelihood that the environment (atmosphere, hydrosphere, lithosphere, and biosphere) may be affected as a result of exposure to one or more stressors (Figure 10). The ecological effects may relate to plants, animals, and ecosystems as a whole, and how humans interact with them. Many stressors do not cause adverse effects when administered into individual subjects, but they cause adverse effects when administered to an ecosystem, possibly due to primary and secondary effects.

Ecological risk assessment addresses the following questions:

- How would a physical barrier affect the marine animal populations in nearby water bodies?
- Can the residential or agricultural application of an insecticide end up harming an endangered species (food chain bioaccumulation effect)?
- What is the risk of introducing a nonnative species to an estuary?

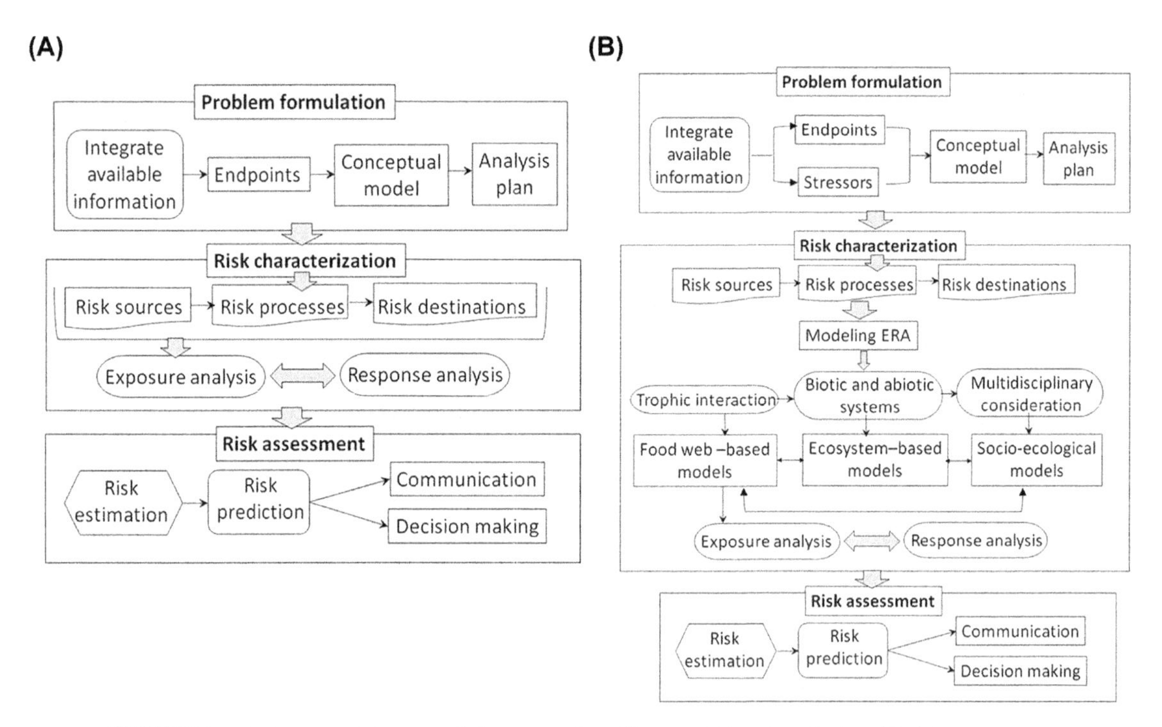

FIGURE 9 An integrated framework for system-based human (A) and ecological (B) risk assessment.

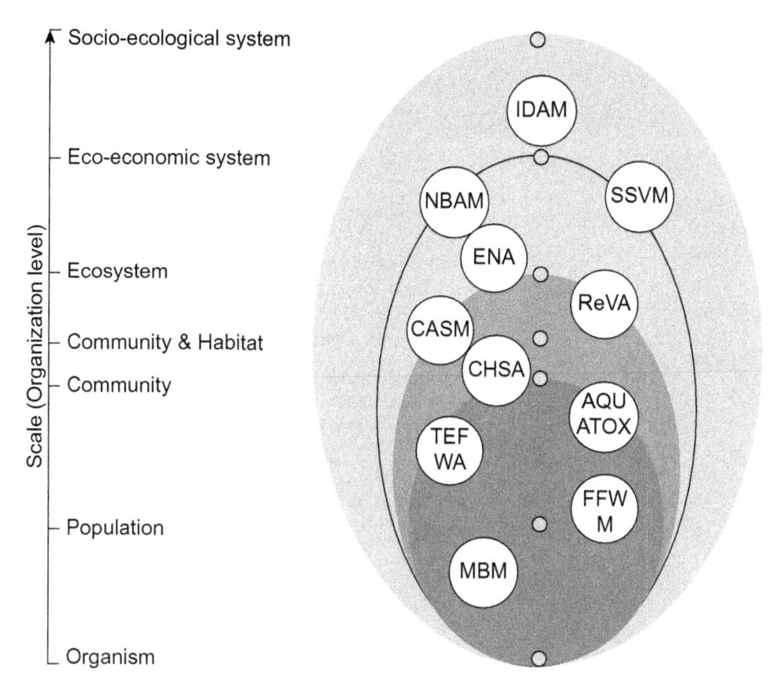

FIGURE 10 System-based models for ecological risk assessment (ERA) at different ecosystem levels. *Abbreviations*: Multimedia box model (MBM) (Mackay, 1991), AQUATOX (USEPA, 2004), Three-phase ecological food web analysis (TEFWA), fugacity-based food-web model (FFWM) (Wang et al., 2010), comprehensive aquatic system model (CASM) (DeAngelis et al., 1989, Bartell et al., 1999), regional vulnerability assessment (ReVA) (Boughton et al., 1999), ecological network analysis (ENA) (Chen and Chen, 2012), communities and habitats status assessment (CHSA), net benefit analysis model (NBAM) (Raghua et al., 2007), socio-ecological systems vulnerability (SSVM) (De Chazal et al., 2008), the integrative Dam assessment model (IDAM) (Brown et al., 2009). *Reprinted from Chen et al. (2013) with permission.*

- How does fertilizer runoff reduce oxygen levels in water bodies, such as bays?
- Are some plants or animals more likely to be susceptible to environmental stressors because of factors such as age, genetics, body size, or differences among species?

Because of these differences, the ecosystem risk assessment is more complex and requires analytical, statistical, and modeling approaches. The basic steps are described in the following sections.

2.2.2.1 PROBLEM FORMULATION

This step defines the problem and presents the plan for analyzing and characterizing risk from specific hazards, such as toxins, pharmaceuticals, or nanoparticles, to specific assessment endpoints (Figure 11; Moore et al., 2004). It includes a review and an understanding of the peer-reviewed literature and the development of a hypothesis. Selection of a toxicological endpoint is based on the potential of the stressor to negatively impact the survival of local populations and, by extension, aquatic community dynamics. The key questions asked are the following (USEPA, 2007):

- Who/what/where is at risk?
- What are the environmental hazards of concern?
- Where do these environmental hazards come from?
- How does exposure occur?
- What does the body do with the environmental hazard and how is this

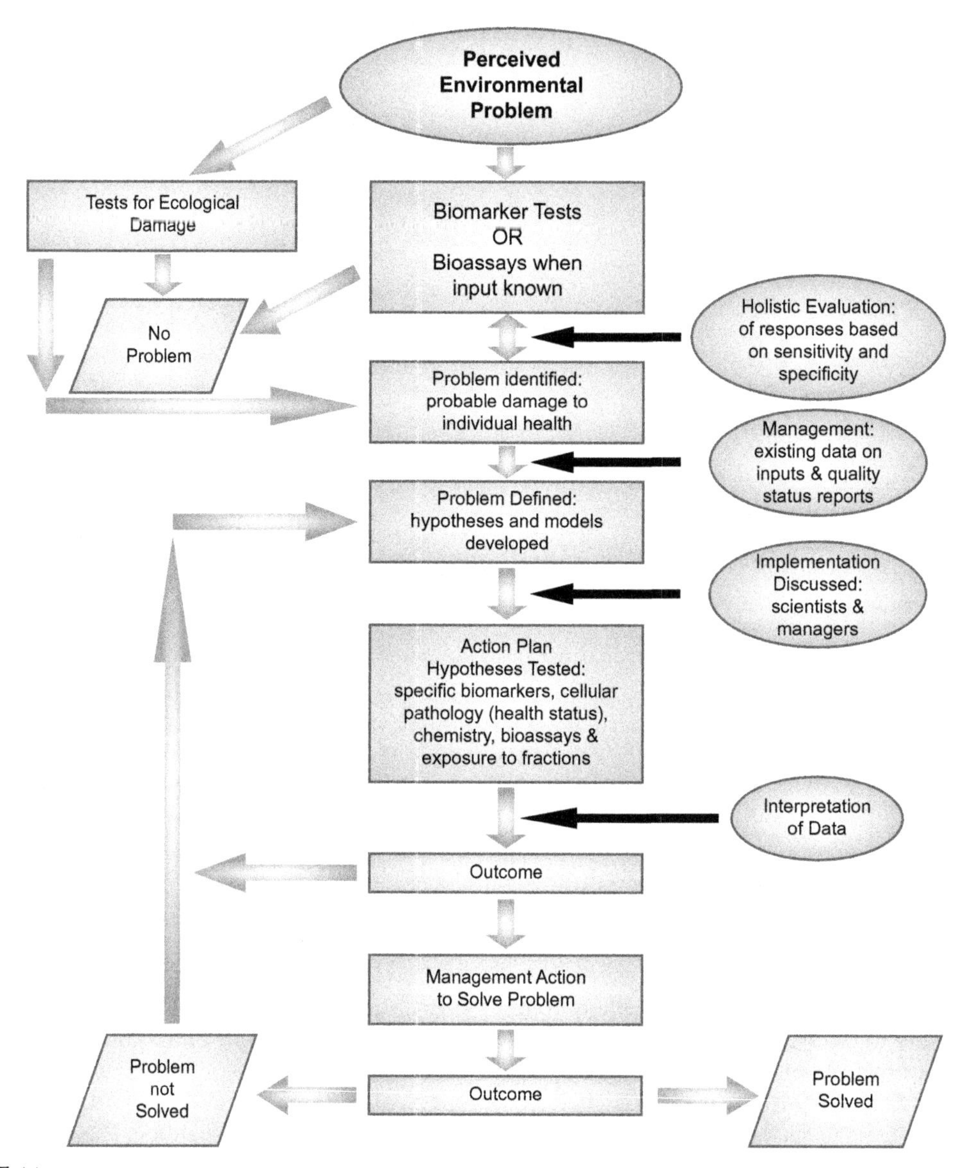

FIGURE 11 Flow diagram describing different stages of ecological and human risk assessment. This include identification of the problem, hazard characterization, risk characterization, and mitigation strategies (US EPA, 2007).

affected by factors such as life stage, genetics, species differences, etc.?
- What are the ecological effects?
- How long does it take for an environmental hazard to cause a toxic effect?

2.2.2.2 EFFECTS CHARACTERIZATION

- *EPA phase1—Problem formulation*: This step defines an assessment endpoint to determine what ecological entity is important to protect. The stressor—response relationships, exposure concentrations, causality, and the relationship between measures of effect are assessed. A key is to develop a predicted no-effect concentration (PNEC) for aquatic chronic toxicity for use during the risk characterization. The EPA generally describes such values as being based on the most sensitive, scientifically acceptable toxicity endpoint.

- *EPA phase 2—Analysis*: The goal of the analysis phase is to provide the ingredients necessary for determining or predicting ecological responses to stressors under exposure conditions of interest. Analysis is performed to determine what plants and animals are exposed, the degree of exposure, and if that level of exposure is likely or not to cause harmful ecological effects. Calculations used may include hazard quotients (HQ, the ratio of chemical contaminant concentration to a selected screening benchmark) to quantify risk and various endpoints to determine the levels of exposure to a stressor (e.g., chemical contaminant) by a selected plant or animal. The endpoint selection is a critical part of the ecosystem risk assessment. Some of the strategies used are listed below (Figure 10):
  - *Organism-level endpoints* that are most directly linked to population-level effects, such as mortality, reproduction, and to a lesser extent, growth (usually most relevant to wildlife risk assessment). (USEPA, cfpub.epa.gov/ncea/cfm/recordisplay.cfm?deid=56830).
  - *Correlated endpoints*: These endpoints are different but closely related (Krewski et al., 2002). In general, these endpoints typically considered in wildlife risk assessments (survival, growth, and reproduction).
  - *Sequential or hierarchal endpoints*: Many endpoints, particularly reproductive endpoints, have sequential or hierarchal relationships. For example, a study may report egg production per breeding pair as well as the proportion of eggs that hatch—these are sequential in terms of their contribution to reproductive output.
  - *Cellular reaction endpoints*: Pathological cellular reaction endpoints, such as lysosomal stability, can provide early-warning distress signals of injurious changes in plants and animals (Galloway et al., 2002; Kohler et al., 2002; Moore et al., 1994; Moore, 2002; Hinton and Lauren, 1990), especially in aquatic organisms (Kohler et al., 1992; Livingstone et al., 2000; Moore, 1990).
- *EPA phase 3—Risk characterization*: This is the final phase of an ecological risk assessment. It is the culmination of all work done during the previous phases. During risk characterization, the results of analysis are used to estimate the risk posed to ecological entities. Then, the risk is described, indicating the overall degree of confidence in the risk estimates, summarizing uncertainties, citing evidence in support of the risk estimates, and interpreting the adversity of ecological effects. Some of the methods commonly used are described below.
  - *Relative risk assessment*: As described above, stressors tend to destabilize the ecosystems. In streams and rivers, a stressor may not be present along the entire length, but it is present in a small region (regional scale). The importance of a regional-aquatic stressor depends both on its regional extent (i.e., how widespread it is) and on the severity of its effects on ecosystems where it is found (Stoddard et al., 2006; Bailey et al., 2004; Reynoldson et al., 1997). The relative risk measures the strength of association between stressor and response variables that can be classified as either *good* (i.e., reference or an area not exposed to the stressor) or *poor* (i.e., different from reference). Table 2 shows the estimated lengths of a stream (km) and sampled site counts (in parentheses) for combinations of good and poor response strength and sediment condition.

TABLE 2 Stressor (Total Nitrogen + Phosphorus) Response Values for a Stream

	Stressor	
Response condition	**Good**	**Poor**
Good	X(i) 22700	Y(i) 3930
Poor	A(i) 27450	B(i) 27680
Total	X+A 50150	Y+B 31610

Data from Sickle et al. (2006).

The relative response (RR) is determined using the following equations:

$$\text{RR (Poor response given poor sediment condition)} = B/(Y+B) \text{ or } 27680/(3930+27680) = 0.876 \quad (1)$$

$$\text{RR (Poor response given good sediment condition)} = A/(X+A) \text{ or } 27450/(22700+27450) = 0.546 \quad (2)$$

The RR ratio $= 0.876/0.546 = 1.6$. This suggests that poor response strength has a greater risk (1.6-fold) of occurring when the stressor conditions are poor rather than when the stressor conditions are good. A ratio of 1 indicates no connection between stressor and response.

- *Hazard quotients (HQs)*: HQs are commonly used to identify risks of contaminants to ecosystem. A typical HQ is determined using Eqn (3), where PNEC is the predicted no-effect concentration or a toxicity reference value (TRV), where the TRV is assumed to represent a safe dose (Hill et al., 2013).

$$HQ = \frac{\text{Measured environmental concentration (MEC)}}{\text{PNEC or TRV}} \quad \text{(Solomon et al., 1996)} \quad (3)$$

An $HQ < 1$ suggests that unacceptable adverse effects on local populations are highly unlikely. Conversely, if $HQ > 1$, the magnitude of potential effects is unknown and further evaluation may be warranted (Barnthouse et al., 2008).

- *Probabilistic risk characterization*: The probabilistic risk quotient compares the exposure and toxicity concentrations of stressors with distributions of probabilistic HQs calculated by a Monte Carlo simulation (10,000 iterations), as described by Solomon et al. (2000).
- *Modelling ecological risk*: As mentioned above, ecosystems are highly complex systems of interactive functional networks that give rise to the whole system function. Environmental contaminants, depending upon the dose, can result in adaptive responses (an appropriate reaction to an environmental response), stress syndromes (an organism's total response to environmental demands or pressures), disease, and changes in functional biodiversity in living systems due to perturbation of one or many component(s) of the interactive networks. Therefore, a reductionist approach may not be appropriate to determine ecosystem toxicity. However, by combining simulation with empirical data, it should be possible to describe the behavior of complex ecosystems. The commonly used models are based on an integration of chemical, molecular, cellular, physiological, and trophic processes, coupled with generic simulation of virtual whole systems, to simulate "virtual organisms". (Bontje, 2010; De Laender, 2007; De Laender, et al., 2008a,b,c,d; Galic et al., 2010; Schreuders et al., 2004; Sharma and Kansal, 2013). Chen et al. (2013) reviewed different models developed for ecological risk assessment and have proposed a holistic approach for ecological risk assessment (ERA) and management at different levels of hierarchical organization, such as food

web-based, ecosystem-based, and socioecological models. The authors propose an integrated framework characterizing problem formulation, risk characterization, and risk assessment for current and future ERA (Figures 5 and 11).

The following articles provide further information regarding different aspects of ecological toxicity: Adger (2006): vulnerability, Brittingham et al. (2014): ecological risk of shale oil exploration, De Lange et al. (2010): ecological vulnerability, Filser et al (2008): ecological theory and ecotoxicology, Hill et al. (2013): wild-life risk assessment, Hiltbrunner et al. (2014): ecological toxicology of expansion of nitrogen-fixing plants in cold-area, Judson et al. (2014): in vitro modeling EPA, Kanwar et al. (2015), Ludwicki et al. (2015): HQ, Laurenson et al. (2014): ethinyl estradiol ecological risk, Li (2014): heavy metals in the Yanghe River, Morrissey et al. (2015): neonicotinoid−global surface water risk, Perrodin et al. (2011): urban ecosystem risk assessment, Sickle et al. (2006): environmental stressors, Tixier et al. (2011): urban storm water. The Appendix section describes in detail the approaches used to control the risk posed by the bulk particles (>100 nm). Readers who are not familiar with bulk-particle risk assessment may wish to review the Appendix section before proceeding to the next section on nanoparticle risk assessment.

3. NANOPARTICLE RISK ASSESSMENT

Nanotechnology has infiltrated almost every aspect of our society including, but not limited to, manufacturing, health care, energy, agriculture, communications, transportation, and electronics (Figure 12). It may account for almost $1 trillion worth of products in the United States alone and will create anywhere from 800,000 to 2 million new jobs (Uldrich and Newberry, 2003).

Weir et al. (2012) have shown that TiO_2 nanoparticles, a common additive in many personal care and other consumer products used by people, were also present in items such as candies, sweets, chewing gums, toothpastes, shampoos, deodorants ,and shaving creams (Figure 13(A) and (B)). Many dental products also contain nanoparticles, as shown in Figure 13(C).

Nano in general, people may consume 0.2−3 mg/kg TiO_2/kg_{BW}/day. After use, nanoparticles from stored nanoparticle-based products may leach and enter the sewage system. Subsequently, they may enter the environment as treated effluent discharged to surface waters or biosolids applied to the agricultural land, incinerated wastes, or landfill solids. Children may have the highest exposure because the TiO_2 content of sweets is higher than other food products. Unfortunately, there are little, if any,

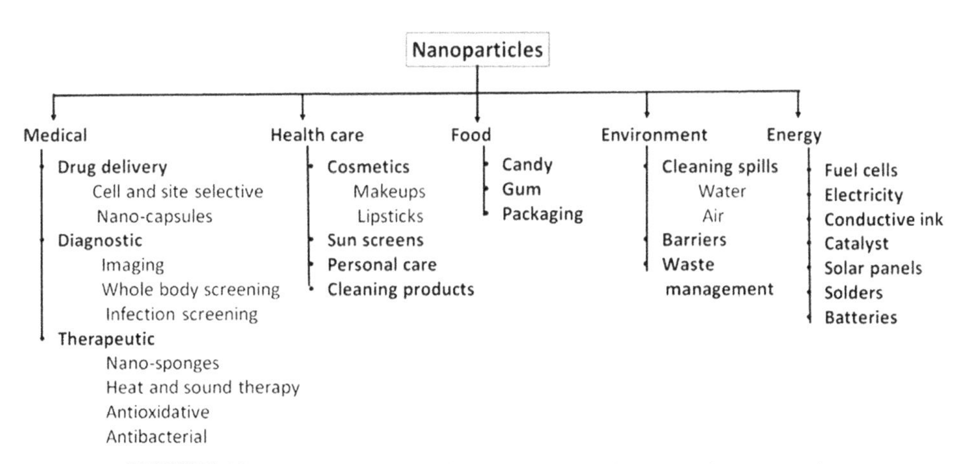

FIGURE 12 The flow diagram shows various applications of nanoparticles.

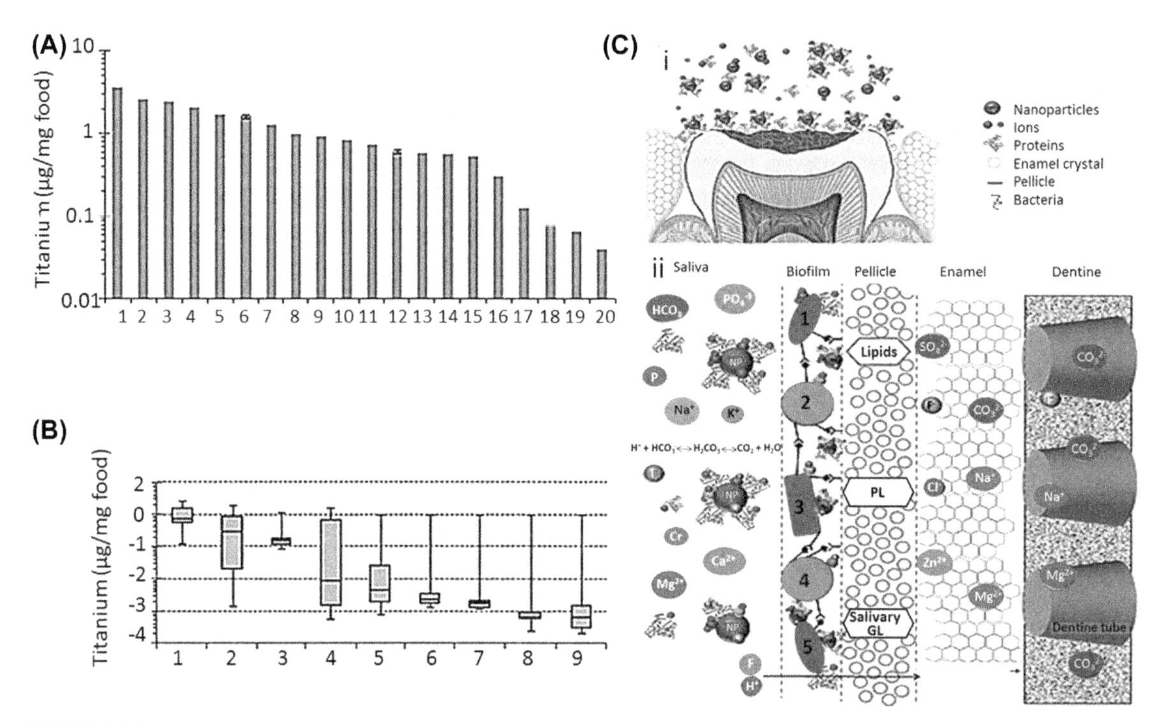

FIGURE 13 (A) Titanium dioxide nanoparticles in food and personal care products. (B) Diagram shows the presence of nanoparticles (isolated particles or agglomerates) in saliva and the structure of dental tissues. (1) The pellicle covers the superficial layer of enamel, and the oral biofilm develops on the pellicle surface. The nanoparticle–ion–protein complexes do not adhere directly to the tooth surfaces, but adhesion occurs either to the pellicle layer or to the developing biofilm. (C) Distribution of nanoparticles in the oral cavity (i), oral biofilm, and dental mineralized tissues (ii). Oral conditions promote particle agglomeration that results in particle sedimentation onto the dental surfaces. The oral biofilm and pellicle act as diffusion/permeation barriers to nanoparticles preventing them from reaching the enamel–pellicle interface. Distribution of certain ions (F^-, Cl^-, SiO_4^{4-}, Zn^{2+}, Na^+, Mg^{2+}, CO_3^{2-}, F^-, Na^+, Mg^{2+}, and CO_3^{2-}) is also shown. i1. Dickinsons Coconut Curd, i2. Mentos Freshmint Gum, i3. Hostess Powdered Donette, i4. Good and Plenty candy, i5. Kool Aid Blue Raspberry, i6. Eclipse Spermint Gum, i7. M&M Chocolate Candy, i8. Albertson Vanilla Pudding, i9. Betty Crocker Whipped Cream Frosting, i10. M&Ms Chocolate with Peanuts, i11. Trident White Peppermint Gum, i12. Jello Banana Cream Pudding, i13. Dentyne Ice Peppermint Gum, i14. Kool Aid Lemonade, i15. Mothers Oatmeal Iced Cookies, i16: Albertson Mini Marsh Mallows, i17: Dentyne Fire Spicy Cinnamon, i18. Vanilla Milkshake Pop Tarts, i19. Mentos Mints, i20. Nestle Original Coffee Creamer; ii1. Chewing Gum (n=5), ii2. Additive (n=10), ii3. Beverage (n=10), ii4. Candy (n=14), ii15: Baked Goods, ii6. Grains, ii7. Dairy (n=17), Sauce (n=6); iii1. *Streptococcus Sobrinus*, iii2. *S Oralis/S mitis*, iii.3. Lactobacillus, iii4. *S Sobrinus/S gordonii*, and iii5. *S sanguinis/S oralis. Reprinted from Weir et al. (2012) with permission.*

oversight or regulation of nanoparticle use in health and food products.

In 2003, President Bush signed the twenty-first Century Nanotechnology Research and Development Act, authorizing $3.7 billion over four years for the emerging science. The bill did not mention any need for improvements in the risk posed by nanoparticles. To address this issue, Dr John Marburger, science advisor, to the president and director of the Office of Science and Technology Policy, issued the following statement: "The risks from nanotechnology do not differ substantially from those of other technology hazards, such as toxicity of new chemicals or new biological materials, or environmental impact. I believe many of these concerns can be addressed with existing regulatory mechanisms. The new nanotechnology act specifies that these risks be investigated, widely discussed, and responsibly addressed."

Over a decade later, although the application of nanoparticles have skyrocketed, the risk assessment and public/environment health issues remain unattended. It is true that a review of the literature reveals extensive publications describing the adverse effects of nanoparticles on human and animal health and ecosystems. However, there are extensive knowledge gaps and uncertainties that have not allowed development of risk-assessment models for nanoparticles. The aim of the following sections is to discuss the progress that has been made in the area of public health and risk assessment of nanoparticles.

3.1 Nanoparticle Toxicology: A Brief Review

3.1.1 Physicochemical Property–Toxicity Relationship

Nanoparticles (1–100 nm diameter) are chemically diverse (metal, metal oxide, carbon, lipids, dendrimers, etc.) groups of particles exhibiting the chemical characteristics of the atoms/molecules they are made of, as well as certain common physicochemical properties that are not found in atoms, molecules (<1 nm), or bulk particles (>100 nm). Some of the key common properties of nanoparticles, not present in bulk particles, are described in the following sections.

3.1.1.1 SIZE-DEPENDENT PROPERTIES AND TOXICITY

At the nanorange (1–100 nm in at least one dimension), a decrease in the nanoparticle's diameter is associated with (1) an increase in its surface-to-volume ratio, surface reactivity (Figure 1), band gap between HOMO and LUMO, and aggregation potency; and (2) a decrease in electronic conductivity and cohesive energy. For bulk particles, the number of atoms/molecules on the surface is minuscule in comparison with the atoms/molecules in the core; thus, their properties are determined by the core atoms that are relatively constant (Figure 14(A)). In nanoparticles, as the particle size decreases, the ratio of surface to core atoms increases. At a certain threshold size (usually 30 nm), there are greater numbers of atoms at the surface than in the core. At or below the threshold size, the surface atoms determine the physiochemical properties of the particle. Thus, a small change in size causes an exponential

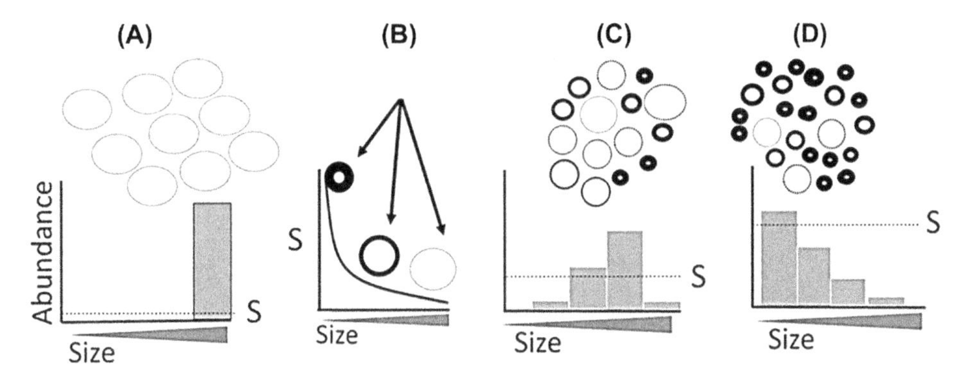

FIGURE 14 Relationship between nanoparticles' diameter and surface activity (the sphere thickness represents surface activity: smaller the particle, greater the surface activity. In the case of specimens containing bulk particles, relatively homogenous size exhibits poor surface activity (dashed line, S), possibly due to a smaller number of atoms at the surface (A: bulk). In the case of specimens containing homogenous nanoparticles, their surface activity (S) will negatively correlate with the particle size (14B: nanoparticles). However, in the case of heterogeneous nanoparticle population (C and D), the one containing a greater proportion of smaller particles (D) has more reactive surface than the one containing a greater proportion of larger particles (C).

change in the physicochemical properties, and their surface activity (S) negatively correlates with the particle size (Figure 14(B)). In heterogeneous groups of nanoparticles (Figure 14(C) and (D)), those containing a greater proportion of smaller particles (Figure 14(D)) have more reactive surfaces than those containing a greater proportion of larger particles (Figure 14(C)).

3.1.1.2 SHAPE-DEPENDENT PROPERTIES AND TOXICITY

Shape is one of the key determinants of nanoparticles' properties, biological effects, and toxicity (Wang et al., 1999; Johnson et al., 2002; Xia et al., 2008). As shown in Figure 15, rod-like shapes of single-crystalline face-centered-cubic (fcc) metal nanocrystals are associated with the presence of predominantly lower density {100} and {110} type facets, whereas octahedral particle shapes are predominantly associated with higher density {111}-type facets. Because different crystallographic orientations of nanoparticle facets are associated with different interfacial energies and bond configurations, their facets have been shown to play a crucial role in determining the chemical activity and properties of nanoparticle systems (Thomas, 2008; Kundu et al., 2009).

3.1.1.3 SURFACE-DEPENDENT PROPERTIES AND TOXICITY

In addition to the shape and size, the nanoparticles' surface also dictates their properties, such as solubility and macromolecule and cell-surface interactions. The nanoparticle surface is highly reactive and catalyzes numerous chemical reactions (Figure 16), including generation of reactive oxygen species (ROS) and, in some cases, neutralization of ROS. Incubation of nanomaterials with cells in media may results in reversible adsorption on their surface (formation of a corona), which further modifies their properties and may induce their entry into the cells by receptor-mediated endocytosis (Khan et al., 2007). To prevent nonspecific interactions, nanoparticles are coated with a neutral ligand such as poly(ethylene glycol) (PEG), which is well known to resist protein adsorption (Xie et al., 2007; Hu et al, 2007; Sant et al., 2008). Hydroxyl-coated quantum dot surfaces have a significant reduction over PEG-coated analogs in nonspecific cellular binding (Kairdolf et al., 2008). Neutral surface with zwitterionic ligands exhibits relatively low cell interactions and uptake (Kim et al., 2008). These studies show that the surface plays a significant role in determining a nanoparticle's properties and toxicity.

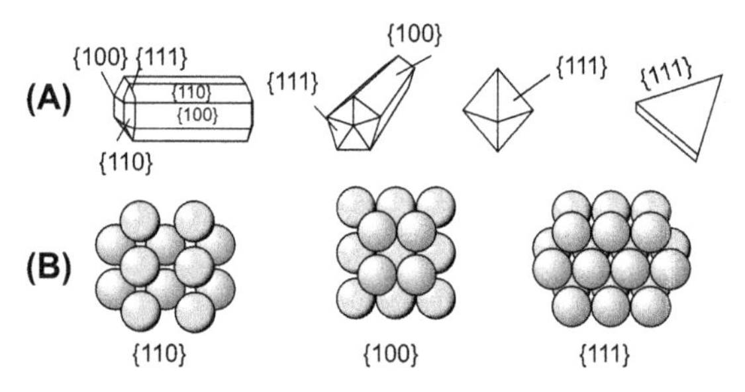

FIGURE 15 Different shape indexes of nanoparticles.

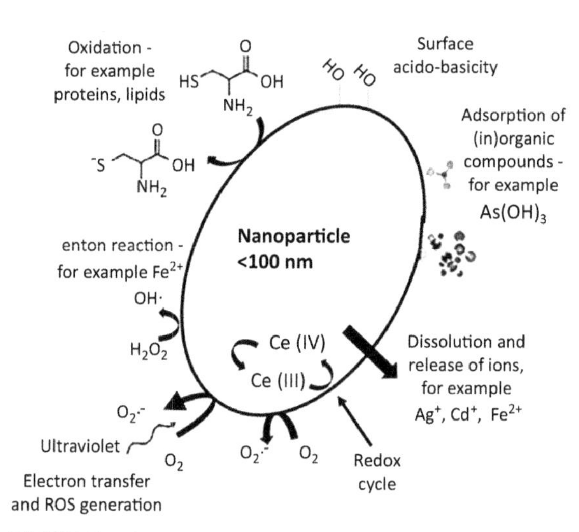

FIGURE 16 Chemical reaction including formation of reactive oxygen species mediated by the nanoparticles' surface.

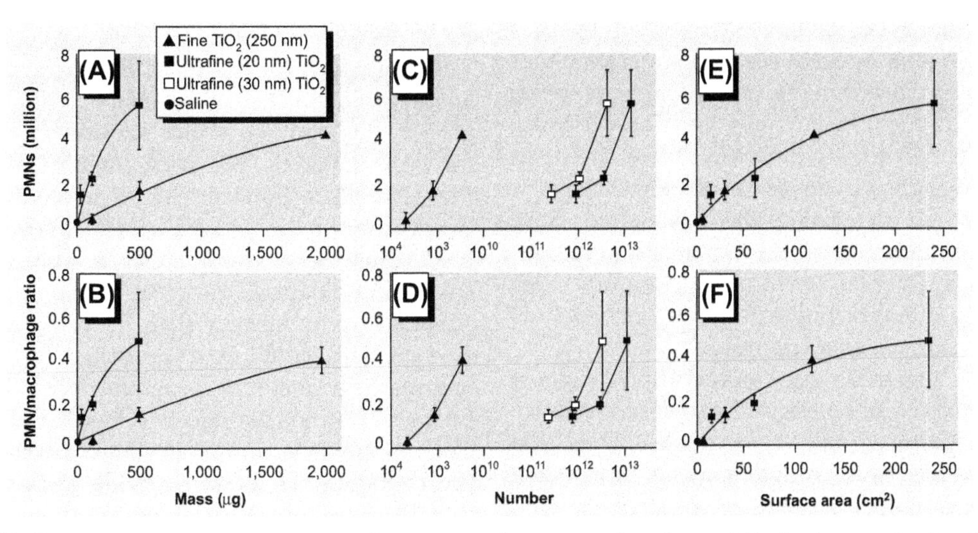

FIGURE 17 Dose metrics for nanoparticles. Inflammatory cell response in lung lavage 24 h after intra-tracheal instillation of fine (~250 nm) and ultrafine (20–30 nm) TiO_2 expressed by different dose metrics [particle mass (A, B), number (C, D), and surface area (E, F)] and different response metrics [number of PMNs (A, C, E) and PMN/macrophage ratio (B, D, F)]. *Reprinted from Oberdörster et al (2007) with permission.*

3.1.1.4 THE DOSE METRIC

A key problem associated with nanoparticles is that the dose in terms of mass per unit body weight (e.g., mg/kg) may not correlate with their adverse effects (Monteiller et al., 2007; Warheit et al., 2007; Oberdörster, 2002). Thus, the risk assessment indices calculated using the dose (in terms of mass)–response curve may not accurately reflect a nanoparticle's risk. Dose in terms of specific surface area or surface activity may be more important (Oberdörster, 2002). As shown in Figure 17, the use of surface area, but not mass or number metric, as the dose metric provided an identical inflammatory response for different-sized particles (Oberdörster et al., 2007). However, Wittmaack (2007) suggested the contrary: particle number, not surface area, is the better dose metric. Wittmaack (2007) also suggested that the mass-based surface area is a better dose metric than Brunauer, Emmett, and Teller (BET) method-determined surface area.

As shown in Figure 18, there appears to be an inverse relationship between particle size and the normalized surface toxicity (S_{pd}, SootL, and SootH). However, PrintexG (PrtxG) diameter was larger (50 nm diameter) than Printex (Prtx) 90 diameter (15 nm diameter), but PrtxG had several-fold greater normalized surface toxicity than Prtx90. Wittmaack (2007) attributed this to

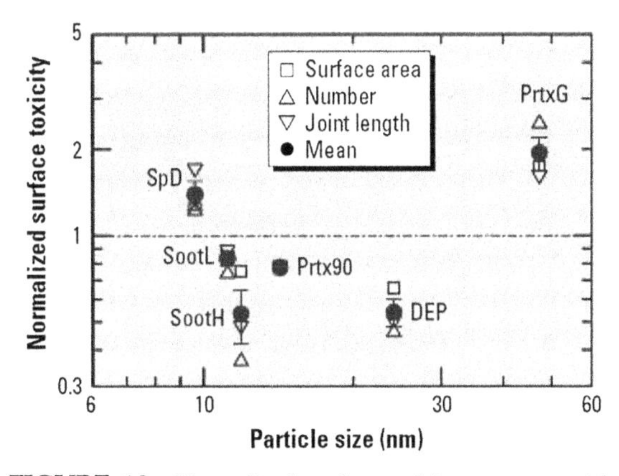

FIGURE 18 Normalized surface toxicity versus particle size. Solid circles are mean ± SD. *Reprinted from Wittmaack (2007) with permission.*

the fact that PrtxG and Prtx90 were synthesized using different procedures. He suggested that identical particles made of the same material but by different techniques are not generally suited for identifying the proper dose metric in studies designed to determine the effect of particle size. To determine the best-suited dose metric, particles of the same material and with sufficiently similar surface toxicity, but with distinctly different physical parameters such as size and surface area, are needed.

Taken together, these observations suggest considerable differences in the surface reactivity of bulk particles and nanoparticles. Thus, the bulk particle toxicity data may be unsuitable for determining nanoparticles' public or environmental health risks.

3.1.2 Mechanisms of Toxicity

In nanoparticles less than 30 nm, the percentage of atoms in the core is insignificant in relation to the number at the surface. Contrarily, for bulk materials, the percentage of atoms at the surface is insignificant in relation to the number of atoms in the core (Nel et al., 2006). Because the surface atoms/molecules are more reactive than those in bulk particles, a nanoparticle's surface is more reactive than the bulk particle's surface. This could be responsible for greater nanoparticle–membrane interactions, leading to their toxicological effects (Oberdörster et al., 2005; Donaldson and Tran, 2002) by inducing the formation of the superoxide radical (O_2^-) and ROS (Figure 19) (Shvedova et al., 2005). ROS accumulation is currently the best-

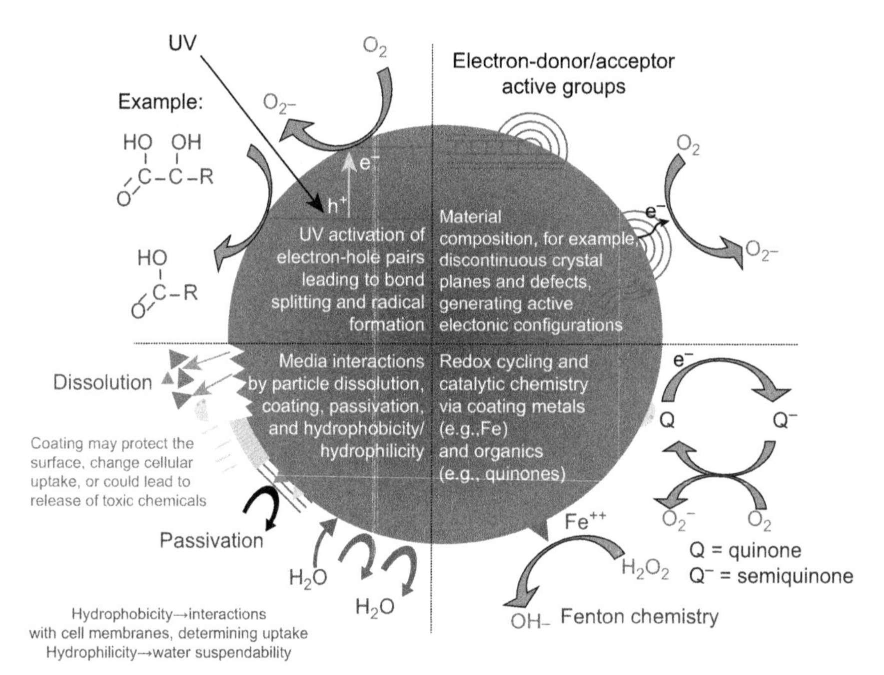

FIGURE 19 This sketch shows possible mechanisms by which nanomaterials interact with biological tissue. Possible roles of the material composition, electronic structure, bonded surface species (e.g., metal-containing), surface coatings (active or passive), and solubility, including the contribution of surface species and coatings and interactions with other environmental factors (e.g., UV activation), have been addressed. Nanoparticle surface can lead to toxicity by inducing the formation of the superoxide radical (O_2^-) and ROS that is currently the best-developed paradigm for nanoparticle toxicity. *Reprinted from Nel et al. (2006).*

developed paradigm for nanoparticle toxicity (Table 3) (Nel et al., 2006).

In vitro (in cell cultures) and in vivo (exposure to animals) exposure have been used in toxicological research studying the health effects of nanoparticles (Choi et al., 2015; El-Sayed et al., 2015; Ng et al., 2015; Smulders et al., 2015; Sultana et al., 2015; Adamcakova-Dodd et al., 2012; Kim et al., 2011; Grassian et al., 2007a,b, Elder et al., 2006; Elder, 2009; Oberdörster et al., 2004, 2000 and Frampton, 2001). Toxicity due to the occupational exposure of nanoparticles has been studied by Cho et al. (2012), McKinney et al. (2012), Murphy et al. (2012), Pacurari et al. (2012), Knuckles et al. (2011), Cho et al (2006), Shvedova et al. (2008), Grassian et al. (2007b), and Donaldson et al. (2006) Kim et al., 2010. Oberdörster et al. (2004, 2000) and Oberdörster (2001) have shown that, irrespective of the mode of exposure, nanoparticles may translocate from one organ to another and the significance of particle surface area concentration is a valid dose metric for nanoparticle toxicity. O'Shaughnessy (2013) generalized nanoparticle toxicity as follows:

1. In the nanorange, smaller particles can be more toxic than larger particles on a per mass basis.
2. Surface reactivity may be associated with particle surface area rather than mass concentration for certain nanoparticles.

TABLE 3 Mechanisms of Action of Nanoparticles

Experimental NM effects	Possible pathophysiological outcomes
ROS generation*	Protein, DNA, and membrane injury,* oxidative stress†
Oxidative stress*	Phase II enzyme induction, inflammation,† mitochondrial perturbation*
Mitochondrial perturbation*	Inner membrane damage,* permeability transition (PT) pore opening,* energy failure,* apoptosis,* apo-necrosis, cytotoxicity
Inflammation*	Tissue infiltration with inflammatory cells,† fibrosis,† granulomas,† atherogenesis,† acute phase protein expression (e.g., C-reactive protein)
Uptake by reticulo-endothelial system*	Asymptomatic sequestration and storage in liver,* spleen, lymph nodes,† possible organ enlargement, and dysfunction
Protein denaturation, degradation*	Loss of enzyme activity,* auto-antigenicity
Nuclear uptake*	DNA damage, nucleoprotein clumping,* autoantigens
Uptake in neuronal tissue*	Brain and peripheral nervous system injury
Perturbation of phagocytic function,* "particle overload," mediator release*	Chronic inflammation,† fibrosis,† granulomas,† interference in clearance of infectious agents†
Endothelial dysfunction, effects on blood clotting*	Atherogenesis,* thrombosis,* stroke, myocardial infarction
Generation of neoantigens, breakdown in immune tolerance	Autoimmunity, adjuvant effects
Altered cell cycle regulation	Proliferation, cell cycle arrest, senescence
DNA damage	Mutagenesis, metaplasia, carcinogenesis

Effects supported by limited experimental evidence are marked with asterisks, while effects supported by limited clinical evidence are marked with daggers.
Reprinted from Nel et al (2006).

3. Nanoparticle inhalation can lead to oxidative stress and a subsequent chain of events that produce adverse health effects.
4. Physicochemical characteristics, especially surface functionalization and dissolution rate (in lung surfactant), are significant determinants of nanoparticle toxicity.

Interaction of nanoparticles with the biological system at the molecular, cellular, organ, and whole body levels is determined by the unique physicochemical properties of nanoparticles (Vincent, 2011). However, further research is needed to characterize human and ecological risk assessment of nanoparticles.

3.1.3 Uncertainties Associated with Nanoparticle Toxicity Data

A review of nanoparticle toxicity data reveals a large number of knowledge gaps or uncertainties that must be addressed when developing a risk assessment. The uncertainties arise at all aspects of the risk-assessment process (Singh, 2015; Kandlikar et al., 2007), as shown in Figure 20 and described below.

3.1.3.1 METHODOLOGICAL UNCERTAINTIES

The surface area of a nanoparticle is determined either mathematically from mass values or experimentally using the BET method. However, surface areas determined mathematically and experimentally do not correlate, causing the first methodological uncertainty ($U1_{np}$). The second uncertainty ($U2_{np}$) is associated with dynamic light scattering (DLS), a commonly used method to measure the size of nanoparticles in aqueous solutions in situ by extrapolating the apparent diffusion coefficients at different scattering angles and the sample concentration to infinite conditions. There are uncertainties in the measurement related to the extrapolating

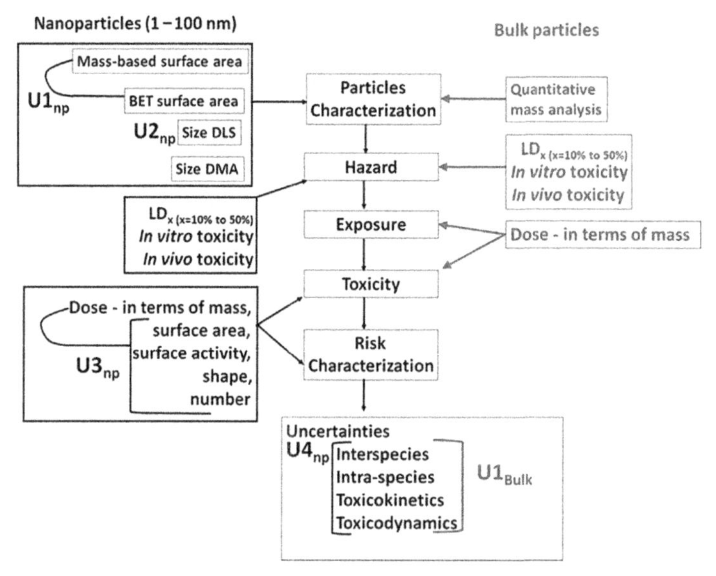

FIGURE 20 Uncertainties associated with bulk particles (red font—gray in print versions, right) and nanoparticles (black font, left). For bulk properties are associated with inter- and intraspecies conversions including toxicokinetic and toxicodynamic data ($U1_{bulk}$) (Dorne and Renwick, 2005). Nanoparticle risk-assessment encounters many levels of uncertainties: $U1_{np}$ deals with surface area determination, $U2_{nm}$ deals with size determination, $U3_{np}$ deals with different dose metrics and $U4_{np}$ is the same as the $U1_{bulk}$. In addition, formation of reversible protein-lipid corona adds to an uncertainty that cannot be resolved at present (see Figure 25).

procedure used, possibly because of the presence of a number of unevaluated factors (Kwon et al., 2011; Takahashi et al., 2008). The DLS-related uncertainty increases as the particle size decreases, reaching >20% for 20-nm nanoparticles. A major uncertainty ($U3_{np}$) is associated with the dose metrics. Earlier studies have mostly used dose as the metric for dose–response studies. However, there is sufficient evidence indicating that surface area or reactivity, not mass, is the preferred index for determining the dose–response relationships. An uncertainty factor is needed to be applied to account for this uncertainty.

3.1.3.2 UNCERTAINTIES ASSOCIATED WITH PARTICLE CHARACTERIZATION

Singh (2015) described the fate of a nanoparticle in a biological system. Bare nanoparticles are highly reactive and, if not stabilized, may aggregate (Figure 21(1)). Functionalization with PEG (Figure 21(2-i)) and/or functional groups (Figure 21(2-ii)) stabilizes the particles and adds desired functionalities to it. In an aqueous solution, functionalized nanoparticles may remain dispersed but interact reversibly with some proteins (Figure 21(3-iii), Singh, 2015). Administration of functionalized particles in an organism may alter the nanoparticles' characteristics by (1) replacing the external protein (the released proteins may adversely affect the organisms) with endogenous proteins (Figure 21(4-v)); (2) removing the functional groups (defunctionalization) from nanoparticles (Figure 21(5, 6, 8, and 9)); and (3) causing protein exchange in or formation of a protein corona around the defunctionalized nanoparticles (Figure 21(10–12)). As the nanoparticles translocate, the composition of the bound proteins changes. At this stage, a nanoparticle's characteristics are determined by the bound proteins and not by the nanoparticles' surface. These changing characteristics and ensuing changes in toxicity are major uncertainties because the current methodologies may not be able to track the changes in a protein corona in vivo.

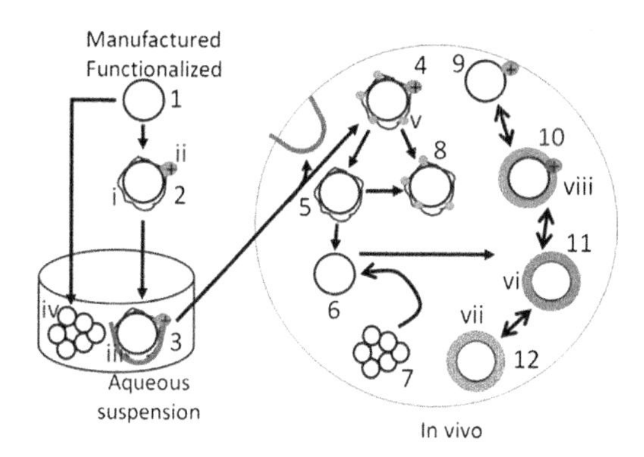

FIGURE 21 Interaction of bare and functionalized nanoparticles with organism/environmental components. (1) Bare nanoparticle. (2) functionalized (PEG (i) and positively charged (ii) group) nanoparticles. (3) Functionalized nanoparticles in aqueous solution where proteins may form corona ((iii) green line—dark gray line in print versions) or bare nanoparticles may form aggregates. (4) When administered in an organism, the media protein dissociated and the intracellular protein for surface corona (v: represents binding of functional groups onto the PEG). (5) Partial defunctionalization of nanoparticles—loss of positively charged functional group. (6) Loss of PEG results in the release of bare nanoparticles. (7) Bare nanoparticles aggregate into larger particles. Under certain circumstances, aggregated nanoparticles may disaggregate. (8) Loss of the positive functional group but presence of PEG and associated proteins. (9) formation of bare nanoparticles with positively charged group. This may not aggregate due to particle repulsion. (10) Formation of protein corona in particle-9. (11 and 12) The composition of protein corona may change as the particles move from one organ to another. At this stage, the corona, not the surface, may determine the particles' properties.

3.1.3.3 UNCERTAINTIES ASSOCIATED WITH AGGLOMERATION

As discussed above, nanoparticles have a tendency to agglomerate and form larger structures of varying sizes, resulting in a reduction in the number of atoms at the surface with a reduction in surface energy and reactivity (Kandlikar et al., 2007). Thus, agglomeration may result in unpredicted changes in toxicity. Because agglomeration is rapid (tens of microseconds to a few milliseconds, Preining, 1998), nanoparticle concentrations can decrease rapidly. Nanoparticle functionalization reduces agglomeration in the body. This may increase the potential for human inhalation exposures, affecting their disposition in the body (Araujo et al. (2008); Oberdörster et al., 1994). Because agglomeration properties of many engineered nanoparticles are unknown, it cannot be accounted for during risk determination.

3.1.3.4 UNCERTAINTIES ASSOCIATED WITH PARTICLE SHAPE

Engineered nanoparticles come in various shapes, such as spheres, dendrimers, nanotubes, nano-clay flakes, fullerenes, capsules, needles, etc. Although the precise effects of shape on nanoparticle toxicity are not known, earlier studies have shown that the particles' shape (i.e., their length and diameter) has a profound effect on toxicity: smaller diameter fibers penetrate deeper into the respiratory tract, while longer fibers are cleared more slowly (Iman and Helton, (1988), Mossman et al., 1990). Ding et al. (2012), using computer modeling, have shown that penetration of nanoparticles through the cell membrane was size and shape dependent (Figure 22). As the size increased, the depth of

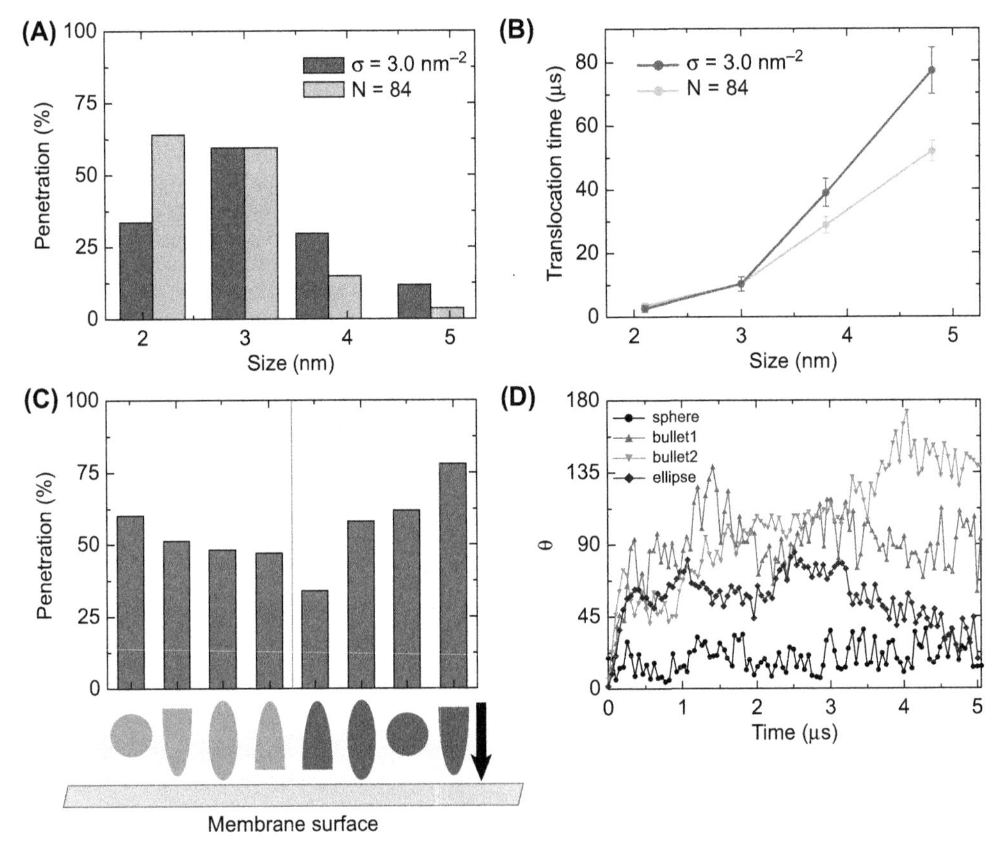

FIGURE 22 Effect of the nanoparticles' size and properties on their penetration efficiency (A) and translocation time (B). Red line (dark gray line in print versions) shows the case of the same ligand surface density, while the green line (gray line in print versions) denotes the same ligand number on the surface. (C) Penetration efficiency of nanoparticles with different shapes and fixed volume (ellipse with half lengths of 1.05, 1.05, and 3.0 nm in three axes; sphere with radius of 1.5 nm; bullet with half lengths of 1.2, 1.2, and 4.5 nm in three axes). The orange (light gray in print versions) nanoparticles can rotate freely, while the rotation of red nanoparticles is fixed during the penetration process. (D) Time evolution of different (orange) nanoparticles' orientations during the penetration processes. *Reprinted from Ding et al. (2012).*

penetration decreased (Figure 22(A)), while the translocation rate increased (Figure 22(B)). Figure 22(C) shows the penetration of differently shaped (flexible rotation and fixed rotation) nanoparticles through a membrane. Except for the sphere, the flexible nanoparticles exhibited greater penetration than fixed-rotation nanoparticles. Bullet-shaped nanoparticles with the flat side facing the membrane exhibited the least penetration, while the flexible bullet with the tip facing the membrane exhibited maximum penetration. Because nanoparticle penetration is a prerequisite in toxicity development, differently shaped nanoparticles may exhibit different toxicity. However, this is a major information gap needing more research.

3.1.3.5 UNCERTAINTIES ASSOCIATED WITH EXPOSURE METRICS

The commonly used metrics for bulk particle exposure are mass per body weight (g/kg_{bw}) for injection or oral exposure, surface area (g/m^2) for dermal exposure, or g/m^3 for inhalation exposure. However, for nanoparticles, surface area (example: (diffusion charge (DC)/μm^2) cm^{-3} or (photoelectric aerosol sensor (PAS)/ ng PAH) m^{-3}) or particle number (particle cm^{-3}) are more appropriate metrics than mass. Ramachandran et al. (2005) evaluated the exposures of three occupational groups (bus garage mechanics, bus drivers, and parking booth attendants) using the following metrics: traditional mass concentration, surface area, and number concentration of elemental carbon. They observed that the three groups had statistically significantly different exposures to elemental carbon:

1. In terms of mass, the highest levels were observed in the bus garage mechanics (almost entirely diesel exhaust) and the lowest levels in the parking ramp booth attendants.
2. In terms of surface area concentrations measured as DC, the average exposures of parking ramp attendants are significantly greater than those of bus garage mechanics, which in turn are significantly greater than those of bus drivers.
3. In terms of PAS measurements, the ramp booth attendants had the highest mean exposures, while the exposures for bus drivers and garage mechanics were comparable.
4. In terms of number concentrations, the exposures of garage mechanics exceeded those of ramp booth attendants by a factor of 5–6.

These observations suggest that, depending on the exposure metric chosen, the three occupational groups have quite different exposure rankings, although it is still unclear which exposure metric is the most relevant to human health. Thus, the choice of the exposure metric will affect the classification of workers into similarly or differently exposed groups in future epidemiological studies.

3.1.3.6 UNCERTAINTIES ASSOCIATED WITH TRANSLOCATION

An important difference between nanoparticles and bulk particles is the ability of nanoparticles to move or translocate to different parts of the body (Figure 23). For example,

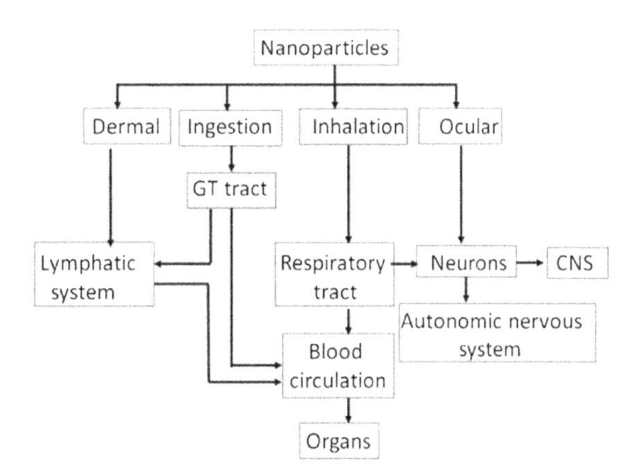

FIGURE 23 The flow diagram showing translocation of nanoparticles from the site of exposure to distant tissues/ organs.

inhaled nanoparticles may translocate from the alveolar spaces into the interstitium (Oberdörster et al., 2005), regional lymph nodes (Oberdörster et al., 1994), and the circulatory system (Nemmar et al., 2002; Kreyling et al., 2002). Inhaled nanoparticles may also translocate to the central nervous system, either through neural pathways in the olfactory system or through the circulatory systems. A critical point is that nanoparticles can translocate to parts of the body that are not accessible to bulk particles. Thus, it is difficult to extrapolate the known laboratory and epidemiological findings for bulk particles to nanoparticles. There is considerable uncertainty regarding the extent of translocation of nanoparticles.

Putting it all together, there are several information gaps and uncertainties that pervade every element of the exposure–response–risk paradigm for nanoparticles. Considerable research is needed to fill these gaps before rigorous, specific guidelines or working practices can be fully developed. Maynard and Kuempel (2005) has proposed that the following issues must be addressed:

- Release of inhalable and respirable particles into the air during manufacturing, handling, storage, or cleanup
- Characteristics of released airborne nanoparticles, such as diameters, nanostructure, surface area, unique surface chemistry, and other size- and structure-related properties, which may lead to differences in hazards when compared to that for the component chemicals
- Characteristics of released aerosol particles that indicate the use of exposure metrics other than mass-based metrics
- Characterization of dose metrics
- The measures that can be taken to characterize and reduce or eliminate exposure

3.2 Current Risk Characterization Strategies

Because nanotoxicology is a developing field, information about nanoparticle toxicity contains substantial knowledge gaps. Therefore, the nanoparticle risk assessment is performed using the following approaches: (1) when the toxicity data are adequate, (2) when the data is limited that requires inclusion of uncertainty factors, and (3) when data is sparse, which requires a pragmatic approach relying on probability-based models and professional judgment (Figure 24).

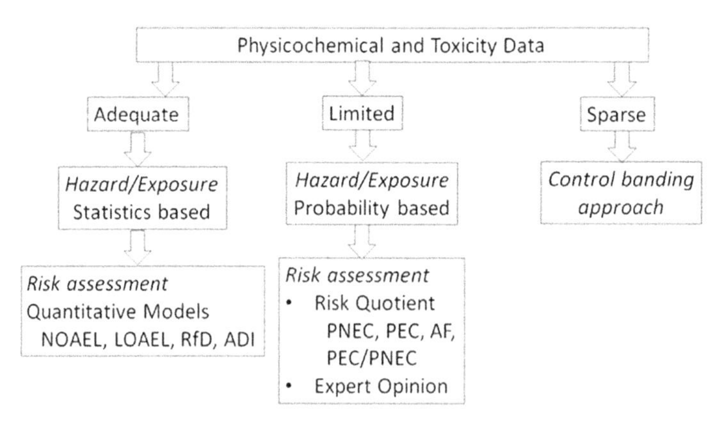

FIGURE 24 Flow diagram showing risk-assessment strategies in case of adequate, limited and spears data.

3.2.1 *Risk Evaluation with Adequate Data (Relatively Few Knowledge Gaps)*

In 2007, the US EPA published a nanotechnology white paper to establish the science needs associated with nanoparticle risk assessment and management (http://www.epa.gov/osainter/pdfs/nanotech/epa-nanotechnology-whitepaper-0207.pdf). Thereafter, considerable changes have been made in the data requirements, as described below. It should be noted that each nanoparticle preparation may need to be considered as an individual toxin.

1. Physicochemical properties of nanoparticles
   a. Size, specific surface area, surface reactivity, and particle enumeration
   b. Surface functionalization
   c. Stability (aggregation and defunctionalization) inside an organism or an ecosystem
2. Environmental characterization
   a. Ecosystem structure (biotic and abiotic)
   b. The food chain and food web from producers to top-level predators
   c. Nanoparticle distribution—seasonal changes
3. Sources, fate, transport, and exposure
   a. Sources from a life-cycle perspective (manufacturing, processing, application, and disposal)
   b. Mode of exposure (ingestion, dermal, inhalation, injection/medicinal use, gills in the case of fish)
   c. Disaggregation and release of free nanoparticles in an organism or the ecosystem
   d. Nanoparticles' fate in an organism (translocation, transformation, storage, corona formation, defunctionalization, excretion and metabolism, detoxification, and activation)
   e. Transport (within an organism or ecosystem and in the environment)
   f. Toxicokinetics and toxicodynamics
4. Hazard characterization
   Toxicology inherent to the functionalized nanoparticle that is being investigated. The indices may include.
   a. Lethal dose$_x$ (LD_x) or lethal concentration$_x$ (LC_x), where x is the percentage of animals that died for a given dose (in terms of appropriate metrics)
   b. Effective dose$_x$ (ED_x) or lethal concentration$_x$ (EC_x), where x is the percentage of animals that died for a given dose (for specific toxicity indices such as reproduction, nervous system, respiratory system, immune system, etc.)
   c. Other in vitro toxicity and/or in vivo toxicity indices
   d. Structural indices, such as ecosystem biota population density
   e. Extinct species
   f. Food chain alteration
   g. Most sensitive species
   h. Resistance
   i. Resilience
   j. Other ecosystem toxicity indices

Appropriate uncertainty factors should be applied in the case of a data gap. In human risk assessment, adequate data are available for CNTs and TiO_2 nanoparticles. Table 4 lists selected studies that describe data for occupational exposure to these nanoparticles. The data include various interspecies scaling factors and/or uncertainty factors. If data are adequate, then the strategies described for bulk particles can be used to determine NOAEL and LOAEL, as well as RfD and ADI values.

3.2.2 *Risk Evaluation with Deficient Data (Relatively Large Knowledge Gap)*

In human risk assessment, occupational exposure can be modeled using NOAEL and/or LOAEL. The NOAEL for occupational TiO_2 exposure was $1-3\ mg/m^3$ Shi et al. (2013) have reported the NOAEL for TiO_2 to be around

TABLE 4 Summary of Aquatic Threshold Concentrations for the Photosynthesis-Inhibiting Herbicides Metribuzin and Metamitron, Derived from Different Risk-Assessment Procedures

	Metribuzin (μg/L)	Metamitron (μg/L)
First-tier Maximum Permissible Concentration NL (1/10 * NOEC for most sensitive standard alga	0.2	8.4
First-tier Maximum Permissible Concentration EU (1/10 * EC_{50} for most sensitive standard alga	1.2	85.2
HC_5 chronic toxicity data for primary producers	1.4	–
HC_{10} chronic toxicity data for primary producers	2.1	–
NOEC enclosure study (single application)	5.6	280
HC_5 acute toxicity data for primary producers	7.4	667
HC_{10} acute toxicity data for primary producers	10.0	763
LOEC enclosure study, functional (single application) (Class 2 effect on community metabolism)	18.0	1120
LOEC enclosure study, structural (single application) (Class 2 effect on community structure)	18.0	>4480

Data derived from Brock et al. (2004).

1.2 mg/m^3 respirable dust in the case of a hypothetical 8-h day, 5-day working week. The American Conference of Governmental Industrial Hygienists has assigned TiO_2 fine particles (total dust) a threshold limit value (TLV) of 10 mg/m^3 as a time-weighted average (TWA) for a normal 8-h workday and a 40 h workweek (Shi et al., 2012). Permissible exposure limit (PEL)-TWA of the Occupational Safety and Health Administration for TiO_2 FPs is 15 mg/m^3 (CGIH, 2001). The United States National Institute for Occupational Safety and Health proposed a recommended exposure limit (REL) for TiO_2 nanoparticles at 0.3 mg/m^3, which was 10 times lower than the REL for TiO_2 FPs (Kitchin et al., 2010; NIOSH, 2011). The New Energy and Industrial Technology Development Organization (NEDO) in Japan has set the acceptable exposure concentration of TiO_2 nanoparticles to be 1.2 mg/m^3 as a TWA for an 8-h workday and a 40-h workweek (Morimoto et al., 2010). Compared to TiO_2, CNTs' NOAEL ranged from 0.1 to 0.185 mg/m^3 (Table 5). Pauluhn (2010) and Aschberger et al. (2011) have calculated SWCNT's estimated occupational exposure in humans to be 50 μg/m^3 and 2 μg/m^3, respectively, using an identical NOAEL value (0.1 mg/m^3). This difference may be because they applied different uncertainty factor (2 and 5, respectively). For MWCNTs, starting from the NOAEL in a 28-day rat intratracheal instillation study, Nakanishi (2011) derived an exposure level of 30 μg/m^3 for a 15-year exposure period.

In cases of environmental exposure, the risk assessment to the ecosystem becomes complicated and, depending upon the situation, different methods may be needed, as described below.

- *Equilibrium partitioning sediment benchmark (ESB) method*: The ESB is based on equilibrium partitioning (EqP) theory that a nanoparticle in sediment partitions between the sediment's organic carbon, the interstitial (pore) water, and the benthic organisms (Figure 25). At

TABLE 5 Examples of Proposed Occupational Exposure Limits (OELs) for Nanomaterials and the Associated Exposure Control Bin

Substance	OEL ($\mu g/m^3$)[a]	Basis	References	Exposure control bin ($\mu g/m^3$)
TiO_2—ultrafine	610[b]	Estimated human-equivalent concentration to rat subchronic estimated NOAEL[c] of 2 mg/m^3 (Bermudez et al., 2004), UF of 3	Gamo, (2011), Nakanishi (2011)	100–1000
TiO_2—ultrafine	300	Working lifetime (45-year) excess risk <1/1000 (95% LCL estimate) based on the benchmark dose model average estimate of particle surface area dose and rat lung tumor response. MOA: secondary genotoxicity from persistent pulmonary inflammation	NIOSH (2011a,b,c)	
Fullerene (C_{60})	390[b]	Estimated human-equivalent concentration to rat NOAEL of 0.12 mg/m^3 (from 28-day inhalation, estimated subchronic equivalent), UF of 9	Shinohara (2011), Nakanishi (2011)	
MWCNT	50	Estimated human-equivalent rat NOAEL of 0.1 mg/m^3 by breathing rate, exposure time, deposition, alveolar macrophage volume, and retention kinetics; no UF	Pauluhn (2010)	10–100
CNT	30[b]	Estimated human-equivalent concentration to rat 28-day NOAEL of 0.065 mg/m^3 for SWCNT and 0.185 for MWCNT. Proposed the lower SWCNT value for all CNT[d], UF of 3	Nakanishi (2011)	
CNT and CNF	7 (draft)	Set at LOQ of the measurement method. Working lifetime (45 year) excess risk >10% (95% LCL estimate) of early-stage pulmonary inflammation or fibrosis in rat or mouse short-term or subchronic studies	NIOSH (2010c)	1–10
MWCNT	1–2	Adjusted rat NOAEL or LOAEL of 0.1 mg/m^3 (Pauluhn, 2010; Ma-Hock et al., 2009, respectively) for exposure day and breathing rate, UF of 25 or 50	Aschberger et al. (2010)	

Abbreviations: TiO_2, titanium dioxide; MOA, mode of action; UF, uncertainty factor.
[a] *8-h TWA concentration.*
[b] *OEL (PL): 15-year period-limited OEL. Rat to human adjustments: breathing rate, exposure time, deposition, and body weight.*
[c] *Bermudez et al. (2004) report responses in rat at 2 mg/m^3, suggesting that it could be interpreted as a LOAEL.*
[d] *An OEL (PL) of 0.08 mg/m^3 was also derived for an MWCNT (44-nm diameter); however, the OEL of 0.03 mg/m^3 for an SWCNT with the largest specific surface area was proposed as the common value for CNT.*
Reprinted from Kuempel et al. (2012).

equilibrium, if the concentration in any one phase is known, then the concentrations in the others can be predicted. The ratio of the concentration in water to the concentration in organic carbon is termed the organic carbon partition coefficient (K_{OC}), which is a constant for each chemical. EqP approach to establishing sediment benchmarks (http://www.epa.gov/nheerl/download_files/publications/dieldrin.pdf) is as follows:

- The concentrations of nonionic particles in sediments, expressed on an organic carbon basis, and in interstitial waters correlate with the observed biological effects on sediment-dwelling organisms across a range of sediments (Figure 25).

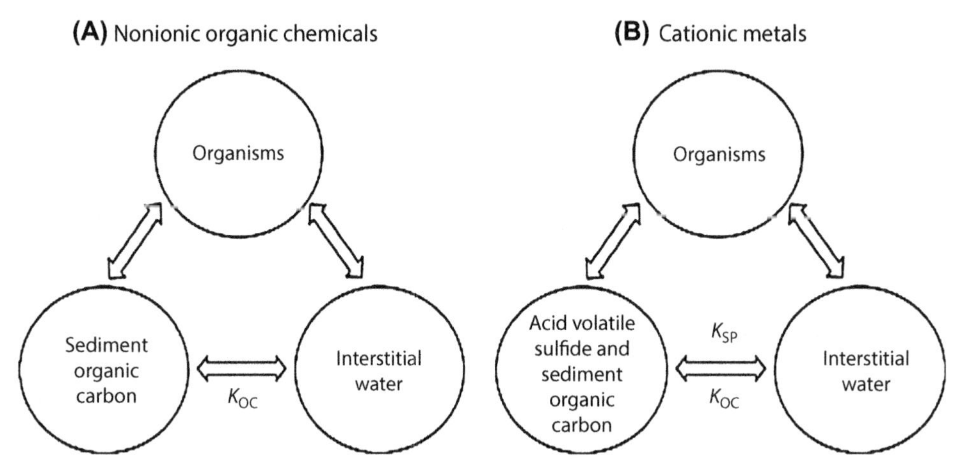

FIGURE 25 Principle of the equilibrium partitioning sediment benchmark for hydrophobic and cationic particles. The concentrations of nonionic particles in sediments expressed on an organic carbon basis and in interstitial waters correlate with the observed biological effects on sediment-dwelling organisms. The concentrations of ionic particles in sediments acid volatile sulfide and sediment organic carbon and in interstitial waters correlate with the observed biological effects on sediment-dwelling organisms across a range of sediments. OC: organic carbon, SP: solubility parameter and K is a constant. *Reprinted from Bondarenko et al (2013).*

- Partitioning models can relate sediment concentrations for nonionic particles on an organic carbon basis to freely dissolved chemical concentrations in interstitial water.
- The distribution of sensitivities to nanoparticles of benthic organisms is similar to that of water column organisms; thus, the currently established Water Quality Criteria WQC, Final Chronic Values FCV or Secondary Chronic Values SCV can be used to define the acceptable effect concentration of a chemical freely dissolved in interstitial water.
  - *EqP of nonionic organic chemicals*: As shown in Figure 25, the freely dissolved interstitial water concentration is predicted using the organic carbon–water partition coefficient and Eqn (4), where C_{OC} is the organic carbon normalized sediment concentration (micrograms per kilogram OC) and C_d is the dissolved concentration (Burgess et al., 2013).

$$C_{OC} = K_{OC} \, C_d \tag{4}$$

$$\text{ESB} = K_{OC} \, \text{FCV} \left(\frac{1}{1000} \right) \tag{5}$$

$$\text{Log } K_{OC} = 0.00028 + OC \, 0.983 \text{ Log } K_{OW} \tag{6}$$

In Eqn (5), ESB is calculated by substituting a known effect concentration (e.g., the FCV or SCV) for the C_d and dividing with 1000. ESB is the organic carbon normalized particulate concentration for natural organic carbon NOCs below which or equal to which adverse effects are unlikely to occur and above which effects may occur. Because K_{OC} is not easily measured, Eqn (3) was developed using K_{OW} (liters per kilogram octanol), which is more easily measured to estimate K_{OC}. For this, Eqn (6) will be used (Di Toro et al., 1991).

  - *EqP of cationic metals*: In sediments, the sulfide conjugates relate most directly to the bioavailable concentration of metals. Sulfide conjugates are a significant sedimentary phase in reduced anoxic sediments, called acid volatile sulfide (AVS) (Di Toro et al., 2005, 1992, 1990). Primarily composed of amorphous FeS, AVS is the particulate sulfide fraction released from the sediment during a weak acid extraction. Cationic particles liberated from the sediment during the extraction, collection, and measurement of AVS are called simultaneously extracted metals (SEMs). SEMs are metals either associated with sulfides or adsorbed to other phases that are in equilibrium with the interstitial water. Toxic metals may outcompete the Fe, as a function of their solubility product (K_{SP}), to form an insoluble and nonbioavailable metal sulfide (e.g., NiS, ZnS, CdS, PbS, CuS, and Ag_2S), as shown in Eqn (7).

$$Cd^{2+}(\text{aqueous}) + \text{FeS (solid)} = \text{CdS (solid)} + Fe^{2+}(\text{aqueous}) \quad (7)$$

In anoxic sediments uncontaminated by heavy metals, the associated metal is usually iron (Fe) or manganese (Mn). The K_{sp} for FeS is $10^{-22.39}$ (with a comparable value for manganese (II) sulfide), so all the sulfide in an uncontaminated sediment will be bound to Fe or Mn. If a toxic heavy metal, such as Cd (K_{sp} equal to $10^{-32.85}$), has K_{sp} values substantially lower than $10^{-22.39}$, the metal binds sulfide with a much higher affinity than does iron or Mn. The EqP-based mechanistic ESBs can be used to predict ecotoxicity:

$$\text{SEM} - \text{AVS} > 0 \quad \text{Possible bioavailable metals} \quad (8)$$

$$\text{SEM} - \text{AVS} < 0 \quad \text{No bioavailable metal} \quad (9)$$

$$\text{SEM} = \text{AVS} + f_{OC}K_{OC}C_d \text{ (Eqn (7)) or } C_d = \frac{\text{SEM} - \text{AVS}}{f_{OC}\,K_{OC}} \quad (10)$$

Equation 10 provides a more accurate measure of the bioavailable metal, which includes interactions with both AVS and organic carbon. Equation 10 can be further rearranged to Eqn (11), which provides the FCV calculation.

$$\frac{\text{SEM} - \text{AVS}}{f_{OC}} = K_{OC}\cdot\text{FCV} \quad (11)$$

Studies have shown that sediment concentrations less than or equal to the ESB values are not expected to result in adverse effects and benthic organisms should be protected. Sediment concentrations above the ESB values may result in adverse effects to benthic organisms.

Calculation of predicted no-effect-concentration (PNEC): The equilibrium partitioning theory can be used to determine $PNEC_{sediment}$ using Eqn (12), in which $RHO_{suspension}$ is the bulk density of wet suspended matter, $K_{suspension\ water}$ is the suspended water matter, and $PNEC_{water}$ is PNEC in water.

$$PNEC_{sediment} = \frac{K_{suspension\ water}}{RHO_{suspension}}\cdot PNEC_{water}\cdot 1000 \quad (12)$$

A problem is that Eqn (12) only considers uptake via water. To account for uptake via ingestion of sediment, the $PEC_{sed}/PNEC_{sed}$ (sed: sediment) ratio is increased by a factor of 10. The results obtained by this method must be verified experimentally. Only a few studies (Mueller and Nowack, 2008; Boxall et al., 2008 and Blaser et al., 2008) have reported predicted environmental nanoparticle concentrations (PEC_{nano}). Mueller and Nowack (2008) found that, although the current concentrations of TiO_2 nanoparticles may currently pose a threat to aquatic organisms, the current concentrations of nano-Ag and CNT may be nonhazardous. Development of probabilistic methods of environmental exposure analysis (Finley and Paustenbach, 1994; McKone and Bogen, 1991; van der Voet and Slob, 2007; Macleod et al., 2007; Gottschalk et al., 2010, 2013; Sun et al., 2014) addressed the inconsistency and variability of model input parameters by using probability distributions based on empirical data, expert judgment, or a combination of these sources. The PEC/PNEC ratio can be used to classify nanoparticle risk assessment (Table 6).

- *Risk classification based on lethal concentration$_{50}$/effective concentration$_{50}$ ($L(E)C_{50}$) values*: Bondarenko et al. (2013) and Kahru and Dabourguier (2010) have classified nanoparticle risk according to the EU Directive 93/67/EEC, based on the lowest median L(E) C_{50} value of key environmental organisms of algae, crustaceans, and fish (CEC, 1996). The lowest median $L(E)C_{50}$ value <1 mg/L = very toxic to aquatic organisms. 1–10 mg/L = toxic to aquatic organisms. 10–100 mg/L = harmful to aquatic organisms. >100 mg/L = not classified (CEC, 1996).

TABLE 6 The Environmental Risk of a Substance Can be Estimated by the PEC/PNEC

Classification	PEC/PNEC
Serious risk	>10
Unacceptable risk	1–10
Acceptable risk	0.01–1
Negligible risk	<0.01

The classifications that are commonly used.

An additional category of *extremely toxic* was applied by Sanderson et al. (2003) and Blaise et al. (2008) and was also employed in the current review. According to the EU Directive 93/67/EEC, the lowest EC_{50} value obtained either in tests with crustaceans, algae, or fish will determine the final hazard class of the chemical compound.

Bondarenko et al. (2013) showed that Ag nanoparticles (target species algae) exhibited the highest toxicity to the crustaceans (L(E) C50 value of 0.01 mg/L), followed by algae, fish, nematodes, bacteria, yeast, various mammalian cells, *Vibrio fischeri*, and protozoa (Figure 26(A); Table 7). Thus, Ag nanoparticles, classified as extremely toxic to crustaceans, exhibit the most toxicity toward nontarget aqueous crustaceans, which are crucial components of the aquatic food web. Similarly to Ag nanoparticles, CuO nanoparticles (target species-bacteria) were most toxic to crustaceans and algae, but at a slightly higher level (2–3 mg CuO/L; Figure 26(C); Table 7). Thus, the target species was not most sensitive toward CuO nanoparticles (MIC >250 mg/L). Unlike CuO nanoparticles, Cu ions were most toxic to the target species of bacteria. Cu ions were more toxic than CuO nanoparticles to all organisms, except for yeast and mammalian cells in vitro (Figure 26(D)). As in the case of Ag and CuO nanoparticles, the toxicity of ZnO nanoparticles to algae ($L(E)C_{50}$ = 0.08 mg/L), crustaceans ($L(E)C_{50}$ = 2.3 mg/L), and fish ($L(E)C_{50}$ = 3.0 mg/L) was higher than that to bacteria (MIC 622 mg/L). Thus, according to the most sensitive organism of the test battery crustaceans–algae–fish, ZnO nanoparticles is classified as *very toxic* to aquatic organisms. A key observation of this study was that the toxicity of ZnO nanoparticles and Zn ions to different organisms was similar (Figure 26(E and F); Table 7), indicating that the toxicity of ZnO nanoparticles may be largely caused by dissolved Zn ions.

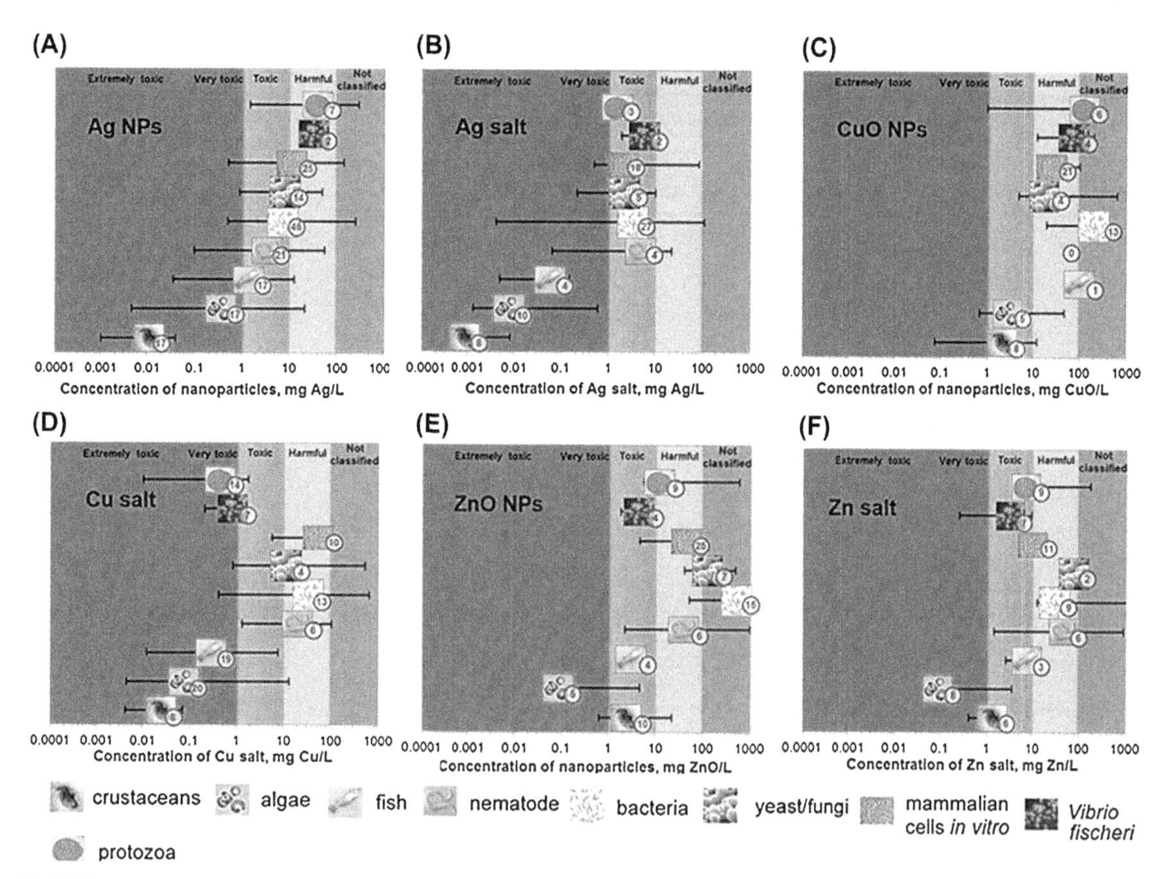

FIGURE 26 Ecotoxicity of CuO, ZnO, and Ag nanoparticles to different organisms. Median $L(E)C_{50}$ values for all other organisms except bacteria (shown MIC for bacteria) ± minimum and maximum values are presented. Different organisms/cells are shown by the respective pictograms. $L(E)C_{50}$ or MIC values on the x-axis are plotted in the logarithmic scales. The classification to hazard categories is explained in Table 7. *Reprinted from Bondarenko et al (2013).*

Kahru and Dabourguier (2010) have shown that ZnO nanoparticles are the most toxic nanoparticles for algae and nematodes (Table 8). C_{60} fullerene was most toxic for bacteria, fish, and ciliates, and Ag nanoparticles were most toxic for the crustaceans (Table 8). However, there was no big difference between the toxicities of nanoparticles and their respective bulk particles, except in the case of CuO. The most toxic particles were Ag nanoparticles, ZnO nanoparticles, C_{60} fullerenes, and CuO nanoparticles. Both TiO_2 nanoparticles and corresponding bulk particles were less toxic but still were classified as harmful. The few available data indicate that carbon nanotubes are probably less toxic than fullerenes and metal-containing nanoparticles. As nanoparticles show harmful effects at very low concentrations, they can lead to disruption of the food chain, leading further to global effects concerning the entire ecosystem. From all the compounds studied, the most toxic was Cu^{2+}. The toxicity of Cu^{2+} varied from 0.02 mg/L (algae) to 45 mg/L (nematodes; Table 9). These values concur with the results reported by Riedel et al. (2008).

TABLE 7 Median L(E)C50 Values for All Organisms except Bacteria for Which Median MIC is Shown for Ag, CuO, and ZnO Nanoparticles (NPs) and their Respective Metal Salts

	Median L(E)C50 or MIC, on a compound basis, mg/L (number of data)			Median L(E)C50 or MIC, on a metal basis, mg metal/L (number of data)		
Group of organisms	**Ag NPs**	**CuO NPs**	**ZnO NPs**	**Ag salt**	**Cu salt**	**Zn salt**
Crustaceans	0.01 (17)	2.1 (8)	2.3 (10)	0.00085 (8)	0.024 (8)	1.3 (6)
Algae	0.36 (17)	2.8 (5)	0.08 (5)	0.0076 (10)	0.07 (20)	0.09 (8)
Fish	1.36 (17)	100 (1)	3.0 (4)	0.058 (4)	0.28 (19)	7.5 (3)
Nematodes	3.34 (21)	Not found (0)	39 (6)	4.8 (4)	19.4 (6)	49 (6)
Bacteria	7.10 (46)	200 (13)	500 (15)	3.3 (27)	32 (13)	30 (9)
Yeast	7.90 (14)	17 (4)	121 (7)	2.16 (5)	11.1 (4)	78 (2)
Mammalian cells in vitro	11.3 (25)	25 (21)	43 (25)	2 (18)	53 (10)	9.8 (11)
Vibrio fischeri[a]	32 (2)	73.6 (4)	4.3 (4)	5.7 (2)	0.78 (7)	3.2 (7)
Protozoa	38 (7)	124 (6)	11.7 (9)	1.5 (3)	0.43 (14)	7 (9)
Lowest L(E)C50, MIC	0.01	2.1	0.08	0.00085	0.024	0.09
Most sensitive organisms	Crustaceans	Crustaceans	Algae	Crustaceans	Crustaceans	Algae
Classification (1)[b]	Very toxic	Toxic	Very toxic	Very toxic	Very toxic	Very toxic
Classification (2)[c]	Extremely toxic	Toxic	Extremely toxic	Extremely toxic	Extremely toxic	Extremely toxic

The L/E/D_{50} values are obtained from Bondarenko et al. (2013). Risk classification 1: determined according to CEC (1996) and risk classification-2 is from Sanderson et al. (2003) and Blaise et al. (2008).

[a] Vibrio fischeri *data are retrieved separately from other bacteria, because V. fischeri was considered as non-target aquatic species.*

[b] *Classification of NPs and their soluble salts to hazard categories adheres to EU-Directive 93/67/EEC (CEC, 1996) and is based on the lowest median L(E) C50 value of the three key environmental organisms: algae, crustaceans, and fish. <1 mg/L = very toxic to aquatic organisms; 1–10 mg/L = toxic to aquatic organisms; 10–100 mg/L = harmful to aquatic organisms; >100 mg/L = not classified.*

[c] *Analogous to the classification of CEC (1996), except that one category is added: <0.1 mg/L = extremely toxic to aquatic organisms.*

Reprinted from Bondarenlo et al. (2013) with permission.

3.2.3 *Risk Evaluation with Scarce Data (Broad Knowledge Gap or Lack of Toxicology Data)*

The Control Banding (CB) framework has been widely adopted by the pharmaceutical industry to assess risks associated with new components. CB has proven to be an effective strategy for controlling worker exposure in the absence of complete relevant information about toxicity and exposure—that is, without Occupational Exposure Limit (OEL). CB has been widely adopted as an alternative option for controlling exposure to nanoparticles. It can be considered as a first step in a complex risk-assessment process that may involve quantitative exposure measurements and accidental risk assessment. The commonly used CB method are Groso's method (Groso et al., 2010), CB NanoTool (Paik et al., 2008), ANSES method (ANSES, 2010), Stoffenmanager Nanoor STM Nano (Van Duuren-Stuurman et al., 2012), and multiple-path particle dosimetry (MPPD). Brief descriptions on these methods are described here:

- *Groso's method*: This method is based on a decision tree approach that helps to manage

TABLE 8 Median L(E)C50 Values for Selected Synthetic Nanoparticles toward Different Groups of Organisms and Classification of Nanoparticles to Different Hazard Categories

		Inorganic nanoparticles[a]				Organic nanoparticles[a]		
		mg TiO_2/L	mg ZnO/L	mg CuO/L	mg Ag/L	mg/L	mg/L	mg/L
	Group of organisms	Nano TiO_2	Nano ZnO	Nano CuO	Nano Ag	SWCNT	MWCNT	C60
1	Crustaceans	67.7 (10)	0.62 (3)	2.65 (2)	**0.040** (1)	15.0 (3)	**8.7** (1)	35.0 (5)
2	Bacteria	603 (4)	20 (3)	71 (2)	7.60 (5)	163 (2)	500 (1)	0.81 (4)
3	Algae	**65.5** (4)	**0.068** (2)	**0.87** (1)	0.23 (2)	**1.04** (1)	NF	100.0 (1)
4	Fish	300 (4)	1.9 (2)	NF	7.1 (1)	NF	NF	1.0 (3)
5	Ciliates	NF	5.4 (1)	156.5 (1)	39.0 (1)	6.8 (1)	NF	**0.25** (1)
6	Nematodes	80.1 (1)	2.24 (1)	NF	NF	NF	NF	NF
7	Yeasts	20000 (1)	121.2 (1)	20.5 (1)	NF	NF	NF	NF
1–7	No. of data	24	13	7	10	7	2	14
1–7	Lowest L(E)C50	65.5	0.068	0.87	0.040	1.04	8.7	0.25
1–7	Most sensitive organisms	Algae	Algae	Algae	Crust.	Algae	Crust.	Ciliates
1–7	Classification (1–7)[b]	Harmful	ET	VT	ET	Toxic	Toxic	VT
1–3	Classification (1–3)[c]	Harmful	ET	VT	ET	Toxic	Toxic	VT

The classification scheme applied adheres to EU-Directive 93/67/EEC classification scheme (CEC, 1996). Evaluation grid applied by Sanderson et al. (2003) and Blaise et al. (2008) was used. Classification is based on the median L(E)C50 value of the most sensitive organism used: <0.1 mg/L = extremely toxic to aquatic organisms; 0.1–1 mg/L = very toxic to aquatic organisms; 1–10 mg/L = toxic to aquatic organisms; 10–100 mg/L = harmful to aquatic organisms; >100 mg/L = non-toxic to aquatic organisms. Lowest median L(E)C50 value for each compound is in bold. In the brackets the number of values that was used for the calculation of the median value is indicated. NF, Not found.

[a] *Used for the calculation of the median values (inorganic NPs + organic NPs = total).*

[b] *Classification of the potential hazard according to the lowest median L(E)C50 values (all test organisms).*

[c] *Classification of the potential hazard according to the lowest median L(E)C50 values (crustaceans, bacteria, and algae).*

Reprinted from Kahru and Dabourguier (2010) with permission.

the risks involving nanoparticles and insufficient data. This method classifies hazard into three hazard classes: nano-1, nano-2, and nano-3 in increasing severity. This tool provides an easy way to assess and compare the level of risk.

- *CB Nanotool*: This method consists of a risk-level (RL) matrix that classifies nanoparticles as lowest risk level (RL1–RL3) and the higher risk level (RL4). Hazard assessment is classified into two groups: physicochemical properties and toxicological properties (carcinogenicity, mutagenicity, and dermal toxicity).
- *Stoffenmanager Nano method*: In this method, the input parameters for the hazard assessment are selected from the Material Safety Data Sheets MSDS and other technical data sheets. Stoffenmanager combines the available hazard information with a qualitative assessment of the potential exposure.
- *Multiple-path particle dosimetry (MPPD)*: This is a widely used and well-accepted model to estimate deposited and retained mass doses in the pulmonary alveolar region of nanomaterial-exposed workers (Oberdörster, 2012). The model calculates deposition and clearance in the respiratory tract (Anjilvel and Asgharian,

TABLE 9 Median L(E)C50 Values for Bulk Materials of ZnO, TiO_2, and CuO as well as for Zn^{2+}, Cu^{2+}, Pentachlorophenol (PCP), and Aniline toward Different Groups of Organisms and Their Classification to Different Hazard Categories

No.	Group of organisms	Reference compounds						
		Bulk TiO_2	Bulk ZnO	Bulk CuO	Zn^{2+}	Cu^{2+}	PCP	Aniline
1	Crustaceans	20000 (3)	0.48 (3)	127.8 (2)	0.192 (4)	0.029 (8)	0.48 (3)	0.30 (3)
2	Bacteria	20000 (1)	20.0 (3)	3758 (1)	3.50 (5)	0.47 (4)	2.9 (4)	135 (2)
3	Algae	**60** (1)	**0.052** (2)	**14.2** (1)	**0.051** (2)	**0.020 (3)**	0.24 (1)	11 (3)
4	Fish	500 (2)	1.8 (2)	NF	36.9 (1)	0.21 (13)	0.23 (5)	49.0 (3)
5	Ciliates	NF	4.9 (1)	1947 (1)	9.8 (4)	1.10 (5)	**0.15** (1)	220 (2)
6	Nematodes	137 (1)	2.2 (1)	NF	111.0 (3)	45.2 (2)	NF[a]	NF[a]
7	Yeasts	20000 (1)	134.4 (1)	1277 (1)	62.6 (2)	8.2 (2)	NF[a]	NF[a]
1–7	No. of data	9	13	6	21	37	14	13
1–7	Lowest L(E)C50	**60**	**0.052**	**14.2**	**0.051**	**0.020**	**0.15**	**0.30**
1–7	Most sensitive organisms	Algae	Algae	Algae	Algae	Algae	Ciliates	Crust
1–7	Classification (1–7)[b,e]	Harmful	ET	Harmful	ET	ET	VT	VT
	Risk phrases (R)[c,d]	NF	R50/R53	NF	R50/R53	R50/R53	R50/R53	R50

The classification scheme adheres to the EU-Directive 93/67/EEC classification scheme (CEC, 1996). Evaluation grid applied by Sanderson et al. (2003) and Blaise et al. (2008) was used. Classification is based on the median L(E)C50 value of the most sensitive organism used: <0.1 mg/L = extremely toxic to aquatic organisms; 0.1–1 mg/L = very toxic to aquatic organisms; 1–10 mg/L = toxic to aquatic organisms; 10–100 mg/L = harmful to aquatic organisms; >100 mg/L = non-toxic to aquatic organisms. Lowest median L(E)C50 value for each compound is in bold. In the brackets the number of values that was used for the calculation of the median value is indicated. PCP—Pentachlorophenol; NF—not found.
[a] *Probably available but were not found in the selection of the literature examined for the current review.*
[b] *Classification of the potential hazard according to the lowest median L(E)C50 values (all test organisms).*
[c] *Classification of the potential hazard according to the lowest median L(E)C50 values (crustaceans, bacteria, and algae).*
[d] *Classification under Directive 67/548/EEC. The classification is carried out according to the lowest effect concentration. Substances are classified as dangerous for the environment and labeled with the symbol N (dangerous for the environment) and an adequate risk phrase (R). R50—very toxic to aquatic organisms; L(E)C50 < 0.1 mg/L; R53—may cause long-term adverse effects in the aquatic environment.*
[e] *Classification under Regulation No. 1272/2008 (European Parliament, 2008).*
Reprinted from Kahru and Dabourguier (2010) with permission.

1995; Asgharian et al., 2001). Figure 27 shows the deposition of Ag or TiO_2 nanoparticles (Figure 27(A)) and CNTs (Figure 27(B)). The highest amount tested for Ag nanoparticles ranges from 1.6 to 500 μg/mL, whereas for TiO_2 nanoparticles, the high-side range is 100–1000 μg/mL or 20–520 μg/cm^2.

- *ANSES method*: This method can be used in any work environment in which nanomaterials are manufactured or used. The ANSES method classifies hazards from the lowest toxicity (CL1) to the highest toxicity (CL5).

The methods listed above provide a simple means to estimate the risk despite high uncertainties. This is important because decisions need to be made about the possible outcomes of nanoparticle exposure. As discussed earlier, engineered nanoparticles may possess such unique properties that extrapolating from other types of studies may introduce, not resolve, uncertainty. Although considerable health effect studies are being conducted on engineered nanoparticles, it may take years or even decades to collect sufficient data to fully and

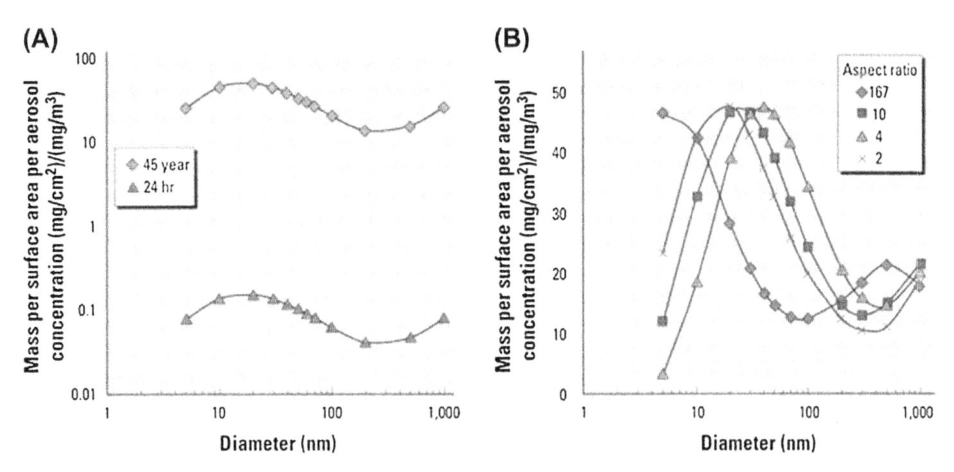

FIGURE 27 Multiple-path particle dosimetry (MPPD) model yields alveolar mass retained per alveolar surface area per inputted aerosol concentration that has been plotted against particle diameter in human lungs. (A) TiO_2 and Ag nanoparticles with exposure durations of 45 years (full working lifetime) and 24 h, and (B) CNTs with aspect ratios of 167, 10, 4, and 2 after 45 years of exposure. Both (A) and (B) are based on a light exercise breathing pattern. *Reprinted from Gangwal et al. (2011) with permission.*

confidently understand the potential for health or environmental risk from them. The methods listed above may provide a starting point to conduct more studies and/or to set up sole regulations.

3.2.4 Expert Judgment in Uncertainty Assessment

As discussed in the previous section, there are deep uncertainties associated with nanoparticle toxicity. The uncertainties pervade every element of the exposure–response–risk paradigm for nanoparticles. When the uncertainties prevent the development of risk assessment, professional or expert judgment can be useful in assessing relevant variables, characterizing model forms, and estimating model parameters.

Expert judgment is also useful in the early stages of a scientific issue when uncertainty is high or undetermined. Expert assessment can be used to structure problems, to indicate key variables, and to examine relationships and influences between variables by building influence diagrams (Morgan and Henrion, 1990), as shown in Figures 28–30. Influence diagrams are very useful devices for structuring problems and can be used quantitatively (if sufficient data are available about the quantitative relationships among variables) or qualitatively (in the absence of such data, they are more useful as qualitative tools and cannot be used for risk assessment). Expert judgment approaches have been used in environmental applications, such as environmental exposure assessment (Hawkins and Evans, 1989; Ramachandran and Vincent, 1999; Ramachandran, 2001; Walker et al., 2001, 2003; Ramachandran et al., 2003) and assessment of global climate change (Morgan and Keith, 1995; Risbey et al., 2000; Risbey and Kandlikar, 2002). Expert judgment may occur in two stages: (1) subjective degrees of belief/agreement with specific statements and (2) relative importance of variables.

- *Subjective degrees of belief/agreement with specific statements*: Experts are given 10 specific statements in all domains shown in Figure 28 (center column). The extent of their agreement with statement will be noted using the Likert

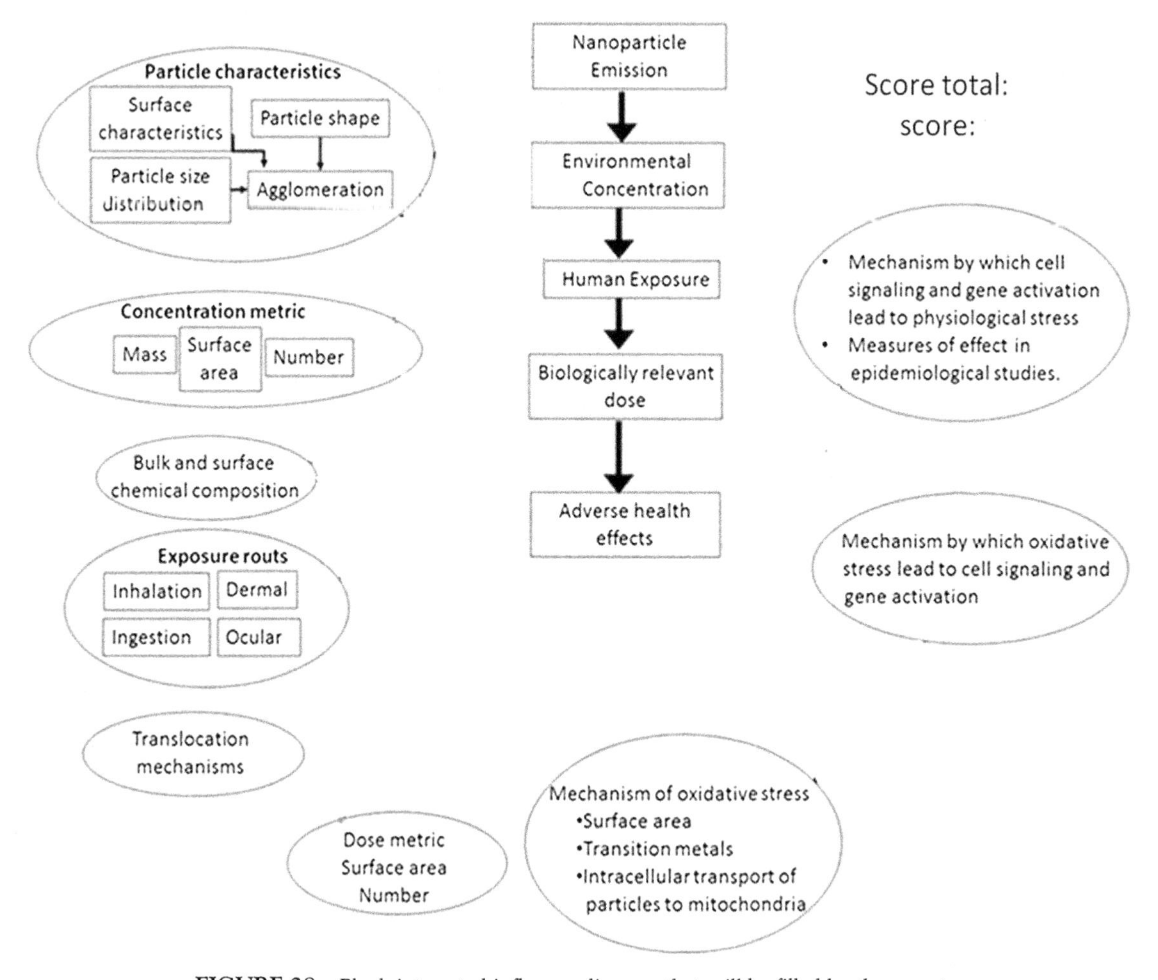

FIGURE 28 Blank-integrated influence diagram that will be filled by the experts.

scale (Wuensch, 2005): strongly agree (5 points), agree (3 points), undecided (0 points), disagree (−3 points), or strongly disagree (−5 points). The statements will be posed as definitive propositions. Some of the examples are provided below:

a. The smaller the nanoparticle, the more reactive their surface is. Answer: agree strongly (5), agree (3), total 8
b. Manufactured nanoparticles (e.g., fullerenes, carbon nanotubes) are more likely to persist in the environment than other nanoparticles (e.g., generated as a byproduct of combustion) due to different agglomeration processes resulting from special coatings on the particles. Answer: undecided (0), disagree (−3), total −3
c. The smaller nanoparticles (<100 nm) are more toxic than the bulk particles (>500 nm). Answer: strongly agree (5), strongly agree (5), total 10
d. A decrease in nanoparticle size is associated with an increase in their agglomeration, resulting in the formation of larger particles. Answer: strongly agree (5), disagree (−3), total 2

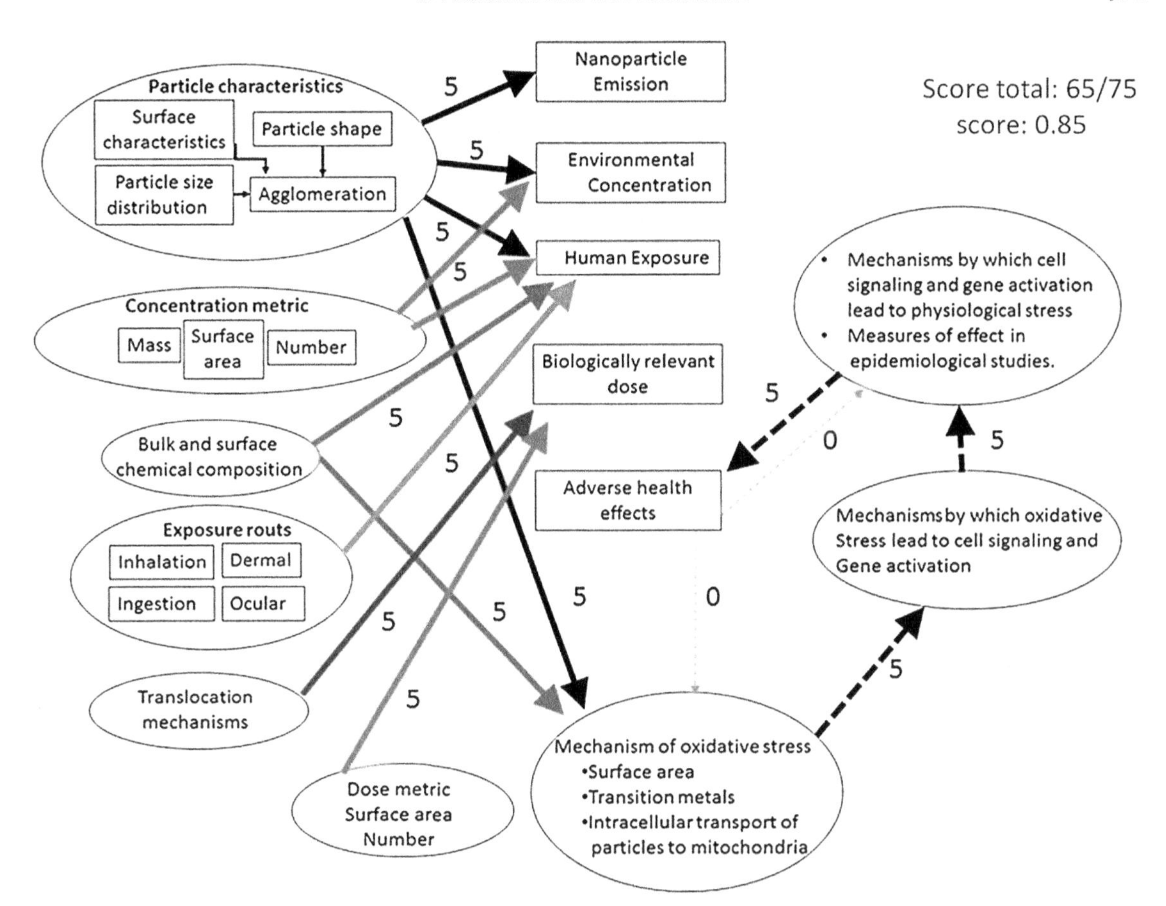

FIGURE 29 Integrated influence diagram completed by expert-1. The line thickness shows the relative importance of the connection. The numerical values indicate the expert's rating.

Possible answers are shown in red font. In this response, the experts' answers match once and differ three times; thus, the uncertainty is 3:4. The survey score is a = 8, b = −3, c = 10, d = 2. Thus, questions a and d may have a higher probability of occurring in the environment than questions b and c (a hypothetical example).

- *Relative importance of the variables*: Experts are shown in Figure 28 (without the connecting lines) and asked to evaluate the importance of specific variables using ranking exercises or by assigning numerical values ranging from 0 (not important or not occurring) to 5 (extremely important or essential). The experts could be advised to award more points to the choice they consider more likely. In this way, a subjective probability measure that reflects the expert's state of belief regarding the choice of exposure metric can be obtained. For example, the experts may assign 1 point (probability = 0.2) to mass concentration as the appropriate exposure metric, 3 points (probability = 0.6) to surface area concentration as the appropriate exposure metric, and 1 point (probability = 0.2) that particle number is the appropriate exposure metric. Eliciting opinions from a large group of experts (20 or

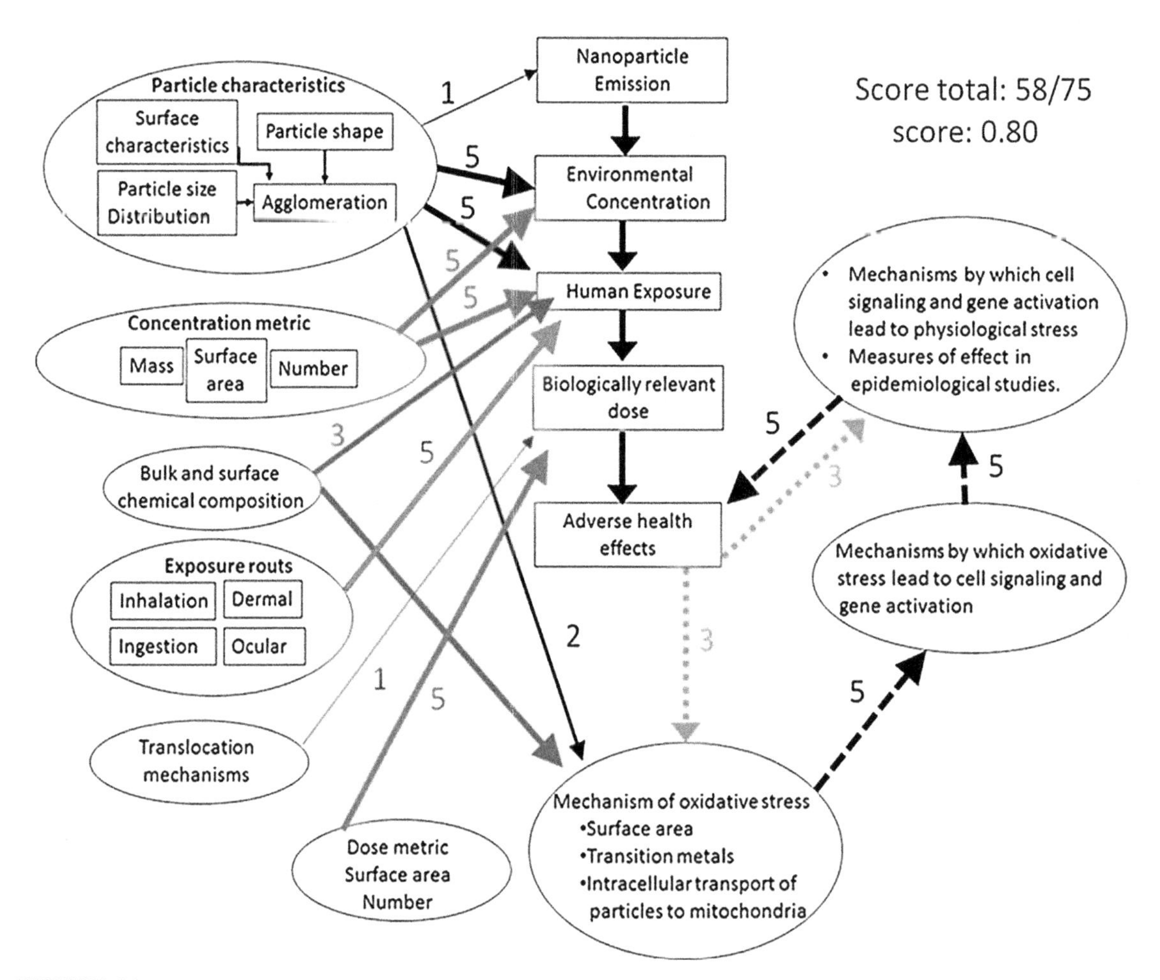

FIGURE 30 Integrated influence diagram completed by expert-2. The line thickness shows relative importance of the connection.

more) would assure that the causal chains and the associated probabilities would be sufficiently populated so as to allow meaningful comparisons across expert responses. Figures 29 and 30 show a hypothetical representation of how consensus and divergence in different elements of the causal chain can be evaluated. The data can be interpreted based on each variable's probability. Figure 31 shows the level of agreement among six experts (Kandlikar et al., 2007). The particle characteristic parameters (black), the concentration metric parameters, and the toxicity parameters are shown on the vertical axis, while the individual expert's responses (individual symbols) in probability ranging from 0 to 1 are shown on horizontal axis. There is unanimity about parameters 3 and 4, general agreement about parameters 1 and 2 (although one expert differs from all others), and there is broad disagreement about parameters 5 and 6. Opinion is evenly divided about parameter N.

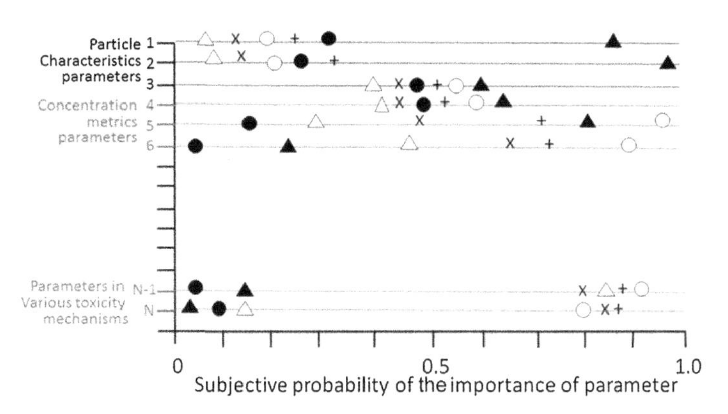

FIGURE 31 An example of the level of agreement among six experts. See the text for detailed description.

- *Interpretation of expert opinions*: Expert opinion, in the absence of necessary data, cannot be used in risk determination. The data can be used as a starting point to further characterize the nanoparticles experimentally. When the survey exhibits considerable variability (parameters 5 and 6), other methods may be used to clarify the risk.

4. GOVERNMENTAL LEGISLATIONS REGARDING NANOPARTICLE RISK

Realizing the risk posed by engineered nanoparticles, many national and international governmental organizations have passed legislations to set specific standards for their manufacture, use, and disposal (Handy and Shaw, 2007; SCENIHR, 2007; Bard et al., 2009). In the following sections, some of the legislation approved in the United States and the European Union are described.

4.1 Nanoparticle Regulation in the United States

The current regulatory framework concerning the manufacture and use of nanoparticles is inadequate to protect Americans from the potential threat posed by nanoparticles. The same physical traits that give nanoparticles their value also make them uniquely dangerous toxins and environmental contaminants. Currently, there is no specific legislation for nanoparticles. Nanoparticle-containing items are regulated by a variety of federal agencies, depending on their intended use (Bashow, 2009).

4.1.1 Environmental Protection Agency

The EPA has authority under the Toxic Substances Control Act (TSCA) to regulate nanoparticles (Bashow, 2012a, b, Toxic substance control act 1976, 15 USA 2605(a), ID. § 2604(a)(1), Id. § 2604(a)(1), Id. § 2603(a)). TSCA allows for the regulation of nanoparticle manufacture and emission into the environment, as could also be done under the Clean Air Act or Clean Water Act (Bashow, 2012a,b). In TSCA, nanoparticles are defined on the basis of their molecular structure. Manufacturers are often free to produce nanoscale materials that exhibit novel properties, but without reporting them to the EPA; they are regulated as an existing substance despite often possessing dramatically different physical properties.

Under section 5(a)(1) of TSCA, a company may not manufacture or import a "new" chemical substance without first completing a premanufacture notification (PMN) review by the EPA

(15 U.S.C. § 2604(a)(1)). Based on its PMN review, the EPA may permit the new substance to be placed on the inventory, allowing it to be manufactured, processed, distributed, or imported without restriction; the agency also may issue an administrative order under TSCA section 5(e) to impose restrictions on the manufacture, use, and/or distribution of the substance, require the submission of more data on the substance, or ban the chemical outright (Section 5(e) order, TSCA).

Given the EPA's lack of success in pursuing new rules under this section of TSCA, it would be wishful thinking to assume that TSCA as it stands can provide sufficient protection against potential pollution from a wide variety of emerging nanoparticles. Furthermore, the relative lack of conclusive information on their safety, coupled with the EPA's difficulties in properly classifying them, may make the administrative burden on the EPA even more insurmountable.

4.1.2 US Food and Drug Administration

The US Food and Drug Administration (FDA) does not make a categorical reference to nanotechnology as inherently safe or harmful. It plans to consider the specific characteristics of individual products. The FDA has issued four guidance documents to encourage manufacturers to consult (not a requirement) with the agency before taking their products to market (the EPA may or may not receive information). The guidance is described on their Website: http://www.fda.gov/NewsEvents/Newsroom/PressAnnouncements/ucm402499.htm.

Final Guidance for Industry: Considering Whether an FDA-Regulated Product Involves the Application of Nanotechnology: This document is intended to help industry and others identify when they should consider potential implications for regulatory status, safety, effectiveness, or public health impact that may arise with the application of nanotechnology in FDA-regulated products.

Final Guidance for Industry: Safety of Nanomaterials in Cosmetics: The guidance describes the FDA's current thinking on the safety assessment of nanomaterials when used in cosmetic products and encourages manufacturers to consult with the FDA on test methods and data needed to support the substantiation of a product's safety.

Final Guidance for Industry: Assessing the Effects of Significant Manufacturing Process Changes, Including Emerging Technologies, on the Safety and Regulatory Status of Food Ingredients and Food Contact Substances, Including Food Ingredients that Are Color Additives: The guidance alerts manufacturers to the potential impact of any significant manufacturing process changes, including changes involving nanotechnology, on the safety and regulatory status of food substances. This guidance also describes considerations for determining whether a significant manufacturing process change for a food substance already on the market affects the identity, safety, or regulatory status of the food substance, potentially warranting a regulatory submission to the FDA.

Guidance for Industry: Use of Nanomaterials in Food for Animals: A guidance addressing issues related to the use of nanotechnology in food ingredients intended for use in food for animals was issued by the Food and Drug Administration in 2015. http://www.fda.gov/ScienceResearch/SpecialTopics/Nanotechnology/ucm401782.htm.

4.2 European Union's Legislations

This section include a review of the EU in developing the regulatory structure on the approval of nanomedicine. The legislations set by the European Union deal with the health and safety regulations designed for the protection of workers (General Directive 89/391/EEC on the improvements in the safety and health of workers at work; Directive 98/24/EC, Directive 2004/37/EC, Directive 1992/92/EC on the protection of the health and safety of workers from the risks related to chemical agents, carcinogens or mutagens, and explosive atmospheres, respectively; Council Directive 89/655/EEC concerning the minimum safety and health requirements for the use of work equipment by workers at work).

In July 2007, a legislation called Registration, Evaluation, Authorization, and Restriction of Chemicals was passed by the EU that notifies nanoparticle hazardous data to general public and businesses. The listed nanoparticles undergo extensive testing and safety evaluations as their concentration may reach a critical limit. Toma-Biano (2012) has presented a detailed overview of the nanoparticle regulation in the EU. Selected important legislations are listed below.

EC 258/1997: The Novel Food Regulation – did not mention nanoparticles.
EC 1223/2009: Nanoparticles in cosmetic industry.
EC 178/2002: Nanoparticles in food. Legislation modified in 2008 to address nanoparticle risk.
EC 1935/2004: Nanomaterials in contact with food must be subject to risk assessment.
EC 282/2008: Recycling of nanomaterials.

Unlike in the United States, where nanoparticles are being regulated under the TSCA, Clean Water Act 1977, and Clean Water Act 1987, the EU has developed legislations specifically for the manufacture, application of, and exposure to nanoparticles.

4.3 Future Perspectives

One problem with current legislations is that they do not require the disclosed use of nanoparticles in consumer goods, although the practice is increasing rapidly. Furthermore, no comprehensive and compulsory danger assessment scheme has been introduced to manage the potential risks posed by nanoparticles to public and environmental health. Lin (2007) has presented a balanced argument for and against stricter nanotechnology legislations. On the pro-stricter-regulation side, there is sufficient, but not compelling, evidence that nanomaterials, already making their way into the marketplace today, are possibly harmful to consumers and the environment, so stronger and new laws are needed to ensure that they are safe. On the business side, it can be argued that more regulation will slow down the pace of business and innovation in nanotechnology. The industry believes that self-regulation will address the health issues. Lin has posed some interesting questions that need to be answered before the debate is settled or a third option is found. The questions are as follows:

1. Are there any interim approaches that the industry can follow?
2. What are the specific steps needed to be taken to strengthen preregulatory planning, methods for testing materials, and toxicology testing?
3. Do we need (paradoxically) a legal basis for ensuring that this greater focus on environmental health and safety (EHS) risks and testing actually occurs, for example, by stipulating that some percentage of all nanotechnology research funds will go toward these areas?
4. How do we know that more and faster study of the EHS aspects of nanotechnology can keep up with the full-throttle research and development (R&D) and commercialization of nanotechnology in not just the US but also abroad?
5. How much more funding is needed for ethics and risk to catch up with R&D?
6. Would the United States (or any other nation that adopts such an interim solution) need to compensate with even more funding if other nations do not focus as much on the risks?
7. If additional funding is warranted, where would that come from? Would it be diverted from other programs that are working on cures for current ills, outside of nanotechnology's risks (which seem to be future risks, as opposed to actually harming people or the environment right now)?

To my knowledge, these and other relevant questions have not yet been addressed, while the nanotechnology regulation debate rages on.

5. CONCLUSIONS

Nanoparticles have a vast impact on society by providing various useful products and solutions to global problems. The beneficial effects of nanoparticles are attributed to their unique physicochemical properties, which are highly dependent on their size, shape, and surface properties. The same unique properties responsible for beneficial effects are also responsible for adversely affecting human health and the natural environment. Therefore, it is vital to comprehend the human and environmental risks that nanoparticles pose. The problem is that comprehensive experimental data is needed to develop risk assessment, management, and communication; unfortunately, this level of information is not yet available for nanoparticles. There are still many gaps in available experimental data devoted to risk assessment of engineered nanoparticles that are already available on the market. Studies have proposed the concept of uncertainty factors to account for the missing information. This chapter described the risk-assessment processes when sufficient data are available and when there are broad knowledge gaps.

In the Introduction section, two questions were posed for greater understanding of nanoparticle toxicity and their risk to general public. Based on the information presented, the questions can be answered as follows.

1. *Do the unique features of nanoparticles pose any inherent threats (different from their bulk counterpart) to humans and the environment?* Earlier studies have shown conflicting results regarding toxicities of nanoparticles and bulk particles. Warheit et al. (2007) have shown that nanoparticles exhibited greater oxidative-toxicity than their larger counterparts and that the toxicity of quartz particles was not dependent on particle size/surface area, but rather the surface reactivity. Similarly, Karlsson et al. (2008a) showed that CuO nanoparticles exhibited greater toxicity from CuO microparticles and other metal oxide nanoparticles and carbon nanotubes. The differential toxicity between CuO nanoparticles and CuO microparticles may be due to differential mitochondrial damage (Karlsson et al., 2008b, 2009). The exceptionally high toxicity of CuO nanoparticles compared to CuO microparticles shows that the CuO nanoparticles may be of specific concern. Because of inherent variability, it is proposed that the nanoparticles' toxicity must be assessed on a case-by-case basis.
2. *Are existing risk-assessment methodologies developed for bulk particles appropriate to assess potential and plausible risks associated with nanoparticles?* The answer to this question, at best, can be equivocal. The EU's SCENIHR (http://copublications.greenfacts.org/en/nanotechnologies//l-2/9-conclusion.htm) has looked into this question and came with following opinions:
   a. Although the existing methods are appropriate to assess many of the hazards associated with the products and processes involving nanoparticles, they may not be sufficient to address all the hazards.
   b. More specifically, the mode of delivery of the nanoparticle to the test system should adequately reflect the exposure scenarios. Additional tests may be needed.
   c. Expressing the dose of exposure in terms of mass alone is not sufficient; it also needs to be expressed in terms of total surface area, number of particles, or a combination of the two.
   d. Also, the existing methods used for environmental exposure assessment are not necessarily appropriate.
   e. Therefore, the current risk-assessment procedures require modification for nanoparticles.

However, SCENIHR's opinion may be optimistic because of the excessive knowledge gaps identified by SCENIHR itself. Listed below are the major gaps in knowledge that need to be filled to allow for satisfactory risk assessments for humans and ecosystems:

a. Mechanisms and the rate at which nanoparticles are released from products and processes.
b. Actual levels of exposure to nanoparticles, both for humans and the environment.
c. Extent to which it is possible to calculate the toxicology of nanoparticles from the knowledge on the same chemicals in larger physical forms.
d. Movement of nanoparticles inside the body and the effects that nanoparticles cause at the cellular level.
e. How target organs respond to different doses of nanoparticles.
f. Exposure and related health effects of workers involved in the production and processing of nanoparticles.
g. Behavior of nanoparticles in the environment, how they are distributed, if they persist in it, and whether they accumulate in different species, including micro-organisms.
h. The effects of nanoparticles on various environmental species.

In addition, the technology used to characterize nanoparticles' risk suffers from excessive uncertainties, as shown in Figure 24. For bulk particles, the uncertainties are associated with inter- and intraspecies variation, toxicokinetics, and toxicodynamics (discussed earlier). However, the uncertainties are those listed above, as well as associated with the analytical methods.

6. APPENDIX: BULK PARTICLES' RISK CHARACTERIZATION

Large numbers of books describing the basic risk characterization of bulk chemicals are available (Pine, 2014; Simon, 2014; Fenton and Neil, 2012; Ostrom and Wilhelmsen, 2012; Theodore and Dupont, 2012; Aven, 2011; Dikshith, 2010; Paustenbach, 2009; Nielsen et al., 2008; Ott et al., 2006; Hester and Harrison, 1998). Therefore, the aim of this section is not to repeat the information already published, but to review selected research papers that discuss different aspects of ecosystems and human health following the exposure to bulk particles, such as insecticides, metals, or nutrients.

1. Downing et al. (2011) have shown the possible role of species diversity in resilience and resistance of the ecosystems. They modeled stability of ecosystem having 30, 20, and 10 species in the presence or absence of a stressor, such as a predator. They reported that diversity increased the resilience of ecosystems. The investigators analyzed a simple model of a diverse (10–30 species) community where each competing species inflicts a small mortality pressure on an introduced predator (Figure 32). When diversity was high (30 species), a single species extinction usually led to a slight decrease in the total biomass of the native community (Figure 32(a1) and (b1)). However, when starting from a lower initial diversity (10 species), a few consecutive species extinctions caused a relatively large biomass loss that ultimately lead to collapse (Figure 32(c1)). They also showed that the predator is unable to invade a high diverse system. With a gradual loss of native species, the introduced predator can escape control and the system collapses into a contrasting, invaded, low-diversity state (Figure 32(c1)). A diverse system has high complementarity gains in resilience, while a diverse system with high-functional redundancy has high resistance. Loss of resilience can display early-warning signals of a collapse, but not loss of resistance.
De Laender et al. (2008d) described the sensitivity of an ecosystem's (consisting of one species of fish as a predator, three species of zooplanktons, and two species of phytoplankton, lines showing the predation

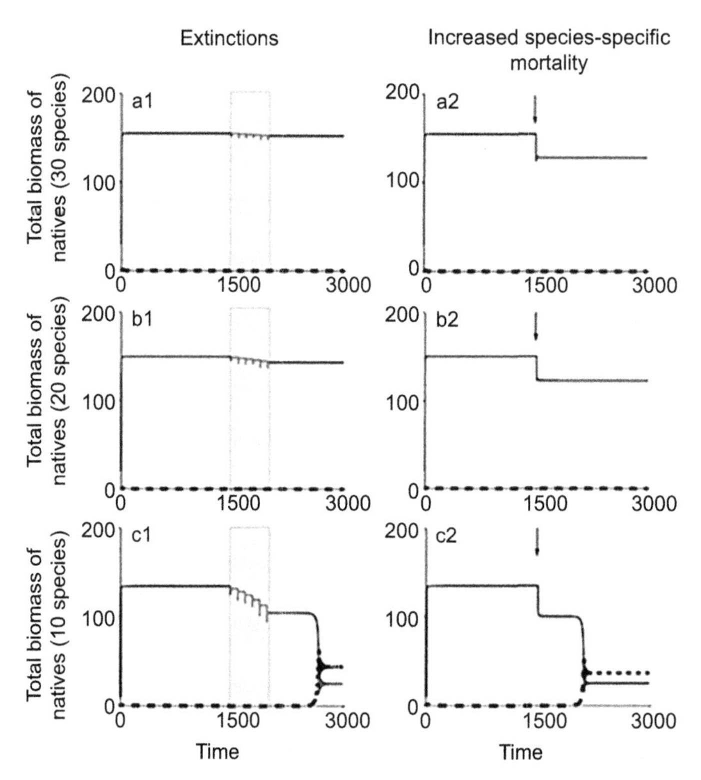

FIGURE 32 Effect of biotic diversity on the stability of aquatic ecosystems. As species diversity decreases (from 30 to 10), the adverse effects of an extinction of a single species on total biomass and species mortality increase in an aquatic ecosystem. (a1 and a2) At high diversity (30 species), extinction of a single species causes a minor fluctuation in total biomass and a slight increase in species-selective mortality. (b1 and b2) When the species size was 20, the biomass decreased slightly. (c1 and c2) When an ecosystem contained 10 species, extinction of a single species may result in the ecosystem collapse. *Reprinted from Downing et al. (2011) with permission.*

ratio, Figure 33(A)) species, structure (composition of producer and consumer populations), or function (energy transfer expressed as biomass flow) to environmental stressors, including chemical toxicants. They hypothesized that the ecosystem's structure NOEC was lower than or equal to the corresponding ecosystem's function NOEC. To test that hypothesis, a hypothetical freshwater ecosystem was exposed to different ($n = 1000$) toxicants and then, using an ecosystem model, calculated NOECs for ecosystem structure (ecosystem structure NOECs) and function (ecosystem function NOECs) for each of 1000 hypothetical toxicants. They observed that, for 979 of these toxicants, the ecosystem structure NOEC was lower than or equal to the ecosystem function NOEC (Figure 33(B)), indicating that the proposed hypothesis may be considered valid. Interestingly, the trend of the relationship between NOEC and trophic level was opposite for the ecosystem structure (NOEC (fish) > NOEC (zooplankton) > NOEC (phytoplankton)) and ecosystem function (NOEC ($PS_{phyto,tot}$) > NOEC ($I_{zoo,tot}$) > NOEC (I_{fish})). The observation that these trends are independent of the toxicant

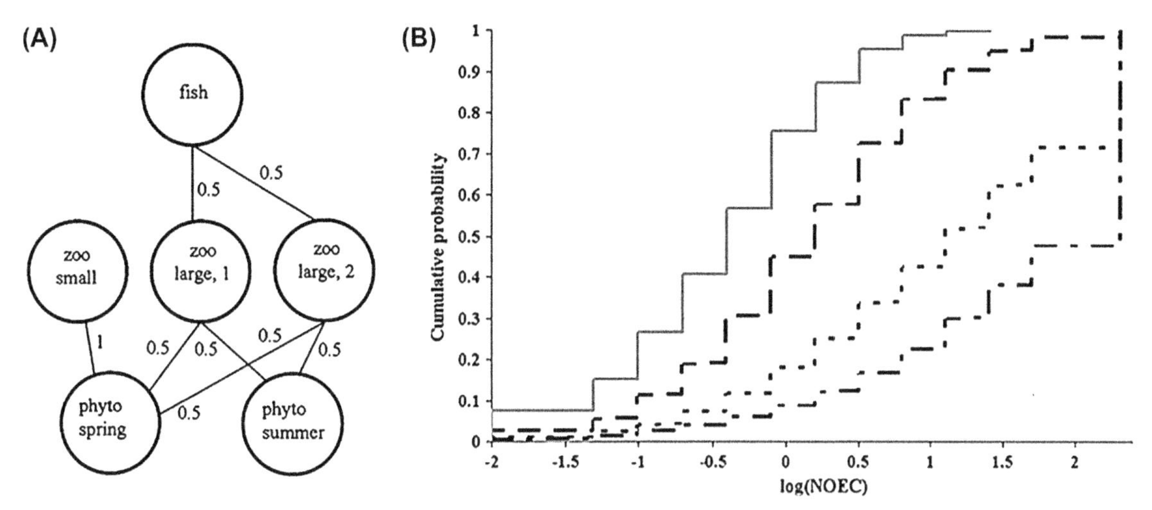

FIGURE 33 (A) An artificial ecosystem consisting of two producer species (one blooming in spring and the other in spring), three primary consumers (small and two large zooplanktons), and a single fish species as zooplankton predator. (B) Cumulative probability distribution: gray line: log(ecosystem structure-NOECs); dashed line: log(I_{fish}-NOEC); dotted line: log($I_{zoo,tot}$-NOEC); dotted and dashed line: log($PS_{phyto,tot}$-NOEC). I_{fish} is ingestion of zooplanktons by fish, $I_{zpp,\ tot}$ is ingestion of total (spring + summer) phytoplankton by zooplankton and $PS_{phyto\text{-}tot}$ is total (spring + summer) photosynthesis activity. *Reprinted from De Laender et al. (2008d) with permission.*

type suggests that these trends may be due to the ecological interactions within the system studied and not due to the toxins' direct effects. For about half of the 1000 toxicants, the structure of lower trophic levels appears to be more sensitive than the structure of higher trophic levels (i.e., fish). The ecosystem's structure NOECs are primarily determined by the sensitivity of the structure of lower trophic levels. In contrast, ecosystem's function NOECs are associated with higher trophic levels (e.g., total ingestion by fish). The function NOECs are more sensitive than functions associated with lower trophic levels (e.g., total photosynthesis by phytoplankton) for 749 toxicants. More studies are needed to experimentally delineate the sensitivities of an ecosystem's structure and function to different toxins.

In a separate study, De Laender et al. (2008b) have applied the species sensitivity distributions (SSDs), defined as statistical distributions which extrapolate single-species toxicity test results to ecosystem effects, in determining whether or not the ecological interactions between populations (as grazing and competition) influence the sensitivity of ecosystems. They plotted the chronic single species against the SSDs (termed conventional SSD or cSSD since the values were calculated in the absence of ecological interactions) for all 1000 hypothetical toxicants. If cSSD = ecoSSD, then the ecosystem interactions do not play a role in ecological effects of a toxicant; however, if cSSD ≠ ecoSSD, then the ecological factors are in play. For 254 of the 1000 hypothetical toxicants, the mean and/or variance of the cSSD are ≠ the ecoSSD. They also developed a hierarchal classification tree approach that indicated that the toxicants that directly affect phytoplankton (i.e., herbicides) may have a higher mean for cSSD than for ecoSSD. Conversely, the means of ecoSSD and cSSD tend to be equal for toxicants directly affecting zooplankton and fish, such as insecticides.

This approach provides an important means to understand mechanisms of the adverse effects of toxicants on an ecosystem.

2. Brock et al. (2004) have compared the toxicity of photosynthesis-inhibiting herbicides metribuzin or metamitron on ecosystem community in their natural habitat (for ecosystem effects due to interspecies ecosystem feedback controls) with the single-species laboratory toxicity tests (direct effects of herbicides without the interspecies controls). Metribuzin and metamitron are comparable inhibitors of the photosynthesis (photosystem I), but metribuzin is two- or three-fold more stable than metamitron in an aqueous environment (Figure 34).
The differential stability makes the herbicides ideal to evaluate their direct and ecological risks. The investigators used three different risk-assessment procedures to achieve this: (1) standard first-tier approach using an assessment factor, (2) the SSD approach based

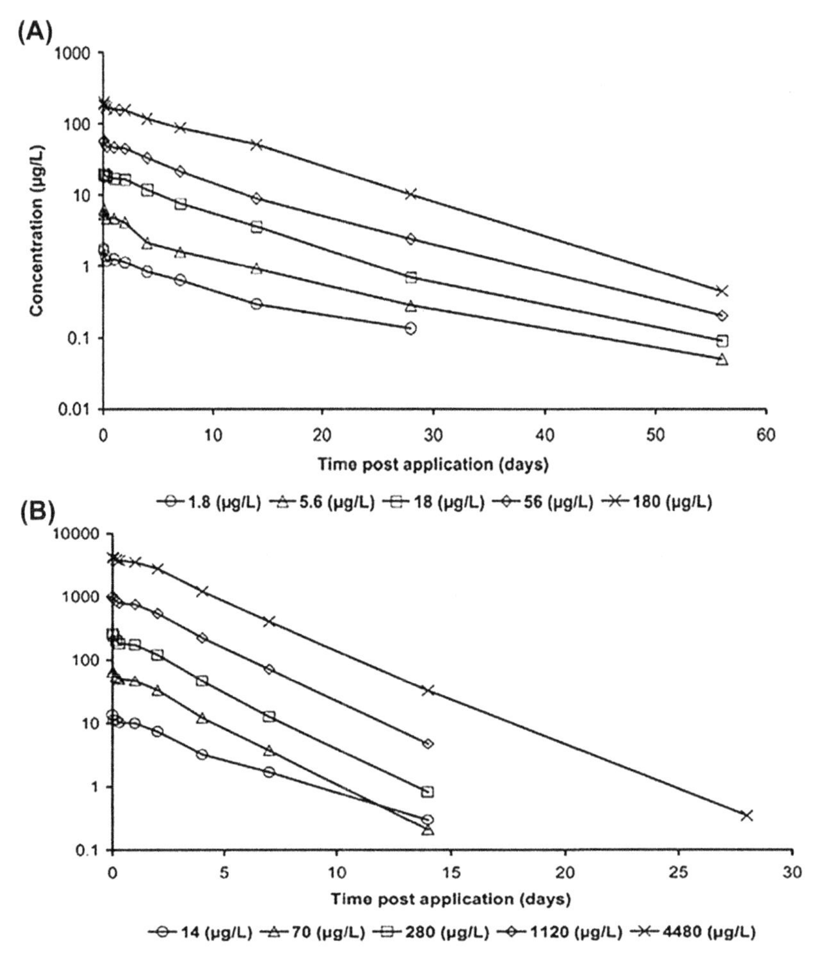

FIGURE 34 Concentration of photosynthesis inhibitor herbicides, metribuzin (A), and metamitron (B), in a water column. Although metribuzin and metamitron are comparable inhibitors of the photosynthesis (photosystem I), metribuzin is two–three folds more stable (thus more toxic) than metamitron in an aqueous environment. *Reprinted from Brock et al. (2004) with permission.*

TABLE 10 Effects and critical threshold levels observed in the enclosures treated with a single dose of the herbicides metribuzin (upper panel) and metamitron (lower panel)

	Nominal exposure concentrations (μg/l)				
	1.8	5.6	18	56	180
Metribuzin					
Community metabolism					
Dissolved oxygen	1	1 (NOEC)	2 ↓ (LOEC)	3 ↓	3 ↓
pH	1	1 (NOEC)	2 ↓ (LOEC)	3 ↓	3 ↓
Community structure					
Phytoplankton	1	1 (NOEC)	2 ↓ (LOEC)	3 ↓	3 ↓
Periphyton (chlorophyll-a)	1	1	1	1 (NOEC)	5 ↑↓ (LOEC)
Macrophytes	1	1	1*	1*	1*
Invertebrates	1	1	1 (NOEC)	3 ↓ (LOEC)	5 ↑↓
	Nominal exposure concentrations (μg/l)				
	14	70	280	1120	4480
Metamitron					
Community metabolism					
Dissolved oxygen	1	1	1 (NOEC)	2 ↓ (LOEC)	3 ↓
pH	1	1	1	1 (NOEC)	3 ↓ (LOEC)
Community structure					
Phytoplankton	1	1	1	1	1
Periphyton (chlorophyll-a)	1	1	1	1	1
Macrophytes	1	1	1	1	1
Invertebrates	1	1	1	1	1

The numbers in the table refer to effect classes described in detail in Brock et al. (2000); 1=no effects; 2=slight effects; 3=clear short-term effects, lasting less than 8 weeks; 5=clear long-term effects, lasting more than 8 weeks). ↓=decrease; ↑↓=decrease and increase depending on species and/or period post treatment. NOEC: no observable effect concentration, LOEL: lowest observable effect concentration.

** A trend towards a smaller final biomass of macrophytes was visible, though the difference was not significant.*

Reprinted from Brock et al. (2004) with permission.

on laboratory tests to calculate the HC_x (the Hazardous Concentration for x% of the species), and (3) the model ecosystem approach using field enclosure studies. Table 10 summarizes aquatic threshold concentrations for metribuzin and metamitron derived using different risk-assessment procedures. This further confirms that metamitron at environmentally relevant concentrations poses no real risk to the ecosystem. Table 10 also shows the treatment-related responses by using the five effect classes described in detail by Brock et al. (2000). A 180 μg/L concentration of the more persistent metribuzin resulted in effect classes ranging from 3 to 5 in all indices, except macrophytes, which showed an effect class of 1. In comparison, a 4480 μg/L concentration of the less persistent metamitron was essentially nontoxic. Long-term treatment-related effects lasting longer than 8 weeks (effect class 5) were found in the 180 μg/L metamitron enclosures. Effect-class 5 comprised responses by the invertebrates and the zooplankters *Chydorus sphaericus* and *Polyarthra remata*. Indirect effects on invertebrates observed in the metribuzin study occurred at a higher concentration (56 μg/L) than the LOEC for direct effects (18 μg/L). For metribuzin, the first-tier (qualitative) aquatic risk-assessment procedure is more conservative when compared with the NOECs (Table 11). The HC_5 and HC_{10} values based on acute toxicity data of primary producers were lower than the LOECs of the most sensitive structural or functional endpoint. This can be used to set maximum permissible concentrations in a cost-effective way. As shown in Table 11, the first-tier approach is the most conservative for both metamitron and metribuzin. The HC_5 and HC_{10} values based on acute EC_{50} values of algae and aquatic vascular plants may be used to derive maximum permissible concentrations for single applications. The HC_x values based on laboratory toxicity tests do not provide information on (1) the recovery potential of sensitive endpoints and (2) indirect effects, the indices important for regulatory decision-making.

3. Malhat et al. (2015) have studied the health hazard assessment of pyridaben (2-*tert*-butyl-5-(4-*tert*-butylbenzylthio)-4-chloropyridazin-3(2H)-one) residues in Egyptian strawberries. Pyridaben is a novel, broad-spectrum acaricide, approved by the Egyptian Pesticides Committee of the Government of Egypt for the control of pests on various crops. The project was commissioned to establish (1) the dose-specific preharvest intervals that are essential to ensure that dissipation of residues is well below the prescribed minimum residue limit (MRL) at harvest time (2) the food safety hazard by evaluating residues of pyridaben in terms of their dietary exposure related to the ADI, and

TABLE 11 Summary of Aquatic Threshold Concentrations for the Photosynthesis-Inhibiting Herbicides Metribuzin and Metamitron, Derived from Different Risk-Assessment Procedures

	Metribuzin (μg/L)	Metamitron (μg/L)
First-tier Maximum Permissible Concentration NL (1/10 * NOEC for the most sensitive standard alga	0.2	8.4
First-tier Maximum Permissible Concentration EU (1/10 * EC_{50} for most sensitive standard alga	1.2	85.2
HC_5 chronic toxicity data for primary producers	1.4	–
HC_{10} chronic toxicity data for primary producers	2.1	–
NOEC enclosure study (single application)	5.6	280
HC_5 acute toxicity data for primary producers	7.4	667
HC_{10} acute toxicity data for primary producers	10.0	763
LOEC enclosure study, functional (single application) (Class 2 effect on community metabolism)	18.0	1120
LOEC enclosure study, structural (single application) (Class 2 effect on community structure)	18.0	>4480

Reprinted from Brock et al (2004) with permission.

(3) and maximum permissible intake. To achieve the aims, the investigators conducted field trials in a randomized block design in farmers' fields that never used pyridaben (information provided by the growers). Field trials were conducted in separate plots each measuring 40 m^2 in three replicates. The end-points measured were pyridaben concentrations in soil and strawberries, and the insecticide dissipation rate. The ADI values were obtained from the literature (WHO, 2005). The following equations were used for hazard characterization:

a. Dissipation kinetics determined as $C_t = C_0^{-kt}$ where C_t is the pesticide concentration at time t, C_0 is the initial applied dose, and k is pesticide dissipation.
b. EED (mg/kg. BW) = [CRL (mg/kg) × FL (mg)]/BW (kg). In this equation, EED is the estimated exposure dose, BW is body weight, CRL is the calculated residue level, FL is food intake, and ADI is acceptable dietary intake (0.005 mg/kg/day).
c. HQ = EED/ADI. In this equation, HQ is the hazard quotient. An HQ value greater than 1 would indicate that the hazard of pesticides in humans is unacceptable. By contrast, an HQ value of less than 1 represents the minimal hazard to humans (Zhang et al., 2009).

The *risk-assessment* phase included determination of the chemical's dissipation, half-life, and the HQ values. At applied doses of 100 and 200 gai/Ha, the C_t values are $0.427e^{-0.262t}$ and $0.7495e^{-264t}$, respectively. The half-life ($t_{1/2}$) for both treatments is approximately 2.3 days for pyridaben on strawberries in an open field. The half-life of this insecticide in oranges range from 11.9 to 24.9 days (Zhang et al., 2005), while the half-life of this insecticide in apples and bananas are 7.3 days (Ma et al., 2007) and 2.8–2.9 days, respectively (Chen et al., 2011). Table 12 shows the EED and HQ values for recommended (100 gai/Ha) and double recommended (200 gai/Ha) doses of the insecticide. The HQ values of pyridaben in strawberries after application of both doses are significantly less than 1 (7.4×10^{-6} to 6.2×10^{-3}). This suggests that the health hazard of pyridaben use in strawberries was negligible to humans. This results for strawberries concur with earlier studies in cabbage (Liu et al., 2014) and apple and citrus (Zhang et al., 2005, 2009 and Ma et al., 2007).

TABLE 12 Hazard Assessment of Pyridaben in Strawberries After Application

	Recommended dosage		Double recommended dosage	
Time after application (days)	**Estimated exposure dose mg/(kg bw day)**	**Hazard quotient (HQ)**	**Estimated exposure dose mg/(kg bw day)**	**Hazard quotient (HQ)**
0	1.9×10^{-5}	3.8×10^{-3}	3.1×10^{-5}	6.2×10^{-3}
1	1.3×10^{-5}	2.6×10^{-3}	2.2×10^{-5}	4.4×10^{-3}
3	3.6×10^{-6}	7.2×10^{-4}	7.4×10^{-6}	1.4×10^{-3}
6	2.3×10^{-6}	4.7×10^{-4}	3.4×10^{-6}	6.8×10^{-4}
9	1.6×10^{-6}	3.3×10^{-4}	3.2×10^{-6}	6.4×10^{-4}
12	6.6×10^{-7}	1.3×10^{-4}	1.7×10^{-6}	3.4×10^{-4}
15	–	–	3.3×10^{-6}	6.6×10^{-5}
21	–	–	–	–

Reprinted from Malhat et al. (2015).

4. Wang et al. (2014) and Silva and Beauvais (2010) have assessed human risk posed by endosulfan (6,7,8,9,10,10-hexachloro-1,5,5a,6,9,9a-hexahydro-6,9-methano-2,4,3-benzodioxathiepin-3-oxide), an organochlorine pesticide consisting of two isomers (I or α (about 70%) and II or β (about 30%)) and a sulfate conjugate (Maier-Bode, 1968; NRCC, 1975). Endosulfan has been shown to induce severe neurotoxicity (Lawrence and Casida, 1984; Cole and Casida, 1986) and reproductive or developmental toxicity (Silva and Beauvais, 2010). A primary risk assessment concern for endosulfan is its prenatal or neonatal effects, resulting in endocrine disruption that may be irreversible. The California Department of Pesticide Regulation and the US EPA have reviewed endosulfan ecotoxicity and have decided that current data do not support the case that endosulfan is a developmental or reproductive toxicant or an endocrine disruptor at doses lower than those that cause other toxic effects (US EPA, 2007). Wang et al. (2014) have reported the risk potential of endosulfan I, II, and sulfate released for a manufacturing facility in China. They have described spatial soil endosulfan concentrations near and far from the field of the endosulfan manufacturing plant, as well as several air, sediment, fish, vegetable, and crop samples taken close to the facility. Figure 35 shows the spatial and vertical distributions of endosulfans (A: endosulfan

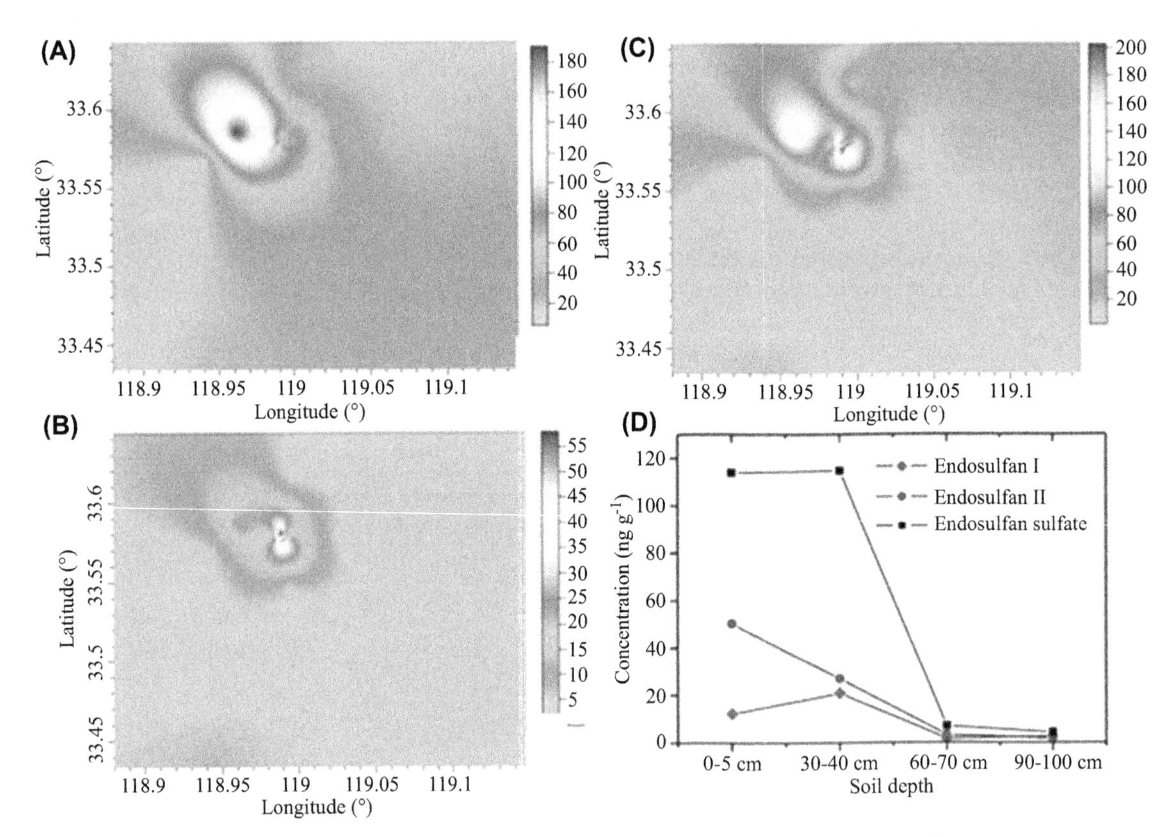

FIGURE 35 (A–C): Spatial distribution of endosulfan I (A), endosulfan II (B), and endosulfan sulfate (C) concentrations in soil surrounding the manufacturing facility. Concentrations of all three compounds apexed at the same location, indicating a common source within the studied geographical grid, which corresponded to the manufacturing facility. (D) Vertical distribution of endosulfans. *Reprinted from Wang et al. (2014) with permission.*

sulfate, B: endosulfan I, and C: endosulfan II) near in the vicinity of the production facility. The highest concentration in the center represents the release source. Figure 35(D) shows the vertical distribution of endosulfan in the soil at different soil depths. The concentrations in vertical soil samples were the highest in the upper layers from 0 to 40 cm and decreased gradually from greater than 20 ng/g in the upper layers to less than 5 ng/g in the lower layers, ranging from 60 to 100 cm. Trace amounts were observed in the deepest layers, indicating that endosulfan can be transported through the leaching of pore water in soil.

Human exposure to endosulfan occurred via dietary intake (estimated exposure dose–dermal, EED_D) (Xing et al., 2011), dermal contact (EED_S) (Minh et al., 2003; Nouwen et al., 2001), and inhalation (EED_I) (Xing et al., 2011) from the manufacturing facility. The dietary intake dose for all people is approximately 3.6×10^{-5} mg/kg/day in which grains, vegetables, and fish accounted for 51%, 43%, and 5% of total dietary intake, respectively (Wang et al., 2014). Dietary intake had the highest contribution. Table 13 shows the HQ, EED, RfD, NOAEL, and UF values for people at work or in residential areas. The manufacturing facility in China is a significant source of endosulfan contamination. The spatial trends of endosulfan sulfate in soil were similar to the total endosulfans, implying that the manufacturing facility plays an important role in the spatial pattern of endosulfan. This study suggests the following:

a. The production facility influenced the distribution of endosulfans in surface soil and is an important source for the immediate surrounding environment.
b. Dietary intake through foodstuffs may account for the main exposure of daily intake for people in this region under the worst-case scenario.
c. The maximum HI values were at least two orders of magnitude less than 1.

Because the hazard indices were at least two orders of magnitude less than 1, no adverse health effects are likely to occur at current exposure levels, and the risk to human health is generally acceptable (Wang et al., 2014).

In conclusion, bulk toxicants, such as pesticides and herbicides, present certain health risks to humans and the ecosystem. For determining

TABLE 13 Hazard Quotient (HQ), Estimated Exposure Dose (EED, mg/kg/day) and Reference Dose (RfD, mg/kg/day) of Total Endosulfans (Endosulfans I and II) for Workers, Residents, and Children via Dietary, Dermal Contact, and Inhalation at Study Sites in the Manufacturing Facility in Huai'an, China

Exposure route	HQ[a]			EED			RfD[b]	*NOAEL*[c]	*UF*[d]			
	Worker	Resident	Children	Worker	Resident	Children			H	S	A	L
Dietary	6.3×10^{-3}	6.3×10^{-3}	6.3×10^{-3}	3.6×10^{-5}	3.6×10^{-5}	3.6×10^{-5}	5.7×10^{-3}	0.57	10	1	10	1
Dermal	1.7×10^{-8}	5.6×10^{-9}	6.1×10^{-9}	9.7×10^{-11}	3.2×10^{-11}	3.5×10^{-11}	5.7×10^{-3}	0.57	10	1	10	1
Inhalation	5.4×10^{-4}	3.2×10^{-4}	4.9×10^{-4}	1.0×10^{-7}	6.3×10^{-8}	9.6×10^{-8}	1.9×10^{-4}	0.019	10	1	10	1

[a] *HQ = EED/RfD.*
[b] *RfD* = NOAEL (experimental dose)/(UF × MF), UF *is standard uncertainty factor.*
[c] *No observed adverse effect levels* (NOAEL *mg/kg/day) via oral, dermal, and inhalation obtained from the repeated dose toxicity.*
[d] *UF = A × H × S × L, A = animal-to-human (interspecies); H = inter-individual (intraspecies); S = subchronic-to-chronic duration; L = LOAEL-to-*NOAEL.
Reprinted from Wang et al. (2014) with permission.

human risk, the dose–response data for different exposure modes are modeled to calculate NOAEL, LOAEL, and RfDs. Risk assessment for ecosystems is a highly complex process because it involves an interaction between (1) the abiotic and biotic process, (2) different producers and consumers, and (3) biotic–environment interactions. The ecosystem conditions may vary rapidly, which may modulate the ecosystem structure and function. Therefore, complex models have been designed to accurately assess an ecosystem's health. Often, it is difficult for the general public to comprehend the model results.

References

Adamcakova-Dodd, A., Stebounova, L.V., O'Shaughnessy, P.T., Kim, J.S., Grassian, V.H., Thorne, P.S., 2012. Murine pulmonary responses after sub-chronic exposure to aluminum oxide-based nanowhiskers. Part. Fibre Toxicol. 9, 22.

Adger, W.N., 2006. Vulnerability. Global Environ. Change 16, 268–281.

Anjilvel, S., Asgharian, B., 1995. A multiple-path model of particle deposition in the rat lung. Fundam. Appl. Toxicol. 28, 41–50.

ANSES, 2010. Development of a Specific Control Banding Tool for Nanomaterials.

Araujo, J.A., Barajas, B., Kleinman, M., Wang, X., Bennett, B.J., Gong, K.W., Navab, M., Harkema, J., Sioutas, C., Lusis, A.J., Nel, A.E., 2008. Ambient particulate pollutants in the ultrafine range promote early atherosclerosis and systemic oxidative stress. Circ. Res. 102, 589–596.

Aschberger, K., Micheletti, C., Sokull-Klüttgen, B., Christensen, F.M., 2011. Analysis of currently available data for characterising the risk of engineered nanomaterials to the environment and human health – lessons learned from four case studies. Environ. Int. 37, 1143–1156.

Aschberger, K., Johnston, H.J., Stone, V., et al., 2010. Review of carbon nanotubes toxicity and exposure—appraisal of human health risk assessment based on open literature. Crit. Rev. Toxicol. 2010 (40), 759–790.

Asgharian, B., Hofman, W., Bergmann, R., 2001. Particle deposition in a multiple-path model of the human lung. Aerosol. Sci. Technol. 34, 332–339.

Aven, T., 2011. Quantitative Risk Assessment: The Scientific Platform. Cambridge University Press.

Bailey, R.C., Norris, R.H., Reynoldson, T.B., 2004. Bioassessment of Freshwater Ecosystems: Using the Reference Condition Approach. Kluwer Academic Publishers, New York.

Bard, D., Mark, D., Mohlmann, C., 2009. Current standardisation for nanotechnology. Nanosafe 2008: international conference on safe production and use of nanomaterials. J. Phys. Conf. Ser. 170, 012036.

Barnthouse, L.W., Munns Jr., W.R., Sorensen, M.T., 2008. In: Population-level Ecological Risk Assessment. Taylor & Francis, Boca Raton (FL), 376 pp.

Bartell, S.M., Lefebvre, G., Kaminski, G., Carreau, M., Campbell, K.R., 1999. An ecosystem model for assessing ecological risks in Québec rivers, lakes, and reservoirs. Ecol. Model. 124, 43–67.

Bashow, 2012a, supra note 13, at 476.

Bashow, 2012b, supra note 13, at 476; see also Clean Air Act of 1963 § 101(a), 42 U.S.C. § 7401(2012).

Bashow, J., 2009. Regulation of nanoparticles: trying to keep pace with a scientific revolution. Nanotechnology L. Bus. 6, 475–476 [12].

Bermudez, E., Mangum, J.B., Wong, B.A., et al., 2004. Pulmonary responses of mice, rats, and hamsters to subchronic inhalation of ultrafine titanium dioxide particles. Toxicol. Sci. 77, 347–357.

Blaise, C., Gagné, F., Férard, J.F., Eullaffroy, P., 2008. Ecotoxicity of selected nano-materials to aquatic organisms. Environ. Toxicol. 23, 591–598.

Blaser, S.A., Scheringer, M., MacLeod, M., Hungerbuehler, K., 2008. Estimation of cumulative aquatic exposure and risk due to silver: contribution of nano-functionalized plastics and textiles. Sci. Total Environ. 390, 396–409 [9].

Bondarenko, O., Juganson, K., Ivask, A., Kasemets, K., Mortimer, M., Kahru, A., 2013. Toxicity of Ag, CuO and ZnO nanoparticles to selected environmentally relevant test organisms and mammalian cells in vitro: a critical review. Arch. Toxicol. 87, 1181–1200.

Bontje, D.M., 2010. Analysis of Toxic Effects and Nutrient Stress in Aquatic Ecosystems (Ph.D. thesis). Vrije Universiteit, Amsterdam, The Netherlands.

Boughton, D.A., Smith, E.R., O'Neill, R.V., 1999. Regional vulnerability: a conceptual framework. Ecosyst. Health 5, 312–322.

Boxall, A.B.A., Chaudhry, Q., Jones, A., Jefferson, B., Watts, C.D., 2008. Current and Future Predicted Environmental Exposure to Engineered Nanoparticles. Central Science Laboratory, Sand Hutton, UK.

Brittingham, M.C., Maloney, K.O., Farag, A.M., Harper, D.D., Bowen, Z.H., 2014. Ecological risks of shale oil and gas development to wildlife, aquatic resources and their habitats. Environ. Sci. Technol. 48, 11034–11047.

Brock, T.C.M., Crum, S.J.H., Deneer, J.W., Heimbach, F., Roijackers, R.M.M., Sinkeldam, J.A., 2004. Comparing aquatic risk assessment methods for the photosynthesis-inhibiting herbicides metribuzin and metamitron. Env. Pollut. 130, 403–426.

Brock, T.C.M., Lahr, J., Van den Brink, P.J., 2000. Ecological Risks of Pesticides in Freshwater Ecosystems. Part 1: Herbicides. Alterra-rapport 088. Wageningen, The Netherlands, 127 pp.

Brown, P.H., Tullos, D., Tilt, B., Magee, D., Wolf, A.T., 2009. Modeling the costs and benefits of dam construction from a multidisciplinary perspective. J. Environ. Manage. 90, S303–S311.

Burgess, R.M., Berry, W.J., Mount, D.R., Di Toro, D.M., 2013. Mechanistic sediment quality guidelines based on contaminant bioavailability: equilibrium partitioning sediment benchmarks. Environ. Toxicol. Chem. 32, 102–114.

CEC, 1996. CEC (Commission of the European Communities) Technical Guidance Document in Support of Commission Directive 93/67/EEC on Risk Assessment for New Notified Substances. Part II, Environmental Risk Assessment. Office for Official Publications of the European Communities, Luxembourg.

CGIH, 2001. Titanium dioxide. In: Documentation of the Threshold Limit Values for Chemical Substances, seventh ed. American Conference of Governmental Industrial Hygienists, Cincinnati, OH.

Chen, S.Q., Chen, B., 2012. Network environ perspective for urban metabolism and carbon emissions: a case study of Vienna, Austria. Environ. Sci. Technol. 46, 4498–4506.

Chen, L., Cai, E., Guo, J., 2011. Dissipation dynamics of residual pyridaben in bananas and soils. Mod. Agrochem. 10, 40–43.

Chen, S., Chen, B., Fath, B.D., February 10 , 2013. Ecological risk assessment on the system scale: a review of state-of-the-art models and future perspectives. Ecol. Model. 250, 25–33.

Cho, W.S., Duffin, R., Howie, S.E., Scotton, C.J., Wallace, W.A., Macnee, W., Bradley, M., Megson, I.L., Donaldson, K., 2006. Progressive severe lung injury by zinc oxide nanoparticles; the role of Zn^{2+} dissolution inside lysosomes. Part. Fibre Toxicol. 8, 27.

Cho, W.S., Duffin, R., Thielbeer, F., Bradley, M., Megson, I.L., Macnee, W., Poland, C.A., Tran, C.L., Donaldson, K., 2012. Zeta potential and solubility to toxic ions as mechanisms of lung inflammation caused by metal/metal oxide nanoparticles. Toxicol. Sci. 126, 469–477.

Choi, J.S., Choi, J.Y., Na, H.B., et al., 2015. Quantitation of oxidative stress gene expression in human cell lines treated with water-dispersible MnO nanoparticles. J. Nanosci. Nanotech. 15, 4126–4135.

Cohrssen, J.J., Covello, V.T., 1989. Risk Analysis: A Guide to Principles and Methods for Analyzing Health and Environmental Risks. Council on Environmental Quality, Washington DC.

Cole, L.M., Casida, J.E., 1986. Polychlorocycloalkane insecticide-induced convulsions in mice in relation to disruption of the GABA-regulated chloride ionophore. Life Sci. 39, 1855–1862.

De Angelis, D.L., Bartell, S.M., Brenkert, A.L., 1989. Effects of nutrient recycling and food-chain length on resilience. Am. Naturalist 134 (5), 778–805.

De Chazal, J., Quétier, F., Lavorel, S., Van Doorn, A., 2008. Including multiple differing stakeholder values into vulnerability assessments of socio-ecological systems. Global Environ. Change 18, 508–520.

De Laender, F., De Schamphelaere, F.A.C., Vanrolleghem, P.A., Janssen, C.R., 2008a. Comparison of different toxic effect sub-models in ecosystem modelling used for ecological effect assessments and water quality standard setting. Ecotox. Environ. Saf. 69, 13–23.

De Laender, F., De Schamphelaere, K.A.C., Vanrolleghem, P.A., Janssen, C.R., 2008b. Do we have to incorporate ecological interactions in the sensitivity assessment of ecosystems? An examination of a theoretical assumption underlying species sensitivity distribution models. Environ. Int. 34, 390–396.

De Laender, F., De Schamphelaere, K.A.C., Vanrolleghem, P.A., Janssen, C.R., 2008c. Validation of an ecosystem modelling approach as a tool for ecological effect assessments. Chemosphere 71, 529–545.

De Laender, F., De Schamphelaere, K.A.C., Vanrolleghem, P.A., Janssen, R.C., 2008d. Is ecosystem structure the target of concern in ecological effect assessments? Water Res. 42, 2395–2402.

De Laender, F., 2007. Predicting Effects of Chemicals on Freshwater Ecosystems: Model Development, Validation and Application (Ph.D. thesis). Ghent University, Ghent, Belgium.

De Lange, H.J., Sala, S., Vighi, M., Faber, J.H., 2010. Ecological vulnerability in risk assessment – a review and perspectives. Sci. Total Environ. 408, 3871–3879.

Di Toro, D.M., Mahony, J.D., Hansen, D.J., Scott, K.J., Carlson, A.R., Ankley, G.T., 1992. Acid volatile sulfide predicts the acute toxicity of cadmium and nickel in sediments. Environ. Sci. Tech. 26, 96–101.

Di Toro, D.M., Mahony, J.D., Hansen, D.J., Scott, K.J., Hicks, M.B., Mayr, S.M., Redmond, M.S., 1990. Toxicity of cadmium in sediments: the role of acid volatile sulfide. Environ. Toxicol. Chem. 9, 1487–1502.

Di Toro, D.M., McGrath, J.A., Hansen, D.J., Berry, W.J., Paquin, P.R., Mathew, R., Wu, K.B., Santore, R.C., 2005. Predicting sediment metal toxicity using a sediment biotic ligand model: methodology and initial application. Environ. Toxicol. Chem. 24, 2410–2427.

Di Toro, D.M., Zarba, C.S., Hansen, D.J., Berry, W.J., Swartz, R.C., Cowan, C.E., Pavlou, S.P., Allen, H.E., Thomas, N.A., Paquin, P.R., 1991. Technical basis for the equilibrium partitioning method for establishing sediment quality criteria. Environ. Toxicol. Chem. 11, 1541–1583.

Dikshith, T.S.S., 2010. Handbook of Chemicals and Safety. CRC Press.

Ding, H-m, Tian, W-d, Ma, Y-q, 2012. Designing nanoparticle translocation through membranes by computer simulations. ACS Nano 6, 1230–1238.

Donaldson, K., Tran, C.L., 2002. Inflammation caused by particles and fibers. Inhal. Toxicol. 14, 5–27.

Donaldson, K., Aitken, R., Tran, L., Stone, V., Duffin, R., Forrest, G., Alexander, A., 2006. Carbon nanotubes: a review of their properties in relation to pulmonary toxicology and workplace safety. Toxicol. Sci. 92, 5–22.

Dorne, J.C.L.M., Renwick, A.G., 2005. The refinement of uncertainty/safety factors in risk assessment by the incorporation of data on toxicokinetic variability in humans. Toxicol. Sci. 86, 20–26.

Downing, A.S., van Nes, E.H., Mooij, W.M., Scheffe, M., 2011. The resilience and resistance of an ecosystem to a collapse of diversity. PLoS One 7, e46135.

ECB, 2003. Technical Guidance Document on Risk Assessment; Part II. European Chemicals Bureau.

Elder, A., Gelein, R., Silva, V., Feikert, T., Opanashuk, L., Carter, L., Potter, R., Maynard, M., Ito, Y., Finkelstein, J., Oberdörster, G., 2006. Translocation of inhaled ultrafine manganese oxide particles to the central nervous system. Environ. Health Perspect. 114, 1172–1178.

Elder, L., 2009. Quantification of uncertainty within and between species and the role of uncertainty factors. Hum. Exp. Toxicol. 28, 115–117.

El-Sayed, Y.S., Ryuhei, S., Atsuto, O., et al., 2015. Carbon black nanoparticle exposure during middle and late fetal development induces immune activation in male offspring mice. Toxicology 327, 53–61.

Fenton, N., Neil, M., 2012. Risk Assessment and Decision Analysis with Bayesian Networks. CRC Press.

Ferenc, S.A., Foran, J.A., 2000. Multiple Stressors in Ecological Risk and Impact Assessment: Approaches to Risk Estimation. SETAC Press, Pensacola, FL.

Filser, J., Koehler, H., Ruf, A., Roombke, J., Prinzing, A., Schaefer, M., 2008. Ecological theory meets soil ecotoxicology: challenge and chance. Basic Appl. Ecol. 9, 346–355.

Finley, B., Paustenbach, D., 1994. The benefits of probabilistic exposure assessment: three case studies air, water, and soil. Risk Anal. 14, 54–73 [10].

Frampton, M.W., 2001. Systemic and cardiovascular effects of airway injury and inflammation: ultrafine particle exposure in humans. Environ. Health Perspect. 109 (Suppl. 4), 529–532.

Galic, N., Hommen, U., Baveco, J.M., van den Brink, P.J., 2010. Potential application of population models in the European ecological risk assessment of chemicals II: review of models and their potential to address environmental protection aims. Integr. Environ. Assess. Manage. 6, 338–360.

Galloway, T.S., Sanger, R.C., Smith, K.L., Fillmann, G., Readman, J.W., Ford, T.E., Depledge, M.H., 2002. Rapid assessment of marine pollution using multiple biomarkers and chemical immunoassays. Environ. Sci. Technol. 36, 2219–2226.

Final report issued on 22 July 2011. In: Gamo, M. (Ed.), 2011. Risk Assessment of Manufactured Nanomaterisl: Titanium Dioxide (TiO_2). New Energy and Industrial Technology Development.

Gangwal, S., Brown, J.S., Wang, A., Houck, K.A., Dix, D.J., Kavlock, R.J., Hubal, E.A.C., 2011. Informing selection of nanomaterial concentrations for ToxCast *in vitro* testing based on occupational exposure potential. Environ. Health Perspect. 119, 1539–1546.

Gottschalk, F., Kost, E., Nowack, B., 2013. Engineered nanomaterials in water and soils: a risk quantification based on probabilistic exposure and effect modeling. Environ. Toxicol. Chem. 32, 1278–1287.

Gottschalk, F., Sonderer, T., Scholz, R.W., Wand, R., Nowack, B., 2010. Possibilities and limitations of modeling environmental exposure by probabilistic material flow analysis. Environ. Toxicol. Chem. 29, 1036–1048.

Grassian, V.H., Adamcakova-Dodd, A., Pettibone, J.M., O'Shaughnessy, P.T., Thorne, P.S., 2007a. Inflammatory response of mice to manufactured titanium dioxide nanoparticles: comparison of size effects through different exposure routes. Nanotoxicology 1, 211–226.

Grassian, V.H., O'Shaughnessy, P.T., Adamcakova-Dodd, A., Pettibone, J.M., Thorne, P.S., 2007b. Inhalation exposure study of titanium dioxide nanoparticles with a primary particle size of 2 to 5 nm. Environ. Health Perspect. 115, 397–402.

Groso, A., Petri-fink, A., Magrez, A., Riediker, M., Meyer, T., 2010. Management of nanomaterials safety in research environment. Part. Fibre Toxicol. 7, 40.

Handy, R.D., Shaw, B.J., 2007. Toxic effects of nanoparticles and nanomaterials: implications for public health, risk management and the public perception of nanotechnology. Health Risk Soc. 9, 125–144.

Hawkins, N.C., Evans, J.S., 1989. Subjective estimation of toluene exposures: a calibration study of industrial hygien-ists. Appl. Ind. Hyg. 4, 61–68.

Hester, R.E., Harrison, R.M., 1998. Risk Assessment and Risk Management. Volume 9 of Issues in Environmental Science and Technology, ISSN: 1350–7583. Royal Society of Chemistry.

Hill, R.A., Pyper, B.J., Lawrence, G.S., Mann, G.S., Allard, P., Mackintosh, C.E., Healey, N., Dwyer, J., Trowell, J., 2013. Using sparse dose–response data for wildlife risk assessment. Integr. Environ. Assess. Manage. 10, 3–11.

Hiltbrunner, E., Aerts, R., Bühlmann, T., Huss-Danell, K., Magnusson, B., Myrold, D.D., Reed, S.C., Sigurdsson, B.D., Korner, C., 2014. Ecological consequences of the expansion of N_2-fixing plants in cold biomes. Oecologia 176, 11–24.

Hinton, D.E., Lauren, D.J., 1990. Liver structural alterations accompanying chronic toxicity in fishes: potential biomarkers of exposure. In: McCarthy, J.F., Shugart, L.K. (Eds.), Biomarkers of Environmental Contamination. Lewis Publishers, Boca Rota, pp. 17–37.

Hu, Y., Xie, J., Tong, Y.W., Wang, C.H., 2007. Effect of PEG conformation and particle size on the cellular uptake efficiency of nanoparticles with the HepG2 cells. J. Control. Release 118, 7–17.

Iman, R.L., Helton, J.C., 1988. An investigation of uncertainty and sensitivity analysis techniques for computer models. Risk Anal. 8, 71–90.

Johnson, C.J., Dujardin, E., Davis, S.A., Murphy, C.J., Mann, S., 2002. Growth and form of gold nanorods prepared by seed-mediated, surfactant-directed synthesis. J. Mater. Chem. 12, 1765–1770.

Judson, R., Houck, K., Martin, M., Knudsen, T., Thomas, R.S., Sipes, N., Shah, I., Wambaugh, J., Crofton, K., 2014. In vitro and modelling approaches to risk assessment from the U.S. Environmental Protection Agency ToxCast program. Basic Clin. Pharm. Toxicol. 115, 69–76.

Kahru, A., Dubourguier, H.-C., 2010. From ecotoxicology to nanoecotoxicology. Toxicology 269, 105–119.

Kairdolf, B.A., Mancini, M.C., Smith, A.M., Nie, S., 2008. Minimizing nonspecific cellular binding of quantum dots with hydroxyl-derivatized surface coatings. Anal. Chem. 80, 3029–3034.

Kandlikar, M., Ramachandran, G., Maynard, M., Murdock, B., Toscano, W.A., 2007. Health risk assessment for nanoparticles: a case for using expert judgment. J. Nanopart Res. 9, 137–156.

Kandlikar, M., Risbey, J., Dessai, S., 2005. Representing and communicating deep uncertainity in climate change assessment. C R Geosci. 2337, 443–455.

Kanwar, P., Bowden, W.B., Greenhalgh, S., 2015. A regional ecological risk assessment of the Kaipara Harbour, New Zealand, using a relative risk model. Hum. Ecol. Risk Assess. Int. J. 21, 1123–1146.

Karlsson, H.L., Cronholm, P., Gustafsson, J., Möller, L., 2008a. Copper oxide nanoparticles are highly toxic: a comparison between metal oxide nanoparticles and carbon nanotubes. Chem. Res. Toxicol. 21, 1726–1732.

Karlsson, H.L., Holgersson, A., Möller, L., 2008b. Mechanisms related to the genotoxicity of particles in the subway and from other sources. Chem. Res. Toxicol. 21, 726–731.

Karlsson, H.L., Gustafsson, J., Cronholm, P., Möller, L., 2009. Size-dependent toxicity of metal oxide particles—a comparison between nano- and micrometer size. Toxicol. Lett. 188, 112–118.

Khan, J.A., Pillai, B., Das, T.K., Singh, Y., Maiti, S., 2007. Molecular effects of uptake of gold nanoparticles in HeLa cells. ChemBioChem 8, 1237.

Kim, B.S., Kim, C.S., Lee, K.M., 2008. The intracellular uptake ability of chitosan-coated Poly (D,L-lactideco-glycolide) nanoparticles. Arch. Pharm. Res. 31, 1050–1054.

Kim, J.S., Adamcakova-Dodd, A., O'Shaughnessy, P.T., Grassian, V.H., Thorne, P.S., 2011. Murine pulmonary responses after sub-chronic exposure to oxide-based nanowhiskers. Part. Fibre Toxicol. 8, 29.

Kim, Y.S., Song, M.Y., Park, J.D., Song, K.S., Ryu, H.R., Chung, Y.H., Chang, H.K., Lee, J.H., Oh, K.H., Kelman, B.J., Hwang, I.K., Yu, I.J., 2010. Subchronic oral toxicity of silver nanoparticles. Part. Fibre Toxicol. 7, 20.

Kitchin, K.T., Prasad, R.Y., Wallace, K., 2010. Oxidative stress studies of six TiO(2) and two CeO_2 nanomaterials: immuno-spin trapping results with DNA. Nanotoxicology 5, 546–556.

Knuckles, T.L., Yi, J., Frazer, D.G., Leonard, H.D., Chen, B.T., Castranova, V., Nurkiewicz, T.R., 2011. Nanoparticle inhalation alters systemic arteriolar vasoreactivity through sympathetic and cyclooxygenase-mediated pathways. Nanotoxicology 6, 724–735.

Kohler, A., Deisemann, H., Lauritzen, B., 1992. Ultrastructural and cytochemical indices of toxic injury in dab liver. Mar. Ecol. Prog. Ser. 91, 141–153.

Kohler, A., Wahl, E., Söffker, K., 2002. Functional and morphological changes of lysosomes as prognostic biomarkers of toxic liver injury in a marine flatfish (*Platichthys flesus* (L)). Environ. Toxicol. Chem. 21, 2434–2444.

Kraus, N.N., Slovic, P., 1988. Taxonomic analysis of perceived risk: modeling individual and group perceptions within homogenous hazard domains. Risk Anal. 8, 435–455.

Krewski, D., Brand, K.P., Burnett, R.T., Zielinski, J.M., 2002. Simplicity vs complexity in the development of risk models for dose-response assessment. Hum. Ecol. Risk Assess. 8, 1355–1374.

Kreyling, W., Semmler, M., Erbe, M.F., Mayer, P., Takenaka, S., Schulz, H., et al., 2002. Translocation of ultrafine insoluble iridium particles from lung epithelium to extrapulmonary organs is size-dependent but very low. J. Toxicol. Environ. Health 65A, 1513–1530.

Kuempel, E.D., Geraci, C.L., Schulte, P.A., 2012. Risk assessment and risk management of nanomaterials in the workplace: translating research to practice. Ann. Occup. Hyg. 56, 491–505.

Kuhlbusch, T.A.J., Asbach, C., Fissan, H., et al., 2011. Nanoparticle exposure at nanotechnology workplaces: a review. Part. Fibre Toxicol. 8, 22.

Kundu, S., Lau, S., Liang, H., 2009. Shape-controlled catalysis by cetyltrimethylammonium bromide terminated gold nanospheres, nanorods, and nanoprisms. J. Phys. Chem. C 13, 5150.

Kwon, S.Y., Kim, Y.-G., Lee, S.H., Moon, J.H., 2011. Uncertainty analysis of measurements of the size of nanoparticles in aqueous solutions using dynamic light scattering. Metrologia 48, 417–425.

Laurenson, J.P., Bloom, R.A., Page, S., Sadrieh, N., 2014. Ethinyl estradiol and other human pharmaceutical estrogens in the aquatic environment: a review of recent risk assessment data. AAPS J. 16, 299–310.

Lawrence, L.J., Casida, J.E., 1984. Interactions of lindane, toxaphene and cyclodienes with brain specific t-butylcyclophosphorothionate receptor. Life Sci. 35, 171–178.

Li, J., 2014. Risk assessment of heavy metals in surface sediments from the Yanghe River, China. Int. J. Environ. Res. Public Health 11, 12441–12453.

Lin, P., 2007. Nanotechnology bound: evaluating the case for more regulation. NanoEthics 1, 105–122.

Liu, C., Lu, D., Wang, Y., 2014. Residue and risk assessment of pyridaben in cabbage. Food Chem. 149, 233–236.

Livingstone, D.R., Chipman, J.K., Lowe, D.M., Minier, C., Mitchelmore, C.L., Moore, M.N., Peters, L.D., Pipe, R.K., 2000. Development of biomarkers to detect the effects of organic pollution on aquatic invertebrates: recent molecular, genotoxic, cellular and immunological studies on the common mussel (*Mytilus edulis* L.) and other mytilids. Int. J. Environ. Pollut. 13, 56–91.

Ludwicki, J.K., Góralczyk, K., Struciński, P., Wojtyniak, B., Rabczenko, D., Toft, G., Lindh, C.H., Jonsson, B.A.G., Lenters, V., Heederik, D., Czaja, K., Hernik, A., Pedersen, H.S., Zvyezday, V., Bonde, J.P., 2015. Hazard quotient profiles used as a risk assessment tool for PFOS and PFOA serum levels in three distinctive European populations. Environ. Int. 74, 112–118.

Ma, H., Zhang, S., Zhang, D., 2007. Residues dynamics of pyridaben in apple and soil. J. Heb-ei Agricul. Sci. 11, 108–110 [13].

Ma-Hock, L., Burkhardt, S., Strauss, V., Gamer, A.O., Wiench, K., van Ravenzwaay, B., Landsiedel, R., 2009. Development of a short-term inhalation test in the rat using nano-titanium dioxide as a model substance. Inhal. Toxicol. 21, 102–118.

Mackay, D., 1991. Multimedia Environmental Fate Models: The Fugacity Approach. Lewis Publ., Chelsea.

Macleod, M., Scheringer, M., Podey, H., Jones, K.C., Hungerbuhler, K., 2007. The origin and significance of short-term variability of semivolatile contaminants in air. Environ. Sci. Technol. 41, 3249–3253 [13].

MacPhail, R.C., Grulke, E.A., Yokel, R.A., 2013. Assessing nanoparticle risk poses prodigious challenges. Wiley Interdis. Rev. Nanomed. Nanobiotech. 5, 374–378.

Maier-Bode, H., 1968. Properties, effect, residues and analytics of the insecticide endosulfan. Residue Rev. 22, 1–44.

Malhat, F.M., El-Mesallamy, A., Assy, M., Madian, W., Loutfy, N.M., Ahmed, M.T., 2015. Health hazard assessment of pyridaben residues in Egyptian strawberries. Hum. Ecol. Risk Assess. 21, 241–249.

Maynard, A.M., Kuempel, E.D., 2005. Airborne nanostructured particles and occupational health. J. Nanopart. Res. 7, 587–614.

McKinney, W., Jackson, M., Sager, T.M., Reynolds, J.S., Chen, B.T., Afshari, A., Krajnak, K., Waugh, S., Johnson, R., Mercer, R.R., Frazer, D.G., Thomas, T.A., Castranova, V., 2012. Pulmonary and cardiovascular responses of rats to inhalation of a commercial antimicrobial spray containing titanium dioxide nanoparticles. Inhal. Toxicol. 24, 447–457.

McKone, T.E., Bogen, K.T., 1991. Predicting the uncertainties in risk assessments. A California groundwater case-study. Environ. Sci. Technol. 25, 1674–1681 [11].

Minh, N.H., Minh, T.B., Watanabe, M., Kunisue, T., Monirith, I., Tanabe, S., et al., 2003. Open dumping site in Asian developing countries: a potential source of polychlorinated dibenzo-p-dioxins and polychlorinated dibenzofurans. Environ. Sci. Technol. 37, 1493–1502.

Monteiller, C., Tran, L., MacNee, W., Faux, S., Jones, A., Miller, B., Donaldson, K., 2007. The pro-inflammatory effects of low-toxicity low-solubility particles, nanoparticles and fine particles, on epithelial cells in vitro: the role of surface area. Occup. Environ. Med. 64, 609–615.

Moore, M.N., Depledge, M.H., Readman, J.W., Leonard, D.R.P., 2004. An integrated biomarker-based strategy for ecotoxicological evaluation of risk in environmental management. Mutat. Res/Fund Mol Mech. Mutagenesis 552, 247–268.

Moore, M.N., Kohler, A., Lowe, D.L., Simpson, M.L., 1994. An integrated approach to cellular biomarkers in fish. In: Fossi, M.C., Leonzio, C. (Eds.), Non-destructive Biomarkers in Vertebrates. Lewis/CRC, Boca Raton, pp. 171–197.

Moore, M.N., 1990. Lysosomal cytochemistry in marine environmental monitoring. Histochem. J. 22, 187–191.

Moore, M.N., 2002. Biocomplexity: the post-genome challenge in ecotoxicology. Aquat. Toxicol. 59, 1–15.

Morgan, M.G., 1993. Risk analysis and management. Sci. Am. 269, 32–41.

Morgan, M.G., Henrion, M., 1992. Uncertainty—Guide to Dealing with Uncertainty in Quantitative Risk and Policy Analysis. Cambridge University Press, Cambridge.

Morgan, M.G., Keith, D., 1995. Subjective judgments by climate experts. Environ. Sci. Technol. 29, 468–476.

Morgan, M.G., Henrion, M., 1990. Uncertainty: A Guide to Dealing with Uncertainty in Quantitative Risk and Policy Analysis. University Press, Cambridge.

Morimoto, Y., Kobayashi, N., Shinohara, N., Myojo, T., Tanaka, I., Nakanishi, J., 2010. Hazard assessments of manufactured nanomaterials. J. Occup. Health 52, 325–334.

Morrissey, C.A., Mineau, P., Devries, J.H., Sanchez-Bayo, F., Liess, M., Cavallaro, M.C., Liber, K., 2015. Neonicotinoid contamination of global surface waters and associated risk to aquatic invertebrates: a review. Environ. Int. 74, 291–303.

Mossman, B.T., Bignon, J., Corn, M., Seaton, A., Gee, J.B.L., 1990. Asbestos: scientific developments and implications for public policy. Science 247, 294–301.

Mueller, N.C., Nowack, B., 2008. Exposure modeling of engineered nanoparticles in the environment. Environ. Sci. Technol. 42, 4447–4453 [7].

Murphy, F.A., Schinwald, A., Poland, C.A., Donaldson, K., 2012. The mechanism of pleural inflammation by long carbon nanotubes: interaction of long fibres with macrophages stimulates them to amplify pro-inflammatory responses in mesothelial cells. Part. Fibre Toxicol. 9, 8.

Nakanishi, J., 2011. Risk Assessment of Manufactured Nanomaterials "Approaches" – Overview of Approaches and Results. Final report issued on 17 August 2011. New Energy and Industrial.

Nel, A., Xia, T., Madler, L., Li, N., 2006. Toxic potential of materials at the nanolevel. Science 311, 622–627.

Nemmar, A., Hoet, P.H.M., Vanquickenborne, B., Dinsdale, D., Thomeer, M., Hoylaerts, M.F., Vanbilloen, H., Mortelmans, L., Nemery, B., 2002. Passage of inhaled particles into the blood circulation in humans. Circulation 105, 411–414.

Nielsen, E., Ostergaard, G., Larsen, J.C., 2008. Toxicological Risk Assessment of Chemicals: A Practical Guide. CRC Press.

Ng, C.-T., Yung, L.Y.L., Swa, H.L.F., et al., 2015. Altered protein expression profile associated with phenotypic changes in lung fibroblasts co-cultured with gold nanoparticle-treated small airway epithelial cells. Biomaterials 39, 31–38.

NIOSH, 2010a. Progress toward Safe Nanotechnology in the Workplace: A Report from the NIOSH Nanotechnology Research Center. U.S. Department of Health and Human Service, Centers for Disease Control and Prevention, National Institute for Occupational Safety and Health, DHHS (NIOSH), Cincinnati, OH. Publication No. 2010-104.

NIOSH, 2010b. Strategic Plan for NIOSH Nanotechnology Research and Guidance Filling the Knowledge Gaps. U.S. Department of Health and Human Service, Centers for Disease Control and Prevention, National Institute for Occupational Safety and Health, DHHS (NIOSH), Cincinnati, OH. Publication No. 2010-105.

NIOSH, 2010c. Occupational Exposure to Carbon Nanotubes and Nanofibers. Draft for public comment. Current intelligence bulletin, NIOSH Docket Number: NIOSH 161-A.

NIOSH, 2011. Occupational exposure to titanium dioxide. In: Current Intelligence Bulletin 63. National Institute for Occupational Safety and Health, Cincinnati.

Nouwen, J., Cornelis, C., De Fr, R., Wevers, M., Viaene, P., Mensink, C., et al., 2001. Health risk assessment of dioxin emissions from municipal waste incinerators: the Neerlandquarter (Wilrijk, Belgium). Chemosphere 43, 909–923.

NRCC, 1975. Endosulfan: Its Effects on Environmental Quality. National Research Council Canada, Environmental Secretariat, Ottawa, ON. Publication #: NRCC 14098.

Technology Development Organization (NEDO) Project (P06041), 2011. Research and Development of Nanoparticle Characterization Methods. National Institute of Advanced Industrial Science and Technology (AIST). Available at. http://www.aist-riss.jp/main/?ml_lang=en.

Oberdörster, G., Ferin, J., Lehnert, B.S., 1994. Correlation between particle size, in vivo particle persistence, and lung injury. Environ. Health Perspect. 102 (Suppl. 5), 173–179.

Oberdörster, G., 2012. Nanotoxicology: in vitro–in vivo dosimetry. Environ. Health Perspect. 120, a13.

Oberdörster, G., Finkelstein, J.N., Johnston, C., Gelein, R., Cox, C., Baggs, R., Elder, A.C., 2000. Acute pulmonary effects of ultrafine particles in rats and mice. Res. Rep. Health Eff. Inst. 96, 75–86.

Oberdörster, G., Oberdörster, E., Oberdörster, J., 2005. Invited review: nanotechnology: an emerging discipline evolving from studies of ultrafine particles. Environ. Health Perspect. 113, 823–839.

Oberdörster, G., Oberdörster, E., Oberdörster, J., 2007. Concepts of nanoparticle dose metric and response metric. Environ. Health Perspect. 115, A290.

Oberdörster, G., Sharp, Z., Atudorei, V., Elder, A., Gelein, R., Kreyling, W., Cox, C., 2004. Translocation of inhaled ultrafine particles to the brain. Inhalation Toxicol. 16, 437–445.

Oberdörster, G., 2001. Pulmonary effects of inhaled ultrafine particles. Int. Arch. Occup. Environ. Health 74, 1–8.

Oberdörster, G., 2002. Significance in parameters in the evaluation of exposure-dose response relationships of inhaled particles. Inhal. Toxicol. 8, 73–89.

Olsen, A.R., 2006. Using relative risk to compare the effects of aquatic stressors at a regional scale. Environ. Manage. 38, 1020–1030.

O'Shaughnessy, P.T., 2013. Occupational health risk to nanoparticulate exposure. Environ. Sci. Process. Impacts 15, 49–62.

Ostrom, L.T., Wilhelmsen, C.A., 2012. Risk Assessment: Tools, Techniques, and Their Applications. Wiley Publishing.

Ott, W.R., Steinemann, A.C., Wallace, L.A., 2006. Exposure Analysis. CRC Press.

Pacurari, M., Qian, Y., Fu, W., Schwegler-Berry, D., Ding, M., Castranova, V., Guo, N.L., 2012. Cell permeability, migration, and reactive oxygen species induced by multiwalled carbon nanotubes in human microvascular endothelial cells. J. Toxicol. Environ. Health Part A 75, 129–147.

Paik, S.Y., Zalk, D.M., Swuste, P., 2008. Application of a pilot control banding tool for risk level assessment and control of nanoparticle exposures. Ann. Occup. Hyg. 52, 419–428.

Pauluhn, J., 2010. Multi-walled carbon nanotubes (Baytubes®): approach for derivation of occupational exposure limit. Regul. Toxicol. Pharmacol. 57, 78–89.

Paustenbach, D.J., 2009. Human and Ecological Risk Assessment: Theory and Practice. Wiley Classics Library.

Perrodin, Y., Boillot, C., Angerville, R., Donguy, G., Emmanuel, E., 2011. Ecological risk assessment of urban and industrial systems: a review. Sci. Total Environ. 409, 5162–5176.

Pine, J.C., 2014. Hazards Analysis: Reducing the Impact of Disasters, Second ed. CRC Press.

Preining, O., 1998. The physical nature of very, very small particles and its impact on their behavior. J. Aerosol. Sci. 29, 481–495.

Raghua, S., Dhileepan, K., Scanlan, J.C., 2007. Predicting risk and benefit a priori in biological control of invasive plant species: a systems modelling approach. Ecol. Model. 208, 247–262.

Ramachandran, G., Paulsen, D., Watts, W., Kittelson, D., 2005. Mass, surface area and number metrics in diesel occupational exposure assessment. J. Environ. Monit. 7, 728–735.

Ramachandran, G., Vincent, J.H., 1999. A Bayesian approach to retrospective exposure assessment. Appl. Occup. Environ. Hyg 14, 547–557.

Ramachandran, G., Banerjee, S., Vincent, J.H., 2003. Expert judgment and occupational hygiene: application to aerosol speciation in the nickel primary production industry. Ann. Occup. Hyg. 47, 461–475.

Ramachandran, G., 2001. Retrospective exposure assessmentusing Bayesian methods. Ann. Occup. Hyg. 45 (8), 651–667.

Renwick, A.G., Lazarus, N.R., 1998. Human variability and noncancer risk assessment—an analysis of the default uncertainty factor. Regul. Toxicol. Pharmacol. 27, 3–20 [5].

Renwick, A.G., Barlow, S.M., Hertz-Piccioto, I., Boobis, A.R., Dybing, E., Edler, L., et al., 2003. Risk characterization of chemicals in food and diet. Final part of Food Safety in Europe (FOSIE): risk assessment of chemicals in food and diet (EC concerted action QLK1-1999-00156). Food Chem. Toxicol. 41, 1211–1271 [2].

Renwick, A.G., 1993. Data derived safety factors for the evaluation of food additives and environmental contaminants. Food Addit. Contam. 10 (3), 275–305.

Reynoldson, T.B., Norris, R.H., Resh, V.H., Day, K.E., Rosenberg, D.M., 1997. The reference condition: a comparison of multimetric and multivariate approaches to assess water-quality impairment using benthic macroinvertebrates. J. North Am. Benthol. Soc. 16, 833–852.

Riedel, G.F., Copper, S.E., Fath, J.B. (Eds.), 2008. Encyclopedia of Ecology. Elsevier, pp. 778–783.

Risbey, J.S., Kandlikar, M., 2002. Expert assessment of uncertainties in detection and attribution of climate change. Bull. Am. Meteorol. Soc. 83, 1317–1326.

Risbey, J.S., Kandlikar, M., Karoly, D.J., 2000. A protocol toarticulate and quantify uncertainties in climate changedetection and attribution. Climate Res. 16, 61–78.

Sanderson, H., Johnson, D.J., Wilson, C.J., Brain, R.A., Solomon, K.R., 2003. Probabilistic hazard assessment of environmentally occurring pharmaceuticals toxicity to fish, daphnids and algae by ECOSAR screening. Toxicol. Lett. 144, 383–395.

Sandman, P.M., Muller, P.M., Johnson, B.S., Weinstein, N.N., 1993. Agency communication, community outrage and perception of risk: three simulation experiments. Risk Anal. 13, 585–592.

Sand, P., 1987. Perceptions of risk. Science 236, 280–284.

Sant, S., Poulin, S., Hildgen, P., 2008. Effect of polymer architecture on surface properties, plasma protein adsorption, and cellular interactions of pegylated nanoparticles. J. Biomed. Mater. Res. Part A 87A, 885–895.

SCENIHR – Scentific Committee on Emerging and Newly Identified Health Risks, March 29, 2007. Opinion on the Appropriateness of the Risk Assessment Methodology in Accordance with the Technical Guidance Documents for New and Existing Substances for Assessing the Risks of Nanomaterials.

Schreuders, P.D., Nagoda, C., Lomander, A., Gipson, G., Rebar, J., Cheng, X., 2004. Creation of a virtual aquatic mesocosm using Stella software. Trans. ASAE 47, 2123–2213.

Section 5(e) orders: Because they are issued pursuant to Section 5(e) of TSCA, these administrative orders are commonly referred to as "section 5(e) orders." [5].

Sharma, D., Kansal, A., 2013. Assessment of river quality models: a review. Rev. Environ. Sci. Biotech. 12, 285–311.

Shi, H., Magaye, R., Castranova, V., Zhao, J., 2012. Titanium dioxide nanoparticles: a review of current toxicological data. Ann. Occup. Hyg. 56, 491–505.

Shi, H., Magaye, R., Castranova, V., Zhao, J., 2013. Titanium dioxide nanoparticles: a review of current toxicological data. Part. Fibre Toxicol. 10, 15.

Final report issued on July 2011. In: Shinohara, N. (Ed.), 2011. Risk Assessment of Manufactured Nanomaterials – Fullerence (C60). New Energy and Industrial Technology Development Organization (NEDO). project (P06041)

"Research and Development of Nanoparticle Characterization Methods." National Institute of Advanced Industrial Science and Technology (AIST). Available at. http://www.aist-riss.jp/main/?ml_lang=en.

Shvedova, A.A., Kisin, E., Murray, A.R., Johnson, V.J., Gorelik, O., Arepalli, S., Hubbs, A.F., Mercer, R.R., Keohavong, P., Sussman, N., Jin, J., Yin, J., Stone, S., Chen, B.T., Deye, G., Maynard, A., Castranova, V., Baron, P.A., Kagan, V.E., 2008. Inhalation vs aspiration of single-walled carbon nanotubes in C57BL/6 mice: inflammation, fibrosis, oxidative stress, and mutagenesis. Am. J. Physiol. Lung Cell. Mol. Physiol. 295, L552–L565.

Shvedova, A.A., Kisin, E.R., Mercer, R., Murray, A.R., Johnson, V.J., Potapovich, A.I., Tyurina, Y.Y., Gorelik, O., Arepalli, S., Schwegler-Berry, D., et al., 2005. Unusual inflammatory and fibrogenic pulmonary responses to single-walled carbon nanotubes in mice. Am. J. Physiol. Lung Cell. Mol. Physiol. 289, L698–L708.

Sickle, J.V., Stoddard, J.L., Paulsen, S.G., Olsen, A.R., 2006. Using relative risk to compare the effects of aquatic stressors at a regional scale. Environ. Manage. 38, 1020–1030.

Silva, M.H., Beauvais, S.L., 2010. Human health risk assessment of endosulfan. I: toxicology and hazard identification. Regul. Toxicol. Pharmacol. 56, 4–17.

Simon, T., 2014. Environmental Risk Assessment: A Toxicological Approach. CRC Press.

Singh, A.K., 2015. Challenges in the medicinal applications of carbon nanotubes (CNTs): toxicity of the central nervous system and safety issues. J. Nanomed. Nanotech. E published 2015.

Slovic, P., 1987. Perception of risk. Science 236, 280–285.

Slovic, P., 1993. Perceived risk, trust and democracy. Risk Anal. 13, 675–682.

Smulders, S., Luyts, K., Brabants, G., Golanski, L., Martens, J., Vanoirbeek, J., Hoet, P.H., 2015. Toxicity of nanoparticles embedded in paints compared to pristine nanoparticles, in vitro study. Toxicol. Lett. 232, 333–339.

Solomon, K., Giesy, J., Jones, P., 2000. Probabilistic risk assessment of agrochemicals in the environment. Crop Prot. 19, 10649–10655.

Solomon, K.R., Baker, D.B., Richards, R.P., et al., 1996. Ecological risk assessment of atrazine in North American surface waters. Environ. Toxicol. Chem. 15, 31–76.

Stoddard, J.L., Larsen, D.P., Hawkins, C.P., Johnson, R.K., Norris, R.H., 2006. Setting expectations for the ecological condition of streams: the concept of reference condition. Ecol. Appl. 16, 1267–1276.

Sultana, S., Djaker, N., Boca-Farcau, S., Salerno, M., Charnaux, N., Astilean, S., Hlawaty, H., de la Chapelle, M.L., 2015. Comparative toxicity evaluation of flower-shaped and spherical gold nanoparticles on human endothelial cells. Nanotechnology 23, 85–94.

Sun, T.Y., Gottschalk, F., Hungerbühler, K., Nowack, B., 2014. Comprehensive probabilistic modelling of environmental emissions of engineered nanomaterials. Environ. Pollut. 185, 69–76.

Takahashi, K., Kato, H., Saito, T., Matsuyama, S., Kinugasa, S., 2008. Precise measurement of the size of nanoparticles by dynamic light scattering with uncertainty analysis. Part. Part. Syst. Charact. 25, 31–38.

Theodore, L., Dupont, R.R., 2012. Environmental Health and Hazard Risk Assessment: Principles and Calculations. CRC Press.

Thomas, J.M., 2008. Heterogeneous catalysis: enigmas, illusions, challenges, realities, and emergent strategies of design. J. Chem. Phys. 128, 182502.

Tixier, G., Lafont, M., Grapentine, L., Rochfort, Q., Marsalek, J., 2011. Ecological risk assessment of urban stormwater ponds: literature review and proposal of a new conceptual approach providing ecological quality goals and the associated bioassessment tools. Ecol. Indic. 11, 1497–1506.

Toma-Biano, A., 2012. European Union regulations on nanomedicine. Bull. Transilvania Univ. Braşov Ser. VII: Soc. Sci. Law 5, 2.

Uldrich, J., Newberry, D., 2003. Next Big Thing Is Really Small: How Nanotechnology Will Change the Future of Your Business. ISBN-13: 9781400046898. Crown Publishing Group.

USEPA, 2004. http://www2.epa.gov/exposure-assessment-models/aquatox.

USEPA, 2007. Endosulfan. Hazard Characterization and Endpoint Selection Reflecting Receipt of a Developmental Neurotoxicity Study and Subchronic Neurotoxicity Study. PC Code: 079401. DP Barcode D338576.

van der Voet, H., Slob, W., 2007. Integration of probabilistic exposure assessment and probabilistic hazard characterization. Risk Anal. 27, 351–365 [12].

Van Duuren-Stuurman, B., Vink, S.R., Verbist, K.J.M., Heussen, H.G.A., Brouwer, D.H., Kroese, D.E.D., Van niftrik, M.F.J., Tielemans, E., Fransman, W., 2012. Stoffenmanager nano version 1.0: a web-based tool for risk prioritization of airborne manufactured nano objects. Ann. Occup. Hyg. 56, 525–541.

Van Eijndhoven, J.C.M., Wetering, R.A.P.M., Worrel, C.W., de Boer, J., Van der Pligt, J., Stallen, P.J.M., 1994. Risk communication in The Netherlands: the monitored introduction of the EC "Post Seveso" directive. Risk Anal. 14, 87–96.

Vincent, C., 2011. Overview of current toxicological knowledge of engineered nanoparticles. J. Occupat. Environ. Med. 53 (Suppl. 6S), S14–S17.

Walker, K.D., Evans, J.S., Macintosh, D., 2001. Use of expert judgment in exposure assessment. Part I. Characterization of personal exposure to benzene. J. Exposure Anal. Environ. Epidemiol. 11, 308–322.

Walker, K.D., Catalano, P., Hammitt, J.K., Evans, J.S., 2003. Use of expert judgment in exposure assessment: part 2. Calibration of expert judgments about personal exposures tobenzene. J. Expo. Anal. Environ. Epidemiol. 13, 1–16.

Wang, B., Yu, G., Jun, H., Wang, T., Hu, T., 2010. Probabilistic ecological risk assessment of DDTs in the Bohai Bay based on a food web bioaccumulation model. Sci. Total Environ. 409, 495–502.

Wang, D.G., Alaee, M., Guo, M.X., Pei, W., Wu, Q., 2014. Concentration, distribution, and human health risk assessment of endosulfan from a manufacturing facility in Huai'an, China. Sci. Total Environ. 491–492, 163–169.

Wang, Z.L., Mohamed, M.B., Link, S., El-Sayed, M.A., 1999. Crystallographic facets and shapes of gold nanorods of different aspect ratios. Surf. Sci. 440, L809–L814.

Warheit, D., Webb, T., Reed, K., Frerichs, S., Sayes, C., 2007. Pulmonary toxicity study in rats with three forms of ultrafine-TiO_2 particles: differential responses related to surface properties. Toxicology 230, 90–104.

Weil, C.S., 1972. Statistics vs safety factors and scientific judgment in the evaluation of safety for man. Tox. Appl. Pharmacol. 21, 454–463.

Weir, A., Westerhoff, P., Fabricius, L., Hristovski, K., von Goetz, N., 2012. Titanium dioxide nanoparticles in food and personal care products. Environ. Sci. Technol. 46, 2242–2250.

Wittmaack, K., 2007. In search of the most relevant parameter for quantifying lung inflammatory response to nanoparticle exposure: particle number, surface area, or what? Environ. Health Perspect. 115, 187–194.

WHO, 1994. International Programme on Chemical Safety: Assessing human health risks of chemicals: Derivation of Guidance values for health-based exposure limits. Environmenal Health Criteria, 170, 73pp. International Programme on Chemical Safety, World Health Organisation, Geneva.

WHO, 1987. Principles for the safety assessment of food additives and contaminants in food. Environmenal Health Criteria. 70, 174pp. International Programme on Chemical Safety, World Health Organisation, Geneva.

WHO, 2005. Chemical Specific Adjustment Factors for Interspecies Differences and Human Variability. Guidance Document for Use of Data in Dose/concentration–response Assessment. WHO, Geneva [6].

Wuensch, K.L., 2005. What Is a Likert Scale? And How Do You Pronounce 'Likert?'. East Carolina University.

Xia, Y., Xiong, Y., Lim, B., Skrabalak, S.E., 2008. Shape-controlled synthesis of metal nanocrystals: simple chemistry meets complex physics? Angew. Chem. Int. 48, 60–103.

Xie, J., Xu, C., Kohler, N., Hou, Y., Sun, S., 2007. Controlled PEGylation of monodisperse Fe_3O_4 nanoparticles for reduced non-specific uptake by macrophage cells. Adv. Mater. 19, 3163–3166.

Xing, G.H., Liang, Y., Chen, L.X., Wu, S.C., Wong, M.H., 2011. Exposure to PCBs, through inhalation, dermal contact and dust ingestion at Taizhou, China – a major site for recycling transformers. Chemosphere 83, 605–611.

Zhang, X., Shang, H., Chen, J., 2005. Studies on the dynamic variation of pyridaben residues in citrus and soil. Southwest China J. Agricul. Sci. 18, 777–781.

Zhang, Z., Li, H., Wu, M., 2009. Residue and risk assessment of chlorothalonil, myclobutanil and pyraclostrobin in greenhouse strawberry. Chin. J. Pestic. Sci. 11, 449–455.

CHAPTER

10

The Past, Present, and the Future of Nanotechnology

OUTLINE

1. INTRODUCTION

Nanoparticles are gradually being incorporated into our daily lives as part of medicine, cosmetics, food, clothing, and electronic goods. A large sector of the economy depends on nanotechnology. At the same time, nanoparticles exhibit unusually high toxicity and pose serious human and ecological health risks. Governments around the world are struggling with the dilemma of choosing between the economy and public environmental health. We are in this situation because the industrial application of nanoparticles is outpacing research on nanoparticle toxicity, which can be used to develop the regulatory policies. Although numerous probabilistic and statistical models are available to determine the risk, more definitive answers are needed to form informed pubic policies. Although a large portion of the chapters in this book are dedicated to the adverse effects of nanoparticles, the aim is not to spread fear, but

Engineered Nanoparticles
http://dx.doi.org/10.1016/B978-0-12-801406-6.00010-8

rather to bring the risks posed by nanoparticles at par with their beneficial effects, which are fueling nanoparticles' application in commercial products.

The three questions listed below are posed to explore possible status of nanotechnology in the future:

1. Do the perceptions of the general public regarding nanotechnology matter?
2. Does high toxicity preclude the use of nanoparticles in medicine?
3. What are the future applications and risks of nanoparticles?

Of these questions, the public perception is most relevant in determining the fate of nanotechnology. The answers to these questions are discussed in this chapter.

2. DO THE PERCEPTIONS OF THE GENERAL PUBLIC REGARDING NANOTECHNOLOGY MATTER?

As mentioned, a clear understanding of risks and benefits will help shape the general public's perceptions of nanomaterials, which is critically important to the future of nanotechnology. In the past, a lack of understanding by the general public has resulted in serious public protests against biotechnology in the United Kingdom and Europe or chemical and nuclear technologies in the United States (Sparks and Shepherd, 1994; Siegrist, 1999), often based on unfounded perceptions rather than scientific facts. The public perception may depend upon the extent of their familiarity with the technology as well as the source of information. People who claim to know much about nanotechnology are substantially more likely to believe its benefits outweigh its risks. However, Kahan et al. (2008) do not support the familiarity hypothesis. They found that half the public has at least some familiarity with nanotechnology, and those who perceive the greater benefits outnumber those who perceive greater risks by 3 to 1. In addition, a large minority of those surveyed (44%) is unsure, suggesting malleability of the risk judgments. According to them, knowledge deficit may correlate with positive perceptions. Given the potential malleability of perceptions, novel methods for understanding future public responses to nanotechnologies will need to be developed.

Siegrist et al. (2007) and Savadori et al. (2004) have suggested that experts and the public generally differ in their perceptions of nanoparticles; however, it is not clear whether the observed differences are due to expertise or to other factors (Rowe and Wright, 2001). People for whom nanotechnology is associated with positive outcomes and those who are not afraid of possible negative side effects of nanotechnology may assess nanotechnology applications more positively than people for whom negative effects outweigh positive effects. Thus, there may be a negative correlation between perceived risks and perceived benefits (Alhakami and Slovic, 1994; Siegrist, 2000; Siegrist and Cvetkovich, 2000). Siegrist et al. (2007) suggest that people who perceive more benefits from nanotechnology applications perceive fewer risks than people for whom nanotechnology applications offer little in the way of benefits. In the US context, male, white, high-income, and well-educated individuals perceive the risks of most hazardous technologies as being much lower than those in all other demographic groups, including white women and all nonwhite men and women (Finucane et al., 2000; Flin et al., 1994). Other people who focus more on risk than benefits (Kraus et al., 1992) are those regarding themselves as vulnerable and subject to injustice (Bord and O'Connor, 1997; Satterfield et al., 2004), wary of science and technology (Poortinga and Pidgeon, 2003; Frewer et al., 1997), skeptical of political authority or expertise (Braman and Kahan, 2001; Kraus et al., 1992) and dose-insensitive (i.e., they see risk as a function of any exposure, however small). A mishandling of risk management that creates distrust of regulatory agencies may also result in the stigmatization (Peters et al. 2004; Gregory et al., 1995) of specific technologies or risks (Siegrist, 2000; Slovic, 1992; Powell and Leiss 1997). Based on these and earlier observations, Satterfield et al. (2009) have hypothesized the following:

- Public response to nanotechnology would parallel other new and unknown technologies, such as biotechnology, and thus early evidence of risk aversion would prevail.
- An increase in knowledge will not result in reduced aversion to risks.
- Judgments about nanotechnology will be highly malleable and subject to persuasion given risk-centric information.
- Contextual, psychometric, and attitudinal predictors of perceived risk from earlier studies can anticipate future perceptions of nanotechnologies.

Therefore, as nanotechnology becomes an integral part of the society, the risk managers and regulation authorities should ensure that the public safety evaluation process is transparent and the results are communicated to the general public. This will help people understand the benefits and risks of nanotechnology.

3. DOES HIGH TOXICITY PRECLUDE NANOPARTICLES FROM BEING USED AS DRUGS?

The answer is hidden in another question: have we stopped using many highly toxic pharmaceuticals that serve an important purpose? As shown in Table 1, the answer is obviously *no*. Even though nanoparticles are considered to be a modern phenomenon, they have been used (and still being used) in indigenous and Chinese medicine. One example is metallic bhasma or

TABLE 1 Currently Approved NTI Drugs

Drug category	Drugs in that category	Treatment use
Cardiac drugs	Digoxin, digitoxin, quinidine, procainamide, amiodarone	Congestive heart failure, angina, arrhythmias
Antibiotics	Aminoglycosides (gentamicin, tobramycin, amikacin), vancomycin, chloramphenicol	Infections with bacteria that are resistant to less toxic antibiotics
Anti-epileptics	Phenobarbital, phenytoin, valproic acid, carbamazepine, ethosuximide, sometimes gabapentin, lamotrigine	Epilepsy, prevention of seizures, sometimes to stabilize moods
Bronchodilators	Theophylline, caffeine	Asthma, chronic obstructive pulmonary disorder (COPD), neonatal apnea
Immunosuppressants	Cyclosporine, tacrolimus, sirolimus, mycophenolate mofetil, azathioprine	Prevent rejection of transplanted organs, autoimmune disorders
Anticancer	All cytotoxic agents	Multiple malignancies
Psychiatric drugs	Lithium, valproic acid, some antidepressants (imipramine, amitriptyline, nortriptyline, doxepin, desipramine)	Bipolar disorder (manic depression), depression
Protease inhibitors	Indinavir, ritonavir, lopinavir, saquinavir, atazanavir, nelfinavi	HIV/AIDS

http://www.palmettogba.com/Palmetto/Providers.nsf/files/Drug_Therapy_Requiring_Intensive_Monitoring_for_Toxicity.pdf/$FIle/Drug_Therapy_Requiring_Intensive_Monitoring_for_Toxicity.pdf

ashes (especially gold bhasma). Metal bhasmas are used for treatment of many diseases, including *pandu* (anemia), *anidra* (insomnia), *apasmara* (convulsions), *mandagni* (poor digestion), and *kustha* (skin diseases) (Arya, 2014). An analysis of commercial gold bhasma revealed peaks of Au, Fe_2O_3, iron sulfide (FeS_2), CuS, and silicon oxide (SiO_2) (Mohaptra and Jha, 2010). Bhowmick et al. (2009) have analyzed the physicochemical properties of another bhasma called Jasada bhasma that contained ZnO nanoparticles (Figure 1). This suggests that significant amounts of synthetic nanoparticles were being used as long ago as 430 BC.

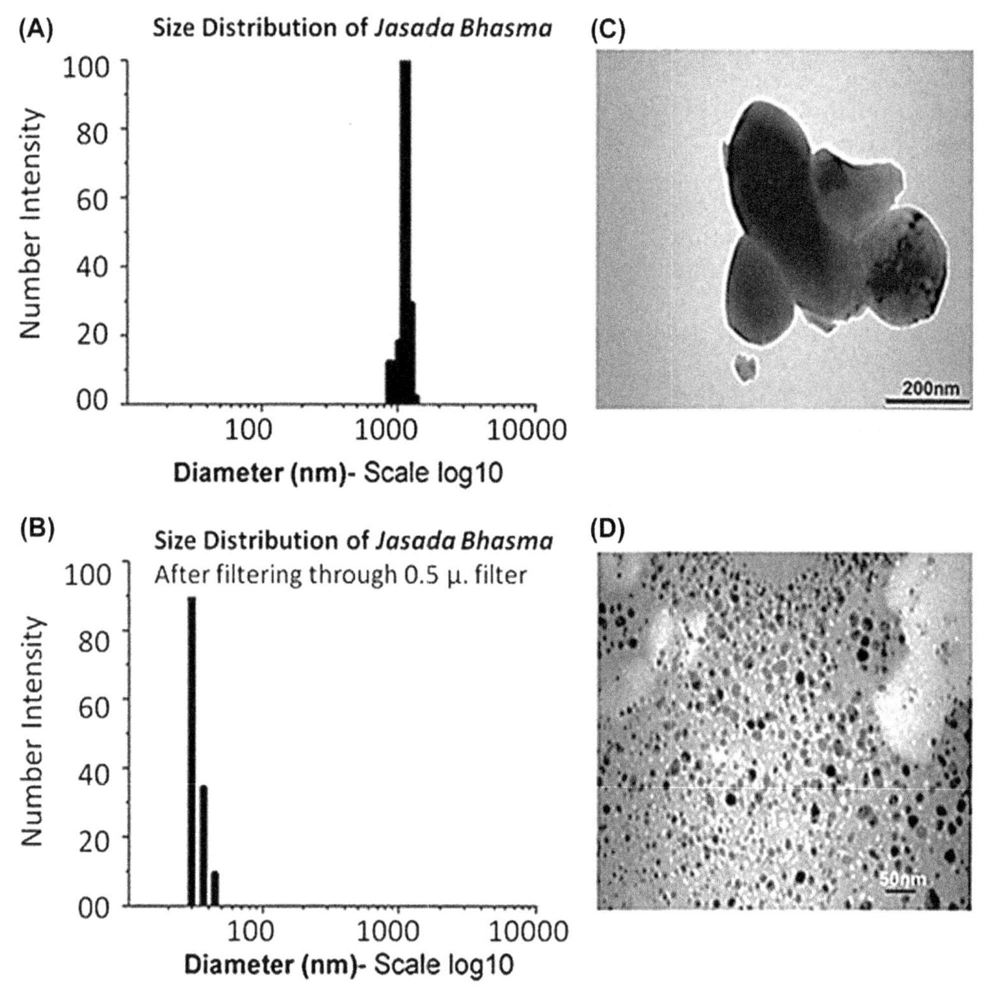

FIGURE 1 Particle size distribution in Jasada Bhasma by Dynamic light scattering (DLS). (A) Bulk particle size distribution; (B) particle size distribution of Jasada Bhasma after filtration through a 0.5-μm filter. Particle size analysis with transmission electron microscope (TEM); (C) TEM photograph of bulk particles of Jasada Bhasma; (D) TEM photograph of Jasada Bhasma particles present in the filtrate after filtration through a 0.5-μm filter. The mean particle size of Jasada Bhasma is ~1 μm. After filtration through a 0.5-μm filter, we can observe a significant number of small particles of 10–25 nm size present in the sample. *Reprinted from Bhowmick et al (2009).*

4. WHAT ARE THE FUTURE APPLICATIONS AND RISKS OF NANOPARTICLES?

Even though people may not understand nanoparticles, these tiny particles are embedded in many existing products that may or may not be so labeled. Some of the novel applications of nanoparticles are listed below.

4.1 Novel Applications

4.1.1 *Treatments of Drug-Resistant Bacteria*

Recently, there has been an increase in drug-resistant bacterial infections of patients undergoing different surgical procedures in hospitals. In addition, there has been an emergence of the development of antibiotics resistance to aquatic microbes. These situations have renewed a great interest in alternative methods of prevention and control of diseases. Swain et al. (2014), Dizaj et al. (2014), Kujda et al. (2015), Chatterjee et al. (2015), and others have shown that metal nanoparticles such as gold, silver, zinc, and copper nanoparticles exhibit potent antimicrobial activity. Dizaj et al. (2014) have proposed that combination therapy with metal nanoparticles might be one of the possible strategies to overcome the current bacterial resistance to the antibacterial agents.

4.1.2 *CNT-Based Stem-Cell-Based Regenerative Therapy*

Huang et al. (2012), by using the analysis of neuronal gene and protein expressions by quantitative polymerase chain reaction and immunostaining, respectively, have shown that neural stem cells plated on CNT ropes are boosted toward differentiated neurons in the early culture stage when compared to conventional tissue culture plates. Furthermore, a set of electrical stimulation parameters (5 mV, 0.5 mA, 25 ms intermittent stimulation) promotes neuronal maturity while also increasing the speed of neurite outgrowth. These results indicate that an electroconductive CNT rope substrate along with electrical stimulation may have a synergistic effect on promoting neurite elongation and boosting effects on the differentiation of the stem cells into mature neuronal cells for therapeutic application in neural regeneration.

4.1.3 *Medicinal Nanodevices*

Currently, nanoparticles are being used to probe cellular movements and molecular changes associated with pathological states (Sandhiya et al., 2009). CNT-based nanodevices can locate and deliver anticancer drugs at the specific tumor site (Torchilin, 2007). Other nanodevices such as respirocytes, microbivores, and probes encapsulated by biologically localized embedding have greater application in the treatment of anemia and infections (Kim et al., 2007; Freitas, 1998). Research is in progress to develop artificial cells, enzymes, and genes using nanotechnology. Many zero-dimensional nanoparticles (0D, all dimensions <100 nm, near-spherical) are used for triggered release of drugs (Willner et al., 2012).

4.1.4 *Energy Storage*

CNTs possess high capacity to store energy (Fabbro et al., 2012; Lu and Lieber, 2006, Yang, 2005) and serve as promising nanoelectronic building blocks for a variety of applications, including sensing (Patolsky et al., 2006b; Stern et al., 2007), photonics (Wang, 2007), energy conversion (Wang, 2012; Tian et al., 2009), and electrical devices capable of forming complex logical functions (Thelander et al., 2006; Yang, 2005; Grumer, 2006; Lin et al., 2009).

4.1.5 *Recording Electrical Signals*

Electrical signals can be recorded with nanostructures, such as Si nanowires, which has several advantages compared to conventional detection techniques with planar field-effect transistors (FETs) and multielectrode arrays (MEAs). Nanodevices, because of their high surface-to-volume ratios, exhibit high sensitivities with signal-to-noise ratios that outperform planar structures (Patolsky et al., 2006b; Stern et al., 2007). Therefore, they can be used as ultrasensitive sensors for various analytes, including single virus particle detection, protein sensing in the femtomolar range, and DNA sequencing using nanowire-based sensors (Patolsky et al., 2004; Xie et al., 2012). Nanostructures can also enhance cellular adhesion and activity, perhaps because their dimensions are closer to those of the subcellular building blocks of biological entities, such as proteins within the cell membrane (Stevens and George, 2005; Sniadecki et al., 2006; Kotov et al., 2009).

4.1.6 Two- and Three-Dimensional Devices

Two-dimensional layered materials, such as the one shown in Figure 2, consist of metal (Ni–Au–Ni) nanowires and carbon-nanotubes (CNTs), having distinct physicochemical properties, that trigger bioelectrocatalytic transformations of alcohols, reversibly and on-demand using magnetic on/off switch (Wang, 2008, Laocharoensuk et al., 2007; Willner and Katz, 2003). The nanowire is functionalized with two enzymes alcohol dehydrogenase (ADH) and alcohol oxidase (AOX) halfway along a CNT-modified amperometric transducer. Switching the orientation of the nanowires from vertical to horizontal—done by a magnetic field—brings the enzyme in contact with the CNT coated electrode, which allows the enzyme—in this case alcohol dehydrogenase (ADH)—to catalyze the transformation of ethanol to acetaldehyde. Regeneration of the enzyme's cofactor (NAD+) by the nanotube surface maintains the catalytic activity, and allows analysis by electrochemical methods. Three-dimensional (3D) self-assembled nanostructures serve a wide range of purposes for such devices, including delivery of molecules of interest, tissue engineering, and nanogenerators for self-sustained biosystems (Matson and Stupp, 2012; Hoare et al., 2011; Dvir et al., 2011; Aida et al., 2012; Wang, 2012). Nanowire-based devices have been used to record extracellular signals from cultured neurons, cardiomyocytes, brain slices, and whole embryonic chicken hearts (Patolsky et al., 2006a; Cohen-Karni et al., 2009; Qing et al., 2010; Timko et al., 2009; Pui et al., 2009). Recent developments in the synthesis of nanowires have enabled a free-standing nanowire-based three-dimensional nanoprobe, which allows the recording of the intracellular electrical activity of cardiomyocytes (Tian et al., 2010).

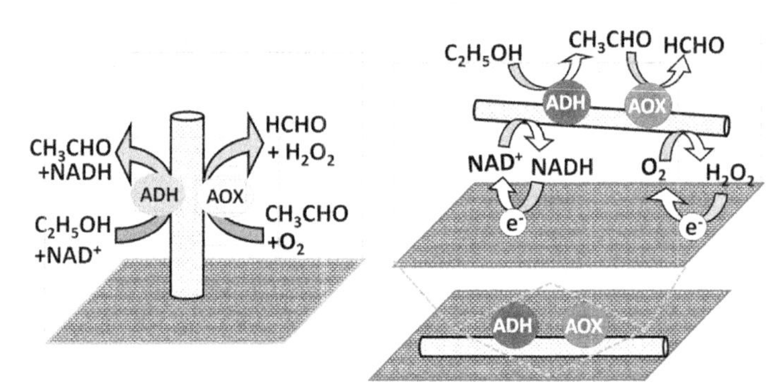

FIGURE 2 Application of carbon nanotubes in parallel on-demand biosensing of ethanol and methanol using Ni–Au–Ni nanowires functionalized with the ADH and AOX, and CNT-promoted electrocatalyic detection of NADH or hydrogen peroxide products. Bi-enzyme alcohol dehydrogenase (ADH) and alcohol oxidase (AOX) functionalized nanowires in the vertical positions, along with the CNT-coated transducer.

4.1.7 *Intracellular Applications*

Various device geometries and synthetic approaches have been developed to internalize electrical devices within cells (Duan et al., 2012; Gao et al., 2012; Robinson et al., 2012). Such nanostructures can be assembled on flexible plastic substrates, allowing them to conform to the shape of organs and tissues and enabling electrical monitoring and mapping of whole organs (Qing et al., 2010; Viventi et al., 2010, 2011; Kim et al., 2011). Laocharoensuk et al. (2007) have described an adaptive nanomaterial bioelectronic system, integrating nanowires (NW) and carbon-nanotubes (CNT), for on-demand bioelectrocatalytic transformations (Figure 2). Magnetic control of bioelectrocatalytic processes that used the attraction and retraction of functionalized magnetic spheres for on-off switching of enzymatic reactions (Willner and Katz, 2003). These devices can have multiple applications, including the screening of enzyme activities, biofuel cells, and bioreactors (Loaiza et al., 2007; Musameh et al., 2002).

4.1.8 *Future Developments*

In the near future (2015–2020), the field will expand to include molecular nanosystems—heterogeneous networks in which molecules and supramolecular structures serve as distinct devices. The proteins inside cells work together this way, but biological systems are water-based and markedly temperature-sensitive, and these molecular nanosystems will be able to operate in a far wider range of environments and should be much faster. Computers and robots could be reduced to extraordinarily small sizes. Medical applications might be as ambitious as new types of genetic therapies and antiaging treatments. New interfaces linking people directly to electronics could change telecommunications.

Over time, nanotechnology should benefit every industrial sector and health-care field. It should also help the environment through more efficient use of resources and better methods of pollution control. Nanotechnology does, however, pose new challenges to risk governance as well. Internationally, more needs to be done to collect the scientific information needed to resolve the ambiguities and to install the proper regulatory oversight. Helping the public to perceive nanotechnology soberly in a big picture that retains human values and quality of life will also be essential for this powerful new discipline to live up to its astonishing potential. But, how do we reconcile toxicity with the beneficial effects of nanoparticles, especially in the face of wide uncertainties? One approach is calculation of the therapeutic index (TI), determined by comparing the dose–response curves for therapeutic effects and the toxic effects (Figure 3). The TI is a comparison between the effective dose (ED_{50}) and the toxic dose (TD_{50}). It is given as a ratio between ED_{50} and the TD_{50}.

- **ED_{50}** is the dose that produces a therapeutic effect in 50% of the population (ED = effective dose)
- **TD_{50}** is the dose that produces a toxic effect in 50% of the population (TD = toxic dose)

For a drug to be safe, there must be at least a twofold gap between the ED_{50} and LD_{50} doses (called a narrow TI or NTI). Thus, nanoparticles may fall into the NTI group. Does this mean that NTI nanoparticles should not be used in treating diseases? As shown in Table 1, there are many NTI drugs currently approved for treatment of a number of diseases. Patients receiving these drugs must be monitored closely for signs of toxicity. The following conditions must be met for using NTI drugs:

1. An alternative safer option is not available.
2. The toxic effects must be scalable, tolerable, and reversible.
3. The monitoring plan must adhere to a previously agreed plan.

In fact, for certain diseases, the only option may be either loss of life or living with the help of a toxic drug and, thus, there may be some

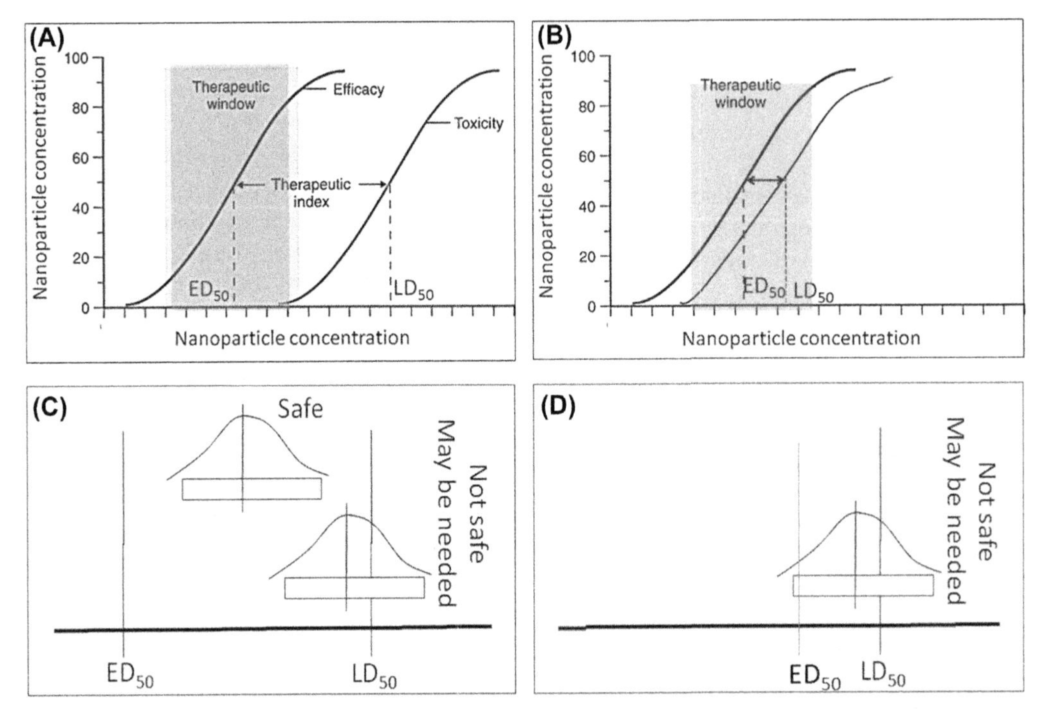

FIGURE 3 The concept of therapeutic index. (A) The dose–response curves for therapeutic and toxic effects. The ED_{50} represents a dose at which 50% of total response occurred, while TD_{50} represents a dose at which 50% of the toxic effect occurred. The ED_{50} and TD_{50} values were folds apart (no toxicity in therapeutic window); (B) the dose–response curves for therapeutic and toxic effects for a drug having close ED_{50} and TD_{50} values, resulting in an overlap of the therapeutic window and toxic effects; (C) and (D) probability distribution for (A) and (B), respectively.

discomfort (e.g., cancer, mental disorders, exposure to drug-resistant bacteria, AIDS). In fact, drug-resistant bacterial infection has become a serious problem. If NTI nanoparticles can kill the drug-resistant bacteria, would the associated risk be acceptable? This is the question that needs to be answered by society.

4.2 Future Risks of Nanoparticles

This is the dawn of nanoparticle application in different aspects of society. It is expected that, within the next few decades, nanotechnology will be closely integrated in every aspect of the life of nearly every person on the planet. With the potential benefits will come unknown risks, as described here (adapted from http://www.futureforall.org/nanotechnology/risks.htm).

The currently used nanoparticles are mostly passive nanoparticles that are small enough to enter the body and penetrate cell membranes. Nanoparticles are used in some of household and beauty products, such as antiaging cosmetics and sunscreen. The highest risk is to the workers in nanotechnology research and manufacturing processes. In the coming years, there will be active nanoparticles with limited ability to perform analytical functions. There may be virtually undetectable surveillance devices that could dramatically increase spying on governments, corporations, and private citizens. The distant future may be the era of molecular manufacturing (the assembly of products one molecule at a time), which means that every product can be made using molecular manufacturing. It is unclear whether this would bring boom or bust to the global economy. Scientists will be able to construct functioning cells and viruses. Untraceable weapons made with nanotechnology could be smaller than an insect with the intelligence of a supercomputer.

The proactive action taken today may go a long way in reducing future risks. There may be many unique or novel applications and risks that may appear in the future, which will shape the future of humanity. However, nanotechnology is not the only risk factor—there are biotechnology, genetic engineering, and nuclear energy risks, as well. The overall risk to the world will be determined by our ability to effectively use these technologies.

References

Aida, T., Meijer, E.W., Stupp, S.I., 2012. Functional supramolecular polymers. Science 335, 813–817.

Alhakami, A.S., Slovic, P., 1994. A psychological study of the inverse relationship between perceived risk and perceived benefit. Risk Anal. 14, 1085–1096.

Arya, R.K., 2014. Characterization of bio-active nanoparticles – bhasma, an Indian Ayurvedic drug. Ind. J. Pharm. Edu. Res. 48, 61–68.

Bhowmick, T.K., Suresh, A.K., Kane, S.G., Joshi, A.C., Bellare, J.R., 2009. Physicochemical characterization of an Indian traditional medicine, Jasada Bhasma: detection of nanoparticles containing non-stoichiometric zinc oxide. J. Nanopart. Res. 11, 655–664.

Bord, R.J., O'Connor, R.E., 1997. The gender gap in environmental attitudes: the case of perceived vulnerability to risk: research on the environment. Soc. Sci. Quart. 78, 830–840.

Braman, D., Kahan, D.M., 2001. More Statistics, Less Persuasion: A Cultural Theory of Gun-risk Perceptions. http://papers.ssrn.com/sol3/papers.cfm?abstract_id=286205.

Chatterjee, T., Chatterjee, B.K., Dipanwita, M., et al., 2015. Antibacterial effect of silver nanoparticles and the modeling of bacterial growth kinetics using a modified Gompertz model. Biochem. Biophys. Acta 1850, 299–306.

Cohen-Karni, T., Timko, B.P., Weiss, L.E., Lieber, C.M., 2009. Flexible electrical recording from cells using nanowire transistor arrays. Proc. Natl. Acad. Sci. USA 106, 7309–7313.

Dizaj, S.M., Lotfipour, F., Barzegar-Jalali, M., Zarrintan, M.H., Adibkia, K., 2014. Antimicrobial activity of the metals and metal oxide nanoparticles. Mater. Sci. Eng. C 44, 278–284.

Duan, X.J., Gao, R.X., Xie, P., Cohen-Karn, T., Qing, Q., Choe, H.S., Tian, B.Z., Jiang, X.C., Lieber, C.M., 2012. Intracellular recordings of action potentials by an extracellular nanoscale field-effect transistor. Nat. Nanotechnol. 7, 174–179.

Dvir, T., Timko, B.P., Brigham, M.D., Naik, S.R., Karajanagi, S.S., Levy, O., Jin, H., Parker, K.K., Langer, R., Kohane, D.S., 2011. Nanowired three-dimensional cardiac patches. Nat. Nanotechnol. 6, 720–725.

Fabbro, A., Villari, A., Laishram, J., Scaini, D., Toma, F.M., Turco, A., Prato, M., Ballerini, L., 2012. Spinal cord explants use carbon nanotube interfaces to enhance neurite outgrowth and to fortify synaptic inputs. ACS Nano 6, 2041–2055.

Finucane, M.L., Slovic, P., Mertz, C.K., Flynn, J., Satterfield, T.A., 2000. Gender, race and perceived risk: the white male effect. Health Risk Soc. 2, 159–172.

Flynn, J., Slovic, P., Mertz, C.K., 1994. Gender, race and perception of environmental health risks. Risk Anal. 14, 1101–1108.

Freitas Jr., R.A., 1998. Exploratory design in medical nanotechnology: a mechanical artificial red cell. Artif. Cells Blood Substit. Immobil. Biotechnol. 26, 411–430.

Frewer, L.J., Howard, C., Shepherd, R., 1997. Public concerns in the United Kingdom about general and specific applications of genetic engineering: risk, benefit and ethics. Sci. Technol. Hum. Val. 22, 98–124.

Gao, R., Strehle, S., Tian, B., Cohen-Karni, T., Xie, P., Duan, X., Qing, Q., Lieber, C.M., 2012. Outside looking in: nanotube transistor intracellular sensors. Nano Lett. 12, 3329–3333.

Gregory, R., Flynn, J., Slovic, P., 1995. Technological stigma. Am. Sci. 83, 220–223.

Gruner, G., 2006. Carbon nanotube transistors for biosensing applications. Anal. Bioanal. Chem. 384, 322–335.

Hoare, T., Timko, B.P., Santamaria, J., Goya, G.F., Irusta, S., Lau, S., Stefanescu, C.F., Lin, D.B., Langer, R., Kohane, D.S., 2011. Magnetically triggered nanocomposite membranes: a versatile platform for triggered drug release. Nano Lett. 11, 1395–1400.

Huang, Y.J., Wu, H.C., Tai, N.H., Wang, T.W., 2012. Carbon nanotube rope with electrical stimulation promotes the differentiation and maturity of neural stem cells. Small 8, 2869–2877.

Kahan, D.M., Slovic, P., Gastil, J., Cohen, G.L., 2008. Cultural cognition of the risks and benefits of nanotechnology. Nat. Nanotech. 4, 87–90.

Kim, G., Huang, S.W., Day, K.C., et al., 2007. Indocyanine-green-embedded PEBBLEs as a contrast agent for photoacoustic imaging. J. Biomed. Opt. 12, 044020.

Kim, D.H., Lu, N.S., Ma, R., Kim, Y.S., Kim, R.H., Wang, S.D., Wu, J., Won, S.M., Tao, H., Islam, A., et al., 2011. Epidermal electronics. Science 333, 838–843.

Kotov, N.A., Winter, J.O., Clements, I.P., Jan, E., Timko, B.P., Campidelli, S., Pathak, S., Mazzatenta, A., Lieber, C.M., Prato, M., et al., 2009. Nanomaterials for neural interfaces. Adv. Mater. 21, 3970–4004.

Kraus, N., Malmfors, T., Slovic, P., 1992. Intuitive toxicology: expert and lay judgments of chemical risks. Risk Anal. 12, 215–232.

Kujda, M., Ocwieja, M., Adamczyk, Z., et al., 2015. Charge stabilized silver nanoparticles applied as antibacterial agents. J. Nanosci. Technol. 15, 3574–3583.

Laocharoensuk, R., Bulbarello, A., Mannino, S., Wang, J., 2007. Adaptive nanowire–nanotube bioelectronic system for on-demand bioelectrocatalytic transformations. The future application and risks of nanoparticles. Chem. Commun. 2007, 3362–3364.

Lin, W.B., Rieter, W.J., Taylor, K.M.L., 2009. Modular synthesis of functional nanoscale coordination polymers. Angew. Chem. Int. Ed. 48, 650–658.

Loaiza, O.A., Laocharoensuk, R., Burdick, J., 2007. Adaptive orientation of multifunctional nanowires for magnetic control of bioelectrocatalytic processe. Angew. Chem. Int. 46, 1508–1511.

Lu, W., Lieber, C.M., 2006. Semiconductor nanowires. J. Phys. D: Appl. Phys. 39, R387–R406.

Matson, J.B., Stupp, S.I., 2012. Self-assembling peptide scaffolds for regenerative medicine. Chem. Commun. 48, 26–33.

Mohaptra, S., Jha, C.B., 2010. Physicochemical characterization of Ayurvedic bhasma (*Swarna makshika bhasma*): an approach to standardization. Int. J. Ayurveda Res. 1, 82–86.

Musameh, M., Wang, J., Merkoci, A., Lin, Y., 2002. Low-potential stable NADH detection at carbon-nanotube-modified glassy carbon electrodes. Electrochem. Commun. 4, 743–746.

Patolsky, F., Zheng, G.F., Hayden, O., Lakadamyali, M., Zhuang, X.W., Lieber, C.M., 2004. Electrical detection of single viruses. Proc. Natl. Acad. Sci. USA 101, 14017–14022.

Patolsky, F., Timko, B.P., Yu, G.H., Fang, Y., Greytak, A.B., Zheng, G.F., Lieber, C.M., 2006a. Detection, stimulation, and inhibition of neuronal signals with high-density nanowire transistor arrays. Science 313, 1100–1104.

Patolsky, F., Zheng, G., Lieber, C.M., 2006b. Nanowire sensors for medicine and the life sciences. Nanomedicine 1, 51–65.

Peters, E.M., Burraston, B., Mertz, C.K., 2004. An emotion-based model of risk perception and stigma susceptibility: cognitive appraisals of emotion, affective reactivity, worldviews and risk perceptions in the generation of technological stigma. Risk Anal. 24, 1349–1367.

Poortinga, W., Pidgeon, N.F., 2003. Exploring the dimensionality of trust in risk regulation. Risk Anal. 23, 961–972.

Powell, D., Leiss, W., 1997. Mad Cows and Mother's Milk: The Perils of Poor Risk Communication. McGill-Queen's Univ. Press.

Pui, T.S., Agarwal, A., Ye, F., Balasubramanian, N., Chen, P., 2009. CMOS-compatible nanowire sensor arrays for detection of cellular bioelectricity. Small 5, 208–212.

Qing, Q., Pal, S.K., Tian, B.Z., Duan, X.J., Timko, B.P., Cohen-Karni, T., Murthy, V.N., Lieber, C.M., 2010. Nanowire transistor arrays for mapping neural circuits in acute brain slices. Proc. Natl. Acad. Sci. USA 107, 1882–1887.

Robinson, J.T., Jorgolli, M., Shalek, A.K., Yoon, M.H., Gertner, R.S., Park, H., 2012. Nanowire electrode arrays as a scalable platform for intracellular interfacing to neuronal circuits. Nat. Nanotechnol. 7, 180–184.

Rowe, G., Wright, G., 2001. Differences in expert and lay judgments of risk: myth or reality? Risk Anal. 21, 341–356.

Sandhiya, S., Dkhar, S.A., Surendiran, A., 2009. Emerging trends of nanomedicine – an overview. Fund. Clin. Pharmacol. 23, 263–269.

Satterfield, T., Kandlikar, M., Beaudrie, C.E.H., Conti, J., Harthorn, B.H., 2009. Anticipating the perceived risk of nanotechnologies. Nat. Nanotech. 4, 752–758.

Satterfield, T.A., Mertz, C.K., Slovic, P., 2004. Discrimination, vulnerability and justice in the face of risk. Risk Anal. 24, 115–129.

Savadori, L., Savio, S., Nicotra, E., Rumiati, R., Finucane, M., Slovic, P., 2004. Expert and public perception of risk from biotechnology. Risk Anal. 24, 1289–1299.

Siegrist, M., 1999. A causal model explaining the perception and acceptance of gene technology. J. Appl. Soc. Psychol. 29, 2093–2106.

Siegrist, M., 2000. The influence of trust and perceptions of risks and benefits on the acceptance of gene technology. Risk Anal. 20, 195–204.

Siegrist, M., Cvetkovich, G., 2000. Perception of hazards: the role of social trust and knowledge. Risk Anal. 20, 713–719.

Siegrist, M., Keller, C., Kastenholz, H., Frey, S., Wiek, S.A., 2007. Laypeople's and experts' perception of nanotechnology hazards. Risk Anal. 27, 2007.

Slovic, P., 1992. In: Krimsky, S., Golding, D. (Eds.), Social Theories of Risk, vol. 1. Praeger, pp. 117–152.

Sniadecki, N., Desai, R.A., Ruiz, S.A., Chen, C.S., 2006. Nanotechnology for cell–substrate interactions. Ann. Biomed. Eng. 34, 59–74.

Sparks, P., Shepherd, R., 1994. Public perceptions of the potential hazards associated with food production and food consumption: an empirical study. Risk Anal. 14, 799–806.

Stern, E., Klemic, J.F., Routenberg, D.A., Wyrembak, P.N., Turner-Evans, D.B., Hamilton, A.D., LaVan, D.A., Fahmy, T.M., Reed, M.A., 2007. Label-free immunodetection with CMOS-compatible semiconducting nanowires. Nature 445, 519–522.

Stevens, M.M., George, J.H., 2005. Exploring and engineering the cell surface interface. Science 310, 1135–1138.

Swain, P., Nayak, S.K., Sasmal, A., Behera, T., Barik, S.K., Swain, S.K., Mishra, S.S., Sen, A.K., Das, J.K., Jayasankar, J.K., 2014. Antimicrobial activity of metal based nanoparticles against microbes associated with diseases in aquaculture. World J. Microbiol. Biotechnol. 30, 2491–2502.

Thelander, C., Agarwal, P., Brongersma, S., Eymery, J., Feiner, L.F., Forchel, A., Scheffler, M., Riess, W., Ohlsson, B.J., Gosele, U., Samuelson, L., 2006. Nanowire-based one-dimensional electronics. Mater. Today 9, 28–35.

Tian, B., Kempa, T.J., Lieber, C.M., 2009. Single nanowire photovoltaics. Chem. Soc. Rev. 38, 16–24.

Tian, B.Z., Cohen-Karni, T., Qing, Q., Duan, X.J., Xie, P., Lieber, C.M., 2010. Three-dimensional, flexible nanoscale field-effect transistors as localized bioprobes. Science 329, 830–834.

Timko, B.P., Cohen-Karni, T., Yu, G.H., Qing, Q., Tian, B.Z., Lieber, C.M., 2009. Electrical recording from hearts with flexible nanowire device array. Nano Lett. 9, 914–918.

Torchilin, V.P., 2007. Targeted pharmaceutical nanocarriers for cancer therapy and imaging. AAPS J. 9, E128–E147.

Viventi, J., Kim, D.H., Moss, J.D., Kim, Y.S., Blanco, J.A., Annetta, N., Hicks, A., Xiao, J.L., Huang, Y.G., Callans, D.J., et al., 2010. A conformal, bio-interfaced class of silicon electronics for mapping cardiac electrophysiology. Sci. Trans. Med. 2, 24ra22.

Viventi, J., Kim, D.H., Vigeland, L., Frechette, E.S., Blanco, J.A., Kim, Y.S., Avrin, A.E., Tiruvadi, V.R., Hwang, S.W., Vanleer, A.C., et al., 2011. Flexible, foldable, actively multiplexed, high-density electrode array for mapping brain activity in vivo. Nat. Neurosci. 14, 1599–1605.

Wang, Z.L., 2007. Novel nanostructures of ZnO for nanoscale photonics, optoelectronics, piezoelectricity, and sensing. Appl. Phys. A 88, 7–15.

Wang, J., 2008. Adaptive nanowires for on-demand control of electrochemical microsystems. Electrolysis 20, 611–615.

Wang, Z.L., 2012. Self-powered nanosensors and nanosystems. Adv. Mater. 24, 280–285.

Wildavsky, A., Dake, K., 1990. Theories of risk perception: who fears what and why? Daedalus 119, 41–60.

Willner, I., Katz, E., 2003. Magnetic control of electrocatalytic and bioelectrocatalytic processes. Angew. Chemi. Int. 42, 4576–4588.

Willner, I., Katz, E., Schroeder, A., Heller, D.A., Winslow, M.M., Dahlman, J.E., Pratt, G.W., Langer, R., Jacks, T., Anderson, D.G., 2012. Treating metastatic cancer with nanotechnology. Nat. Rev. Cancer 12, 39–50.

Xie, P., Xiong, Q.H., Fang, Y., Qing, Q., Lieber, C.M., 2012. Local electrical potential detection of DNA by nanowire–nanopore sensors. Nat. Nanotechnol. 7, 119–125.

Yang, P.D., 2005. The chemistry and physics of semiconductor nanowires. MRS Bull. 30, 85–91.

Index

Note: Page numbers followed by "f" and "t" indicate figures and tables respectively.

N

O

Q

U

V

W

X

Y

Z

Printed in the United States
By Bookmasters